Student Solutions Manual

To Accompany

Mathematical Applications for the Management, Life, and Social Sciences

Eighth Edition

Ronald J. Harshbarger
University of South Carolina Beaufort

James J. Reynolds
Clarion University of Pennsylvania

HOUGHTON MIFFLIN COMPANY   BOSTON   NEW YORK

Publisher: Richard Stratton
Development Editor: Maureen Ross
Editorial Associate: Allison Carol Lewis
Senior Project Editor: Nancy Blodget
Editorial Assistant: Sarah Driver
Manufacturing Coordinator: Eddloune Labrissiere
Senior Marketing Manager: Jennifer Jones
Marketing Associate: Mary Legere

Printed in the U.S.A

ISBN 13: 978-0-618-67692-7
ISBN 10: 0-618-67692-9

1 2 3 4 5 6 7 8 9-POO-10 09 08 07 06

STUDENT SOLUTIONS MANUAL

Table of Contents

Chapter 0: Algebraic Concepts

Exercise 0.1

1. $x \in \{x, y, z, a\}$

3. $12 \in \{1, 2, 3, 4, \ldots\}$

5. $6 \notin \{1, 2, 3, 4, 5\}$

7. $\{1, 2, 3, 4, 5, 6, 7\}$

9. $\{x : x$ is a natural number greater than 2 and less than 8$\}$

11. Yes. Every element of A is an element of B.

13. No. $c \in A$ but $c \notin B$.

15. $D \subseteq C$ since every element of D is an element of C.

17. $D \subseteq A$ since every element of D is an element of A.

19. $A \subseteq B$ and $B \subseteq A$. (Also $A = B$.)

21. $A \subseteq B$ and $B \subseteq A$. Thus, $A = B$.

23. $D \neq E$ because $4 \in E$ and $4 \notin D$.

25. A and B are disjoint since they have no elements in common. B and D are disjoint since they have no elements in common. C and D are disjoint.

27. $A \cap B = \{4, 6\}$ since 4 and 6 are elements of each set.

29. $A \cap B = \varnothing$ since they have no common elements.

31. $A \cup B = \{1, 2, 3, 4, 5\}$

33. $A \cup B = \{1, 2, 3, 4\}$ or $A \cup B = B$.

For problems 35 - 45, we have
$$U = \{1, 2, 3, \ldots, 9, 10\}.$$

35. $A' = \{4, 6, 9, 10\}$ since these are the only elements in U that are not elements of A.

37. $B' = \{1, 2, 5, 6, 7, 9\}$
    $A \cap B' = \{1, 2, 5, 7\}$

39. $A \cup B = \{1, 2, 3, 4, 5, 7, 8, 10\}$
    $(A \cup B)' = \{6, 9\}$

41. $A' = \{4, 6, 9, 10\}$
    $B' = \{1, 2, 5, 6, 7, 9\}$
    $A' \cup B' = \{1, 2, 4, 5, 6, 7, 9, 10\}$

43. $B' = \{1, 2, 5, 6, 7, 9\}$, $C' = \{1, 3, 5, 7, 9\}$
    $A \cap B' = \{1, 2, 3, 5, 7, 8\} \cap \{1, 2, 5, 6, 7, 9\}$
    $\qquad = \{1, 2, 5, 7\}$
    $(A \cap B') \cup C' = \{1, 2, 3, 5, 7, 9\}$

45. $B' = \{1, 2, 5, 6, 7, 9\}$,
    $A \cap B' = \{1, 2, 3, 5, 7, 8\} \cap \{1, 2, 5, 6, 7, 9\}$
    $\qquad = \{1, 2, 5, 7\}$
    $(A \cap B')' \cap C$
    $\qquad = \{3, 4, 6, 8, 9, 10\} \cap \{2, 4, 6, 8, 10\}$
    $\qquad = \{4, 6, 8, 10\}$

47. $A - B = \{1, 3, 7, 9\} - \{3, 5, 8, 9\} = \{1, 7\}$

49. $A - B = \{2, 1, 5\} - \{1, 2, 3, 4, 5, 6\} = \varnothing$

51. **a.** $L = \{99, 2000, 2001\}$ $H = \{97, 98, 99, 2000, 2001, 2002\}$
    $C = \{95, 2001, 2002\}$
    b. $L \subseteq H$
    c. C' is the years when the percentage change (from low to high) was 35% or less.
    d. $H' = \{95, 96\}$ $C' = \{96, 97, 98, 99, 2000\}$
    $H' \cup C' = \{95, 96, 97, 98, 99, 2000\}$ $H' \cup C'$ are the years when the high was less than or equal to 75,000 or the percentage gain was less than or equal to 35%.
    e. $L' = \{95, 96, 97, 98, 2002\}$ $L' \cap C = \{95, 2002\}$ $L' \cap C$ are the years when the low was less than or equal to 75,000 and the percentage gain was more than 35%.

53. **a.** From the table, there are 100 white Republicans and 30 non-white Republicans who favor national health care, for a total of 130.
    b. From the table, there are $350 + 40$ Republicans, and $250 + 200$ Democrats who favor national health care, for a total of 840.
    c. From the table, there are 350 white Republicans, and 150 white Democrats and 20 non-whites who oppose national health care, for a total of 520.

55. a. The key to solving this problem is to work from "the inside out". There are 40 aides in $E \cap F$. This leaves $65 - 40 = 25$ aides who speak English but do not speak French. Also we have $60 - 40 = 20$ aides who speak French but do not speak English. Thus there are $40 + 25 + 20 = 85$ aides who speak English or French. This means there are 15 aides who do not speak English or French.

b. From the Venn diagram $E \cap F$ has 40 aides.

c. From the Venn diagram $E \cup F$ has 85 aides.

d. From the Venn diagram $E \cap F'$ has 25 aides.

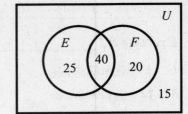

57. Since 12 students take M and E but not FA, and 15 take M and E, 3 take all three classes. Since 9 students take M and FA and we have already counted 3, there are 6 taking M and FA which are not taking E. Since 4 students take E and FA and we have already counted 3, there is only 1 taking E and FA but not taking M also. Since 20 students take E and we already have 16 enrolled in E, this leaves 4 taking only E. Since 42 students take FA and we already have 10 enrolled in FA, this leaves 32 taking only FA. Since 38 students take M and we already have 21 enrolled in M, this leaves 17 taking only M.

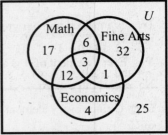

a. In the union of the 3 courses we have $17 + 12 + 3 + 6 + 32 + 1 + 4 = 75$ students enrolled. Thus, there are $100 - 75 = 25$ students who are not enrolled in any of these courses.

b. In $M \cup E$ we have $17 + 12 + 3 + 6 + 1 + 4 = 43$ enrolled.

c. We have $17 + 32 + 4 = 53$ students enrolled in exactly one of the courses.

59. (a) and (b)

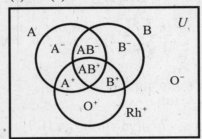

Exercise 0.2

1. a. Note that $-\dfrac{\pi}{10} = \pi \cdot \left(-\dfrac{1}{10}\right)$, where π is irrational and $-\dfrac{1}{10}$ is rational. The product of a rational number and an irrational number is an irrational number.

b. -9 is rational and an integer.

c. $\dfrac{9}{3} = \dfrac{3}{1} = 3$. This is a natural number, an integer, and a rational number.

d. Division by zero is meaningless.

3. a. Commutative

b. Distributive

c. Multiplicative identity

5. $-14 < -3$

7. $0.333 < \dfrac{1}{3}\left(\dfrac{1}{3} = 0.3333\cdots\right)$

9. $8 > 2$

11. $-3^2 + 10 \cdot 2 = -3^2 + 20 = -9 + 20 = 11$

13. $\dfrac{4 + 2^2}{2} = \dfrac{4 + 4}{2} = \dfrac{8}{2} = 4$

15. $\dfrac{16 - (-4)}{8 - (-2)} = \dfrac{16 + 4}{8 + 2} = \dfrac{20}{10} = 2$

17. $\dfrac{|5 - 2| - |-7|}{|5 - 2|} = \dfrac{|3| - |-7|}{|3|} = \dfrac{3 - 7}{3} = -\dfrac{4}{3}$

19. $\dfrac{(-3)^2 - 2 \cdot 3 + 6}{4 - 2^2 + 3} = \dfrac{9 - 6 + 6}{4 - 4 + 3} = \dfrac{9}{3} = 3$

21. $\dfrac{-4^2+5-2\cdot3}{5-4^2}=\dfrac{-16+5-6}{5-16}=\dfrac{-17}{-11}=\dfrac{17}{11}$

23. The entire line

25. (1, 3]; half-open interval

27. (2, 10); open interval

29. $-3\le x<5$

31. $x>4$

33. $(-\infty,4)\cap(-3,\infty)=(-3,4)$

35. $x>4$ and $x\ge0=(4,\infty)$

37. $[0,\infty)\cup[-1,5]=[-1,\infty)$

39. $(-\infty,0)\cup(7,\infty)$

41. -0.0000385850105

43. 9122.387471

45. $\dfrac{2500}{[(1.1^6)-1]}=\dfrac{2500}{0.771561}=3240.184509$

47. a. $\$300.00+\$788.91=\$1088.91$

b. $0.25[1088.91-0.05(1088.91)]=\$258.62$

c.   Retirement: $0.05(1088.91)=\$54.45$

State tax = Retirement = $\$54.45$

Local tax = $0.01(1088.91)=\$10.89$

Federal tax = $0.25(1088.91-54.45)=\$258.62$

Soc. Sec. tax = $0.0765(1088.91)=\underline{\$83.30}$

Total Withholding = $\$461.71$

Take-home pay = $1088.91-461.71$
   = $\$627.20$

49. a.   1996:

$C=1.2393(16)+4.2594=\$24.09$

$C=0.0232(16)^2+0.7984(16)+5.8271$
   $=\$24.54$

1997:

$C=1.2393(17)+4.2594=\$25.33$

$C=0.0232(17)^2+0.7984(17)+5.8271$
   $=\$26.10$

The second formula is more accurate.

b. $C=0.0232(28)^2+0.7984(28)+5.8271$
   $=\$46.37$

51. a. $\$70,351\le I\le\$146,750$

$\$146,751\le I\le\$319,100$

$I>\$319,100$

b. $T=715+0.15(29,050-7150)=\$4000$

$T=4000+0.25(70,350-29,050)$

$=\$14,325$

c. $[4000,\ 14,325]$

Exercise 0.3

1. $(-4)^4=(-4)(-4)(-4)(-4)=256$

3. $-2^6=-1\cdot2\cdot2\cdot2\cdot2\cdot2\cdot2=-64$

5. $3^{-2}=\dfrac{1}{3^2}=\dfrac{1}{9}$

7. $-\left(\dfrac{3}{2}\right)^2=(-1)\left(\dfrac{3}{2}\right)\left(\dfrac{3}{2}\right)=-\dfrac{9}{4}$

9. $6^5\cdot6^3=6^{5+3}=6^8$

11. $\dfrac{10^8}{10^9}=10^{8-9}=10^{-1}=\dfrac{1}{10}$

13. $\left(3^3\right)^3=3^{3\cdot3}=3^9$

15. $\left(\dfrac{2}{3}\right)^{-2}=\left(\dfrac{3}{2}\right)^2=\dfrac{9}{4}$

17. $(x^2)^{-3}=x^{2(-3)}=x^{-6}=\dfrac{1}{x^6}$

19. $xy^{-2}z^0=x\cdot\dfrac{1}{y^2}\cdot1=\dfrac{x}{y^2}$

21. $x^3\cdot x^4=x^{3+4}=x^7$

23. $x^{-5}\cdot x^3=x^{-5+3}=x^{-2}=\dfrac{1}{x^2}$

25. $\dfrac{x^8}{x^4}=x^{8-4}=x^4$

27. $\dfrac{y^5}{y^{-7}} = y^{5-(-7)} = y^{12}$

29. $(x^4)^3 = x^{3\cdot4} = x^{12}$

31. $(xy)^2 = x^2 y^2$

33. $\left(\dfrac{2}{x^5}\right)^4 = \dfrac{2^4}{(x^5)^4} = \dfrac{16}{x^{5\cdot4}} = \dfrac{16}{x^{20}}$

35. $\left(2x^{-2}y\right)^{-4} = 2^{-4}x^8 y^{-4} = \dfrac{x^8}{16y^4}$

37. $\left(-8a^{-3}b^2\right)\left(2a^5 b^{-4}\right) = -16a^{-3+5}b^{2-4}$

$$= -16a^2 b^{-2} = -\dfrac{16a^2}{b^2}$$

39. $2x^{-2} \div x^{-1}y^2 = \dfrac{2}{x^2} \div \dfrac{y^2}{x} = \dfrac{2}{x^2} \cdot \dfrac{x}{y^2} = \dfrac{2}{xy^2}$

41. $\left(\dfrac{x^3}{y^{-2}}\right)^{-3} = \dfrac{x^{-9}}{y^6} = \dfrac{1}{x^9} \cdot \dfrac{1}{y^6} = \dfrac{1}{x^9 y^6}$

43. $\left(\dfrac{a^{-2}b^{-1}c^{-4}}{a^4 b^{-3}c^0}\right)^{-3} = \left(\dfrac{b^2}{a^6 c^4}\right)^{-3} = \left(\dfrac{a^6 c^4}{b^2}\right)^3$

$$= \dfrac{a^{18}c^{12}}{b^6}$$

45. a. $\dfrac{2x^{-2}}{(2x)^2} = 2 \cdot \dfrac{1}{x^2} \cdot \dfrac{1}{(2x)^2} = 2 \cdot \dfrac{1}{x^2} \cdot \dfrac{1}{4x^2} = \dfrac{1}{2x^4}$

b. $\dfrac{(2x)^{-2}}{(2x)^2} = \dfrac{1}{(2x)^2} \cdot \dfrac{1}{(2x)^2} = \dfrac{1}{4x^2} \cdot \dfrac{1}{4x^2} = \dfrac{1}{16x^4}$

c. $\dfrac{2x^{-2}}{2x^2} = 2 \cdot \dfrac{1}{x^2} \cdot \dfrac{1}{2x^2} = \dfrac{1}{x^4}$

d. $\dfrac{2x^{-2}}{(2x)^{-2}} = 2 \cdot \dfrac{1}{x^2} \cdot (2x)^2 = 2 \cdot \dfrac{1}{x^2} \cdot 4x^2 = 8$

47. $\dfrac{1}{x} = x^{-1}$

49. $(2x)^3 = 2^3 x^3 = 8x^3$

51. $\dfrac{1}{(4x^2)} = \dfrac{1}{4} \cdot \dfrac{1}{x^2} = \dfrac{1}{4}x^{-2}$

53. $\left(\dfrac{-x}{2}\right)^3 = \dfrac{-x^3}{2^3} = -\dfrac{1}{8}x^3$

55. 1.2 $\boxed{y^x}$ 4 $\boxed{=}$ 2.0736

57. 1.5 $\boxed{y^x}$ -5 $\boxed{=}$ 0.1316872428

59. $P = 1200$, $i = 0.12$, $n = 5$

$S = P(1+i)^n$

$\quad = 1200(1+0.12)^5$

$\quad = 1200(1.12)^5$

$\quad = \$2114.81$

$I = S - P = 2114.81 - 1200 = \$914.81$

61. $P = 5000$, $i = 0.115$, $n = 6$

$S = P(1+i)^n$

$\quad = 5000(1+0.115)^6$

$\quad = 5000(1.115)^6$

$\quad = \$9607.70$

$I = S - P = 9607.70 - 5000 = \$4607.70$

63. $S = 15{,}000$, $n = 6$, $i = 0.115$

$P = S(1+i)^{-n}$

$\quad = 15{,}000(1+0.115)^{-6}$

$\quad = 15{,}000(1.115)^{-6}$

$\quad = \$7806.24$

65. $T = 257.8(1.0693)^t$

	Year	1970	1980	1990	1999
a.	t-value	20	30	40	49
b.	Tax	\$984.63	\$1924.29	\$3760.67	\$6873.24

c. $T = 257.8(1.0683)^{60} = \$14{,}363.39$

d. 2010 corresponds to $t = 60$, $T(60) = 257.8(1.0693)^{60} \approx \$14{,}363.4$ (seems high)

67. $y = \dfrac{1209}{1+6.46(1.17)^{-t}}$

a.

Year	1980	1990	2003
t – value	0	10	23
Predicted number of endangered species	162	516	1029

b. Year 2000: $t = 20$; $y = \dfrac{1209}{1+6.46(1.17)^{-20}} \approx 945$

Year 2010: $t = 30$; $y = \dfrac{1209}{1+6.46(1.17)^{-30}} \approx 1143$

Increase between 2000 and 2010 is $1143 - 945 = 198$ species

c. There are only a limited number of species and hopefully environmental protection would also contribute.

d. Same answer as c. To find the upper limit, which is 1209, compute y for large t-values:

Year	2050	2100	2200
t – value	70	120	220
Predicted number of endangered species	1208.9	1209	1209

69. $H = 29.57(1.102)^t$

a. $t = 10$ corresponds to 1970

b. $H = 29.57(1.102)^{10} \approx \$78.1$ billion

c. $H = 29.57(1.102)^{33} \approx \$729.2$ billion

d. $H = 29.57(1.102)^{50} \approx \$3801.3$ billion

Exercise 0.4

1. a. Since $\left(\dfrac{16}{3}\right)^2 = \dfrac{256}{9}$ we have $\sqrt{\dfrac{256}{9}} = \dfrac{16}{3}$.

b. $\sqrt{1.44} = 1.2$

3. a. $16^{3/4} = \left(\sqrt[4]{16}\right)^3 = 2^3 = 8$

b. $(-16)^{-3/2} = \left(\sqrt{-16}\right)^{-3}$ The square root of a negative number is not real.

5. $\left(\dfrac{8}{27}\right)^{-2/3} = \left(\dfrac{27}{8}\right)^{2/3} = \left(\sqrt[3]{\dfrac{27}{8}}\right)^2 = \left(\dfrac{3}{2}\right)^2 = \dfrac{9}{4}$

7. a. $8^{2/3} = \left(\sqrt[3]{8}\right)^2 = 2^2 = 4$

b. $(-8)^{-2/3} = \dfrac{1}{(-8)^{2/3}}$

$= \dfrac{1}{\left(\sqrt[3]{-8}\right)^2} = \dfrac{1}{(-2)^2} = \dfrac{1}{4}$

9. $\sqrt[9]{(6.12)^4} = (6.12)^{4/9} \approx 2.2370$

11. $\sqrt{m^3} = m^{3/2}$

13. $\sqrt[4]{m^2 n^5} = \left(m^2 n^5\right)^{1/4} = m^{2/4} n^{5/4} = m^{1/2} n^{5/4}$

15. $x^{7/6} = \sqrt[6]{x^7}$

17. $-\left(\dfrac{1}{4}\right) x^{-5/4} = -\dfrac{1}{4} \cdot \dfrac{1}{x^{5/4}} = \dfrac{-1}{4\sqrt[4]{x^5}}$

19. $y^{1/4} \cdot y^{1/2} = y^{(1/4)+(1/2)} = y^{3/4}$

21. $z^{3/4} \cdot z^4 = z^{(3/4)+(16/4)} = z^{19/4}$

23. $y^{-3/2} \cdot y^{-1} = y^{(-3/2)-(2/2)} = y^{-5/2} = \dfrac{1}{y^{5/2}}$

25. $\dfrac{x^{1/3}}{x^{-2/3}} = x^{(1/3)-(-2/3)} = x^{3/3} = x$

27. $\dfrac{y^{-5/2}}{y^{-2/5}} = y^{(-5/2)-(-2/5)}$

$\qquad\qquad = y^{(-25/10)+(4/10)} = y^{-21/10} = \dfrac{1}{y^{21/10}}$

29. $(x^{2/3})^{3/4} = x^{(2/3)(3/4)} = x^{2/4} = x^{1/2}$

31. $(x^{-1/2})^2 = x^{-1} = \dfrac{1}{x}$

33. $\sqrt{64x^4} = 8x^2$

35. $\sqrt{128x^4y^5} = \sqrt{64x^4y^4 \cdot 2y}$

$\qquad\qquad = \sqrt{64} \cdot \sqrt{x^4} \cdot \sqrt{y^4} \cdot \sqrt{2y}$

$\qquad\qquad = 8x^2y^2\sqrt{2y}$

37. $\sqrt[3]{40x^8y^5} = \sqrt[3]{8x^6y^3 \cdot 5x^2y^2}$

$\qquad\qquad = \sqrt[3]{8} \cdot \sqrt[3]{x^6} \cdot \sqrt[3]{y^3} \cdot \sqrt[3]{5x^2y^2}$

$\qquad\qquad = 2x^2y\sqrt[3]{5x^2y^2}$

39. $\sqrt{12x^3y} \cdot \sqrt{3x^2y} = \sqrt{36x^5y^2}$

$\qquad\qquad = \sqrt{36} \cdot \sqrt{x^5} \cdot \sqrt{y^2}$

$\qquad\qquad = 6x^2y\sqrt{x}$

41. $\sqrt{63x^5y^3} \cdot \sqrt{28x^2y} = \sqrt{9x^4y^2 \cdot 7xy} \cdot \sqrt{4x^2 \cdot 7y}$

$\qquad\qquad = 3x^2y\sqrt{7xy} \cdot 2x\sqrt{7y}$

$\qquad\qquad = 42x^3y^2\sqrt{x}$

43. $\dfrac{\sqrt{12x^3y^{12}}}{\sqrt{27xy^2}} = \sqrt{\dfrac{4x^2y^{10}}{9}} = \dfrac{2xy^5}{3}$

45. $\dfrac{\sqrt[4]{32a^9b^5}}{\sqrt[4]{162a^{17}}} = \sqrt[4]{\dfrac{16b^4}{81a^8} \cdot \dfrac{b}{1}} = \dfrac{2b}{3a^2}\sqrt[4]{b}$

47. $(A^9)^x = A^{9x}$

$\quad A^{9x} = A^1$

$\quad 9x = 1$

$\quad x = \dfrac{1}{9}$

49. $\left(\sqrt[7]{R}\right)^x = R^{x/7}$

$\quad R^{x/7} = R^1$

$\quad \dfrac{x}{7} = 1$

$\quad x = 7$

51. $\sqrt{\dfrac{2}{3}} \cdot \dfrac{\sqrt{3}}{\sqrt{3}} = \dfrac{\sqrt{2} \cdot \sqrt{3}}{\sqrt{3} \cdot \sqrt{3}} = \dfrac{\sqrt{6}}{3}$

53. $\dfrac{\sqrt{m^2x}}{\sqrt{mx^2}} = \sqrt{\dfrac{m}{x}} = \dfrac{\sqrt{m} \cdot \sqrt{x}}{\sqrt{x} \cdot \sqrt{x}} = \dfrac{\sqrt{mx}}{x}$

55. $\dfrac{\sqrt[3]{m^2x}}{\sqrt[3]{mx^5}} = \sqrt[3]{\dfrac{m}{x^4}} = \dfrac{\sqrt[3]{m}}{\sqrt[3]{x^3} \cdot \sqrt[3]{x}} \cdot \dfrac{\sqrt[3]{x^2}}{\sqrt[3]{x^2}}$

$\qquad\qquad = \dfrac{\sqrt[3]{mx^2}}{x\sqrt[3]{x^3}} = \dfrac{\sqrt[3]{mx^2}}{x^2}$

57. $\dfrac{-2}{3\sqrt[3]{x^2}} = \dfrac{-2}{3} \cdot \dfrac{1}{x^{2/3}} = -\dfrac{2}{3}x^{-2/3}$

59. $3x\sqrt{x} = 3x \cdot x^{1/2} = 3x^{3/2}$

61. $\dfrac{3}{2}x^{1/2} = \dfrac{3}{2}\sqrt{x}$

63. $\dfrac{1}{2}x^{-1/2} = \dfrac{1}{2} \cdot \dfrac{1}{x^{1/2}} = \dfrac{1}{2\sqrt{x}}$

65. a. $R = 8.5 = \dfrac{17}{2} \quad I = 10^{17/2} = \sqrt{10^{17}}$

    b. $I = \sqrt{10^{17}} = 10^{8.5} = 316{,}227{,}766$

    c. $\dfrac{I_{06}}{I_{89}} = \dfrac{10^{8.25}}{10^{7.10}} = 10^{1.15} \approx 14.125$

67. a. $S = 1000\sqrt{\left(1+\dfrac{r}{100}\right)^5}$

    b. $S = 1000\sqrt{\left(1+\dfrac{6.6}{100}\right)^5} \approx \$1173.26$

69. a. $P = 0.924t^{13/100} = 0.92\sqrt[100]{t^{13}}$

    b.

Year	t	Population
2005	5	1.139
2010	10	1.246
2045	45	1.516
2050	50	1.537

Change from 2005 to 2010: 9.39%

Change from 2045 to 2050: 1.39%

By 2045 and 2050 the population is much larger than earlier in the 21st century, and there is a limited number of people that any land can support, both in terms of space and food.

71. $k = 25$, $t = 10$, $q_0 = 98$

$$q = q_0\left(2^{-t/k}\right)$$
$$= 98\left(2^{-10/25}\right)$$
$$= 98\left(2^{-2/5}\right)$$
$$\approx 74 \text{ kg}$$

73. $P = P_0(2.5)^{ht}$
$$= 30,000(2.5)^{0.03(10)}$$
$$= 30,000(2.5)^{0.3}$$
$$\approx 39,491$$

75. a. $N = 500(0.02)^{(0.7)^t}$; at $t = 0$ we have $(0.7)^0 = 1$. Thus, $N = 500(0.02)^1 = 10$.

b. $N = 500(0.02)^{(0.7)^5} = 500(0.02)^{0.16807} = 259$

Exercise 0.5

1. $10 - 3x - x^2$
   a. The largest exponent is 2. The degree of the polynomial is 2.
   b. The coefficient of x^2 is -1.
   c. The constant term is 10.
   d. It is a polynomial of one variable x.

3. $7x^2y - 14xy^3z$
   a. The sum of the exponents in each term is 3 and 5, respectively. The degree of the polynomial is 5.
   b. The coefficient of xy^3z is -14.
   c. The constant term is zero.
   d. It is a polynomial of three variables; x, y, and z.

5. $2x^5 - 3x^2 - 5$
   a. $a_n x^n$ means $2 = a_5$.
   b. $a_3 = 0$ (Term is $0x^3$)
   c. $-3 = a_2$
   d. $a_0 = -5$, the constant term.

7. $4x - x^2$
   When $x = -2$,
   $$4x - x^2 = 4(-2) - (-2)^2 = -8 - 4 = -12$$

9. $\dfrac{2x - y}{x^2 - 2y}$
   When $x = -5$ and $y = -3$,
   $$\frac{2(-5) - (-3)}{(-5)^2 - 2(-3)} = \frac{-10 + 3}{25 + 6} = -\frac{7}{31}.$$

11. $1.98(74.7) - 1.09(1 - 0.80)(74.7 - 58) - 56.8$
   $$= 147.906 - 3.6406 - 56.8 = 87.4654$$

13. $\left(16pq - 7p^2\right) + \left(5pq + 5p^2\right) = 21pq - 2p^2$

15. $\left(4m^2 - 3n^2 + 5\right) - \left(3m^2 + 4n^2 + 8\right)$
   $$= 4m^2 - 3n^2 + 5 - 3m^2 - 4n^2 - 8$$
   $$= m^2 - 7n^2 - 3$$

17. $-\left[8 - 4(q + 5) + q\right] = -\left[8 - 4q - 20 + q\right]$
   $$= -\left[-12 - 3q\right] = 12 + 3q$$

19. $x^2 - \left[x - \left(x^2 - 1\right) + 1 - \left(1 - x^2\right)\right] + x$
   $$= x^2 - \left[x - x^2 + 1 + 1 - 1 + x^2\right] + x$$
   $$= x^2 - (x + 1) + x = x^2 - x - 1 + x$$
   $$= x^2 - 1$$

21. $\left(5x^3\right)\left(7x^2\right) = 35x^{3+2} = 35x^5$

23. $\left(39r^3s^2\right) \div \left(13r^2s\right) = 3r^{3-2}s^{2-1} = 3rs$

25. $ax^2\left(2x^2 + ax + ab\right) = 2ax^4 + a^2x^3 + a^2bx^2$

27. $(3y + 4)(2y - 3) = 6y^2 - 9y + 8y - 12$
   $$= 6y^2 - y - 12$$

29. $6\left(1 - 2x^2\right)\left(2 - x^2\right)$
   $$= 6\left(2 - x^2 - 4x^2 + 2x^4\right)$$
   $$= 6\left(2 - 5x^2 + 2x^4\right)$$
   $$= 12x^4 - 30x^2 + 12$$

31. $(4x + 3)^2 = 16x^2 + 2(4x)(3) + 9 = 16x^2 + 24x + 9$

33. $\left(x^2 - \dfrac{1}{2}\right)^2 = x^4 + 2(x^2)\left(-\dfrac{1}{2}\right) + \left(-\dfrac{1}{2}\right)^2$

$\qquad\qquad = x^4 - x^2 + \dfrac{1}{4}$

35. $9(2x+1)(2x-1) = 9\left[(2x)^2 - 1^2\right]$

$\qquad = 9\left[4x^2 - 1\right] = 36x^2 - 9$

37. $(0.1 - 4x)(0.1 + 4x) = (0.1)^2 - (4x)^2$

$\qquad\qquad\qquad\qquad = 0.01 - 16x^2$

39. $(0.1x - 2)(x + 0.05) = 0.1x^2 + 0.005x - 2x - 0.10$

$\qquad\qquad\qquad\qquad = 0.1x^2 - 1.995x - 0.10$

41.
$$
\begin{array}{r}
x^2 + 2x + 4 \\
x - 2 \\
\hline
-2x^2 - 4x - 8 \\
x^3 + 2x^2 + 4x \\
\hline
x^3 \qquad\qquad -8
\end{array}
$$

43.
$$
\begin{array}{r}
x^5 - 2x^3 + 5 \\
x^3 + 5x \\
\hline
5x^6 - 10x^4 \qquad + 25x \\
x^8 - 2x^6 \qquad + 5x^3 \\
\hline
x^8 + 3x^6 - 10x^4 + 5x^3 + 25x
\end{array}
$$

45. a. $(3x-2)^2 - 3x - 2(3x-2) + 5$

$\qquad = 9x^2 - 12x + 4 - 3x - 6x + 4 + 5$

$\qquad = 9x^2 - 21x + 13$

    b. $(3x-2)^2 - (3x-2)(3x-2) + 5$

$\qquad = (3x-2)^2 - (3x-2)^2 + 5 = 5$

47. $\dfrac{18m^2n + 6m^3n + 12m^4n^2}{6m^2n}$

$= \dfrac{18m^2n}{6m^2n} + \dfrac{6m^3n}{6m^2n} + \dfrac{12m^4n^2}{6m^2n}$

$= 3 + m + 2m^2n$

49. $\dfrac{24x^8y^4 + 15x^5y - 6x^7y}{9x^5y^2}$

$= \dfrac{24x^8y^4}{9x^5y^2} + \dfrac{15x^5y}{9x^5y^2} - \dfrac{6x^7y}{9x^5y^2}$

$= \dfrac{8x^3y^2}{3} + \dfrac{5}{3y} - \dfrac{2x^2}{3y}$

51. $(x+1)^3 = x^3 + 3(x^2)(1) + 3(x)(1)^2 + 1^3$

$\qquad\qquad = x^3 + 3x^2 + 3x + 1$

53. $(2x-3)^3 = (2x)^3 - 3(2x)^2(3) + 3(2x)(3)^2 - 3^3$

$\qquad\qquad = 8x^3 - 36x^2 + 54x - 27$

55.
$$
\begin{array}{r}
x^2 - 2x + 5 \\
x+2\ \overline{)\ x^3 \qquad\quad + x - 1} \\
x^3 + 2x^2 \\
\hline
-2x^2 + \ x - 1 \\
-2x^2 - 4x \\
\hline
5x - \ 1 \\
5x + 10 \\
\hline
-11
\end{array}
$$

Quotient: $x^2 - 2x + 5 - \dfrac{11}{x+2}$

57.
$$
\begin{array}{r}
x^2 + 3x - 1 \\
x^2+1\ \overline{)\ x^4 + 3x^3 \qquad\quad - x + 1} \\
x^4 \qquad\ + x^2 \\
\hline
3x^3 - x^2 - x + 1 \\
3x^3 \qquad + 3x \\
\hline
-x^2 - 4x + 1 \\
-x^2 \qquad - 1 \\
\hline
-4x + 2
\end{array}
$$

Quotient: $x^2 + 3x - 1 + \dfrac{-4x+2}{x^2+1}$

59. $x^{1/2}\left(x^{1/2} + 2x^{3/2}\right) = x^{2/2} + 2x^{4/2} = x + 2x^2$

61. $\left(x^{1/2} + 1\right)\left(x^{1/2} - 2\right) = x - 2x^{1/2} + x^{1/2} - 2 = x - x^{1/2} - 2$

63. $\left(\sqrt{x} + 3\right)\left(\sqrt{x} - 3\right) = \left(\sqrt{x}\right)^2 - (3)^2 = x - 9$

65. $(2x+1)^{1/2}\left[(2x+1)^{3/2} - (2x+1)^{-1/2}\right] = (2x+1)^2 - (2x+1)^0 = 4x^2 + 4x + 1 - 1 = 4x^2 + 4x$

67. $R = 55x$

69. a. $C = 99.95 + 0.21x$

    b. $C = 99.95 + 0.21(132) = \$127.67$

71. a. $4000 - x$

b. $0.10x$

c. $0.08(4000 - x)$

d. $0.10x + 0.08(4000 - x)$ or $320 + 0.02x$

73. $V = x(15 - 2x)(10 - 2x)$

75. a. Lengths decrease by one. Thus
$(A3) = A2 - 1$.

b. Width = 50 − Length

c. $(B2) = 50 - A2$

d. $(C2) = A2 \cdot B2$

e. Your spreadsheet will give Length = 33, Width = 17.

77. $P = 5.74t + 14.61$

a.

Year	t – value	Internet Households	Year	t – value	Internet Households
1995	0	43.31	2003	8	89.23
1996	1	49.05	2004	9	94.97
1997	2	54.79	2005	10	100.71
1998	3	60.53	2006	11	106.45
1999	4	66.27	2007	12	112.19
2000	5	72.01	2008	13	117.93
2001	6	77.75	2009	14	123.67
2002	7	83.49	2010	15	129.41

b. In 2001, P is greater than 75%.

c. The formula is no longer valid when $P > 100\%$, so the year 2005.

Exercise 0.6

1. $9ab - 12a^2b + 18b^2 = 3b\left(3a - 4a^2 + 6b\right)$

3. $4x^2 + 8xy^2 + 2xy^3 = 2x\left(2x + 4y^2 + y^3\right)$

5. $\left(7x^3 - 14x^2\right) + (2x - 4) = 7x^2(x - 2) + 2(x - 2)$
$$= (x - 2)\left(7x^2 + 2\right)$$

7. $6x - 6m + xy - my = (6x - 6m) + (xy - my)$
$$= 6(x - m) + y(x - m)$$
$$= (x - m)(6 + y)$$

9. $x^2 + 8x + 12 = (x + 6)(x + 2)$

11. $x^2 - x - 6 = (x - 3)(x + 2)$

13. $7x^2 - 10x - 8$
$7x^2 \cdot 8 = 56x^2$
The factors $-14x$ and $+4x$ give a sum of $-10x$.
$7x^2 - 10x - 8 = 7x^2 - 14x + 4x - 8$
$$= 7x(x - 2) + 4(x - 2)$$
$$= (x - 2)(7x + 4)$$

15. $x^2 - 10x + 25 = x^2 - 2 \cdot 5x + 5^2 = (x - 5)^2$

17. $49a^2 - 144b^2 = (7a)^2 - (12b)^2$
$$= (7a + 12b)(7a - 12b)$$

19. a. $9x^2 + 21x - 8$
$9x^2(-8) = -72x^2$
The factors $24x$ and $-3x$ give a sum of $21x$.
$9x^2 + 21x - 8 = 9x^2 + 24x - 3x - 8$
$$= 3x(3x + 8) - 1(3x + 8)$$
$$= (3x + 8)(3x - 1)$$

b. $9x^2 + 22x + 8$
$9x^2 \cdot 8 = 72x^2$
The factors $18x$ and $4x$ give a sum of $22x$.
$9x^2 + 22x + 8 = 9x^2 + 18x + 4x + 8$
$$= 9x(x + 2) + 4(x + 2)$$
$$= (x + 2)(9x + 4)$$

21. $4x^2 - x = x(4x - 1)$

23. $x^3 + 4x^2 - 5x - 20 = x^2(x+4) - 5(x+4)$
$$= (x+4)(x^2-5)$$

25. $2x^2 - 8x - 42 = 2(x^2 - 4x - 21) = 2(x-7)(x+3)$

27. $2x^3 - 8x^2 + 8x = 2x(x^2 - 4x + 4)$
$$= 2x(x^2 - 2\cdot 2x + 2^2)$$
$$= 2x(x-2)^2$$

29. $2x^2 + x - 6$
$2x^2 \cdot (-6) = -12x^2$
The factors $4x$ and $-3x$ give a sum of x.
$2x^2 + x - 6 = 2x^2 + 4x - 3x - 6$
$$= 2x(x+2) - 3(x+2)$$
$$= (2x-3)(x+2)$$

31. $3x^2 + 3x - 36 = 3(x^2 + x - 12) = 3(x+4)(x-3)$

33. $2x^3 - 8x = 2x(x^2 - 4) = 2x(x+2)(x-2)$

35. $10x^2 + 19x + 6$
$10x^2 \cdot 6 = 60x^2$
The factors $4x$ and $15x$ give a sum of $19x$.
$10x^2 + 19x + 6 = 10x^2 + 4x + 15x + 6$
$$= 2x(5x+2) + 3(5x+2)$$
$$= (5x+2)(2x+3)$$

37. $9 - 47x + 10x^2$
$9 \cdot 10x^2 = 90x^2$
The factors $-45x$ and $-2x$ give a sum of $-47x$.
$9 - 47x + 10x^2 = 9 - 45x - 2x + 10x^2$
$$= 9(1-5x) - 2x(1-5x)$$
$$= (1-5x)(9-2x)$$
$$\text{or } (5x-1)(2x-9)$$

39. $y^4 - 16x^4 = (y^2)^2 - (4x^2)^2$
$$= (y^2 - 4x^2)(y^2 + 4x^2)$$
$$= (y-2x)(y+2x)(y^2 + 4x^2)$$

41. $x^4 - 8x^2 + 16 = (x^2)^2 - 2\cdot 4x^2 + 4^2 = (x^2 - 4)^2$
$$= [(x-2)(x+2)]^2$$
$$= (x-2)^2(x+2)^2$$

43. $4x^4 - 5x^2 + 1 = (4x^2 - 1)(x^2 - 1)$
$$= (2x+1)(2x-1)(x+1)(x-1)$$

45. $x^3 + 3x^2 + 3x + 1 = (x+1)^3$

47. $x^3 - 12x^2 + 48x - 64 = x^3 - 3(4x^2) + 3(16x) - 4^3$
$$= x^3 - 3x^2(4) + 3x(4)^2 - 4^3$$
$$= (x-4)^3$$

49. $x^3 - 64 = x^3 - 4^3 = (x-4)(x^2 + 4x + 16)$

51. $27 + 8x^3 = 3^3 + (2x)^3 = (3+2x)(9 - 6x + 4x^2)$

53. $x^{3/2} + x^{1/2} = x^{1/2}(x^{2/2} + 1) = x^{1/2}(x+1)$
$? = (x+1)$

55. $x^{-3} + x^{-2} = x^{-3}(1 + x^1) = x^{-3}(1+x)$
$? = 1 + x$

57. $(-x^3 + x)(3 - x^2)^{-1/2} + 2x(3 - x^2)^{1/2}$
$$= (3 - x^2)^{-1/2}\left[(-x^3 + x) + 2x(3 - x^2)^{2/2}\right]$$
$$= (3 - x^2)^{-1/2}\left[-x^3 + x + 6x - 2x^3\right]$$
$$= (3 - x^2)^{-1/2}\left[7x - 3x^3\right]$$
$? = 7x - 3x^3$

59. $P + Prt = P(1+rt)$

61. $S = cm - m^2 = m(c-m)$

63. a. In the form px we have $p(10{,}000 - 100p)$.
$x = 10{,}000 - 100p$
b. If $p = 38$, then
$x = 10{,}000 - 100\cdot 38 = 6200$.

65. a. $R = x(300 - x)$
b. $P = 300 - x$

Exercise 0.7

1. $\dfrac{18x^3 y^3}{9x^3 z} = \dfrac{2x^3 y^3}{x^3 z} = \dfrac{2y^3}{z}$

3. $\dfrac{x - 3y}{3x - 9y} = \dfrac{1(x-3y)}{3(x-3y)} = \dfrac{1}{3}$

5. $\dfrac{x^2-2x+1}{x^2-4x+3}=\dfrac{(x-1)(x-1)}{(x-3)(x-1)}=\dfrac{x-1}{x-3}$

7. $\dfrac{6x^3y^3-15x^2y}{3x^2y^2+9x^2y}=\dfrac{3x^2y(2xy^2-5)}{3x^2y(y+3)}=\dfrac{2xy^2-5}{y+3}$

9. $\dfrac{6x^3}{8y^3}\cdot\dfrac{16x}{9y^2}\cdot\dfrac{15y^4}{x^3}=\dfrac{6}{y^3}\cdot\dfrac{2x}{9y^2}\cdot\dfrac{15y^4}{1}$

$\qquad\quad=\dfrac{2}{1}\cdot\dfrac{2x}{3y^2}\cdot\dfrac{15y}{1}$

$\qquad\quad=\dfrac{2}{1}\cdot\dfrac{2x}{y}\cdot\dfrac{5}{1}=\dfrac{20x}{y}$

11. $\dfrac{8x-16}{x-3}\cdot\dfrac{4x-12}{3x-6}=\dfrac{8(x-2)}{x-3}\cdot\dfrac{4(x-3)}{3(x-2)}=\dfrac{8\cdot4}{3}=\dfrac{32}{3}$

13. $\dfrac{x^2+7x+12}{3x^2+13x+4}\cdot\dfrac{9x+3}{1}$

$\qquad=\dfrac{(x+4)(x+3)}{(3x+1)(x+4)}\cdot\dfrac{3(3x+1)}{1}$

$\qquad=3(x+3)=3x+9$

15. $\dfrac{x^2-x-2}{2x^2-8}\cdot\dfrac{18-2x^2}{x^2-5x+4}\cdot\dfrac{x^2-2x-8}{x^2-6x+9}$

$\qquad=\dfrac{(x-2)(x+1)}{2(x^2-4)}\cdot\dfrac{-2(x^2-9)}{(x-4)(x-1)}\cdot\dfrac{(x-4)(x+2)}{(x-3)(x-3)}$

$\qquad=\dfrac{(x-2)(x+1)}{2(x-2)(x+2)}\cdot\dfrac{-2(x-3)(x+3)}{(x-1)}\cdot\dfrac{(x+2)}{(x-3)(x-3)}$

$\qquad=-\dfrac{(x+1)(x+3)}{(x-1)(x-3)}$

17. $\dfrac{15ac^2}{7bd}\div\dfrac{4a}{14b^2d}=\dfrac{15ac^2}{7bd}\cdot\dfrac{14b^2d}{4a}$

$\qquad=\dfrac{15c^2}{1}\cdot\dfrac{2b}{4}=\dfrac{15bc^2}{2}$

19. $\dfrac{y^2-2y+1}{7y^2-7y}\div\dfrac{y^2-4y+3}{35y^2}$

$\qquad=\dfrac{y^2-2y+1}{7y(y-1)}\cdot\dfrac{35y^2}{y^2-4y+3}$

$\qquad=\dfrac{(y-1)(y-1)}{7y(y-1)}\cdot\dfrac{35y^2}{(y-3)(y-1)}=\dfrac{5y}{y-3}$

21. $\dfrac{x^2-x-6}{1}\div\dfrac{9-x^2}{x^2-3x}$

$\qquad=\dfrac{x^2-x-6}{1}\cdot\dfrac{x^2-3x}{-1(x^2-9)}$

$\qquad=\dfrac{(x-3)(x+2)}{1}\cdot\dfrac{x(x-3)}{-1(x-3)(x+3)}$

$\qquad=\dfrac{-x(x-3)(x+2)}{x+3}$

23. $\dfrac{2x}{x^2-x-2}-\dfrac{x+2}{x^2-x-2}=\dfrac{2x-x-2}{(x-2)(x+1)}$

$\qquad=\dfrac{x-2}{(x-2)(x+1)}$

$\qquad=\dfrac{1}{x+1}$

25. $\dfrac{a}{a-2}-\dfrac{a-2}{a}=\dfrac{a}{a-2}\cdot\dfrac{a}{a}-\dfrac{a-2}{a}\cdot\dfrac{a-2}{a-2}$

$\qquad=\dfrac{a^2-(a^2-4a+4)}{a(a-2)}$

$\qquad=\dfrac{4a-4}{a(a-2)}$

$\qquad=\dfrac{4(a-1)}{a(a-2)}$

27. $\dfrac{x}{x+1}-x+1=\dfrac{x}{x+1}-\dfrac{x}{1}\cdot\dfrac{x+1}{x+1}+\dfrac{1}{1}\cdot\dfrac{x+1}{x+1}$

$\qquad=\dfrac{x-x^2-x+x+1}{x+1}$

$\qquad=\dfrac{-x^2+x+1}{x+1}$

29. $\dfrac{4a}{3x+6}+\dfrac{5a^2}{4x+8}=\dfrac{4a}{3(x+2)}+\dfrac{5a^2}{4(x+2)}$

$\qquad=\dfrac{4a}{3(x+2)}\cdot\dfrac{4}{4}+\dfrac{5a^2}{4(x+2)}\cdot\dfrac{3}{3}$

$\qquad=\dfrac{16a+15a^2}{12(x+2)}$

31. $\dfrac{3x-1}{2x-4}+\dfrac{4x}{3x-6}-\dfrac{x-4}{5x-10}=\dfrac{3x-1}{2(x-2)}+\dfrac{4x}{3(x-2)}-\dfrac{x-4}{5(x-2)}$

$\qquad\qquad = \dfrac{3x-1}{2(x-2)}\cdot\dfrac{3\cdot5}{3\cdot5}+\dfrac{4x}{3(x-2)}\cdot\dfrac{2\cdot5}{2\cdot5}-\dfrac{(x-4)}{5(x-2)}\cdot\dfrac{3\cdot2}{3\cdot2}$

$\qquad\qquad = \dfrac{(45x-15)+40x-6x+24}{30(x-2)}$

$\qquad\qquad = \dfrac{79x+9}{30(x-2)}$

33. $\dfrac{x}{x^2-4}+\dfrac{4}{x^2-x-2}-\dfrac{x-2}{x^2+3x+2}=\dfrac{x}{(x+2)(x-2)}+\dfrac{4}{(x-2)(x+1)}-\dfrac{x-2}{(x+2)(x+1)}$

$\qquad\qquad = \dfrac{x}{(x+2)(x-2)}\cdot\dfrac{x+1}{x+1}+\dfrac{4}{(x-2)(x+1)}\cdot\dfrac{x+2}{x+2}-\dfrac{x-2}{(x+2)(x+1)}\cdot\dfrac{x-2}{x-2}$

$\qquad\qquad = \dfrac{(x^2+x)+(4x+8)-(x^2-4x+4)}{(x+2)(x+1)(x-2)}=\dfrac{9x+4}{(x+2)(x+1)(x-2)}$

35. $\dfrac{-x^3+x}{\sqrt{3-x^2}}+\dfrac{2x\sqrt{3-x^2}}{1}=\dfrac{-x^3+x}{\sqrt{3-x^2}}+\dfrac{2x\sqrt{3-x^2}}{1}\cdot\dfrac{\sqrt{3-x^2}}{\sqrt{3-x^2}}=\dfrac{-x^3+x+2x(3-x^2)}{\sqrt{3-x^2}}$

$\qquad = \dfrac{-x^3+x+6x-2x^3}{\sqrt{3-x^2}}=\dfrac{7x-3x^3}{\sqrt{3-x^2}}$

37. $\dfrac{\frac{3}{1}-\frac{2}{3}}{\frac{14}{1}}\cdot\dfrac{3}{3}=\dfrac{9-2}{14(3)}=\dfrac{7}{14(3)}=\dfrac{1}{6}$

39. $\dfrac{x+y}{\frac{1}{x}+\frac{1}{y}}=\dfrac{(x+y)}{\frac{1}{x}+\frac{1}{y}}\cdot\dfrac{xy}{xy}=\dfrac{xy(x+y)}{y+x}=xy$

41. $\dfrac{2-\frac{1}{x}}{2x-\frac{3x}{x+1}}=\dfrac{\frac{2}{1}-\frac{1}{x}}{\frac{2x}{1}-\frac{3x}{x+1}}\cdot\dfrac{x(x+1)}{x(x+1)}$

$\qquad = \dfrac{2x(x+1)-1(x+1)}{2x^2(x+1)-3x(x)}$

$\qquad = \dfrac{2x^2+x-1}{2x^3-x^2}$

$\qquad = \dfrac{(2x-1)(x+1)}{x^2(2x-1)}=\dfrac{x+1}{x^2}$

43. $\dfrac{\sqrt{a}-\frac{b}{\sqrt{a}}}{a-b}=\dfrac{\frac{\sqrt{a}}{1}-\frac{b}{\sqrt{a}}}{\frac{a-b}{1}}\cdot\dfrac{\sqrt{a}}{\sqrt{a}}=\dfrac{a-b}{\sqrt{a}(a-b)}$

$\qquad = \dfrac{1}{\sqrt{a}}$ or $\dfrac{\sqrt{a}}{a}$

45. $\dfrac{\sqrt{x^2+9}-\frac{13}{\sqrt{x^2+9}}}{x^2-x-6}$

$\qquad = \dfrac{\frac{\sqrt{x^2+9}}{1}-\frac{13}{\sqrt{x^2+9}}}{(x-3)(x+2)}\cdot\dfrac{\sqrt{x^2+9}}{\sqrt{x^2+9}}$

$\qquad = \dfrac{x^2+9-13}{(x-3)(x+2)\sqrt{x^2+9}}$

$\qquad = \dfrac{(x-2)(x+2)}{(x-3)(x+2)\sqrt{x^2+9}}$

$\qquad = \dfrac{x-2}{(x-3)\sqrt{x^2+9}}$

47. a. $(2^{-2}-3^{-1})^{-1}=\left(\dfrac{1}{2^2}-\dfrac{1}{3}\right)^{-1}=\left(-\dfrac{1}{12}\right)^{-1}=-12$

b. $(2^{-1}+3^{-1})^2=\left(\dfrac{1}{2}+\dfrac{1}{3}\right)^2=\left(\dfrac{5}{6}\right)^2=\dfrac{25}{36}$

Hint: Work inside ( ) first when adding or subtracting is involved.

49. $\dfrac{2a^{-1}-b^{-1}}{(ab)^{-1}}=\dfrac{\frac{2}{a}-\frac{1}{b}}{\frac{1}{ab}}\cdot\dfrac{ab}{ab}=\dfrac{2b-a}{1}$ or $2b-a$

51. $\dfrac{1-\sqrt{x}}{1+\sqrt{x}}=\dfrac{1-\sqrt{x}}{1+\sqrt{x}}\cdot\dfrac{1-\sqrt{x}}{1-\sqrt{x}}=\dfrac{1-2\sqrt{x}+x}{1-x}$

53. $\dfrac{\sqrt{x+h}-\sqrt{x}}{h} = \dfrac{\sqrt{x+h}-\sqrt{x}}{h} \cdot \dfrac{\sqrt{x+h}+\sqrt{x}}{\sqrt{x+h}+\sqrt{x}}$

$= \dfrac{(x+h)-(x)}{h\left(\sqrt{x+h}+\sqrt{x}\right)} = \dfrac{h}{h\left(\sqrt{x+h}+\sqrt{x}\right)}$

$= \dfrac{1}{\sqrt{x+h}+\sqrt{x}}$

55. $\dfrac{1}{a}+\dfrac{1}{b}+\dfrac{1}{c} = \dfrac{1}{a}\cdot\dfrac{bc}{bc}+\dfrac{1}{b}\cdot\dfrac{ac}{ac}+\dfrac{1}{c}\cdot\dfrac{ab}{ab}$

$= \dfrac{bc+ac+ab}{abc}$

57. a. Avg. cost $= \dfrac{4000}{x}+\dfrac{55}{1}+\dfrac{0.1x}{1}$

$= \dfrac{4000+55x+0.1x^2}{x}$

b. Total cost $=$ (Avg. cost)(number of units)

$= 4000+55x+0.1x^2$

59. $SV = 1+\dfrac{3}{t+3}-\dfrac{18}{(t+3)^2}$

$= \dfrac{(t+3)^2+3(t+3)-18}{(t+3)^2}$

$= \dfrac{t^2+6t+9+3t+9-18}{(t+3)^2}$

$= \dfrac{t^2+9t}{(t+3)^2}$

Review Exercises

1. $B = \{1, 2, 3, 4, 5, 6, 7, 8\}$. Since every element of A is also an element of B, A is a subset of B.

2. No. $3 \notin \{x : x > 3\}$

3. A and B are not disjoint since each set contains the element 1.

4. $A = \{1, 2, 3, 9\}$ $B' = \{2, 4, 9\}$
   $A \cup B' = \{1, 2, 3, 4, 9\}$

5. $\{4, 5, 6, 7, 8, 10\} \cap \{1, 3, 5, 6, 7, 8, 10\}$
   $= \{5, 6, 7, 8, 10\}$

6. $A = \{1, 2, 3, 9\}$ $B = \{1, 3, 5, 6, 7, 8, 10\}$
   $A' = \{4, 5, 6, 7, 8, 10\}$
   $A' \cap B = \{5, 6, 7, 8, 10\}$
   $(A' \cap B)' = \{1, 2, 3, 4, 9\}$

7. $\{4, 5, 6, 7, 8, 10\} \cup \{2, 4, 9\}$
   $= \{2, 4, 5, 6, 7, 8, 9, 10\}$
   $(A' \cup B')' = \{2, 4, 5, 6, 7, 8, 9, 10\}' = \{1, 3\}$
   $A \cap B = \{1, 2, 3, 9\} \cap \{1, 3, 5, 6, 7, 8, 10\}$
   $= \{1, 3\}$ Yes.

8. a. $6 + \dfrac{1}{3} = \dfrac{1}{3} + 6$ illustrates the commutative property of addition.

   b. $2(3 \cdot 4) = (2 \cdot 3)4$ illustrates the associative property of multiplication.

   c. $\dfrac{1}{3}(6+9) = 2 + 3$ illustrates the distributive property.

9. a. irrational
   b. rational, integer
   c. meaningless

10. a. $\pi > 3.14$
    b. $-100 < 0.1$
    c. $-3 > -12$

11. $|5 - 11| = |-6| = -(-6) = 6$

12. $44 \div 2 \cdot 11 - 10^2 = 22 \cdot 11 - 100 = 242 - 100 = 142$

13. $(-3)^2 - (-1)^3 = 9 - (-1) = 10$

14. $\dfrac{(3)(2)(15) - (5)(8)}{(4)(10)} = \dfrac{90 - 40}{40} = \dfrac{50}{40} = \dfrac{5}{4}$

15. $2 - [3 - (2 - |-3|)] + 11 = 2 - [3 - (2 - 3)] + 11$
    $= 2 - [3 - (-1)] + 11$
    $= 2 - [3 + 1] + 11 = 2 - 4 + 11 = 9$

16. $-4^2 - (-4)^2 + 3 = -16 - 16 + 3 = -32 + 3 = -29$

17. $\dfrac{4 + 3^2}{4} = \dfrac{4 + 9}{4} = \dfrac{13}{4}$

18. $\dfrac{(-2.91)^5}{\sqrt{3.29^5}} \approx \dfrac{-208.6724}{19.6331} \approx -10.6286$

19. a. $[0, 5]$, closed

    b. $[-3, 7)$, half open

    c. $(-4, 0)$ open

20. a. $(-1, 16)$
    $-1 < x < 16$

    b. $[-12, 8]$
    $-12 \le x \le 8$

    c. $x < -1$

21. a. $\left(\dfrac{3}{8}\right)^0 = 1$

    b. $2^3 \cdot 2^{-5} = 2^{-2} = \dfrac{1}{2^2} = \dfrac{1}{4}$

    c. $\dfrac{4^9}{4^3} = 4^6 = 4096$

    d. $\left(\dfrac{1}{7}\right)^3 \left(\dfrac{1}{7}\right)^{-4} = \left(\dfrac{1}{7}\right)^{-1} = 7$

22. a. $x^5 \cdot x^{-7} = x^{5+(-7)} = x^{-2} = \dfrac{1}{x^2}$

    b. $\dfrac{x^8}{x^{-2}} = x^{8-(-2)} = x^{10}$

    c. $(x^3)^3 = x^{3 \cdot 3} = x^9$

    d. $(y^4)^{-2} = y^{(4)(-2)} = y^{-8} = \dfrac{1}{y^8}$

    e. $(-y^{-3})^{-2} = y^{(-3)(-2)} = y^6$

There are other correct methods of working problems 23–28.

23. $\dfrac{-(2xy^2)^{-2}}{(3x^{-2}y^{-3})^2} = \dfrac{(-1)(2)^{-2}x^{-2}y^{-4}}{3^2x^{-4}y^{-6}} = \dfrac{(-1)x^4y^6}{2^2 \cdot 3^2 x^2 y^4} = -\dfrac{x^2y^2}{36}$

24. $\left(\dfrac{2}{3}x^2y^{-4}\right)^{-2} = \left(\dfrac{2}{3}\right)^{-2}(x^2)^{-2}(y^{-4})^{-2} = \left(\dfrac{3}{2}\right)^2(x^{-4})(y^8) = \left(\dfrac{9}{4}\right)\left(\dfrac{1}{x^4}\right)(y^8) = \dfrac{9y^8}{4x^4}$

25. $\left(\dfrac{x^{-2}}{2y^{-1}}\right)^2 = \left(\dfrac{y}{2x^2}\right)^2 = \dfrac{y^2}{4x^4}$

26. $\dfrac{(-x^4y^{-2}z^2)^0}{-(x^4y^{-2}z^2)^{-2}} = \dfrac{1}{-(x^4)^{-2}(y^{-2})^{-2}(z^2)^{-2}} = \dfrac{1}{-x^{-8}y^4z^{-4}} = \dfrac{-x^8z^4}{y^4}$

27. $\left(\dfrac{x^{-3}y^4z^{-2}}{3x^{-2}y^{-3}z^{-3}}\right)^{-1} = \left(\dfrac{y^{4-(-3)}z^{-2-(-3)}}{3x^{-2-(-3)}}\right)^{-1} = \left(\dfrac{y^7z}{3x}\right)^{-1} = \dfrac{3x}{y^7z}$

28. $\left(\dfrac{x}{2y}\right)\left(\dfrac{y}{x^2}\right)^{-2} = \left(\dfrac{x}{2y}\right)\left(\dfrac{x^2}{y}\right)^2 = \left(\dfrac{x}{2y}\right)\left(\dfrac{(x^2)^2}{y^2}\right) = \left(\dfrac{x}{2y}\right)\left(\dfrac{x^4}{y^2}\right) = \dfrac{x^5}{2y^3}$

29. a. $-\sqrt[3]{-64} = -\sqrt[3]{(-4)^3} = -(-4) = 4$

    b. $\sqrt{\dfrac{4}{49}} = \sqrt{\dfrac{2^2}{7^2}} = \dfrac{2}{7}$

    c. $\sqrt[3]{1.9487171} = 1.1$

30. a. $\sqrt{x} = x^{1/2}$

    b. $\sqrt[3]{x^2} = x^{2/3}$

    c. $1/\sqrt[4]{x} = \dfrac{1}{x^{1/4}} = x^{-1/4}$

31. a. $x^{2/3} = \sqrt[3]{x^2}$

    b. $x^{-1/2} = \dfrac{1}{\sqrt{x}} = \dfrac{\sqrt{x}}{x}$

    c. $-x^{3/2} = -x\sqrt{x}$

32. a. $\dfrac{5xy}{\sqrt{2x}} \cdot \dfrac{\sqrt{2x}}{\sqrt{2x}} = \dfrac{5xy\sqrt{2x}}{2x} = \dfrac{5y\sqrt{2x}}{2}$

    b. $\dfrac{y}{x\sqrt[3]{xy^2}} \cdot \dfrac{\sqrt[3]{x^2y}}{\sqrt[3]{x^2y}} = \dfrac{y\sqrt[3]{x^2y}}{x\sqrt[3]{x^3y^3}} = \dfrac{y\sqrt[3]{x^2y}}{x(xy)}$

        $= \dfrac{y\sqrt[3]{x^2y}}{x^2y} = \dfrac{\sqrt[3]{x^2y}}{x^2}$

33. $x^{1/2} \cdot x^{1/3} = x^{(3/6)+(2/6)} = x^{5/6}$

34. $\dfrac{y^{-3/4}}{y^{-7/4}} = y^{-3/4-(-7/4)} = y^{4/4} = y$

35. $x^4 \cdot x^{1/4} = x^{(16/4)+(1/4)} = x^{17/4}$

36. $\dfrac{1}{x^{-4/3} \cdot x^{-7/3}} = \dfrac{1}{x^{-11/3}} = x^{11/3}$

37. $(x^{4/5})^{1/2} = x^{(4/5)(1/2)} = x^{2/5}$

38. $(x^{1/2}y^2)^4 = (x^{1/2})^4(y^2)^4 = x^2y^8$

39. $\sqrt{12x^3y^5} = \sqrt{4x^2y^4 \cdot 3xy} = 2xy^2\sqrt{3xy}$

40. $\sqrt{1250x^6y^9} = \sqrt{625x^6y^8 \cdot 2y} = 25x^3y^4\sqrt{2y}$

41. $\sqrt[3]{24x^4y^4} \cdot \sqrt[3]{45x^4y^{10}}$

    $= \sqrt[3]{8x^3y^3 \cdot 3xy} \cdot \sqrt[3]{9x^3y^9 \cdot 5xy}$

    $= 2xy\sqrt[3]{3xy} \cdot xy^3\sqrt[3]{9 \cdot 5xy}$

    $= 2x^2y^4\sqrt[3]{27 \cdot 5x^2y^2}$

    $= 6x^2y^4\sqrt[3]{5x^2y^2}$

42. $\sqrt{16a^2b^3} \cdot \sqrt{8a^3b^5} = \sqrt{128a^5b^8}$

              $= \sqrt{64a^4b^8 \cdot 2a}$

              $= 8a^2b^4\sqrt{2a}$

43. $\dfrac{\sqrt{52x^3y^6}}{\sqrt{13xy^4}} = \sqrt{4x^2y^2} = 2xy$

44. $\dfrac{\sqrt{32x^4y^3}}{\sqrt{6xy^{10}}} = \sqrt{\dfrac{16x^3}{3y^7}} = \dfrac{4x\sqrt{x}}{y^3\sqrt{3y}} \cdot \dfrac{\sqrt{3y}}{\sqrt{3y}} = \dfrac{4x\sqrt{3xy}}{3y^4}$

45. $(3x+5)-(4x+7)=3x+5-4x-7=-x-2$

46. $x(1-x)+x[x-(2+x)]=x-x^2+x(-2)=-x^2-x$

47. $(3x^3-4xy-3)+(5xy+x^3+4y-1)$
$=4x^3+xy+4y-4$

48. $(4xy^3)(6x^4y^2)=24x^{1+4}y^{3+2}=24x^5y^5$

49. $(3x-4)(x-1)=3x^2-3x-4x+4=3x^2-7x+4$

50. $(3x-1)(x+2)=3x^2+6x-x-2=3x^2+5x-2$

51. $(4x+1)(x-2)=4x^2-8x+x-2=4x^2-7x-2$

52. $(3x-7)(2x+1)=6x^2+3x-14x-7$
$=6x^2-11x-7$

53. $(2x-3)^2=(2x)^2-2(2x)(3)+3^2=4x^2-12x+9$

54. $(4x+3)(4x-3)=16x^2-9$
Difference of two squares

55.
$$\begin{array}{r} x^2+x-3 \\ 2x^2+1 \\ \hline x^2+x-3 \\ 2x^4+2x^3-6x^2 \\ \hline 2x^4+2x^3-5x^2+x-3 \end{array}$$

56. $(2x-1)^3=8x^3-12x^2+6x-1$ Binomial cubed

57.
$$\begin{array}{r} x^2+xy+y^2 \\ x-y \\ \hline -x^2y-xy^2-y^3 \\ x^3+x^2y+xy^2 \\ \hline x^3\qquad\qquad -y^3 \end{array}$$ Difference of two cubes

58. $\dfrac{4x^2y-3x^3y^3-6x^4y^2}{2x^2y^2}=\dfrac{2}{y}-\dfrac{3xy}{2}-3x^2$

59. $x^2+1\overline{\smash{\big)}\,3x^4+2x^3-x+4}$
$$\begin{array}{r} 3x^2+2x-3 \\ 3x^4+3x^2 \\ \hline 2x^3-3x^2-x+4 \\ 2x^3+2x \\ \hline -3x^2-3x+4 \\ -3x^2-3 \\ \hline -3x+7 \end{array}$$
Quotient is $3x^2+2x-3+\dfrac{7-3x}{x^2+1}$.

60. $x-3\overline{\smash{\big)}\,x^4-4x^3+5x^2+x+7}$
$$\begin{array}{r} x^3-x^2+2x+7 \\ x^4-3x^3 \\ \hline -x^3+5x^2 \\ -x^3+3x^2 \\ \hline 2x^2+x \\ 2x^2-6x \\ \hline 7x \\ 7x-21 \\ \hline 21 \end{array}$$
Quotient is $x^3-x^2+2x+7+\dfrac{21}{x-3}$.

61. $x^{4/3}(x^{2/3}-x^{-1/3})=x^{6/3}-x^{3/3}=x^2-x$

62. $\left(\sqrt{x}+\sqrt{a-x}\right)\left(\sqrt{x}-\sqrt{a-x}\right)=\left(\sqrt{x}\right)^2-\left(\sqrt{a-x}\right)^2$
$=x-(a-x)$
$=x-a+x$
$=2x-a$

63. $2x^4-x^3=x^3(2x-1)$

64. $4(x^2+1)^2-2(x^2+1)^3=2(x^2+1)^2[2-(x^2+1)]$
$=2(x^2+1)^2(2-x^2-1)$
$=2(x^2+1)^2(1-x^2)$
$=2(x^2+1)^2(1+x)(1-x)$

65. $4x^2-4x+1=(2x)^2-2(2x)+1^2=(2x-1)^2$

66. $16-9x^2=(4+3x)(4-3x)$

67. $2x^4-8x^2=2x^2(x^2-4)=2x^2(x+2)(x-2)$

68. $x^2-4x-21=(x-7)(x+3)$

69. $3x^2-x-2=(3x+2)(x-1)$

70. $12x^2-23x-24$
Two expressions whose product is
$12x^2(-24)=-288x^2$ and whose sum is
$-23x$ are $-32x$ and $9x$. So,
$12x^2-23x-24=12x^2+9x-32x-24$
$=3x(4x+3)-8(4x+3)$
$=(4x+3)(3x-8)$.

71. $16x^4-72x^2+81=(4x^2)^2-2(4x^2\cdot 9)+9^2$
$=(4x^2-9)^2$
$=[(2x+3)(2x-3)]^2$
$=(2x+3)^2(2x-3)^2$

72. $x^{-2/3} + x^{-4/3} = x^{-4/3}(?)$

$x^{-2/3} + x^{-4/3} = x^{-4/3}(x^{2/3}+1)$

$? = x^{2/3}+1$

73. a. $\dfrac{2x}{2x+4} = \dfrac{2x}{2(x+2)} = \dfrac{x}{x+2}$

   b. $\dfrac{4x^2y^3 - 6x^3y^4}{2x^2y^2 - 3xy^3} = \dfrac{2x^2y^3(2-3xy)}{xy^2(2x-3y)}$

$= \dfrac{2xy(2-3xy)}{2x-3y}$

74. $\dfrac{x^2-4x}{x^2+4} \cdot \dfrac{x^4-16}{x^4-16x^2} = \dfrac{x(x-4)}{x^2+4} \cdot \dfrac{(x^2-4)(x^2+4)}{x^2(x^2-16)}$

$= \dfrac{(x-4)(x+2)(x-2)}{x(x-4)(x+4)}$

$= \dfrac{(x+2)(x-2)}{x(x+4)}$

75. $\dfrac{x^2+6x+9}{x^2-7x+12} \cdot \dfrac{x^2-3x-4}{x^2+4x+3}$

$= \dfrac{(x+3)(x+3)}{(x-4)(x-3)} \cdot \dfrac{(x-4)(x+1)}{(x+3)(x+1)} = \dfrac{x+3}{x-3}$

76. $\dfrac{x^4-2x^3}{3x^2-x-2} \div \dfrac{x(x^2-4)}{9x^2-4}$

$= \dfrac{x^3(x-2)}{(3x+2)(x-1)} \cdot \dfrac{(3x+2)(3x-2)}{x(x+2)(x-2)}$

$= \dfrac{x^2(3x-2)}{(x-1)(x+2)}$

77. $1 + \dfrac{3}{2x} - \dfrac{1}{6x^2} = \dfrac{1}{1} \cdot \dfrac{6x^2}{6x^2} + \dfrac{3}{2x} \cdot \dfrac{3x}{3x} - \dfrac{1}{6x^2}$

$= \dfrac{6x^2+9x-1}{6x^2}$

78. $\dfrac{1}{x-2} - \dfrac{x-2}{4} = \dfrac{1 \cdot 4 - (x-2)(x-2)}{4(x-2)} = \dfrac{4x-x^2}{4(x-2)}$

79. $\dfrac{x+2}{x(x-1)} - \dfrac{x^2+4}{(x-1)(x-1)} + \dfrac{1}{1}$

$= \dfrac{(x+2)(x-1) - (x^2+4)x + x(x-1)(x-1)}{x(x-1)(x-1)}$

$= \dfrac{x^2+x-2-x^3-4x+x^3-2x^2+x}{x(x-1)^2}$

$= \dfrac{-(x^2+2x+2)}{x(x-1)^2}$

80. $\dfrac{x-1}{x^2-x-2} - \dfrac{x}{x^2-2x-3} + \dfrac{1}{x-2} = \dfrac{x-1}{(x-2)(x+1)} - \dfrac{x}{(x-3)(x+1)} + \dfrac{1}{x-2}$

$= \dfrac{(x-1)(x-3)}{(x-2)(x+1)(x-3)} - \dfrac{x(x-2)}{(x-2)(x+1)(x-3)} + \dfrac{(x+1)(x-3)}{(x-2)(x+1)(x-3)} = \dfrac{x^2-4x+3-x^2+2x+x^2-2x-3}{(x-2)(x+1)(x-3)}$

$= \dfrac{x^2-4x}{(x-2)(x+1)(x-3)} = \dfrac{x(x-4)}{(x-2)(x+1)(x-3)}$

81. $\dfrac{\frac{x-1}{1} - \frac{x-1}{x}}{\frac{1}{x-1}+1} \cdot \dfrac{x(x-1)}{x(x-1)} = \dfrac{x(x-1)^2 - (x-1)^2}{x+x(x-1)} = \dfrac{(x-1)^2(x-1)}{x^2} = \dfrac{(x-1)^3}{x^2}$

82. $\dfrac{x^{-2}-x^{-1}}{x^{-2}+x^{-1}} = \dfrac{\frac{1}{x^2}-\frac{1}{x}}{\frac{1}{x^2}+\frac{1}{x}} \cdot \dfrac{x^2}{x^2} = \dfrac{1-x}{1+x}$

83. $\dfrac{3x-3}{\sqrt{x}-1} \cdot \dfrac{\sqrt{x}+1}{\sqrt{x}+1} = \dfrac{3(x-1)(\sqrt{x}+1)}{x-1} = 3(\sqrt{x}+1)$

84. $\dfrac{\sqrt{x}-\sqrt{x-4}}{2} \cdot \dfrac{\sqrt{x}+\sqrt{x-4}}{\sqrt{x}+\sqrt{x-4}} = \dfrac{x-(x-4)}{2(\sqrt{x}+\sqrt{x-4})} = \dfrac{x-x+4}{2(\sqrt{x}+\sqrt{x-4})} = \dfrac{4}{2(\sqrt{x}+\sqrt{x-4})} = \dfrac{2}{\sqrt{x}+\sqrt{x-4}}$

85. a. *R*: Recognized
    C: Involved
    E: Exercised

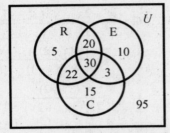

Numbered statement indicates solution for that question.
1. 30
2. $50 - 30 = 20$
3. $52 - 30 = 22$
4. $30 + 22 + 20 + \underline{5} = 77$
5. $37 - 22 = 15$
6. $77 + 15 + \underline{3} = 95$

b. $200 - (95 + 5 + 22 + 30 + 20 + 3 + 15) = 10$ So, 10 exercised only.

c. $63 + 70 - (3 + 30) = 100$ So, 100 exercised or were involved in the community.

86. $S = 100\left[\dfrac{(1.0075)^n - 1}{0.0075}\right]$

a. $S(36) = 100\left[\dfrac{(1.0075)^{36} - 1}{0.0075}\right]$

$\approx 100\left[\dfrac{0.30865}{0.0075}\right]$

$\approx \$4115.27$

b. $S(240) = 100\left[\dfrac{(1.0075)^{240} - 1}{0.0075}\right]$

$\approx 100\left[\dfrac{5.00915}{0.0075}\right]$

$\approx \$66,788.69$

87. a. $R = 10,000\left[\dfrac{0.0065}{1 - (1.0065)^{-n}}\right]$

$= 10,000\left[\dfrac{0.0065}{1 - \frac{1}{1.0065^n}}\right] \cdot \dfrac{1.0065^n}{1.0065^n}$

$= 10,000\left[\dfrac{0.0065(1.0065)^n}{1.0065^n - 1}\right]$

$= \dfrac{65(1.0065)^n}{1.0065^n - 1}$

b. $R = 10,000\left[\dfrac{0.0065}{1 - (1.0065)^{-48}}\right] \approx 243.19$

$R = \dfrac{65(1.0065)^{48}}{1.0065^{48} - 1} \approx 243.19$

88. $S = kA^{1/3}$

a. $S = k\sqrt[3]{A}$

b. Let S_1 be the number of species on 20,000 acres. Then $S_1 = k\sqrt[3]{20,000}$. Let S_2 be the number of species on 45,000 acres. Then

$S_2 = k\sqrt[3]{45000}$

$= \sqrt[3]{2.25 \cdot 20,000}$

$= \sqrt[3]{2.25} \cdot k\sqrt[3]{20,000}$

$= \sqrt[3]{2.25} \cdot S_1$

$S_2 \approx 1.31 S_1$

89. a. $C = \dfrac{540,000}{100 - p} - \dfrac{5400}{1}$

$= \dfrac{540,000 - 5400(100 - p)}{100 - p}$

$= \dfrac{5400p}{100 - p}$

b. If $p = 0$, $C = \dfrac{5400(0)}{100 - 0} = \dfrac{0}{100} = 0$.
The cost of removing no pollution is zero.

c. $C = \dfrac{5400(98)}{100 - 98} = \$264,600$

d. The formula is not defined when $p = 100$. We are dividing by zero. The cost increases as p approaches 100. It is cost prohibitive to remove all of the pollution.

Chapter Test

1. a. $A = \{6, 8\}$ $B' = \{3, 4, 6\}$
   $A \cup B' = \{3, 4, 6, 8\}$

   b. $\{3, 4\}$, $\{3, 6\}$, and $\{4, 6\}$ are disjoint from B.

   c. $\{6\}$, and $\{8\}$ are non-empty subsets of A.

2. $(4 - 2^3)^2 - 3^4 \cdot 0^{15} + 12 \div 3 + 1 = (-4)^2 - 0 + 4 + 1$
$$= 16 + 4 + 1 = 21$$

3. a. $x^4 \cdot x^4 = x^8$

   b. $x^0 = 1$, if $x \neq 0$

   c. $\sqrt{x} = x^{1/2}$

   d. $(x^{-5})^2 = x^{-10}$ or $\dfrac{1}{x^{10}}$

   e. $a^{27} \div a^{-3} = a^{27-(-3)} = a^{30}$

   f. $x^{1/2} \cdot x^{1/3} = x^{5/6}$

   g. $\dfrac{1}{\sqrt[3]{x^2}} = \dfrac{1}{x^{2/3}}$

   h. $\dfrac{1}{x^3} = x^{-3}$

4. a. $x^{1/5} = \sqrt[5]{x}$

   b. $x^{-3/4} = \sqrt[4]{x^{-3}}$ or $\left(\sqrt[4]{x}\right)^{-3}$ or $\dfrac{1}{\sqrt[4]{x^3}}$

5. a. $x^{-5} = \dfrac{1}{x^5}$

   b. $\left(\dfrac{x^{-8}y^2}{x^{-1}}\right)^{-3} = \dfrac{x^{24}y^{-6}}{x^3} = \dfrac{x^{21}}{y^6}$

6. a. $\dfrac{x}{\sqrt{5x}} \cdot \dfrac{\sqrt{5x}}{\sqrt{5x}} = \dfrac{x\sqrt{5x}}{5x} = \dfrac{\sqrt{5x}}{5}$

   b. $\sqrt{24a^2b} \cdot \sqrt{a^3b^4} = 2a\sqrt{6b} \cdot ab^2\sqrt{a}$
$$= 2a^2b^2\sqrt{6ab}$$

   c. $\dfrac{1-\sqrt{x}}{1+\sqrt{x}} \cdot \dfrac{1-\sqrt{x}}{1-\sqrt{x}} = \dfrac{1-2\sqrt{x}+x}{1-x}$

7. $2x^3 - 7x^5 - 5x - 8$

   a. Degree is 5.
   b. Constant is -8.
   c. Coefficient of x is -5.

8. In interval notation, $(-2, \infty) \cap (-\infty, 3] = (-2, 3]$

9. a. $8x^3 - 2x^2 = 2x^2(4x - 1)$

   b. $x^2 - 10x - 24 = (x - 12)(x + 2)$

   c. $6x^2 - 13x + 6 = (2x - 3)(3x - 2)$

   d. $2x^3 - 32x^5 = 2x^3(1 - 16x^2)$
$$= 2x^3(1 - 4x)(1 + 4x)$$

10. A quadratic polynomial has degree two. (c) is the quadratic.
$$4 - x - x^2 = 4 - (-3) - (-3)^2$$
$$= 4 + 3 - 9$$
$$= -2, \text{ when } x = -3$$

11.
$$
\begin{array}{r}
2x + 1 \\
x^2 - 1 \overline{)2x^3 + x^2 - 7} \\
\underline{2x^3 - 2x} \\
x^2 + 2x - 7 \\
\underline{x^2 - 1} \\
2x - 6
\end{array}
$$

Quotient: $2x + 1 + \dfrac{2x - 6}{x^2 - 1}$

12. a. $4y - 5(9 - 3y) = 4y - 45 + 15y = 19y - 45$

   b. $-3t^2(2t^4 - 3t^7) = -6t^6 + 9t^9$

   c.
$$
\begin{array}{r}
x^2 - 5x + 2 \\
4x - 1 \\
\hline
-x^2 + 5x - 2 \\
4x^3 - 20x^2 + 8x \\
\hline
4x^3 - 21x^2 + 13x - 2
\end{array}
$$

   d. $(6x - 1)(2 - 3x) = 12x - 18x^2 - 2 + 3x$
$$= -18x^2 + 15x - 2$$

   e. $(2m - 7)^2 = 4m^2 - 28m + 49$

   f. $\dfrac{x^6}{x^2 - 9} \cdot \dfrac{x - 3}{3x^2} = \dfrac{x^4}{(x+3)(x-3)} \cdot \dfrac{(x-3)}{3}$
$$= \dfrac{x^4}{3(x+3)}$$

   g. $\dfrac{x^4}{9} \div \dfrac{9x^3}{x^6} = \dfrac{x^4}{9} \cdot \dfrac{x^6}{9x^3} = \dfrac{x^7}{81}$

   h. $\dfrac{4}{x - 8} - \dfrac{x - 2}{x - 8} = \dfrac{4 - x + 2}{x - 8} = \dfrac{6 - x}{x - 8}$

i. $\dfrac{x-1}{x^2-2x-3}-\dfrac{3}{x^2-3x}$

$=\dfrac{x-1}{(x-3)(x+1)}-\dfrac{3}{x(x-3)}$

$=\dfrac{x(x-1)-3(x+1)}{x(x-3)(x+1)}$

$=\dfrac{x^2-x-3x-3}{x(x-3)(x+1)}$

$=\dfrac{x^2-4x-3}{x(x-3)(x+1)}$

13. $\dfrac{\frac{1}{x}-\frac{1}{y}}{\frac{1}{x}+y}\cdot\dfrac{xy}{xy}=\dfrac{y-x}{y+xy^2}$ or $\dfrac{y-x}{y(1+xy)}$

14. Construct a Venn diagram:

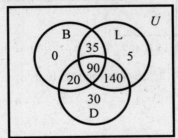

a. 0 students ate only breakfast.
b. $320-145=175$
175 students skipped breakfast.

15. $S=1000\left(1+\dfrac{0.08}{4}\right)^{4x}=1000\left(1+\dfrac{0.08}{4}\right)^{4(20)}$

$=1000(1+0.02)^{80}=1000(1.02)^{80}\approx 4875.44$

In 20 years, the future value will be about $4875.44.

Chapter 1: Linear Equations and Functions

Exercise 1.1

1.
$$4x - 7 = 8x + 2$$
$$4x - 7 + 7 - 8x = 8x + 2 + 7 - 8x$$
$$-4x = 9$$
$$x = -\frac{9}{4}$$

3.
$$x + 8 = 8(x + 1)$$
$$x + 8 = 8x + 8$$
$$x - 8x = 8 - 8$$
$$-7x = 0$$
$$x = 0$$

5.
$$-\frac{3x}{4} = 24$$
$$-3x = 4(24) = 96$$
$$x = -32$$

7.
$$2(x - 7) = 5(x + 3) - x$$
$$2x - 14 = 5x + 15 - x$$
$$2x - 5x + x = 15 + 14$$
$$-2x = 29$$
$$x = -\frac{29}{2}$$

9.
$$\frac{5x}{2} - 4 = \frac{2x - 7}{6}$$
$$6\left(\frac{5x}{2} - 4\right) = 6\left(\frac{2x - 7}{6}\right)$$
$$15x - 24 = 2x - 7$$
$$15x - 2x = 24 - 7$$
$$13x = 17$$
$$x = \frac{17}{13}$$

11.
$$x + \frac{1}{3} = 2\left(x - \frac{2}{3}\right) - 6x$$
$$x + \frac{1}{3} = 2x - \frac{4}{3} - 6x$$
$$3x + 1 = 6x - 4 - 18x$$
$$3x + 18x - 6x = -4 - 1$$
$$15x = -5$$
$$x = \frac{-5}{15} = -\frac{1}{3}$$

13.
$$(5x)\left(\frac{33 - x}{5x}\right) = 5x(2)$$
$$33 - x = 10x$$
$$-x - 10x = -33$$
$$-11x = -33$$
$$x = 3$$
Check: $\dfrac{33 - 3}{5(3)} \overset{?}{=} 2$
$$\frac{30}{15} \overset{?}{=} 2$$
$$2 = 2$$
$x = 3$ is the solution.

15.
$$\frac{2x}{2x + 5} = \frac{2}{3} - \frac{5}{2(2x + 5)}$$
Multiply each term by $6(2x + 5)$.
$$12x = (8x + 20) - 15$$
$$12x - 8x = 20 - 15$$
$$4x = 5 \text{ or } x = \frac{5}{4}$$
Check: $\dfrac{2\left(\frac{5}{4}\right)}{2\left(\frac{5}{4}\right) + 5} \overset{?}{=} \dfrac{2}{3} - \dfrac{5}{4\left(\frac{5}{4}\right) + 10}$
$$\frac{10}{10 + 20} \overset{?}{=} \frac{2}{3} - \frac{5}{15}$$
$$\frac{10}{30} = \frac{1}{3} \text{ and } \frac{2}{3} - \frac{5}{15} = \frac{1}{3}$$
$x = \dfrac{5}{4}$ is the solution.

17.
$$\frac{2x}{x - 1} + \frac{1}{3} = \frac{5}{6} + \frac{2}{x - 1}$$
$$\frac{2x - 2}{x - 1} + \frac{1}{3} = \frac{5}{6}$$
$$\frac{2(x - 1)}{x - 1} + \frac{1}{3} = \frac{5}{6}$$
$$2 + \frac{1}{3} \neq \frac{5}{6}$$
There is no solution.

19.
$$3.259x - 8.638 = -3.8(8.625x + 4.917)$$
$$3.259x - 8.638 = -32.775x - 18.6846$$
$$3.259x + 32.775x = 8.638 - 18.6846$$
$$36.034x = -10.0466$$
$$x = \frac{-10.0466}{36.034} \approx -0.279$$

21.
$$0.000316x + 9.18 = 2.1(3.1 - 0.0029x) - 4.68$$
$$0.000316x + 9.18 = 6.51 - 0.00609x - 4.68$$
$$0.000316x + 0.00609x = 6.51 - 4.68 - 9.18$$
$$0.006406x = -7.35$$
$$x = \frac{-7.35}{0.006406} \approx -1147.362$$

23.
$$3x - 4y = 15$$
$$-4y = -3x + 15$$
$$y = \frac{-3x}{-4} + \frac{15}{-4}$$
$$y = \frac{3}{4}x - \frac{15}{4}$$

25.
$$2\left(9x + \frac{3}{2}y\right) = 2(11)$$
$$18x + 3y = 22$$
$$3y = -18x + 22$$
$$y = -6x + \frac{22}{3}$$

27.
$$S = P + Prt$$
$$Prt = S - P$$
$$t = \frac{S - P}{Pr}$$

29.
$$3(x - 1) < 2x - 1$$
$$3x - 3 < 2x - 1$$
$$x - 3 < -1$$
$$x < 2$$

31.
$$1 - 2x > 9$$
$$-2x > 8$$
$$\left(-\tfrac{1}{2}\right)(-2x) > 8\left(-\tfrac{1}{2}\right)$$
$$x < -4$$

33.
$$\frac{3(x - 1)}{2} \leq x - 2$$
$$3(x - 1) \leq 2(x - 2)$$
$$3x - 3 \leq 2x - 4$$
$$x - 3 \leq -4$$
$$x \leq -1$$

35.
$$2(x - 1) - 3 > 4x + 1$$
$$2x - 2 - 3 > 4x + 1$$
$$2x - 5 > 4x + 1$$
$$-2x - 5 > 1$$
$$-2x > 6$$
$$\left(-\tfrac{1}{2}\right)(-2x) > 6\left(-\tfrac{1}{2}\right)$$
$$x < -3$$

(number line: -6 -5 -4 -3 -2 -1 0, closed point at -3 with ray to left)

37.
$$\frac{-3x}{2} > 3 - x$$
$$-3x > 2(3 - x)$$
$$-3x > 6 - 2x$$
$$-x > 6$$
$$(-1)(-x) > 6(-1)$$
$$x < -6$$

(number line: -12 -10 -8 -6 -4 -2 0, open point at -6 with ray to left)

39.
$$12\left(\frac{3x}{4} - \frac{1}{6}\right) < 12\left(x - \frac{2(x - 1)}{3}\right)$$
$$9x - 2 < 12x - 8(x - 1)$$
$$9x - 2 < 12x - 8x + 8$$
$$9x - 2 < 4x + 8$$
$$5x - 2 < 8$$
$$5x < 10$$
$$\left(\tfrac{1}{5}\right)(5x) < \left(\tfrac{1}{5}\right)(10)$$
$$x < 2$$

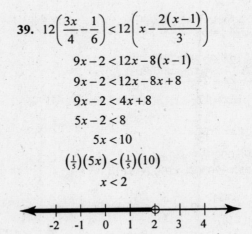

(number line: -2 -1 0 1 2 3 4, open point at 2 with ray to left)

41.
$$y = 648,000 - 1800x$$
$$387,000 = 648,000 - 1800x$$
$$1800x = 648,000 - 387,000 = 261,000$$
$$x = \frac{261,000}{1800} = 145 \text{ months}$$

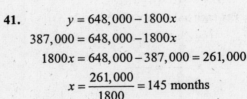

43. $\dfrac{I}{175.393} + 0.663 = r$

$\dfrac{I}{175.393} + 0.663 = 19.8$

$\dfrac{I}{175.393} = 19.8 - 0.663 = 19.137$

$I = 19.137(175.393)$

$I = \$3356.50$

45. $R = C$ for breakeven point

$20x = 2x + 7920$

$18x = 7920$

$x = 440$ packs or 220,000 CD's

47. $170,500 = 5.76x$

$x = \dfrac{170,500}{5.76} = \$29,600$

49. $959C - 1000I = 456.8$

$959C - 1000(75) = 456.8$

$959C - 75000 = 456.8$

$959C = 75456.8$

$C = 78.68\%$

51. $\dfrac{93 + 69 + 89 + 97 + FE + FE}{6} = 90$

$2FE + 348 = 540$

$2FE = 192$

$FE = 96$

A 96 is the lowest grade that can be earned on the final.

53. x = amount in safe fund

$120,000 - x$ = amount in risky fund

Yield: $0.09x + 0.13(120,000 - x) = 12,000$

$0.09x + 15,600 - 0.13x = 12,000$

$-0.04x = -3600$

$x = 90,000$

$x = \$90,000$ in 9% fund

$120,000 - 90000 = \$30,000$ in 13% fund.

55. Reduced salary: $2000 - 0.10(2000) = \$1800$

Increased salary: $1800 + 0.20(1800) = \$2160$

$160 = R\%$ of 2000

$R = \dfrac{160}{2000} = \dfrac{8}{100}$

$\$160$ is an 8% increase.

57. $40x > 20x + 1600$

$20x > 1600$

$x > 80$

59. $695 + 5.75t \le 900$

$5.75t \le 205$

$t \le 35.65$

He could buy 35 tapes.

61. a. $2005 - 1995 = 10$

b. $6.205 + 11.23t > 150$

$11.23t > 143.795$

$t > 12.8$

c. 2008

63. $A = 90.2 + 41.3h$

a. $90.2 + 41.3h \ge 110$

$41.3h \ge 19.8$

$h \ge 0.48$

b. $90.2 + 41.3h < 100$

$41.3h < 9.8$

$h < 0.24$

Exercise 1.2

1. a. For each value of x there is only one y.

b. $D = \{-7, -1, 0, 3, 4.2, 9, 11, 14, 18, 22\}$

$R = \{0, 1, 5, 9, 11, 22, 35, 60\}$

3. $f(0) = 1$, $f(11) = 35$

5. This is a function, since for each x there is only one y. $D = \{1, 2, 3, 8, 9\}$, $R = \{-4, 5, 16\}$

7. The vertical line test shows that graph (a) is a function of x, and that graph (b) is not a function of x.

9. If $y = 3x^3$, then y is a function of x.

11. If $y^2 = 3x$, then y is not a function. If, for example, $x = 3$, there are two possible values for y.

13. $R(x) = 8x - 10$
  a. $R(0) = 8(0) - 10 = -10$
  b. $R(2) = 8(2) - 10 = 6$
  c. $R(-3) = 8(-3) - 10 = -34$
  d. $R(1.6) = 8(1.6) - 10 = 2.8$

15. $C(x) = 4x^2 - 3$
  a. $C(0) = 4(0)^2 - 3 = -3$
  b. $C(-1) = 4(-1)^2 - 3 = 1$
  c. $C(-2) = 4(-2)^2 - 3 = 13$
  d. $C\left(-\dfrac{3}{2}\right) = 4\left(-\dfrac{3}{2}\right)^2 - 3 = 6$

17. $f(x) = x^3 - \dfrac{4}{x}$
  a. $f\left(-\dfrac{1}{2}\right) = \left(-\dfrac{1}{2}\right)^3 - \dfrac{4}{-\frac{1}{2}} = -\dfrac{1}{8} + 8 = \dfrac{63}{8}$
  b. $f(2) = 2^3 - \dfrac{4}{2} = 8 - 2 = 6$
  c. $f(-2) = (-2)^3 - \dfrac{4}{-2} = -8 + 2 = -6$

19. $f(x) = 1 + x + x^2$
  a. $f(2 + 1) = f(3) = 1 + 3 + 3^2 = 13$
  $f(2) + f(1) = 7 + 3 = 10$
  $f(2) + f(1) \neq f(2 + 1)$
  b. $f(x + h) = 1 + (x + h) + (x + h)^2$
  c. No. This is equivalent to a.
  d. $f(x) + h = 1 + x + x^2 + h$
  No. $f(x + h) \neq f(x) + h$
  e. $f(x + h) = 1 + (x + h) + (x + h)^2$
  $\qquad = 1 + x + h + x^2 + 2xh + h^2$
  $f(x) = 1 + x + x^2$
  $f(x + h) - f(x) = h + 2xh + h^2$
  $\qquad = h(1 + 2x + h)$
  $\dfrac{f(x + h) - f(x)}{h} = 1 + 2x + h$

21. $f(x) = x - 2x^2$
  a. $f(x + h) = (x + h) - 2(x + h)^2$
  $\qquad = -2x^2 - 4xh - 2h^2 + x + h$
  b. $f(x + h) - f(x)$
  $\qquad = (x + h) - 2(x + h)^2 - (x - 2x^2)$
  $\qquad = x + h - 2x^2 - 4xh - 2h^2 - x + 2x^2$
  $\qquad = h - 4xh - 2h^2$
  $\dfrac{f(x + h) - f(x)}{h} = \dfrac{h - 4xh - 2h^2}{h}$
  $\qquad = 1 - 4x - 2h$

23. Since $(9, 10)$ and $(5, 6)$ are points on the graph:
  a. $f(9) = 10$
  b. $f(5) = 6$

25. a. The ordered pair (a, b) satisfies the equation. Thus $b = a^2 - 4a$.
  b. The coordinates of $Q = (1, -3)$. Since the point is on the curve, the coordinates satisfy the equation.
  c. The coordinates of $R = (3, -3)$. They satisfy the equation.
  d. The x values are 0 and 4. These values are also solutions of $x^2 - 4x = 0$.

27. $y = x^2 + 4$
  There is no division by zero or square roots. Domain is all the reals, i.e., $\{x : x \in \text{Reals}\}$. Since $x^2 \geq 0$, $x^2 + 4 \geq 4$, the range is reals ≥ 4 or $\{y : y \geq 4\}$.

29. $y = \sqrt{x + 4}$
  There is no division by zero. To get a real number y, we must have $x + 4 \geq 0$ or $x \geq -4$. Domain: $x \geq -4$. The square root is always nonnegative. Thus, the range is $\{y : y \in \text{reals}, y \geq 0\}$.

31. $D : \{x : x \geq 1, x \neq 2\}$

33. $D : \{x : -7 \leq x \leq 7\}$

35. $f(x) = 3x$, $g(x) = x^3$
  a. $(f + g)(x) = 3x + x^3$
  b. $(f - g)(x) = 3x - x^3$
  c. $(f \cdot g)(x) = 3x \cdot x^3 = 3x^4$
  d. $\left(\dfrac{f}{g}\right)(x) = \dfrac{3x}{x^3} = \dfrac{3}{x^2}$

37. $f(x) = \sqrt{2x}$, $g(x) = x^2$

   a. $(f+g)(x) = \sqrt{2x} + x^2$

   b. $(f-g)(x) = \sqrt{2x} - x^2$

   c. $(f \cdot g)(x) = \sqrt{2x} \cdot x^2 = x^2\sqrt{2x}$

   d. $\left(\dfrac{f}{g}\right)(x) = \dfrac{\sqrt{2x}}{x^2}$

39. $f(x) = (x-1)^3$, $g(x) = 1 - 2x$

   a. $(f \circ g)(x) = f(1-2x) = (1-2x-1)^3 = -8x^3$

   b. $(g \circ f)(x) = g((x-1)^3) = 1 - 2(x-1)^3$

   c. $f(f(x)) = f((x-1)^3) = [(x-1)^3 - 1]^3$

   d. $(f \cdot f)(x) = (x-1)^3 \cdot (x-1)^3 = (x-1)^6$

      $\left[(f \cdot f)(x) \neq f(f(x))\right]$

41. $f(x) = 2\sqrt{x}$, $g(x) = x^4 + 5$

   a. $(f \circ g)(x) = f(x^4 + 5) = 2\sqrt{x^4 + 5}$

   b. $(g \circ f)(x) = g(2\sqrt{x}) = (2\sqrt{x})^4 + 5$

           $= 16x^2 + 5$

   c. $f(f(x)) = f(2\sqrt{x}) = 2\sqrt{2\sqrt{x}}$

   d. $(f \cdot f)(x) = 2\sqrt{x} \cdot 2\sqrt{x} = 4x$

      $\left[(f \cdot f)(x) \neq f(f(x))\right]$

43. a. $f(20) = 103{,}000$ means it will take 20 years to pay off a debt of \$103,000 (at \$800 per month and 7.5% compounded monthly.)

   b. $f(5+5) = f(10) = 69{,}000$;

      $f(5) + f(5) = 80{,}000$; No.

   c. It will take 15 years to pay off the debt, i.e., $89{,}000 = f(15)$.

45. a. $f(1950) = 16.5$ means that in 1950 there were 16.5 workers supporting each person receiving Social Security benefits.

   b. $f(1990) = 3.4$

   c. The points based on known data must be the same and those based on projections might be the same.

   d. Domain: $1950 \leq t \leq 2050$

      Range: $1.9 \leq n \leq 16.5$

47. a. $f(95) = 1{,}000{,}000$, $g(95) = 600{,}000$

   b. $f(100) = 1{,}200{,}000$. This was the number of prisoners in 2000.

   c. $g(90) = 500{,}000$. This was the number of parolees in 1990.

   d. $(f-g)(100) = 1{,}200{,}000 - 700{,}000$

      $= 500{,}000$. There were this many more in prison than were out on parole.

   e. $(f-g)(93) = 900{,}000 - 600{,}000 = 300{,}000$

      $(f-g)(98) = 1{,}100{,}000 - 600{,}000 = 500{,}000$

      $(f-g)(98)$ is greater. Possible reason is increased prison capacity but parolee level is constant.

49. a. Since the wind speed cannot be negative, $s \geq 0$.

   b. $f(10) = 45.694 + 1.75(10) - 29.26\sqrt{10} = -29.33$

      At a temperature of $-5°$F and a wind speed of 10 mph, the temperature feels like $-29.33°$F.

   c. $f(0) = 45.694$, but $f(0)$ should equal the air temperature, $-5°$F.

51. $C(x) = 300x + 0.1x^2 + 1200$

   a. $C(10) = 300(10) + 0.1(10)^2 + 1200$

      $= 3000 + 0.1(100) + 1200$

      $= 3000 + 10 + 1200$

      $= \$4210$

   b. The value $C(100)$ is the total cost of producing 100 units.

   c. $C(100) = 300(100) + 0.1(100)^2 + 1200$

      $= \$32{,}200$

53. $C(p) = \dfrac{7300p}{100 - p}$

   a. Domain: $\{p : 0 \leq p < 100\}$

   b. $C(45) = \dfrac{7300(45)}{100 - 45} = \dfrac{328{,}500}{55} = \$5972.73$

   c. $C(90) = \dfrac{7300(90)}{100 - 90} = \dfrac{657{,}000}{10} = \$65{,}700$

   d. $C(99) = \dfrac{7300(99)}{100 - 99} = \dfrac{722{,}700}{1} = \$722{,}700$

   e. $C(99.6) = \dfrac{7300(99.6)}{100 - 99.6} = \dfrac{727{,}080}{0.4} = \$1{,}817{,}700$

   In each case above, to remove $p\%$ of the particulate pollution would cost $C(p)$.

55. a. A is a function of x.

b. $A(2) = 2(50 - 2) = 96$ sq ft
$A(30) = 30(50 - 30) = 600$ sq ft

c. For the problem to have meaning we have $0 < x < 50$.

57. a. $P(q(t)) = P(1000 + 10t)$

$$= 180(1000 + 10t) - \frac{(1000 + 10t)^2}{100} - 200$$

$$= 169,800 + 1600t - t^2$$

b. $q(15) = 1000 + 10(15) = 1150$
$P(q(15)) = \$193,575$

59. $R = f(C)$ $C = g(A)$

a. $(f \circ g)(x) = f(C) = R$

b. $(g \circ f)(x)$ is not defined.

c. A is the independent variable and R is the dependent variable. Revenue depends on money spent for advertising.

61. length $= x$ width $= y$ $L = 2x + 2y$

$$1600 = xy \text{ or } y = \frac{1600}{x}$$

$$L = 2x + 2\left(\frac{1600}{x}\right) = 2x + \frac{3200}{x}$$

63. Revenue $=$ (no. of people)(price per person)
Example: $R = 30 \times 10$

$$R = 31 \times 9.80$$

$$R = 32 \times 9.60$$

Solution: $R = (30 + x)(10 - 0.20x)$

Exercise 1.3

1. $3x + 4y = 12$

x-intercept: $y = 0$ then $x = 4$.
y-intercept: $x = 0$ then $y = 3$.

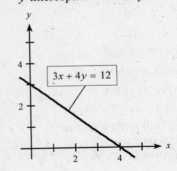

3. $5x - 8y = 60$

x-intercept: $y = 0$ then $x = 12$.
y-intercept: $x = 0$ then $y = -7.5$.

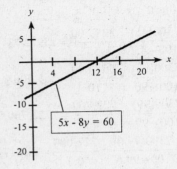

5. $3x + 2y = 0$

x-intercept: $y = 0$ then $x = 0$
Likewise, y-intercept is $y = 0$.

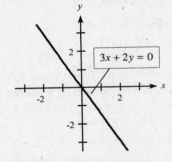

7. $(22, 11)$ and $(15, -17)$

$$m = \frac{y_2 - y_1}{x_2 - x_1} = \frac{-17 - 11}{15 - 22} = \frac{-28}{-7} = 4$$

9. $(3, 2)$ and $(-1, 2)$

$$m = \frac{y_2 - y_1}{x_2 - x_1} = \frac{2 - 2}{3 - (-1)} = \frac{0}{4} = 0$$

11. A horizontal line has a slope of 0.

13. $(3, 2)$ and $(-1, 2)$

The rate of change is equivalent to the slope of the line.

$$m = \frac{y_2 - y_1}{x_2 - x_1} = \frac{2 - 2}{-1 - 3} = \frac{0}{-4} = 0$$

15. a. Slope is negative.

b. Slope is undefined.

17. $y = \frac{7}{3}x - \frac{1}{4}$, $m = \frac{7}{3}$, $b = -\frac{1}{4}$

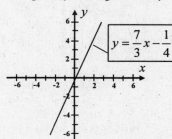

19. $y = 3$ or $y = 0x + 3$, $m = 0$, $b = 3$

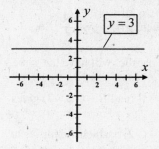

21. $x = -8$
Slope is undefined.
There is no y-intercept.

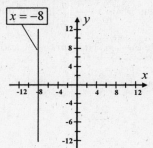

23. $2x + 3y = 6$ or $y = -\frac{2}{3}x + 2$, $m = -\frac{2}{3}$, $b = 2$.

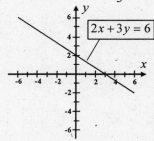

25. $m = \frac{1}{2}$, $b = -3$

$y = \frac{1}{2}x - 3$

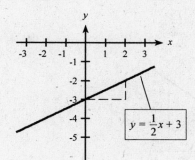

27. $m = -2$, $b = \frac{1}{2}$

$y = -2x + \frac{1}{2}$

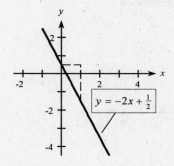

29. $P(2, 0)$, $m = -5$
$y - 0 = -5(x - 2)$
$y = -5x + 10$

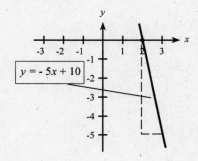

31. $P(-1, 4)$, $m = \dfrac{3}{4}$

$$y - 4 = \frac{3}{4}(x - (-1))$$

$$y = \frac{3}{4}x + \frac{19}{4}$$

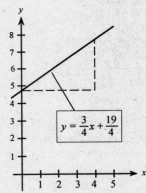

$y = \frac{3}{4}x + \frac{19}{4}$

33. $P(-1, 1)$, m is undefined

$x = -1$

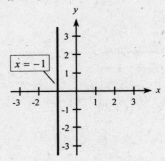

$x = -1$

35. $P_1 = (3, 2)$, $P_2 = (-1, -6)$

$$m = \frac{-6 - 2}{-1 - 3} = 2$$

$$y - 2 = 2(x - 3)$$

$$y = 2x - 4$$

37. $P_1 = (7, 3)$, $P_2 = (-6, 2)$

$$m = \frac{2 - 3}{-6 - 7} = \frac{-1}{-13} = \frac{1}{13}$$

$$y - 3 = \frac{1}{13}(x - 7)$$

$$y = \frac{1}{13}x - \frac{7}{13} + 3$$

$$y = \frac{1}{13}x + \frac{32}{13} \text{ or } -x + 13y = 32$$

39. $3x + 2y = 6$ $\qquad$ $2x - 3y = 6$

$\quad y = -\dfrac{3}{2}x + 3$ $\qquad$ $y = \dfrac{2}{3}x - 2$

Lines are perpendicular since $\left(-\dfrac{3}{2}\right)\left(\dfrac{2}{3}\right) = -1$.

41. $6x - 4y = 12$ $\qquad$ $3x - 2y = 6$

$\quad y = \dfrac{6}{4}x - \dfrac{12}{4}$ $\qquad$ $y = \dfrac{3}{2}x - 3$

or $y = \dfrac{3}{2}x - 3$

Lines are the same.

43. If $3x + 5y = 11$, then $y = -\dfrac{3}{5}x + \dfrac{11}{5}$. So,

$m = -\dfrac{3}{5}$. A line parallel will have the same

slope. Thus, $m = -\dfrac{3}{5}$ and $P = (-2, -7)$ gives

$y - (-7) = -\dfrac{3}{5}(x - (-2))$ which simplifies to

$y = -\dfrac{3}{5}x - \dfrac{41}{5}$.

45. If $5x - 6y = 4$, then $y = \dfrac{5}{6}x - \dfrac{4}{6}$. Slope of the

perpendicular line is $-\dfrac{6}{5}$. Thus $m = -\dfrac{6}{5}$ and

$P = (3, 1)$ gives $y - 1 = -\dfrac{6}{5}(x - 3)$ which

simplifies to $y = -\dfrac{6}{5}x + \dfrac{23}{5}$.

47. a.

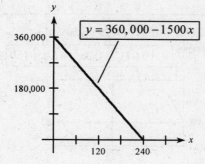

$y = 360,000 - 1500x$

b. $0 = 360,000 - 1500x$

$x = \dfrac{360000}{1500} = 240$ months

In 240 months, the building will be completely depreciated.

c.   (60, 270,000) means that after 60 months the value of the building will be $270,000.

49. $y = 5.74x + 14.61$

a. $m = 5.74,\ b = 14.61$

b.   In 1995, when $x = 0$, 14.61% of the U.S. population had Internet.

c.   The percentage of the U.S. population with Internet is changing at the rate of 5.74% per year.

51. $M(x) = -0.762x + 85.284$

a. $m = -0.762\quad b = 85.284$

b.   In 1950, 85% of the unmarried women became married.

c.   The annual rate of change is -0.762%. For each passing year the percent of unmarried women who get married decreases by 0.762%.

53. $F = 0.551M + 6.762$

a. $m = 0.551$

b.   For each $1 increase in male earnings, the female's earning increases only by $0.551.

c. $F(40) = 0.551(40) + 6.762$

$= 28.802$ thousands

$= \$28,802$

55. $y = 0.0838x + 4.95$ Both units are in dollars.

57. a. $m = \dfrac{61.90}{100} = 0.619$ and $(676, 527)$

$B - 527 = 0.6190(W - 676)$

$B = 0.6190W + 108.556$

b. $B(850) = 0.6190(850) + 108.556$

$= \$634.71$

59. a. $(1980, 82.4)$ and $(2002, 179.9)$

$m = \dfrac{179.9 - 82.4}{2002 - 1980} = \dfrac{97.5}{22} = 4.43$

$y - 82.4 = 4.43(x - 1980)$

$y = 4.43x - 8689$

b.   The consumer price index for urban consumers increases at the rate of $4.43 per year.

61. (x, p) is the reference. (0, 85000) is one point.

$m = \dfrac{-1700}{1} = -1700$

$p - 85,000 = -1700(x - 0)$ or

$p = -1700x + 85,000$

63. (t, R) is the ordered pair.

$P_1 = \left(\dfrac{7}{2}, 11\right),\ P_2 = (6, 19)$

$m = \dfrac{19 - 11}{6 - \frac{7}{2}} = \dfrac{8}{\frac{5}{2}} = \dfrac{16}{5} = 3.2$

$R - 19 = 3.2(t - 6)$ or $R = 3.2t - 19.2 + 19$ or

$R = 3.2t - 0.2$

65. $P_1 = (200, 25)\ P_2 = (250, 49)$

$m = \dfrac{49 - 25}{250 - 200} = \dfrac{24}{50} = 0.48$

$y - 25 = 0.48(x - 200)$ or $y = 0.48x - 71$

Exercise 1.4

1.

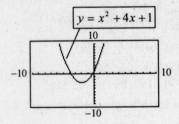

$y = x^2 + 4x + 1$

3.

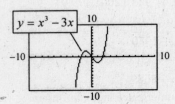

$y = x^3 - 3x$

5.

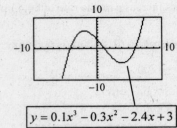

$y = 0.1x^3 - 0.3x^2 - 2.4x + 3$

7.

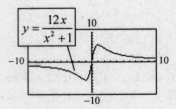

$y = \dfrac{12x}{x^2 + 1}$

9.

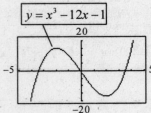

$y = x^3 - 12x - 1$

11.

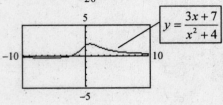

$y = \dfrac{3x + 7}{x^2 + 4}$

13. $y = 0.01x^3 + 0.3x^2 - 72x + 150$

a.

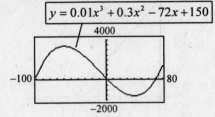

$y = 0.01x^3 + 0.3x^2 - 72x + 150$

b.

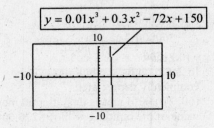

$y = 0.01x^3 + 0.3x^2 - 72x + 150$

15. a.

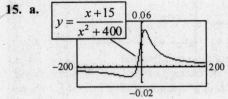

$y = \dfrac{x + 15}{x^2 + 400}$

b.   Standard Window

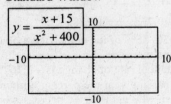

$y = \dfrac{x + 15}{x^2 + 400}$

17. a. y-intercept $= -0.03$

x-intercept: $0.001x = 0.03$

$x = 30$

b.

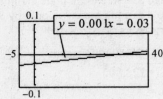

$y = 0.001x - 0.03$

19 – 21. **Complete graphs can be seen with different windows. A hint is to look at the equation and try to determine the max and/or min of y. Also, find the x-intercepts.**

19. $y = -0.15(x - 10.2)^2 + 10$

There is no min.
Max value of $y = 10$.
x-intercepts:

$$0 = -0.15(x - 10.2)^2 + 10$$

$$(x - 10.2)^2 = \frac{10}{0.15} = 66.66$$

$$x - 10.2 = \pm\sqrt{66.66} \approx \pm 8$$

$x = 10.2 \pm 8$ or $x = 2.2$ or 18.2

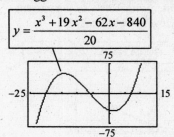

21. If $x = 0$, $y = -42$.
A suggested window is shown below.

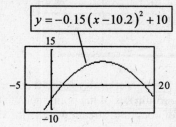

23. $4x - y = 8$

$$y = 4x - 8$$

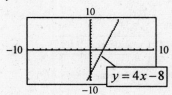

25. $4x^2 + 2y = 5$

$$2y = -4x^2 + 5$$

$$y = -2x^2 + \frac{5}{2}$$

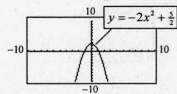

27. $x^2 + 2y = 6$

$$2y = -x^2 + 6$$

$$y = -\frac{1}{2}x^2 + 3$$

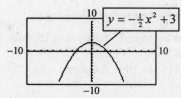

29. $f(x) = x^3 - 3x^2 + 2$

$$f(-2) = (-2)^3 - 3(-2)^2 + 2 = -8 - 12 + 2 = -18$$

$$f\left(\frac{3}{4}\right) = \left(\frac{3}{4}\right)^3 - 3\left(\frac{3}{4}\right)^2 + 2 = 0.734375$$

Use your graphing calculator, and evaluate the function at these two points. If either of your answers differ, can you explain the difference?

31. As x gets large, y approaches 12.
When $x = 0$, $y = -12$, x intercepts at ± 1.

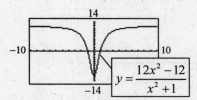

33. $y = \dfrac{x^2 - x - 6}{x^2 + 5x + 6} = \dfrac{(x - 3)(x + 2)}{(x + 3)(x + 2)}$

What happens to y as x approaches -3? -2?

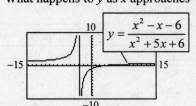

35. $6x - 21 = 0$

$$6x = 21$$

$$x = \frac{21}{6} = \frac{7}{2}$$

37. $x^2 - 3x - 10 = 0$

$$(x - 5)(x + 2) = 0$$

$$x = -2 \text{ or } 5$$

39. *Find the zeros* and *find the x-intercepts* are equivalent statements. Use a graphing calculator's **TRACE** or **ZERO**.
a.-b. Graphing calculator approximation is $x = -1.1098$, 8.1098.

41. a.

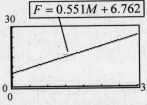

$$F = 0.551M + 6.762$$

b. If males average $50,000, then females earn $34,312.

c. $F(62.5) = 41.1995$ thousand $= $41,200$

43. a.

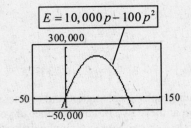

$$E = 10,000p - 100p^2$$

b. $E \geq 0$ when $0 \leq p \leq 100$.

45. a.

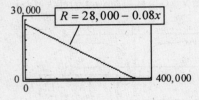

$$R = 28,000 - 0.08x$$

b. The rate is -0.08. As more people become aware of the product, there are fewer to learn about it.

47. a. x-min $= 0$, x-max $= 120$

b. y-min $= 0$, y-max $= 70$

c.

$$p = 0.00008x^3 - 0.0128x^2 - 0.0032x + 66.6$$

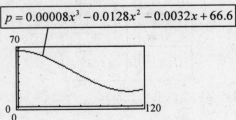

d.

$$p = 0.00008x^3 - 0.0128x^2 - 0.0032x + 66.6$$

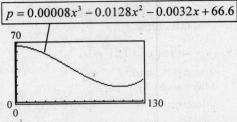

e. The percentage decreases before 2000 and increases after 2000.

49. a.

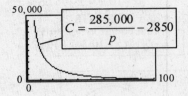

$$C = \frac{285,000}{p} - 2850$$

b. Near $p = 0$, cost grows without bound.

c. The coordinates of the point mean that the cost of obtaining stream water with 1% of the current pollution levels would cost $282,150.

d. The p-intercept means that the cost of stream water with 100% of the current pollution levels would cost $0.

51. a.

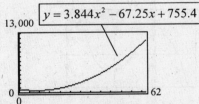

$$y = 3.844x^2 - 67.25x + 755.4$$

b. Increasing. The per capita federal tax burden is increasing.

Exercise 1.5

1. Solution: $(-1, -2)$

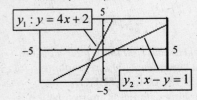

$y_1 : y = 4x + 2$

$y_2 : x - y = 1$

3. Infinitely many solutions.

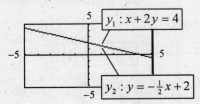

$y_1 : x + 2y = 4$

$y_2 : y = -\frac{1}{2}x + 2$

5.   Solution: $(2,2)$

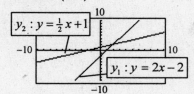

7.   Solution: No solution since the graphs do not intersect.

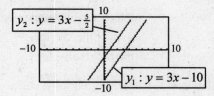

9. $3x - 2y = 6$

$4y = 8$ Solve for y. $y = \frac{8}{4} = 2$

Substitute for this variable in $3x - 2(2) = 6$

first equation and solve for $3x = 6 + 4 = 10$

the other variable. $x = \frac{10}{3}$

The solution of the system is $x = \dfrac{10}{3}$ and $y = 2$, or $\left(\dfrac{10}{3}, 2\right)$.

11. $2x - y = 2$

$3x + 4y = 6$ Solve for y. $y = 2x - 2$

Substitute for this variable in $3x + 4(2x - 2) = 6$

second equation and solve for $3x + 8x - 8 = 6$

the other variable. $11x = 14$

$x = 14/11$

Solve for y: $y = 2\left(\dfrac{14}{11}\right) - 2 = \dfrac{6}{11}$

The solution of the system is $x = \dfrac{14}{11}$ and $y = \dfrac{6}{11}$, or $\left(\dfrac{14}{11}, \dfrac{6}{11}\right)$.

13. $3x + 4y = 1$ Multiply 1st equation by 3. $9x + 12y = 3$

$2x - 3y = 12$ Multiply 2nd equation by 4. $\underline{8x - 12y = 48}$

Add the two equations. $17x \quad\quad = 51$

Solve for the variable. $x = 3$

Substitute for this variable in $3(3) + 4y = 1$

either original equation and $4y = -8$

solve for the other variable. $y = -2$

The solution of the system is $x = 3$ and $y = -2$, or $(3, -2)$.

15. $-4x + 3y = -5$ Multiply first equation by 3. $-12x + 9y = -15$

$3x - 2y = 4$ Multiply second equation by 4. $\underline{12x - 8y = 16}$

Add the two equations. $y = 1$

Substitute for this variable in $-4x + 3(1) = -5$

either original equation and $-4x = -8$

solve for the other variable. $x = 2$

The solution of the system is $x = 2$ and $y = 1$.

17. $0.2x - 0.3y = 4$ $0.20x - 0.3y = 4$

   $2.3x - \ \ y = 1.2$ Multiply 2nd equation by 0.3. $\underline{0.69x - 0.3y = 0.36}$

   Subtract the two equations. $-0.49x \qquad = 3.64$

   Solve for the variable. $x = -\frac{52}{7}$

   Substitute, solve for y. $y = -\frac{128}{7}$

   The solution of the system is $x = -\dfrac{52}{7}$ and $y = -\dfrac{128}{7}$, or $\left(-\dfrac{52}{7}, -\dfrac{128}{7}\right)$.

19. $\frac{5}{2}x - \frac{7}{2}y = -1$ Multiply first equation by 6. $15x - 21y = -6$

   $8x + 3y = 11$ Multiply second equation by 7. $\underline{56x + 21y = 77}$

   Add the two equations. $71x \qquad = 71$

   Substitute for this variable in $x = 1$

   either original equation and $8(1) + 3y = 11$

   solve for the other variable. $3y = 3$

   $y = 1$

   The solution of the system is $x = 1$ and $y = 1$, or $(1, 1)$.

21. $4x + 6y = 4$ $4x + 6y = 4$

   $2x + 3y = 2$ Multiply second equation by -2. $\underline{-4x - 6y = -4}$

   Add the two equations: $0 = 0$

   There are infinitely many solutions. The system is dependent. Solve for one of the variables in terms of the

   remaining variable: $y = \dfrac{2}{3} - \dfrac{2}{3}x$. Then a general solution is $\left(c, \dfrac{2}{3} - \dfrac{2}{3}c\right)$, where any value of c will give a

   particular solution.

23–25 Use the standard window and graph each equation. Use the TRACE or INTERSECT feature to find
   the solution.

23. $\begin{cases} y = 8 - \dfrac{3x}{2} \\ y = \dfrac{3x}{4} - 1 \end{cases}$

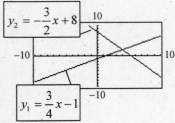

Solution: $(4, 2)$

25. $\begin{cases} y_1 : 5x + 3y = -2 \\ y_2 : 3x + 7y = 4 \end{cases}$

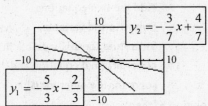

Solution: $(-1, 1)$

27. Eq. 1 $x + 2y + z = 2$ Steps 1, 2, and 3 of the systematic

     Eq. 2 $-y + 3z = 8$ procedure are completed.

     Eq. 3 $2z = 10$ Step 4: $z = 5$

        From Eq. 2 $-y + 3(5) = 8$ or $y = 7$

        From Eq. 1 $x + 2(7) + 5 = 2$ or $x = -17$

   The solution is $x = -17,\ y = 7,\ z = 5$.

29. Eq. 1 $x - y - 8z = 0$ Steps 1 and 2 of the systematic

     Eq. 2 $y + 4z = 8$ procedure are completed.

     Eq. 3 $3y + 14z = 22$

        Step 3: $(-3) \times$ Eq 2 added to Eq. 3 gives $2z = -2$ or $z = -1$.

        From Eq. 2 $y + 4(-1) = 8$ or $y = 12$

        From Eq. 1 $x - 12 - 8(-1) = 0$ or $x = 4$

   The solution is $x = 4,\ y = 12,\ z = -1$ or $(4,\ 12,\ -1)$.

31. Eq. 1 $x + 4y - 2z = 9$ Step 1 is completed.

     Eq. 2 $x + 5y + 2z = -2$

     Eq. 3 $x + 4y - 28z = 22$

   Step 2:

         $x + 4y - 2z = 9$ Eq. 1

   Eq. 4 $y + 4z = -11$ $(-1) \times$ Eq. 1 added to Eq. 2

   Eq. 5 $-26z = 13$ $(-1) \times$ Eq. 1 added to Eq. 3

   Step 3 is also completed.

   Step 4: $z = -\dfrac{1}{2}$ from Eq. 5.

        From Eq. 4 $y + 4\left(-\dfrac{1}{2}\right) = -11$ or $y = -9$

        From Eq. 1 $x + 4(-9) - 2\left(-\dfrac{1}{2}\right) = 9$ or $x = 44$

   The solution is $x = 44,\ y = -9,\ z = -\dfrac{1}{2}$ or $\left(44,\ -9,\ -\dfrac{1}{2}\right)$.

33. $f(x) = 40.74x + 742.65,\ h(x) = 47.93x + 725$

   $40.74x + 742.65 = 47.93x + 725$

         $17.65 = 7.19x$

           $x = 2.455$ during 1998

   $h(2.455) = f(2.455) = \$842.67$ billion

35. a. $x + y = 1800$ Total number of tickets

b. $20x =$ revenue from \$20 tickets

c. $30y =$ revenue from \$30 tickets

d. $20x + 30y = 42,000$ Total Revenue

e.   Multiply equation from part (a) by -20.

$$-20x - 20y = -36000$$
$$\underline{20x + 30y = 42000}$$
$$10y = 6000$$
$$y = 600$$

Substitution into equation from part (a) gives $x = 1200$.
Sell 1200 of the \$20 tickets and 600 of the \$30 tickets.

37.   x = amount of safe investment.
y = amount of risky investment.
$x + y = 145,600$ Total amount invested
$0.1x + 0.18y = 20,000$ Income from investments
The solution is the solution of the above system of equations.

$$x + \ \ y = 145,600$$
$$\underline{x + 1.8y = 200,000} \quad (10) \times \text{ second equation}$$
$$0.8y = 54,400 \quad \text{Subtract equations}$$
$$y = 68,000 \quad \text{Solve for } y \text{ or amount of risky investment.}$$

Substituting $y = 68,000$ into one of the original equations we have $x + 68,000 = 145,600$ or $x = \$77,600$.
Solution: Put \$77,600 in a safe investment and \$68,000 in a risky investment.

39. $x =$ amount invested at 10%.
$y =$ amount invested at 12%.
$x + y = 235,000$ Total amount invested
$0.10x + 0.12y = 235,000$ Investment income
Solve the system of equation:

$$x + \ \ y = 235,000$$
$$\underline{x + 1.2y = 25,500} \quad (10) \times \text{ second equation}$$
$$0.2y = 20,000 \quad \text{Subtract equations}$$
$$y = 100,000 \quad \text{Solve for } y$$

Substituting into the first equation we have $x + 100,000 = 235,000$ or $x = 135,000$.
Thus, \$135,000 is invested at 10% and \$100,000 is invested at 12%.

41. A = ounces of substance A.

B = ounces of substance B.

Required ratio $\dfrac{A}{B} = \dfrac{3}{5}$ gives $5A - 3B = 0$.

Required nutrition is $5\%A + 12\%B = 100\%$. This gives $5A + 12B = 100$.

The % notation can be trouble. Be careful! Now we can solve the system.

$$5A - 3B = 0$$
$$\underline{5A + 12B = 100}$$
$$\qquad\quad 15B = 100 \qquad \text{Subtract first equation from second.}$$
$$B = \dfrac{100}{15} = \dfrac{20}{3}$$

Substituting into the original equation gives $5A - 3\left(\dfrac{20}{3}\right) = 0$ or $A = 4$.

The solution is 4 ounces of substance A and $6\dfrac{2}{3}$ ounces of substance B.

43. x = population of species A.

y = population of species B.

$2x + y = 10{,}600$ \qquad units of first nutrient

$3x + 4y = 19{,}650$ \qquad units of second nutrient

$$8x + 4y = 42{,}400 \quad (4) \times \text{ first equation}$$
$$\underline{3x + 4y = 19{,}650}$$
$$5x \quad\;\; = 22{,}750 \qquad \text{Subtract}$$
$$x \quad\;\; = 4550 \qquad \text{Solve for } x$$

Substituting $x = 4550$ into an original equation we have $2(4550) + y = 10{,}600$.

So, $y = 1500$. Solution is 4550 of species A and 1500 of species B.

45. x = amount of 20% concentration.

y = amount of 5% concentration.

$x + y = 10$ \qquad amount of solution

$0.20x + 0.05y = 0.155(10)$ \qquad concentration of medicine

Solving this system of equations:

$$x + \quad y = 10$$
$$\underline{x + 0.25y = 7.75} \quad (5) \times \text{ second equation}$$
$$0.75y = 2.25 \quad \text{Subtract equations}$$
$$y = 3 \qquad \text{Solve for } y$$

Substituting into the first equation we have $x + 3 = 10$ or $x = 7$.

The solution is 3 cc of 5% concentration and 7 cc of 20% concentration.

47. x = number of \$20 tickets.

y = number of \$30 tickets.

$x + y = 16{,}000$ \qquad total number of tickets

$20x + 30y = 380{,}000$ \qquad total revenue

To solve the system of equations multiply Eq. 1 by 30.

$$30x + 30y = 480{,}000$$
$$\underline{20x + 30y = 380{,}000}$$
$$10x \quad\;\; = 100{,}000 \text{ or } x = 10{,}000$$

Substituting into the first equation gives $y = 6000$.

Sell 10,000 tickets for \$20 each and 6000 tickets for \$30 each.

49. $x =$ amount of 20% solution to be added.
$0.20x =$ concentration of nutrient in 20% solution.
$0.02(100) = 2$ is the concentration of nutrient in 2% solution.

$$0.20x + 2 = 0.10(x + 100)$$

$$0.20x + 2 = 0.10x + 10$$

$$0.1x = 8 \text{ or } x = 80 \text{ cc of 20% solution is needed.}$$

51. $x =$ ounces of substance A,
$y =$ ounces of substance B, and
$z =$ ounces of substance C.

$5x + 15y + 12z = 100$ Nutrition requirements

$x = z$ Digestive restrictions

$y = \frac{1}{5}z$ Digestive restrictions

Since both x and y are in terms of z, we can substitute in the first equation and solve for z.
So, $5z + 3z + 12z = 100$ or $20z = 100$. So, $z = 5$. Now, since $x = z$, we have $x = 5$.

Since $y = \frac{1}{5}z$, we have $y = 1$. The solution is 5 ounces of substance A, 1 ounce of substance B, and

5 ounces of substance C.

53. $A =$ number of A type clients.
$B =$ number of B type clients.
$C =$ number of C type clients.

$A + B + C = 500$ Total clients
$200A + 500B + 300C = 150{,}000$ Counseling costs
$300A + 200B + 100C = 100{,}000$ Food and shelter

To find the solution we must solve the system of equations.

Eq. 1 $A + B + C = 500$

Eq. 2 $2A + 5B + 3C = 1500$ Original equation divided by 100

Eq. 3 $3A + 2B + C = 1000$ Original equation divided by 100

          $A + B + C = 500$ Eq. 1

Eq. 4 $3B + C = 500$ $(-2)\times$ Eq. 1 added to Eq. 2

Eq. 5 $-B - 2C = -500$ $(-3)\times$ Eq. 1 added to Eq. 3

         $A + B + C = 500$ Eq. 1

            $3B + C = 500$ Eq. 4

          $-\frac{5}{3}C = \frac{-1000}{3}$ $\frac{1}{3}\times$ Eq. 4 added to Eq. 5

             $C = \frac{1000}{3} \cdot \frac{3}{5} = 200$

Substituting $C = 200$ into Eq. 4 gives $3B + 200 = 500$ or $3B = 300$. So, $B = 100$.
Substituting $C = 200$ and $B = 100$ into Eq. 1 gives $A + 100 + 200 = 500$. So, $A = 200$.
Thus, the solution is 200 type A clients, 100 type B clients, and 200 type C clients.

Exercise 1.6

1. **a.** $P(x) = R(x) - C(x)$

$$= 34x - (17x + 3400)$$

$$= 17x - 3400$$

b. $P(300) = 17(300) - 3400 = \$1700$

3. **a.** $P(x) = R(x) - C(x)$

$$= 80x - (43x + 1850)$$

$$= 37x - 1850$$

b. $P(30) = 37(30) - 1850 = -\$740$

The total costs are more than the revenue.

c. $P(x) = 0$ or $37x - 1850 = 0$

So, $x = \dfrac{1850}{37} = 50$ units is the break-even point.

5. $C(x) = 5x + 250$

a. $m = 5$, C-intercept: 250

b. $\overline{MC} = 5$ means that each additional unit produced costs \$5.

c. Slope = marginal cost. C-intercept = fixed costs.

d. \$5, \$5 ($\overline{MC} = 5$ at every point)

7. $R = 27x$

a. $m = 27$

b. 27; each additional unit sold yields \$27 in revenue.

c. In each case, one more unit yields \$27.

9. $R(x) = 27x$, $C(x) = 5x + 250$

a. $P(x) = 27x - (5x + 250) = 22x - 250$

b. $m = 22$

c. Marginal profit is 22.

d. Each additional unit sold gives a profit of \$22. To maximize profit sell all that you can produce. Note that this is not always true.

11. (x, P) is the correct form.

$P_1 = (200, 3100)$

$P_2 = (250, 6000)$

$m = \dfrac{6000 - 3100}{250 - 200} = 58$

$P - 3100 = 58(x - 200)$ or $P = 58x - 8500$

The marginal profit is 58.

13. **a.** $TC = 35H + 6600$

b. $TR = 60H$

c. $P = R - C$

$$= 60H - (35H + 6600)$$

$$= 25H - 6600$$

d. $C(200) = 35(200) + 6600$

$$= \$13,600 \text{ cost of 200 helmets}$$

$R(200) = 60(200)$

$$= \$12,000 \text{ revenue from 200 helmets}$$

$P(200) = R(200) - C(200)$

$$= \$12,000 - 13,600$$

$$= -\$1600 \text{ loss from 200 helmets}$$

e. $C(300) = 35(300) + 6600$

$$= \$17,100 \text{ cost of 300 helmets}$$

$R(300) = 60(300)$

$$= \$18,000 \text{ revenue from 300 helmets}$$

$P(300) = R(300) - C(300)$

$$= 18,000 - 17,100$$

$$= \$900 \text{ profit from 300 helmets}$$

f. The marginal profit is \$25. Each additional helmet sold gives a profit of \$25.

15. **a.** The revenue function is the graph that passes through the origin.

b. At a production of zero the fixed costs are \$2000.

c. From the graph, the break-even point is 400 units and \$3000 in revenue or costs.

d. Marginal cost $= \dfrac{3000 - 2000}{400 - 0} = 2.5$

Marginal revenue $= \dfrac{3000 - 0}{400 - 0} = 7.5$

17. $R(x) = C(x) = 85x = 35x + 1650$ or $50x = 1650$ or $x = 33$.

Thus, 33 necklaces must be sold to break even.

19. **a.** $R(x) = 12x$, $C(x) = 8x + 1600$

b. $R(x) = C(x)$ if $12x = 8x + 1600$ or $4x = 1600$ or $x = 400$.

It takes 400 units to break even.

21. **a.** $P(x) = R(x) - C(x)$

$$= 12x - (8x + 1600)$$

$$= 4x - 1600$$

b. By setting $P(x) = 0$ we get $x = 400$. Same as 19(b).

23. a. $TC = 4.50x + 1045$

  b. $TR = 10x$

  c. $P = R - C$

$$= 10x - (4.50x + 1045)$$

$$= 5.50x - 1045$$

  d.   Breakeven also means $P = 0$.

$$5.50x - 1045 = 0$$

$$5.50x = 1{,}045$$

$$x = 190 \text{ units to break even}$$

25. a. $R(x) = 54.90x$

  b. $P_1 = (2000, \ 50000)$

$$P_2 = (800, \ 32120)$$

$$m = \frac{32{,}120 - 50{,}000}{800 - 2000} = \frac{-17{,}880}{-1200} = 14.90$$

$$y - 50{,}000 = 14.90(x - 2000) \text{ or}$$

$$y = 14.90x + 20{,}200 = C(x)$$

  c.   From $54.90x = 14.90x + 20{,}200$ we have $x = 505$ units to break even.

27. a.

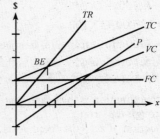

  b.   *TR* starts at the origin and intersects *TC* at the break-even (*BE*). *FC* is a horizontal line from the vertical intercept of *TC*. *VC* starts at the origin and is parallel to *TC*.

29.   If price increases, then the demand for the product decreases.

31. a.   If $p = \$100$, then $q = 600$ (approximately).

  b.   If $p = \$100$, then $q = 300$.

  c.   There is a shortage since more is demanded.

33.   Demand: $2p + 5q = 200$

$$2(60) + 5q = 200$$

$$5q = 80$$

$$q = 16$$

Supply: $p - 2q = 10$

$$60 - 2q = 10$$

$$2q = 50$$

$$q = 25$$

There will be a surplus of 9 units at a price of $\$60.00$.

35.   Remember that (q, p) is the correct form.

$$P_1 = (240, \ 900)$$

$$P_2 = (315, \ 850)$$

$$m = \frac{850 - 900}{315 - 240} = -\frac{50}{75} = -\frac{2}{3}$$

Note: $m < 0$ for demand equations.

$$p - 900 = -\frac{2}{3}(q - 240) \text{ or}$$

$$p = -\frac{2}{3}q + 1060$$

37. (q, p) is the correct form.

$$P_1 = (10000, \ 1.50)$$

$$P_2 = (5000, \ 1.00)$$

$$m = \frac{1 - 1.50}{5000 - 10000} = \frac{-0.50}{-5000} = 0.0001$$

Note: $m > 0$ for supply equations.

$$p - 1 = 0.0001(q - 5000) \text{ or } p = 0.0001q + 0.5$$

39. a.   The decreasing function is the demand curve. The increasing function is the supply curve.

  b.   Reading the graph, we have equilibrium at $q = 30$ and $p = 25$.

41. a.   Reading the graph, at $p = 20$ we have 20 units supplied.

  b.   Reading the graph, at $p = 20$ we have 40 units demanded.

  c.   At $p = 20$ there is a shortage of 20 units.

43.   By observing the graph in the figure, we see that a price below the equilibrium price results in a shortage.

45. $-\frac{1}{2}q+28=\frac{1}{3}q+\frac{34}{3}$ Required condition.

$-3q+168=2q+68$ Multiply both sides by 6 to simplify.

$-5q=-100$

$q=20$

Substituting into one of the original equations gives $p=-\frac{1}{2}(20)+28=18.$

Thus, the equilibrium point is $(q,p)=(20,18).$

47. $-4q+220=15q+30$ Required condition.

$190=19q$

$q=10$ Solve for q.

Substituting $q=10$ into one of the original equations gives $p=180.$
Thus, the equilibrium point is $(q,p)=(10,180).$

49. Demand: $(80,350)$ and $(120,300)$ are two points. $m=\dfrac{350-300}{80-120}=-\dfrac{5}{4}$

$p-p_1=m(q-q_1)$ or $p-300=-\dfrac{5}{4}(q-120)$ or $p=-\dfrac{5}{4}q+450$

Supply: $(60,280)$ and $(140,370)$ are two points. $m=\dfrac{280-370}{60-140}=\dfrac{9}{8}$

$p-p_1=m(q-q_1)$ or $p-280=\dfrac{9}{8}(q-60)$ or $p=\dfrac{9}{8}q+212.5$

Now, set these two equations for p equal to each other and solve for q.

$\frac{9}{8}q+212.5=-\frac{5}{4}q+450$ Required for equilibrium.

$9q+1700=-10q+3600$ Multiply both sides by 8 to simplify.

$19q=1900$

$q=100$

Substituting $q=100$ into one of the original equations gives $p=325.$
Thus, the equilibrium point is $(q,p)=(100,325).$

51. a. Reading the graph, we have that the tax is \$15.
b. From the graph, the original equilibrium was $(100,100).$
c. From the graph, the new equilibrium is $(50,110).$
d. The supplier suffers because the increased price reduces the demand.

53. New supply price: $p=15q+30+38=15q+68$
$15q+68=-4q+220$ Required condition
$19q=152$
$q=8$
Substituting $q=8$ into one of the original equations gives $p=188.$
Thus, the new equilibrium point is $(q,p)=(8,188).$

55. New supply price: $p=\dfrac{q}{20}+10+5=\dfrac{q}{20}+15$

$\frac{q}{20}+15=-\frac{q}{20}+65$ Required condition

$q+300=-q+1300$

$2q=1000$

$q=500$ Thus, $p=\frac{500}{20}+15=40.$

The new equilibrium point is $(500,40).$

57. Demand: $p = \dfrac{-q + 2100}{60}$　　Supply: $p = \dfrac{q + 540}{120}$

New supply: $p = \dfrac{q + 540}{120} + \dfrac{1}{2} = \dfrac{q + 540}{120} + \dfrac{60}{120} = \dfrac{q + 600}{120}$

$\dfrac{q+600}{120} = \dfrac{-q+2100}{60}$　　　　Required condition

$q + 600 = -2q + 4200$　　　Multiply both sides by 120

$3q = 3600$

$q = 1200$　　　　Thus, $p = \dfrac{1200 + 600}{120} = 15$.

The new equilibrium quantity is 1200.
The new equilibrium price is $15.

Review Exercises

For this set of exercises we will not give reasons for any steps or list any formulas.

1. $3x - 8 = 23$

$3x = 31$

$x = \dfrac{31}{3}$

2. $2x - 8 = 3x + 5$

$-x = 13$

$x = -13$

3. $\dfrac{6x + 3}{6} = \dfrac{5(x - 2)}{9}$

$18\left(\dfrac{6x + 3}{6}\right) = 18\left(\dfrac{5(x - 2)}{9}\right)$

$3(6x + 3) = 10(x - 2)$

$18x + 9 = 10x - 20$

$8x = -29$

$x = -\dfrac{29}{8}$

4. $2x + \dfrac{1}{2} = \dfrac{x}{2} + \dfrac{1}{3}$

$12x + 3 = 3x + 2$

$9x = -1$

$x = -\dfrac{1}{9}$

5. $\dfrac{6}{3x - 5} = \dfrac{6}{2x + 3}$

$6(2x + 3) = 6(3x - 5)$

$2x + 3 = 3x - 5$

$3 + 5 = 3x - 2x$

$x = 8$

6. $\dfrac{2x + 5}{x + 7} = \dfrac{1}{3} + \dfrac{x - 11}{2(x + 7)}$

$6(2x + 5) = 2(x + 7) + 3(x - 11)$

$12x + 30 = 2x + 14 + 3x - 33$

$12x - 2x - 3x = 14 - 33 - 30$

$7x = -49$

$x = -7$

There is no solution since we have division by zero when $x = -7$.

7. $3y - 6 = -2x - 10$

$3y = -2x - 4$

$y = \dfrac{-2x - 4}{3}$

$y = -\dfrac{2}{3}x - \dfrac{4}{3}$

8. $3x - 9 \le 4(3 - x)$

$3x - 9 \le 12 - 4x$

$7x \le 21$

$x \le 3$

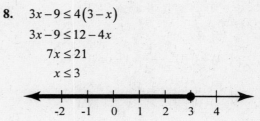

9. $\dfrac{2}{5}x \le x + 4$

$5\left(\dfrac{2}{5}x\right) \le 5(x + 4)$

$2x \le 5x + 20$

$-3x \le 20$

$x \ge -\dfrac{20}{3}$

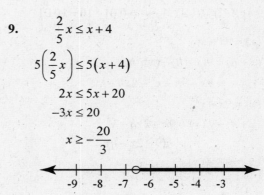

10. $5x + 1 \ge \dfrac{2}{3}(x - 6)$

$3(5x + 1) \ge 3 \cdot \dfrac{2}{3}(x - 6)$

$15x + 3 \ge 2(x - 6)$

$15x + 3 \ge 2x - 12$

$13x \ge -15$

$x \ge -\dfrac{15}{13}$

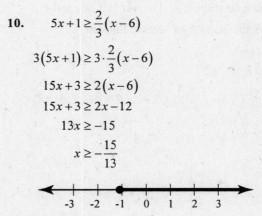

11. Yes.

12. $y^2 = 9x$, is not a function of x. If $x = 1$, then $y = \pm 3$.

13. Yes.

14. $y = \sqrt{9-x}$

Domain: $9 - x \geq 0$ or $9 \geq x$ or $x \leq 9$.

Range: Positive square root means $y \geq 0$.

15. $f(x) = x^2 + 4x + 5$

   a. $f(-3) = (-3)^2 + 4(-3) + 5 = 9 - 12 + 5 = 2$

   b. $f(4) = (4)^2 + 4(4) + 5 = 16 + 16 + 5 = 37$

   c. $f\left(\dfrac{1}{2}\right) = \left(\dfrac{1}{2}\right)^2 + 4\left(\dfrac{1}{2}\right) + 5 = \dfrac{1}{4} + 2 + 5 = \dfrac{29}{4}$

16. $g(x) = x^2 + \dfrac{1}{x}$

   a. $g(-1) = (-1)^2 + \dfrac{1}{-1} = 1 - 1 = 0$

   b. $g\left(\dfrac{1}{2}\right) = \left(\dfrac{1}{2}\right)^2 + \dfrac{1}{\frac{1}{2}} = \dfrac{1}{4} + 2 = 2\dfrac{1}{4}$

   c. $g(0.1) = (0.1)^2 + \dfrac{1}{0.1} = 0.01 + 10 = 10.01$

17. $f(x) = 9x - x^2$

$f(x+h) = 9(x+h) - (x+h)^2$

$\qquad\quad = 9x + 9h - x^2 - 2xh - h^2$

$f(x) = 9x - x^2$

$f(x+h) - f(x) = 9h - 2xh - h^2$

$\qquad\qquad\qquad = h(9 - 2x - h)$

$\dfrac{f(x+h) - f(x)}{h} = 9 - 2x - h$

18. y is a function of x. (Use vertical line test.)

19. No, the graph fails vertical line test.

20. $f(2) = 4$

21. $x = 0, x = 4$

22. a. $D = \{-2, -1, 0, 1, 3, 4\}$, $R = \{-3, 2, 4, 7, 8\}$

   b. $f(4) = 7$

   c. $f(x) = 2$ if $x = -1, 3$

   d.

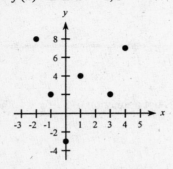

   e. No. For $y = 2$, there are two values of x.

23. $f(x) = 3x + 5$, $g(x) = x^2$

   a. $(f+g)x = (3x+5) + x^2 = x^2 + 3x + 5$

   b. $\left(\dfrac{f}{g}\right)x = \dfrac{3x+5}{x^2}$ or $\dfrac{3x}{x^2} + \dfrac{5}{x^2} = \dfrac{3}{x} + \dfrac{5}{x^2}$

   c. $f(g(x)) = f(x^2) = 3x^2 + 5$

   d. $(f \circ f)x = f(3x+5)$

$\qquad\qquad\quad = 3(3x+5) + 5$

$\qquad\qquad\quad = 9x + 20$

24. $5x + 2y = 10$

x-intercept: If $y = 0$, $x = 2$

y-intercept: If $x = 0$, $y = 5$

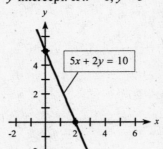

25. $6x + 5y = 9$

x-intercept: If $y = 0$, $x = \dfrac{9}{6} = \dfrac{3}{2}$

y-intercept: If $x = 0$ or $y = \dfrac{9}{5}$

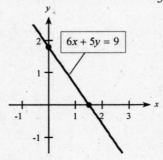

26. $x = -2$

x-intercept: $x = -2$

There is no y-intercept.

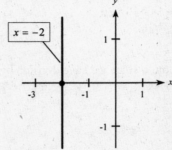

27. $P_1(2, -1)$; $P_2(-1, -4)$

$$m = \frac{-4 - (-1)}{-1 - 2} = \frac{-3}{-3} = 1$$

28. $(-3.8, -7.16)$ and $(-3.8, 1.16)$

$$m = \frac{-7.16 - 1.16}{-3.8 - (-3.8)} = \frac{-8.32}{0}$$

Slope is undefined.

29. $2x + 5y = 10$

$$y = -\frac{2}{5}x + 2, \quad m = -\frac{2}{5}, \quad b = 2$$

30. $x = -\frac{3}{4}y + \frac{3}{2}$ or $y = -\frac{4}{3}x + 2$

$$m = -\frac{4}{3}, \quad b = 2$$

31. $m = 4$, $b = 2$, $y = 4x + 2$

32. $m = -\frac{1}{2}$, $b = 3$, $y = -\frac{1}{2}x + 3$

33. $P = (-2, 1)$, $m = \frac{2}{5}$

$$y - 1 = \frac{2}{5}(x + 2) \text{ or } y = \frac{2}{5}x + \frac{9}{5}$$

34. $(-2, 7)$ and $(6, -4)$

$$m = \frac{-4 - 7}{6 - (-2)} = \frac{-11}{8}$$

$$y - 7 = \frac{-11}{8}(x - (-2)) \text{ or}$$

$$y = \frac{-11}{8}x + \frac{17}{4}$$

35. $P_1(-1, 8)$; $P_2(-1, -1)$

The line is vertical since the x-coordinates are the same. Equation: $x = -1$

36. Parallel to $y = 4x - 6$ means $m = 4$.

$y - 6 = 4(x - 1)$ or $y = 4x + 2$

37. $P(-1, 2)$; $\perp$ to $3x + 4y = 12$

or

$$y = -\frac{3}{4}x + 3$$

$$m = \frac{4}{3}$$

$$y - 2 = \frac{4}{3}(x + 1) \text{ or}$$

$$y = \frac{4}{3}x + \frac{10}{3}$$

38. $x^2 + y - 2x - 3 = 0$

$$y = -x^2 + 2x + 3$$

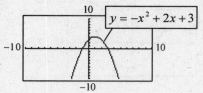

39.

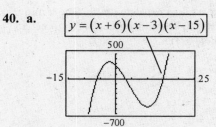

40. a.

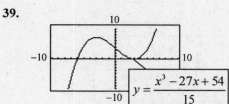

b.

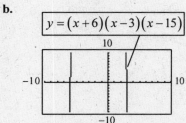

c. The graph in (a) shows the complete graph. The graph in (b) shows a piece that rises toward the high point and a piece between the high and low points.

41. $y = x^2 - x - 42$ is a parabola opening upward.

a.

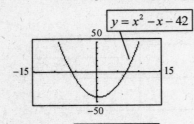

b.

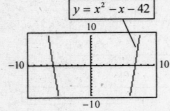

c. (a) shows the complete graph. The y-min is too large in absolute value for (b) to get a complete graph.

42. $y = \dfrac{\sqrt{x+3}}{x}$

$x \neq 0;\ x + 3 \geq 0$ or $x \geq -3$;

Domain: $x \neq 0,\ x \geq -3$

43. Trace approximates $x = -7.2749,\ x = 0.2749$

44.
$$4x - 2y = 6$$
$$3x + 3y = 9$$
Then,
$$12x - 6y = 18$$
$$\underline{6x + 6y = 18}$$
$$18x \qquad = 36$$
$$x = 2$$
$$4(2) - 2y = 6$$
$$y = 1$$

Solution: (2, 1)

45.
$$2x + y = 19$$
$$x - 2y = 12$$
Then,
$$4x + 2y = 38$$
$$\underline{x - 2y = 12}$$
$$5x \qquad = 50$$
$$x = 10$$
$$2(10) + y = 19$$
$$y = -1$$

Solution: (10, −1)

46.
$$3x + 2y = 5$$
$$2x - 3y = 12$$
Then,
$$9x + 6y = 15$$
$$\underline{4x - 6y = 24}$$
$$13x \qquad = 39$$
$$x = 3$$
$$3(3) + 2y = 5$$
$$2y = -4$$
$$y = -2$$

Solution: (3, −2)

47.
$$6x + 3y = 1$$
$$y = -2x + 1$$
$$6x + 3(-2x + 1) = 1$$
$$6x - 6x + 3 = 1$$
$$3 = 1$$

No solution.

48.
$$4x - 3y = 253 \qquad 8x - 6y = 506 \qquad 4(10) - 3y = 253$$
$$13x + 2y = -12 \qquad \underline{39x + 6y = -36} \qquad -3y = 213$$
$$47x \qquad = 470 \qquad y = -71$$
$$x \qquad = 10$$

Solution: (10, −71)

49. $x + 2y + 3z = 5$ Steps 1 and 2: Nothing to be done.

$\qquad y + 11z = 21$ Step 3: $x + 2y + 3z = 5$

$\qquad 5y + 9z = 13$ $\qquad\qquad\qquad y + 11z = 21$

$\qquad\qquad\qquad\qquad\qquad\qquad\qquad\qquad -46z = -92$

$\qquad\qquad\qquad$ Step 4: $z = 2$ $\qquad\qquad$ $y + 11(2) = 21$ $x + 2(-1) + 3(2) = 5$

$\qquad\qquad\qquad\qquad\qquad\qquad\qquad\qquad\qquad\qquad y = -1$ $\qquad\qquad\qquad x = 1$

$\qquad\qquad\qquad$ Solution is $x = 1$, $y = -1$, $z = 2$.

50. $\quad x + y - z = 12$ $\qquad\qquad$ Thus $z = 9$

$\qquad\quad 2y - 3z = -7$ $\qquad\qquad$ $2y - 27 = -7$

$\quad 3x + 3y - 7z = 0$ $\qquad\qquad$ $2y = 20$ or $y = 10$

$\qquad\quad x + y - z = 12$ $\qquad\qquad$ $x + 10 - 9 = 12$

$\qquad\qquad 2y - 3z = -7$ $\qquad\qquad$ $x = 11$

$\qquad\qquad\quad -4z = -36$ $\qquad\qquad$ Solution: $(11, 10, 9)$

51. a. 1997

    b. $x = 2007 - 1980 = 27$

    c. $\quad 328 = 10.06x + 16.73$

$\qquad\quad 311.27 = 10.06x$

$\qquad\qquad\quad x = 30.94$ in 2011

52. Student has total points of $91 + 82 + 88 + 50 + 42 + 42 = 395$.

Total of possible points is $300 + 150 + 200 = 650$.

To earn an A students need at least $0.9(650) = 585$ points.

Student must earn $585 - 395 = 190$ points on the final. This is the same as 95%.

53. Diesel: $C = 0.24x + 38,000$

Gas: $C = 0.30x + 35,600$

$\quad 0.24x + 38,000 = 0.30x + 35,600$

$\qquad\qquad 0.06x = 2400$

$\qquad\qquad\quad x = 40,000$

Costs are equal at 40,000 miles. A truck is used more than 40,000 miles in 5 years. Buy the diesel.

54. a. Yes

    b. No

    c. $f(300) = 4$

55. a. $f(80) = 565.44$

    b. The monthly payment on a \$70,000 loan is \$494.75.

56. $P(x) = 180x - \dfrac{x^2}{100} - 200$ $x = q(t) = 1000 + 10t$

    a. $(P \circ q)(t) = P(1000 + 10t) = 180(1000 + 10t) - \dfrac{(1000 + 10t)^2}{100} - 200$

    b. $x = q(15) = 1000 + 10(15) = 1150$ units produced

$\qquad P(1150) = 180(1150) - \dfrac{(1150)^2}{100} - 200 = \$193,575$

57. $W(L) = kL^3$, $L(t) = 50 - \dfrac{(t-20)^2}{10}$, $0 \le t \le 20$

$(W \circ L)(t) = W\left(50 - \dfrac{(t-20)^2}{10}\right) = 0.02\left(50 - \dfrac{(t-20)^2}{10}\right)^3$

58. a.

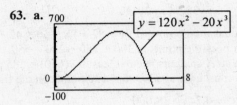

$d = \dfrac{t}{4.8}$

b.   (9.6, 2) means that the thunderstorm is two miles away if the flash and thunder are 9.6 seconds apart.

59.

$H_c = 90 - T_a$

Using $y = mx + b$, $F = \dfrac{9}{5}C + 32$.

60. a. (x, P) is the required form.

$P_1 = (200, 3100)$, $P_2 = (250, 6000)$

$m = \dfrac{6000 - 3100}{250 - 200} = \dfrac{2900}{50} = 58$

$P - 3100 = 58(x - 200)$ or

$P(x) = 58x - 8500$

b.   For each additional unit sold the profit increases by \$58.

61. $p = 0.219x + 2.75$

a.   Yes.

b. $m = 0.219$, p-intercept is 2.75

c.   In 1950 (the year that corresponds to $x = 0$), the total cost of health care in the U.S. was 2.75% of the GDP.

d.   The total cost of health care in the U.S. as a percentage of the GDP increases by 0.219% per year.

62. $(C, F): (0, 32)$ and $(100, 212)$

$m = \dfrac{212 - 32}{100 - 0} = \dfrac{180}{100} = \dfrac{9}{5}$

63. a.

$y = 120x^2 - 20x^3$

b.   Algebraically, $y \ge 0$ if

$120x^2 - 20x^3 = 20x^2(x - 6) \ge 0$.

Answer: $0 \le x \le 6$

64. a. $v^2 = 1960(h + 10)$

$h + 10 = \dfrac{v^2}{1960}$

$h = \dfrac{v^2}{1960} - 10$

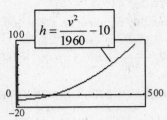

$h = \dfrac{v^2}{1960} - 10$

b. $h(210) = \dfrac{210^2}{1960} - 10 = 12.5$ cm

65. x = amount of safer investment and
y = amount of other investment.
$$x + \quad y = 150000$$
$$0.095x + 0.11y = 15000$$
Solving the system:
$$0.11x + 0.11y = 16500$$
$$\underline{0.095x + 0.11y = 15000}$$
$$0.015x \qquad = 1500$$
$$x \qquad = 100000$$
Then $y = 50000$. Thus, invest \$100,000 at 9.5% and \$50,000 at 11%.

66. x = liters of 20% solution
y = liters of 70% solution
$$x + \quad y = 4$$
$$0.2x + 0.7y = 1.4$$
$$x + \quad y = 4$$
$$\underline{x + 3.5y = 7}$$
$$2.5y = 3 \qquad y = 1.2$$
$$x + \quad 1.2 = 4$$
$$x = 2.8$$
Answer: 2.8 liters of 20%, 1.2 of 70%.

67. $S: p = 4q + 5,\ D: p = -2q + 81$

a.
$$S: 53 = 4q + 5 \qquad D: 53 = -2q + 81$$
$$4q = 48 \qquad\qquad 2q = 28$$
$$q = 12 \qquad\qquad q = 14$$

b. Demand is greater.
There is a shortfall.

c. Price is likely to increase.

68. a. – c.

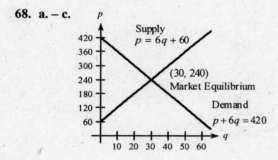

69. $C(x) = 38.80x + 4500,\ R(x) = 61.30x$
a. Marginal cost is \$38.80.
b. Marginal revenue is \$61.30.
c. Marginal profit is $61.30 - 38.80 = \$22.50$.
d. $61.30x = 38.80x + 4500$
$$22.50x = 4500$$
$$x = 200 \text{ units to break even.}$$

70. $FC = \$1500,\ VC = \$22$ per unit, $R = \$52$ per unit
a. $C(x) = 22x + 1500$
b. $R(x) = 52x$
c. $P = R - C = 30x - 1500$
d. $\overline{MC} = 22$
e. $\overline{MR} = 52$
f. $\overline{MP} = 30$
g. Break even means $30x - 1500 = 0$ or $x = 50$.

71. Supply: $\quad m = \dfrac{100 - 200}{200 - 400} = \dfrac{1}{2}$ Demand: $\quad m = \dfrac{200 - 0}{200 - 600} = -\dfrac{1}{2}$

$$p - 100 = \dfrac{1}{2}(q - 200) \qquad\qquad p - 0 = -\dfrac{1}{2}(q - 600)$$

$$p = \dfrac{1}{2}q \qquad\qquad\qquad\qquad p = -\dfrac{1}{2}q + 300$$

So, $\dfrac{1}{2}q = -\dfrac{1}{2}q + 300$ or $q = 300$. The equilibrium price is $p = \dfrac{1}{2}(300) = \$150$.

72. New supply equation: $\quad p = \dfrac{q}{10} + 8 + 2 = \dfrac{q}{10} + 10$

Demand:

$$p = \dfrac{-q + 1500}{10} = -\dfrac{q}{10} + 150$$

$$\dfrac{q}{10} + 10 = -\dfrac{q}{10} + 150$$

$$\dfrac{2q}{10} = 140 \text{ or } q = 700$$

$$p = \dfrac{700}{10} + 10 = 80$$

Solution: (700, 80)

Chapter Test

1. $4x - 3 = \dfrac{x}{2} + 6$

$8x - 6 = x + 12$

$7x = 18$

$x = \dfrac{18}{7}$

2. $\dfrac{3}{x} + 4 = \dfrac{4x}{x+1}$

$3(x+1) + 4x(x+1) = 4x(x)$

$3x + 3 + 4x^2 + 4x = 4x^2$

$7x = -3$

$x = -\dfrac{3}{7}$

3. $\dfrac{3x-1}{4x-9} = \dfrac{5}{7}$

$7(3x-1) = 5(4x-9)$

$21x - 7 = 20x - 45$

$x = -38$

4. $f(x) = 7 + 5x - 2x^2$

$f(x+h) = 7 + 5(x+h) - 2(x+h)^2$

$\qquad = 7 + 5x + 5h - 2x^2 - 4xh - 2h^2$

$f(x) = 7 + 5x - 2x^2$

$f(x+h) - f(x) = 5h - 4xh - 2h^2$

$\dfrac{f(x+h) - f(x)}{h} = 5 - 4x - 2h$

5. $1 + \dfrac{2}{3}t \le 3t + 22$

$3\left(1 + \dfrac{2}{3}t\right) \le 3(3t + 22)$

$3 + 2t \le 9t + 66$

$-7t \le 63$

$t \ge -9$

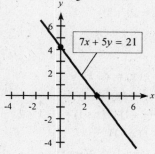

6. $5x - 6y = 30$

x-intercept: 6

y-intercept: -5

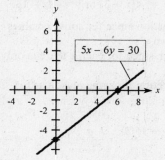

7. $7x + 5y = 21$

x-intercept: 3

y-intercept: $\dfrac{21}{5}$

8. $f(x) = \sqrt{4x + 16}$

a. $4x + 16 \ge 0$

$4x \ge -16$

Domain: $x \ge -4$; Range: $y \ge 0$

For range, note square root is positive.

b. $f(3) = \sqrt{12 + 16} = 2\sqrt{7}$

c. $f(5) = \sqrt{20 + 16} = 6$

9. $(-1, 2)$ and $(3, -4)$

$m = \dfrac{-4 - 2}{3 - (-1)} = \dfrac{-6}{4} = \dfrac{-3}{2}$

$y - 2 = \dfrac{-3}{2}(x - (-1))$

$y = \dfrac{-3}{2}x + \dfrac{1}{2}$

10. $5x + 4y = 15$

$y = -\dfrac{5}{4}x + \dfrac{15}{4}$

$m = -\dfrac{5}{4},\ b = \dfrac{15}{4}$

11. Point $(-3, -1)$

   a. Undefined slope means vertical line. $x = -3$

   b. $\perp$ to $y = \frac{1}{4}x + 2$ means $m = -4$.

     Thus, $y + 1 = -4(x + 3)$ or $y = -4x - 13$.

12. a. is not a function since for some x-values there are two y's.

   b. is a function since for each x there is only one y.

   c. is not a function for same reason as (a).

13. $3x + 2y = -2$

   $4x + 5y = 2$

   $12x + 8y = -8$

   $\underline{12x + 15y = 6}$

       $-7y = -14$

         $y = 2$

   $3x + 2(2) = -2$

       $3x = -6$

        $x = -2$

   Solution: $(-2, 2)$

14. $f(x) = 5x^2 - 3x$, $g(x) = x + 1$

   a. $(fg)(x) = (5x^2 - 3x)(x + 1)$

   b. $g(g(x)) = g(x + 1) = (x + 1) + 1 = x + 2$

   c. $(f \circ g)(x) = f(x + 1)$

           $= 5(x + 1)^2 - 3(x + 1)$

           $= 5x^2 + 10x + 5 - 3x - 3$

           $= 5x^2 + 7x + 2$

15. $R(x) = 38x$, $C(x) = 30x + 1200$

   a. $\overline{MC} = \$30$

   b. $P(x) = 38x - (30x + 1200)$

           $= 8x - 1200$

   c. Break-even means $P(x) = 0$.

     $8x = 1200$ or $x = 150$ units

   d. $\overline{MP} = \$8$. Each additional unit sold increases the profit by $8.

16. a. $R(x) = 50x$

   b. $C(100) = 10(100) + 18000$

           $= \$19,000$

     It costs $19,000 to make 100 units.

   c. $50x = 10x + 18000$

     $40x = 18000$

       $x = 450$ units

17. $S : p = 5q + 1500$, $D : p = -3q + 3100$

   $5q + 1500 = -3q + 3100$

   $8q = 1600$ or $q = 200$

   $p(200) = 5(200) + 1500 = \$2500$

18. $y = 360000 - 1500x$

   a. $b = 360,000$

     The original value is $360,000.

   b. $m = -1500$.

     The building is depreciating $1500 each month.

19. $x =$ number of reservations

   $0.90x = 360$

       $x = 400$

   Accept 400 reservations.

20. $x =$ amount invested at 9%

   $y =$ amount invested at 6%

   $x + y = 20000$ Amount

   $0.09x + 0.06y = 1560$ Interest

   $0.09x + 0.09y = 1800$

   $\underline{0.09x + 0.06y = 1560}$

         $0.03y = 240$

            $y = \$8000$

   Invest $8000 at 6% and $12000 at 9%

Chapter 2: Special Functions

Exercise 2.1

All problems must be in the form of $ax^2 + bx + c = 0$ before solutions can be found.

The Quadratic formula, $x = \dfrac{-b \pm \sqrt{b^2 - 4ac}}{2a}$, is used when factoring is difficult or not possible.

1. $2x^2 + 3 = x^2 - 2x + 4$

$x^2 + 2x - 1 = 0$

3. $(y+1)(y+2) = 4$

$y^2 + 3y + 2 = 4$

$y^2 + 3y - 2 = 0$

5. $9 - 4x^2 = 0$

$(3 + 2x)(3 - 2x) = 0$

$3 + 2x = 0$ or $3 - 2x = 0$

Solution: $x = -\dfrac{3}{2}, \dfrac{3}{2}$

7. $x = x^2$

$x^2 - x = 0$

$x(x - 1) = 0$

Solution: $x = 0, 1$

Never divide by a variable. A root is lost if you divide.

9. $x^2 + 5x = 21 + x$

$x^2 + 4x - 21 = 0$

$(x + 7)(x - 3) = 0$

$x + 7 = 0$ or $x - 3 = 0$

Solution: $x = -7, 3$

11. $4t^2 - 4t + 1 = 0$

$(2t - 1)(2t - 1) = 0$

$2t - 1 = 0$

Solution: $t = \dfrac{1}{2}$

13. $\dfrac{w^2}{8} - \dfrac{w}{2} - 4 = 0$

$w^2 - 4w - 32 = 0$

$(w - 8)(w + 4) = 0$

$w - 8 = 0$ or $w + 4 = 0$

Solution: $w = 8, -4$

15. $(x - 1)(x + 5) = 7$

$x^2 + 4x - 5 = 7$

$x^2 + 4x - 12 = 0$

$(x + 6)(x - 2) = 0$

Solution: $x = -6, 2$

17. $x + \dfrac{8}{x} = 9$

$x^2 + 8 = 9x$

$x^2 - 9x + 8 = 0$

$(x - 8)(x - 1) = 0$

Solution: $x = 1, 8$

19. $\dfrac{x}{x-1} = 2x + \dfrac{1}{x-1}$

$x = (2x^2 - 2x) + 1$

$2x^2 - 3x + 1 = 0$

$(2x - 1)(x - 1) = 0$

Solution: $x = \dfrac{1}{2}$

1 is not a root since division by zero is not defined.

21. a. $x^2 - 4x - 4 = 0$

$a = 1, b = -4, c = -4$

$x = \dfrac{-(-4) \pm \sqrt{(-4)^2 - 4(1)(-4)}}{2(1)}$

$= \dfrac{4 \pm \sqrt{32}}{2} = \dfrac{4 \pm 4\sqrt{2}}{2} = 2 \pm 2\sqrt{2}$

b. Since $\sqrt{2} \approx 1.414$, the solutions are approximately 4.83, −0.83.

23. $2w^2 + w + 1 = 0$

$a = 2, b = 1, c = 1$

$w = \dfrac{-1 \pm \sqrt{1 - 8}}{4} = \dfrac{-1 \pm \sqrt{-7}}{4}$

There are no real solutions.

25. a. $16z^2 + 16z - 21 = 0$

$a = 16, b = 16, c = -21$

$$z = \frac{-16 \pm \sqrt{256 + 1344}}{32}$$

$$= \frac{-16 \pm 40}{32} = \frac{3}{4} \text{ or } -\frac{7}{4}$$

b. $0.75, -1.75$

27. a. $5x^2 = 2x + 6$ or $5x^2 - 2x - 6 = 0$

$a = 5, b = -2, c = -6$

$$x = \frac{2 \pm \sqrt{4 + 120}}{10} = \frac{1 \pm \sqrt{31}}{5}$$

b. $\dfrac{1 \pm \sqrt{31}}{5} \approx \dfrac{1 \pm 5.57}{5} \approx 1.31, -0.91$

29. $y^2 = 7$

$y = \pm\sqrt{7}$

31. $y^2 + 9 = 0$

$y^2 + 3^2 = 0$

The sum of two squares cannot be factored.
There are no real solutions.

33. $(x + 4)^2 = 25$

$x + 4 = \pm 5$

$x = -4 \pm 5$

Solution: $x = 1, -9$

35. $(x + 8)^2 + 3(x + 8) + 2 = 0$

$[(x + 8) + 2][(x + 8) + 1] = 0$

$(x + 8) + 2 = 0$ or $(x + 8) + 1 = 0$

Solution: $x = -10, -9$

37. $21x + 70 - 7x^2 = 0$

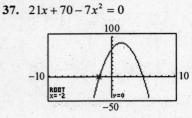

Divide by -7 and rearrange.

$x^2 - 3x - 10 = 0$

$(x - 5)(x + 2) = 0$

Solution: $x = -2, 5$

39. $300 - 2x - 0.01x^2 = 0$

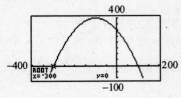

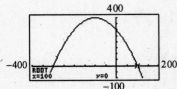

$a = -0.01, b = -2, c = 300$

$$x = \frac{2 \pm \sqrt{4 + 12}}{-0.02} = \frac{2 \pm 4}{-0.02}$$

$$= -300 \text{ or } 100$$

41. $25.6x^2 - 16.1x - 1.1 = 0$

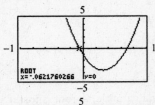

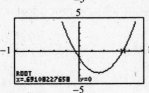

$a = 25.6, b = -16.1, c = -1.1$

$$x = \frac{16.1 \pm \sqrt{259.21 + 112.64}}{51.2}$$

$$= \frac{16.1 \pm \sqrt{371.85}}{51.2}$$

$$\approx 0.69 \text{ or } -0.06$$

43. $P = -x^2 + 90x - 200$

$1200 = -x^2 + 90x - 200$

$0 = x^2 - 90x + 1400$

$0 = (x - 20)(x - 70)$

A profit of \$1200 is earned at $x = 20$ units or $x = 70$ units of production.

45. a.
$$P = -18x^2 + 6400x - 400$$
$$61,800 = -18x^2 + 6400x - 400$$
$$18x^2 - 6400x + 62,200 = 0$$
Factoring appears difficult, so let us apply the quadratic formula.
$$x = \frac{6400 \pm \sqrt{6400^2 - 4(18)(62,200)}}{36}$$
$$= \frac{6400 \pm \sqrt{36,481,600}}{36}$$
$$= \frac{6400 \pm 6040}{36} = 10 \text{ or } 345.56$$
So, a profit of \$61,800 is earned for 10 units or for 345.56 units.
Yes. Maximum profit occurs at vertex as seen using the graphing calculator.

47.
$$5 = 100 + 96t - 16t^2$$
$$100 = 100 + 96t - 16t^2$$
$$0 = 96t - 16t^2 = 16t(6 - t)$$
The ball is 100 feet high 6 seconds later.

49. $p = 25 - 0.01s^2$
   a.
$$0 = 25 - 0.01s^2$$
$$= (5 + 0.1s)(5 - 0.1s)$$
$$p = 0 \text{ if } 5 - 0.1s = 0 \text{ or } s = 50.$$
   b. $s \geq 0$. $p = 0$ means there is no particulate pollution.

51.
$$t = 0.001(0.872x^2 + 8.308x + 711.311)$$
$$9.03 = 0.001(0.872x^2 + 8.308x + 711.311)$$
$$9030 = 0.872x^2 + 8.308x + 711.311$$
$$0 = 0.872x^2 + 8.308x - 8318.689$$
$$x = \frac{-8.308 \pm \sqrt{(8.308)^2 - 4(0.872)(8318.689)}}{2(0.872)}$$
$$x = 95.793 \text{ or } -99.587$$
The positive answer is the one that makes sense here, 95.8 mph.

53. $y = 0.034x^2 - 0.044x + 12.642$
   a. $y = 0.034(7)^2 - 0.044(7) + 12.642 = 14$

b.
$$14 = 0.034x^2 - 0.044x + 12.642$$
$$0 = 0.034x^2 - 0.044x - 1.358$$
$$x = \frac{0.044 \pm \sqrt{(-.044)^2 - 4(.034)(-1.358)}}{2(0.034)}$$
$$x = -5.71 \text{ or } x = 7$$
The year 1997.

55.
$$p = f(t) = 0.121t^2 - 4.180t + 79.620$$
$$44.4 = 0.121t^2 - 4.180t + 79.620$$
$$0 = 0.121t^2 - 4.180t + 35.220$$
Use the graphing calculator and vary the range. This will give $x = 19.97$ or $x = 14.58$. In 2000 the percentage will reach 44.4, and it also reached that percentage in 1995.

57. $C = \text{cost}$
$$\text{profit} = \text{selling price} - \text{cost} = 144 - C$$
$$\text{percentage of profit} = \frac{144 - C}{C} \cdot 100$$
$$C = 100\left(\frac{144 - C}{C}\right)$$
$$C^2 + 100C - 14,400 = 0$$
$$(C - 80)(C + 180) = 0$$
The store paid \$80.

59.
$$y = 11.786x^2 - 142.214x + 493$$
$$812 = 11.786x^2 - 142.214x + 493$$
$$0 = 11.786x^2 - 142.214x - 319$$
Vary the range on the graphing calculator to see where the graph crosses the x-axis. This is approximately $x = 14$. In 2004 there will be 812 million users.

61. $K^2 = 16v + 4$
   a. $K^2 = 16(20) + 4 = 324$
$$K = 18$$
   b. $K^2 = 16(60) + 4 = 964$
$$K \approx 31$$
   c. Speed triples but K changes only by a factor of 1.72.

Exercise 2.2

1. $y = \dfrac{1}{2}x^2 + x$

   a. $x = \dfrac{-b}{2a} = \dfrac{-1}{2(1/2)} = -1$

$$y = \dfrac{1}{2}(-1)^2 + (-1) = -\dfrac{1}{2}$$

Vertex is at $\left(-1, -\dfrac{1}{2}\right)$.

   b. $a > 0$, so vertex is a minimum.

   c. -1

   d. $-\dfrac{1}{2}$

3. $y = 8 + 2x - x^2$

   a. $x = \dfrac{-b}{2a} = \dfrac{-2}{2(-1)} = 1$

$$y = 8 + 2(1) - (1)^2 = 9$$

Vertex is at $(1, 9)$.

   b. $a < 0$, so vertex is a maximum.

   c. 1

   d. 9

5. $f(x) = 6x - x^2$

$$x = \dfrac{-b}{2a} = \dfrac{-6}{-2} = 3.$$

$$f(3) = 6(3) - (3)^2 = 9$$

Vertex is at $(3, 9)$.

$a < 0$, so vertex is a maximum.

3

9

7. $y = -\dfrac{1}{4}x^2 + x$

Vertex is a maximum point since $a < 0$.

V: $x = \dfrac{-b}{2a} = \dfrac{-1}{2(-1/4)} = 2$

$$y = -\dfrac{1}{4}(2)^2 + 2 = 1$$

Zeros: $-\dfrac{1}{4}x^2 + x = 0$

$$x\left(-\dfrac{1}{4}x + 1\right) = 0$$

$$x = 0, 4$$

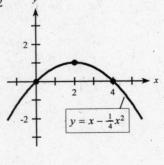

9. $y = x^2 + 4x + 4$

Vertex is a minimum point since $a > 0$.

V: $x = \dfrac{-b}{2a} = \dfrac{-4}{2(1)} = -2$

$$y = (-2)^2 + 4(-2) + 4 = 0$$

Zeros: $x^2 + 4x + 4 = (x + 2)(x + 2) = 0$

$$x = -2$$

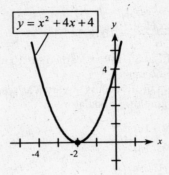

11. $y = \dfrac{1}{2}x^2 + x - 3$

Vertex is a minimum point since $a > 0$.

V: $x = \dfrac{-b}{2a} = \dfrac{-1}{2(1/2)} = -1$

$$y = \dfrac{1}{2}(-1)^2 + (-1) - 3 = -\dfrac{7}{2}$$

Zeros: $\dfrac{1}{2}x^2 + x - 3 = 0 \;\rightarrow\; x^2 + 2x - 6 = 0$

$$x = \dfrac{-2 \pm \sqrt{4 + 24}}{2} = \dfrac{-2 \pm 2\sqrt{7}}{2} = -1 \pm \sqrt{7}$$

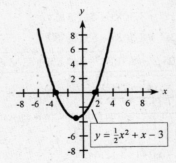

13. $y = (x-3)^2 + 1$

   a. Graph is shifted 3 units to the right and 1 unit up.

   b.

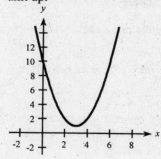

15. $y = (x-10)^2 + 12$

   a. Graph is shifted 10 units to the right and 12 units up.

   b.

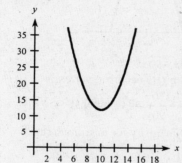

17. $y = \dfrac{1}{2}x^2 - x - \dfrac{15}{2}$

   V: $x = \dfrac{-b}{2a} = \dfrac{-(-1)}{2(1/2)} = 1$

   $y = \dfrac{1}{2}(1)^2 - 1 - \dfrac{15}{2} = -8$

   Zeros: $x^2 - 2x - 15 = (x-5)(x+3) = 0$

         $x = 5, -3$

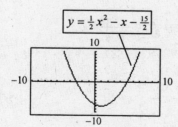

19. $y = \dfrac{1}{4}x^2 + 3x + 12$

   V: $x = \dfrac{-b}{2a} = \dfrac{-3}{2\left(\dfrac{1}{4}\right)} = -6$

   $y = \dfrac{1}{4}(-6)^2 + 3(-6) + 12 = 3$

   Zeros: $x^2 + 12x + 48 = 0$

      $b^2 - 4ac = 144 - 192 < 0$

      There are no zeros.

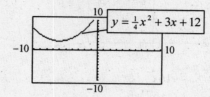

21. $f(x) = y = -5x - x^2$

   Average Rate of Change $= \dfrac{f(1) - f(-1)}{1 - (-1)}$

                     $= \dfrac{-6 - 4}{2} = -\dfrac{10}{2} = -5$

23. $y = 63 + 0.2x - 0.01x^2$

   V: $x = \dfrac{-0.2}{-0.02} = 10$

      $y = 63 + 2 - 1 = 64$

   Zeros: $x^2 - 20x - 6300 = (x-90)(x+70) = 0$

         $x = 90, -70$

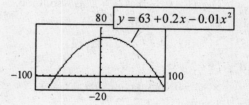

25. $y = 0.0001x^2 - 0.01$

$\quad$ V: $x = \dfrac{-0}{2(0.0001)} = 0$

$\qquad y = 0 - 0.01 = -0.01$

$\quad$ Zeros: $0.0001x^2 - 0.01 = 0.01(0.01x^2 - 1) = 0$

$\qquad\qquad 0.01(0.1x + 1)(0.1x - 1) = 0$

$\qquad\qquad x = -10, 10$

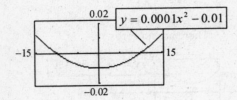

27. $f(x) = 8x^2 - 16x - 16$

$\quad$ **a.** $x = \dfrac{-b}{2a} = \dfrac{16}{16} = 1$ and $f(1) = -24$

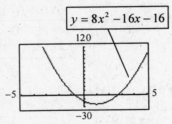

$\quad$ **b.** Graphical approximation gives
$\qquad x = -0.73,\ 2.73$

29. $f(x) = 3x^2 - 8x + 4$

$\quad$ **a.** The TRACE gives $x = 2$ as a solution.

$\quad$ **b.** $(x - 2)$ is a factor.

$\quad$ **c.** $3x^2 - 8x + 4 = (x - 2)(3x - 2)$

$\quad$ **d.** $(x - 2)(3x - 2) = 0$

$\qquad x - 2 = 0$ or $3x - 2 = 0$

$\qquad$ Solution is $x = 2,\ 2/3$.

31. $P = -0.1x^2 + 16x - 100$

$\quad$ The vertex coordinates are the answers to the questions.

$\quad$ **a.** $a = -0.1,\ b = 16$

$\qquad x = \dfrac{-b}{2a} = \dfrac{-16}{-0.2} = 80$

$\qquad$ Profit is maximized at a production level of 80 units.

$\quad$ **b.** $P(80) = -0.1(80)^2 + 16(80) - 100 = \$540$

$\qquad$ is the maximum profit.

33. $Y = 800x - x^2$

$\quad$ Opens down so maximum Y is at vertex.

$\quad$ V: $x = \dfrac{-800}{-2} = 400$

$\quad$ Maximum yield occurs at $x = 400$ trees.

35. $S = 1000x - x^2$

$\quad$ Maximum sensitivity occurs at vertex.

$\quad$ V: $x = \dfrac{-1000}{-2} = 500$

$\quad$ The dosage for maximum sensitivity is 500.

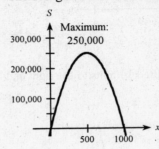

37. $R = 270x - 90x^2$

$\quad$ Maximum rate occurs at vertex.

$\quad$ V: $x = \dfrac{-270}{2(-90)} = \dfrac{3}{2}$ (lumens)

$\qquad$ is the intensity for maximum rate.

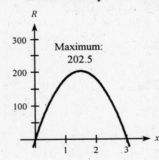

39. **a.** $y = -0.0013x^2 + x + 10$

$\qquad$ V: $x = \dfrac{-1}{-0.0026} = 384.62$;

$\qquad y = -0.0013(384.62)^2 + 384.62 + 10$

$\qquad\quad = 202.31$

$\quad$ **b.** $y = -\dfrac{1}{81}x^2 + \dfrac{4}{3}x + 10$

$\qquad$ V: $x = \dfrac{\frac{-4}{3}}{\frac{-2}{81}} = 54$;

$\qquad y = -\dfrac{1}{81}(54)^2 + \dfrac{4}{3}(54) + 10 = 46$

$\quad$ Projectile **a.** goes $202.31 - 46 = 156.31$ feet higher.

41. a. From *b* to *c*. The average rate of change is the same as the slope of the segment. The segment from *b* to *c* is steeper.
  b. Needs to satisfy $d > b$ to make the segment from *a* to *d* have a greater slope.

43. a.

No. of Apts	Rent	Total Revenue
50	$600	$30,000
49	$620	$30,380
48	$640	$30,720

  b. Revenue increases $720
  c. $R = (50 - x)(600 + 20x)$
  d. $R = -20x^2 + 400x + 30,000$

  R is maximized at $x = \dfrac{-400}{2(-20)} = 10$.

  Rent would be $600 + $200 = $800.

45. a. A quadratic function or parabola.
  b. $a < 0$ because the graph opens downward.
  c. The vertex occurs after 2004 (or when $x > 0$), so $-\frac{b}{2a} > 0$. Hence with $a < 0$ we must have $b > 0$. The value $c = f(0)$ or the y-value during 2004 which is positive.

47. a.

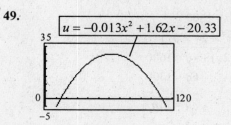

  b. The shape appears to be quadratic.
  c. The graph is of the form $y = ax^2 + 26.7$.
    Using $a \le 1$ gives $y = 0.78x^2 + 26.7$.
  d. $y = 0.78(40)^2 + 26.7 = 1275 million

49.

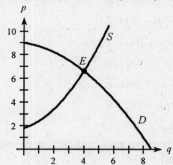

$u = -0.013x^2 + 1.62x - 20.33$

51. a. 1914 and 2010
    2011 (% is negative).

Exercise 2.3

1. a. Supply: $p = \frac{1}{4}q^2 + 10$ (see below)
  b. Demand: $p = 86 - 6q - 3q^2$ (see below)

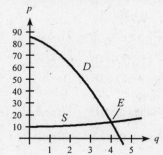

3. $\dfrac{1}{4}q^2 + 10 = 86 - 6q - 3q^2$

$q^2 + 40 = 344 - 24q - 12q^2$

$0 = 13q^2 + 24q - 304$

$0 = (q - 4)(13q + 76)$

$q = 4$ must be positive.

$p = \dfrac{1}{4}(4)^2 + 10 = 14$

E: (4, 14)

5. a. Supply: $p = 0.2q^2 + 0.4q + 1.8$ (see below)
  b. Demand: $p = 9 - 0.2q - 0.1q^2$ (see below)

  c. See E on graph.
  d. $0.2q^2 + 0.4q + 1.8 = 9 - 0.2q - 0.1q^2$

  $0.3q^2 + 0.6q - 7.2 = 0$

  $q^2 + 2q - 24 = 0$

  $(q + 6)(q - 4) = 0$

  $q = 4$ positive root only

  $p = 0.2(4)^2 + 0.4(4) + 1.8 = 6.60$

  E: (4, 6.60)

7. $p = q^2 + 8q + 16$

$p = -3q^2 + 6q + 436$

$q^2 + 8q + 16 = -3q^2 + 6q + 436$

$4q^2 + 2q - 420 = 0$

$2q^2 + q - 210 = 0$

$(2q + 21)(q - 10) = 0$

$q = 10$

$p = 10^2 + 8(10) + 16 = 196$

E: (10, 196)

9. $p^2 + 4q = 1600$

$300 - p^2 + 2q = 0$

$(300 + 2q) + 4q = 1600$

$6q = 1300$

$q = 216\frac{2}{3}$

$p^2 + 4\left(\frac{1300}{6}\right) = 1600$ or $p^2 = 733.33$ or $p = 27.08$

E: $\left(216\frac{2}{3}, 27.08\right)$

11. $p - q = 10$ or $q = p - 10$

$q(2p - 10) = 2100$

$q = \frac{2100}{2p - 10}$

$p - 10 = \frac{2100}{2p - 10}$

$(p - 10)(2p - 10) = 2100$

$2p^2 - 30p + 100 = 2100$

$2p^2 - 30p - 2000 = 0$

$p^2 - 15p - 1000 = 0$

$(p - 40)(p + 25) = 0$

$p = 40$ or $p = -25$

(only the positive answer makes sense here)

$q = 40 - 10 = 30$

$E : (30, 40)$

13. $2p - q - 10 = 0$

$(p + 10)(q + 30) = 7200$

So, $(p + 10)(2p - 10 + 30) = 7200$

$p^2 + 20p + 100 = 3600$

$p^2 + 20p - 3500 = 0$

$(p + 70)(p - 50) = 0$

$p = 50$

$q = 2(50) - 10 = 90$

E: $(q, p) = (90, 50)$

15. $p = \frac{1}{2}q + 5 + 22 = \frac{1}{2}q + 27$

So, $\left(\frac{1}{2}q + 27 + 10\right)(q + 30) = 7200$

$(q + 74)(q + 30) = 14,400$

$q^2 + 104q - 12,180 = 0$

$(q + 174)(q - 70) = 0$

$p = \frac{1}{2}(70) + 27 = 62$

E: (70, 62)

17. $C(x) = x^2 + 40x + 2000$

$R(x) = 130x$

$x^2 + 40x + 2000 = 130x$

$x^2 - 90x + 2000 = 0$

$(x - 40)(x - 50) = 0$

$x = 40$ or $x = 50$

Break-even values are at $x = 40$ and 50 units.

19. $C(x) = 15,000 + 35x + 0.1x^2$

$R(x) = 385x - 0.9x^2$

$15,000 + 35x + 0.1x^2 = 385x - 0.9x^2$

$x^2 - 350x + 15,000 = 0$

$(x - 300)(x - 50) = 0$

$x = 300$ or $x = 50$

Break-even values are at $x = 50$ and 300 units.

21. $C(x) = 150 + x + 0.09x^2$

$R(x) = 12.5x - 0.01x^2$

$x < 75$

$150 + x + 0.09x^2 = 12.5x - 0.01x^2$

$0.1x^2 - 11.5x + 150 = 0$

$x^2 - 115x + 1500 = 0$

$(x - 100)(x - 15) = 0$

$x = 100$ or $x = 15$

Break-even value is at $x = 15$ units since we were given that $x < 75$.

23. $R(x) = 385x - 0.9x^2$

$a = -0.9, b = 385$

Maximum revenue is at the vertex.

V: $x = \dfrac{-385}{-1.8} = 213.89$ or 214 total units

$R(214) = 385(214) - 0.9(214)^2 = \$41,173.60$

25. $R(x) = x(175 - 0.50x) = 175x - 0.5x^2$

$a = -0.50, b = 175$

Revenue is a maximum at $x = \dfrac{-175}{-1} = 175.$

Price that will maximize revenue is
$p = 175 - 87.50 = \$87.50.$

27. $P(x) = -x^2 + 110x - 1000$

Maximum profit is at the vertex or when

$x = \dfrac{-110}{-2} = 55.$

$P(55) = \$2025.$

29. a.

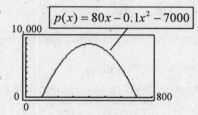

$p(x) = 80x - 0.1x^2 - 7000$

    b. $(400, 9000)$ is the maximum

    c. positive

    d. negative

    e. closer to 0

31. $R(x) = 385x - 0.9x^2$

$C(x) = 15,000 + 35x + 0.1x^2$

  a. $P(x) = 385x - 0.9x^2 - (15,000 + 35x + 0.1x^2)$

          $= -x^2 + 350x - 15,000$

     At the vertex we have $x = \dfrac{-350}{-2} = 175.$

     So, $P(175) = \$15,625.$

  b. No. More units are required to maximize revenue.

  c. The break-even values and zeros of $P(x)$ are the same.

33. a. $C(x) = 28,000 + \left(\dfrac{2}{5}x + 222\right)x = \dfrac{2}{5}x^2 + 222x + 28,000$

      $R(x) = \left(1250 - \dfrac{3}{5}x\right)x = 1250x - \dfrac{3}{5}x^2$

      (The key is "per unit x.")

               $R(x) = C(x)$

         $1250x - \dfrac{3}{5}x^2 = \dfrac{2}{5}x^2 + 222x + 28,000$

     $x^2 - 1028x + 28,000 = 0$

      $(x - 1000)(x - 28) = 0$

     Break-even values are at $x = 28$ and $x = 1000.$

  b. Maximum revenue occurs at $x = \dfrac{-1250}{-\dfrac{6}{5}} = 1042$ (rounded).

     $R(1042) = \$651,041.60$ is the maximum revenue.

  c. $P(x) = 1250x - \dfrac{3}{5}x^2 - \left(\dfrac{2}{5}x^2 + 222x + 28,000\right) = -x^2 + 1028x - 28,000$

     Maximum profit is at $x = \dfrac{-1028}{-2} = 514.$ $P(514) = \$236,196$ is the maximum profit.

  d. Price that will maximize profit is $p = 1250 - \dfrac{3}{5}(514) = \$941.60.$

35. $R(t) = 0.253t^2 - 4.03t + 76.84$

    a.   Minimum occurs at vertex where $t = \dfrac{-(-4.03)}{2(0.253)} = 7.96 \approx 8$. The minimum revenue is in 2000.

        $R(8) = \$60.79$ million .

    b.   The data shows that the minimum revenue was in 1999.

    c.

    d.   Except for 1998, the model and the data are a good fit.

37. a. $P = R - C = (-0.031t^2 + 0.776t + 0.179) - (-0.012t^2 + 0.492t + 0.725) = -0.019t^2 + 0.284t - 0.546$

    b.   V: $t = \dfrac{-0.284}{-0.038} = 7.47 \approx 7$ years or 1999

    c.

    d.   The model projects decreasing profits. Except for 2004, the data supports the model.

    e.   Revenue must be increased and/or costs must be decreased.

Exercise 2.4

Problems 1–27 Helpful methods:

   a.   Set $x = 0$.

   b.   Set $y = 0$.

   c.   As the denominator gets close to zero does y increase without bound?

   d.   As x gets large what happens to y?

1. j

3. b

5. f

7. g

9. i

11. k

13. cubic

15. $y = x^3 - x = x(x+1)(x-1) : \text{b}$

17. $y = 16x^2 - x^4 = x^2(4+x)(4-x) : \text{g}$

19. $y = x^2 + 7x = x(x+7) : \text{a}$

21. $y = \dfrac{x-3}{x+1} : \text{d}$

23.

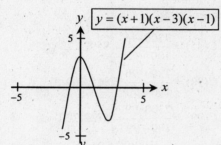

$$y = (x+1)(x-3)(x-1)$$

25.

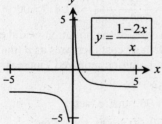

$$y = \frac{1-2x}{x}$$

27.

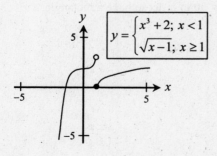

$$y = \begin{cases} x^3 + 2; & x < 1 \\ \sqrt{x-1}; & x \ge 1 \end{cases}$$

29. $F(x) = \dfrac{x^2 - 1}{x}$

$$F\left(-\frac{1}{3}\right) = \frac{\frac{1}{9} - 1}{-\frac{1}{3}} = \frac{8}{3}$$

$$F(10) = \frac{100 - 1}{10} = \frac{99}{10}$$

$$F(0.001) = \frac{0.000001 - 1}{0.001} = \frac{-0.999999}{0.001}$$

$$= -999.999$$

$F(0)$ is not defined–division by zero.

31. $f(x) = x^{3/2}$

   a. $f(16) = (\sqrt{16})^3 = 64$

   b. $f(1) = (\sqrt{1})^3 = 1$

   c. $f(100) = (\sqrt{100})^3 = 1000$

   d. $f(0.09) = (\sqrt{0.09})^3 = 0.027$

33. $k(x) = \begin{cases} 2 & \text{if } x < 0 \\ x+4 & \text{if } 0 \le x < 1 \\ 1-x & \text{if } x \ge 1 \end{cases}$

   a. $k(-5) = 2$ since $x < 0$.

   b. $k(0) = 0 + 4 = 4$

   c. $k(1) = 1 - 1 = 0$

   d. $k(-0.001) = 2$ since $x < 0$.

35. $y = 1.6x^2 - 0.1x^4$

a.

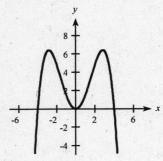

b. polynomial
c. no asymptotes
d. turning points at $x = 0$ and approximately $x = -2.8$ and $x = 2.8$

37. $y = \dfrac{2x+4}{x+1}$

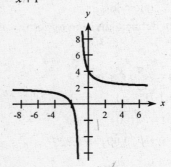

rational
vertical: $x = -1$
   horizontal: $y = 2$
no turning points

39. $f(x) = \begin{cases} -x & \text{if } x < 0 \\ 5x & \text{if } x \geq 0 \end{cases}$

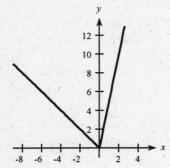

piecewise
no asymptotes
turning point at $x = 0$.

41. $V = V(x) = x^2(108 - 4x)$

$V(10) = 100(68) = 6800$ cubic inches
$\quad V(20) = 400(28) = 11{,}200$ cubic inches
$108 - 4x > 0$
$\quad -4x > -108$
$\quad\quad 0 < x < 27$

43. $f(x) = 105.095x^{1.5307}$

This is a power function graph turning upward.

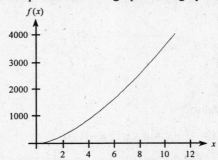

c. From the graph, assets reached $4000 billion in 10.7 years, or late 2000.
d. No, investments declined after 9/11/01.

45. $C(p) = \dfrac{7300p}{100 - p}$

a. $0 \leq p < 100$

b. $C(45) = \dfrac{7300 \cdot 45}{100 - 45} = \$5972.73$

c. $C(90) = \dfrac{7300 \cdot 90}{100 - 90} = \$65{,}700$

d. $C(99) = \dfrac{7300 \cdot 99}{100 - 99} = \$722{,}700$

e. $C(99.6) = \dfrac{7300(99.6)}{100 - 99.6} = \$1{,}817{,}700$

f. To remove p% of the pollution would cost $C(p)$. Note how cost increases as p (the percent of pollution removed) increases.

47. $A = A(x) = x(50 - x)$

$A(2) = 2 \cdot 48 = 96$ square feet

$\quad A(30) = 30 \cdot 20 = 600$ square feet

$0 < x < 50$ in order to have a rectangle.

49. a. $P(t) = \begin{cases} 1.965t - 5.65 & 5 \le t \le 20 \\ 0.095t^2 - 2.925t + 54.15 & t > 20 \end{cases}$

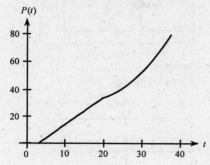

From the graph, $P(20) = \$33.65$ billion .

From the graph,

$P(38) = \$80.18$ billion (approx.)

51. a. $P(x) = \begin{cases} 0.37 & \text{if } 0 < x \le 1 \\ 0.60 & \text{if } 1 < x \le 2 \\ 0.83 & \text{if } 2 < x \le 3 \\ 1.06 & \text{if } 3 < x \le 4 \end{cases}$

The postage for a 1.2 ounce letter is $0.60.

$D = \{ x : 0 < x \le 4 \}$, $R = \{ 0.37, 0.60, 0.83, 1.06 \}$

2 ounces cost is $0.60, 2.01 ounces cost is $0.83

$p = \dfrac{200}{2 + 0.1x}$

b.

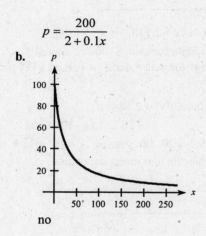

no

$C(x) = 30(x - 1) + \dfrac{3000}{x + 10}$

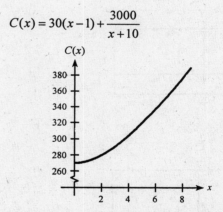

A turning point indicates a minimum or
maximum cost.

This is the fixed cost of production.

Exercise 2.5

1. linear   The points are in a straight line.

3. quadratic   The points appear to fit a parabola.

5. quartic   The graph crosses the x-axis four
times. Also there are three bends.

7. quadratic   There is one bend. A parabola is the
best fit.

9. $y = 2x - 3$ is the best fit.

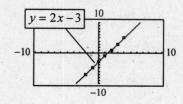

11. $y = 2x^2 - 1.5x - 4$ is the best fit.

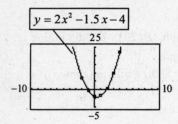

13. $y = x^3 - x^2 - 3x - 4$ is the best fit.

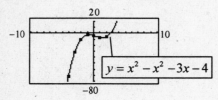

15. $y = 2x^{1/2}$ is the best fit.

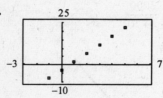

17. a.

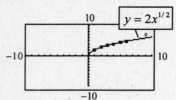

   b. linear

   c. $y = 5x - 3$

19. a.

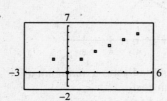

   b. quadratic

   c. $y = 0.0959x^2 + 0.4656x + 1.4758$

21. a.

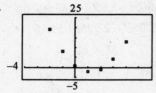

   b. quadratic

   c. $y = 2x^2 - 5x + 1$

23. a.

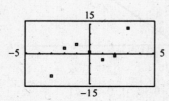

   b. cubic

   c. $y = x^3 - 5x + 1$

25. a.

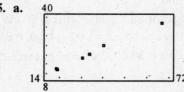

   b. $y = 0.518x + 2.775$

   c. $m = 0.518$; for each \$1000 of annual earnings for males, females earned \$518 per year.

27. a. $y = -0.0620x^3 + 2.5848x^2$
$$- 28.2315x - 103.9594$$

   b. $x = 20.3 \approx 20$; the year for $x = 19$, which is 1999 has the maximum data value.

29. a.

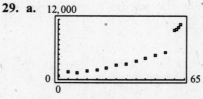

   b. $y = 149.931x - 683.033$

   c. $y = 0.1156x^3 - 7.6370x^2$
$$+ 186.4609x + 597.9660$$

   d.

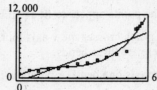

The cubic model is the better fit.

31. a. $y_1 = 7.888x^2 + 117.803x - 421.039$,

$y_2 = -1.974x^3 + 33.575x^2$
$$- 106.491x - 205.453$$

b. The cubic model is a much better fit to the data.
cubic: $y \approx -912.0$
quadratic: $y \approx -428.7$

33. a. quadratic or power

b. quadratic: $y = 1.0275x^2 - 29.598x + 299.31$

power: $y = 0.05193x^{2.5679}$

c. quadratic (it predicts \$1392.2 billion)

d. The year 2010 corresponds to $x = 60$.
quadratic model gives \$2152.4 billion.

35. a. $y = 0.3488x^{4.208}$

b. 2005 corresponds to $x = 25$
$\approx 266,139$ thousand subscribers

c. It is nearly the entire population of the U.S.

Review Exercises

1. $3x^2 + 10x = 5x$

$3x^2 + 5x = 0$

$x(3x + 5) = 0$

$x = 0$ or $x = -\dfrac{5}{3}$

2. $4x - 3x^2 = 0$

$x(4 - 3x) = 0$

$x = 0$ or $x = \dfrac{4}{3}$

3. $x^2 + 5x + 6 = 0$

$(x + 3)(x + 2) = 0$

$x = -3$ or $x = -2$

4. $11 - 10x - 2x^2 = 0$

$a = -2,\ b = -10,\ c = 11$

$x = \dfrac{10 \pm \sqrt{100 + 88}}{-4} = \dfrac{-5 \pm \sqrt{47}}{2}$

5. $(x - 1)(x + 3) = -8$

$x^2 + 2x - 3 = -8$

$x^2 + 2x + 5 = 0$

$b^2 - 4ac < 0$

No real solution

6. $4x^2 = 3$

$x^2 = \dfrac{3}{4}$

$x = \pm\sqrt{\dfrac{3}{4}} = \pm\dfrac{\sqrt{3}}{2}$

7. $20x^2 + 3x = 20 - 15x^2$

$35x^2 + 3x - 20 = 0$

$(7x - 5)(5x + 4) = 0$

$x = \dfrac{5}{7}$ or $x = -\dfrac{4}{5}$

8. $8x^2 + 8x = 1 - 8x^2$

$16x^2 + 8x - 1 = 0$

$a = 16,\ b = 8,\ c = -1$

$x = \dfrac{-8 \pm \sqrt{64 + 64}}{32} = \dfrac{-1 \pm \sqrt{2}}{4}$

9. $0.02x^2 - 2.07x + 7 = 0$

$a = 0.02,\ b = -2.07,\ c = 7$

$x = \dfrac{2.07 \pm \sqrt{4.2849 - 0.56}}{0.04} = \dfrac{2.07 \pm 1.93}{0.04}$

$= 100$ or 3.5

10. $46.3x - 117 - 0.5x^2 = 0$

$a = -0.5,\ b = 46.3,\ c = -117$

$x = \dfrac{-46.3 \pm \sqrt{2143.69 + (-234)}}{-1} = \dfrac{-46.3 \pm 43.7}{-1}$

$= 90$ or 2.6

11. $4z^2 + 25 = 0$

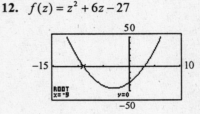

$4z^2 + 5^2 = 0$

The sum of 2 squares cannot be factored. There are no real solutions.

12. $f(z) = z^2 + 6z - 27$

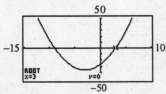

From the graph, the zeros are –9 and 3.

Algebraic solution:

$z(z + 6) = 27$

$z^2 + 6z - 27 = 0$

$(z + 9)(z - 3) = 0$

$z = -9$ or $z = 3$

13. $3x^2 - 18x - 48 = 0$

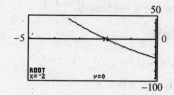

$$3(x^2 - 6x - 16) = 0$$
$$3(x - 8)(x + 2) = 0$$
$$x = -2, \ x = 8$$

14. $f(x) = 3x^2 - 6x - 9$

$$3x^2 - 6x - 9 = 0$$
$$3\left(x^2 - 2x - 3\right) = 0$$
$$3(x - 3)(x + 1) = 0$$
$$x = 3, \ x = -1$$

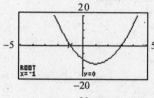

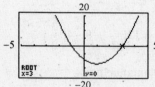

15. $x^2 + ax + b = 0$

To apply the quadratic formula we have "a" = 1, "b" = a, and "c" = b.

$$x = \frac{-a \pm \sqrt{a^2 - 4b}}{2}$$

16. $xr^2 - 4ar - x^2c = 0$

To solve for r, use the quadratic formula with "a" = x, "b" = $-4a$, and "c" = $-x^2c$.

$$r = \frac{4a \pm \sqrt{16a^2 + 4x(x^2c)}}{2x} = \frac{4a \pm \sqrt{16a^2 + 4x^3c}}{2x}$$

$$= \frac{4a \pm 2\sqrt{4a^2 + x^3c}}{2x} = \frac{2a \pm \sqrt{4a^2 + x^3c}}{x}$$

17. $-0.002x^2 - 14.1x + 23.1 = 0$

$$x = \frac{14.1 \pm \sqrt{198.81 + 0.1848}}{-0.004} = \frac{14.1 \pm 14.107}{-0.004}$$

$$= -7051.64, \ 1.64, \ \text{or} \ 1.75 \ \text{(using 14.107)}$$

18. $1.03x^2 + 2.02x - 1.015 = 0$

$a = 1.03, \ b = 2.02, \ c = -1.015$

$$x = \frac{-2.02 \pm \sqrt{4.0804 + 4.1818}}{2.06} = \frac{-2.02 \pm 2.87}{2.06}$$

$$= -2.38 \ \text{or} \ 0.41$$

19. $y = \dfrac{1}{2}x^2 + 2x$

$a > 0$, thus vertex is a minimum.

$$V: \ x = \frac{-2}{2\left(\dfrac{1}{2}\right)} = -2$$

$$y = \frac{1}{2}(-2)^2 + 2(-2) = -2$$

Zeros: $\dfrac{1}{2}x^2 + 2x = 0$

$$x\left(\frac{1}{2}x + 2\right) = 0$$

$$x = 0, -4$$

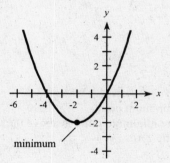

20. $y = 4 - \dfrac{1}{4}x^2$

V: x-coordinate = 0
  y-coordinate = 4
  (0, 4) is a maximum point
Zeros are $x = \pm 4$.

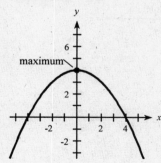

21. $y = 6 + x - x^2$

$a < 0$, thus vertex is a maximum.

V: $x = \dfrac{-1}{2(-1)} = \dfrac{1}{2}$

$y = 6 + \dfrac{1}{2} - \left(\dfrac{1}{2}\right)^2 = \dfrac{25}{4}$

Zeros: $6 + x - x^2 = 0$

$(3 - x)(2 + x) = 0$

$x = -2, 3$

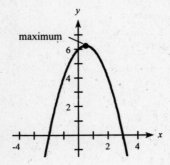

22. $y = x^2 - 4x + 5$

V: x-coordinate $= \dfrac{4}{2} = 2$

y-coordinate $= 2^2 - 4(2) + 5 = 1$

$(2, 1)$ is a minimum point.

Zeros: Since the minimum point is above the x-axis, there are no zeros.

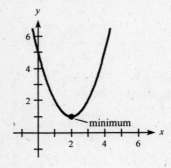

23. $y = x^2 + 6x + 9$

$a > 0$, thus vertex is a minimum.

V: $x = \dfrac{-6}{2(1)} = -3$

$y = (-3)^2 + 6(-3) + 9 = 0$

Zeros: $x^2 + 6x + 9 = 0$

$(x + 3)(x + 3) = 0$

$x = -3$

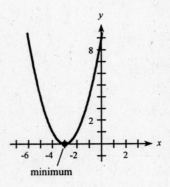

24. $y = 12x - 9 - 4x^2$

V: x-coordinate $= -\dfrac{12}{-8} = \dfrac{3}{2}$

y-coordinate $= 12\left(\dfrac{3}{2}\right) - 9 - 4\left(\dfrac{3}{2}\right)^2 = 0$

$\left(\dfrac{3}{2}, 0\right)$ is a maximum point.

Zeros: From the vertex we have that $x = \dfrac{3}{2}$ is the only zero.

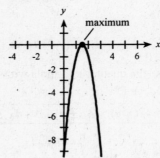

25. $y = \dfrac{1}{3}x^2 - 3$

V: $(0, -3)$

Zeros: $\dfrac{1}{3}x^2 - 3 = 0$

$$x^2 = 9$$
$$x = \pm 3$$

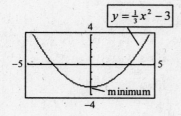

26. $y = \dfrac{1}{2}x^2 + 2$

Vertex: $(0, 2) \leftarrow$ minimum

No zeros.

The graph using x-min $= -4$ y-min $= 0$

$\qquad\qquad\qquad x$-max $= 4$ y-max $= 6$

is shown below.

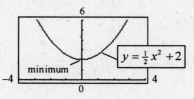

27. $y = x^2 + 2x + 5$

V: $(-1, 4)$

There are no real zeros.

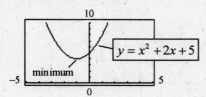

28. $y = -10 + 7x - x^2$

Vertex: $\left(\dfrac{7}{2}, \dfrac{9}{4}\right) \leftarrow$ maximum

Zeros: $x^2 - 7x + 10 = 0$

$\qquad (x - 5)(x - 2) = 0$

$\qquad\quad x = 5$ or $x = 2$

Graph using x-min $= 0$ y-min $= -5$

$\qquad\qquad\quad x$-max $= 8$ y-max $= 5$

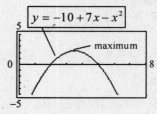

29. $y = 20x - 0.1x^2$

Zeros: $x(20 - 0.1x) = 0$

$\qquad\quad x = 0, 200$

(This is an alternative method of getting the vertex.)

The x-coordinate of the vertex is halfway between the zeros.

V: $(100, 1000)$

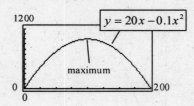

30. $y = 50 - 1.5x + 0.01x^2$

Vertex: $(75, -6.25) \leftarrow$ minimum

Zeros: $\qquad 0.01x^2 - 1.5x + 50 = 0$

$\qquad\quad 0.01(x^2 - 150x + 5000) = 0$

$\qquad\qquad 0.01(x - 50)(x - 100) = 0$

$\qquad\qquad\quad x = 50$ or $x = 100$

Graph using x-min $= 0$ y-min $= -10$

$\qquad\qquad\quad x$-max $= 125$ y-max $= 10$

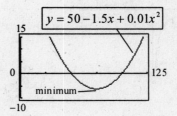

31. $\dfrac{f(50)-f(30)}{50-30} = \dfrac{2500-2100}{20} = \dfrac{400}{20} = 20$

32. $\dfrac{f(50)-f(0)}{50-0} = \dfrac{1022-22}{50} = \dfrac{1000}{50} = 20$

33. a. The vertex is halfway between the zeros. So, the vertex is $\left(1, -4\dfrac{1}{2}\right)$.

    b. The zeros are where the graph crosses the x-axis. $x = -2, 4$.

    c. The graph matches B.

34. From the graph,
    a. Vertex is $(0, 49)$
    b. Zeros are $x = \pm 7$.
    c. Matches with D.

35. a. The vertex is halfway between the zeros. So, the vertex is $(7, 24.5)$.
    b. Zeros are $x = 0, 14$.
    c. The graph matches A.

36. From the graph,
    a. Vertex is $(-1, 9)$.
    b. Zeros are $x = -4$ and $x = 2$.
    c. Matches with C.

37. a. $f(x) = x^2$

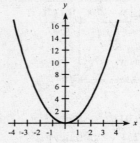

    b. $f(x) = \dfrac{1}{x}$

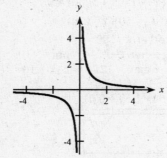

c. $f(x) = x^{1/4}$

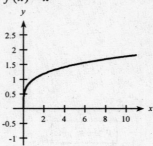

38. $f(x) = \begin{cases} -x^2 & \text{if } x \le 0 \\ \dfrac{1}{x} & \text{if } x > 0 \end{cases}$

    a. $f(0) = -(0^2) = 0$

    b. $f(0.0001) = \dfrac{1}{0.0001} = 10,000$

    c. $f(-5) = -(-5)^2 = -25$

    d. $f(10) = \dfrac{1}{10} = 0.1$

39. $f(x) = \begin{cases} x & \text{if } x \le 1 \\ 3x-2 & \text{if } x > 1 \end{cases}$

    a. $f(-2) = -2$
    b. $f(0) = 0$
    c. $f(1) = 1$
    d. $f(2) = 3 \cdot 2 - 2 = 4$

40. $f(x) = \begin{cases} x & \text{if } x \le 1 \\ 3x-2 & \text{if } x > 1 \end{cases}$

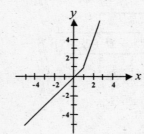

41. a. $f(x) = (x-2)^2$

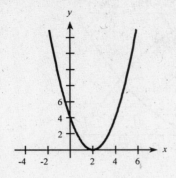

b. $f(x) = (x+1)^3$

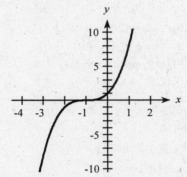

42. $y = x^3 + 3x^2 - 9x$

Using x-min $= -10$, x-max $= 10$, y-min $= -10$, y-max $= 35$, the turning points are at $x = -3$ and 1.

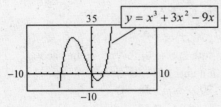

43. $y = x^3 - 9x$

Using x-min $= -4.7$, x-max $= 4.7$, y-min $= -15$, y-max $= 15$, the turning points are at $x = \pm 1.732$.

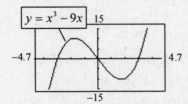

Note: Your turning points in 42–43. may vary depending on your scale.

44. $y = \dfrac{1}{x-2}$

There is a vertical asymptote $x = 2$.
There is a horizontal asymptote $y = 0$.

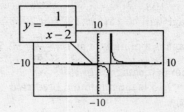

45. $y = \dfrac{2x-1}{x+3} = \dfrac{2 - \dfrac{1}{x}}{1 + \dfrac{3}{x}}$

Vertical asymptote is $x = -3$.
Horizontal asymptote is $y = 2$.

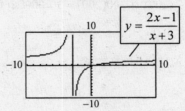

46. a.   Left to reader.
  b. $y = -2.179x + 159.857$ is a good fit to the data.
  c. $y = -0.082x^2 - 0.214x + 153.310$ is a slightly better fit.

47. a.   Left to reader.
  b. $y = 2.141x + 34.391$ is a good fit to the data.
  c. $y = -0.045x^2 + 3.607x + 26.610$ is a slightly better fit.

48. $S = 96 + 32t - 16t^2$
  a. $16(6 + 2t - t^2) = 0$
    $$t = \frac{-2 \pm \sqrt{4 + 24}}{-2}$$
    $t \approx -1.65$ or $t \approx 3.65$
  b. $t \geq 0$ Use $t = 3.65$
  c.   After 3.65 seconds

49. $P(x) = -0.10x^2 + 82x - 1600$
    $(-0.10x + 80)(x - 20) = 0$
    Break-even at $x = 20, 800$

50. $f(t) = 0.0241904t^2 - 4.47459t + 216.074$

a. $0.0241904t^2 - 4.47459t + (216.074 - 15) = 0$

Using the calculator and the quadratic formula.

$t = 76.9$ or 108

Percentage will be 15% in 1977 or 2008.

b. Percentage is a minimum at $t = \dfrac{-b}{2a} = 92.5$

Minimum percentage is in 1993.

$f(92.5) = 9.15$ is the minimum predicted percentage.

51. $A = -\dfrac{3}{4}x^2 + 300x$

a. V: $x = \dfrac{-300}{-\dfrac{3}{2}} = 200$ ft

b. $A = -\dfrac{3}{4}(200)^2 + 300(200) = 30,000$ sq ft

52. $p = 2q^2 + 4q + 6$

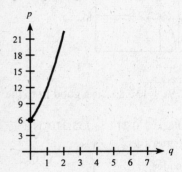

53.

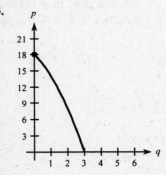

54. a.

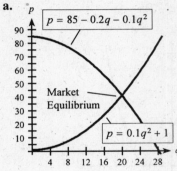

b. $\qquad 0.1q^2 + 1 = 85 - 0.2q - 0.1q^2$

$0.2q^2 + 0.2q - 84 = 0$

$0.2(q^2 + q - 420) = 0$

$0.2(q - 20)(q + 21) = 0$

$q = 20$ (only positive value)

$p = 0.1(20)^2 + 1 = 41$

55. $p = q^2 + 300$

$p = -q + 410$

$q^2 + 300 = -q + 410$

$q^2 + q - 110 = 0$

$(q + 11)(q - 10) = 0$

$q = 10$

$p = -10 + 410 = 400$

So, E: (10, 400).

56. D: $p^2 + 5q = 200 \rightarrow p^2 = 200 - 5q$

S: $40 - p^2 + 3q = 0$

Substitute $200 - 5q$ for p^2 in the second equation and solve for q.

$40 - (200 - 5q) + 3q = 0$

$-160 = -8q$

$q = 20$

$p^2 = 200 - 5(20)$

$p^2 = 100$ or $p = 10$

57. $R(x) = 100x - 0.4x^2$

$C(x) = 1760 + 8x + 0.6x^2$

$100x - 0.4x^2 = 1760 + 8x + 0.6x^2$

$x^2 - 92x + 1760 = 0$

$x = \dfrac{92 \pm \sqrt{1424}}{2} = 46 \pm 2\sqrt{89} \approx 64.87, 27.13$

$(\sqrt{1424} = \sqrt{16 \cdot 89})$

58. $C(x) = 900 + 25x$

$\quad R(x) = 100x - x^2$

$\quad\quad 900 + 25x = 100x - x^2$

$\quad\quad x^2 - 75x + 900 = 0$

$\quad\quad (x - 60)(x - 15) = 0$

$\quad\quad x = 60$ or $x = 15$

$\quad\quad R(60) = 2400;\ R(15) = 1275$

$\quad\quad (60, 2400)$ and $(15, 1275)$

59. $R(x) = 100x - x^2$

$\quad$ V: $x = \dfrac{-100}{-2} = 50$

$\quad R(50) = 100(50) - 50^2$

$\quad\quad\quad = \$2500$ max revenue

$\quad P(x) = (100x - x^2) - (900 + 25x)$

$\quad\quad\quad = -x^2 + 75x - 900$

$\quad$ V: $x = \dfrac{-75}{-2} = 37.5$

$\quad P(37.5) = \$506.25$ max profit

60. $P(x) = 1.3x - 0.01x^2 - 30$

$\quad$ x-coordinate of the vertex $= \dfrac{1.3}{0.02} = 65$

$\quad P(65) = 1.3(65) - 0.01(65)^2 - 30 = 12.25 \leftarrow$ max

$\quad$ Break-even points:

$\quad 0 = 1.3x - 0.01x^2 - 30$

$\quad 0 = -0.01(x^2 - 130x + 3000)$

$\quad 0 = -0.01(x - 30)(x - 100)$

$\quad x = 30$ or $x = 100$

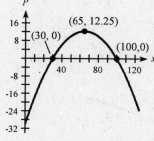

61. $P(x) = (50x - 0.2x^2) - (360 + 10x + 0.2x^2)$

$\quad\quad\quad = -0.4x^2 + 40x - 360$

$\quad$ V: $x = \dfrac{-40}{-0.8} = 50$ units for maximum profit.

$\quad P(50) = -0.4(50)^2 + 40(50) - 360$

$\quad\quad\quad = \$640$ maximum profit.

62. a. $C(x) = 15,000 + (140 + 0.04x)x$

$\quad\quad\quad\quad = 15,000 + 140x + 0.04x^2$

$\quad\quad R(x) = (300 - 0.06x)x$

$\quad\quad\quad\quad = 300x - 0.06x^2$

$\quad$ **b.** $\quad 15,000 + 140x + 0.04x^2 = 300x - 0.06x^2$

$\quad\quad\quad 0.10x^2 - 160x + 15,000 = 0$

$\quad\quad\quad 0.1(x^2 - 1600x + 150,000) = 0$

$\quad\quad\quad\quad 0.1(x - 100)(x - 1500) = 0$

$\quad\quad\quad x = 100$ or $x = 1500$

$\quad$ **c.** Maximum revenue:

$\quad\quad\quad$ x-coordinate: $-\dfrac{300}{-0.12} = 2500$

$\quad$ **d.** $P(x) = R(x) - C(x)$

$\quad\quad\quad\quad = -0.10x^2 + 160x - 15,000$

$\quad\quad\quad$ x-coordinate of max $= -\dfrac{160}{-0.20} = 800$

$\quad$ **e.** $P(2500) = \$240,000$ loss

$\quad\quad\quad P(800) = \$49,000$ profit

63. $H(t) = 0.099t^{1.969}$

$\quad$ **a.** Power function

$\quad$ **b.** $H(25) = 0.099(25)^{1.969} = 55.998845$

$\quad\quad\quad$ or $5,599,885$ cases.

$\quad\quad\quad H(t)$ is in hundreds of thousands.

$\quad$ **c.** $H(15) = 0.099(15)^{1.969} = 20.481364$

$\quad\quad\quad$ There will be $2,048,136$ cases in 1995.

64. $y = 120x^2 - 20x^3$, for $x \geq 0$.

$\quad$ **a.**

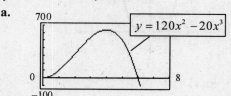

$\quad$ **b.** $y = 20x^2(6 - x)$.

$\quad\quad\quad$ Domain: $0 \leq x \leq 6$

65. $C(p) = \dfrac{4800p}{100 - p}$

$\quad$ **a.** Rational function

$\quad$ **b.** Domain: $0 \leq p < 100$

$\quad$ **c.** $C(0) = 0$ means there is no cost if no pollution is removed.

$\quad$ **d.** $C(99) = \dfrac{4800(99)}{100 - 99} = \$475,200$

66. $C(x) = \begin{cases} 1.557x & 0 \le x \le 100 \\ 155.7 + 1.04(x-100) & 100 < x \le 1000 \\ 1091.7 + 0.689(x-1000) & x > 1000 \end{cases}$

a. $C(12) = 1.557(12) = \$18.68$

b. $C(825) = 155.70 + 1.04(825 - 100) =$
$\$909.70$

67. a. The best fit is the power function.
$y = 17.3969x^{0.5094}$

b.

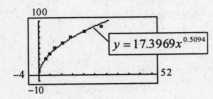

c. $f(5) = 17.3969(5)^{0.5094} = 39.5$ mph

d. Use the TRACE KEY. It will take 18. 3 seconds.

68. a. The best fit is the quadratic function
$g = 1.8155x^2 - 11.1607x + 29.1845$.

b. For $2000 - 2010$ $x = 11$.
$C(11) = 126$ million cu. yards.

69. a. $y = 6.2992x^{1.6683}$

b. $y = 3.3416x^2 - 86.7630x + 981.8941$

c. They are both good fits, but the quadratic model fits better.

Chapter Test

1. **a.** $f(x) = x^4$

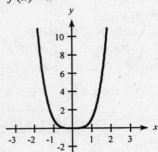

b. $g(x) = |x|$

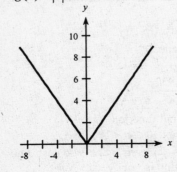

c. $h(x) = -1$

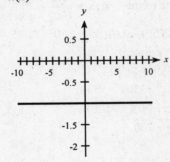

d. $k(x) = \sqrt{x}$

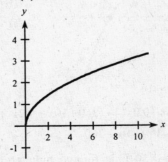

2. figure b is the graph for $b > 1$.
figure a is the graph for $0 < b < 1$.

3. $f(x) = ax^2 + bx + c$ and $a < 0$ is a parabola opening downward.

4. **a.** $f(x) = (x+1)^2 - 1$

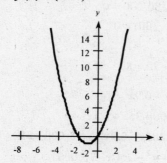

b. $f(x) = (x-2)^3 + 1$

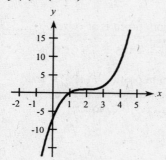

5. $f(x) = x^3 - 4x^2 = x^2(x-4)$.

a. and **b.** are the cubic choices. $f(x) < 0$ if $0 < x < 4$. Answer: b

6. $f(x) = \begin{cases} 8x + \dfrac{1}{x} & \text{if } x < 0 \\ 4 & \text{if } 0 \le x \le 2 \\ 6 - x & \text{if } x > 2 \end{cases}$

a. $f(16) = 6 - 16 = -10$

b. $f(-2) = 8(-2) + \dfrac{1}{-2} = -16\dfrac{1}{2}$

c. $f(13) = 6 - 13 = -7$

7. $g(x) = \begin{cases} x^2 & \text{if } x \le 1 \\ 4 - x & \text{if } x > 1 \end{cases}$

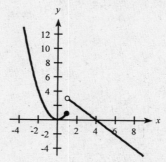

8. $f(x) = 21 - 4x - x^2 = (7+x)(3-x)$

Vertex: $x = \dfrac{-b}{2a} = \dfrac{-(-4)}{2(-1)} = -2$

Point: $(-2, 25)$

Zeros: $f(x) = 0$ at $x = -7$ or 3.

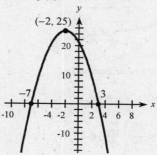

9.
$$3x^2 + 2 = 7x$$
$$3x^2 - 7x + 2 = 0$$
$$(3x - 1)(x - 2) = 0$$
$$3x - 1 = 0 \text{ or } x - 2 = 0$$
$$x = \frac{1}{3}, 2$$

10. $2x^2 + 6x - 9 = 0$

$$x = \frac{-6 \pm \sqrt{36 + 72}}{4} = \frac{-6 \pm 6\sqrt{3}}{4} = \frac{-3 \pm 3\sqrt{3}}{2}$$

11. $\left(\dfrac{1}{x} + 2x = \dfrac{1}{3} + \dfrac{x+1}{x} \right) 3x$

$$3 + 6x^2 = x + 3x + 3$$
$$6x^2 - 4x = 0$$
$$2x(3x - 2) = 0$$
$$x = \frac{2}{3} \text{ is the only solution.}$$

12. $g(x) = \dfrac{3(x-4)}{x+2}$

Vertical asymptote at $x = -2$.

$g(4) = 0$

Answer: c

13. $f(x) = \dfrac{8}{2x-10}$

Horizontal: $y = 0$

Vertical: $2x - 10 = 0$

$\qquad\qquad 2x = 10$

$\qquad\qquad\ \ x = 5$

14. $\dfrac{f(40)-f(10)}{40-10} = \dfrac{320-(-940)}{30} = \dfrac{1260}{30} = 42$

15. a. quartic
  b. cubic

16. a. $f(x) = -0.3577x + 19.9227$

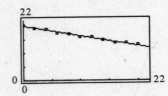

  b. $f(40) = 5.6$

  c. $f(x) = 0$ if $x = \dfrac{19.9227}{0.3577} \approx 55.7$

17. S: $p = \dfrac{1}{6}q + 30$

  D: $p = \dfrac{30,000}{q} - 20$

$\left(\dfrac{1}{6}q + 30 = \dfrac{30,000}{q} - 20 \right) 6q$

$q^2 + 180q = 180,000 - 120q$

$q^2 + 300q - 180,000 = 0$

$(q + 600)(q - 300) = 0$

$E_q : q = 300$

$E_p : p = 50 + 30 = 80$

18. $R(x) = 285x - 0.9x^2$

  $C(x) = 15,000 + 35x + 0.1x^2$

  a. $P(x) = 285x - 0.9x^2 - (15,000 + 35x + 0.1x^2)$

  $\qquad\quad = -x^2 + 250x - 15,000$

  $\qquad\quad = (100 - x)(x - 150)$

  b. Maximum profit is at vertex.

  $\qquad x = \dfrac{-250}{2(-1)} = 125$

  Maximum profit $= P(125) = \$625$

  c. Break-even means $P(x) = 0$.
  From **a.**, $x = 100, 150$.

19. $WC = f(s) = \begin{cases} 0 & \text{if } 0 \le s \le 4 \\ 91.4 - 7.46738(5.81 + 3.7\sqrt{s} - 0.25s) & \text{if } 4 < s \le 45 \\ -55 & \text{if } s > 45 \end{cases}$

  b. Use middle rule for $s = 15$. $f(15) = -31$
  If the air temperature is 0°F and the wind speed is 15 mph, the air temperature feels like -31°F. In winter, the T.V weather report usually gives the wind chill temperature.
  c. $f(48) = -55$ (last rule, $s > 45$)

20. a.

  b. quadratic
  c. $y = -0.313x^2 + 4.034x + 22.797$
  d. 2002 corresponds to $x = 12$, $y(12) \approx 26.1\%$

Chapter 3: Matrices

Exercise 3.1 _____

1. Matrix B has 3 rows.

3. $-F = \begin{bmatrix} -1 & -2 & -3 \\ 1 & 0 & -1 \\ -2 & 3 & 4 \end{bmatrix}$

5. $A, C, D, F,$ and Z are square.

7. $a_{23} = 1$

9. $A^T = \begin{bmatrix} 1 & 3 & 4 \\ 0 & 2 & 0 \\ 2 & 1 & 3 \end{bmatrix}$

11. $M + (-M) = \begin{bmatrix} 0 & 0 & 0 \\ 0 & 0 & 0 \\ 0 & 0 & 0 \end{bmatrix}$

13. $C + D = \begin{bmatrix} 5+4 & 3+2 \\ 1+3 & 2+5 \end{bmatrix} = \begin{bmatrix} 9 & 5 \\ 4 & 7 \end{bmatrix}$

15. $A - F = \begin{bmatrix} 1-1 & 0-2 & 2-3 \\ 3+1 & 2-0 & 1-1 \\ 4-2 & 0+3 & 3+4 \end{bmatrix} = \begin{bmatrix} 0 & -2 & -1 \\ 4 & 2 & 0 \\ 2 & 3 & 7 \end{bmatrix}$

17. $A + A^T = \begin{bmatrix} 1 & 0 & 2 \\ 3 & 2 & 1 \\ 4 & 0 & 3 \end{bmatrix} + \begin{bmatrix} 1 & 3 & 4 \\ 0 & 2 & 0 \\ 2 & 1 & 3 \end{bmatrix} = \begin{bmatrix} 2 & 3 & 6 \\ 3 & 4 & 1 \\ 6 & 1 & 6 \end{bmatrix}$

19. Because the orders are different, $D - E$ is undefined.

21. $3B = 3 \cdot \begin{bmatrix} 1 & 1 & 3 & 0 \\ 4 & 2 & 1 & 1 \\ 3 & 2 & 0 & 1 \end{bmatrix} = \begin{bmatrix} 3 & 3 & 9 & 0 \\ 12 & 6 & 3 & 3 \\ 9 & 6 & 0 & 3 \end{bmatrix}$

23. $4C + 2D = \begin{bmatrix} 20 & 12 \\ 4 & 8 \end{bmatrix} + \begin{bmatrix} 8 & 4 \\ 6 & 10 \end{bmatrix} = \begin{bmatrix} 28 & 16 \\ 10 & 18 \end{bmatrix}$

25. Because the orders are different, $2A - 3B$ is undefined.

27. $\begin{bmatrix} x & 1 & 0 \\ 0 & y & z \\ w & 2 & 1 \end{bmatrix} = \begin{bmatrix} 3 & 1 & 0 \\ 0 & 2 & 3 \\ 4 & 2 & 1 \end{bmatrix}$

By setting corresponding elements equal we have $x = 3, y = 2, z = 3, w = 4.$

29. $\begin{bmatrix} x & 3 & 2x-1 \\ y & 4 & 4y \end{bmatrix} = \begin{bmatrix} 2x-4 & z & 7 \\ 1 & w+1 & 3y+1 \end{bmatrix}$

$2x - 4 = x$ or $x = 4, z = 3, y = 1,$
$w + 1 = 4$ or $w = 3.$

31. $\begin{bmatrix} 3x & 3y \\ 3y & 3z \end{bmatrix} + \begin{bmatrix} 4x & -2y \\ 6y & -8z \end{bmatrix} = \begin{bmatrix} 14 & 4-y \\ 18 & 15 \end{bmatrix}$

Setting corresponding elements equal gives the following:

$3x + 4x = 14$ $3y - 2y = 4 - y$
$7x = 14$ $2y = 4$
$x = 2$ $y = 2$

$3y + 6y = 18$ $3z - 8z = 15$
$9y = 18$ $-5z = 15$
$y = 2$ $z = -3$

33. a. $A = \begin{bmatrix} 63 & 78 & 14 & 10 & 70 \\ 9 & 14 & 22 & 8 & 44 \end{bmatrix}$

$B = \begin{bmatrix} 251 & 175 & 64 & 8 & 11 \\ 17 & 6 & 15 & 1 & 0 \end{bmatrix}$

b. $A + B = \begin{bmatrix} 314 & 253 & 78 & 18 & 81 \\ 26 & 20 & 37 & 9 & 44 \end{bmatrix}$

c. $B - A = \begin{bmatrix} 188 & 97 & 50 & -2 & -59 \\ 8 & -8 & -7 & -7 & -44 \end{bmatrix}$

The negative entries tell us how many more endangered or threatened species are in the US.

35. a. Capital Expenditures + Gross Operating Costs = $\begin{bmatrix} 11041.7 & 8978.4 & 6461.0 \\ 8739.8 & 9159.6 & 6877.3 \\ 9798.1 & 9086.7 & 6448.4 \\ 9696.6 & 8926.7 & 6109.5 \end{bmatrix}$

b. R3, C1 was the most expensive

37. a. $A + B = \begin{bmatrix} 825 & 580 & 1560 \\ 810 & 650 & 350 \end{bmatrix}$ **b.** $B - A = \begin{bmatrix} -75 & 20 & -140 \\ 10 & -50 & 50 \end{bmatrix}$

39.

$A = \begin{array}{cc} M & F \\ \begin{bmatrix} 54.4 & 55.6 \\ 62.1 & 66.6 \\ 67.4 & 74.1 \\ 70.7 & 78.1 \\ 72.9 & 79.4 \\ 75.0 & 80.2 \end{bmatrix} \end{array}$ $B = \begin{array}{cc} M & F \\ \begin{bmatrix} 45.5 & 45.2 \\ 51.5 & 54.9 \\ 61.1 & 67.4 \\ 63.8 & 72.5 \\ 64.5 & 73.6 \\ 68.6 & 75.5 \end{bmatrix} \end{array}$ $A - B = \begin{array}{cc} M & F \\ \begin{bmatrix} 8.9 & 10.4 \\ 10.6 & 11.7 \\ 6.3 & 6.7 \\ 6.9 & 5.6 \\ 8.4 & 5.8 \\ 6.4 & 4.7 \end{bmatrix} \end{array}$

41. a. $\begin{bmatrix} 342 & 376 \\ 493 & 603 \\ 520 & 731 \\ 565 & 777 \\ 505 & 738 \\ 378 & 537 \end{bmatrix}$

b. $1.2 \cdot \begin{bmatrix} 342 & 376 \\ 493 & 603 \\ 520 & 731 \\ 565 & 777 \\ 505 & 738 \\ 378 & 537 \end{bmatrix} = \begin{bmatrix} 410.4 & 451.2 \\ 591.6 & 723.6 \\ 624 & 877.2 \\ 678 & 932.4 \\ 606 & 885.6 \\ 453.6 & 644.4 \end{bmatrix}$

43. a. $1.05A = \begin{bmatrix} 46.20 & 84 & 210 & 10.50 \\ 42 & 84 & 42 & 0 \\ 58.80 & 147 & 94.50 & 0 \\ 31.50 & 147 & 42 & 21 \\ 42 & 0 & 210 & 10.50 \end{bmatrix}$

b. $1.10A = \begin{bmatrix} 48.40 & 88 & 220 & 11 \\ 44 & 88 & 44 & 0 \\ 61.60 & 154 & 99 & 0 \\ 33 & 154 & 44 & 22 \\ 44 & 0 & 220 & 11 \end{bmatrix}$

45. a. $\begin{bmatrix} 0 & 1 & 1 & 0 \\ 1 & 0 & 1 & 1 \\ 0 & 0 & 0 & 1 \\ 0 & 1 & 1 & 0 \end{bmatrix}$

b. $\begin{bmatrix} 0 & 1 & 1 & 1 \\ 1 & 0 & 1 & 1 \\ 0 & 1 & 0 & 1 \\ 1 & 1 & 1 & 0 \end{bmatrix}$

c. $A + B = \begin{bmatrix} 0 & 2 & 2 & 1 \\ 2 & 0 & 2 & 2 \\ 0 & 1 & 0 & 2 \\ 1 & 2 & 2 & 0 \end{bmatrix}$

Sum in Row 1 = 5, Row 2 = 6, Row 3 = 3, Row 4 = 5. Person 2 is the most active.

47. a. $A + B = \begin{bmatrix} 50+30 & 30+45 \\ 36+22 & 44+62 \end{bmatrix} = \begin{bmatrix} 80 & 75 \\ 58 & 106 \end{bmatrix}$

b. $A+B+C = \begin{bmatrix} 80+96 & 75+52 \\ 58+81 & 106+37 \end{bmatrix}$

$= \begin{bmatrix} 176 & 127 \\ 139 & 143 \end{bmatrix}$

c. $A - D = \begin{bmatrix} 50-40 & 30-26 \\ 36-29 & 44-42 \end{bmatrix} = \begin{bmatrix} 10 & 4 \\ 7 & 2 \end{bmatrix}$

d. $B - D = \begin{bmatrix} 30-40 & 45-26 \\ 22-29 & 62-42 \end{bmatrix} = \begin{bmatrix} -10 & 19 \\ -7 & 20 \end{bmatrix}$

Shortage; take from inventory.

49. a. Row 7: 3, 4, 5, 6
b. Row 1: 1

51. For each row: (C1 + C2 + C3 + C4) ÷ 4 = average efficiency

W1: 0.9625   W2: 0.9375   W3: 0.9125

W4: 0.8875   W5: 0.85   W6: 0.875

W7: 0.90   W8: 0.925   W9: 0.95

W5 is least efficient. He works best at center 5.

Exercise 3.2

1. $\begin{bmatrix} 1 & 2 & 3 \end{bmatrix} \begin{bmatrix} 4 \\ 5 \\ 6 \end{bmatrix} = \begin{bmatrix} 1 \cdot 4 + 2 \cdot 5 + 3 \cdot 6 \end{bmatrix} = \begin{bmatrix} 32 \end{bmatrix}$ (not 32)

3. $\begin{bmatrix} 1 & 2 \end{bmatrix} \begin{bmatrix} 3 & 5 \\ 4 & 6 \end{bmatrix} = \begin{bmatrix} 1 \cdot 3 + 2 \cdot 4 & 1 \cdot 5 + 2 \cdot 6 \end{bmatrix}$

$= \begin{bmatrix} 11 & 17 \end{bmatrix}$

5. $CD = \begin{bmatrix} 5 & 3 \\ 1 & 2 \end{bmatrix} \cdot \begin{bmatrix} 4 & 2 \\ 3 & 5 \end{bmatrix}$

$= \begin{bmatrix} 20+9 & 10+15 \\ 4+6 & 2+10 \end{bmatrix}$

$= \begin{bmatrix} 29 & 25 \\ 10 & 12 \end{bmatrix}$

7. $DC = \begin{bmatrix} 4 & 2 \\ 3 & 5 \end{bmatrix} \cdot \begin{bmatrix} 5 & 3 \\ 1 & 2 \end{bmatrix}$

$= \begin{bmatrix} 20+2 & 12+4 \\ 15+5 & 9+10 \end{bmatrix}$

$= \begin{bmatrix} 22 & 16 \\ 20 & 19 \end{bmatrix}$

Note $CD \neq DC$

9. $AB = \begin{bmatrix} 1 & 0 & 2 \\ 3 & 2 & 1 \\ 4 & 0 & 3 \end{bmatrix} \begin{bmatrix} 1 & 1 & 3 & 0 \\ 4 & 2 & 1 & 1 \\ 3 & 2 & 0 & 1 \end{bmatrix} = \begin{bmatrix} 1+0+6 & 1+0+4 & 3+0+0 & 0+0+2 \\ 3+8+3 & 3+4+2 & 9+2+0 & 0+2+1 \\ 4+0+9 & 4+0+6 & 12+0+0 & 0+0+3 \end{bmatrix} = \begin{bmatrix} 7 & 5 & 3 & 2 \\ 14 & 9 & 11 & 3 \\ 13 & 10 & 12 & 3 \end{bmatrix}$

11. BA is undefined.

13. $EB = \begin{bmatrix} 1 & 0 & 4 \\ 5 & 1 & 0 \end{bmatrix} \begin{bmatrix} 1 & 1 & 3 & 0 \\ 4 & 2 & 1 & 1 \\ 3 & 2 & 0 & 1 \end{bmatrix} = \begin{bmatrix} 1+0+12 & 1+0+8 & 3+0+0 & 0+0+4 \\ 5+4+0 & 5+2+0 & 15+1+0 & 0+1+0 \end{bmatrix} = \begin{bmatrix} 13 & 9 & 3 & 4 \\ 9 & 7 & 16 & 1 \end{bmatrix}$

15. $EA^T = \begin{bmatrix} 1 & 0 & 4 \\ 5 & 1 & 0 \end{bmatrix} \begin{bmatrix} 1 & 3 & 4 \\ 0 & 2 & 0 \\ 2 & 1 & 3 \end{bmatrix} = \begin{bmatrix} 1+0+8 & 3+0+4 & 4+0+12 \\ 5+0+0 & 15+2+0 & 20+0+0 \end{bmatrix} = \begin{bmatrix} 9 & 7 & 16 \\ 5 & 17 & 20 \end{bmatrix}$

17. $A^2 = \begin{bmatrix} 1 & 0 & 2 \\ 3 & 2 & 1 \\ 4 & 0 & 3 \end{bmatrix} \begin{bmatrix} 1 & 0 & 2 \\ 3 & 2 & 1 \\ 4 & 0 & 3 \end{bmatrix} = \begin{bmatrix} 1+0+8 & 0+0+0 & 2+0+6 \\ 3+6+4 & 0+4+0 & 6+2+3 \\ 4+0+12 & 0+0+0 & 8+0+9 \end{bmatrix} = \begin{bmatrix} 9 & 0 & 8 \\ 13 & 4 & 11 \\ 16 & 0 & 17 \end{bmatrix}$

19. $C^3 = \begin{bmatrix} 5 & 3 \\ 1 & 2 \end{bmatrix} \cdot \left(\begin{bmatrix} 5 & 3 \\ 1 & 2 \end{bmatrix} \begin{bmatrix} 5 & 3 \\ 1 & 2 \end{bmatrix} \right) = \begin{bmatrix} 5 & 3 \\ 1 & 2 \end{bmatrix} \cdot \begin{bmatrix} 25+3 & 15+6 \\ 5+2 & 3+4 \end{bmatrix}$

$= \begin{bmatrix} 5 & 3 \\ 1 & 2 \end{bmatrix} \cdot \begin{bmatrix} 28 & 21 \\ 7 & 7 \end{bmatrix} = \begin{bmatrix} 140+21 & 105+21 \\ 28+14 & 21+14 \end{bmatrix} = \begin{bmatrix} 161 & 126 \\ 42 & 35 \end{bmatrix}$

21. $AA^T = \begin{bmatrix} 1 & 0 & 2 \\ 3 & 2 & 1 \\ 4 & 0 & 3 \end{bmatrix}\begin{bmatrix} 1 & 3 & 4 \\ 0 & 2 & 0 \\ 2 & 1 & 3 \end{bmatrix} = \begin{bmatrix} 5 & 5 & 10 \\ 5 & 14 & 15 \\ 10 & 15 & 25 \end{bmatrix}$

$(AA^T)^T = \begin{bmatrix} 5 & 5 & 10 \\ 5 & 14 & 15 \\ 10 & 15 & 25 \end{bmatrix}$ $A^T A = \begin{bmatrix} 1 & 3 & 4 \\ 0 & 2 & 0 \\ 2 & 1 & 3 \end{bmatrix}\begin{bmatrix} 1 & 0 & 2 \\ 3 & 2 & 1 \\ 4 & 0 & 3 \end{bmatrix} = \begin{bmatrix} 26 & 6 & 17 \\ 6 & 4 & 2 \\ 17 & 2 & 14 \end{bmatrix}$

No, they are not equal.

23. No. See problems 5 and 7.

25. $AB = \begin{bmatrix} 2 & 5 & 4 \\ 1 & 4 & 3 \\ 1 & -3 & -2 \end{bmatrix}\begin{bmatrix} -1 & 2 & 1 \\ -5 & 8 & 2 \\ -7 & 11 & -3 \end{bmatrix} = \begin{bmatrix} -2-25-28 & 4+40+44 & 2+10-12 \\ -1-20-21 & 2+32+33 & 1+8-9 \\ -1+15+14 & 2-24-22 & 1-6+6 \end{bmatrix} = \begin{bmatrix} -55 & 88 & 0 \\ -42 & 67 & 0 \\ 28 & -44 & 1 \end{bmatrix}$

27. $CD = \begin{bmatrix} 3 & 0 & 4 \\ 1 & 7 & -1 \\ 3 & 0 & 4 \end{bmatrix}\begin{bmatrix} 4 & 4 & -8 \\ -1 & -1 & 2 \\ -3 & -3 & 6 \end{bmatrix} = \begin{bmatrix} 12+0-12 & 12+0-12 & -24+0+24 \\ 4-7+3 & 4-7+3 & -8+14-6 \\ 12+0-12 & 12+0-12 & -24+0+24 \end{bmatrix} = \begin{bmatrix} 0 & 0 & 0 \\ 0 & 0 & 0 \\ 0 & 0 & 0 \end{bmatrix}$

29. $F^2 = \begin{bmatrix} 0 & 1 & 2 \\ 0 & 0 & -4 \\ 0 & 0 & 0 \end{bmatrix}\begin{bmatrix} 0 & 1 & 2 \\ 0 & 0 & -4 \\ 0 & 0 & 0 \end{bmatrix} = \begin{bmatrix} 0+0+0 & 0+0+0 & 0-4+0 \\ 0+0+0 & 0+0+0 & 0+0+0 \\ 0+0+0 & 0+0+0 & 0+0+0 \end{bmatrix} = \begin{bmatrix} 0 & 0 & -4 \\ 0 & 0 & 0 \\ 0 & 0 & 0 \end{bmatrix}$

$F^3 = \begin{bmatrix} 0 & 0 & -4 \\ 0 & 0 & 0 \\ 0 & 0 & 0 \end{bmatrix}\begin{bmatrix} 0 & 1 & 2 \\ 0 & 0 & -4 \\ 0 & 0 & 0 \end{bmatrix} = \begin{bmatrix} 0+0+0 & 0+0+0 & 0+0+0 \\ 0+0+0 & 0+0+0 & 0+0+0 \\ 0+0+0 & 0+0+0 & 0+0+0 \end{bmatrix} = \begin{bmatrix} 0 & 0 & 0 \\ 0 & 0 & 0 \\ 0 & 0 & 0 \end{bmatrix}$

31. $AI = A$ See Example 6.

33. $Z(CI) = ZC = Z$ $(ZC)I = ZI = Z$
Product of zero matrix and A is the zero matrix.

35. Problems 35 and 37 show that a product can be "zero" and neither of the factors is zero.

37. a. We will use $\dfrac{1}{ad-bc}AB$ and $\dfrac{1}{ad-bc}BA$.

$AB = \begin{bmatrix} a & b \\ c & d \end{bmatrix}\begin{bmatrix} d & -b \\ -c & a \end{bmatrix} = \begin{bmatrix} ad-bc & -ab+ba \\ cd-dc & -bc+ad \end{bmatrix} = \begin{bmatrix} ad-bc & 0 \\ 0 & ad-bc \end{bmatrix}$

$BA = \begin{bmatrix} d & -b \\ -c & a \end{bmatrix}\begin{bmatrix} a & b \\ c & d \end{bmatrix} = \begin{bmatrix} ad-bc & bd-bd \\ -ac+ac & -bc+ad \end{bmatrix} = \begin{bmatrix} ad-bc & 0 \\ 0 & ad-bc \end{bmatrix}$

$\dfrac{1}{ad-bc}AB = \dfrac{1}{ad-bc}BA = \begin{bmatrix} 1 & 0 \\ 0 & 1 \end{bmatrix}.$

b. For B to exist, we must have $ad-bc \neq 0$.

39. $\begin{bmatrix} 1 & 2 & 1 \\ 3 & 4 & -2 \\ 2 & 0 & -1 \end{bmatrix}\begin{bmatrix} 2 \\ -1 \\ 2 \end{bmatrix} = \begin{bmatrix} 2-2+2 \\ 6-4-4 \\ 4+0-2 \end{bmatrix} = \begin{bmatrix} 2 \\ -2 \\ 2 \end{bmatrix}$

These values are the solution.

41.
$$\begin{bmatrix} 1 & 1 & 2 \\ 4 & 0 & 1 \\ 2 & 1 & 1 \end{bmatrix} \begin{bmatrix} 1 \\ 2 \\ 1 \end{bmatrix} = \begin{bmatrix} 1+2+2 \\ 4+0+1 \\ 2+2+1 \end{bmatrix} = \begin{bmatrix} 5 \\ 5 \\ 5 \end{bmatrix}$$
These values are the solution.

43. The graphing calculator should show 1's on the main diagonal and 0's everywhere else (1 may be 0.9999 and 0 may be 0.00001).

45.
$$\begin{bmatrix} 0.88 & 0 \\ 0 & 0.85 \end{bmatrix} \begin{bmatrix} 25,000 & 28,000 \\ 36,000 & 42,000 \end{bmatrix} = \begin{bmatrix} 22,000 & 24,640 \\ 30,600 & 35,700 \end{bmatrix}$$
The product is the price the dealer pays for the cars

47. a.
$$A = \begin{bmatrix} 2703 & 2378 \\ 2383 & 2723 \\ 4376 & 5114 \\ 1443 & 1581 \\ 3152 & 2490 \\ 1289 & 1245 \\ 916 & 824 \end{bmatrix} \qquad B^T = \begin{bmatrix} 0.961 & 0.481 & 0.794 & 0.99 & 0.773 & 0.658 & 0.895 \\ 1.31 & 0.478 & 0.868 & 1.18 & 1.07 & 0.642 & 0.963 \end{bmatrix}$$

b. $\frac{1}{12} AB^T$ is a 7×7 matrix.

c. $\frac{1}{12}\left[(2703)(0.961) + (2378)(1.31) \right] \approx \$476$ million gives the total value of production for Alabama in 1995 and 2000. The other states are likewise represented on the diagonal entries.

49. TC is the only product that is defined.
$$\begin{bmatrix} 50 & 30 & 5 \\ 40 & 60 & 30 \\ 45 & 40 & 60 \end{bmatrix} \begin{bmatrix} 20 \\ 9 \\ 6 \end{bmatrix} = \begin{bmatrix} 1300 \\ 1520 \\ 1620 \end{bmatrix} \begin{matrix} \text{Teens} \\ \text{Single} \\ \text{Married} \end{matrix}$$

51. T h e   blank   d i e   blank   i s   blank   c a s t
20 8 5   27   4 9 5   27   9 19   27   3 1 19 20
$$\begin{bmatrix} 5 & 9 \\ 6 & 11 \end{bmatrix} \begin{bmatrix} 20 & 5 & 4 & 5 & 9 & 27 & 1 & 20 \\ 8 & 27 & 9 & 27 & 19 & 3 & 19 & 27 \end{bmatrix} = \begin{bmatrix} 172 & 268 & 101 & 268 & 216 & 162 & 176 & 343 \\ 208 & 327 & 123 & 327 & 263 & 195 & 215 & 417 \end{bmatrix}$$
So, 172, 208, 268, 327, 101, 123, 268, 327, 216, 263, 162, 195, 176, 215, 343, 417 is the message sent.

53. We need one-half the sum of the two matrices. This represents Male and Female per 100,000.
Answer:
$$\begin{bmatrix} 2137 & 84 & 41.5 & 128.5 & 158.5 & 317 & 738 \\ 1285.5 & 64 & 30.5 & 115 & 136 & 229 & 590 \\ 969.5 & 46.5 & 24 & 98 & 139 & 224 & 476.5 \\ 809 & 44.5 & 22.5 & 99 & 141.5 & 240.5 & 455.5 \\ 687.5 & 33 & 17.5 & 80 & 104.5 & 203.5 & 430.5 \end{bmatrix}$$

55. a. $B = \begin{bmatrix} 0.7 & 8.5 & 10.2 & 1.1 & 5.6 & 3.6 \\ 0.5 & 0.2 & 6.1 & 1.3 & 0.2 & 1.0 \\ 2.2 & 0.4 & 8.8 & 1.2 & 1.2 & 4.8 \\ 251.8 & 63.4 & 81.6 & 35.2 & 54.3 & 144.2 \\ 30.0 & 1.0 & 1.0 & 1.0 & 1.0 & 1.0 \\ 788.9 & 0 & 0 & 0 & 0 & 0 \end{bmatrix}$

b. $A = \begin{bmatrix} 1.11 & 0 & 0 & 0 & 0 & 0 \\ 0 & 0.95 & 0 & 0 & 0 & 0 \\ 0 & 0 & 1.11 & 0 & 0 & 0 \\ 0 & 0 & 0 & 1.11 & 0 & 0 \\ 0 & 0 & 0 & 0 & 0.95 & 0 \\ 0 & 0 & 0 & 0 & 0 & 0.95 \end{bmatrix}$

57. To get new efficiencies, multiply each entry by 0.9 to reflect the 10% decrease.

$$\begin{bmatrix}
0.900 & 0.855 & 0.855 & 0.855 & 0.855 & 0.765 & 0.765 & 0.765 & 0.765 \\
0.765 & 0.900 & 0.855 & 0.855 & 0.855 & 0.855 & 0.765 & 0.765 & 0.765 \\
0.765 & 0.765 & 0.900 & 0.855 & 0.855 & 0.855 & 0.855 & 0.765 & 0.765 \\
0.765 & 0.765 & 0.765 & 0.900 & 0.855 & 0.855 & 0.855 & 0.855 & 0.765 \\
0.765 & 0.765 & 0.765 & 0.765 & 0.900 & 0.855 & 0.855 & 0.855 & 0.855 \\
0.855 & 0.765 & 0.765 & 0.765 & 0.765 & 0.900 & 0.855 & 0.855 & 0.855 \\
0.855 & 0.855 & 0.765 & 0.765 & 0.765 & 0.765 & 0.900 & 0.855 & 0.855 \\
0.855 & 0.855 & 0.855 & 0.765 & 0.765 & 0.765 & 0.765 & 0.900 & 0.855 \\
0.855 & 0.855 & 0.855 & 0.855 & 0.765 & 0.765 & 0.765 & 0.765 & 0.900
\end{bmatrix}$$

Exercise 3.3

1. $\begin{bmatrix} 1 & -2 & -1 & | & -7 \\ 3 & 1 & 2 & | & 0 \\ 4 & 2 & 2 & | & 1 \end{bmatrix}$ $\begin{matrix} -3R_1 + R_2 \to R_2 \\ \to \end{matrix}$

$\begin{bmatrix} 1 & -2 & -1 & | & -7 \\ 0 & 7 & 5 & | & 21 \\ 4 & 2 & 2 & | & 1 \end{bmatrix}$

3. $\begin{bmatrix} 1 & -3 & 4 & | & 2 \\ 2 & 0 & 2 & | & 1 \\ 1 & 2 & 1 & | & 1 \end{bmatrix}$

5. $x = 2,\ y = \dfrac{1}{2},\ z = -5$

7. $x = 18,\ y = 10,\ z = 0$

9. $\begin{bmatrix} 1 & 1 & 2 & | & -1 \\ 0 & 3 & 1 & | & 7 \\ 0 & -2 & 4 & | & 0 \end{bmatrix}$ $\begin{matrix} \to \\ R_3 + R_2 \to R_2 \end{matrix}$ $\begin{bmatrix} 1 & 1 & 2 & | & -1 \\ 0 & 1 & 5 & | & 7 \\ 0 & -2 & 4 & | & 0 \end{bmatrix}$ $\begin{matrix} -R_2 + R_1 \to R_1 \\ \to \\ 2R_2 + R_3 \to R_3 \end{matrix}$

$\begin{bmatrix} 1 & 0 & -3 & | & -8 \\ 0 & 1 & 5 & | & 7 \\ 0 & 0 & 14 & | & 14 \end{bmatrix}$ $\begin{matrix} \to \\ \frac{1}{14}R_3 \to R_3 \end{matrix}$ $\begin{bmatrix} 1 & 0 & -3 & | & -8 \\ 0 & 1 & 5 & | & 7 \\ 0 & 0 & 1 & | & 1 \end{bmatrix}$ $\begin{matrix} 3R_3 + R_1 \to R_1 \\ -5R_3 + R_2 \to R_2 \end{matrix}$ $\begin{bmatrix} 1 & 0 & 0 & | & -5 \\ 0 & 1 & 0 & | & 2 \\ 0 & 0 & 1 & | & 1 \end{bmatrix}$

Solution: $x = -5,\ y = 2,\ z = 1$

11. $\begin{bmatrix} 1 & 2 & 5 & | & -4 \\ 2 & -2 & 4 & | & -2 \\ 0 & 1 & -3 & | & 7 \end{bmatrix}$ $\begin{array}{c} \rightarrow \\ -2R_1 + R_2 \rightarrow R_2 \end{array}$ $\begin{bmatrix} 1 & 2 & 5 & | & -4 \\ 0 & -6 & -6 & | & 6 \\ 0 & 1 & -3 & | & 7 \end{bmatrix}$ $\begin{array}{c} R_2 \leftrightarrow R_3 \\ \rightarrow \end{array}$ $\begin{bmatrix} 1 & 2 & 5 & | & -4 \\ 0 & 1 & -3 & | & 7 \\ 0 & -6 & -6 & | & 6 \end{bmatrix}$ $\begin{array}{c} -2R_2 + R_1 \rightarrow R_1 \\ \rightarrow \\ 6R_2 + R_3 \rightarrow R_3 \end{array}$

$\begin{bmatrix} 1 & 0 & 11 & | & -18 \\ 0 & 1 & -3 & | & 7 \\ 0 & 0 & -24 & | & 48 \end{bmatrix}$ $\begin{array}{c} \rightarrow \\ -\frac{1}{24}R_3 \rightarrow R_3 \end{array}$ $\begin{bmatrix} 1 & 0 & 11 & | & -18 \\ 0 & 1 & -3 & | & 7 \\ 0 & 0 & 1 & | & -2 \end{bmatrix}$ $\begin{array}{c} -11R_3 + R_1 \rightarrow R_1 \\ 3R_3 + R_2 \rightarrow R_2 \\ \rightarrow \end{array}$ $\begin{bmatrix} 1 & 0 & 0 & | & 4 \\ 0 & 1 & 0 & | & 1 \\ 0 & 0 & 1 & | & -2 \end{bmatrix}$

Solution: $x = 4$, $y = 1$, $z = -2$

13. $\begin{bmatrix} 1 & 1 & -1 & | & 0 \\ 1 & 2 & 3 & | & -5 \\ 2 & -1 & -13 & | & 17 \end{bmatrix}$ $\begin{array}{c} \rightarrow \\ -R_1 + R_2 \rightarrow R_2 \\ -2R_1 + R_3 \rightarrow R_3 \end{array}$ $\begin{bmatrix} 1 & 1 & -1 & | & 0 \\ 0 & 1 & 4 & | & -5 \\ 0 & -3 & -11 & | & 17 \end{bmatrix}$ $\begin{array}{c} -R_2 + R_1 \rightarrow R_1 \\ \rightarrow \\ 3R_2 + R_3 \rightarrow R_3 \end{array}$

$\begin{bmatrix} 1 & 0 & -5 & | & 5 \\ 0 & 1 & 4 & | & -5 \\ 0 & 0 & 1 & | & 2 \end{bmatrix}$ $\begin{array}{c} 5R_3 + R_1 \rightarrow R_1 \\ -4R_3 + R_2 \rightarrow R_2 \\ \rightarrow \end{array}$ $\begin{bmatrix} 1 & 0 & 0 & | & 15 \\ 0 & 1 & 0 & | & -13 \\ 0 & 0 & 1 & | & 2 \end{bmatrix}$

Solution: $x = 15$, $y = -13$, $z = 2$

15. $\begin{bmatrix} 2 & -6 & -12 & | & 6 \\ 3 & -10 & -20 & | & 5 \\ 2 & 0 & -17 & | & -4 \end{bmatrix}$ $\begin{array}{c} \frac{1}{2}R_1 \rightarrow R_1 \\ \rightarrow \end{array}$ $\begin{bmatrix} 1 & -3 & -6 & | & 3 \\ 3 & -10 & -20 & | & 5 \\ 2 & 0 & -17 & | & -4 \end{bmatrix}$ $\begin{array}{c} \rightarrow \\ -3R_1 + R_2 \rightarrow R_2 \\ -2R_1 + R_3 \rightarrow R_3 \end{array}$

$\begin{bmatrix} 1 & -3 & -6 & | & 3 \\ 0 & -1 & -2 & | & -4 \\ 0 & 6 & -5 & | & -10 \end{bmatrix}$ $\begin{array}{c} \rightarrow \\ -R_2 \rightarrow R_2 \end{array}$ $\begin{bmatrix} 1 & -3 & -6 & | & 3 \\ 0 & 1 & 2 & | & 4 \\ 0 & 6 & -5 & | & -10 \end{bmatrix}$ $\begin{array}{c} 3R_2 + R_1 \rightarrow R_1 \\ \rightarrow \\ -6R_2 + R_3 \rightarrow R_3 \end{array}$

$\begin{bmatrix} 1 & 0 & 0 & | & 15 \\ 0 & 1 & 2 & | & 4 \\ 0 & 0 & -17 & | & -34 \end{bmatrix}$ $\begin{array}{c} \rightarrow \\ -\frac{1}{17}R_3 \rightarrow R_3 \end{array}$ $\begin{bmatrix} 1 & 0 & 0 & | & 15 \\ 0 & 1 & 2 & | & 4 \\ 0 & 0 & 1 & | & 2 \end{bmatrix}$ $\begin{array}{c} \rightarrow \\ -2R_3 + R_2 \rightarrow R_2 \end{array}$ $\begin{bmatrix} 1 & 0 & 0 & | & 15 \\ 0 & 1 & 0 & | & 0 \\ 0 & 0 & 1 & | & 2 \end{bmatrix}$

Solution: $x = 15$, $y = 0$, $z = 2$

17. $\begin{bmatrix} 1 & -3 & 3 & | & 7 \\ 1 & 2 & -1 & | & -2 \\ 3 & 2 & 4 & | & 5 \end{bmatrix}$ $\begin{array}{c} \rightarrow \\ -R_1 + R_2 \rightarrow R_2 \\ -3R_1 + R_3 \rightarrow R_3 \end{array}$ $\begin{bmatrix} 1 & -3 & 3 & | & 7 \\ 0 & 5 & -4 & | & -9 \\ 0 & 11 & -5 & | & -16 \end{bmatrix}$ $\begin{array}{c} \rightarrow \\ -2R_2 + R_3 \rightarrow R_3 \end{array}$

$\begin{bmatrix} 1 & -3 & 3 & | & 7 \\ 0 & 5 & -4 & | & -9 \\ 0 & 1 & 3 & | & 2 \end{bmatrix}$ $\begin{array}{c} R_2 \leftrightarrow R_3 \\ \rightarrow \end{array}$ $\begin{bmatrix} 1 & -3 & 3 & | & 7 \\ 0 & 1 & 3 & | & 2 \\ 0 & 5 & -4 & | & -9 \end{bmatrix}$ $\begin{array}{c} 3R_2 + R_1 \rightarrow R_1 \\ \rightarrow \\ -5R_2 + R_3 \rightarrow R_3 \end{array}$

$\begin{bmatrix} 1 & 0 & 12 & | & 13 \\ 0 & 1 & 3 & | & 2 \\ 0 & 0 & -19 & | & -19 \end{bmatrix}$ $\begin{array}{c} \rightarrow \\ -\frac{1}{19}R_3 \rightarrow R_3 \end{array}$ $\begin{bmatrix} 1 & 0 & 12 & | & 13 \\ 0 & 1 & 3 & | & 2 \\ 0 & 0 & 1 & | & 1 \end{bmatrix}$ $\begin{array}{c} -12R_3 + R_1 \rightarrow R_1 \\ -3R_3 + R_2 \rightarrow R_2 \\ \rightarrow \end{array}$ $\begin{bmatrix} 1 & 0 & 0 & | & 1 \\ 0 & 1 & 0 & | & -1 \\ 0 & 0 & 1 & | & 1 \end{bmatrix}$

Solution: $x = 1$, $y = -1$, $z = 1$

19.
$$\begin{bmatrix} 1 & 3 & 2 & 2 & | & 3 \\ 1 & 1 & 3 & 0 & | & 4 \\ 2 & 0 & 2 & -3 & | & 4 \\ 1 & -3 & 0 & 0 & | & 1 \end{bmatrix} \begin{array}{l} \rightarrow \\ -R_1 + R_2 \rightarrow R_2 \\ -2R_1 + R_3 \rightarrow R_3 \\ -R_1 + R_4 \rightarrow R_4 \end{array}$$

$$\begin{bmatrix} 1 & 3 & 2 & 2 & | & 3 \\ 0 & -2 & 1 & -2 & | & 1 \\ 0 & -6 & -2 & -7 & | & -2 \\ 0 & -6 & -2 & -2 & | & -2 \end{bmatrix} \begin{array}{l} \text{*See note} \\ -3R_4 + R_3 \rightarrow R_3 \\ \rightarrow \end{array}$$

$$\begin{bmatrix} 1 & 3 & 2 & 2 & | & 3 \\ 0 & -2 & 1 & -2 & | & 1 \\ 0 & 12 & 4 & -1 & | & 4 \\ 0 & -6 & -2 & -2 & | & -2 \end{bmatrix} \begin{array}{l} \rightarrow \\ -R_3 \rightarrow R_3 \\ \text{then } R_3 \leftrightarrow R_4 \end{array}$$

$$\begin{bmatrix} 1 & 3 & 2 & 2 & | & 3 \\ 0 & -2 & 1 & -2 & | & 1 \\ 0 & -6 & -2 & -2 & | & -2 \\ 0 & -12 & -4 & 1 & | & -4 \end{bmatrix} \begin{array}{l} -2R_4 + R_1 \rightarrow R_1 \\ 2R_4 + R_2 \rightarrow R_2 \\ 2R_4 + R_3 \rightarrow R_3 \\ \rightarrow \end{array}$$

$$\begin{bmatrix} 1 & 27 & 10 & 0 & | & 11 \\ 0 & -26 & -7 & 0 & | & -7 \\ 0 & -30 & -10 & 0 & | & -10 \\ 0 & -12 & -4 & 1 & | & -4 \end{bmatrix} \begin{array}{l} \rightarrow \\ -3R_2 \rightarrow R_2 \\ \\ -2R_3 \rightarrow R_3 \end{array}$$

$$\begin{bmatrix} 1 & 27 & 10 & 0 & | & 11 \\ 0 & 78 & 21 & 0 & | & 21 \\ 0 & 60 & 20 & 0 & | & 20 \\ 0 & -12 & -4 & 1 & | & -4 \end{bmatrix} \begin{array}{l} \rightarrow \\ -R_3 + R_2 \rightarrow R_2 \end{array}$$

$$\begin{bmatrix} 1 & 27 & 10 & 0 & | & 11 \\ 0 & 18 & 1 & 0 & | & 1 \\ 0 & 60 & 20 & 0 & | & 20 \\ 0 & -12 & -4 & 1 & | & -4 \end{bmatrix} \begin{array}{l} -10R_2 + R_1 \rightarrow R_1 \\ \rightarrow \\ -20R_2 + R_3 \rightarrow R_3 \\ 4R_2 + R_4 \rightarrow R_4 \end{array}$$

$$\begin{bmatrix} 1 & -153 & 0 & 0 & | & 1 \\ 0 & 18 & 1 & 0 & | & 1 \\ 0 & -300 & 0 & 0 & | & 0 \\ 0 & 60 & 0 & 1 & | & 0 \end{bmatrix} \begin{array}{l} \rightarrow \\ -\frac{1}{300} R_3 \rightarrow R_3 \\ \text{then } R_2 \leftrightarrow R_3 \end{array}$$

$$\begin{bmatrix} 1 & -153 & 0 & 0 & | & 1 \\ 0 & 1 & 0 & 0 & | & 0 \\ 0 & 18 & 1 & 0 & | & 1 \\ 0 & 60 & 0 & 1 & | & 0 \end{bmatrix} \begin{array}{l} 153R_2 + R_1 \rightarrow R_1 \\ \rightarrow \\ -18R_2 + R_3 \rightarrow R_3 \\ -60R_2 + R_4 \rightarrow R_4 \end{array}$$

$$\begin{bmatrix} 1 & 0 & 0 & 0 & | & 1 \\ 0 & 1 & 0 & 0 & | & 0 \\ 0 & 0 & 1 & 0 & | & 1 \\ 0 & 0 & 0 & 1 & | & 0 \end{bmatrix}$$

Solution: $x_1 = 1, x_2 = 0, x_3 = 1, x_4 = 0$

*Note: The author's step by step method always works. However, it is often necessary to introduce fractions. This problem has been worked "backwards" to illustrate an alternate method and avoid fractions.

21.
$$\begin{bmatrix} 1 & -2 & 3 & 1 & | & -2 \\ 1 & -3 & 1 & -1 & | & -7 \\ 1 & -1 & 0 & 0 & | & -2 \\ 1 & 0 & 1 & 1 & | & 2 \end{bmatrix} \begin{array}{l} \rightarrow \\ -R_1 + R_2 \rightarrow R_2 \\ -R_1 + R_3 \rightarrow R_3 \\ -R_1 + R_4 \rightarrow R_4 \end{array}$$

$$\begin{bmatrix} 1 & -2 & 3 & 1 & | & -2 \\ 0 & -1 & -2 & -2 & | & -5 \\ 0 & 1 & -3 & -1 & | & 0 \\ 0 & 2 & -2 & 0 & | & 4 \end{bmatrix} \begin{array}{l} \rightarrow \\ -1R_2 \rightarrow R_2 \end{array}$$

$$\begin{bmatrix} 1 & -2 & 3 & 1 & | & -2 \\ 0 & 1 & 2 & 2 & | & 5 \\ 0 & 1 & -3 & -1 & | & 0 \\ 0 & 2 & -2 & 0 & | & 4 \end{bmatrix} \begin{array}{l} 2R_2 + R_1 \rightarrow R_1 \\ \rightarrow \\ -R_2 + R_3 \rightarrow R_3 \\ -2R_2 + R_4 \rightarrow R_4 \end{array}$$

$$\begin{bmatrix} 1 & 0 & 7 & 5 & | & 8 \\ 0 & 1 & 2 & 2 & | & 5 \\ 0 & 0 & -5 & -3 & | & -5 \\ 0 & 0 & -6 & -4 & | & -6 \end{bmatrix} \begin{array}{l} \rightarrow \\ -\frac{1}{5} R_3 \rightarrow R_3 \end{array}$$

$$\begin{bmatrix} 1 & 0 & 7 & 5 & | & 8 \\ 0 & 1 & 2 & 2 & | & 5 \\ 0 & 0 & 1 & 3/5 & | & 1 \\ 0 & 0 & -6 & -4 & | & -6 \end{bmatrix} \begin{array}{l} -7R_3 + R_1 \rightarrow R_1 \\ -2R_3 + R_2 \rightarrow R_2 \\ \rightarrow \\ 6R_3 + R_4 \rightarrow R_4 \end{array}$$

$$\begin{bmatrix} 1 & 0 & 0 & 4/5 & | & 1 \\ 0 & 1 & 0 & 4/5 & | & 3 \\ 0 & 0 & 1 & 3/5 & | & 1 \\ 0 & 0 & 0 & 2/5 & | & 0 \end{bmatrix} \begin{array}{l} \text{Now take } \frac{5}{2} R_4 \\ \text{and get 0's} \\ \text{in column 4;} \\ \text{Result is} \end{array} \begin{bmatrix} 1 & 0 & 0 & 0 & | & 1 \\ 0 & 1 & 0 & 0 & | & 3 \\ 0 & 0 & 1 & 0 & | & 1 \\ 0 & 0 & 0 & 1 & | & 0 \end{bmatrix}$$

Solution: $x = 1, \ y = 3, \ z = 1, \ w = 0$

23. There is no solution since the last row of the reduced matrix says $0x + 0y + 0z = 1$, which is not possible.

25. From first row: $x - \dfrac{2}{3}z = \dfrac{11}{3}$

From second row: $y + \dfrac{1}{3}z = -\dfrac{1}{3}$

General solution: $x = \dfrac{11}{3} + \dfrac{2}{3}z = \dfrac{11 + 2z}{3}$, $y = -\dfrac{1}{3} - \dfrac{1}{3}z = \dfrac{-1-z}{3}$ for any real number z.

27. If the bottom row of the reduced matrix has all zeros on the left side of the augment and a number other than zero on the right side, the system has no solutions.

29.
$$\begin{bmatrix} 1 & 1 & 1 & | & 0 \\ 2 & -1 & -1 & | & 0 \\ -1 & 2 & 2 & | & 0 \end{bmatrix} \quad \begin{matrix} \rightarrow \\ -2R_1 + R_2 \rightarrow R_2 \\ R_1 + R_3 \rightarrow R_3 \end{matrix} \quad \begin{bmatrix} 1 & 1 & 1 & | & 0 \\ 0 & -3 & -3 & | & 0 \\ 0 & 3 & 3 & | & 0 \end{bmatrix} \quad \begin{matrix} \rightarrow \\ R_2 + R_3 \rightarrow R_3 \end{matrix}$$

$$\begin{bmatrix} 1 & 1 & 1 & | & 0 \\ 0 & -3 & -3 & | & 0 \\ 0 & 0 & 0 & | & 0 \end{bmatrix} \quad \begin{matrix} \rightarrow \\ -\frac{1}{3}R_2 \rightarrow R_2 \end{matrix} \quad \begin{bmatrix} 1 & 1 & 1 & | & 0 \\ 0 & 1 & 1 & | & 0 \\ 0 & 0 & 0 & | & 0 \end{bmatrix} \quad \begin{matrix} -R_2 + R_1 \rightarrow R_1 \\ \rightarrow \end{matrix} \quad \begin{bmatrix} 1 & 0 & 0 & | & 0 \\ 0 & 1 & 1 & | & 0 \\ 0 & 0 & 0 & | & 0 \end{bmatrix}$$

General solution: $x = 0$, $y = -z$

31.
$$\begin{bmatrix} 3 & 2 & 1 & | & 0 \\ 1 & 1 & 2 & | & 1 \\ 2 & 1 & -1 & | & -1 \end{bmatrix} \quad \begin{matrix} R_1 \leftrightarrow R_2 \\ \rightarrow \end{matrix} \quad \begin{bmatrix} 1 & 1 & 2 & | & 1 \\ 3 & 2 & 1 & | & 0 \\ 2 & 1 & -1 & | & -1 \end{bmatrix} \quad \begin{matrix} \rightarrow \\ -3R_1 + R_2 \rightarrow R_2 \\ -2R_1 + R_3 \rightarrow R_3 \end{matrix}$$

$$\begin{bmatrix} 1 & 1 & 2 & | & 1 \\ 0 & -1 & -5 & | & -3 \\ 0 & -1 & -5 & | & -3 \end{bmatrix} \quad \begin{matrix} \rightarrow \\ -R_2 + R_3 \rightarrow R_3 \end{matrix} \quad \begin{bmatrix} 1 & 1 & 2 & | & 1 \\ 0 & -1 & -5 & | & -3 \\ 0 & 0 & 0 & | & 0 \end{bmatrix} \quad \begin{matrix} R_2 + R_1 \rightarrow R_1 \\ \rightarrow \end{matrix}$$

$$\begin{bmatrix} 1 & 0 & -3 & | & -2 \\ 0 & -1 & -5 & | & -3 \\ 0 & 0 & 0 & | & 0 \end{bmatrix} \quad \begin{matrix} \rightarrow \\ -R_2 \rightarrow R_2 \end{matrix} \quad \begin{bmatrix} 1 & 0 & -3 & | & -2 \\ 0 & 1 & 5 & | & 3 \\ 0 & 0 & 0 & | & 0 \end{bmatrix}$$

General solution: $x = 3z - 2$, $y = 3 - 5z$

33.
$$\begin{bmatrix} 2 & 2 & 1 & | & 2 \\ 1 & -2 & 2 & | & 1 \\ -1 & 2 & -2 & | & -1 \end{bmatrix} \quad \begin{matrix} R_1 \leftrightarrow R_2 \\ \rightarrow \end{matrix} \quad \begin{bmatrix} 1 & -2 & 2 & | & 1 \\ 2 & 2 & 1 & | & 2 \\ -1 & 2 & -2 & | & -1 \end{bmatrix} \quad \begin{matrix} \rightarrow \\ -2R_1 + R_2 \rightarrow R_2 \\ R_1 + R_3 \rightarrow R_3 \end{matrix}$$

$$\begin{bmatrix} 1 & -2 & 2 & | & 1 \\ 0 & 6 & -3 & | & 0 \\ 0 & 0 & 0 & | & 0 \end{bmatrix} \quad \begin{matrix} \rightarrow \\ \frac{1}{6}R_2 \rightarrow R_2 \end{matrix} \quad \begin{bmatrix} 1 & -2 & 2 & | & 1 \\ 0 & 1 & -1/2 & | & 0 \\ 0 & 0 & 0 & | & 0 \end{bmatrix} \quad \begin{matrix} 2R_2 + R_1 \rightarrow R_1 \\ \rightarrow \end{matrix} \quad \begin{bmatrix} 1 & 0 & 1 & | & 1 \\ 0 & 1 & -1/2 & | & 0 \\ 0 & 0 & 0 & | & 0 \end{bmatrix}$$

General solution: $x = 1 - z$, $y = \dfrac{1}{2}z$

35.
$$\begin{bmatrix} 2 & -5 & 1 & | & -9 \\ 1 & 4 & -6 & | & 2 \\ 3 & -4 & -2 & | & -10 \end{bmatrix} \quad \begin{matrix} R_1 \leftrightarrow R_2 \\ \rightarrow \end{matrix} \quad \begin{bmatrix} 1 & 4 & -6 & | & 2 \\ 2 & -5 & 1 & | & -9 \\ 3 & -4 & -2 & | & -10 \end{bmatrix} \quad \begin{matrix} \rightarrow \\ -2R_1 + R_2 \rightarrow R_2 \\ -3R_1 + R_3 \rightarrow R_3 \end{matrix} \quad \begin{bmatrix} 1 & 4 & -6 & | & 2 \\ 0 & -13 & 13 & | & -13 \\ 0 & -16 & 16 & | & -16 \end{bmatrix} \quad \begin{matrix} \rightarrow \\ -\frac{1}{13}R_2 \rightarrow R_2 \end{matrix}$$

$$\begin{bmatrix} 1 & 4 & -6 & | & 2 \\ 0 & 1 & -1 & | & 1 \\ 0 & -16 & 16 & | & -16 \end{bmatrix} \quad \begin{matrix} -4R_2 + R_1 \rightarrow R_1 \\ \rightarrow \\ 16R_2 + R_3 \rightarrow R_3 \end{matrix} \quad \begin{bmatrix} 1 & 0 & -2 & | & -2 \\ 0 & 1 & -1 & | & 1 \\ 0 & 0 & 0 & | & 0 \end{bmatrix}$$

General solution: $x = 2z - 2$, $y = z + 1$

37.
$$\begin{bmatrix} 1 & 1 & 1 & | & 3 \\ 1 & -1 & 1 & | & 4 \end{bmatrix} \quad \xrightarrow{\;\;} \atop -R_1 + R_2 \to R_2 \quad \begin{bmatrix} 1 & 1 & 1 & | & 3 \\ 0 & -2 & 0 & | & 1 \end{bmatrix} \quad \xrightarrow{\;\;} \atop -\frac{1}{2}R_2 \to R_2$$

$$\begin{bmatrix} 1 & 1 & 1 & | & 3 \\ 0 & 1 & 0 & | & -1/2 \end{bmatrix} \quad \xrightarrow[\;\;]{-R_2 + R_1 \to R_1} \quad \begin{bmatrix} 1 & 0 & 1 & | & 7/2 \\ 0 & 1 & 0 & | & -1/2 \end{bmatrix}$$

General Solution: $x = \dfrac{7}{2} - z$, $y = -\dfrac{1}{2}$, $z = z$

39.
$$\begin{bmatrix} 3 & -2 & 5 & | & 14 \\ 2 & -3 & 4 & | & 8 \end{bmatrix} \quad \xrightarrow[\;\;]{-R_2 + R_1 \to R_1} \quad \begin{bmatrix} 1 & 1 & 1 & | & 6 \\ 2 & -3 & 4 & | & 8 \end{bmatrix} \quad \xrightarrow{\;\;} \atop -2R_1 + R_2 \to R_2$$

$$\begin{bmatrix} 1 & 1 & 1 & | & 6 \\ 0 & -5 & 2 & | & -4 \end{bmatrix} \quad \xrightarrow[\;\;]{-\frac{1}{5}R_2 \to R_2} \quad \begin{bmatrix} 1 & 1 & 1 & | & 6 \\ 0 & 1 & -\frac{2}{5} & | & \frac{4}{5} \end{bmatrix} \quad \xrightarrow{\;\;} \atop -R_2 + R_1 \to R_1 \quad \begin{bmatrix} 1 & 0 & \frac{7}{5} & | & \frac{26}{5} \\ 0 & 1 & -\frac{2}{5} & | & \frac{4}{5} \end{bmatrix}$$

General Solution: $x = \dfrac{26}{5} - \dfrac{7}{5}z$, $y = \dfrac{4}{5} + \dfrac{2}{5}z$

41.
$$\begin{bmatrix} -0.6 & 0.1 & 0.3 & | & 0 \\ 0.4 & -0.7 & 0.2 & | & 0 \\ 0.2 & 0.6 & -0.5 & | & 0 \end{bmatrix} \quad \begin{matrix} 10R_1 \to R_1 \\ 10R_2 \to R_2 \\ 10R_3 \to R_3 \end{matrix} \quad \begin{bmatrix} -6 & 1 & 3 & | & 0 \\ 4 & -7 & 2 & | & 0 \\ 2 & 6 & -5 & | & 0 \end{bmatrix} \quad \begin{matrix} -\frac{1}{6}R_1 \to R_1 \\ -2R_3 + R_2 \to R_2 \\ \to \end{matrix}$$

$$\begin{bmatrix} 1 & -\frac{1}{6} & -\frac{1}{2} & | & 0 \\ 0 & -19 & 12 & | & 0 \\ 2 & 6 & -5 & | & 0 \end{bmatrix} \quad \begin{matrix} \to \\ -\frac{1}{19}R_2 \to R_2 \\ -2R_1 + R_3 \to R_3 \end{matrix} \quad \begin{bmatrix} 1 & -\frac{1}{6} & -\frac{1}{2} & | & 0 \\ 0 & 1 & -\frac{12}{19} & | & 0 \\ 0 & \frac{19}{3} & -4 & | & 0 \end{bmatrix} \quad \begin{matrix} \to \\ \\ -\frac{19}{3}R_2 + R_3 \to R_3 \end{matrix}$$

$$\begin{bmatrix} 1 & -\frac{1}{6} & -\frac{1}{2} & | & 0 \\ 0 & 1 & -\frac{12}{19} & | & 0 \\ 0 & 0 & 0 & | & 0 \end{bmatrix} \quad \begin{matrix} \frac{1}{6}R_2 + R_1 \to R_1 \\ \\ \end{matrix} \quad \begin{bmatrix} 1 & 0 & -\frac{23}{38} & | & 0 \\ 0 & 1 & -\frac{12}{19} & | & 0 \\ 0 & 0 & 0 & | & 0 \end{bmatrix}$$

General solution: $x_1 = \dfrac{23}{38}x_3$, $x_2 = \dfrac{12}{19}x_3$

43.
$$\begin{bmatrix} 3 & 2 & 1 & -1 & | & 3 \\ 1 & -1 & -2 & 2 & | & 2 \\ 2 & 3 & -1 & 1 & | & 1 \\ -1 & 1 & 2 & -2 & | & -2 \end{bmatrix} \quad \begin{matrix} R_1 \leftrightarrow R_2 \\ \to \end{matrix} \quad \begin{bmatrix} 1 & -1 & -2 & 2 & | & 2 \\ 3 & 2 & 1 & -1 & | & 3 \\ 2 & 3 & -1 & 1 & | & 1 \\ -1 & 1 & 2 & -2 & | & -2 \end{bmatrix} \quad \begin{matrix} \to \\ -3R_1 + R_2 \to R_2 \\ -2R_1 + R_3 \to R_3 \\ R_1 + R_4 \to R_4 \end{matrix}$$

$$\begin{bmatrix} 1 & -1 & -2 & 2 & | & 2 \\ 0 & 5 & 7 & -7 & | & -3 \\ 0 & 5 & 3 & -3 & | & -3 \\ 0 & 0 & 0 & 0 & | & 0 \end{bmatrix} \quad \begin{matrix} \to \\ \\ -R_2 + R_3 \to R_3 \end{matrix} \quad \begin{bmatrix} 1 & -1 & -2 & 2 & | & 2 \\ 0 & 5 & 7 & -7 & | & -3 \\ 0 & 0 & -4 & 4 & | & 0 \\ 0 & 0 & 0 & 0 & | & 0 \end{bmatrix} \quad \begin{matrix} \to \\ \frac{1}{5}R_2 \to R_2 \\ -\frac{1}{4}R_3 \to R_3 \end{matrix}$$

$$\begin{bmatrix} 1 & -1 & -2 & 2 & | & 2 \\ 0 & 1 & \frac{7}{5} & -\frac{7}{5} & | & -\frac{3}{5} \\ 0 & 0 & 1 & -1 & | & 0 \\ 0 & 0 & 0 & 0 & | & 0 \end{bmatrix} \quad \begin{matrix} R_2 + R_1 \to R_1 \\ -\frac{7}{5}R_3 + R_2 \to R_2 \\ \to \end{matrix} \quad \begin{bmatrix} 1 & 0 & -\frac{3}{5} & \frac{3}{5} & | & \frac{7}{5} \\ 0 & 1 & 0 & 0 & | & -\frac{3}{5} \\ 0 & 0 & 1 & -1 & | & 0 \\ 0 & 0 & 0 & 0 & | & 0 \end{bmatrix} \quad \begin{matrix} \frac{3}{5}R_3 + R_1 \to R_1 \\ \to \end{matrix} \quad \begin{bmatrix} 1 & 0 & 0 & 0 & | & \frac{7}{5} \\ 0 & 1 & 0 & 0 & | & -\frac{3}{5} \\ 0 & 0 & 1 & -1 & | & 0 \\ 0 & 0 & 0 & 0 & | & 0 \end{bmatrix}$$

General solution: $x = \frac{7}{5}$, $y = -\frac{3}{5}$, $z = w$

45. $\begin{bmatrix} 1 & 2 & -1 & 1 & 3 \\ 1 & 3 & 4 & 1 & -2 \\ 2 & 5 & 2 & 2 & 1 \\ 2 & 3 & -6 & 2 & 3 \end{bmatrix}$ $\begin{array}{c} \rightarrow \\ -R_1 + R_2 \rightarrow R_2 \\ -2R_1 + R_3 \rightarrow R_3 \\ -2R_1 + R_4 \rightarrow R_4 \end{array}$ $\begin{bmatrix} 1 & 2 & -1 & 1 & 3 \\ 0 & 1 & 5 & 0 & -5 \\ 0 & 1 & 4 & 0 & -5 \\ 0 & -1 & -4 & 0 & -3 \end{bmatrix}$ $\begin{array}{c} -2R_2 + R_1 \rightarrow R_1 \\ \rightarrow \\ -R_2 + R_3 \rightarrow R_3 \\ R_2 + R_4 \rightarrow R_4 \end{array}$ $\begin{bmatrix} 1 & 0 & -11 & 1 & 13 \\ 0 & 1 & 5 & 0 & -5 \\ 0 & 0 & -1 & 0 & 0 \\ 0 & 0 & 1 & 0 & -8 \end{bmatrix}$

The last two rows state that $x_3 = 0$ and $x_3 = -8$. Thus, there is no solution.

47. $\begin{bmatrix} 1 & 3 & 4 & -1 & 2 & 1 \\ 1 & -1 & -2 & 1 & 0 & 3 \\ 1 & 2 & 3 & 1 & 4 & 0 \\ 2 & 2 & 2 & 0 & 2 & 4 \end{bmatrix}$ $\begin{array}{c} \rightarrow \\ R_1 - R_2 \rightarrow R_2 \\ R_1 - R_3 \rightarrow R_3 \\ -2R_1 + R_4 \rightarrow R_4 \end{array}$ $\begin{bmatrix} 1 & 3 & 4 & -1 & 2 & 1 \\ 0 & 4 & 6 & -2 & 2 & -2 \\ 0 & 1 & 1 & -2 & -2 & 1 \\ 0 & -4 & -6 & 2 & -2 & 2 \end{bmatrix}$ $R_2 \leftrightarrow R_3$

$\begin{bmatrix} 1 & 3 & 4 & -1 & 2 & 1 \\ 0 & 1 & 1 & -2 & -2 & 1 \\ 0 & 4 & 6 & -2 & 2 & -2 \\ 0 & -4 & -6 & 2 & -2 & 2 \end{bmatrix}$ $\begin{array}{c} -3R_2 + R_1 \rightarrow R_1 \\ \rightarrow \\ -4R_2 + R_3 \rightarrow R_3 \\ 4R_2 + R_4 \rightarrow R_4 \end{array}$ $\begin{bmatrix} 1 & 0 & 1 & 5 & 8 & -2 \\ 0 & 1 & 1 & -2 & -2 & 1 \\ 0 & 0 & 2 & 6 & 10 & -6 \\ 0 & 0 & -2 & -6 & -10 & 6 \end{bmatrix}$ $\begin{array}{c} \rightarrow \\ \rightarrow \\ \frac{1}{2}R_3 \rightarrow R_3 \\ R_4 + R_3 \rightarrow R_4 \end{array}$

$\begin{bmatrix} 1 & 0 & 1 & 5 & 8 & -2 \\ 0 & 1 & 1 & -2 & -2 & 1 \\ 0 & 0 & 1 & 3 & 5 & -3 \\ 0 & 0 & 0 & 0 & 0 & 0 \end{bmatrix}$ $\begin{array}{c} -R_3 + R_1 \rightarrow R_1 \\ -R_3 + R_2 \rightarrow R_2 \\ \rightarrow \end{array}$ $\begin{bmatrix} 1 & 0 & 0 & 2 & 3 & 1 \\ 0 & 1 & 0 & -5 & -7 & 4 \\ 0 & 0 & 1 & 3 & 5 & -3 \\ 0 & 0 & 0 & 0 & 0 & 0 \end{bmatrix}$

There are infinitely many solutions.
General solution: $x_1 = 1 - 2x_4 - 3x_5$, $x_2 = 4 + 5x_4 + 7x_5$, $x_3 = -3 - 3x_4 - 5x_5$

49. $\begin{bmatrix} a_1 & b_1 & c_1 \\ a_2 & b_2 & c_2 \end{bmatrix}$ $\begin{array}{c} \frac{1}{a_1}R_1 \rightarrow R_1 \\ \rightarrow \end{array}$ $\begin{bmatrix} 1 & \dfrac{b_1}{a_1} & \dfrac{c_1}{a_1} \\ a_2 & b_2 & c_2 \end{bmatrix}$ $\begin{array}{c} \rightarrow \\ -a_2R_1 + R_2 \rightarrow R_2 \end{array}$ $\begin{bmatrix} 1 & \dfrac{b_1}{a_1} & \dfrac{c_1}{a_1} \\ 0 & \dfrac{b_2a_1 - b_1a_2}{a_1} & \dfrac{c_2a_1 - c_1a_2}{a_1} \end{bmatrix}$

$\begin{array}{c} -\dfrac{b_1}{a_1} \cdot \dfrac{a_1}{b_2a_1 - b_1a_2}R_2 + R_1 \rightarrow R_1 \\ \rightarrow \end{array}$ $\begin{bmatrix} 1 & 0 & \dfrac{b_2c_1 - c_2b_1}{a_1b_2 - a_2b_1} \\ 0 & \dfrac{b_2a_1 - b_1a_2}{a_1} & \dfrac{c_2a_1 - c_1a_2}{a_1} \end{bmatrix}$ Thus, $x = \dfrac{b_2c_1 - c_2b_1}{a_1b_2 - a_2b_1}$.

Fractions and messy computations cannot always be avoided.

51. a. Let $x = 12\%$ investment amount, $y = 10\%$ investment amount, and $z = 8\%$ investment amount.

$$x + y + z = 235,000 \quad \text{Total investment}$$
$$0.12x + 0.10y + 0.08z = 22,500 \quad \text{Investment income}$$
$$2x \quad - \quad z = 0 \quad\quad 2x = z$$

$$\begin{bmatrix} 1 & 1 & 1 & | & 235,000 \\ 0.12 & 0.10 & 0.08 & | & 22,500 \\ 2 & 0 & -1 & | & 0 \end{bmatrix} \begin{matrix} \rightarrow \\ 100R_2 \rightarrow R_2 \\ \rightarrow \end{matrix} \begin{bmatrix} 1 & 1 & 1 & | & 235,000 \\ 12 & 10 & 8 & | & 2,250,000 \\ 2 & 0 & -1 & | & 0 \end{bmatrix} \begin{matrix} \rightarrow \\ -12R_1 + R_2 \rightarrow R_2 \\ -2R_1 + R_3 \rightarrow R_3 \end{matrix}$$

$$\begin{bmatrix} 1 & 1 & 1 & | & 235,000 \\ 0 & -2 & -4 & | & -570,000 \\ 0 & -2 & -3 & | & -470,000 \end{bmatrix} \begin{matrix} \rightarrow \\ -\frac{1}{2}R_2 \rightarrow R_2 \\ \rightarrow \end{matrix} \begin{bmatrix} 1 & 1 & 1 & | & 235,000 \\ 0 & 1 & 2 & | & 285,000 \\ 0 & -2 & -3 & | & -470,000 \end{bmatrix} \begin{matrix} -R_2 + R_1 \rightarrow R_1 \\ \rightarrow \\ 2R_2 + R_3 \rightarrow R_3 \end{matrix}$$

$$\begin{bmatrix} 1 & 0 & -1 & | & -50,000 \\ 0 & 1 & 2 & | & 285,000 \\ 0 & 0 & 1 & | & 100,000 \end{bmatrix} \begin{matrix} R_3 + R_1 \rightarrow R_1 \\ -2R_3 + R_2 \rightarrow R_2 \\ \rightarrow \end{matrix} \begin{bmatrix} 1 & 0 & 0 & | & 50,000 \\ 0 & 1 & 0 & | & 85,000 \\ 0 & 0 & 1 & | & 100,000 \end{bmatrix}$$

Invest \$50,000 at 12%, \$85,000 at 10% and \$100,000 at 8%.

b. Income: $0.12(50,000) = \$6000$, $0.10(85,000) = \$8,500$, and $0.08(100,000) = \$8,000$

53. Let $x =$ cups of Beef. Let $y =$ cups of Sirloin Burger.
Fat: $4x + 9y = 80$, Cholesterol: $25x + 20y = 210$

Use the result for x in #51. $x = \dfrac{20(80) - 210(9)}{4(20) - 25(9)} = \dfrac{-290}{-145} = 2$

Then $4(2) + 9y = 80$ gives $y = 8$. Use 2 cups of Beef and 8 cups of Sirloin Burger.

55. Let $x =$ ounces of AF, $y =$ ounces of FP, $z =$ ounces of NMG
We start directly with the augmented matrix.

$$\begin{matrix} \text{Calories} \\ \text{Fat} \\ \text{Carbohydrates} \end{matrix} \begin{bmatrix} 50 & 108 & 127 & | & 443 \\ 0 & 0.1 & 5.5 & | & 5.7 \\ 22 & 25.7 & 18 & | & 113.4 \end{bmatrix} \begin{matrix} \frac{1}{50}R_1 \rightarrow R_1 \\ 10R_2 \rightarrow R_2 \\ \rightarrow \end{matrix} \begin{bmatrix} 1 & 2.16 & 2.54 & | & 8.86 \\ 0 & 1 & 55 & | & 57 \\ 22 & 25.7 & 18 & | & 113.4 \end{bmatrix} \begin{matrix} \rightarrow \\ \\ -22R_1 + R_3 \rightarrow R_3 \end{matrix}$$

$$\begin{bmatrix} 1 & 2.16 & 2.54 & | & 8.86 \\ 0 & 1 & 55 & | & 57 \\ 0 & -21.82 & -37.88 & | & -81.52 \end{bmatrix} \begin{matrix} -2.16R_2 + R_1 \rightarrow R_1 \\ \rightarrow \\ 21.82R_2 + R_3 \rightarrow R_3 \end{matrix} \begin{bmatrix} 1 & 0 & -116.26 & | & -114.26 \\ 0 & 1 & 55 & | & 57 \\ 0 & 0 & 1162.22 & | & 1162.22 \end{bmatrix} \begin{matrix} \rightarrow \\ \\ \frac{1}{1162.22}R_3 \rightarrow R_3 \end{matrix}$$

$$\begin{bmatrix} 1 & 0 & -116.26 & | & -114.26 \\ 0 & 1 & 55 & | & 57 \\ 0 & 0 & 1 & | & 1 \end{bmatrix} \begin{matrix} 116.26R_3 + R_1 \rightarrow R_1 \\ -55R_3 + R_2 \rightarrow R_2 \\ \rightarrow \end{matrix} \begin{bmatrix} 1 & 0 & 0 & | & 2 \\ 0 & 1 & 0 & | & 2 \\ 0 & 0 & 1 & | & 1 \end{bmatrix}$$

Use 2 ounces of AF, 2 ounces of FP, and 1 ounce of NMG.

57. $x = \#$ porfolio I $\quad 2x + 4y + 2z = 12$
$y = \#$ portfolio II $\quad x + 2y + 2z = 6$
$z = \#$ portfolio III $\quad 3y + 3z = 6$

$$\begin{bmatrix} 2 & 4 & 2 & | & 12 \\ 1 & 2 & 2 & | & 6 \\ 0 & 3 & 3 & | & 6 \end{bmatrix} \begin{matrix} R_1 \rightarrow R_2 \\ \rightarrow \end{matrix}$$

$$\begin{bmatrix} 1 & 2 & 2 & | & 6 \\ 2 & 4 & 2 & | & 12 \\ 0 & 3 & 3 & | & 6 \end{bmatrix} \begin{matrix} \rightarrow \\ -2R_1 + R_2 \rightarrow R_2 \\ \rightarrow \end{matrix} \begin{bmatrix} 1 & 2 & 2 & | & 6 \\ 0 & 0 & -2 & | & 0 \\ 0 & 3 & 3 & | & 6 \end{bmatrix} \begin{matrix} \rightarrow \\ -\frac{1}{2}R_2 \rightarrow R_2 \\ \frac{1}{3}R_3 \rightarrow R_3 \end{matrix} \begin{bmatrix} 1 & 2 & 2 & | & 6 \\ 0 & 0 & 1 & | & 0 \\ 0 & 1 & 1 & | & 2 \end{bmatrix} \begin{matrix} -2R_2 + R_1 \rightarrow R_1 \\ \rightarrow \\ -R_2 + R_3 \rightarrow R_3 \end{matrix}$$

$$\begin{bmatrix} 1 & 2 & 0 & | & 6 \\ 0 & 0 & 1 & | & 0 \\ 0 & 1 & 0 & | & 2 \end{bmatrix} \begin{matrix} R_2 \leftrightarrow R_3 \\ \rightarrow \end{matrix} \begin{bmatrix} 1 & 2 & 0 & | & 6 \\ 0 & 1 & 0 & | & 2 \\ 0 & 0 & 1 & | & 0 \end{bmatrix} \begin{matrix} -2R_2 + R_1 \rightarrow R_1 \\ \rightarrow \end{matrix} \begin{bmatrix} 1 & 0 & 0 & | & 2 \\ 0 & 1 & 0 & | & 2 \\ 0 & 0 & 1 & | & 0 \end{bmatrix} \begin{matrix} \text{2 units each} \\ \text{of portfolio I} \\ \text{and portfolio II.} \end{matrix}$$

59. $0.1S + 3.4M + 2.2B = 12.1$ iron needed $\begin{bmatrix} 0.1 & 3.4 & 2.2 & | & 12.1 \\ 8.5 & 22 & 10 & | & 97 \\ 1 & 20 & 12 & | & 70 \end{bmatrix} \begin{matrix} 10R_1 \to R_1 \\ \to \\ {} \end{matrix}$

     $8.5S + 22M + 10B = 97$ protein needed

     $1S + 20M + 12B = 70$ carbohydrates needed

$\begin{bmatrix} 1 & 34 & 22 & | & 121 \\ 8.5 & 22 & 10 & | & 97 \\ 1 & 20 & 12 & | & 70 \end{bmatrix} \begin{matrix} \to \\ -8.5R_1 + R_2 \to R_2 \\ -R_1 + R_3 \to R_3 \end{matrix} \begin{bmatrix} 1 & 34 & 22 & | & 121 \\ 0 & -267 & -177 & | & -931.5 \\ 0 & -14 & -10 & | & -51 \end{bmatrix} \begin{matrix} \to \\ -19R_3 + R_2 \to R_2 \\ {} \end{matrix}$

$\begin{bmatrix} 1 & 34 & 22 & | & 121 \\ 0 & -1 & 13 & | & 37.5 \\ 0 & -14 & -10 & | & -51 \end{bmatrix} \begin{matrix} 34R_2 + R_1 \to R_1 \\ \to \\ -14R_2 + R_3 \to R_3 \end{matrix} \begin{bmatrix} 1 & 0 & 464 & | & 1396 \\ 0 & -1 & 13 & | & 37.5 \\ 0 & 0 & -192 & | & -576 \end{bmatrix} \begin{matrix} \to \\ -\frac{1}{192}R_3 \to R_3 \\ {} \end{matrix}$

$\begin{bmatrix} 1 & 0 & 464 & | & 1396 \\ 0 & -1 & 13 & | & 37.5 \\ 0 & 0 & 1 & | & 3 \end{bmatrix} \begin{matrix} -464R_3 + R_1 \to R_1 \\ -13R_3 + R_2 \to R_2 \\ \to \end{matrix} \begin{bmatrix} 1 & 0 & 0 & | & 4 \\ 0 & -1 & 0 & | & -1.5 \\ 0 & 0 & 1 & | & 3 \end{bmatrix} \begin{matrix} -R_2 \to R_2 \\ \to \\ {} \end{matrix} \begin{bmatrix} 1 & 0 & 0 & | & 4 \\ 0 & 1 & 0 & | & 1.5 \\ 0 & 0 & 1 & | & 3 \end{bmatrix}$

4 glasses of milk, 1.5 quarter-pound servings of meat, and 3 2-slice servings of bread. Note that this is 3/8 pounds of meat and 6 slices of bread.

61. x = Type I bags; y = Type II bags; z = Type III bags; w = Type IV bags

$\begin{bmatrix} 5 & 5 & 10 & 5 & | & 10{,}000 \\ 10 & 5 & 30 & 10 & | & 20{,}000 \\ 5 & 15 & 10 & 25 & | & 20{,}000 \end{bmatrix} \begin{matrix} \frac{1}{5}R_1 \to R_1 \\ \frac{1}{5}R_2 \to R_2 \\ \frac{1}{5}R_3 \to R_3 \end{matrix} \begin{bmatrix} 1 & 1 & 2 & 1 & | & 2000 \\ 2 & 1 & 6 & 2 & | & 4000 \\ 1 & 3 & 2 & 5 & | & 4000 \end{bmatrix} \begin{matrix} \to \\ -2R_1 + R_2 \to R_2 \\ -R_1 + R_3 \to R_3 \end{matrix}$

$\begin{bmatrix} 1 & 1 & 2 & 1 & | & 2000 \\ 0 & -1 & 2 & 0 & | & 0 \\ 0 & 2 & 0 & 4 & | & 2000 \end{bmatrix} \begin{matrix} R_2 + R_1 \to R_1 \\ \to \\ 2R_2 + R_3 \to R_3 \end{matrix} \begin{bmatrix} 1 & 0 & 4 & 1 & | & 2000 \\ 0 & -1 & 2 & 0 & | & 0 \\ 0 & 0 & 4 & 4 & | & 2000 \end{bmatrix} \begin{matrix} \to \\ -R_2 \to R_2 \\ \frac{1}{4}R_3 \to R_3 \end{matrix}$

$\begin{bmatrix} 1 & 0 & 4 & 1 & | & 2000 \\ 0 & 1 & -2 & 0 & | & 0 \\ 0 & 0 & 1 & 1 & | & 500 \end{bmatrix} \begin{matrix} -4R_3 + R_1 \to R_1 \\ 2R_3 + R_2 \to R_2 \\ \to \end{matrix} \begin{bmatrix} 1 & 0 & 0 & -3 & | & 0 \\ 0 & 1 & 0 & 2 & | & 1000 \\ 0 & 0 & 1 & 1 & | & 500 \end{bmatrix}$

Solution: $x = 3w$, $y = 1000 - 2w$, $z = 500 - w$, where w is any non-negative amount ≤ 500.

63.

Nutrient	Species			
	x_1	x_2	x_3	
A	1	2	2	5100 units
B	1	0	3	6900 units
C	2	2	5	12,000 units

$\begin{bmatrix} 1 & 2 & 2 & | & 5100 \\ 1 & 0 & 3 & | & 6900 \\ 2 & 2 & 5 & | & 12{,}000 \end{bmatrix} \begin{matrix} \to \\ -R_1 + R_2 \to R_2 \\ -2R_1 + R_3 \to R_3 \end{matrix} \begin{bmatrix} 1 & 2 & 2 & | & 5100 \\ 0 & -2 & 1 & | & 1800 \\ 0 & -2 & 1 & | & 1800 \end{bmatrix} \begin{matrix} R_2 + R_1 \to R_1 \\ \to \\ -R_2 + R_3 \to R_3 \end{matrix} \begin{bmatrix} 1 & 0 & 3 & | & 6900 \\ 0 & -2 & 1 & | & 1800 \\ 0 & 0 & 0 & | & 0 \end{bmatrix}$

Species $x_1 = 6900 - 3x_3$ and species $x_2 = -900 + \frac{1}{2}x_3$, with any amount of x_3 between 1800 and 2300.

65. a.   Let $x =$ computer shares, $y =$ utility shares, and $z =$ retail shares.

$30x + 44y + 26z = 392,000$ Total Cost

$6x + 6y + 2.4z = 0.15(392,000)$ Total Growth

$$\begin{bmatrix} 30 & 44 & 26 & | & 392,000 \\ 6 & 6 & 2.4 & | & 58,800 \end{bmatrix} \; R_1 \leftrightarrow \tfrac{1}{6}R_2 \; \begin{bmatrix} 1 & 1 & 0.4 & | & 9,800 \\ 30 & 44 & 26 & | & 392,000 \end{bmatrix} \; \begin{array}{c} \rightarrow \\ -30R_1 + R_2 \rightarrow R_2 \end{array}$$

$$\begin{bmatrix} 1 & 1 & 0.4 & | & 9,800 \\ 0 & 14 & 14 & | & 98,000 \end{bmatrix} \; \tfrac{1}{14}R_2 \rightarrow R_2 \; \begin{bmatrix} 1 & 1 & 0.4 & | & 9,800 \\ 0 & 1 & 1 & | & 7,000 \end{bmatrix} \; -R_2 + R_1 \rightarrow R_1 \; \begin{bmatrix} 1 & 0 & -0.6 & | & 2,800 \\ 0 & 1 & 1 & | & 7,000 \end{bmatrix}$$

General solution: $x = 2,800 + 0.6z$ and $y = 7000 - z$

b.   If $z = 1000$, then $x = 3400$ and $y = 6000$. So 3400 shares computer stock and 6000 shares utility stock.

c.   If they buy no shares of retail stock ($z = 0$), then $x = 2,800$ and $y = 7000$. So, the minimum number of shares of computer stock is 2,800 and they would also buy 7,000 shares of utility stock.

d.   The maximum number of computer shares depends on the maximum value of z. $y = 7000 - z$, so z cannot be greater than 7000. If $z = 7000$, then $x = 2800 + 0.6(7000) = 7000$ and $y = 0$. So they would purchase 7,000 computer shares and 7,000 retail shares.

67. $x =$ units of Portfolio I $2x + 4y + 2z = 16$ common stock blocks

$y =$ units of Portfolio II $x + 2y + z = 8$ municipal bond blocks

$z =$ units of Portfolio III $3y + 3z = 6$ preferred stock blocks

$$\begin{bmatrix} 2 & 4 & 2 & | & 16 \\ 1 & 2 & 1 & | & 8 \\ 0 & 3 & 3 & | & 6 \end{bmatrix} \; \begin{array}{c} R_1 \leftrightarrow R_2 \\ \rightarrow \end{array}$$

$$\begin{bmatrix} 1 & 2 & 1 & | & 8 \\ 2 & 4 & 2 & | & 16 \\ 0 & 3 & 3 & | & 6 \end{bmatrix} \; \begin{array}{c} \rightarrow \\ -2R_1 + R_2 \rightarrow R_2 \\ \tfrac{1}{3}R_3 \rightarrow R_3 \end{array} \; \begin{bmatrix} 1 & 2 & 1 & | & 8 \\ 0 & 0 & 0 & | & 0 \\ 0 & 1 & 1 & | & 2 \end{bmatrix} \; \begin{array}{c} -2R_3 + R_1 \rightarrow R_1 \\ \rightarrow \end{array} \; \begin{bmatrix} 1 & 0 & -1 & | & 4 \\ 0 & 0 & 0 & | & 0 \\ 0 & 1 & 1 & | & 2 \end{bmatrix}$$

Thus, $x = z + 4$ and $y = -z + 2$. Note that $z = 0, 1, 2$ only.

Possible Offerings

I	II	III
4	2	0
5	1	1
6	0	2

Exercise 3.4

1. The product is the 3×3 identity matrix.

3. $\begin{bmatrix} 1 & 2 & 1 \\ 0 & 0 & 3 \\ 1 & 0 & 1 \end{bmatrix} \begin{bmatrix} 0 & -1/3 & 1 \\ 1/2 & 0 & -1/2 \\ 0 & 1/3 & 0 \end{bmatrix} = \begin{bmatrix} 1 & 0 & 0 \\ 0 & 1 & 0 \\ 0 & 0 & 1 \end{bmatrix}$

Yes, $B = A^{-1}$.

5. $A^{-1} = \dfrac{1}{(4)(2) - 7(1)} \begin{bmatrix} 2 & -7 \\ -1 & 4 \end{bmatrix} = \begin{bmatrix} 2 & -7 \\ -1 & 4 \end{bmatrix}$

7. $ad - bc = 2(2) - (-4)(-1) = 0$, so no inverse exists.

9. $A^{-1} = \dfrac{1}{2(5) - 2(4)} \begin{bmatrix} 5 & -2 \\ -4 & 2 \end{bmatrix} = \begin{bmatrix} 5/2 & -1 \\ -2 & 1 \end{bmatrix}$

11. $A^{-1} = \dfrac{1}{4(1) - 7(2)} \begin{bmatrix} 1 & -7 \\ -2 & 4 \end{bmatrix} = \begin{bmatrix} -1/10 & 7/10 \\ 1/5 & -2/5 \end{bmatrix}$

13. $\left[\begin{array}{ccc|ccc} 3 & 0 & 0 & 1 & 0 & 0 \\ 0 & 3 & 0 & 0 & 1 & 0 \\ 0 & 0 & 3 & 0 & 0 & 1 \end{array}\right] \begin{array}{c} \frac{1}{3}R_1 \to R_1 \\ \frac{1}{3}R_2 \to R_2 \\ \frac{1}{3}R_3 \to R_3 \end{array}$

$\left[\begin{array}{ccc|ccc} 1 & 0 & 0 & \frac{1}{3} & 0 & 0 \\ 0 & 1 & 0 & 0 & \frac{1}{3} & 0 \\ 0 & 0 & 1 & 0 & 0 & \frac{1}{3} \end{array}\right]$ $A^{-1} : \begin{bmatrix} \frac{1}{3} & 0 & 0 \\ 0 & \frac{1}{3} & 0 \\ 0 & 0 & \frac{1}{3} \end{bmatrix}$

15. $\left[\begin{array}{ccc|ccc} 0 & 1 & 0 & 1 & 0 & 0 \\ 1 & 1 & 0 & 0 & 1 & 0 \\ 0 & 1 & 1 & 0 & 0 & 1 \end{array}\right] \begin{array}{c} R_1 \leftrightarrow R_2 \\ \to \end{array}$

$\left[\begin{array}{ccc|ccc} 1 & 1 & 0 & 0 & 1 & 0 \\ 0 & 1 & 0 & 1 & 0 & 0 \\ 0 & 1 & 1 & 0 & 0 & 1 \end{array}\right] \begin{array}{c} -R_2 + R_1 \to R_1 \\ \to \\ -R_2 + R_3 \to R_3 \end{array}$

$\left[\begin{array}{ccc|ccc} 1 & 0 & 0 & -1 & 1 & 0 \\ 0 & 1 & 0 & 1 & 0 & 0 \\ 0 & 0 & 1 & -1 & 0 & 1 \end{array}\right]$

Inverse $= \begin{bmatrix} -1 & 1 & 0 \\ 1 & 0 & 0 \\ -1 & 0 & 1 \end{bmatrix}$

$AA^{-1} = \begin{bmatrix} 3 & 0 & 0 \\ 0 & 3 & 0 \\ 0 & 0 & 3 \end{bmatrix} \begin{bmatrix} \frac{1}{3} & 0 & 0 \\ 0 & \frac{1}{3} & 0 \\ 0 & 0 & \frac{1}{3} \end{bmatrix}$

$= \begin{bmatrix} 1 & 0 & 0 \\ 0 & 1 & 0 \\ 0 & 0 & 1 \end{bmatrix}$

17. $\left[\begin{array}{ccc|ccc} 3 & 1 & 2 & 1 & 0 & 0 \\ 1 & 2 & 3 & 0 & 1 & 0 \\ 1 & 1 & 1 & 0 & 0 & 1 \end{array}\right] \begin{array}{c} R_1 \leftrightarrow R_3 \\ \to \end{array}$

$\left[\begin{array}{ccc|ccc} 1 & 1 & 1 & 0 & 0 & 1 \\ 1 & 2 & 3 & 0 & 1 & 0 \\ 3 & 1 & 2 & 1 & 0 & 0 \end{array}\right] \begin{array}{c} \to \\ -R_1 + R_2 \to R_2 \\ -3R_1 + R_3 \to R_3 \end{array}$

$\left[\begin{array}{ccc|ccc} 1 & 1 & 1 & 0 & 0 & 1 \\ 0 & 1 & 2 & 0 & 1 & -1 \\ 0 & -2 & -1 & 1 & 0 & -3 \end{array}\right] \begin{array}{c} -R_2 + R_1 \to R_1 \\ \to \\ 2R_2 + R_3 \to R_3 \end{array}$

$\left[\begin{array}{ccc|ccc} 1 & 0 & -1 & 0 & -1 & 2 \\ 0 & 1 & 2 & 0 & 1 & -1 \\ 0 & 0 & 3 & 1 & 2 & -5 \end{array}\right] \begin{array}{c} \to \\ \frac{1}{3}R_3 \to R_3 \end{array}$

$\left[\begin{array}{ccc|ccc} 1 & 0 & -1 & 0 & -1 & 2 \\ 0 & 1 & 2 & 0 & 1 & -1 \\ 0 & 0 & 1 & 1/3 & 2/3 & -5/3 \end{array}\right] \begin{array}{c} R_3 + R_1 \to R_1 \\ -2R_3 + R_2 \to R_2 \\ \to \end{array}$

$\left[\begin{array}{ccc|ccc} 1 & 0 & 0 & 1/3 & -1/3 & 1/3 \\ 0 & 1 & 0 & -2/3 & -1/3 & 7/3 \\ 0 & 0 & 1 & 1/3 & 2/3 & -5/3 \end{array}\right]$

Inverse $= \begin{bmatrix} 1/3 & -1/3 & 1/3 \\ -2/3 & -1/3 & 7/3 \\ 1/3 & 2/3 & -5/3 \end{bmatrix}$

19. $\left[\begin{array}{ccc|ccc} 1 & 3 & 5 & 1 & 0 & 0 \\ -1 & -1 & 2 & 0 & 1 & 0 \\ 1 & 5 & 12 & 0 & 0 & 1 \end{array}\right] \begin{array}{c} \to \\ R_1 + R_2 \to R_2 \\ -R_1 + R_3 \to R_3 \end{array} \left[\begin{array}{ccc|ccc} 1 & 3 & 5 & 1 & 0 & 0 \\ 0 & 2 & 7 & 1 & 1 & 0 \\ 0 & 2 & 7 & -1 & 0 & 1 \end{array}\right] \begin{array}{c} \to \\ -R_2 + R_3 \to R_3 \end{array} \left[\begin{array}{ccc|ccc} 1 & 3 & 5 & 1 & 0 & 0 \\ 0 & 2 & 7 & 1 & 1 & 0 \\ 0 & 0 & 0 & -2 & -1 & 1 \end{array}\right]$

There is no inverse since there is a row of 0's in the original matrix.

21. $\begin{bmatrix} 1 & 2 & 4 & | & 1 & 0 & 0 \\ 1 & -1 & -3 & | & 0 & 1 & 0 \\ 2 & 1 & 1 & | & 0 & 0 & 1 \end{bmatrix}$ $\begin{matrix} \\ -R_1 + R_2 \to R_2 \\ -2R_1 + R_3 \to R_3 \end{matrix}$ $\begin{bmatrix} 1 & 2 & 4 & | & 1 & 0 & 0 \\ 0 & -3 & -7 & | & -1 & 1 & 0 \\ 0 & -3 & -7 & | & -2 & 0 & 1 \end{bmatrix}$ $\begin{matrix} \to \\ \\ -R_2 + R_3 \to R_3 \end{matrix}$ $\begin{bmatrix} 1 & 2 & 4 & | & 1 & 0 & 0 \\ 0 & -3 & -7 & | & -1 & 1 & 0 \\ 0 & 0 & 0 & | & -1 & -1 & 1 \end{bmatrix}$

There is no inverse since there is a row of 0's in the original matrix.

23. $C^{-1} = \begin{bmatrix} 2 & 2 & 0 & 2 & 2 \\ 1 & 0 & 2 & 2 & 1 \\ 0 & 1 & 0 & 2 & 1 \\ 2 & 0 & 2 & 2 & 1 \\ 1 & 0 & 0 & 0 & 2 \end{bmatrix}$ Use a graphing calculator to find the inverse.

25. $AX = \begin{bmatrix} 3 \\ 2 \end{bmatrix}$ $\quad X = A^{-1}\begin{bmatrix} 3 \\ 2 \end{bmatrix} = \begin{bmatrix} 3 & 2 \\ 1 & 1 \end{bmatrix}\begin{bmatrix} 3 \\ 2 \end{bmatrix} = \begin{bmatrix} 13 \\ 5 \end{bmatrix}$

27. $AX = \begin{bmatrix} 3 \\ -1 \\ 2 \end{bmatrix}$ $\quad X = A^{-1}\begin{bmatrix} 3 \\ -1 \\ 2 \end{bmatrix} = \begin{bmatrix} 3 & 2 & 1 \\ 1 & 1 & 2 \\ 1 & 2 & 1 \end{bmatrix} \cdot \begin{bmatrix} 3 \\ -1 \\ 2 \end{bmatrix} = \begin{bmatrix} 9 \\ 6 \\ 3 \end{bmatrix}$

29. $A \cdot \begin{bmatrix} x \\ y \\ z \end{bmatrix} = \begin{bmatrix} 1 \\ 2 \\ 3 \end{bmatrix}$ $\quad \begin{bmatrix} x \\ y \\ z \end{bmatrix} = A^{-1}\begin{bmatrix} 1 \\ 2 \\ 3 \end{bmatrix} = \begin{bmatrix} -1 & 1 & 0 \\ 1 & 0 & 0 \\ -1 & 0 & 1 \end{bmatrix} \cdot \begin{bmatrix} 1 \\ 2 \\ 3 \end{bmatrix} = \begin{bmatrix} 1 \\ 1 \\ 2 \end{bmatrix}$ $\quad \begin{matrix} x = 1 \\ \text{So, } y = 1. \\ z = 2 \end{matrix}$

31. $A = \begin{bmatrix} 1 & 2 \\ 3 & 4 \end{bmatrix}$, $A^{-1} = \dfrac{1}{-2}\begin{bmatrix} 4 & -2 \\ -3 & 1 \end{bmatrix} = \begin{bmatrix} -2 & 1 \\ 3/2 & -1/2 \end{bmatrix}$; $\begin{bmatrix} x \\ y \end{bmatrix} = A^{-1}\begin{bmatrix} 4 \\ 10 \end{bmatrix} = \begin{bmatrix} -2 & 1 \\ 3/2 & -1/2 \end{bmatrix}\begin{bmatrix} 4 \\ 10 \end{bmatrix} = \begin{bmatrix} 2 \\ 1 \end{bmatrix}$

So, $x = 2$, $y = 1$.

33. $A = \begin{bmatrix} 2 & 1 \\ 3 & 1 \end{bmatrix}$, $A^{-1} = \dfrac{1}{-1}\begin{bmatrix} 1 & -1 \\ -3 & 2 \end{bmatrix} = \begin{bmatrix} -1 & 1 \\ 3 & -2 \end{bmatrix}$; $\begin{bmatrix} x \\ y \end{bmatrix} = A^{-1}\begin{bmatrix} 4 \\ 5 \end{bmatrix} = \begin{bmatrix} -1 & 1 \\ 3 & -2 \end{bmatrix}\begin{bmatrix} 4 \\ 5 \end{bmatrix} = \begin{bmatrix} 1 \\ 2 \end{bmatrix}$

So, $x = 1$, $y = 2$.

35. $\begin{bmatrix} 1 & 1 & 1 & | & 1 & 0 & 0 \\ 2 & 1 & 1 & | & 0 & 1 & 0 \\ 2 & 2 & 1 & | & 0 & 0 & 1 \end{bmatrix}$ $\begin{matrix} \to \\ -2R_1 + R_2 \to R_2 \\ -2R_1 + R_3 \to R_3 \end{matrix}$ $\begin{bmatrix} 1 & 1 & 1 & | & 1 & 0 & 0 \\ 0 & -1 & -1 & | & -2 & 1 & 0 \\ 0 & 0 & -1 & | & -2 & 0 & 1 \end{bmatrix}$ $\begin{matrix} \to \\ -R_2 \to R_2 \\ -R_3 \to R_3 \end{matrix}$

$\begin{bmatrix} 1 & 1 & 1 & | & 1 & 0 & 0 \\ 0 & 1 & 1 & | & 2 & -1 & 0 \\ 0 & 0 & 1 & | & 2 & 0 & -1 \end{bmatrix}$ $\begin{matrix} -R_2 + R_1 \to R_1 \\ \to \\ \end{matrix}$ $\begin{bmatrix} 1 & 0 & 0 & | & -1 & 1 & 0 \\ 0 & 1 & 1 & | & 2 & -1 & 0 \\ 0 & 0 & 1 & | & 2 & 0 & -1 \end{bmatrix}$ $\begin{matrix} \to \\ -R_3 + R_2 \to R_2 \end{matrix}$

$\begin{bmatrix} 1 & 0 & 0 & | & -1 & 1 & 0 \\ 0 & 1 & 0 & | & 0 & -1 & 1 \\ 0 & 0 & 1 & | & 2 & 0 & -1 \end{bmatrix}$ $\qquad \begin{bmatrix} x \\ y \\ z \end{bmatrix} = \begin{bmatrix} -1 & 1 & 0 \\ 0 & -1 & 1 \\ 2 & 0 & -1 \end{bmatrix}\begin{bmatrix} 3 \\ 4 \\ 5 \end{bmatrix} = \begin{bmatrix} 1 \\ 1 \\ 1 \end{bmatrix}$

So, $x = 1$, $y = 1$, $z = 1$.

37. $\begin{bmatrix} 1 & 1 & 2 & | & 1 & 0 & 0 \\ 2 & 1 & 1 & | & 0 & 1 & 0 \\ 2 & 2 & 1 & | & 0 & 0 & 1 \end{bmatrix}$ $\begin{array}{c} \rightarrow \\ -2R_1 + R_2 \rightarrow R_2 \\ -2R_1 + R_3 \rightarrow R_3 \end{array}$ $\begin{bmatrix} 1 & 1 & 2 & | & 1 & 0 & 0 \\ 0 & -1 & -3 & | & -2 & 1 & 0 \\ 0 & 0 & -3 & | & -2 & 0 & 1 \end{bmatrix}$ $\begin{array}{c} R_2 + R_1 \rightarrow R_1 \\ \rightarrow \\ -\frac{1}{3}R_3 \rightarrow R_3 \end{array}$

$\begin{bmatrix} 1 & 0 & -1 & | & -1 & 1 & 0 \\ 0 & -1 & -3 & | & -2 & 1 & 0 \\ 0 & 0 & 1 & | & 2/3 & 0 & -1/3 \end{bmatrix}$ $\begin{array}{c} R_3 + R_1 \rightarrow R_1 \\ \rightarrow \\ 3R_3 + R_2 \rightarrow R_2 \end{array}$ $\begin{bmatrix} 1 & 0 & 0 & | & -1/3 & 1 & -1/3 \\ 0 & -1 & 0 & | & 0 & 1 & -1 \\ 0 & 0 & 1 & | & 2/3 & 0 & -1/3 \end{bmatrix}$ $\begin{array}{c} -R_2 \rightarrow R_2 \\ \rightarrow \end{array}$

$\begin{bmatrix} 1 & 0 & 0 & | & -1/3 & 1 & -1/3 \\ 0 & 1 & 0 & | & 0 & -1 & 1 \\ 0 & 0 & 1 & | & 2/3 & 0 & -1/3 \end{bmatrix}$ $\begin{bmatrix} x \\ y \\ z \end{bmatrix} = \begin{bmatrix} -1/3 & 1 & -1/3 \\ 0 & -1 & 1 \\ 2/3 & 0 & -1/3 \end{bmatrix}\begin{bmatrix} 8 \\ 7 \\ 10 \end{bmatrix} = \begin{bmatrix} 1 \\ 3 \\ 2 \end{bmatrix}$

So, $x = 1, \ y = 3, \ z = 2$ ..

39. The graphing calculator yields

$A^{-1} = \begin{bmatrix} 3 & 2 & 2 & 4 & 3 \\ 1 & 0 & 2 & 2 & 1 \\ 0.5 & 1 & 1 & 3 & 1.5 \\ 2 & 0 & 2 & 2 & 1 \\ 1 & 0 & 0 & 0 & 2 \end{bmatrix}; \quad X = A^{-1}\begin{bmatrix} 0.7 \\ -1.6 \\ 1.275 \\ 1.15 \\ -0.15 \end{bmatrix} = \begin{bmatrix} 5.6 \\ 5.4 \\ 3.25 \\ 6.1 \\ 0.4 \end{bmatrix}$

41. $\begin{vmatrix} 1 & 2 \\ 3 & 4 \end{vmatrix} = 1(4) - 3(2) = -2$

43. $\begin{vmatrix} 3 & -1 \\ 2 & 4 \end{vmatrix} = 3(4) - 2(-1) = 14$

45. Using technology, $\begin{vmatrix} 3 & 2 & 1 \\ -1 & 0 & 2 \\ 0 & 1 & 1 \end{vmatrix} = -5$

47. Using technology, $\begin{vmatrix} 0 & 1 & 2 \\ 3 & 1 & 1 \\ 4 & -1 & 3 \end{vmatrix} = -19$

49. Since det $\begin{bmatrix} 2 & 3 \\ -1 & 4 \end{bmatrix} = 11 \neq 0$, $\begin{bmatrix} 2 & 3 \\ -1 & 4 \end{bmatrix}$ has an inverse.

51. Since det $\begin{bmatrix} 1 & 3 & -2 \\ 2 & -1 & 5 \\ 3 & 2 & 3 \end{bmatrix} = 0$, $\begin{bmatrix} 1 & 3 & -2 \\ 2 & -1 & 5 \\ 3 & 2 & 3 \end{bmatrix}$ does not have an inverse.

53. $A^{-1} = \begin{bmatrix} 11 & -9 \\ -6 & 5 \end{bmatrix}$, $\begin{bmatrix} 11 & -9 \\ -6 & 5 \end{bmatrix} \cdot \begin{bmatrix} 49 & 133 & 270 & 313 \\ 59 & 161 & 327 & 381 \end{bmatrix} = \begin{bmatrix} 8 & 14 & 27 & 14 \\ 1 & 7 & 15 & 27 \end{bmatrix}$

8 1 14 7 27 15 14 27

H A N G   O N

55. $A^{-1} = \begin{bmatrix} 1 & 1 & -2 \\ 2 & 1 & -3 \\ -3 & -2 & 6 \end{bmatrix}$, $\begin{bmatrix} 1 & 1 & -2 \\ 2 & 1 & -3 \\ -3 & -2 & 6 \end{bmatrix} \cdot \begin{bmatrix} 47 & 28 & 63 & 56 & 17 \\ 22 & 87 & 66 & 44 & 14 \\ 34 & 46 & 55 & 43 & 15 \end{bmatrix} = \begin{bmatrix} 1 & 23 & 19 & 14 & 1 \\ 14 & 5 & 27 & 27 & 3 \\ 19 & 18 & 9 & 2 & 11 \end{bmatrix}$

1 14 19 23 5 18 19 27 9 14 27 2 1 3 11

A N S W E R S   I N   B A C K

57. $\begin{bmatrix} 1900 \\ 1700 \end{bmatrix} = \begin{bmatrix} 2/3 & 1/4 \\ 1/3 & 3/4 \end{bmatrix} \begin{bmatrix} x_0 \\ y_0 \end{bmatrix}$ For ease in calculations, use a graphing calculator.

$\begin{bmatrix} x_0 \\ y_0 \end{bmatrix} = \begin{bmatrix} 2/3 & 1/4 \\ 1/3 & 3/4 \end{bmatrix}^{-1} \begin{bmatrix} 1900 \\ 1700 \end{bmatrix} = \begin{bmatrix} 9/5 & -3/5 \\ -4/5 & 8/5 \end{bmatrix} \begin{bmatrix} 1900 \\ 1700 \end{bmatrix} = \begin{bmatrix} 2400 \\ 1200 \end{bmatrix}$

59. Medication A is given every 4 hours, or 6 times per day. Medication B is given 2 times per day.

The ratio of the dosage of A to the dosage of B is always 5 to 8: $\dfrac{A}{B} = \dfrac{5}{8}$ or, rearranging, $8A - 5B = 0$.

a. For Patient I, the total dosage is 50.6 mg per day: $6A + 2B = 50.6$, or rearranging, $3A + B = 25.3$

We need to solve: $\begin{bmatrix} 3 & 1 \\ 8 & -5 \end{bmatrix} \begin{bmatrix} A \\ B \end{bmatrix} = \begin{bmatrix} 25.3 \\ 0 \end{bmatrix}$ The inverse of $\begin{bmatrix} 3 & 1 \\ 8 & -5 \end{bmatrix}$ is $-\dfrac{1}{23} \begin{bmatrix} -5 & -1 \\ -8 & 3 \end{bmatrix} = \begin{bmatrix} 5/23 & 1/23 \\ 8/23 & -3/23 \end{bmatrix}$.

Patient 1: $\begin{bmatrix} A \\ B \end{bmatrix} = \begin{bmatrix} 5/23 & 1/23 \\ 8/23 & -3/23 \end{bmatrix} \begin{bmatrix} 25.3 \\ 0.0 \end{bmatrix} = \begin{bmatrix} 5.5 \\ 8.8 \end{bmatrix}$ 5.5 mg of A 8.8 mg of B

b. For Patient II, the total dosage is 92.0 mg per day: $6A + 2B = 92.0$, or rearranging, $3A + B = 46.0$

We need to solve: $\begin{bmatrix} 3 & 1 \\ 8 & -5 \end{bmatrix} \begin{bmatrix} A \\ B \end{bmatrix} = \begin{bmatrix} 46.0 \\ 0 \end{bmatrix}$

Using the inverse found in part a:

Patient II: $\begin{bmatrix} A \\ B \end{bmatrix} = \begin{bmatrix} 5/23 & 1/23 \\ 8/23 & -3/23 \end{bmatrix} \begin{bmatrix} 46.0 \\ 0 \end{bmatrix} = \begin{bmatrix} 10 \\ 16 \end{bmatrix}$ 10 mg of A 16 mg of B

61. x = amount invested at 10%.
y = amount invested at 18%.
$x + y = 145,600$ Total investment
$0.10x + 0.18y = 20,000$ Total income

$\begin{bmatrix} x \\ y \end{bmatrix} = \begin{bmatrix} 1 & 1 \\ 0.10 & 0.18 \end{bmatrix}^{-1} \begin{bmatrix} 145,600 \\ 20,000 \end{bmatrix} = \begin{bmatrix} 77,600 \\ 68,000 \end{bmatrix}$

$77,600 at 10% and $68,000 at 18%
Note - Use the graphing utility to find the inverse, then multiply.

63. Use the graphing utility to find the inverse.

$\begin{bmatrix} \text{Deluxe} \\ \text{Premium} \\ \text{Ultimate} \end{bmatrix} = \begin{bmatrix} 1.6 & 2 & 2.4 \\ 2 & 3 & 4 \\ 0.5 & 0.5 & 1 \end{bmatrix}^{-1} \begin{bmatrix} 96 \\ 156 \\ 37 \end{bmatrix} = \begin{bmatrix} 2.5 & -2 & 2 \\ 0 & 1 & -4 \\ -1.25 & 0.5 & 2 \end{bmatrix} \begin{bmatrix} 96 \\ 156 \\ 37 \end{bmatrix} = \begin{bmatrix} 2 \\ 8 \\ 32 \end{bmatrix}$

Produce 2 deluxe, 8 premium, and 32 ultimate models.

65. $x = 8\%$ investment, $y = 10\%$ investment, $z = 6\%$ investment
$x + y + z = 1,000,000$ Total investment
$0.08x + 0.10y + 0.06z = 86,000$ Investment income
$y = x + z$ Third condition
Use the graphing utility to find the inverse.

$\begin{bmatrix} x \\ y \\ z \end{bmatrix} = \begin{bmatrix} 1 & 1 & 1 \\ 0.08 & 0.10 & 0.06 \\ 1 & -1 & 1 \end{bmatrix}^{-1} \begin{bmatrix} 1,000,000 \\ 86,000 \\ 0 \end{bmatrix} = \begin{bmatrix} -4 & 50 & 1 \\ 0.5 & 0 & -0.5 \\ 4.5 & -50 & -0.5 \end{bmatrix} \begin{bmatrix} 1,000,000 \\ 86,000 \\ 0 \end{bmatrix} = \begin{bmatrix} 300,000 \\ 500,000 \\ 200,000 \end{bmatrix}$

67. a.

$$\begin{bmatrix} a & b & c \\ d & e & f \\ g & h & i \end{bmatrix}\begin{bmatrix} n_t \\ n_{t+1} \\ n_{t+2} \end{bmatrix} = \begin{bmatrix} n_{t+1} \\ n_{t+2} \\ n_t + n_{t+1} + n_{t+2} \end{bmatrix}$$

$an_t + bn_{t+1} + cn_{t+2} = n_{t+1}$, etc.

implies $a = 0$, $b = 1$, $c = 0$, $d = 0$, $e = 0$, $f = 1$, $g = h = i = 1$

$$M = \begin{bmatrix} 0 & 1 & 0 \\ 0 & 0 & 1 \\ 1 & 1 & 1 \end{bmatrix}$$

b. $M^{-1} = \begin{bmatrix} -1 & -1 & 1 \\ 1 & 0 & 0 \\ 0 & 1 & 0 \end{bmatrix}$ $\begin{bmatrix} n_t \\ n_{t+1} \\ n_{t+2} \end{bmatrix} = M^{-1}\begin{bmatrix} 191 \\ 346 \\ 645 \end{bmatrix} = \begin{bmatrix} 108 \\ 191 \\ 346 \end{bmatrix}$

There were 108 visitors on the day before there were 191 visitors.

69. a. See exercise 67 where M was found to be $\begin{bmatrix} 0 & 1 & 0 \\ 0 & 0 & 1 \\ 1 & 1 & 1 \end{bmatrix}$.

b. $M^{-1} = \begin{bmatrix} -1 & -1 & 1 \\ 1 & 0 & 0 \\ 0 & 1 & 0 \end{bmatrix}$

At the end of 3 successive 1-hour periods, the value for N was $N = \begin{bmatrix} 200 \\ 370 \\ 600 \end{bmatrix}$.

$$M^{-1}N = \begin{bmatrix} -1 & -1 & 1 \\ 1 & 0 & 0 \\ 0 & 1 & 0 \end{bmatrix}\begin{bmatrix} 200 \\ 370 \\ 600 \end{bmatrix} = \begin{bmatrix} 30 \\ 200 \\ 370 \end{bmatrix}$$

So, the population was 30 one hour before it was 200.

Exercise 3.5

1. a. Row 3, column 2 = 0.15   100(0.15) = 15
   b. Row 4, column 1 = 0.10   40(0.10) = 4

3. 1000(0.008) = 8.

5. 1000(0.040) = 40.

7. Most dependent would be the largest entry on the main diagonal. Raw materials is the most self dependent. Likewise, Fuels is least dependent.

9. The largest entries in Row 2 give the industries most affected by a rise in raw material cost. These are Raw materials, Manufacturing, and Service industries.

11. $10M = 50$ $M = 5$ units of manufacturing
   $8A - 4(5) = 60$ $A = 10$ units of agriculture
   $U - 2(10) - 1(5) = 80$ $U \doteq 105$ units of utilities

13. $D = \begin{bmatrix} 96 \\ 8 \end{bmatrix}$. $(I - A)X = D$ or $\begin{bmatrix} 0.5 & -0.1 \\ -0.1 & 0.7 \end{bmatrix}\begin{bmatrix} P \\ M \end{bmatrix} = \begin{bmatrix} 96 \\ 8 \end{bmatrix}$

$\begin{bmatrix} P \\ M \end{bmatrix} = \begin{bmatrix} 0.5 & -0.1 \\ -0.1 & 0.7 \end{bmatrix}^{-1} \cdot \begin{bmatrix} 96 \\ 8 \end{bmatrix} = \frac{1}{0.34}\begin{bmatrix} 0.7 & 0.1 \\ 0.1 & 0.5 \end{bmatrix}\begin{bmatrix} 96 \\ 8 \end{bmatrix} = \frac{1}{0.34}\begin{bmatrix} 68 \\ 13.6 \end{bmatrix} = \begin{bmatrix} 200 \\ 40 \end{bmatrix}$

$P = 200$ units of farm products and $M = 40$ units of machinery.

15. $D = \begin{bmatrix} 0 \\ 610 \end{bmatrix}$. $(I-A)X = D$ or $\begin{bmatrix} 0.7 & -0.1 \\ -0.2 & 0.9 \end{bmatrix}\begin{bmatrix} AP \\ OP \end{bmatrix} = \begin{bmatrix} 0 \\ 610 \end{bmatrix}$

$\begin{bmatrix} AP \\ OP \end{bmatrix} = \begin{bmatrix} 0.7 & -0.1 \\ -0.2 & 0.9 \end{bmatrix}^{-1} \cdot \begin{bmatrix} 0 \\ 610 \end{bmatrix} = \frac{1}{0.61}\begin{bmatrix} 0.9 & 0.1 \\ 0.2 & 0.7 \end{bmatrix}\begin{bmatrix} 0 \\ 610 \end{bmatrix} = \begin{bmatrix} 100 \\ 700 \end{bmatrix}$ Ag. Products
Oil Products

17. $D = \begin{bmatrix} 80 \\ 180 \end{bmatrix}$. $(I-A)X = D$ or $\begin{bmatrix} 0.7 & -0.15 \\ -0.3 & 0.6 \end{bmatrix}\begin{bmatrix} U \\ M \end{bmatrix} = \begin{bmatrix} 80 \\ 180 \end{bmatrix}$

$\begin{bmatrix} U \\ M \end{bmatrix} = \begin{bmatrix} 0.7 & -0.15 \\ -0.3 & 0.6 \end{bmatrix}^{-1} \cdot \begin{bmatrix} 80 \\ 180 \end{bmatrix} = \frac{1}{0.375}\begin{bmatrix} 0.6 & 0.15 \\ 0.3 & 0.7 \end{bmatrix}\begin{bmatrix} 80 \\ 180 \end{bmatrix} = \begin{bmatrix} 200 \\ 400 \end{bmatrix}$ Utility
Manufacturing

19. $D = \begin{bmatrix} 36 \\ 278 \end{bmatrix}$. $(I-A)X = D$ or $\begin{bmatrix} 0.8 & -0.4 \\ -0.3 & 0.7 \end{bmatrix}\begin{bmatrix} \text{MINE} \\ \text{MFG} \end{bmatrix} = \begin{bmatrix} 36 \\ 278 \end{bmatrix}$

$\begin{bmatrix} \text{MINE} \\ \text{MFG} \end{bmatrix} = \begin{bmatrix} 0.8 & -0.4 \\ -0.3 & 0.7 \end{bmatrix}^{-1} \cdot \begin{bmatrix} 36 \\ 278 \end{bmatrix} = \frac{1}{0.44}\begin{bmatrix} 0.7 & 0.4 \\ 0.3 & 0.8 \end{bmatrix}\begin{bmatrix} 36 \\ 278 \end{bmatrix} = \begin{bmatrix} 310 \\ 530 \end{bmatrix}$ Mining
Manufacturing

21. a. Let E = gross production for electronics and C = gross production for computers.

$\begin{matrix} & E & C \end{matrix}$

$A = \begin{bmatrix} 0.3 & 0.2 \\ 0.6 & 0.2 \end{bmatrix}$

b. $D = \begin{bmatrix} 648 \\ 16 \end{bmatrix} \quad x = \begin{bmatrix} E \\ C \end{bmatrix}$

$X = (I-A)^{-1} D$

$I - A = \begin{bmatrix} 1 & 0 \\ 0 & 1 \end{bmatrix} - \begin{bmatrix} 0.3 & 0.2 \\ 0.6 & 0.2 \end{bmatrix} = \begin{bmatrix} 0.7 & -0.8 \\ -0.6 & 0.8 \end{bmatrix}$

$(I-A)^{-1} = \frac{1}{0.56-0.48}\begin{bmatrix} 0.8 & 0.8 \\ 0.6 & 0.7 \end{bmatrix} = \begin{bmatrix} 1.8182 & 0.4545 \\ 1.3636 & 1.5909 \end{bmatrix}$

$X = (I-A)^{-1}D = \begin{bmatrix} 1.8182 & 0.4545 \\ 1.3636 & 1.5909 \end{bmatrix}\begin{bmatrix} 648 \\ 16 \end{bmatrix} = \begin{bmatrix} 1185 \\ 909 \end{bmatrix}$

Gross production for electronics is 1185 units and for computers is 909 units.

23. $D = \begin{bmatrix} 20 \\ 1090 \end{bmatrix}$. $(I-A)X = D$ or $\begin{bmatrix} 0.7 & -0.04 \\ -0.35 & 0.9 \end{bmatrix}\begin{bmatrix} \text{Fish} \\ \text{Oil} \end{bmatrix} = \begin{bmatrix} 20 \\ 1090 \end{bmatrix}$

$\begin{bmatrix} \text{Fish} \\ \text{Oil} \end{bmatrix} = \begin{bmatrix} 0.7 & -0.04 \\ -0.35 & 0.9 \end{bmatrix}^{-1} \cdot \begin{bmatrix} 20 \\ 1090 \end{bmatrix} = \frac{1}{0.616}\begin{bmatrix} 0.9 & 0.04 \\ 0.35 & 0.7 \end{bmatrix}\begin{bmatrix} 20 \\ 1090 \end{bmatrix} = \begin{bmatrix} 100 \\ 1250 \end{bmatrix}$ Fishing
Oil

25. $\begin{bmatrix} x \\ y \end{bmatrix} = \begin{bmatrix} 0 & 0.05 \\ 0.1 & 0 \end{bmatrix}\begin{bmatrix} x \\ y \end{bmatrix} + \begin{bmatrix} 20,400 \\ 9900 \end{bmatrix}$ or $\begin{bmatrix} 1 & -0.05 \\ -0.1 & 1 \end{bmatrix}\begin{bmatrix} x \\ y \end{bmatrix} = \begin{bmatrix} 20,400 \\ 9900 \end{bmatrix}$

$\begin{bmatrix} x \\ y \end{bmatrix} = \frac{1}{1-0.005}\begin{bmatrix} 1 & 0.05 \\ 0.1 & 1 \end{bmatrix}\begin{bmatrix} 20,400 \\ 9900 \end{bmatrix} = \frac{1}{0.995}\begin{bmatrix} 20,895 \\ 11,940 \end{bmatrix} = \begin{bmatrix} 21,000 \\ 12,000 \end{bmatrix}$

Costs for development are $21,000. Costs for promotional are $12,000.

27. $\begin{bmatrix} E \\ C \end{bmatrix} = \begin{bmatrix} 11,750 \\ 10,000 \end{bmatrix} + \begin{bmatrix} 0 & 0.25 \\ 0.2 & 0 \end{bmatrix}\begin{bmatrix} E \\ C \end{bmatrix}$

$\begin{bmatrix} 1 & -0.25 \\ -0.2 & 1 \end{bmatrix}\begin{bmatrix} E \\ C \end{bmatrix} = \begin{bmatrix} 11,750 \\ 10,000 \end{bmatrix}$ or $\begin{bmatrix} E \\ C \end{bmatrix} = \frac{1}{1-0.05}\begin{bmatrix} 1 & 0.25 \\ 0.2 & 1 \end{bmatrix}\begin{bmatrix} 11,750 \\ 10,000 \end{bmatrix} = \begin{bmatrix} 15,000 \\ 13,000 \end{bmatrix}$

Engineering costs are $15,000. Computer costs are $13,000.

29. $D = \begin{bmatrix} 110 \\ 50 \\ 50 \end{bmatrix}$. $(I-A)X = D$ or $X = (I-A)^{-1} \cdot D$

$$X = \begin{bmatrix} 0.5 & -0.1 & -0.1 \\ -0.3 & 0.5 & -0.2 \\ -0.1 & -0.3 & 0.6 \end{bmatrix}^{-1} \cdot \begin{bmatrix} 110 \\ 50 \\ 50 \end{bmatrix} = \begin{bmatrix} 2.79 & 1.05 & 0.81 \\ 2.33 & 3.37 & 1.51 \\ 1.63 & 1.86 & 2.56 \end{bmatrix} \cdot \begin{bmatrix} 110 \\ 50 \\ 50 \end{bmatrix} = \begin{bmatrix} 400 \\ 500 \\ 400 \end{bmatrix}$$

400 units of fishing output, 500 units of agricultural goods and 400 units of mining goods are needed.

31. $D = \begin{bmatrix} 100 \\ 272 \\ 200 \end{bmatrix}$. $(I-A)X = D$ or $\begin{bmatrix} 0.4 & -0.2 & -0.2 & | & 100 \\ -0.1 & 0.6 & -0.5 & | & 272 \\ -0.1 & -0.2 & 0.8 & | & 200 \end{bmatrix}$ is the required augmented matrix.

By reducing this to $\begin{bmatrix} 1 & 0 & 0 & | & 1240 \\ 0 & 1 & 0 & | & 1260 \\ 0 & 0 & 1 & | & 720 \end{bmatrix}$ we have 1240 electronics, 1260 steel, and 720 autos.

33. $D = \begin{bmatrix} 24 \\ 62 \\ 32 \end{bmatrix}$. $(I-A)X = D$ or $X = (I-A)^{-1} \cdot D$

$$X = \begin{bmatrix} 0.6 & -0.1 & -0.1 \\ -0.2 & 0.5 & -0.2 \\ -0.2 & -0.1 & 0.7 \end{bmatrix}^{-1} \cdot \begin{bmatrix} 24 \\ 62 \\ 32 \end{bmatrix} = \begin{bmatrix} 1.964 & 0.476 & 0.417 \\ 1.071 & 2.381 & 0.833 \\ 0.714 & 0.476 & 1.667 \end{bmatrix} \cdot \begin{bmatrix} 24 \\ 62 \\ 32 \end{bmatrix} = \begin{bmatrix} 90 \\ 200 \\ 100 \end{bmatrix} = \begin{bmatrix} S \\ M \\ A \end{bmatrix}$$

35.
$$\begin{aligned} 0.5P + 0.1M + 0.2H &= P \\ 0.1P + 0.3M + 0.0H &= M \\ 0.4P + 0.6M + 0.8H &= H \end{aligned} \quad \text{or} \quad \begin{aligned} -0.5P + 0.1M + 0.2H &= 0 \\ 0.1P - 0.7M + 0.0H &= 0 \\ 0.4P + 0.6M - 0.2H &= 0 \end{aligned}$$

$$\begin{bmatrix} 1 & -7 & 0 & | & 0 \\ -5 & 1 & 2 & | & 0 \\ 4 & 6 & -2 & | & 0 \end{bmatrix} \begin{matrix} \\ 5R_1 + R_2 \rightarrow R_2 \\ -4R_1 + R_3 \rightarrow R_3 \end{matrix} \rightarrow \begin{bmatrix} 1 & -7 & 0 & | & 0 \\ 0 & -34 & 2 & | & 0 \\ 0 & 34 & -2 & | & 0 \end{bmatrix} \begin{matrix} \\ \\ R_2 + R_3 \rightarrow R_3 \end{matrix} \rightarrow$$

$$\begin{bmatrix} 1 & -7 & 0 & | & 0 \\ 0 & -34 & 2 & | & 0 \\ 0 & 0 & 0 & | & 0 \end{bmatrix} \begin{matrix} \\ -\frac{1}{34}R_2 \rightarrow R_2 \end{matrix} \rightarrow \begin{bmatrix} 1 & -7 & 0 & | & 0 \\ 0 & 1 & -1/17 & | & 0 \\ 0 & 0 & 0 & | & 0 \end{bmatrix} \begin{matrix} 7R_2 + R_1 \rightarrow R_1 \\ \\ \end{matrix} \rightarrow \begin{bmatrix} 1 & 0 & -7/17 & | & 0 \\ 0 & 1 & -1/17 & | & 0 \\ 0 & 0 & 0 & | & 0 \end{bmatrix}$$

So, Farm Products $= \dfrac{7}{17}$ Households and Farm Machinery $= \dfrac{1}{17}$ Households.

37.
$$\begin{aligned} 0.4G + 0.2I + 0.2H &= G \\ 0.2G + 0.3I + 0.3H &= I \\ 0.4G + 0.5I + 0.5H &= H \end{aligned} \quad \text{or} \quad \begin{aligned} -0.6G + 0.2I + 0.2H &= 0 \\ 0.2G - 0.7I + 0.3H &= 0 \\ 0.4G + 0.5I - 0.5H &= 0 \end{aligned}$$

$$\begin{bmatrix} 2 & -7 & 3 & | & 0 \\ -6 & 2 & 2 & | & 0 \\ 4 & 5 & -5 & | & 0 \end{bmatrix} \begin{matrix} \\ 3R_1 + R_2 \rightarrow R_2 \\ -2R_1 + R_3 \rightarrow R_3 \end{matrix} \begin{bmatrix} 2 & -7 & 3 & | & 0 \\ 0 & -19 & 11 & | & 0 \\ 0 & 19 & -11 & | & 0 \end{bmatrix} \begin{matrix} \rightarrow \\ \\ R_2 + R_3 \rightarrow R_3 \end{matrix} \begin{bmatrix} 2 & -7 & 3 & | & 0 \\ 0 & -19 & 11 & | & 0 \\ 0 & 0 & 0 & | & 0 \end{bmatrix} \begin{matrix} \\ -\frac{1}{19}R_2 \rightarrow R_2 \\ \rightarrow \end{matrix}$$

$$\begin{bmatrix} 2 & -7 & 3 & | & 0 \\ 0 & 1 & -11/19 & | & 0 \\ 0 & 0 & 0 & | & 0 \end{bmatrix} \begin{matrix} 7R_2 + R_1 \rightarrow R_1 \\ \\ \rightarrow \end{matrix} \begin{bmatrix} 2 & 0 & -20/19 & | & 0 \\ 0 & 1 & -11/19 & | & 0 \\ 0 & 0 & 0 & | & 0 \end{bmatrix} \begin{matrix} \frac{1}{2}R_1 \rightarrow R_1 \\ \\ \rightarrow \end{matrix} \begin{bmatrix} 1 & 0 & -10/19 & | & 0 \\ 0 & 1 & -11/19 & | & 0 \\ 0 & 0 & 0 & | & 0 \end{bmatrix}$$

Government $= \frac{10}{19}$ Households. Industry $= \frac{11}{19}$ Households.

39.
$$0.5M + 0.4U + 0.3H = M$$
$$0.4M + 0.5U + 0.3H = U \quad \text{or}$$
$$0.1M + 0.1U + 0.4H = H$$

$$-0.5M + 0.4U + 0.3H = 0$$
$$0.4M - 0.5U + 0.3H = 0$$
$$0.1M + 0.1U - 0.6H = 0$$

$$\begin{bmatrix} 1 & 1 & -6 & | & 0 \\ 4 & -5 & 3 & | & 0 \\ -5 & 4 & 3 & | & 0 \end{bmatrix} \begin{array}{l} -4R_1 + R_2 \to R_2 \\ 5R_1 + R_3 \to R_3 \end{array} \begin{bmatrix} 1 & 1 & -6 & | & 0 \\ 0 & -9 & 27 & | & 0 \\ 0 & 9 & -27 & | & 0 \end{bmatrix} \begin{array}{c} \to \\ \\ R_2 + R_3 \to R_3 \end{array} \begin{bmatrix} 1 & 1 & -6 & | & 0 \\ 0 & -9 & 27 & | & 0 \\ 0 & 0 & 0 & | & 0 \end{bmatrix} \begin{array}{c} \to \\ -\frac{1}{9}R_2 \to R_2 \end{array}$$

$$\begin{bmatrix} 1 & 1 & -6 & | & 0 \\ 0 & 1 & -3 & | & 0 \\ 0 & 0 & 0 & | & 0 \end{bmatrix} \begin{array}{c} -R_2 + R_1 \to R_1 \\ \to \end{array} \begin{bmatrix} 1 & 0 & -3 & | & 0 \\ 0 & 1 & -3 & | & 0 \\ 0 & 0 & 0 & | & 0 \end{bmatrix}$$

Manufacturing = 3 Households and Utilities = 3 Households.

41.
$$\begin{bmatrix} 1 & 0 & 0 & 0 & 0 & 0 & | & 24 \\ -4 & 1 & 0 & 0 & 0 & 0 & | & 0 \\ -1 & 0 & 1 & 0 & 0 & 0 & | & 0 \\ 0 & -1 & -1 & 1 & 0 & 0 & | & 0 \\ 0 & -4 & -4 & 0 & 1 & 0 & | & 12 \\ -20 & -24 & -24 & 0 & 0 & 1 & | & 96 \end{bmatrix} \begin{array}{l} \to \\ 4R_1 + R_2 \to R_2 \\ R_1 + R_3 \to R_3 \\ \\ \\ 20R_1 + R_6 \to R_6 \end{array} \begin{bmatrix} 1 & 0 & 0 & 0 & 0 & 0 & | & 24 \\ 0 & 1 & 0 & 0 & 0 & 0 & | & 96 \\ 0 & 0 & 1 & 0 & 0 & 0 & | & 24 \\ 0 & -1 & -1 & 1 & 0 & 0 & | & 0 \\ 0 & -4 & -4 & 0 & 1 & 0 & | & 12 \\ 0 & -24 & -24 & 0 & 0 & 1 & | & 576 \end{bmatrix} \begin{array}{l} \to \\ \\ \\ R_2 + R_4 \to R_4 \\ 4R_2 + R_5 \to R_5 \\ 24R_2 + R_6 \to R_6 \end{array}$$

$$\begin{bmatrix} 1 & 0 & 0 & 0 & 0 & 0 & | & 24 \\ 0 & 1 & 0 & 0 & 0 & 0 & | & 96 \\ 0 & 0 & 1 & 0 & 0 & 0 & | & 24 \\ 0 & 0 & -1 & 1 & 0 & 0 & | & 96 \\ 0 & 0 & -4 & 0 & 1 & 0 & | & 396 \\ 0 & 0 & -24 & 0 & 0 & 1 & | & 2880 \end{bmatrix} \begin{array}{l} \to \\ \\ \\ R_3 + R_4 \to R_4 \\ 4R_3 + R_5 \to R_5 \\ 24R_3 + R_6 \to R_6 \end{array} \begin{bmatrix} 1 & 0 & 0 & 0 & 0 & 0 & | & 24 \\ 0 & 1 & 0 & 0 & 0 & 0 & | & 96 \\ 0 & 0 & 1 & 0 & 0 & 0 & | & 24 \\ 0 & 0 & 0 & 1 & 0 & 0 & | & 120 \\ 0 & 0 & 0 & 0 & 1 & 0 & | & 492 \\ 0 & 0 & 0 & 0 & 0 & 1 & | & 3456 \end{bmatrix}$$

Therefore, 120 sheets, 492 braces, and 3456 bolts are required to fill the order.

43. $D = \begin{bmatrix} 10 \\ 0 \\ 0 \\ 6 \\ 0 \\ 6 \\ 100 \end{bmatrix}$ $X = \begin{bmatrix} x_1 \\ x_2 \\ x_3 \\ x_4 \\ x_5 \\ x_6 \\ x_7 \end{bmatrix} = \begin{bmatrix} \text{total sawhorses} \\ \text{total tops} \\ \text{total leg pairs} \\ \text{total 2} \times 4\text{s} \\ \text{total braces} \\ \text{total clamps} \\ \text{total nails} \end{bmatrix}$ Solve $(I - A)X = D$

$$\left[\begin{array}{ccccccc|c} 1 & 0 & 0 & 0 & 0 & 0 & 0 & 10 \\ -1 & 1 & 0 & 0 & 0 & 0 & 0 & 0 \\ -2 & 0 & 1 & 0 & 0 & 0 & 0 & 0 \\ 0 & -1 & -2 & 1 & 0 & 0 & 0 & 6 \\ 0 & 0 & -1 & 0 & 1 & 0 & 0 & 0 \\ 0 & 0 & -1 & 0 & 0 & 1 & 0 & 6 \\ -4 & 0 & -8 & 0 & 0 & 0 & 1 & 100 \end{array}\right] \begin{array}{l} \\ R_1 + R_2 \to R_2 \\ 2R_1 + R_3 \to R_3 \\ \\ \to \\ \\ 4R_1 + R_7 \to R_7 \end{array} \left[\begin{array}{ccccccc|c} 1 & 0 & 0 & 0 & 0 & 0 & 0 & 10 \\ 0 & 1 & 0 & 0 & 0 & 0 & 0 & 10 \\ 0 & 0 & 1 & 0 & 0 & 0 & 0 & 20 \\ 0 & -1 & -2 & 1 & 0 & 0 & 0 & 6 \\ 0 & 0 & -1 & 0 & 1 & 0 & 0 & 0 \\ 0 & 0 & -1 & 0 & 0 & 1 & 0 & 6 \\ 0 & 0 & -8 & 0 & 0 & 0 & 1 & 140 \end{array}\right] \begin{array}{l} \\ \to \\ \\ R_2 + R_4 \to R_4 \end{array}$$

$$\left[\begin{array}{ccccccc|c} 1 & 0 & 0 & 0 & 0 & 0 & 0 & 10 \\ 0 & 1 & 0 & 0 & 0 & 0 & 0 & 10 \\ 0 & 0 & 1 & 0 & 0 & 0 & 0 & 20 \\ 0 & 0 & -2 & 1 & 0 & 0 & 0 & 16 \\ 0 & 0 & -1 & 0 & 1 & 0 & 0 & 0 \\ 0 & 0 & -1 & 0 & 0 & 0 & 10 & 6 \\ 0 & 0 & -8 & 0 & 0 & 0 & 1 & 140 \end{array}\right] \begin{array}{l} \\ \\ \\ 2R_3 + R_4 \to R_4 \\ R_3 + R_5 \to R_5 \\ R_3 + R_6 \to R_6 \\ 8R_3 + R_7 \to R_7 \end{array} \left[\begin{array}{ccccccc|c} 1 & 0 & 0 & 0 & 0 & 0 & 0 & 10 \\ 0 & 1 & 0 & 0 & 0 & 0 & 0 & 10 \\ 0 & 0 & 1 & 0 & 0 & 0 & 0 & 20 \\ 0 & 0 & 0 & 1 & 0 & 0 & 0 & 56 \\ 0 & 0 & 0 & 0 & 1 & 0 & 0 & 20 \\ 0 & 0 & 0 & 0 & 0 & 1 & 0 & 26 \\ 0 & 0 & 0 & 0 & 0 & 0 & 1 & 300 \end{array}\right]$$

To fill the order, 56 2×4s , 20 braces, 26 clamps, and 300 nails.

Review Exercises

1. $a_{12} = 4$

2. $b_{23} = 0$

3. A and B

4. None

5. D, F, G, I

6. The negative of B is $\begin{bmatrix} -2 & 5 & 11 & -8 \\ -4 & 0 & 0 & -4 \\ 2 & 2 & -1 & -9 \end{bmatrix}$.

7. Zero matrix

8. Two matrices can be added if they have the same order.

9. $A + B$

$$= \begin{bmatrix} 4 & 4 & 2 & -5 \\ 6 & 3 & -1 & 0 \\ 0 & 0 & -3 & 5 \end{bmatrix} + \begin{bmatrix} 2 & -5 & -11 & 8 \\ 4 & 0 & 0 & 4 \\ -2 & -2 & 1 & 9 \end{bmatrix}$$

$$= \begin{bmatrix} 6 & -1 & -9 & 3 \\ 10 & 3 & -1 & 4 \\ -2 & -2 & -2 & 14 \end{bmatrix}$$

10. $C - E = \begin{bmatrix} 4 & -2 \\ 5 & 0 \\ 6 & 0 \\ 1 & 3 \end{bmatrix} - \begin{bmatrix} 1 & 1 \\ 1 & 1 \\ 4 & 6 \\ 0 & 5 \end{bmatrix} = \begin{bmatrix} 3 & -3 \\ 4 & -1 \\ 2 & -6 \\ 1 & -2 \end{bmatrix}$

11. $D^T - I = \begin{bmatrix} 3 & 1 \\ 5 & 2 \end{bmatrix} - \begin{bmatrix} 1 & 0 \\ 0 & 1 \end{bmatrix} = \begin{bmatrix} 2 & 1 \\ 5 & 1 \end{bmatrix}$

12. $3C = 3 \begin{bmatrix} 4 & -2 \\ 5 & 0 \\ 6 & 0 \\ 1 & 3 \end{bmatrix} = \begin{bmatrix} 12 & -6 \\ 15 & 0 \\ 18 & 0 \\ 3 & 9 \end{bmatrix}$

13. $4I = \begin{bmatrix} 4 & 0 \\ 0 & 4 \end{bmatrix}$

14. $-2F = -2 \begin{bmatrix} -1 & 6 \\ 4 & 11 \end{bmatrix} = \begin{bmatrix} 2 & -12 \\ -8 & -22 \end{bmatrix}$

15. $4D - 3I = \begin{bmatrix} 12 & 20 \\ 4 & 8 \end{bmatrix} - \begin{bmatrix} 3 & 0 \\ 0 & 3 \end{bmatrix} = \begin{bmatrix} 9 & 20 \\ 4 & 5 \end{bmatrix}$

16. $F + 2D = \begin{bmatrix} -1 & 6 \\ 4 & 11 \end{bmatrix} + \begin{bmatrix} 6 & 10 \\ 2 & 4 \end{bmatrix} = \begin{bmatrix} 5 & 16 \\ 6 & 15 \end{bmatrix}$

17. $3A - 5B$

$$= \begin{bmatrix} 12 & 12 & 6 & -15 \\ 18 & 9 & -3 & 0 \\ 0 & 0 & -9 & 15 \end{bmatrix} - \begin{bmatrix} 10 & -25 & -55 & 40 \\ 20 & 0 & 0 & 20 \\ -10 & -10 & 5 & 45 \end{bmatrix}$$

$$= \begin{bmatrix} 2 & 37 & 61 & -55 \\ -2 & 9 & -3 & -20 \\ 10 & 10 & -14 & -30 \end{bmatrix}$$

18. $AC = \begin{bmatrix} 4 & 4 & 2 & -5 \\ 6 & 3 & -1 & 0 \\ 0 & 0 & -3 & 5 \end{bmatrix} \begin{bmatrix} 4 & -2 \\ 5 & 0 \\ 6 & 0 \\ 1 & 3 \end{bmatrix} = \begin{bmatrix} 43 & -23 \\ 33 & -12 \\ -13 & 15 \end{bmatrix}$

19. $CD = \begin{bmatrix} 4 & -2 \\ 5 & 0 \\ 6 & 0 \\ 1 & 3 \end{bmatrix} \begin{bmatrix} 3 & 5 \\ 1 & 2 \end{bmatrix} = \begin{bmatrix} 10 & 16 \\ 15 & 25 \\ 18 & 30 \\ 6 & 11 \end{bmatrix}$

20. $DF = \begin{bmatrix} 3 & 5 \\ 1 & 2 \end{bmatrix} \begin{bmatrix} -1 & 6 \\ 4 & 11 \end{bmatrix} = \begin{bmatrix} 17 & 73 \\ 7 & 28 \end{bmatrix}$

21. $FD = \begin{bmatrix} -1 & 6 \\ 4 & 11 \end{bmatrix} \begin{bmatrix} 3 & 5 \\ 1 & 2 \end{bmatrix} = \begin{bmatrix} 3 & 7 \\ 23 & 42 \end{bmatrix}$

22. $FI = F$

23. $IF = F$

24. $DG^T = \begin{bmatrix} 3 & 5 \\ 1 & 2 \end{bmatrix} \begin{bmatrix} 2 & -1 \\ -5 & 3 \end{bmatrix} = \begin{bmatrix} -19 & 12 \\ -8 & 5 \end{bmatrix}$

25. $DG = \begin{bmatrix} 3 & 5 \\ 1 & 2 \end{bmatrix} \begin{bmatrix} 2 & -5 \\ -1 & 3 \end{bmatrix} = \begin{bmatrix} 1 & 0 \\ 0 & 1 \end{bmatrix} = I$

So, $(DG)F = F$.

26. **a.** Infinitely many solutions. The general solution is: $x = 6 + 2z$ and $y = 7 - 3z$.

   b. Answers will vary depending on the choice of z.
   If we let $z = 0$, then $y = 6 + 2(0) = 6$ and $y = 7 - 3(0) = 7$.

27. a. No solution. If translated into equation form, the last row would be $0x + 0y + 0z = 1$. Zero cannot equal one, so there is no solution.

28. a. Unique solution.
    b. $x = 0,\ y = -10,\ z = 14$

29. $\begin{bmatrix} 1 & 1 & 2 & | & 5 \\ 4 & 0 & 1 & | & 5 \\ 2 & 1 & 1 & | & 5 \end{bmatrix}$ $\begin{array}{c} \rightarrow \\ -4R_1 + R_2 \rightarrow R_2 \\ -2R_1 + R_3 \rightarrow R_3 \end{array}$ $\begin{bmatrix} 1 & 1 & 2 & | & 5 \\ 0 & -4 & -7 & | & -15 \\ 0 & -1 & -3 & | & -5 \end{bmatrix}$ $\begin{array}{c} R_2 \leftrightarrow R_3 \\ \rightarrow \end{array}$ $\begin{bmatrix} 1 & 1 & 2 & | & 5 \\ 0 & -1 & -3 & | & -5 \\ 0 & -4 & -7 & | & -15 \end{bmatrix}$ $\begin{array}{c} \rightarrow \\ -R_2 \rightarrow R_2 \end{array}$

$\begin{bmatrix} 1 & 1 & 2 & | & 5 \\ 0 & 1 & 3 & | & 5 \\ 0 & -4 & -7 & | & -15 \end{bmatrix}$ $\begin{array}{c} -R_2 + R_1 \rightarrow R_1 \\ \rightarrow \\ 4R_2 + R_3 \rightarrow R_3 \end{array}$ $\begin{bmatrix} 1 & 0 & -1 & | & 0 \\ 0 & 1 & 3 & | & 5 \\ 0 & 0 & 5 & | & 5 \end{bmatrix}$ $\begin{array}{c} \rightarrow \\ \frac{1}{5}R_3 \rightarrow R_3 \end{array}$ $\begin{bmatrix} 1 & 0 & -1 & | & 0 \\ 0 & 1 & 3 & | & 5 \\ 0 & 0 & 1 & | & 1 \end{bmatrix}$ $\begin{array}{c} R_3 + R_1 \rightarrow R_1 \\ -3R_3 + R_2 \rightarrow R_2 \\ \rightarrow \end{array}$

$\begin{bmatrix} 1 & 0 & 0 & | & 1 \\ 0 & 1 & 0 & | & 2 \\ 0 & 0 & 1 & | & 1 \end{bmatrix}$ The solution is $(1, 2, 1)$.

30. $\begin{bmatrix} 1 & -2 & | & 4 \\ -3 & 10 & | & 24 \end{bmatrix}$ $\begin{array}{c} \rightarrow \\ 3R_1 + R_2 \rightarrow R_2 \end{array}$ $\begin{bmatrix} 1 & -2 & | & 4 \\ 0 & 4 & | & 36 \end{bmatrix}$ $\begin{array}{c} \rightarrow \\ \frac{1}{4}R_2 \rightarrow R_2 \end{array}$ $\begin{bmatrix} 1 & -2 & | & 4 \\ 0 & 1 & | & 9 \end{bmatrix}$ $\begin{array}{c} 2R_2 + R_1 \rightarrow R_1 \\ \rightarrow \end{array}$ $\begin{bmatrix} 1 & 0 & | & 22 \\ 0 & 1 & | & 9 \end{bmatrix}$

Solution: $x = 22, y = 9$

31. $\begin{bmatrix} 1 & 1 & 1 & | & 4 \\ 3 & 4 & -1 & | & -1 \\ 2 & -1 & 3 & | & 3 \end{bmatrix}$ $\begin{array}{c} \rightarrow \\ -3R_1 + R_2 \rightarrow R_2 \\ -2R_1 + R_3 \rightarrow R_3 \end{array}$ $\begin{bmatrix} 1 & 1 & 1 & | & 4 \\ 0 & 1 & -4 & | & -13 \\ 0 & -3 & 1 & | & -5 \end{bmatrix}$ $\begin{array}{c} -R_2 + R_1 \rightarrow R_1 \\ \rightarrow \\ 3R_2 + R_3 \rightarrow R_3 \end{array}$

$\begin{bmatrix} 1 & 0 & 5 & | & 17 \\ 0 & 1 & -4 & | & -13 \\ 0 & 0 & -11 & | & -44 \end{bmatrix}$ $\begin{array}{c} \rightarrow \\ -\frac{1}{11}R_3 \rightarrow R_3 \end{array}$ $\begin{bmatrix} 1 & 0 & 5 & | & 17 \\ 0 & 1 & -4 & | & -13 \\ 0 & 0 & 1 & | & 4 \end{bmatrix}$ $\begin{array}{c} -5R_3 + R_1 \rightarrow R_1 \\ 4R_3 + R_2 \rightarrow R_2 \\ \rightarrow \end{array}$ $\begin{bmatrix} 1 & 0 & 0 & | & -3 \\ 0 & 1 & 0 & | & 3 \\ 0 & 0 & 1 & | & 4 \end{bmatrix}$

Solution: $x = -3,\ y = 3,\ z = 4$

32. $\begin{bmatrix} -1 & 1 & 1 & | & 3 \\ 3 & 0 & -1 & | & 1 \\ 2 & -3 & -4 & | & -2 \end{bmatrix}$ $\begin{array}{c} \rightarrow \\ 3R_1 + R_2 \rightarrow R_2 \\ 2R_1 + R_3 \rightarrow R_3 \end{array}$ $\begin{bmatrix} -1 & 1 & 1 & | & 3 \\ 0 & 3 & 2 & | & 10 \\ 0 & -1 & -2 & | & 4 \end{bmatrix}$ $\begin{array}{c} R_3 + R_1 \rightarrow R_1 \\ 3R_3 + R_2 \rightarrow R_2 \\ \rightarrow \end{array}$ $\begin{bmatrix} -1 & 0 & -1 & | & 7 \\ 0 & 0 & -4 & | & 22 \\ 0 & -1 & -2 & | & 4 \end{bmatrix}$ $\begin{array}{c} -R_1 \rightarrow R_1 \\ \rightarrow \\ R_2 \leftrightarrow R_3 \end{array}$

$\begin{bmatrix} 1 & 0 & 1 & | & -7 \\ 0 & -1 & -2 & | & 4 \\ 0 & 0 & -4 & | & 22 \end{bmatrix}$ $\begin{array}{c} \rightarrow \\ -R_2 \rightarrow R_2 \\ -\frac{1}{4}R_3 \rightarrow R_3 \end{array}$ $\begin{bmatrix} 1 & 0 & 1 & | & -7 \\ 0 & 1 & 2 & | & -4 \\ 0 & 0 & 1 & | & -11/2 \end{bmatrix}$ $\begin{array}{c} -R_3 + R_1 \rightarrow R_1 \\ -2R_3 + R_2 \rightarrow R_2 \\ \rightarrow \end{array}$ $\begin{bmatrix} 1 & 0 & 0 & | & -3/2 \\ 0 & 1 & 0 & | & 7 \\ 0 & 0 & 1 & | & -11/2 \end{bmatrix}$

Solution: $x = -\dfrac{3}{2}, y = 7, z = -\dfrac{11}{2}$

33. $\begin{bmatrix} 1 & 1 & -2 & | & 5 \\ 3 & 2 & 5 & | & 10 \\ -2 & -3 & 15 & | & 2 \end{bmatrix}$ $\begin{array}{c} \rightarrow \\ -3R_1 + R_2 \rightarrow R_2 \\ 2R_1 + R_3 \rightarrow R_3 \end{array}$ $\begin{bmatrix} 1 & 1 & -2 & | & 5 \\ 0 & -1 & 11 & | & -5 \\ 0 & -1 & 11 & | & 12 \end{bmatrix}$ $\begin{array}{c} \rightarrow \\ -R_2 + R_3 \rightarrow R_3 \end{array}$ $\begin{bmatrix} 1 & 1 & -2 & | & 5 \\ 0 & -1 & 11 & | & -5 \\ 0 & 0 & 0 & | & 17 \end{bmatrix}$

No solution.

34.
$$\begin{bmatrix} 1 & -1 & 0 & | & 3 \\ 1 & 1 & 4 & | & 1 \\ 2 & -3 & -2 & | & 7 \end{bmatrix} \begin{array}{c} \\ -R_1 + R_2 \to R_2 \\ -2R_1 + R_3 \to R_3 \end{array} \xrightarrow{} \begin{bmatrix} 1 & -1 & 0 & | & 3 \\ 0 & 2 & 4 & | & -2 \\ 0 & -1 & -2 & | & 1 \end{bmatrix} \tfrac{1}{2}R_2 \to R_2$$

$$\begin{bmatrix} 1 & -1 & 0 & | & 3 \\ 0 & 1 & 2 & | & -1 \\ 0 & -1 & -2 & | & 1 \end{bmatrix} \begin{array}{c} R_2 + R_1 \to R_1 \\ \xrightarrow{} \\ R_2 + R_3 \to R_3 \end{array} \begin{bmatrix} 1 & 0 & 2 & | & 2 \\ 0 & 1 & 2 & | & -1 \\ 0 & 0 & 0 & | & 0 \end{bmatrix}$$

There are infinitely many solutions. The general solution is $x = 2 - 2z, \; y = -1 - 2z$.

35.
$$\begin{bmatrix} 1 & -3 & 1 & | & 4 \\ 2 & -5 & -1 & | & 6 \end{bmatrix} \begin{array}{c} \\ -2R_1 + R_2 \to R_2 \end{array} \begin{bmatrix} 1 & -3 & 1 & | & 4 \\ 0 & 1 & -3 & | & -2 \end{bmatrix} \begin{array}{c} 3R_2 + R_1 \to R_1 \\ \xrightarrow{} \end{array} \begin{bmatrix} 1 & 0 & -8 & | & -2 \\ 0 & 1 & -3 & | & -2 \end{bmatrix}$$

There are infinitely many solutions. The general solution is $x = -2 + 8y$ and $y = -2 + 3z$.

36.
$$\begin{bmatrix} 1 & 1 & 1 & 1 & | & 3 \\ 1 & -2 & 1 & -4 & | & -5 \\ 1 & 0 & -1 & 1 & | & 0 \\ 0 & 1 & 1 & 1 & | & 2 \end{bmatrix} \begin{array}{c} \\ -R_1 + R_2 \to R_2 \\ -R_1 + R_3 \to R_3 \\ \\ \end{array} \begin{bmatrix} 1 & 1 & 1 & 1 & | & 3 \\ 0 & -3 & 0 & -5 & | & -8 \\ 0 & -1 & -2 & 0 & | & -3 \\ 0 & 1 & 1 & 1 & | & 2 \end{bmatrix} \begin{array}{c} R_2 \leftrightarrow R_4 \\ \xrightarrow{} \end{array}$$

$$\begin{bmatrix} 1 & 1 & 1 & 1 & | & 3 \\ 0 & 1 & 1 & 1 & | & 2 \\ 0 & -1 & -2 & 0 & | & -3 \\ 0 & -3 & 0 & -5 & | & -8 \end{bmatrix} \begin{array}{c} -R_2 + R_1 \to R_1 \\ \xrightarrow{} \\ R_2 + R_3 \to R_3 \\ 3R_2 + R_4 \to R_4 \end{array} \begin{bmatrix} 1 & 0 & 0 & 0 & | & 1 \\ 0 & 1 & 1 & 1 & | & 2 \\ 0 & 0 & -1 & 1 & | & -1 \\ 0 & 0 & 3 & -2 & | & -2 \end{bmatrix} \begin{array}{c} \\ -R_3 \to R_3 \end{array} \xrightarrow{}$$

$$\begin{bmatrix} 1 & 0 & 0 & 0 & | & 1 \\ 0 & 1 & 1 & 1 & | & 2 \\ 0 & 0 & 1 & -1 & | & 1 \\ 0 & 0 & 3 & -2 & | & -2 \end{bmatrix} \begin{array}{c} \\ -R_3 + R_2 \to R_2 \\ \\ -3R_3 + R_4 \to R_4 \end{array} \begin{bmatrix} 1 & 0 & 0 & 0 & | & 1 \\ 0 & 1 & 0 & 2 & | & 1 \\ 0 & 0 & 1 & -1 & | & 1 \\ 0 & 0 & 0 & 1 & | & -5 \end{bmatrix} \begin{array}{c} \\ -2R_4 + R_2 \to R_2 \\ R_4 + R_3 \to R_3 \end{array} \xrightarrow{} \begin{bmatrix} 1 & 0 & 0 & 0 & | & 1 \\ 0 & 1 & 0 & 0 & | & 11 \\ 0 & 0 & 1 & 0 & | & -4 \\ 0 & 0 & 0 & 1 & | & -5 \end{bmatrix}$$

Solution: $x_1 = 1, \; x_2 = 11, \; x_3 = -4, \; x_4 = -5$

37. $DG = \begin{bmatrix} 3 & 5 \\ 1 & 2 \end{bmatrix} \begin{bmatrix} 2 & -5 \\ -1 & 3 \end{bmatrix} = \begin{bmatrix} 1 & 0 \\ 0 & 1 \end{bmatrix}$ Yes, D and G are inverse matrices.

38. $\begin{bmatrix} 7 & -1 \\ -10 & 2 \end{bmatrix}^{-1} = \dfrac{1}{14 - 10} \begin{bmatrix} 2 & 1 \\ 10 & 7 \end{bmatrix} = \begin{bmatrix} 1/2 & 1/4 \\ 5/2 & 7/4 \end{bmatrix}$

39.
$$\begin{bmatrix} 1 & 0 & 2 & | & 1 & 0 & 0 \\ 3 & 4 & -1 & | & 0 & 1 & 0 \\ 1 & 1 & 0 & | & 0 & 0 & 1 \end{bmatrix} \begin{array}{c} \\ -3R_1 + R_2 \to R_2 \\ -R_1 + R_3 \to R_3 \end{array} \xrightarrow{} \begin{bmatrix} 1 & 0 & 2 & | & 1 & 0 & 0 \\ 0 & 4 & -7 & | & -3 & 1 & 0 \\ 0 & 1 & -2 & | & -1 & 0 & 1 \end{bmatrix} \begin{array}{c} R_2 \leftrightarrow R_3 \\ \xrightarrow{} \end{array}$$

$$\begin{bmatrix} 1 & 0 & 2 & | & 1 & 0 & 0 \\ 0 & 1 & -2 & | & -1 & 0 & 1 \\ 0 & 4 & -7 & | & -3 & 1 & 0 \end{bmatrix} \begin{array}{c} \\ \\ -4R_2 + R_3 \to R_3 \end{array} \xrightarrow{} \begin{bmatrix} 1 & 0 & 2 & | & 1 & 0 & 0 \\ 0 & 1 & -2 & | & -1 & 0 & 1 \\ 0 & 0 & 1 & | & 1 & 1 & -4 \end{bmatrix} \begin{array}{c} -2R_3 + R_1 \to R_1 \\ 2R_3 + R_2 \to R_2 \\ \xrightarrow{} \end{array}$$

$$\begin{bmatrix} 1 & 0 & 0 & | & -1 & -2 & 8 \\ 0 & 1 & 0 & | & 1 & 2 & -7 \\ 0 & 0 & 1 & | & 1 & 1 & -4 \end{bmatrix} \qquad \text{Answer: } \begin{bmatrix} -1 & -2 & 8 \\ 1 & 2 & -7 \\ 1 & 1 & -4 \end{bmatrix}$$

40. $\begin{bmatrix} 3 & 3 & 2 & | & 1 & 0 & 0 \\ -1 & 4 & 2 & | & 0 & 1 & 0 \\ 2 & 5 & 3 & | & 0 & 0 & 1 \end{bmatrix}$ $\begin{matrix} 2R_2 + R_1 \to R_1 \\ \to \\ 2R_2 + R_3 \to R_3 \end{matrix}$ $\begin{bmatrix} 1 & 11 & 6 & | & 1 & 2 & 0 \\ -1 & 4 & 2 & | & 0 & 1 & 0 \\ 0 & 13 & 7 & | & 0 & 2 & 1 \end{bmatrix}$ $\begin{matrix} \to \\ R_1 + R_2 \to R_2 \end{matrix}$

$\begin{bmatrix} 1 & 11 & 6 & | & 1 & 2 & 0 \\ 0 & 15 & 8 & | & 1 & 3 & 0 \\ 0 & 13 & 7 & | & 0 & 2 & 1 \end{bmatrix}$ $\begin{matrix} \to \\ -R_3 + R_2 \to R_2 \end{matrix}$ $\begin{bmatrix} 1 & 11 & 6 & | & 1 & 2 & 0 \\ 0 & 2 & 1 & | & 1 & 1 & -1 \\ 0 & 13 & 7 & | & 0 & 2 & 1 \end{bmatrix}$ $\begin{matrix} -6R_2 + R_1 \to R_1 \\ \to \\ -7R_2 + R_3 \to R_3 \end{matrix}$

$\begin{bmatrix} 1 & -1 & 0 & | & -5 & -4 & 6 \\ 0 & 2 & 1 & | & 1 & 1 & -1 \\ 0 & -1 & 0 & | & -7 & -5 & 8 \end{bmatrix}$ $\begin{matrix} R_2 \leftrightarrow R_3 \\ \to \\ (-1) \text{ new } R_2 \end{matrix}$ $\begin{bmatrix} 1 & -1 & 0 & | & -5 & -4 & 6 \\ 0 & 1 & 0 & | & 7 & 5 & -8 \\ 0 & 2 & 1 & | & 1 & 1 & -1 \end{bmatrix}$ $\begin{matrix} R_2 + R_1 \to R_1 \\ \to \\ -2R_2 + R_3 \to R_3 \end{matrix}$

$\begin{bmatrix} 1 & 0 & 0 & | & 2 & 1 & -2 \\ 0 & 1 & 0 & | & 7 & 5 & -8 \\ 0 & 0 & 1 & | & -13 & -9 & 15 \end{bmatrix}$ Answer: $\begin{bmatrix} 2 & 1 & -2 \\ 7 & 5 & -8 \\ -13 & -9 & 15 \end{bmatrix}$

41. Use the inverse of problem 35.

$\begin{bmatrix} x \\ y \\ z \end{bmatrix} = \begin{bmatrix} 1 & 0 & 2 \\ 3 & 4 & -1 \\ 1 & 1 & 0 \end{bmatrix}^{-1} \begin{bmatrix} 5 \\ 2 \\ -3 \end{bmatrix} = \begin{bmatrix} -1 & -2 & 8 \\ 1 & 2 & -7 \\ 1 & 1 & -4 \end{bmatrix} \begin{bmatrix} 5 \\ 2 \\ -3 \end{bmatrix} = \begin{bmatrix} -33 \\ 30 \\ 19 \end{bmatrix}$

42. Use the inverse of problem 36.

$\begin{bmatrix} x \\ y \\ z \end{bmatrix} = \begin{bmatrix} 3 & 3 & 2 \\ -1 & 4 & 2 \\ 2 & 5 & 3 \end{bmatrix}^{-1} \begin{bmatrix} 1 \\ -10 \\ -6 \end{bmatrix} = \begin{bmatrix} 2 & 1 & -2 \\ 7 & 5 & -8 \\ -13 & -9 & 15 \end{bmatrix} \begin{bmatrix} 1 \\ -10 \\ -6 \end{bmatrix} = \begin{bmatrix} 4 \\ 5 \\ -13 \end{bmatrix}$

43. $\begin{bmatrix} 1 & 3 & 1 & | & 1 & 0 & 0 \\ 1 & 4 & 3 & | & 0 & 1 & 0 \\ 2 & -1 & -11 & | & 0 & 0 & 1 \end{bmatrix}$ $\begin{matrix} \to \\ -R_1 + R_2 \to R_2 \\ -2R_1 + R_3 \to R_3 \end{matrix}$ $\begin{bmatrix} 1 & 3 & 1 & | & 1 & 0 & 0 \\ 0 & 1 & 2 & | & -1 & 1 & 0 \\ 0 & -7 & -13 & | & -2 & 0 & 1 \end{bmatrix}$ $\begin{matrix} -3R_2 + R_1 \to R_1 \\ \to \\ 7R_2 + R_3 \to R_3 \end{matrix}$

$\begin{bmatrix} 1 & 0 & -5 & | & 4 & -3 & 0 \\ 0 & 1 & 2 & | & -1 & 1 & 0 \\ 0 & 0 & 1 & | & -9 & 7 & 1 \end{bmatrix}$ $\begin{matrix} 5R_3 + R_1 \to R_1 \\ -2R_3 + R_2 \to R_2 \\ \to \end{matrix}$ $\begin{bmatrix} 1 & 0 & 0 & | & -41 & 32 & 5 \\ 0 & 1 & 0 & | & 17 & -13 & -2 \\ 0 & 0 & 1 & | & -9 & 7 & 1 \end{bmatrix}$

$\begin{bmatrix} x \\ y \\ z \end{bmatrix} = \begin{bmatrix} -41 & 32 & 5 \\ 17 & -13 & -2 \\ -9 & 7 & 1 \end{bmatrix} \begin{bmatrix} 0 \\ 2 \\ -12 \end{bmatrix} = \begin{bmatrix} 4 \\ -2 \\ 2 \end{bmatrix}$

44. The determinant of the matrix is 0. The matrix does not have an inverse.

45. $\begin{vmatrix} 4 & 4 \\ -2 & 2 \end{vmatrix} = (4)(2) - (-2)(4) = 8 - (-8) = 16$

Since the determinant does not equal zero, this matrix has an inverse.

46. Using technology, $\begin{vmatrix} 1 & 2 & 3 \\ 4 & -1 & 8 \\ 6 & 3 & 14 \end{vmatrix} = 0$

Since the determinant equals zero, the matrix does not have an inverse.

47. Total production $= N + M = \begin{bmatrix} 250 & 140 \\ 480 & 700 \end{bmatrix}$

48. $M + P - S = \begin{bmatrix} 1030 & 800 \\ 700 & 1200 \end{bmatrix}$

49. June: $\begin{bmatrix} 100 & 0 \\ 0 & 120 \end{bmatrix} \cdot M = \begin{matrix} & A & B \\ & \begin{bmatrix} 15,000 & 8000 \\ 33,600 & 36,000 \end{bmatrix} \\ & \overline{48,600} \ \overline{44,000} \end{matrix}$ July: $\begin{bmatrix} 100 & 0 \\ 0 & 120 \end{bmatrix} \cdot N = \begin{matrix} & A & B \\ & \begin{bmatrix} 10,000 & 6000 \\ 24,000 & 48,000 \end{bmatrix} \\ & \overline{34,000} \ \overline{54,000} \end{matrix}$

   a. Production was higher at Plant A in June.
   b. Production was higher at Plant B in July.

50.
$$\begin{matrix} & \text{S} & \text{M} & \text{L} \\ \text{R} & \begin{bmatrix} 25 & 40 & 45 \\ \text{H} \ 10 & 10 & 10 \end{bmatrix} \end{matrix} \begin{matrix} & \text{M} & \text{W} \\ & \begin{bmatrix} 1 & 14 \\ 12 & 10 \\ 8 & 3 \end{bmatrix} \begin{matrix} \text{S} \\ \text{M} \\ \text{L} \end{matrix} \end{matrix} = \begin{matrix} & \text{M} & \text{W} \\ & \begin{bmatrix} 865 & 885 \\ 210 & 270 \end{bmatrix} \begin{matrix} \text{R} \\ \text{H} \end{matrix} \end{matrix}$$

BA gives costs of robes and hoods for men and women.

51. $\begin{bmatrix} 25 & 40 & 45 \\ 10 & 10 & 10 \end{bmatrix} \begin{bmatrix} 1 & 14 \\ 12 & 10 \\ 8 & 3 \end{bmatrix} = \begin{bmatrix} 865 & 885 \\ 210 & 270 \end{bmatrix}$

$\begin{bmatrix} 865 & 885 \\ 210 & 270 \end{bmatrix} \begin{bmatrix} 1 \\ 1 \end{bmatrix} = \begin{bmatrix} 1750 \\ 480 \end{bmatrix} \begin{matrix} \text{R} \\ \text{H} \end{matrix}$

The cost of new robes is $1750.
The cost of new hoods is $480.

52. a. $\begin{bmatrix} 30 & 20 & 10 \\ 20 & 10 & 20 \end{bmatrix} \begin{bmatrix} 300 & 280 \\ 150 & 100 \\ 150 & 200 \end{bmatrix} = \begin{bmatrix} 9000 + 3000 + 1500 & 8400 + 2000 + 2000 \\ 6000 + 1500 + 3000 & 5600 + 1000 + 4000 \end{bmatrix} = \begin{bmatrix} 13,500 & 12,400 \\ 10,500 & 10,600 \end{bmatrix}$

   b. Column 1 is Ace's price and column 2 is Kink's price. Dept. A buys from Kink and Dept. B buys from Ace.

53. a. $W = \begin{bmatrix} 0.20 & 0.30 & 0.50 \end{bmatrix}$

   b. $R = \begin{bmatrix} 0.013469 \\ 0.013543 \\ 0.006504 \end{bmatrix}$

   c. $WR = 0.0100087$
   d. The historical return is the same as the estimated expected monthly return, i.e., 1%.

54. $x =$ fast food shares; $y =$ software shares; $z =$ pharmaceutical shares

$$50x + 20y + 80z = 50{,}000$$

$$0.115(50x) + 0.15(20y) + 0.10(80z) = 0.12(50{,}000) \quad \rightarrow \quad x = 2z$$

Rearranged and simplified: $\quad x - 2z = 0$

$$5x + 2y + 8z = 5000$$

$$5.75x + 3y + 8z = 6000$$

$$\begin{bmatrix} 1 & 0 & -2 & | & 0 \\ 5 & 2 & 8 & | & 5000 \\ 5.75 & 3 & 8 & | & 6000 \end{bmatrix} \begin{array}{c} \rightarrow \\ -5R_1 + R_2 \rightarrow R_2 \\ -5.75R_1 + R_3 \rightarrow R_3 \end{array} \begin{bmatrix} 1 & 0 & -2 & | & 0 \\ 0 & 2 & 18 & | & 5000 \\ 0 & 3 & 19.5 & | & 6000 \end{bmatrix} \begin{array}{c} \rightarrow \\ \frac{1}{2}R_2 \rightarrow R_2 \end{array} \begin{bmatrix} 1 & 0 & -2 & | & 0 \\ 0 & 1 & 9 & | & 2500 \\ 0 & 3 & 19.5 & | & 6000 \end{bmatrix}$$

$$\begin{array}{c} \rightarrow \\ \\ -3R_2 + R_3 \rightarrow R_3 \end{array} \begin{bmatrix} 1 & 0 & -2 & | & 0 \\ 0 & 1 & 9 & | & 2500 \\ 0 & 0 & -7.5 & | & -1500 \end{bmatrix} \begin{array}{c} \rightarrow \\ \\ -\frac{2}{15}R_3 \rightarrow R_3 \end{array} \begin{bmatrix} 1 & 0 & -2 & | & 0 \\ 0 & 1 & 9 & | & 2500 \\ 0 & 0 & 1 & | & 200 \end{bmatrix} \begin{array}{c} 2R_3 + R_1 \rightarrow R_1 \\ -9R_3 + R_2 \rightarrow R_2 \\ \rightarrow \end{array} \begin{bmatrix} 1 & 0 & 0 & | & 400 \\ 0 & 1 & 0 & | & 700 \\ 0 & 0 & 1 & | & 200 \end{bmatrix}$$

Buy 400 shares of fast food company, 700 shares of software company, and 200 shares of pharmaceutical company.

55. $\quad A + B + 2C = 2000 \qquad$ Units of I

$\quad 3A + 4B + 10C = 8000 \qquad$ Units of II

$\quad A + 2B + 6C = 4000 \qquad$ Units of III

a. $\begin{bmatrix} 1 & 1 & 2 & | & 2000 \\ 3 & 4 & 10 & | & 8000 \\ 1 & 2 & 6 & | & 4000 \end{bmatrix} \begin{array}{c} \rightarrow \\ -3R_1 + R_2 \rightarrow R_2 \\ -R_1 + R_3 \rightarrow R_3 \end{array} \begin{bmatrix} 1 & 1 & 2 & | & 2000 \\ 0 & 1 & 4 & | & 2000 \\ 0 & 1 & 4 & | & 2000 \end{bmatrix} \begin{array}{c} -R_2 + R_1 \rightarrow R_1 \\ \rightarrow \\ -R_2 + R_3 \rightarrow R_3 \end{array} \begin{bmatrix} 1 & 0 & -2 & | & 0 \\ 0 & 1 & 4 & | & 2000 \\ 0 & 0 & 0 & | & 0 \end{bmatrix}$

Solution: $A = 2C$, $B = 2000 - 4C$

b. $\quad 500 = 2C$, so $C = 250$

When $C = 250$, $B = 2000 - 4(250) = 1000$.

Yes, it is possible to support 500 type A slugs; there would be 1000 type B slugs and 250 type C slugs.

c. $\quad 0 = 2000 - 4C$ so $C = 500$

When $C = 500$, $A = 2(500) = 1000$.

The maximum number of type A slugs is 1000, when there are 500 type C slugs and no type B slugs.

56. $\begin{bmatrix} 100 & 100 & 100 & | & 1100 \\ 150 & 20 & 350 & | & 1930 \\ 20 & 65 & 35 & | & 460 \end{bmatrix} \begin{array}{c} \frac{1}{100}R_1 \rightarrow R_1 \\ \frac{1}{10}R_2 \rightarrow R_2 \\ \frac{1}{5}R_3 \rightarrow R_3 \end{array} \begin{bmatrix} 1 & 1 & 1 & | & 11 \\ 15 & 2 & 35 & | & 193 \\ 4 & 13 & 7 & | & 92 \end{bmatrix} \begin{array}{c} \rightarrow \\ -15R_1 + R_2 \rightarrow R_2 \\ -4R_1 + R_3 \rightarrow R_3 \end{array} \begin{bmatrix} 1 & 1 & 1 & | & 11 \\ 0 & -13 & 20 & | & 28 \\ 0 & 9 & 3 & | & 48 \end{bmatrix} \begin{array}{c} \rightarrow \\ \\ \frac{1}{3}R_3 \rightarrow R_3 \end{array}$

$\begin{bmatrix} 1 & 1 & 1 & | & 11 \\ 0 & -13 & 20 & | & 28 \\ 0 & 3 & 1 & | & 16 \end{bmatrix} \begin{array}{c} \rightarrow \\ 4R_3 + R_2 \rightarrow R_2 \\ \end{array} \begin{bmatrix} 1 & 1 & 1 & | & 11 \\ 0 & -1 & 24 & | & 92 \\ 0 & 3 & 1 & | & 16 \end{bmatrix} \begin{array}{c} R_2 + R_1 \rightarrow R_1 \\ \rightarrow \\ 3R_2 + R_3 \rightarrow R_3 \end{array} \begin{bmatrix} 1 & 0 & 25 & | & 103 \\ 0 & -1 & 24 & | & 92 \\ 0 & 0 & 73 & | & 292 \end{bmatrix} \begin{array}{c} \rightarrow \\ -R_2 \rightarrow R_2 \\ \frac{1}{73}R_3 \rightarrow R_3 \end{array}$

$\begin{bmatrix} 1 & 0 & 25 & | & 103 \\ 0 & 1 & -24 & | & -92 \\ 0 & 0 & 1 & | & 4 \end{bmatrix} \begin{array}{c} -25R_3 + R_1 \rightarrow R_1 \\ 24R_3 + R_2 \rightarrow R_2 \\ \rightarrow \end{array} \begin{bmatrix} 1 & 0 & 0 & | & 3 \\ 0 & 1 & 0 & | & 4 \\ 0 & 0 & 1 & | & 4 \end{bmatrix}$

Use 3 passenger, 4 transport, and 4 jumbo.

57. $D = \begin{bmatrix} 4720 \\ 40 \end{bmatrix} \quad (I - A)X = D$ or $\begin{bmatrix} 0.8 & -0.1 \\ -0.1 & 0.8 \end{bmatrix} \begin{bmatrix} S \\ A \end{bmatrix} = \begin{bmatrix} 4720 \\ 40 \end{bmatrix}$

$\begin{bmatrix} S \\ A \end{bmatrix} = \begin{bmatrix} 0.8 & -0.1 \\ -0.1 & 0.8 \end{bmatrix}^{-1} \begin{bmatrix} 4720 \\ 40 \end{bmatrix} = \frac{1}{0.63} \begin{bmatrix} 0.8 & 0.1 \\ 0.1 & 0.8 \end{bmatrix} \begin{bmatrix} 4720 \\ 40 \end{bmatrix} = \begin{bmatrix} 6000 \\ 800 \end{bmatrix}$

58. $D = \begin{bmatrix} 850 \\ 275 \end{bmatrix}$ $(I-A)X = D$ or $\begin{bmatrix} 0.9 & -0.1 \\ -0.2 & 0.95 \end{bmatrix}\begin{bmatrix} S \\ C \end{bmatrix} = \begin{bmatrix} 850 \\ 275 \end{bmatrix}$

$\begin{bmatrix} S \\ C \end{bmatrix} = \begin{bmatrix} 0.9 & -0.1 \\ -0.2 & 0.95 \end{bmatrix}^{-1}\begin{bmatrix} 850 \\ 275 \end{bmatrix} = \dfrac{1}{0.835}\begin{bmatrix} 0.95 & 0.1 \\ 0.2 & 0.9 \end{bmatrix}\begin{bmatrix} 850 \\ 275 \end{bmatrix} = \begin{bmatrix} 1000 \\ 500 \end{bmatrix}$

59. $(I-A)X = D$ or $\begin{bmatrix} 0.6 & -0.4 & -0.2 \\ -0.2 & 0.6 & -0.2 \\ -0.1 & -0.2 & 0.6 \end{bmatrix}\begin{bmatrix} \text{Min} \\ \text{Mfg} \\ \text{Fuel} \end{bmatrix} = \begin{bmatrix} 8 \\ 40 \\ 140 \end{bmatrix}$

$\begin{bmatrix} 6 & -4 & -2 & | & 80 \\ -2 & 6 & -2 & | & 400 \\ -1 & -2 & 6 & | & 1400 \end{bmatrix} \begin{array}{c} -R_3 \leftrightarrow R_1 \end{array} \begin{bmatrix} 1 & 2 & -6 & | & -1400 \\ -2 & 6 & -2 & | & 400 \\ 6 & -4 & -2 & | & 80 \end{bmatrix} \begin{array}{c} \rightarrow \\ 2R_1 + R_2 \rightarrow R_2 \\ -6R_1 + R_3 \rightarrow R_3 \end{array} \begin{bmatrix} 1 & 2 & -6 & | & -1400 \\ 0 & 10 & -14 & | & -2400 \\ 0 & -16 & 34 & | & 8480 \end{bmatrix}$

$\begin{array}{c} \frac{1}{10}R_2 \rightarrow R_2 \end{array} \begin{bmatrix} 1 & 2 & -6 & | & -1400 \\ 0 & 1 & -1.4 & | & -240 \\ 0 & -16 & 34 & | & 8480 \end{bmatrix} \begin{array}{c} -2R_2 + R_1 \rightarrow R_1 \\ \rightarrow \\ 16R_2 + R_3 \rightarrow R_3 \end{array} \begin{bmatrix} 1 & 0 & -3.2 & | & -920 \\ 0 & 1 & -1.4 & | & -240 \\ 0 & 0 & 11.6 & | & 4640 \end{bmatrix} \begin{array}{c} \frac{1}{11.6}R_3 \rightarrow R_3 \end{array}$

$\begin{bmatrix} 1 & 0 & -3.2 & | & -920 \\ 0 & 1 & -1.4 & | & -240 \\ 0 & 0 & 1 & | & 400 \end{bmatrix} \begin{array}{c} 3.2R_3 + R_1 \rightarrow R_1 \\ 1.4R_3 + R_2 \rightarrow R_2 \\ \rightarrow \end{array} \begin{bmatrix} 1 & 0 & 0 & | & 360 \\ 0 & 1 & 0 & | & 320 \\ 0 & 0 & 1 & | & 400 \end{bmatrix}$

360 units of mined goods, 320 units of manufactured goods, and 400 units of fuel.

60. A closed Leontief model must be solved by the Gauss-Jordan elimination method. A graphing calculator is strongly suggested.

$G = \dfrac{64}{93}H, \ A = \dfrac{59}{93}H, \ M = \dfrac{40}{93}H$

Chapter Test

1. $A^T + B = \begin{bmatrix} 1 & 3 & 4 \\ -2 & 2 & 1 \end{bmatrix} + \begin{bmatrix} 2 & -2 & 1 \\ 3 & 1 & 5 \end{bmatrix} = \begin{bmatrix} 3 & 1 & 5 \\ 1 & 3 & 6 \end{bmatrix}$

2. $B - C = \begin{bmatrix} 2 & -2 & 1 \\ 3 & 1 & 5 \end{bmatrix} - \begin{bmatrix} 3 & -4 & -1 \\ 2 & 2 & -1 \end{bmatrix} = \begin{bmatrix} -1 & 2 & 2 \\ 1 & -1 & 6 \end{bmatrix}$

3. $CD = \begin{bmatrix} 3 & -4 & -1 \\ 2 & 2 & -1 \end{bmatrix} \begin{bmatrix} 1 & 2 & 4 \\ 3 & 5 & 41 \\ 3 & 2 & 3 \end{bmatrix} = \begin{bmatrix} -12 & -16 & -155 \\ 5 & 12 & 87 \end{bmatrix}$

4. $DA = \begin{bmatrix} 1 & 2 & 4 \\ 3 & 5 & 41 \\ 3 & 2 & 3 \end{bmatrix} \begin{bmatrix} 1 & -2 \\ 3 & 2 \\ 4 & 1 \end{bmatrix} = \begin{bmatrix} 23 & 6 \\ 182 & 45 \\ 21 & 1 \end{bmatrix}$

5. $BA = \begin{bmatrix} 2 & -2 & 1 \\ 3 & 1 & 5 \end{bmatrix} \begin{bmatrix} 1 & -2 \\ 3 & 2 \\ 4 & 1 \end{bmatrix} = \begin{bmatrix} 0 & -7 \\ 26 & 1 \end{bmatrix}$

6. $ABD = \begin{bmatrix} 1 & -2 \\ 3 & 2 \\ 4 & 1 \end{bmatrix} \begin{bmatrix} 2 & -2 & 1 \\ 3 & 1 & 5 \end{bmatrix} \begin{bmatrix} 1 & 2 & 4 \\ 3 & 5 & 41 \\ 3 & 2 & 3 \end{bmatrix} = \begin{bmatrix} -4 & -4 & -9 \\ 12 & -4 & 13 \\ 11 & -7 & 9 \end{bmatrix} \begin{bmatrix} 1 & 2 & 4 \\ 3 & 5 & 41 \\ 3 & 2 & 3 \end{bmatrix} = \begin{bmatrix} -43 & -46 & -207 \\ 39 & 30 & -77 \\ 17 & 5 & -216 \end{bmatrix}$

7. $\begin{bmatrix} 1 & 3 \\ 2 & 4 \end{bmatrix}^{-1} = \frac{1}{-2} \begin{bmatrix} 4 & -3 \\ -2 & 1 \end{bmatrix} = \begin{bmatrix} -2 & 3/2 \\ 1 & -1/2 \end{bmatrix}$

8. $\begin{bmatrix} 1 & 2 & 4 & | & 1 & 0 & 0 \\ 1 & 2 & 2 & | & 0 & 1 & 0 \\ 1 & 1 & 4 & | & 0 & 0 & 1 \end{bmatrix} \begin{matrix} \\ -R_1 + R_2 \rightarrow R_2 \\ -R_1 + R_3 \rightarrow R_3 \end{matrix}$

$\begin{bmatrix} 1 & 2 & 4 & | & 1 & 0 & 0 \\ 0 & 0 & -2 & | & -1 & 1 & 0 \\ 0 & -1 & 0 & | & -1 & 0 & 1 \end{bmatrix} \begin{matrix} \rightarrow \\ -R_3 \rightarrow R_3 \\ R_2 \leftrightarrow \text{new } R_3 \end{matrix}$

$\begin{bmatrix} 1 & 2 & 4 & | & 1 & 0 & 0 \\ 0 & 1 & 0 & | & 1 & 0 & -1 \\ 0 & 0 & -2 & | & -1 & 1 & 0 \end{bmatrix} \begin{matrix} -2R_2 + R_1 \rightarrow R_1 \\ \rightarrow \\ -\frac{1}{2}R_3 \rightarrow R_3 \end{matrix}$

$\begin{bmatrix} 1 & 0 & 4 & | & -1 & 0 & 2 \\ 0 & 1 & 0 & | & 1 & 0 & -1 \\ 0 & 0 & 1 & | & 1/2 & -1/2 & 0 \end{bmatrix} \begin{matrix} -4R_3 + R_1 \rightarrow R_1 \\ \rightarrow \end{matrix}$

$\begin{bmatrix} 1 & 0 & 0 & | & -3 & 2 & 2 \\ 0 & 1 & 0 & | & 1 & 0 & -1 \\ 0 & 0 & 1 & | & 1/2 & -1/2 & 0 \end{bmatrix}$

$\text{Inverse} = \begin{bmatrix} -3 & 2 & 2 \\ 1 & 0 & -1 \\ 1/2 & -1/2 & 0 \end{bmatrix}$

9. $AX = B$

$A^{-1}AX = A^{-1}B$

$X = A^{-1}B$

$X = \begin{bmatrix} 1 & 2 & 0 \\ 3 & 1 & 2 \\ 4 & 1 & 1 \end{bmatrix} \begin{bmatrix} 3 \\ 1 \\ 2 \end{bmatrix} = \begin{bmatrix} 5 \\ 14 \\ 15 \end{bmatrix}$

10. Begin with the augmented matrix.

$$\begin{bmatrix} 1 & -1 & 2 & | & 4 \\ 1 & 4 & 1 & | & 4 \\ 2 & 2 & 4 & | & 10 \end{bmatrix} \begin{array}{c} \\ \\ \frac{1}{2}R_3 \to R_3 \end{array} \to \begin{bmatrix} 1 & -1 & 2 & | & 4 \\ 1 & 4 & 1 & | & 4 \\ 1 & 1 & 2 & | & 5 \end{bmatrix} \begin{array}{c} \\ -R_1+R_2 \to R_2 \\ -R_1+R_3 \to R_3 \end{array} \to \begin{bmatrix} 1 & -1 & 2 & | & 4 \\ 0 & 5 & -1 & | & 0 \\ 0 & 2 & 0 & | & 1 \end{bmatrix} \begin{array}{c} \\ \frac{1}{2}R_3 \to R_3 \\ R_2 \leftrightarrow \text{new } R_3 \end{array} \to$$

$$\begin{bmatrix} 1 & -1 & 2 & | & 4 \\ 0 & 1 & 0 & | & 1/2 \\ 0 & 5 & -1 & | & 0 \end{bmatrix} \begin{array}{c} R_2+R_1 \to R_1 \\ \to \\ -5R_2+R_3 \to R_3 \end{array} \begin{bmatrix} 1 & 0 & 2 & | & 9/2 \\ 0 & 1 & 0 & | & 1/2 \\ 0 & 0 & -1 & | & -5/2 \end{bmatrix} \begin{array}{c} 2R_3+R_1 \to R_1 \\ \to \\ -R_3 \to R_3 \end{array} \begin{bmatrix} 1 & 0 & 0 & | & -1/2 \\ 0 & 1 & 0 & | & 1/2 \\ 0 & 0 & 1 & | & 5/2 \end{bmatrix}$$

Solution: $x = -\dfrac{1}{2}$, $y = \dfrac{1}{2}$, $z = \dfrac{5}{2}$

11. Begin with the augmented matrix.

$$\begin{bmatrix} 1 & -1 & 2 & | & 4 \\ 1 & 4 & 1 & | & 4 \\ 2 & 3 & 3 & | & 8 \end{bmatrix} \begin{array}{c} \\ -R_1+R_2 \to R_2 \\ -2R_1+R_3 \to R_3 \end{array} \to \begin{bmatrix} 1 & -1 & 2 & | & 4 \\ 0 & 5 & -1 & | & 0 \\ 0 & 5 & -1 & | & 0 \end{bmatrix} \begin{array}{c} \\ \frac{1}{5}R_2 \to R_2 \text{ then} \\ -5R_2+R_3 \to R_3 \end{array} \to \begin{bmatrix} 1 & -1 & 2 & | & 4 \\ 0 & 1 & -1/5 & | & 0 \\ 0 & 0 & 0 & | & 0 \end{bmatrix} \begin{array}{c} R_2+R_1 \to R_1 \\ \to \end{array}$$

$$\begin{bmatrix} 1 & 0 & 9/5 & | & 4 \\ 0 & 1 & -1/5 & | & 0 \\ 0 & 0 & 0 & | & 0 \end{bmatrix} \quad \text{Solution: } x = 4 - \dfrac{9}{5}z, \ y = \dfrac{1}{5}z, \ z = z$$

12. Begin with augmented matrix.

$$\begin{bmatrix} 1 & -1 & 3 & | & 4 \\ 1 & 5 & 2 & | & 3 \\ 2 & 4 & 5 & | & 8 \end{bmatrix} \begin{array}{c} \\ -R_1+R_2 \to R_2 \\ -2R_1+R_3 \to R_3 \end{array} \to \begin{bmatrix} 1 & -1 & 3 & | & 4 \\ 0 & 6 & -1 & | & -1 \\ 0 & 6 & -1 & | & 0 \end{bmatrix} \begin{array}{c} \\ -R_3+R_2 \to R_2 \end{array} \to \begin{bmatrix} 1 & -1 & 3 & | & 4 \\ 0 & 0 & 0 & | & -1 \\ 0 & 6 & -1 & | & 0 \end{bmatrix}$$

There is no solution. In R_2 we have $0 = -1$.

13. Your calculator will give, in decimal notation, A^{-1}.

$$A^{-1} = \begin{bmatrix} 1 & 2 & -1 & 0 \\ -4/3 & -2/3 & 1 & -2/3 \\ 2/3 & -2/3 & 0 & 1/3 \\ 1 & 1 & -1 & 1 \end{bmatrix}; \quad \begin{bmatrix} x \\ y \\ z \\ w \end{bmatrix} = A^{-1} \cdot \begin{bmatrix} 4 \\ 4 \\ 10 \\ 0 \end{bmatrix} = \begin{bmatrix} 2 \\ 2 \\ 0 \\ -2 \end{bmatrix}$$

14. Your calculator will say that the coefficient matrix has no inverse. We will use the Gauss-Jordan method, give the matrices, and ask you to use the same steps with your calculator.

$$\begin{bmatrix} 1 & -1 & 2 & -1 & | & 4 \\ 1 & 4 & 1 & 1 & | & 4 \\ 2 & 2 & 4 & 2 & | & 10 \\ 0 & -1 & 1 & 2 & | & 2 \end{bmatrix} \begin{array}{c} \\ -R_1+R_2 \to R_2 \\ -2R_1+R_3 \to R_3 \\ \end{array} \begin{bmatrix} 1 & -1 & 2 & -1 & | & 4 \\ 0 & 5 & -1 & 2 & | & 0 \\ 0 & 4 & 0 & 4 & | & 2 \\ 0 & -1 & 1 & 2 & | & 2 \end{bmatrix} \begin{array}{c} -R_4+R_1 \to R_1 \\ 5R_4+R_2 \to R_2 \\ 4R_4+R_3 \to R_3 \\ \to \end{array} \begin{bmatrix} 1 & 0 & 1 & -3 & | & 2 \\ 0 & 0 & 4 & 12 & | & 10 \\ 0 & 0 & 4 & 12 & | & 10 \\ 0 & -1 & 1 & 2 & | & 2 \end{bmatrix}$$

$$\begin{array}{c} \to \\ -R_2+R_3 \to R_3 \\ -R_4 \to R_4 \end{array} \begin{bmatrix} 1 & 0 & 1 & -3 & | & 2 \\ 0 & 0 & 4 & 12 & | & 10 \\ 0 & 0 & 0 & 0 & | & 0 \\ 0 & 1 & -1 & -2 & | & -2 \end{bmatrix} \begin{array}{l} \text{1. Interchange } R_3 \text{ and } R_4. \\ \text{2. Interchange } R_2 \text{ and } R_3. \\ \text{3. } \frac{1}{4}R_3 \to R_3 \end{array} \begin{bmatrix} 1 & 0 & 1 & -3 & | & 2 \\ 0 & 1 & -1 & -2 & | & -2 \\ 0 & 0 & 1 & 3 & | & 5/2 \\ 0 & 0 & 0 & 0 & | & 0 \end{bmatrix} \begin{array}{c} -R_3+R_1 \to R_1 \\ R_3+R_2 \to R_2 \\ \to \end{array}$$

$$\begin{bmatrix} 1 & 0 & 0 & -6 & | & -1/2 \\ 0 & 1 & 0 & 1 & | & 1/2 \\ 0 & 0 & 1 & 3 & | & 5/2 \\ 0 & 0 & 0 & 0 & | & 0 \end{bmatrix} \quad \text{Solution: } x = 6w - \dfrac{1}{2}, \ y = -w + \dfrac{1}{2}, \ z = -3w + \dfrac{5}{2}$$

15. a. $H = \$10,000;\quad B = 75,000 - 3(10,000) = \$45,000;\quad E = 20,000 + 2(10,000) = \$40,000$

  b. $0 to $25,000

  c. $E = 20,000 + 2(0) = \$20,000;\quad H = \$0;\quad B = 75,000 - 3(0) = \$75,000$

16. a. $AB = \begin{bmatrix} 0.08 & 0.22 & 0.12 \\ 0.10 & 0.08 & 0.19 \\ 0.05 & 0.07 & 0.09 \\ 0.10 & 0.26 & 0.15 \\ 0.12 & 0.04 & 0.24 \end{bmatrix}$

  b. Plant type 1:
  0.08, 0.22, 0.12 are consumed by carnivores 1, 2, 3, respectively.

  c. Plant 5 by 1; Plant 4 by 2; Plant 5 by 3.

17. a. $\begin{bmatrix} 1000 & 4000 & 2000 & 1000 \end{bmatrix}$

  b. $\begin{bmatrix} 1000 & 4000 & 2000 & 1000 \end{bmatrix} \begin{bmatrix} 10 & 5 & 0 & 0 \\ 5 & 0 & 20 & 10 \\ 5 & 20 & 0 & 10 \\ 5 & 10 & 10 & 10 \end{bmatrix} = \begin{bmatrix} 45,000 & 55,000 & 90,000 & 70,000 \end{bmatrix}$

  c. $\begin{bmatrix} 5 \\ 3 \\ 4 \\ 4 \end{bmatrix}$

  d. $\begin{bmatrix} 45,000 & 55,000 & 90,000 & 70,000 \end{bmatrix} \begin{bmatrix} 5 \\ 3 \\ 4 \\ 4 \end{bmatrix} = \begin{bmatrix} 1,030,000 \end{bmatrix}$

  e. $\begin{bmatrix} 10 & 5 & 0 & 0 \\ 5 & 0 & 20 & 10 \\ 5 & 20 & 0 & 10 \\ 5 & 10 & 10 & 10 \end{bmatrix} \begin{bmatrix} 5 \\ 3 \\ 4 \\ 4 \end{bmatrix} = \begin{bmatrix} 65 \\ 145 \\ 125 \\ 135 \end{bmatrix}$

18. a.
$\begin{array}{cccccccccccc} 12 & 5 & 19 & 19 & 27 & 9 & 19 & 27 & 13 & 15 & 18 & 5 \\ L & E & S & S & & I & S & & M & O & R & E \end{array} = \begin{bmatrix} 12 & 19 & 27 & 19 & 13 & 18 \\ 5 & 19 & 9 & 27 & 15 & 5 \end{bmatrix}$

Encoding: $\begin{bmatrix} 8 & 5 \\ 3 & 2 \end{bmatrix} \begin{bmatrix} 12 & 19 & 27 & 19 & 13 & 18 \\ 5 & 19 & 9 & 27 & 15 & 5 \end{bmatrix} = \begin{bmatrix} 121 & 247 & 261 & 287 & 179 & 169 \\ 46 & 95 & 99 & 111 & 69 & 64 \end{bmatrix}$

Coded message: 121, 46, 247, 95, 261, 99, 287, 111, 179, 69, 169, 64

  b. $A^{-1} = \dfrac{1}{16-15} \begin{bmatrix} 2 & -5 \\ -3 & 8 \end{bmatrix} = \begin{bmatrix} 2 & -5 \\ -3 & 8 \end{bmatrix}$

$\begin{bmatrix} 2 & -5 \\ -3 & 8 \end{bmatrix} \begin{bmatrix} 138 & 140 & 255 & 141 & 201 & 287 \\ 54 & 53 & 99 & 54 & 76 & 111 \end{bmatrix} = \begin{bmatrix} 6 & 15 & 15 & 12 & 22 & 19 \\ 18 & 4 & 27 & 9 & 5 & 27 \end{bmatrix}$

$\begin{array}{cccccccccccc} 6 & 18 & 15 & 4 & 15 & 27 & 12 & 9 & 22 & 5 & 19 & 27 \\ F & R & O & D & O & & L & I & V & E & S & \end{array}$

19. Using the calculator and solving with the inverse is more efficient. This method shows the inefficiency of the Gauss-Jordan method.

x = Growth shares $30x = 100y + 50z$
y = Blue-chip shares $4.6x + 11y + 5z = 0.13(120{,}000)$
z = Utility shares $30x + 100y + 50z = 120{,}000$

$$\begin{bmatrix} 30 & -100 & -50 & | & 0 \\ 4.6 & 11 & 5 & | & 15{,}600 \\ 30 & 100 & 50 & | & 120{,}000 \end{bmatrix} \begin{array}{c} \\ \\ -R_1 + R_3 \to R_3 \end{array} \to \begin{bmatrix} 30 & -100 & -50 & | & 0 \\ 4.6 & 11 & 5 & | & 15{,}600 \\ 0 & 200 & 100 & | & 120{,}000 \end{bmatrix} \begin{array}{c} \frac{1}{30}R_1 \to R_1 \\ \to \\ \end{array}$$

$$\begin{bmatrix} 1 & -10/3 & -5/3 & | & 0 \\ 4.6 & 11 & 5 & | & 15{,}600 \\ 0 & 200 & 100 & | & 120{,}000 \end{bmatrix} \begin{array}{c} \\ -4.6R_1 + R_2 \to R_2 \\ \end{array} \to \begin{bmatrix} 1 & -10/3 & -5/3 & | & 0 \\ 0 & 79/3 & 38/3 & | & 15{,}600 \\ 0 & 200 & 100 & | & 120{,}000 \end{bmatrix} \begin{array}{c} R_2 \leftrightarrow R_3 \\ \to \\ \end{array}$$

$$\begin{bmatrix} 1 & -10/3 & -5/3 & | & 0 \\ 0 & 200 & 100 & | & 120{,}000 \\ 0 & 79/3 & 38/3 & | & 15{,}600 \end{bmatrix} \begin{array}{c} \\ \frac{1}{200}R_2 \to R_2 \\ \end{array} \to \begin{bmatrix} 1 & -10/3 & -5/3 & | & 0 \\ 0 & 1 & 1/2 & | & 600 \\ 0 & 79/3 & 38/3 & | & 15{,}600 \end{bmatrix} \begin{array}{c} \frac{10}{3}R_2 + R_1 \to R_1 \\ \to \\ -\frac{79}{3}R_2 + R_3 \to R_3 \end{array}$$

$$\begin{bmatrix} 1 & 0 & 0 & | & 2000 \\ 0 & 1 & 1/2 & | & 600 \\ 0 & 0 & -1/2 & | & -200 \end{bmatrix} \begin{array}{c} \\ \\ -2R_3 \to R_3 \end{array} \to \begin{bmatrix} 1 & 0 & 0 & | & 2000 \\ 0 & 1 & 1/2 & | & 600 \\ 0 & 0 & 1 & | & 400 \end{bmatrix} \begin{array}{c} \\ -\frac{1}{2}R_3 + R_2 \to R_2 \\ \end{array} \to \begin{bmatrix} 1 & 0 & 0 & | & 2000 \\ 0 & 1 & 0 & | & 400 \\ 0 & 0 & 1 & | & 400 \end{bmatrix}$$

Solution: $x = 2000,\ y = 400,\ z = 400$

20. The technological equation is $(I - A)X = D$.

$$I - A = \begin{bmatrix} 0.6 & -0.2 \\ -0.1 & 0.7 \end{bmatrix}; \qquad (I - A)^{-1} = \frac{1}{0.42 - 0.02}\begin{bmatrix} 0.7 & 0.2 \\ 0.1 & 0.6 \end{bmatrix}$$

$$\begin{bmatrix} Ag \\ M \end{bmatrix} = \frac{1}{0.40}\begin{bmatrix} 0.7 & 0.2 \\ 0.1 & 0.6 \end{bmatrix}\begin{bmatrix} 140 \\ 140 \end{bmatrix} = \frac{1}{0.4}\begin{bmatrix} 126 \\ 98 \end{bmatrix} = \begin{bmatrix} 315 \\ 245 \end{bmatrix}$$

21. The technological equation is $(I - A)X = 0$.
The augmented matrix to be solved is

$$\begin{bmatrix} 0.6 & -0.3 & -0.4 & | & 0 \\ -0.2 & 0.6 & -0.2 & | & 0 \\ -0.4 & -0.3 & 0.6 & | & 0 \end{bmatrix} \begin{array}{c} 10R_1 \to R_1 \\ 5R_2 \to R_2 \\ 10R_3 \to R_3 \end{array} \begin{bmatrix} 6 & -3 & -4 & | & 0 \\ -1 & 3 & -1 & | & 0 \\ -4 & -3 & 6 & | & 0 \end{bmatrix} \begin{array}{c} 6R_2 + R_1 \to R_1 \\ \to \\ -4R_2 + R_3 \to R_3 \end{array}$$

$$\begin{bmatrix} 0 & 15 & -10 & | & 0 \\ -1 & 3 & -1 & | & 0 \\ 0 & -15 & 10 & | & 0 \end{bmatrix} \begin{array}{c} \\ -R_2 \to R_2 \\ R_1 + R_3 \to R_3 \end{array} \to \begin{bmatrix} 0 & 15 & -10 & | & 0 \\ 1 & -3 & 1 & | & 0 \\ 0 & 0 & 0 & | & 0 \end{bmatrix} \begin{array}{c} \frac{1}{15}R_1 \text{ and} \\ \text{then interchange} \to \\ R_1 \text{ and } R_2 \end{array}$$

$$\begin{bmatrix} 1 & -3 & 1 & | & 0 \\ 0 & 1 & -2/3 & | & 0 \\ 0 & 0 & 0 & | & 0 \end{bmatrix} \begin{array}{c} 3R_2 + R_1 \to R_1 \\ \to \\ \end{array} \begin{bmatrix} 1 & 0 & -1 & | & 0 \\ 0 & 1 & -2/3 & | & 0 \\ 0 & 0 & 0 & | & 0 \end{bmatrix}$$

Profit = Households   Non-profit = $\dfrac{2}{3}$ Households

22.

	Ag	M	F	S
Ag	0.2	0.1	0.1	0.1
Mach	0.3	0.2	0.2	0.2
Fuel	0.2	0.2	0.3	0.3
Steel	0.1	0.4	0.2	0.2

23. The technological equation is $(I - A)X = D$.

$$(I - A) = \begin{bmatrix} 0.8 & -0.1 & -0.1 & -0.1 \\ -0.3 & 0.8 & -0.2 & -0.2 \\ -0.2 & -0.2 & 0.7 & -0.3 \\ -0.1 & -0.4 & -0.2 & 0.8 \end{bmatrix} \qquad X = (I - A)^{-1} \begin{bmatrix} 1700 \\ 1900 \\ 900 \\ 300 \end{bmatrix}$$

Then $X = \begin{bmatrix} 1.74 & 0.68 & 0.62 & 0.62 \\ 1.30 & 2.30 & 1.18 & 1.18 \\ 1.39 & 1.55 & 2.50 & 1.50 \\ 1.22 & 1.62 & 1.29 & 2.29 \end{bmatrix} \begin{bmatrix} 1700 \\ 1900 \\ 900 \\ 300 \end{bmatrix} = \begin{bmatrix} 5000 \\ 8000 \\ 8000 \\ 7000 \end{bmatrix}$

Use the graphing utility to find $(I - A)^{-1}$.

24. The Gauss-Jordan method must be used since the equation is $(I - A)X = 0$. We begin with the augmented matrix and will use fractions. The student is encouraged to solve with the graphing calculator and work with decimals.

$$\begin{bmatrix} 3/4 & -1/4 & -3/10 & -1/10 & | & 0 \\ -3/10 & 3/4 & -2/10 & -4/10 & | & 0 \\ -3/20 & -2/10 & 9/10 & -3/10 & | & 0 \\ -3/10 & -3/10 & -4/10 & 8/10 & | & 0 \end{bmatrix} \quad \begin{matrix} \to \\ -4R_1 + R_2 \to R_2 \\ -3R_1 + R_3 \to R_3 \\ 8R_1 + R_4 \to R_4 \end{matrix} \quad \begin{bmatrix} 3/4 & -1/4 & -3/10 & -1/10 & | & 0 \\ -33/10 & 7/4 & 1 & 0 & | & 0 \\ -48/20 & 11/20 & 18/10 & 0 & | & 0 \\ 57/10 & -23/10 & -28/10 & 0 & | & 0 \end{bmatrix}$$

$$\begin{matrix} \frac{3}{10}R_2 + R_1 \to R_1 \\ \to \\ -\frac{18}{10}R_2 + R_3 \to R_3 \\ \frac{28}{10}R_2 + R_4 \to R_4 \end{matrix} \begin{bmatrix} -24/100 & 11/40 & 0 & -1/10 & | & 0 \\ -33/10 & 7/4 & 1 & 0 & | & 0 \\ 354/100 & -104/40 & 0 & 0 & | & 0 \\ -354/100 & 104/40 & 0 & 0 & | & 0 \end{bmatrix} \quad \begin{matrix} \to \\ \\ \\ R_3 + R_4 \to R_4 \end{matrix} \begin{bmatrix} -24/100 & 11/40 & 0 & -1/10 & | & 0 \\ -33/10 & 7/4 & 1 & 0 & | & 0 \\ 354/100 & -104/40 & 0 & 0 & | & 0 \\ 0 & 0 & 0 & 0 & | & 0 \end{bmatrix}$$

$$\begin{matrix} \to \\ \\ -\frac{40}{104}R_3 \to R_3 \\ \\ \end{matrix} \begin{bmatrix} -24/100 & 11/40 & 0 & -1/10 & | & 0 \\ -33/10 & 7/4 & 1 & 0 & | & 0 \\ -177/130 & 1 & 0 & 0 & | & 0 \\ 0 & 0 & 0 & 0 & | & 0 \end{bmatrix} \quad \begin{matrix} -\frac{11}{40}R_3 + R_1 \to R_1 \\ -\frac{7}{4}R_3 + R_2 \to R_2 \\ \to \\ \\ \end{matrix} \begin{bmatrix} 699/5200 & 0 & 0 & -1/10 & | & 0 \\ -477/520 & 0 & 1 & 0 & | & 0 \\ -177/130 & 1 & 0 & 0 & | & 0 \\ 0 & 0 & 0 & 0 & | & 0 \end{bmatrix}$$

Our solution yields: $\text{Hhold} = \dfrac{699}{520} \text{Ag}$, $\text{Fuel} = \dfrac{477}{520} \text{Ag}$, $\text{Steel} = \dfrac{177}{130} \text{Ag}$

Now we must solve in terms of Hholds to yield the desired form.

$$\text{Ag} = \frac{520}{699} \text{Hhold}$$

$$\text{Steel} = \frac{177}{130} \left(\frac{520}{699} \text{Hhold} \right) = \frac{236}{233} \text{Hhold}$$

$$\text{Fuel} = \frac{477}{520} \left(\frac{520}{699} \text{Hhold} \right) = \frac{159}{233} \text{Hhold}.$$

This was a difficult problem. The authors' step by step method always works. Sometimes it is not the shortest method.

Chapter 4: Inequalities and Linear Programming

Exercise 4.1

1. $y \le 2x - 1$

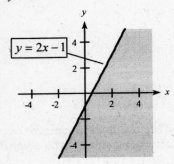

3. $\dfrac{x}{2} + \dfrac{y}{4} < 1$

$2x + y < 4$

$\quad y < -2x + 4$

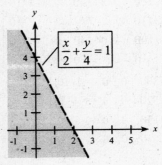

5. $0.4x \ge 0.8$

$\quad x \ge 2$

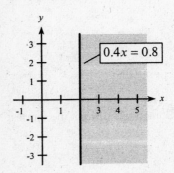

7. a.

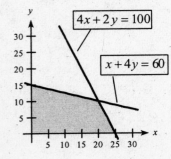

b. From the graph we read intercept corners (0, 15), (0, 0), (25, 0).

$$4x + 2y = 100 \qquad 8x + 4y = 200$$
$$x + 4y = 60 \qquad \underline{x + 4y = 60}$$
$$\qquad\qquad\qquad 7x \quad = 140$$
$$\qquad\qquad\qquad x \quad = 20$$

$20 + 4y = 60$ or $\; y = 10$

Other corner is (20, 10).

9. a.

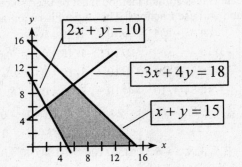

b. From the graph we read intercept corners (5,0) and (15,0).

$$-3x + 4y = 18 \qquad -3x + 4y = 18$$
$$\underline{2x + y = 10} \qquad \underline{-8x - 4y = -40}$$
$$\qquad\qquad\qquad\qquad -11x = -22$$
$$\qquad\qquad\qquad\qquad x = 2$$

$-3(2) + 4y = 18 \;\rightarrow\; 4y = 24 \;\rightarrow\; y = 6$

Corner is (2, 6).

$$-3x + 4y = 18 \qquad -3x + 4y = 18$$
$$\underline{x + y = 15} \qquad \underline{3x + 3y = 45}$$
$$\qquad\qquad\qquad\qquad 7y = 63$$
$$\qquad\qquad\qquad\qquad y = 9$$

$x + 9 = 15 \;\rightarrow\; x = 6 \qquad$ Corner is (6, 9)

11. a.

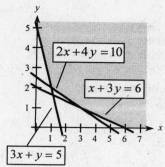

b. From the graph we read intercept corners
(0, 5), (6, 0).

$$3x + y = 5 \qquad 12x + 4y = 20$$
$$2x + 4y = 10 \qquad \underline{2x + 4y = 10}$$
$$ \qquad 10x = 10$$
$$ \qquad x = 1$$
$$3(1) + y = 5 \text{ or } y = 2$$
Corner is (1, 2).

$$2x + 4y = 10 \qquad 2x + 4y = 10$$
$$x + 3y = 6 \qquad \underline{2x + 6y = 12}$$
$$ \qquad 2y = 2$$
$$ \qquad y = 1$$
$$x + 3(1) = 6 \text{ or } x = 3$$
Corner is (3, 1)

13. $\begin{cases} y < 2x \\ y > x - 1 \end{cases}$

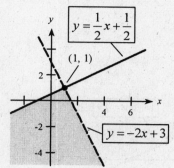

15. $\begin{cases} y < -2x + 3 \\ y \le \dfrac{1}{2}x + \dfrac{1}{2} \end{cases}$

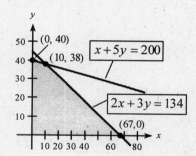

17. $\begin{cases} y \le -\dfrac{1}{5}x + 40 \\ y \le -\dfrac{2}{3}x + \dfrac{134}{3} \\ x \ge 0, \ y \ge 0 \end{cases}$

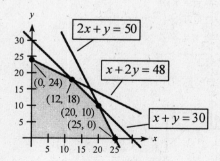

19. $\begin{cases} y \le -\dfrac{1}{2}x + 24 \\ y \le -x + 30 \\ y \le -2x + 50 \\ x \ge 0, \ y \ge 0 \end{cases}$

21. $\begin{cases} y \geq -\dfrac{1}{2}x + \dfrac{19}{2} \\[2mm] y \geq -\dfrac{3}{2}x + \dfrac{29}{2} \\[2mm] x \geq 0, y \geq 0 \end{cases}$

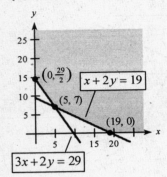

23. $\begin{cases} y \geq -\dfrac{1}{3}x + 1 \\[2mm] y \geq -\dfrac{2}{3}x + \dfrac{5}{3} \\[2mm] y \geq -2x + 3 \\[2mm] x \geq 0, y \geq 0 \end{cases}$

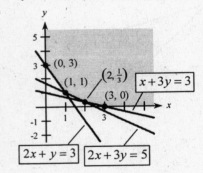

25. $\begin{cases} y \geq -\dfrac{1}{2}x + 10 \\[2mm] y \leq \dfrac{3}{2}x + 2 \\[2mm] x \geq 12 \\[2mm] x \geq 0, \; y \geq 0 \end{cases}$

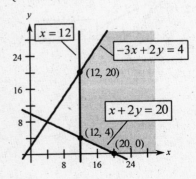

27. Let x = number of deluxe models.
Let y = number of economy models.

a. $x, y \geq 0$ $3x + 2y \leq 24$ $0.5x + y \leq 8$

b.   Points of Intersection:

$$3x + 2y = 24$$
$$\underline{2(0.5x + y = 8)}$$
$$2x \quad = 8$$
$$x \quad = 4$$

$0.5(4) + y = 8 \text{ or } y = 6$

Feasible corners are

$(0,0), (8,0), (4,6),$

and $(0,8)$.

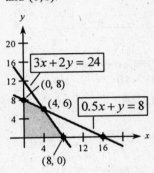

29. Let x = number of cord type models and
y = number of cordless models.

a. $x + y \leq 300$ Packing department

$2x + 4y \leq 800$ Manufacturing

$x \geq 0$

$y \geq 0$

b.

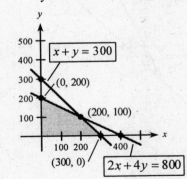

31. Let x = minutes of business/investment program commercials
Let y = minutes of sporting events commercials

   a. $7x + 2y \geq 30$ women reached

        $4x + 12y \geq 28$ men reached

            $x \geq 0, \quad y \geq 0$

   b.

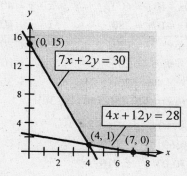

33. x = radio minutes
y = television minutes.

   a. $x + y \geq 80$ Advertising time

      $0.006x + 0.09y \geq 2.16$ People reached

             $x \geq 0$

             $y \geq 0$

   (Selecting a good scale is the most difficult part of the problem.)

   b.

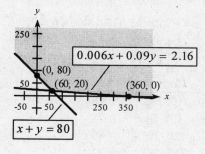

35. x = pounds of regular meat
y = pounds of all beef

   a. $0.18x + 0.75y \leq 1020$ (beef)

        $0.20x + 0.20y \geq 500$ (spices)

               $0.30x \leq 600$ (pork)

                $x, y \geq 0$

   b.

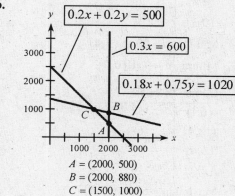

      $A = (2000, 500)$
      $B = (2000, 880)$
      $C = (1500, 1000)$

Exercise 4.2

1.

Corner	$C = 2x + 3y$
$(0,0)$	0
$(7,0)$	14
$(6,2)$	18
$(4,4)$	20
$(0,3)$	9

Maximum = 20 at $(4,4)$

Minimum = 0 at $(0,0)$

3.

Corner	$C = 5x + 2y$
$(6,1)$	32
$(12,4)$	68
$(6,4)$	38
$(2,2)$	14

Maximum = 68 at $(12,4)$.

Minimum = 14 at $(2,2)$.

5.

Corner	$f = 3x + 4y$
$(0,6)$	24
$(1,3)$	15
$(2,1)$	10
$(4,0)$	12

Maximum: None

Minimum = 10 at $(2,1)$.

7. $x + 5y = 100$ If $x = 0$, then $y = 20$.

$2x + y = 40$ If $y = 0$, then $x = 20$.

$8x + 5y = 170$

$10x + 5y = 200$

$\underline{\quad 8x + 5y = 170 \quad}$

$\quad 2x \quad = 30$

$\quad\quad x \quad = 15$

$\quad\quad\quad y = 10$

$\quad\quad\quad x + 5y = 100$

$\quad\quad\quad \underline{8x + 5y = 170}$

$\quad\quad\quad 7x \quad\quad = 70$

$\quad\quad\quad\quad x \quad = 10$

$\quad\quad\quad\quad\quad y = 18$

Corners: (15, 10) and (10, 18)

Corner	$f = 3x + 2y$
$(0,0)$	0
$(0,20)$	40
$(20,0)$	60
$(15,10)$	65
$(10,18)$	66

Maximum = 66 at $(10,18)$

9. $3x + y = 60$

$4x + 10y = 280$

$x + y = 40$

If $x = 0$, then $y = 60$

If $y = 0$, then $x = 70$

$10x + 10y = 400$ $x + y = 40$

$\underline{4x + 10y = 280}$ $\underline{3x + y = 60}$

$\quad 6x \quad = 120$ $\quad 2x \quad = 20$

$\quad\quad x \quad = 20$ $\quad\quad x = 10$

$\quad\quad\quad y = 20$ $\quad\quad\quad y = 30$

Corners: (20, 20) and (10, 30)

Corner	$g = 3x + 2y$
$(0,60)$	120
$(70,0)$	210
$(20,20)$	100
$(10,30)$	90

Minimum = 90 at $(10,30)$.

11. Refer to Problem 19 in Section 4.1, to obtain the corners (0, 24) and (25, 0).

$2x + y = 50$ (20, 10) is a point

$\underline{x + y = 30}$ of intersection.

$\quad x \quad = 20$

$x + 2y = 48$ (12, 18) is a feasible

$\underline{x + y = 30}$ corner.

$\quad\quad y = 18$

Corner	$f = 3x + 2y$
$(0, 24)$	48
$(25, 0)$	75
$(20, 10)$	80
$(12, 18)$	72

Maximum = 80 at (20, 10).

13. Refer to Problem 23 in Section 4.1 to obtain the corners (3, 0) and (0, 3).

$$2x + y = 3 \qquad x + 3y = 3$$
$$\underline{2x + 3y = 5} \qquad \underline{2x + 3y = 5}$$
$$2y = 2 \qquad\qquad x = 2$$
$$y = 1$$

Corner: (1, 1) Corner: $\left(2, \frac{1}{3}\right)$

Corner	$g = 12x + 48y$
(3, 0)	36
(0, 3)	144
(1, 1)	60
$\left(2, \frac{1}{3}\right)$	40

Minimum = 36 at (3, 0).

15. The graph has enough accuracy to read the feasible corners from the graph.

Corner	$f = 3x + 4y$
(0, 4)	16
(2, 4)	22
(4, 2)	20
(5, 0)	15
(0, 0)	0

Maximum = 22 at (2, 4).

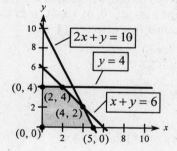

17. The feasible corners can be read directly from the graph.

Corner	$f = 2x + 6y$
(0, 5)	30
(3, 4)	30
(5, 2)	22
(6, 0)	12
(0, 0)	0

Maximum = 30 at any point on line from (0, 5) to (3, 4)

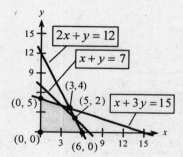

19. The feasible corners can be read directly from the graph.

Corner	$g = 7x + 6y$
(9, 0)	63
(2, 3)	32
(0, 8)	48

Minimum = 32 at (3, 2).

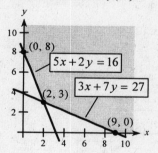

21.

$$3x + 2y = 12 \qquad 3x + 2y = 12$$
$$\underline{4x + y = 11} \qquad \underline{8x + 2y = 22}$$
$$\qquad\qquad\qquad\quad 5x = 10$$
$$\qquad\qquad\qquad\quad x = 2,\ y = 3$$

(2, 3) is a feasible corner.

Corner	$g = 3x + y$
(4, 0)	12
(0, 11)	11
(2, 3)	9

Minimum = 9 at (2, 3).

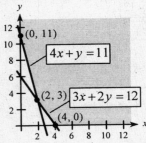

23. The feasible corners can be read directly from the graph.

Corner	$f = x + 2y$
(0, 4)	8
(2, 4)	10
(4, 0)	4

Maximum = 10 at (2, 4).

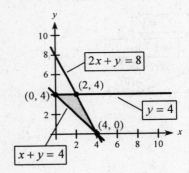

25.

$$x + y = 100 \qquad\quad x + y = 100$$
$$\underline{-x + y = 20} \qquad -2x + 3y = 30$$
$$\quad 2y = 120$$
$$\quad\ y = 60 \qquad\quad 2x + 2y = 200$$
$$\quad\ x = 40 \qquad\ \underline{-2x + 3y = 30}$$
$$\qquad\qquad\qquad\qquad\quad 5y = 230$$
$$\qquad\qquad\qquad\qquad\quad\ y = 46$$
$$\qquad\qquad\qquad\qquad\quad\ x = 54$$

(40, 60) and (54, 46) are the feasible corners.

Corner	$g = 40x + 25y$	
(40, 60)	3100	← minimum
(54, 46)	3310	

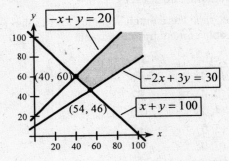

27. From the graph of Problem 31, Section 4.2, the corners along the axes are (0, 0), (0, 8), and (8, 0). The other corner (see graph) is (4, 6).

Corner	$P = 15x + 12y$
(0, 8)	96
(8, 0)	120
(4, 6)	132
(0, 0)	0

Maximum profit of $132 at (4, 6).

29. From the graph of Problem 29, Section 4.1, the corners are (0, 200), (0, 0), (300, 0), and (200, 100).

Corner	$S = 22.50x + 45.00y$
(0, 200)	9000
(300, 0)	6750
(200, 100)	9,000
(0, 0)	0

Maximum sales of $9,000 occurs two ways: 0 corded and 200 cordless or 200 corded and 100 cordless

31. From the graph of Problem 33, Section 4.1, we have corners (0, 80), (60, 20), and (360, 0).

Corner	$C = 100x + 500y$
(0, 80)	40,000
(360, 0)	36,000
(60, 20)	16,000

Minimum cost =$16,000 with 60 minutes of radio time and 20 minutes of TV time.

33. x = number of inkjet printers
y = number of laser printers
Maximize $P = 40x + 60y$

$x + y \leq 70$ Capacity

$x + 3y \leq 120$ Labor

$x, y \geq 0$

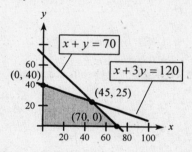

From the graph, we see that (0, 40) and (70, 0) are corners. Solve for the other corner.

$$x + 3y = 120 \qquad x + 3y = 120$$
$$\underline{x + y = 70} \qquad \underline{-x - y = -70}$$
$$\qquad\qquad\qquad 2y = 50$$
$$\qquad\qquad\qquad y = 25$$

$x + 25 = 70 \;\rightarrow\; x = 45$ Corner: (45, 25)

Corner	$P = 40x + 60y$
(0, 40)	$2400
(45, 25)	$3300
(70, 0)	$2800

The maximum profit occurs with 45 inkjet printers and 25 laser printers.

35. x = number of bass, y = number of trout
Maximize $f = x + y$

$2x + 5y \leq 800$ Food A

$4x + 2y \leq 800$ Food B

$x, y \geq 0$

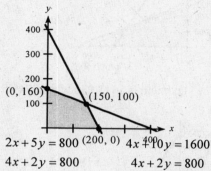

$2x + 5y = 800$ (200, 0) $4x + 10y = 1600$

$4x + 2y = 800$ $\underline{4x + 2y = 800}$

$\qquad\qquad\qquad\qquad\qquad 8y = 800$

$\qquad\qquad\qquad\qquad\qquad y = 100$

$\qquad\qquad\qquad\qquad\qquad x = 150$

Corner	$f = x + y$
(0, 160)	160
(150, 100)	250
(200, 0)	200

Maximum number of fish is 250 with 150 bass and 100 trout.

37. a.

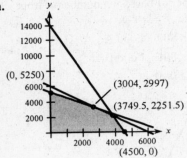

Constraints: $9x + 3y \le 40{,}500$
$x + y \le 6001$
$\frac{3}{4}x + y \le 250$
$x \ge 0,\ y \ge 0$

Corner	$P = 60x + 40y$
$(0, 5250)$	$\$210{,}000$
$(4500, 0)$	$\$270{,}000$
$(3749.5, 2251.5)$	$\$315{,}030$
$(3004, 2997)$	$\$300{,}120$

The maximum is $\$315{,}030$ when 3749.5 acres of corn are planted and 2251.5 acres of soybeans are planted.

b.

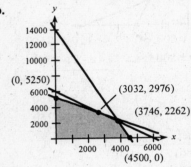

Constraints: $9x + 3y \le 40{,}500$
$x + y \le 6008$
$\frac{3}{4}x + y \le 250$
$x \ge 0,\ y \ge 0$

Corner	$P = 60x + 40y$
$(0, 5250)$	$\$210{,}000$
$(4500, 0)$	$\$270{,}000$
$(3746, 2262)$	$\$315{,}240$
$(3032, 2976)$	$\$300{,}960$

The maximum is $\$315{,}240$ when 3746 acres of corn are planted and 2260 acres of soybeans are planted.

c. $\$315{,}240 - \$315{,}000 = \$240$, $\$240 / 8 = \$30$ per acre

39. Let x = days to keep Factory 1 open, y = days to keep Factory 2 open
Minimize cost: $f = 10{,}000x + 20{,}000y$
Constraints: $80x + 20y \ge 1600$ Revised Constraints: $4x + y \ge 80$
$10x + 10y \ge 500$ $x + y \ge 50$
$20x + 70y \ge 2000$ $2x + 7y \ge 200$
$x \ge 0,\ y \ge 0$ $x \ge 0,\ y \ge 0$

Solving: $\begin{cases} 4x + y = 80 \\ x + y = 50 \end{cases}$ Solving: $\begin{cases} x + y = 50 \\ 2x + 7y = 200 \end{cases}$

$\begin{cases} 4x + y = 80 \\ -x - y = -50 \end{cases}$ $\begin{cases} -2x + -2y = -100 \\ 2x + 7y = 200 \end{cases}$

$3x = 30$ $5y = 100$

$x = 10,\ y = 40$ $y = 20,\ x = 30$

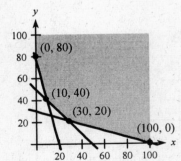

Corners	$f = 10{,}000x + 20{,}000y$
$(0, 80)$	$\$1{,}600{,}000$
$(10, 40)$	$\$900{,}000$
$(30, 20)$	$\$700{,}000$
$(100, 0)$	$\$1{,}000{,}000$

$\leftarrow$ Minimum cost is $\$700{,}000.$

41. x = days at location I, y = days at location II

Objective function: $F = 500x + 800y$

Constraints: $x, y \geq 0$

$$10x + 20y \geq 2000 \quad \text{Deluxe}$$
$$20x + 50y \geq 4200 \quad \text{Better}$$
$$13x + 6y \geq 1200 \quad \text{Standard}$$

$x + 2y = 200 \qquad 13x + 6y = 1200$

$2x + 5y = 420 \qquad\quad x + 2y = 200$

$A(160, 20) \quad B(60, 70)$

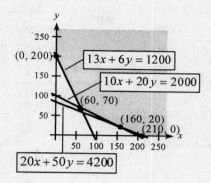

Corner	$F = 500x + 800y$
(160, 20)	96,000
(60, 70)	86,000
(0, 200)	160,000
(210, 0)	105,000

Minimum costs of $86,000 occurs with 60 days at location I and 70 days at location II.

43. See the graph for Problem 35, in Section 4.1, From the graph, feasible corners are (2000, 500), (2000, 880), and (1500, 1000).

Corner	$P = 0.4x + 0.6y$
(1500, 1000)	1200
(2000, 500)	1100
(2000, 880)	1328

Maximum profit = $1328 with 2000 lbs of regular and 880 lbs of all beef.

45. Let x = number from P to B; let y = number from P to Y.

Then $35 - x$ = number from E to B; $40 - y$ = number from E to Y.

Objective function: $C = 18x + 20(35 - x) + 22y + 25(40 - y) = 1700 - 2x - 3y$

Constraints: $x, y \geq 0$ P inventory

$$x + y \leq 60 \qquad \text{or}$$
$$(35 - x) + (40 - y) \leq 30 \qquad \text{E inventory}$$
$$x + y \geq 45$$
$$y \leq 40 \qquad \text{maximum order size}$$
$$x \leq 35 \qquad \text{maximum order size}$$

Corner	$C = 1700 - 2x - 3y$
(5, 40)	1570
(20, 40)	1540
(35, 25)	1555
(35, 10)	1600

The minimum cost is $1540.

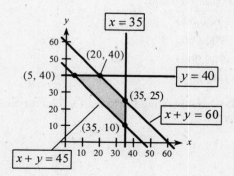

Ship From/To	B	Y
P	20	40
E	15	0

47. x = satellite branches, y = full service

Objective function: $R = 10,000x + 18,000y$

Constraints: $100,000x + 140,000y \le 2,980,000$ Construction costs $\leftrightarrow 10x + 14y \le 298$

$3x + 6y \le 120$ Employees

$x + y \le 25$ Number of branches

$x, y \ge 0$

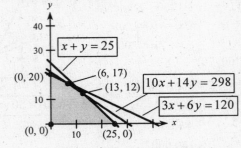

a.

$$10x + 14y = 298 \quad 15x + 21y = 447$$
$$3x + 6y = 120 \quad 15x + 30y = 600$$
$$\overline{ 9y = 153}$$
$$y = 17$$
$$x = 6$$

$$10x + 14y = 298 \quad 10x + 14y = 298$$
$$x + y = 25 \quad 10x + 10y = 250$$
$$\overline{ 4y = 48}$$
$$y = 12$$
$$x = 13$$

Corner	$R = 1000(10x + 18y)$
$(13,12)$	$346,000$
$(6,17)$	$366,000$
$(25,0)$	$250,000$
$(0,20)$	$360,000$

Maximum revenue of $366,000 with 6 satellite and 17 full service branches

b. Branches: Used 23 of 25 possible; 2 not used (slack)

New employees: hired 120 of 120 possible; 0 not hired (slack)

Budget: used all $2.98 million; $0 not used (slack)

c. Additional new employees and additional budget. These items are completely used in the current optimal solution; more could change and improve the optimal solution.

d. Additional branches. The current optimal solution does not use all those allotted; more would just add to the extras.

49. x = days for Factory 1 to operate, y = days for Factory 2 to operate

Objective function: $F = 10,000x + 20,000y$

Constraints: $x, y \ge 0$

$80x + 20y \ge 1600 \quad$ #1

$10x + 10y \ge 500 \quad$ #2

$20x + 70y \ge 2000 \quad$ #3

From the graph (10, 40) and (30, 20) are the two feasible corners not on the coordinate axes.

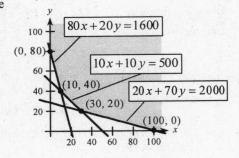

Corner	$F = 10,000x + 20,000y$
$(0, 80)$	$1,600,000$
$(10, 40)$	$900,000$
$(30, 20)$	$700,000$
$(100, 0)$	$1,000,000$

Minimum cost is $700,000 when operating Factory 1 for 30 days and Factory 2 for 20.

Exercise 4.3

1. $3x + 5y + s_1 = 15 \qquad 3x + 6y + s_2 = 20$

3.

$$\begin{array}{ccccc} x & y & s_1 & s_2 & f \\ \left(\begin{array}{ccccc|c} 1 & 5 & 1 & 0 & 0 & 200 \\ 2 & 3 & 0 & 1 & 0 & 134 \\ \hline -4 & -9 & 0 & 0 & 1 & 0 \end{array}\right) \end{array}$$

5.

$$\begin{array}{ccccccc} x & y & z & s_1 & s_2 & s_3 & f \\ \left(\begin{array}{ccccccc|c} 2 & 7 & 9 & 1 & 0 & 0 & 0 & 100 \\ 6 & 5 & 1 & 0 & 1 & 0 & 0 & 145 \\ 1 & 2 & 7 & 0 & 0 & 1 & 0 & 90 \\ \hline -2 & -5 & -2 & 0 & 0 & 0 & 1 & 0 \end{array}\right) \end{array}$$

7. The number of slack variables is equal to the number of constraints (not including the ≥ 0 constraints).

9. **a.** $x_1 = 0$, $x_2 = 0$, $s_1 = 200$, $s_2 = 400$, $s_3 = 350$, $f = 0$

    b. not complete

    c. -7 is the most negative indicator

$$R_1 : \frac{200}{27} \approx 7.41 \qquad R_2 : \frac{400}{51} \approx 7.84 \qquad R_3 : \frac{350}{27} \approx 12.96 \qquad \text{Thus, pivot on 27 in } R_1 C_2.$$

Row Operations: $\dfrac{1}{27} R_1 \to R_1$, then $-51 R_1 + R_2 \to R_2$, $-27 R_1 + R_3 \to R_3$, $7 R_1 + R_4 \to R_4$

11. **a.** $x_1 = 0$, $x_2 = 15$, $s_1 = 12$, $s_2 = 0$, $f = 15$

    b. not complete

    c. -4 is the most negative indicator

$$R_1 : \frac{12}{2} = 6 \qquad R_2 = \frac{15}{3} = 5 \qquad \text{Thus, pivot on 3 in } R_2 C_1.$$

Row Operations: $\dfrac{1}{3} R_2 \to R_2$, then $-2 R_2 + R_1 \to R_1$, $4 R_2 + R_3 \to R_3$

13. **a.** $x_1 = 4$, $x_2 = 0$, $x_3 = 11$, $s_1 = 6$, $s_2 = 0$, $s_3 = 0$, $f = 525$

    b. Because there are no negative indicators, the solution is complete.

15. **a.** $x_1 = 0$, $x_2 = 0$, $x_3 = 12$, $s_1 = 4$, $s_2 = 6$, $s_3 = 0$, $f = 150$

    b. not complete

    c. The pivot could either be in column 1 or 2.

    If you choose column 1:

$$R_1 : \frac{12}{4} = 3 \qquad R_2 : \frac{4}{2} = 2 \qquad \text{Thus, pivot on the 2 in } R_2 C_1.$$

    If you choose column 2:

$$R_1 : \frac{12}{4} = 3 \qquad R_2 : \frac{4}{4} = 1 \qquad \text{Thus, pivot on 4 in } R_2 C_2.$$

    For row operations, assume we use the 4 in $R_2 C_2$:

$$\frac{1}{4} R_2 \to R_2, \text{ then } -4 R_2 + R_1 \to R_1, \ 11 R_2 + R_3 \to R_3, \ 3 R_2 + R_4 \to R_4$$

17. **a.** $x_1 = 0$, $x_2 = 0$, $x_3 = 12$, $s_1 = 5$, $s_2 = 0$, $s_3 = 6$, $f = 120$

    b. no solution is possible because no pivot can be found

19. f is maximized at 20 when $x = 11$ and $y = 9$.

21. f is maximized at 525 when $y = 14$, $z = 11$, and $x = 0$.

23. f is maximized at 100 when $x = 50$ and $y = 10$. Since multiple solutions are possible, pivot on 2 in $R_3 C_5$.

25. $\begin{pmatrix} 14 & \boxed{7} & 1 & 0 & 0 & 35 \\ 5 & 5 & 0 & 1 & 0 & 50 \\ -3 & -10 & 0 & 0 & 1 & 0 \end{pmatrix} \begin{array}{l} \frac{1}{7}R_1 \rightarrow R_1 \\ \rightarrow \end{array}$

-10 is the most negative. Pivot on the 7.

$\begin{bmatrix} 2 & 1 & \frac{1}{7} & 0 & 0 & 5 \\ 5 & 5 & 0 & 1 & 0 & 50 \\ -3 & -10 & 0 & 0 & 1 & 0 \end{bmatrix} \begin{array}{l} \rightarrow \\ -5R_1 + R_2 \rightarrow R_2 \\ 10R_1 + R_3 \rightarrow R_3 \end{array} \begin{bmatrix} 2 & 1 & \frac{1}{7} & 0 & 0 & 5 \\ -5 & 0 & -\frac{5}{7} & 1 & 0 & 25 \\ 17 & 0 & \frac{10}{7} & 0 & 1 & 50 \end{bmatrix}$

All indicators are ≥ 0. The maximum is 50 when $y = 5$ and $x = 0$.

27. $\begin{bmatrix} 1 & \boxed{2} & 1 & 0 & 0 & 10 \\ 1 & 1 & 0 & 1 & 0 & 7 \\ -2 & -3 & 0 & 0 & 1 & 0 \end{bmatrix} \begin{array}{l} \frac{1}{2}R_1 \rightarrow R_1 \end{array} \begin{bmatrix} \frac{1}{2} & 1 & \frac{1}{2} & 0 & 0 & 5 \\ 1 & 1 & 0 & 1 & 0 & 7 \\ -2 & -3 & 0 & 0 & 1 & 0 \end{bmatrix} \begin{array}{l} \rightarrow \\ -R_1 + R_2 \rightarrow R_2 \\ 3R_1 + R_3 \rightarrow R_3 \end{array}$

$\begin{bmatrix} \frac{1}{2} & 1 & \frac{1}{2} & 0 & 0 & 5 \\ \boxed{\frac{1}{2}} & 0 & -\frac{1}{2} & 1 & 0 & 2 \\ -\frac{1}{2} & 0 & \frac{3}{2} & 0 & 1 & 15 \end{bmatrix} 2R_2 \rightarrow R_2 \begin{bmatrix} \frac{1}{2} & 1 & \frac{1}{2} & 0 & 0 & 5 \\ 1 & 0 & -1 & 2 & 0 & 4 \\ -\frac{1}{2} & 0 & \frac{3}{2} & 0 & 1 & 15 \end{bmatrix} \begin{array}{l} -\frac{1}{2}R_2 + R_1 \rightarrow R_1 \\ \rightarrow \\ \frac{1}{2}R_2 + R_3 \rightarrow R_3 \end{array} \begin{bmatrix} 0 & 1 & 1 & -1 & 0 & 3 \\ 1 & 0 & -1 & 2 & 0 & 4 \\ 0 & 0 & 1 & 1 & 1 & 17 \end{bmatrix}$

Maximum is 17 at $x = 4$, $y = 3$.

29. $\begin{bmatrix} -1 & 1 & 1 & 0 & 0 & 0 & 2 \\ 1 & 2 & 0 & 1 & 0 & 0 & 10 \\ \boxed{3} & 1 & 0 & 0 & 1 & 0 & 15 \\ -2 & -1 & 0 & 0 & 0 & 1 & 0 \end{bmatrix} \begin{array}{l} \frac{2}{(-1)} < 0 \\ \frac{10}{1} = 10 \quad \rightarrow \\ \frac{15}{3} = 5 \quad \frac{1}{3}R_3 \rightarrow R_3 \end{array} \begin{bmatrix} -1 & 1 & 1 & 0 & 0 & 0 & 2 \\ 1 & 2 & 0 & 1 & 0 & 0 & 10 \\ 1 & \frac{1}{3} & 0 & 0 & \frac{1}{3} & 0 & 5 \\ -2 & -1 & 0 & 0 & 0 & 1 & 0 \end{bmatrix} \begin{array}{l} R_3 + R_1 \rightarrow R_1 \\ -R_3 + R_2 \rightarrow R_2 \\ \rightarrow \\ 2R_3 + R_4 \rightarrow R_4 \end{array}$

$\begin{bmatrix} 0 & \frac{4}{3} & 1 & 0 & \frac{1}{3} & 0 & 7 \\ 0 & \boxed{\frac{5}{3}} & 0 & 1 & -\frac{1}{3} & 0 & 5 \\ 1 & \frac{1}{3} & 0 & 0 & \frac{1}{3} & 0 & 5 \\ 0 & -\frac{1}{3} & 0 & 0 & \frac{2}{3} & 1 & 10 \end{bmatrix} \begin{array}{l} 7 \div \left(\frac{4}{3}\right) = \frac{21}{4} \quad \rightarrow \\ 5 \div \left(\frac{5}{3}\right) = 3 \quad \frac{3}{5}R_2 \rightarrow R_2 \\ 5 \div \left(\frac{1}{3}\right) = 15 \end{array} \begin{bmatrix} 0 & \frac{4}{3} & 1 & 0 & \frac{1}{3} & 0 & 7 \\ 0 & 1 & 0 & \frac{3}{5} & -\frac{1}{5} & 0 & 3 \\ 1 & \frac{1}{3} & 0 & 0 & \frac{1}{3} & 0 & 5 \\ 0 & -\frac{1}{3} & 0 & 0 & \frac{2}{3} & 1 & 10 \end{bmatrix} \begin{array}{l} -\frac{4}{3}R_2 + R_1 \rightarrow R_1 \\ \rightarrow \\ -\frac{1}{3}R_2 + R_3 \rightarrow R_3 \\ \frac{1}{3}R_2 + R_4 \rightarrow R_4 \end{array}$

$\begin{bmatrix} 0 & 0 & 1 & -\frac{4}{5} & \frac{3}{5} & 0 & 3 \\ 0 & 1 & 0 & \frac{3}{5} & -\frac{1}{5} & 0 & 3 \\ 1 & 0 & 0 & -\frac{1}{5} & \frac{2}{5} & 0 & 4 \\ 0 & 0 & 0 & \frac{1}{5} & \frac{3}{5} & 1 & 11 \end{bmatrix}$

Maximum is 11 at $x = 4$, $y = 3$.

31.
$$\left[\begin{array}{ccccccc|c} 3 & 5 & 4 & 1 & 0 & 0 & 0 & 30 \\ 3 & \boxed{2} & 0 & 0 & 1 & 0 & 0 & 4 \\ 1 & 2 & 0 & 0 & 0 & 1 & 0 & 8 \\ -7 & -10 & -4 & 0 & 0 & 0 & 1 & 0 \end{array}\right] \begin{array}{c} \\ \tfrac{1}{2}R_2 \to R_2 \\ \to \\ \\ \end{array} \left[\begin{array}{ccccccc|c} 3 & 5 & 4 & 1 & 0 & 0 & 0 & 30 \\ \tfrac{3}{2} & 1 & 0 & 0 & \tfrac{1}{2} & 0 & 0 & 2 \\ 1 & 2 & 0 & 0 & 0 & 1 & 0 & 8 \\ -7 & -10 & -4 & 0 & 0 & 0 & 1 & 0 \end{array}\right] \begin{array}{c} -5R_2 + R_1 \to R_1 \\ \to \\ -2R_2 + R_3 \to R_3 \\ 10R_2 + R_4 \to R_4 \end{array}$$

$$\left[\begin{array}{ccccccc|c} -\tfrac{9}{2} & 0 & \boxed{4} & 1 & -\tfrac{5}{2} & 0 & 0 & 20 \\ \tfrac{3}{2} & 1 & 0 & 0 & \tfrac{1}{2} & 0 & 0 & 2 \\ -2 & 0 & 0 & 0 & -1 & 1 & 0 & 4 \\ 8 & 0 & -4 & 0 & 5 & 0 & 1 & 20 \end{array}\right] \begin{array}{c} \tfrac{1}{4}R_1 \to R_1 \\ \\ \to \\ \\ \end{array} \left[\begin{array}{ccccccc|c} -\tfrac{9}{8} & 0 & 1 & \tfrac{1}{4} & -\tfrac{5}{8} & 0 & 0 & 5 \\ \tfrac{3}{2} & 1 & 0 & 0 & \tfrac{1}{2} & 0 & 0 & 2 \\ -2 & 0 & 0 & 0 & -1 & 1 & 0 & 4 \\ 8 & 0 & -4 & 0 & 5 & 0 & 1 & 20 \end{array}\right] \begin{array}{c} \\ \to \\ \\ 4R_1 + R_4 \to R_4 \end{array}$$

$$\left[\begin{array}{ccccccc|c} -\tfrac{9}{8} & 0 & 1 & \tfrac{1}{4} & -\tfrac{5}{8} & 0 & 0 & 5 \\ \tfrac{3}{2} & 1 & 0 & 0 & \tfrac{1}{2} & 0 & 0 & 2 \\ -2 & 0 & 0 & 0 & -1 & 1 & 0 & 4 \\ \tfrac{7}{2} & 0 & 0 & 1 & \tfrac{5}{2} & 0 & 1 & 40 \end{array}\right]$$

Maximum is 40 at $x = 0$, $y = 2$, $z = 5$.

33.
$$\left[\begin{array}{ccccccc|c} 1 & 0 & 1 & 1 & 0 & 0 & 0 & 40 \\ \boxed{1} & 1 & 0 & 0 & 1 & 0 & 0 & 30 \\ 0 & 1 & 1 & 0 & 0 & 1 & 0 & 40 \\ -20 & -12 & -12 & 0 & 0 & 0 & 1 & 0 \end{array}\right] \begin{array}{c} -R_2 + R_1 \to R_1 \\ \to \\ \\ 20R_2 + R_4 \to R_4 \end{array} \left[\begin{array}{ccccccc|c} 0 & -1 & \boxed{1} & 1 & -1 & 0 & 0 & 10 \\ 1 & 1 & 0 & 0 & 1 & 0 & 0 & 30 \\ 0 & 1 & 1 & 0 & 0 & 1 & 0 & 40 \\ 0 & 8 & -12 & 0 & 20 & 0 & 1 & 600 \end{array}\right] \begin{array}{c} \\ \to \\ -R_1 + R_3 \to R_3 \\ 12R_1 + R_4 \to R_4 \end{array}$$

$$\left[\begin{array}{ccccccc|c} 0 & -1 & 1 & 1 & -1 & 0 & 0 & 10 \\ 1 & 1 & 0 & 0 & 1 & 0 & 0 & 30 \\ 0 & \boxed{2} & 0 & -1 & 1 & 1 & 0 & 30 \\ 0 & -4 & 0 & 12 & 8 & 0 & 1 & 720 \end{array}\right] \begin{array}{c} \\ \to \\ \tfrac{1}{2}R_3 \to R_3 \\ \\ \end{array} \left[\begin{array}{ccccccc|c} 0 & -1 & 1 & 1 & -1 & 0 & 0 & 10 \\ 1 & 1 & 0 & 0 & 1 & 0 & 0 & 30 \\ 0 & 1 & 0 & -\tfrac{1}{2} & \tfrac{1}{2} & \tfrac{1}{2} & 0 & 15 \\ 0 & -4 & 0 & 12 & 8 & 0 & 1 & 720 \end{array}\right] \begin{array}{c} R_3 + R_1 \to R_1 \\ -R_3 + R_2 \to R_2 \\ \to \\ 4R_3 + R_4 \to R_4 \end{array}$$

$$\left[\begin{array}{ccccccc|c} 0 & 0 & 1 & \tfrac{1}{2} & -\tfrac{1}{2} & \tfrac{1}{2} & 0 & 25 \\ 1 & 0 & 0 & \tfrac{1}{2} & \tfrac{1}{2} & -\tfrac{1}{2} & 0 & 15 \\ 0 & 1 & 0 & -\tfrac{1}{2} & \tfrac{1}{2} & \tfrac{1}{2} & 0 & 15 \\ 0 & 0 & 0 & 10 & 10 & 2 & 1 & 780 \end{array}\right]$$

Maximum is 780 at $x = 15$, $y = 15$, $z = 25$.

35.
$$\left[\begin{array}{ccccccc|c} 2 & 1 & 1 & 1 & 0 & 0 & 0 & 40 \\ \boxed{1} & 2 & 0 & 0 & 1 & 0 & 0 & 10 \\ 0 & 1 & 3 & 0 & 0 & 1 & 0 & 80 \\ -10 & -8 & -5 & 0 & 0 & 0 & 1 & 0 \end{array}\right] \begin{array}{c} -2R_2 + R_1 \to R_1 \\ \to \\ \\ 10R_2 + R_4 \to R_4 \end{array} \left[\begin{array}{ccccccc|c} 0 & -3 & \boxed{1} & 1 & -2 & 0 & 0 & 20 \\ 1 & 2 & 0 & 0 & 1 & 0 & 0 & 10 \\ 0 & 1 & 3 & 0 & 0 & 1 & 0 & 80 \\ 0 & 12 & -5 & 0 & 10 & 0 & 1 & 100 \end{array}\right] \begin{array}{c} \\ \to \\ -3R_1 + R_3 \to R_3 \\ 5R_1 + R_4 \to R_4 \end{array}$$

$$\left[\begin{array}{ccccccc|c} 0 & -3 & 1 & 1 & -2 & 0 & 0 & 20 \\ 1 & 2 & 0 & 0 & 1 & 0 & 0 & 10 \\ 0 & \boxed{10} & 0 & -3 & 6 & 1 & 0 & 20 \\ 0 & -3 & 0 & 5 & 0 & 0 & 1 & 200 \end{array}\right] \begin{array}{c} \\ \to \\ \tfrac{1}{10}R_3 \to R_3 \\ \\ \end{array} \left[\begin{array}{ccccccc|c} 0 & -3 & 1 & 1 & -2 & 0 & 0 & 20 \\ 1 & 2 & 0 & 0 & 1 & 0 & 0 & 10 \\ 0 & 1 & 0 & -\tfrac{3}{10} & \tfrac{3}{5} & \tfrac{1}{10} & 0 & 2 \\ 0 & -3 & 0 & 5 & 0 & 0 & 1 & 200 \end{array}\right] \begin{array}{c} 3R_3 + R_1 \to R_1 \\ -2R_3 + R_2 \to R_2 \\ \to \\ 3R_3 + R_4 \to R_4 \end{array}$$

$$\left[\begin{array}{ccccccc|c} 0 & 0 & 1 & \tfrac{1}{10} & -\tfrac{1}{5} & \tfrac{3}{10} & 0 & 26 \\ 1 & 0 & 0 & \tfrac{3}{5} & -\tfrac{1}{5} & -\tfrac{1}{5} & 0 & 6 \\ 0 & 1 & 0 & -\tfrac{3}{10} & \tfrac{3}{5} & \tfrac{1}{10} & 0 & 2 \\ 0 & 0 & 0 & \tfrac{41}{10} & \tfrac{9}{5} & \tfrac{3}{10} & 1 & 206 \end{array}\right]$$

Maximum is 206 at $x = 6$, $y = 2$, $z = 26$.

37.
$$\left[\begin{array}{cccccccc|c} 12 & 16 & 20 & 8 & 1 & 0 & 0 & 0 & 880 \\ 5 & \boxed{10} & 10 & 5 & 0 & 1 & 0 & 0 & 460 \\ 10 & -5 & 8 & 5 & 0 & 0 & 1 & 0 & 280 \\ \hline -24 & -30 & -18 & -18 & 0 & 0 & 0 & 1 & 0 \end{array}\right]$$

$\frac{1}{10}R_2 \to R_2$ then

$-16R_2 + R_1 \to R_1$

$5R_2 + R_3 \to R_3$

$30R_2 + R_4 \to R_4$

$$\left[\begin{array}{cccccccc|c} \boxed{4} & 0 & 4 & 0 & 1 & -1.6 & 0 & 0 & 144 \\ 0.5 & 1 & 1 & 0.5 & 0 & 0.1 & 0 & 0 & 46 \\ 12.5 & 0 & 13 & 7.5 & 0 & 0.5 & 1 & 0 & 510 \\ \hline -9 & 0 & 12 & -3 & 0 & 3 & 0 & 1 & 1380 \end{array}\right]$$

$\frac{1}{4}R_1 \to R_1$ then

$-0.5R_1 + R_2 \to R_2$

$-12.5R_1 + R_3 \to R_3$

$9R_1 + R_4 \to R_4$

$$\left[\begin{array}{cccccccc|c} 1 & 0 & 1 & 0 & 0.25 & -0.4 & 0 & 0 & 36 \\ 0 & 1 & 0.5 & 0.5 & -0.125 & 0.3 & 0 & 0 & 28 \\ 0 & 0 & 0.5 & \boxed{7.5} & -3.125 & 5.5 & 1 & 0 & 60 \\ \hline 0 & 0 & 21 & -3 & 2.25 & -0.6 & 0 & 1 & 1704 \end{array}\right]$$

$-0.5R_3 + R_2 \to R_2$

$\frac{2}{15}R_3 \to R_3$ then

$3R_3 + R_4 \to R_4$

$$\left[\begin{array}{cccccccc|c} 1 & 0 & 1 & 0 & 0.25 & -0.4 & 0 & 0 & 36 \\ 0 & 1 & 0.4\overline{6} & 0 & 0.08\overline{3} & -0.0\overline{6} & -0.0\overline{6} & 0 & 24 \\ 0 & 0 & 0.0\overline{6} & 1 & -0.41\overline{6} & 0.7\overline{3} & 0.1\overline{3} & 0 & 8 \\ \hline 0 & 0 & 21.2 & 0 & 1 & 1.6 & 0.4 & 1 & 1728 \end{array}\right]$$

Solution: Maximum = 1728 at $x_1 = 36$, $x_2 = 24$, $x_3 = 0$, $x_4 = 8$.

39.
$$\left[\begin{array}{ccccc|c} 1 & \frac{1}{2} & 1 & 0 & 0 & 16 \\ 0 & \boxed{\frac{1}{2}} & -1 & 1 & 0 & 8 \\ \hline 0 & 0 & 2 & 0 & 1 & 32 \end{array}\right] \xrightarrow{2R_2 \to R_2} \left[\begin{array}{ccccc|c} 1 & \frac{1}{2} & 1 & 0 & 0 & 16 \\ 0 & 1 & -2 & 2 & 0 & 16 \\ \hline 0 & 0 & 2 & 0 & 1 & 32 \end{array}\right] \xrightarrow[\longrightarrow]{-\frac{1}{2}R_2 + R_1 \to R_1} \left[\begin{array}{ccccc|c} 1 & 0 & 2 & -1 & 0 & 8 \\ 0 & 1 & -2 & 2 & 0 & 16 \\ \hline 0 & 0 & 2 & 0 & 1 & 32 \end{array}\right]$$

The maximum is 32 at $x = 8$, $y = 16$.

41.
$$\left[\begin{array}{ccccc|c} \boxed{1} & -10 & 1 & 0 & 0 & 10 \\ -1 & 1 & 0 & 1 & 0 & 40 \\ \hline -3 & -2 & 0 & 0 & 1 & 0 \end{array}\right] \xrightarrow[3R_1 + R_3 \to R_3]{R_1 + R_2 \to R_2} \left[\begin{array}{ccccc|c} 1 & -10 & 1 & 0 & 0 & 10 \\ 0 & -9 & 1 & 1 & 0 & 50 \\ \hline 0 & -32 & 3 & 0 & 1 & 30 \end{array}\right]$$

No nonnegative quotient exists. There is no solution.

43.
$$\left[\begin{array}{ccccc|c} 2 & 1 & 1 & 0 & 0 & 120 \\ 1 & \boxed{4} & 0 & 1 & 0 & 200 \\ \hline -3 & -12 & 0 & 0 & 1 & 0 \end{array}\right] \xrightarrow{\frac{1}{4}R_2 \to R_2} \left[\begin{array}{ccccc|c} 2 & 1 & 1 & 0 & 0 & 120 \\ \frac{1}{4} & 1 & 0 & \frac{1}{4} & 0 & 50 \\ \hline -3 & -12 & 0 & 0 & 1 & 0 \end{array}\right] \begin{array}{c} -R_2 + R_1 \to R_1 \\ \longrightarrow \\ 12R_2 + R_3 \to R_3 \end{array}$$

$$\left[\begin{array}{ccccc|c} \boxed{\frac{7}{4}} & 0 & 1 & -\frac{1}{4} & 0 & 70 \\ \frac{1}{4} & 1 & 0 & \frac{1}{4} & 0 & 50 \\ \hline 0 & 0 & 0 & 3 & 1 & 600 \end{array}\right] \xrightarrow{\frac{4}{7}R_1 \to R_1}$$

Maximum is 600 at $x = 0$, $y = 50$. Continuing to find a second solution we have:

$$\left[\begin{array}{ccccc|c} 1 & 0 & \frac{4}{7} & -\frac{1}{7} & 0 & 40 \\ \boxed{\frac{1}{4}} & 1 & 0 & \frac{1}{4} & 0 & 50 \\ \hline 0 & 0 & 0 & 3 & 1 & 600 \end{array}\right] \xrightarrow{-\frac{1}{4}R_1 + R_2 \to R_2} \left[\begin{array}{ccccc|c} 1 & 0 & \frac{4}{7} & -\frac{1}{7} & 0 & 40 \\ 0 & 1 & -\frac{1}{7} & \frac{2}{7} & 0 & 40 \\ \hline 0 & 0 & 0 & 3 & 1 & 600 \end{array}\right]$$

Maximum is 600 at $x = 40$, $y = 40$.

45. Let x = number of inkjet printers, y = number of laser printers.

Maximize $f = 40x + 60y$

Constraints: $x + y \leq 60$, $x + 3y \leq 120$

a.
$$\left[\begin{array}{ccccc|c} 1 & 1 & 1 & 0 & 0 & 60 \\ 1 & \boxed{3} & 0 & 1 & 0 & 120 \\ \hline -40 & -60 & 0 & 0 & 1 & 0 \end{array}\right] \quad \frac{1}{3}R_2 \to R_2 \quad \left[\begin{array}{ccccc|c} 1 & 1 & 1 & 0 & 0 & 60 \\ \frac{1}{3} & 1 & 0 & \frac{1}{3} & 0 & 40 \\ \hline -40 & -60 & 0 & 0 & 1 & 0 \end{array}\right] \quad \begin{array}{c} -R_2 + R_1 \to R_1 \\ \to \\ 60R_2 + R_3 \to R_3 \end{array}$$

$$\left[\begin{array}{ccccc|c} \boxed{\frac{2}{3}} & 0 & 1 & -\frac{1}{3} & 0 & 20 \\ \frac{1}{3} & 1 & 0 & \frac{1}{3} & 0 & 40 \\ \hline -20 & 0 & 0 & 20 & 1 & 2400 \end{array}\right] \quad \frac{3}{2}R_1 \to R_1 \quad \left[\begin{array}{ccccc|c} 1 & 0 & \frac{3}{2} & -\frac{1}{2} & 0 & 30 \\ \frac{1}{3} & 1 & 0 & \frac{1}{3} & 0 & 40 \\ \hline -20 & 0 & 0 & 20 & 1 & 2400 \end{array}\right] \quad \begin{array}{c} \to \\ -\frac{1}{3}R_1 + R_2 \to R_2 \\ 20R_1 + R_3 \to R_3 \end{array}$$

$$\left[\begin{array}{ccccc|c} 1 & 0 & \frac{3}{2} & -\frac{1}{2} & 0 & 30 \\ 0 & 1 & -\frac{1}{2} & \frac{1}{2} & 0 & 30 \\ \hline 0 & 0 & 30 & 10 & 1 & 3000 \end{array}\right]$$

b. Maximum profit is $3000 with 30 ink jet and 30 laser printers.

47. Let x = number of pairs of style #891 jeans, y = number of pairs of style #917 jeans

Maximize: $f = 6x + 7.5y$

Constraints: $10x + 10y \leq 7500$, $20x + 30y \leq 19500$, $x \geq 0$, $y \geq 0$

$$\left[\begin{array}{ccccc|c} 10 & 10 & 1 & 0 & 0 & 7500 \\ 20 & \boxed{30} & 0 & 1 & 0 & 19500 \\ \hline -6 & -7.5 & 0 & 0 & 1 & 0 \end{array}\right] \quad \frac{1}{30}R_2 \to R_2 \text{ then} \quad \begin{array}{c} -10R_2 + R_1 \to R_1 \\ \to \\ 7.5R_2 + R_3 \to R_3 \end{array}$$

$$\left[\begin{array}{ccccc|c} \boxed{3.\overline{3}} & 0 & 1 & -0.\overline{3} & 0 & 1000 \\ 0.\overline{6} & 1 & 0 & 0.0\overline{3} & 0 & 650 \\ \hline -1 & 0 & 0 & 0.25 & 1 & 4875 \end{array}\right] \quad \frac{3}{10}R_1 \to R_1 \text{ then} \quad \begin{array}{c} \to \\ -0.\overline{6}R_1 + R_2 \to R_2 \\ R_1 + R_3 \to R_3 \end{array}$$

$$\left[\begin{array}{ccccc|c} 1 & 0 & 0.3 & -0.1 & 0 & 300 \\ 0 & 1 & -0.2 & 0.1 & 0 & 450 \\ \hline 0 & 0 & 0.3 & 0.15 & 1 & 5175 \end{array}\right] \quad \text{Solution: } x = 300, \ y = 450, \ f = 5175$$

300 pairs of style #891 jeans and 450 pairs of style #917 jeans for a maximum profit of $5175

49. Let x = number of axles, y = number of wheels.

Maximize: $f = 300x + 300y$

Constraints: $4x + 3y \leq 50$, $3x + 5y \leq 43$, $x \geq 0$, $y \geq 0$

$$\begin{bmatrix} 4 & 3 & 1 & 0 & 0 & | & 50 \\ 3 & \boxed{5} & 0 & 1 & 0 & | & 43 \\ \hline -300 & -300 & 0 & 0 & 1 & | & 0 \end{bmatrix} \quad \frac{1}{5}R_2 \to R_2 \text{ then}$$

$$\begin{array}{c} -3R_2 + R_1 \to R_1 \\ \to \\ 300R_2 + R_3 \to R_3 \end{array}$$

$$\begin{bmatrix} \boxed{\frac{11}{5}} & 0 & 1 & -\frac{3}{5} & 0 & | & \frac{121}{5} \\ \frac{3}{5} & 1 & 0 & \frac{1}{5} & 0 & | & \frac{43}{5} \\ \hline -120 & 0 & 0 & 60 & 1 & | & 2580 \end{bmatrix} \quad \frac{5}{11}R_1 \to R_1 \text{ then}$$

$$\begin{array}{c} \to \\ -\frac{3}{5}R_1 + R_2 \to R_2 \\ 120R_1 + R_3 \to R_3 \end{array}$$

$$\begin{bmatrix} 1 & 0 & \frac{5}{11} & -\frac{3}{11} & 0 & | & 11 \\ 0 & 1 & -\frac{3}{11} & \frac{4}{11} & 0 & | & 2 \\ \hline 0 & 0 & \frac{600}{11} & \frac{300}{11} & 1 & | & 3900 \end{bmatrix}$$

Maximum profit is \$3900 when 11 axles and 2 wheels are produced.

51. Let t = crates of tomatoes and p = crates of peaches. Objective function: $f = t + 2p$

Constraints: $\frac{1}{2}t + \frac{5}{4}p \leq 2500$, $60t + 50p \leq 120{,}000$, $t \geq 0$, $p \geq 0$

$$\begin{bmatrix} \frac{1}{2} & \boxed{\frac{5}{4}} & 1 & 0 & 0 & | & 2500 \\ 60 & 50 & 0 & 1 & 0 & | & 120{,}000 \\ \hline -1 & -2 & 0 & 0 & 1 & | & 0 \end{bmatrix} \quad \frac{4}{5}R_1 \to R_1 \text{ then}$$

$$\begin{array}{c} \to \\ -50R_1 + R_2 \to R_2 \\ 2R_1 + R_3 \to R_3 \end{array}$$

$$\begin{bmatrix} \frac{2}{5} & 1 & \frac{4}{5} & 0 & 0 & | & 2000 \\ \boxed{40} & 0 & -40 & 1 & 0 & | & 20{,}000 \\ \hline -\frac{1}{5} & 0 & \frac{8}{5} & 0 & 1 & | & 4000 \end{bmatrix} \quad \frac{1}{40}R_2 \to R_2 \text{ then}$$

$$\begin{array}{c} -\frac{2}{5}R_2 + R_1 \to R_1 \\ \to \\ \frac{1}{5}R_2 + R_3 \to R_3 \end{array}$$

$$\begin{bmatrix} 0 & 1 & \frac{6}{5} & -0.01 & 0 & | & 1800 \\ 1 & 0 & -1 & \frac{1}{40} & 0 & | & 500 \\ \hline 0 & 0 & \frac{7}{5} & 0.005 & 1 & | & 4100 \end{bmatrix}$$

There is a maximum profit of \$4100 obtained from 500 crates of tomatoes and 1800 crates of peaches.

53. Let x = newspaper ads and y = radio ads. Objective function: $f = 6000x + 8000y$

Constraints: $100x + 300y \leq 6000$, $x \leq 21$, $y \leq 28$

$$\begin{bmatrix} 1 & \boxed{3} & 1 & 0 & 0 & 0 & | & 60 \\ 1 & 0 & 0 & 1 & 0 & 0 & | & 21 \\ 0 & 1 & 0 & 0 & 1 & 0 & | & 28 \\ \hline -6000 & -8000 & 0 & 0 & 0 & 1 & | & 0 \end{bmatrix} \quad \frac{1}{3}R_1 \to R_1 \text{ then}$$

$$\begin{array}{c} \to \\ \to \\ -R_1 + R_3 \to R_3 \\ 8000R_1 + R_4 \to R_4 \end{array}$$

$$\begin{bmatrix} \frac{1}{3} & 1 & \frac{1}{3} & 0 & 0 & 0 & | & 20 \\ \boxed{1} & 0 & 0 & 1 & 0 & 0 & | & 21 \\ -\frac{1}{3} & 0 & -\frac{1}{3} & 0 & 1 & 0 & | & 8 \\ \hline -\frac{10{,}000}{3} & 0 & \frac{8{,}000}{3} & 0 & 0 & 1 & | & 160{,}000 \end{bmatrix}$$

$$\begin{array}{c} -\frac{1}{3}R_2 + R_1 \to R_1 \\ \to \\ \frac{1}{3}R_2 + R_3 \to R_3 \\ \frac{10{,}000}{3}R_2 + R_4 \to R_4 \end{array}$$

$$\begin{bmatrix} 0 & 1 & \frac{1}{3} & -\frac{1}{3} & 0 & 0 & | & 13 \\ 1 & 0 & 0 & 1 & 0 & 0 & | & 21 \\ 0 & 0 & -\frac{1}{3} & \frac{1}{3} & 1 & 0 & | & 15 \\ \hline 0 & 0 & -\frac{8000}{3} & \frac{10{,}000}{3} & 0 & 1 & | & 230{,}000 \end{bmatrix}$$

Maximum exposure occurs with 21 newspaper ads and 13 radio ads. The maximum number of exposures is 230,000.

55. Let x = number of ad packages in medium 1
y = number of ad packages in medium 2
z = number of ad packages in medium 3

Maximize: $f = 3100x + 2000y + 2400z$

Constraints: $10x + 4y + 5z \le 200, \ x \le 18, \ y \le 10, \ z \le 12$

$$
\left[\begin{array}{cccccccc|c}
10 & 4 & 5 & 1 & 0 & 0 & 0 & 0 & 200 \\
\boxed{1} & 0 & 0 & 0 & 1 & 0 & 0 & 0 & 18 \\
0 & 1 & 0 & 0 & 0 & 1 & 0 & 0 & 10 \\
0 & 0 & 1 & 0 & 0 & 0 & 1 & 0 & 12 \\
\hline
-3100 & -2000 & -2400 & 0 & 0 & 0 & 0 & 1 & 0
\end{array}\right]
\begin{array}{l}
-10R_2 + R_1 \to R_1 \\[4pt]
\\[4pt]
\to \\[4pt]
\\[4pt]
3100R_2 + R_5 \to R_5
\end{array}
$$

$$
\left[\begin{array}{cccccccc|c}
0 & 4 & \boxed{5} & 1 & -10 & 0 & 0 & 0 & 20 \\
1 & 0 & 0 & 0 & 1 & 0 & 0 & 0 & 18 \\
0 & 1 & 0 & 0 & 0 & 1 & 0 & 0 & 10 \\
0 & 0 & 1 & 0 & 0 & 0 & 1 & 0 & 12 \\
\hline
0 & -2000 & -2400 & 0 & 3100 & 0 & 0 & 1 & 55800
\end{array}\right]
\begin{array}{l}
\frac{1}{5}R_1 \to R_1 \ \text{then} \quad \to \\[4pt]
\to \\[4pt]
\to \\[4pt]
-R_1 + R_4 \to R_4 \\[4pt]
2400R_1 + R_5 \to R_5
\end{array}
$$

$$
\left[\begin{array}{cccccccc|c}
0 & \frac{4}{5} & 1 & \frac{1}{5} & -2 & 0 & 0 & 0 & 4 \\
1 & 0 & 0 & 0 & 1 & 0 & 0 & 0 & 18 \\
0 & 1 & 0 & 0 & 0 & 1 & 0 & 0 & 10 \\
0 & -\frac{4}{5} & 0 & -\frac{1}{5} & \boxed{2} & 0 & 1 & 0 & 8 \\
\hline
0 & -80 & 0 & 480 & -1700 & 0 & 0 & 1 & 65400
\end{array}\right]
\begin{array}{l}
2R_4 + R_1 \to R_1 \\[4pt]
-R_4 + R_2 \to R_2 \\[4pt]
\to \\[4pt]
\frac{1}{2}R_4 \to R_4 \ \text{then} \quad \to \\[4pt]
1700R_4 + R_5 \to R_5
\end{array}
$$

$$
\left[\begin{array}{cccccccc|c}
0 & 0 & 1 & 0 & 0 & 0 & 1 & 0 & 12 \\
1 & \frac{2}{5} & 0 & \frac{1}{10} & 0 & 0 & -\frac{1}{2} & 0 & 14 \\
0 & \boxed{1} & 0 & 0 & 0 & 1 & 0 & 0 & 10 \\
0 & -\frac{2}{5} & 0 & -\frac{1}{10} & 1 & 0 & \frac{1}{2} & 0 & 4 \\
\hline
0 & -760 & 0 & 310 & 0 & 0 & 850 & 1 & 72{,}200
\end{array}\right]
\begin{array}{l}
-\frac{2}{5}R_3 + R_2 \to R_2 \\[4pt]
\to \\[4pt]
\frac{2}{5}R_3 + R_4 \to R_4 \\[4pt]
760R_3 + R_5 \to R_5
\end{array}
$$

$$
\left[\begin{array}{cccccccc|c}
0 & 0 & 1 & 0 & 0 & 0 & 1 & 0 & 12 \\
1 & 0 & 0 & \frac{1}{10} & 0 & -\frac{2}{5} & -\frac{1}{2} & 0 & 10 \\
0 & 1 & 0 & 0 & 0 & 1 & 0 & 0 & 10 \\
0 & 0 & 0 & -\frac{1}{10} & 1 & \frac{2}{5} & \frac{1}{2} & 0 & 8 \\
\hline
0 & 0 & 0 & 310 & 0 & 760 & 850 & 1 & 79{,}800
\end{array}\right]
$$

Purchase 10 each of medium I and II and purchase 12 units of medium III for a maximum of 79,800 ad exposures.

57. Objective function: $f = 30A + 9B + 15C$

Constraints: $9A + 3B + 0.5C \le 477$, $5A + 4B \le 350$, $3A + 2C \le 150$, $B \le 20$, $A \ge 0$, $B \ge 0$, $C \ge 0$

$$\begin{bmatrix} 9 & 3 & \frac{1}{2} & 1 & 0 & 0 & 0 & 0 & 477 \\ 5 & 4 & 0 & 0 & 1 & 0 & 0 & 0 & 350 \\ \boxed{3} & 0 & 2 & 0 & 0 & 1 & 0 & 0 & 150 \\ 0 & 1 & 0 & 0 & 0 & 0 & 1 & 0 & 20 \\ \hline -30 & -9 & -15 & 0 & 0 & 0 & 0 & 1 & 0 \end{bmatrix}$$

$\frac{1}{3}R_3 \to R_3$ then

$-9R_3 + R_1 \to R_1$
$-5R_3 + R_2 \to R_2$
$\to$
$\to$
$30R_3 + R_5 \to R_5$

$$\begin{bmatrix} 0 & \boxed{3} & -\frac{11}{2} & 1 & 0 & -3 & 0 & 0 & 27 \\ 0 & 4 & -\frac{10}{3} & 0 & 1 & -\frac{5}{3} & 0 & 0 & 100 \\ 1 & 0 & \frac{2}{3} & 0 & 0 & \frac{1}{3} & 0 & 0 & 50 \\ 0 & 1 & 0 & 0 & 0 & 0 & 1 & 0 & 20 \\ \hline 0 & -9 & 5 & 0 & 0 & 10 & 0 & 1 & 1500 \end{bmatrix}$$

$\frac{1}{3}R_1 \to R_1$ then

$-4R_1 + R_2 \to R_2$
$\to$
$-R_1 + R_4 \to R_4$
$9R_1 + R_5 \to R_5$

$$\begin{bmatrix} 0 & 1 & -\frac{11}{6} & \frac{1}{3} & 0 & -1 & 0 & 0 & 9 \\ 0 & 0 & 4 & -\frac{4}{3} & 1 & \frac{7}{3} & 0 & 0 & 64 \\ 1 & 0 & \frac{2}{3} & 0 & 0 & \frac{1}{3} & 0 & 0 & 50 \\ 0 & 0 & \boxed{\frac{11}{6}} & -\frac{1}{3} & 0 & 1 & 1 & 0 & 11 \\ \hline 0 & 0 & -\frac{23}{6} & 3 & 0 & 1 & 0 & 1 & 1581 \end{bmatrix}$$

$\frac{6}{11}R_4 \to R_4$ then

$\frac{11}{6}R_4 + R_1 \to R_1$
$-4R_4 + R_2 \to R_2$
$-\frac{2}{3}R_4 + R_3 \to R_3$
$\to$
$\frac{23}{2}R_4 + R_5 \to R_5$

$$\begin{bmatrix} 0 & 1 & 0 & 0 & 0 & 0 & 1 & 0 & 20 \\ 0 & 0 & 0 & -\frac{20}{33} & 1 & \frac{5}{33} & -\frac{24}{11} & 0 & 40 \\ 1 & 0 & 0 & \frac{4}{33} & 0 & -\frac{1}{33} & -\frac{4}{33} & 0 & 46 \\ 0 & 0 & 1 & -\frac{2}{11} & 0 & \frac{6}{11} & \frac{6}{11} & 0 & 6 \\ \hline 0 & 0 & 0 & \frac{10}{11} & 0 & \frac{80}{11} & \frac{69}{11} & 1 & 1650 \end{bmatrix}$$

The maximum profit is \$1650 when 46 of A, 20 of B, and 6 of C are produced.

59. Objective function: $f = 3x_1 + 4x_2 + 8x_3$

Constraints: $5x_1 + 12x_2 + 24x_3 \le 90,000$, $1x_1 + 2x_2 + 4x_3 \le 12,000$, $1x_1 + 1x_2 + 1x_3 \le 9,000$, $x_1, x_2, x_3 \ge 0$

$$\begin{bmatrix} 5 & 12 & 24 & 1 & 0 & 0 & 0 & 90,000 \\ 1 & 2 & \boxed{4} & 0 & 1 & 0 & 0 & 12,000 \\ 1 & 1 & 1 & 0 & 0 & 1 & 0 & 9,000 \\ \hline -3 & -4 & -8 & 0 & 0 & 0 & 1 & 0 \end{bmatrix}$$

$\frac{1}{4}R_2 \to R_2$ then

$-24R_2 + R_1 \to R_1$
$\to$
$-1R_2 + R_3 \to R_3$
$8R_2 + R_4 \to R_4$

$$\begin{bmatrix} -1 & 0 & 0 & 1 & -6 & 0 & 0 & 18,000 \\ \frac{1}{4} & \frac{1}{2} & 1 & 0 & \frac{1}{4} & 0 & 0 & 3,000 \\ \boxed{\frac{3}{4}} & \frac{1}{2} & 0 & 0 & -\frac{1}{4} & 1 & 0 & 6,000 \\ \hline -1 & 0 & 0 & 0 & 2 & 0 & 1 & 24,000 \end{bmatrix}$$

$\frac{4}{3}R_3 \to R_3$ then

$R_3 + R_1 \to R_1$
$-\frac{1}{4}R_3 + R_2 \to R_2$
$\to$
$R_3 + R_4 \to R_4$

$$\begin{bmatrix} 0 & \frac{2}{3} & 0 & 1 & -\frac{19}{3} & \frac{4}{3} & 0 & 26,000 \\ 0 & \frac{1}{3} & 1 & 0 & \frac{1}{3} & -\frac{1}{3} & 0 & 1,000 \\ 1 & \frac{2}{3} & 0 & 0 & -\frac{1}{3} & \frac{4}{3} & 0 & 8,000 \\ \hline 0 & \frac{2}{3} & 0 & 0 & \frac{5}{3} & \frac{4}{3} & 1 & 32,000 \end{bmatrix}$$

Maximum profit is \$32,000 with 8000 x_1 units and 1000 x_3 units.

61. Let x_1 = number of one-bedroom apartments, x_2 = number of two-bedroom apartments, and x_3 = number of three-bedroom apartments

Objective function: $f = 500x_1 + 800x_2 + 1150x_3$

Constraints: $x_1 + 2x_2 + 3x_3 \leq 250, \ x_1 \leq 26, \ x_2 \leq 40, \ x_3 \leq 60$

$$
\begin{bmatrix}
1 & 2 & 3 & 1 & 0 & 0 & 0 & 0 & 250 \\
1 & 0 & 0 & 0 & 1 & 0 & 0 & 0 & 26 \\
0 & 1 & 0 & 0 & 0 & 1 & 0 & 0 & 40 \\
0 & 0 & \boxed{1} & 0 & 0 & 0 & 1 & 0 & 60 \\
-500 & -800 & -1150 & 0 & 0 & 0 & 0 & 1 & 0
\end{bmatrix}
\begin{array}{l}
-3R_4 + R_1 \to R_1 \\
\to \\
\to \\
\to \\
1150R_4 + R_5 \to R_5
\end{array}
$$

$$
\begin{bmatrix}
1 & \boxed{2} & 0 & 1 & 0 & 0 & -\frac{3}{2} & 0 & 35 \\
1 & 0 & 0 & 0 & 1 & 0 & 0 & 0 & 26 \\
0 & 1 & 0 & 0 & 0 & 1 & 0 & 0 & 40 \\
0 & 0 & 1 & 0 & 0 & 0 & 1 & 0 & 60 \\
-500 & -800 & 0 & 0 & 0 & 0 & 1150 & 1 & 69{,}000
\end{bmatrix}
\begin{array}{l}
\frac{1}{2}R_1 \to R_1 \text{ then} \qquad \to \\
\to \\
-R_1 + R_3 \to R_3 \\
\to \\
800R_1 + R_5 \to R_5
\end{array}
$$

$$
\begin{bmatrix}
\frac{1}{2} & 1 & 0 & \frac{1}{2} & 0 & 0 & -\frac{3}{2} & 0 & 35 \\
\boxed{1} & 0 & 0 & 0 & 1 & 0 & 0 & 0 & 26 \\
-\frac{1}{2} & 0 & 0 & -\frac{1}{2} & 0 & 1 & \frac{3}{2} & 0 & 5 \\
0 & 0 & 1 & 0 & 0 & 0 & 1 & 0 & 60 \\
-100 & 0 & 0 & 400 & 0 & 0 & -50 & 1 & 97{,}000
\end{bmatrix}
\begin{array}{l}
-\frac{1}{2}R_2 + R_1 \to R_1 \\
\to \\
\frac{1}{2}R_2 + R_3 \to R_3 \\
\to \\
100R_2 + R_5 \to R_5
\end{array}
$$

$$
\begin{bmatrix}
0 & 1 & 0 & \frac{1}{2} & -\frac{1}{2} & 0 & -\frac{3}{2} & 0 & 22 \\
1 & 0 & 0 & 0 & 1 & 0 & 0 & 0 & 26 \\
0 & 0 & 0 & -\frac{1}{2} & \frac{1}{2} & 1 & \boxed{\frac{3}{2}} & 0 & 18 \\
0 & 0 & 1 & 0 & 0 & 0 & 1 & 0 & 60 \\
0 & 0 & 0 & 400 & 100 & 0 & -50 & 1 & 99{,}600
\end{bmatrix}
\begin{array}{l}
\frac{3}{2}R_3 + R_1 \to R_1 \\
\to \\
\frac{2}{3}R_3 \to R_3 \text{ then} \qquad \to \\
-R_3 + R_4 \to R_4 \\
50R_3 + R_5 \to R_5
\end{array}
$$

$$
\begin{bmatrix}
0 & 1 & 0 & 0 & 0 & 1 & 0 & 0 & 40 \\
1 & 0 & 0 & 0 & 1 & 0 & 0 & 0 & 26 \\
0 & 0 & 0 & -\frac{1}{3} & \frac{1}{3} & \frac{2}{3} & 1 & 0 & 12 \\
0 & 0 & 1 & \frac{1}{3} & -\frac{1}{3} & -\frac{2}{3} & 0 & 0 & 48 \\
0 & 0 & 0 & \frac{1150}{3} & \frac{350}{3} & \frac{100}{3} & 0 & 1 & 100{,}200
\end{bmatrix}
$$

a-b. Maximum revenue is \$100,200 by renting 26 1-BR, 40 2-BR, and 48 3-BR apartments.

63. Let x_1 = number of 19-inch portable TVs, x_2 = number of 27-inch table models, x_3 = number of 35-inch table models, x_4 = number of 35-inch floor models

Maximize: $f = 46x_1 + 60x_2 + 75x_3 + 100x_4$

Constraints: $x_1 + x_2 + x_3 + x_4 \le 200$

$\qquad\qquad x_1 \le 40$

$\qquad\qquad 7x_1 + 10x_2 + 12x_3 + 15x_4 \le 2000$

$\qquad\qquad 2x_1 + 2x_2 + 4x_3 + 5x_4 \le 500$

$\qquad\qquad x_1 \ge 0,\ x_2 \ge 0,\ x_3 \ge 0,\ x_4 \ge 0$

$$
\left[\begin{array}{ccccccccc|c}
1 & 1 & 1 & 1 & 1 & 0 & 0 & 0 & 0 & 200 \\
1 & 0 & 0 & 0 & 0 & 1 & 0 & 0 & 0 & 40 \\
7 & 10 & 12 & 15 & 0 & 0 & 1 & 0 & 0 & 2000 \\
2 & 2 & 4 & \boxed{5} & 0 & 0 & 0 & 1 & 0 & 500 \\
\hline
-46 & -60 & -75 & -100 & 0 & 0 & 0 & 0 & 1 & 0
\end{array}\right]
$$

$-R_4 + R_1 \to R_1$

$\to$

$-15R_4 + R_3 \to R_3$

$\tfrac{1}{5}R_4 \to R_4$ then $\qquad \to$

$100R_4 + R_5 \to R_5$

$$
\left[\begin{array}{ccccccccc|c}
0.6 & 0.6 & 0.2 & 0 & 1 & 0 & 0 & -0.2 & 0 & 100 \\
1 & 0 & 0 & 0 & 0 & 1 & 0 & 0 & 0 & 40 \\
1 & \boxed{4} & 0 & 0 & 0 & 0 & 1 & -3 & 0 & 500 \\
0.4 & 0.4 & 0.8 & 1 & 0 & 0 & 0 & 0.2 & 0 & 100 \\
\hline
-6 & -20 & 5 & 0 & 0 & 0 & 0 & 20 & 1 & 10{,}000
\end{array}\right]
$$

$-0.6R_3 + R_1 \to R_1$

$\to$

$\tfrac{1}{4}R_3 \to R_3$ then $\qquad \to$

$-0.4R_3 + R_4 \to R_4$

$100R_4 + R_5 \to R_5$

$$
\left[\begin{array}{ccccccccc|c}
0.45 & 0 & 0.2 & 0 & 1 & 0 & -0.15 & 0.25 & 0 & 25 \\
\boxed{1} & 0 & 0 & 0 & 0 & 1 & 0 & 0 & 0 & 40 \\
0.25 & 1 & 0 & 0 & 0 & 0 & 0.25 & -0.75 & 0 & 125 \\
0.3 & 0 & 0.8 & 1 & 0 & 0 & -0.1 & 0.5 & 0 & 50 \\
\hline
-1 & 0 & 5 & 0 & 0 & 0 & 5 & 5 & 1 & 12{,}500
\end{array}\right]
$$

$-0.45R_2 + R_1 \to R_1$

$\to$

$-0.25R_2 + R_3 \to R_3$

$-0.3R_2 + R_4 \to R_4$

$R_2 + R_5 \to R_5$

$$
\left[\begin{array}{ccccccccc|c}
0 & 0 & 0.2 & 0 & 1 & -0.45 & -0.15 & 0.25 & 0 & 7 \\
1 & 0 & 0 & 0 & 0 & 1 & 0 & 0 & 0 & 40 \\
0 & 1 & 0 & 0 & 0 & -0.25 & 0.25 & -0.75 & 0 & 115 \\
0 & 0 & 0.8 & 1 & 0 & -0.3 & -0.1 & 0.5 & 0 & 38 \\
\hline
0 & 0 & 5 & 0 & 0 & 1 & 5 & 5 & 1 & 12{,}540
\end{array}\right]
$$

They should make 40 of the 19-inch portable TVs, 115 of the 27-inch table models, and 38 of the 35-inch floor models for a maximum profit of $12,540 this week.

Exercise 4.4

1. a. $\begin{bmatrix} 5 & 2 & | & 10 \\ 1 & 2 & | & 6 \\ \hline 3 & 1 & | & g \end{bmatrix}$ transpose: $\begin{bmatrix} 5 & 1 & | & 3 \\ 2 & 2 & | & 1 \\ \hline 10 & 6 & | & g \end{bmatrix}$

b. From the transpose in (a):

Maximize $f = 10x_1 + 6x_2$

Constraints: $5x_1 + x_2 \le 3$

$2x_1 + 2x_2 \le 1$

$x_1, x_2 > 0$

3. a. $\begin{bmatrix} 1 & 1 & | & 9 \\ 1 & 3 & | & 15 \\ \hline 5 & 2 & | & g \end{bmatrix}$ transpose: $\begin{bmatrix} 1 & 1 & | & 5 \\ 1 & 3 & | & 2 \\ \hline 9 & 15 & | & g \end{bmatrix}$

b. From the transpose in (a):

Maximize $f = 9x_1 + 15x_2$

Constraints: $x_1 + x_2 \le 5$

$x_1 + 3x_2 \le 2$

$x_1, \ x_2 \ge 0$

5. a. For the minimization problem, the last entries in the columns corresponding to the slack variables are the solution.

Minimum: $g = 252$; $y_1 = 8$, $y_2 = 2$, $y_3 = 0$

b. Maximum: $f = 252$; $x_1 = 5$, $x_2 = 0$, $x_3 = 9$

7. $\begin{bmatrix} 2 & 1 & | & 11 \\ 1 & 3 & | & 11 \\ 1 & 4 & | & 16 \\ \hline 2 & 10 & | & g \end{bmatrix}$ transpose: $\begin{bmatrix} 2 & 1 & 1 & | & 2 \\ 1 & 3 & 4 & | & 10 \\ \hline 11 & 11 & 16 & | & g \end{bmatrix}$

Maximize: $f = 11x_1 + 11x_2 + 16x_3$

Constraints: $2x_1 + x_2 + x_3 \le 2$

$x_1 + 3x_2 + 4x_3 \le 10$

$\begin{bmatrix} 2 & 1 & \boxed{1} & 1 & 0 & 0 & | & 2 \\ 1 & 3 & 4 & 0 & 1 & 0 & | & 10 \\ \hline -11 & -11 & -16 & 0 & 0 & 1 & | & 0 \end{bmatrix}$ $\begin{array}{c} \rightarrow \\ -4R_1 + R_2 \rightarrow R_2 \\ 16R_1 + R_3 \rightarrow R_3 \end{array}$ $\begin{bmatrix} 2 & 1 & 1 & 1 & 0 & 0 & | & 2 \\ -7 & -1 & 0 & -4 & 1 & 0 & | & 2 \\ \hline 21 & 5 & 0 & 16 & 0 & 1 & | & 32 \end{bmatrix}$

Primal: $y_1 = 16$, $y_2 = 0$, $g = 32$ (min)

Dual: $x_1 = 0$, $x_2 = 0$, $x_3 = 2$, $f = 32$ (max)

9.
$$\begin{bmatrix} 4 & 1 & | & 11 \\ 3 & 2 & | & 12 \\ 3 & 1 & | & 6 \\ \hline 3 & 1 & | & g \end{bmatrix} \quad \text{transpose:} \quad \begin{bmatrix} 4 & 3 & 3 & | & 3 \\ 1 & 2 & 1 & | & 1 \\ \hline 11 & 12 & 6 & | & g \end{bmatrix}$$

Maximize $f = 11x_1 + 12x_2 + 6x_3$

Constraints: $4x_1 + 3x_2 + 3x_3 \leq 3$, $x_1 + 2x_2 + x_3 \leq 1$, $x_1, x_2, x_3 \geq 0$

$$\begin{bmatrix} 4 & 3 & 3 & 1 & 0 & 0 & | & 3 \\ 1 & \boxed{2} & 1 & 0 & 1 & 0 & | & 1 \\ \hline -11 & -12 & -6 & 0 & 0 & 1 & | & 0 \end{bmatrix} \quad \begin{array}{l} -3R_2 + R_1 \to R_1 \\ \frac{1}{2}R_2 \to R_2 \quad \text{then} \quad \to \\ 12R_2 + R_3 \to R_3 \end{array}$$

$$\begin{bmatrix} \boxed{\frac{5}{2}} & 0 & \frac{3}{2} & 1 & -\frac{3}{2} & 0 & | & \frac{3}{2} \\ \frac{1}{2} & 1 & \frac{1}{2} & 0 & \frac{1}{2} & 0 & | & \frac{1}{2} \\ \hline -5 & 0 & 0 & 0 & 6 & 1 & | & 6 \end{bmatrix} \quad \begin{array}{l} \frac{2}{5}R_1 \to R_1 \quad \text{then} \qquad \to \\ -\frac{1}{2}R_1 + R_2 \to R_2 \\ 5R_1 + R_3 \to R_3 \end{array} \quad \begin{bmatrix} 1 & 0 & \frac{3}{5} & \frac{2}{5} & -\frac{3}{5} & 0 & | & \frac{3}{5} \\ 0 & 1 & \frac{1}{5} & -\frac{1}{5} & \frac{4}{5} & 0 & | & \frac{1}{5} \\ \hline 0 & 0 & 3 & 2 & 3 & 1 & | & 9 \end{bmatrix}$$

Primal: $y_1 = 2$, $y_2 = 3$, $g = 9$ (min)   Dual: $x_1 = \dfrac{3}{5}$, $x_2 = \dfrac{1}{5}$, $x_3 = 0$, $f = 9$ (max)

11.
$$\begin{bmatrix} 1 & 1 & 1 & | & 3 \\ 0 & 1 & 2 & | & 2 \\ 1 & 0 & 0 & | & 2 \\ \hline 8 & 7 & 12 & | & g \end{bmatrix} \quad \text{transpose:} \quad \begin{bmatrix} 1 & 0 & 1 & | & 8 \\ 1 & 1 & 0 & | & 7 \\ 1 & 2 & 0 & | & 12 \\ \hline 3 & 2 & 2 & | & g \end{bmatrix}$$

Maximize: $f = 3x_1 + 2x_2 + 2x_3$

Constraints: $x_1 + x_3 \leq 8$, $x_1 + x_2 \leq 7$, $x_1 + 2x_2 \leq 12$

$$\begin{bmatrix} 1 & 0 & 1 & 1 & 0 & 0 & 0 & | & 8 \\ \boxed{1} & 1 & 0 & 0 & 1 & 0 & 0 & | & 7 \\ 1 & 2 & 0 & 0 & 0 & 1 & 0 & | & 12 \\ \hline -3 & -2 & -2 & 0 & 0 & 0 & 1 & | & 0 \end{bmatrix} \quad \begin{array}{l} -R_2 + R_1 \to R_1 \\ \to \\ -R_2 + R_3 \to R_3 \\ 3R_2 + R_4 \to R_4 \end{array} \quad \begin{bmatrix} 0 & -1 & \boxed{1} & 1 & -1 & 0 & 0 & | & 1 \\ 1 & 1 & 0 & 0 & 1 & 0 & 0 & | & 7 \\ 0 & 1 & 0 & 0 & -1 & 1 & 0 & | & 5 \\ \hline 0 & 1 & -2 & 0 & 3 & 0 & 1 & | & 21 \end{bmatrix} \quad \begin{array}{l} \to \\ \\ 2R_1 + R_4 \to R_4 \end{array}$$

$$\begin{bmatrix} 0 & -1 & 1 & 1 & -1 & 0 & 0 & | & 1 \\ 1 & 1 & 0 & 0 & 1 & 0 & 0 & | & 7 \\ 0 & \boxed{1} & 0 & 0 & -1 & 1 & 0 & | & 5 \\ \hline 0 & -1 & 0 & 2 & 1 & 0 & 1 & | & 23 \end{bmatrix} \quad \begin{array}{l} R_3 + R_1 \to R_1 \\ -R_3 + R_2 \to R_2 \\ \to \\ R_3 + R_4 \to R_4 \end{array} \quad \begin{bmatrix} 0 & 0 & 1 & 1 & -2 & 1 & 0 & | & 6 \\ 1 & 0 & 0 & 0 & 2 & -1 & 0 & | & 2 \\ 0 & 1 & 0 & 0 & -1 & 1 & 0 & | & 5 \\ \hline 0 & 0 & 2 & 0 & 1 & 1 & | & 28 \end{bmatrix}$$

Minimum: $g = 28$ when $x = 2$, $y = 0$, $z = 1$

13.
$$\begin{bmatrix} 1 & 3 & 0 & 1 \\ 4 & 6 & 1 & 3 \\ 0 & 4 & 1 & 1 \\ \hline 12 & 48 & 8 & g \end{bmatrix} \quad \text{transpose:} \quad \begin{bmatrix} 1 & 4 & 0 & 12 \\ 3 & 6 & 4 & 48 \\ 0 & 1 & 1 & 8 \\ \hline 1 & 3 & 1 & g \end{bmatrix}$$

Maximize: $f = x_1 + 3x_2 + x_3$

Constraints: $x_1 + 4x_2 \le 12, \ 3x_1 + 6x_2 + 4x_3 \le 48, \ x_2 + x_3 \le 8, \ x_1, x_2, x_3 \ge 0$

$$\begin{bmatrix} 1 & \boxed{4} & 0 & 1 & 0 & 0 & 0 & 12 \\ 3 & 6 & 4 & 0 & 1 & 0 & 0 & 48 \\ 0 & 1 & 1 & 0 & 0 & 1 & 0 & 8 \\ \hline -1 & -3 & -1 & 0 & 0 & 0 & 1 & 0 \end{bmatrix} \quad \begin{array}{l} \tfrac{1}{4}R_1 \to R_1 \ \text{ then} \qquad \to \\ \qquad -6R_1 + R_2 \to R_2 \\ \qquad -R_1 + R_3 \to R_3 \\ \qquad 3R_1 + R_4 \to R_4 \end{array}$$

$$\begin{bmatrix} \tfrac{1}{4} & 1 & 0 & \tfrac{1}{4} & 0 & 0 & 0 & 3 \\ \tfrac{3}{2} & 0 & 4 & -\tfrac{3}{2} & 1 & 0 & 0 & 30 \\ -\tfrac{1}{4} & 0 & \boxed{1} & -\tfrac{1}{4} & 0 & 1 & 0 & 5 \\ \hline -\tfrac{1}{4} & 0 & -1 & \tfrac{3}{4} & 0 & 0 & 1 & 9 \end{bmatrix} \quad \begin{array}{l} -4R_3 + R_2 \to R_2 \\ \qquad \to \\ R_3 + R_4 \to R_4 \end{array}$$

$$\begin{bmatrix} \tfrac{1}{4} & 1 & 0 & \tfrac{1}{4} & 0 & 0 & 0 & 3 \\ \boxed{\tfrac{5}{2}} & 0 & 0 & -\tfrac{1}{2} & 1 & -4 & 0 & 10 \\ -\tfrac{1}{4} & 0 & 1 & -\tfrac{1}{4} & 0 & 1 & 0 & 5 \\ \hline -\tfrac{1}{2} & 0 & 0 & \tfrac{1}{2} & 0 & 1 & 1 & 14 \end{bmatrix} \quad \tfrac{2}{5}R_2 \to R_2 \ \text{ then}$$

$$\begin{array}{l} -\tfrac{1}{4}R_2 + R_1 \to R_1 \\ \qquad \to \\ \tfrac{1}{4}R_2 + R_3 \to R_3 \\ \tfrac{1}{2}R_2 + R_4 \to R_4 \end{array} \quad \begin{bmatrix} 0 & 1 & 0 & \tfrac{3}{10} & -\tfrac{1}{10} & \tfrac{2}{5} & 0 & 2 \\ 1 & 0 & 0 & -\tfrac{1}{5} & \tfrac{2}{5} & -\tfrac{8}{5} & 0 & 4 \\ 0 & 0 & 1 & -\tfrac{3}{10} & \tfrac{1}{10} & \tfrac{3}{5} & 0 & 6 \\ \hline 0 & 0 & 0 & \tfrac{2}{5} & \tfrac{1}{5} & \tfrac{1}{5} & 1 & 16 \end{bmatrix}$$

Minimum: $g = 16$ when $y_1 = \dfrac{2}{5}, y_2 = \dfrac{1}{5},$ and $y_3 = \dfrac{1}{5}.$

15. a. Minimize: $g = 120y_1 + 50y_2$

Constraints: $3y_1 + y_2 \ge 40, \quad 2y_1 + y_2 \ge 20$

b.
$$\begin{bmatrix} \boxed{3} & 2 & 1 & 0 & 0 & 120 \\ 1 & 1 & 0 & 1 & 0 & 50 \\ \hline -40 & -20 & 0 & 0 & 1 & 0 \end{bmatrix} \quad \begin{array}{l} \tfrac{1}{3}R_1 \to R_1 \\ \qquad \to \end{array}$$

$$\begin{bmatrix} \boxed{1} & \tfrac{2}{3} & \tfrac{1}{3} & 0 & 0 & 40 \\ 1 & 1 & 0 & 1 & 0 & 50 \\ \hline -40 & -20 & 0 & 0 & 1 & 0 \end{bmatrix} \quad \begin{array}{l} \to \\ -R_1 + R_2 \to R_2 \\ 40R_1 + R_3 \to R_3 \end{array} \quad \begin{bmatrix} 1 & \tfrac{2}{3} & \tfrac{1}{3} & 0 & 0 & 40 \\ 0 & \tfrac{1}{3} & -\tfrac{1}{3} & 1 & 0 & 10 \\ \hline 0 & \tfrac{20}{3} & \tfrac{40}{3} & 0 & 1 & 1600 \end{bmatrix}$$

Primal: $f = 1600$ (max) at $x_1 = 40, x_2 = 0$ \qquad Dual: $g = 1600$ (min) at $y_1 = \dfrac{40}{3}, y_2 = 0$

17.

$$\begin{bmatrix} 1 & 2 & 1 & | & 16 \\ 1 & 5 & 2 & | & 18 \\ 2 & 5 & 3 & | & 38 \\ \hline 40 & 90 & 30 & | & g \end{bmatrix} \quad \text{transpose:} \quad \begin{bmatrix} 1 & 1 & 2 & | & 40 \\ 2 & 5 & 5 & | & 90 \\ 1 & 2 & 3 & | & 30 \\ \hline 16 & 18 & 38 & | & g \end{bmatrix}$$

$$\begin{bmatrix} 1 & 1 & 2 & 1 & 0 & 0 & 0 & | & 40 \\ 2 & 5 & 5 & 0 & 1 & 0 & 0 & | & 90 \\ 1 & 2 & \boxed{3} & 0 & 0 & 1 & 0 & | & 30 \\ \hline -16 & -18 & -38 & 0 & 0 & 0 & 1 & | & 0 \end{bmatrix} \quad \tfrac{1}{3}R_3 \to R_3 \text{ then}$$

$$\begin{array}{l} -2R_3 + R_1 \to R_1 \\ -5R_3 + R_2 \to R_2 \\ \qquad \to \\ 38R_3 + R_4 \to R_4 \end{array}$$

$$\begin{bmatrix} \tfrac{1}{3} & -\tfrac{1}{3} & 0 & 1 & 0 & -\tfrac{2}{3} & 0 & | & 20 \\ \tfrac{1}{3} & \tfrac{5}{3} & 0 & 0 & 1 & -\tfrac{5}{3} & 0 & | & 40 \\ \boxed{\tfrac{1}{3}} & \tfrac{2}{3} & 1 & 0 & 0 & \tfrac{1}{3} & 0 & | & 10 \\ \hline -\tfrac{10}{3} & \tfrac{22}{3} & 0 & 0 & 0 & \tfrac{38}{3} & 1 & | & 380 \end{bmatrix} \quad 3R_3 \to R_3 \text{ then}$$

$$\begin{array}{l} -\tfrac{1}{3}R_3 + R_1 \to R_1 \\ -\tfrac{1}{3}R_3 + R_2 \to R_2 \\ \qquad \to \\ \tfrac{10}{3}R_3 + R_4 \to R_4 \end{array} \quad \begin{bmatrix} 0 & -1 & -1 & 1 & 0 & -1 & 0 & | & 10 \\ 0 & 1 & -1 & 0 & 1 & -2 & 0 & | & 30 \\ 1 & 2 & 3 & 0 & 0 & 1 & 0 & | & 30 \\ \hline 0 & 14 & 10 & 0 & 0 & 16 & 1 & | & 480 \end{bmatrix}$$

Minimum is 480 at $y_1 = 0$, $y_2 = 0$, $y_3 = 16$.

19.

$$\begin{bmatrix} 40 & 20 & 15 & 50 & | & 50 \\ 3 & 2 & 0 & 5 & | & 6 \\ 2 & 2 & 4 & 4 & | & 10 \\ 2 & 4 & 1 & 5 & | & 8 \\ \hline 50 & 20 & 30 & 80 & | & g \end{bmatrix} \quad \text{transpose:} \quad \begin{bmatrix} 40 & 3 & 2 & 2 & | & 50 \\ 20 & 2 & 2 & 4 & | & 20 \\ 15 & 0 & 4 & 1 & | & 30 \\ 50 & 5 & 4 & 5 & | & 80 \\ \hline 50 & 6 & 10 & 8 & | & g \end{bmatrix}$$

$$\begin{bmatrix} 40 & 3 & 2 & 2 & 1 & 0 & 0 & 0 & 0 & | & 50 \\ \boxed{20} & 2 & 2 & 4 & 0 & 1 & 0 & 0 & 0 & | & 20 \\ 15 & 0 & 4 & 1 & 0 & 0 & 1 & 0 & 0 & | & 30 \\ 50 & 5 & 4 & 5 & 0 & 0 & 0 & 1 & 0 & | & 80 \\ \hline -50 & -6 & -10 & -8 & 0 & 0 & 0 & 0 & 1 & | & 0 \end{bmatrix} \quad \tfrac{1}{20}R_2 \to R_2 \text{ then}$$

$$\begin{array}{l} -40R_2 + R_1 \to R_1 \\ \qquad \to \\ -15R_2 + R_3 \to R_3 \\ -50R_2 + R_4 \to R_4 \\ 50R_2 + R_5 \to R_5 \end{array}$$

$$\begin{bmatrix} 0 & -1 & -2 & -6 & 1 & -2 & 0 & 0 & 0 & | & 10 \\ 1 & 0.1 & 0.1 & 0.2 & 0 & 0.05 & 0 & 0 & 0 & | & 1 \\ 0 & -1.5 & \boxed{2.5} & -2 & 0 & -0.75 & 1 & 0 & 0 & | & 15 \\ 0 & 0 & -1 & -5 & 0 & -2.5 & 0 & 1 & 0 & | & 30 \\ \hline 0 & -1 & -5 & 2 & 0 & 2.5 & 0 & 0 & 1 & | & 50 \end{bmatrix} \quad \tfrac{2}{5}R_3 \to R_3 \text{ then}$$

$$\begin{array}{l} 2R_3 + R_1 \to R_1 \\ -0.1R_3 + R_2 \to R_2 \\ \qquad \to \\ R_3 + R_4 \to R_4 \\ 5R_3 + R_5 \to R_5 \end{array}$$

$$\begin{bmatrix} 0 & -2.2 & 0 & -7.6 & 1 & -2.6 & 0.8 & 0 & 0 & | & 22 \\ 1 & \boxed{0.16} & 0 & 0.28 & 0 & 0.08 & -0.04 & 0 & 0 & | & 0.4 \\ 0 & -0.6 & 1 & -0.8 & 0 & -0.3 & 0.4 & 0 & 0 & | & 6 \\ 0 & -0.6 & 0 & -5.8 & 0 & -2.8 & 0.4 & 1 & 0 & | & 36 \\ \hline 0 & -4 & 0 & -2 & 0 & 1 & 2 & 0 & 1 & | & 80 \end{bmatrix} \quad \tfrac{25}{4}R_2 \to R_2 \text{ then}$$

$$\begin{array}{l} 2.2R_2 + R_1 \to R_1 \\ \qquad \to \\ 0.6R_2 + R_3 \to R_3 \\ 0.6R_2 + R_4 \to R_4 \\ 4R_2 + R_5 \to R_5 \end{array}$$

$$\begin{bmatrix} 13.75 & 0 & 0 & -3.75 & 1 & -1.5 & 0.25 & 0 & 0 & | & 27.5 \\ 6.25 & 1 & 0 & 1.75 & 0 & 0.5 & -0.25 & 0 & 0 & | & 2.5 \\ 3.75 & 0 & 1 & 0.25 & 0 & 0 & 0.25 & 0 & 0 & | & 7.5 \\ 3.75 & 0 & 0 & -4.75 & 0 & -2.5 & 0.25 & 1 & 0 & | & 37.5 \\ \hline 25 & 0 & 0 & 5 & 0 & 3 & 1 & 0 & 1 & | & 90 \end{bmatrix}$$

Minimum is 90 at $y_1 = 0$, $y_2 = 3$, $y_3 = 1$, $y_4 = 0$.

21. Let $y_1 =$ hours per week to operate the Atlanta plant, $y_2 =$ hours per week to operate the Ft. Worth plant

Minimize: $g = 700y_1 + 2100y_2$

Constraints: $160y_1 + 800y_2 \geq 64000$, $200y_1 + 200y_2 \geq 40000$, $y_1 \geq 0$, $y_2 \geq 0$

$$\begin{bmatrix} 160 & 800 & | & 64000 \\ 200 & 200 & | & 40000 \\ \hline 700 & 2100 & | & g \end{bmatrix} \quad \text{transpose:} \quad \begin{bmatrix} 160 & 200 & | & 700 \\ 800 & 200 & | & 2100 \\ \hline 64000 & 40000 & | & g \end{bmatrix}$$

$$\begin{bmatrix} 160 & 200 & 1 & 0 & 0 & | & 700 \\ \boxed{800} & 200 & 0 & 1 & 0 & | & 2100 \\ \hline -64000 & -40000 & 0 & 0 & 1 & | & 0 \end{bmatrix} \quad \frac{1}{800}R_2 \to R_2 \text{ then}$$

$$-160R_2 + R_1 \to R_1$$
$$\to$$
$$64000R_2 + R_3 \to R_3$$

$$\begin{bmatrix} 0 & \boxed{160} & 1 & -0.2 & 0 & | & 280 \\ 1 & 0.25 & 0 & 0.00125 & 0 & | & 2.625 \\ \hline 0 & -24000 & 0 & 80 & 1 & | & 168000 \end{bmatrix} \quad \frac{1}{160}R_1 \to R_1 \text{ then}$$

$$\to$$
$$-0.25R_1 + R_2 \to R_2$$
$$24000R_1 + R_3 \to R_3$$

$$\begin{bmatrix} 0 & 1 & 0.00625 & -0.00125 & 0 & | & 1.75 \\ 1 & 0 & -0.0015625 & 0.0015625 & 0 & | & 2.1875 \\ \hline 0 & 0 & 150 & 50 & 1 & | & 210,000 \end{bmatrix}$$

Minimum cost is \$210,000 per week, when the Atlanta plant is operated for 150 hours and the Ft. Worth plant is operated for 50 hours.

23. $x =$ hours for Line 1, $y =$ hours for Line 2

Minimize: $C = 200x + 400y$

Constraints: $30x + 150y \geq 270$, $40x + 40y \geq 200$

Equivalent constraints: $x + 5y \geq 9$, $x + y \geq 5$

$$\begin{bmatrix} 1 & 5 & | & 9 \\ 1 & 1 & | & 5 \\ \hline 200 & 400 & | & g \end{bmatrix} \quad \text{transpose:} \quad \begin{bmatrix} 1 & 1 & | & 200 \\ 5 & 1 & | & 400 \\ \hline 9 & 5 & | & g \end{bmatrix}$$

$$\begin{bmatrix} 1 & 1 & 1 & 0 & 0 & | & 200 \\ \boxed{5} & 1 & 0 & 1 & 0 & | & 400 \\ \hline -9 & -5 & 0 & 0 & 1 & | & 0 \end{bmatrix} \quad \frac{1}{5}R_2 \to R_2 \text{ then}$$

$$-R_2 + R_1 \to R_1$$
$$\to$$
$$9R_2 + R_3 \to R_3$$

$$\begin{bmatrix} 0 & \boxed{\frac{4}{5}} & 1 & -\frac{1}{5} & 0 & | & 120 \\ 1 & \frac{1}{5} & 0 & \frac{1}{5} & 0 & | & 80 \\ \hline 0 & -\frac{16}{5} & 0 & \frac{9}{5} & 1 & | & 720 \end{bmatrix} \quad -\frac{5}{4}R_1 \to R_1 \text{ then}$$

$$\to$$
$$-\frac{1}{5}R_1 + R_2 \to R_2$$
$$\frac{16}{5}R_1 + R_3 \to R_3$$

$$\begin{bmatrix} 0 & 1 & \frac{5}{4} & -\frac{1}{4} & 0 & | & 150 \\ 1 & 0 & -\frac{1}{4} & \frac{1}{4} & 0 & | & 50 \\ \hline 0 & 0 & 4 & 1 & 1 & | & 1200 \end{bmatrix}$$

The minimum cost is \$1200 for 4 hours on Line 1, and 1 hour on Line 2.

25. Minimize: $\text{Cost} = 1000A + 3000B + 4000C$

Constraints: $200A + 200B + 400C \geq 2000 \rightarrow A + B + 2C \geq 10$

$\qquad\qquad\quad 100A + 200B + 100C \geq 1200 \rightarrow A + 2B + C \geq 12$

$$\begin{bmatrix} 1 & 1 & 2 & | & 10 \\ 1 & 2 & 1 & | & 12 \\ \hline 1000 & 3000 & 4000 & | & g \end{bmatrix} \quad \text{transpose:} \quad \begin{bmatrix} 1 & 1 & | & 1000 \\ 1 & 2 & | & 3000 \\ 2 & 1 & | & 4000 \\ \hline 10 & 12 & | & g \end{bmatrix}$$

$$\begin{bmatrix} 1 & \boxed{1} & 1 & 0 & 0 & 0 & | & 1000 \\ 1 & 2 & 0 & 1 & 0 & 0 & | & 3000 \\ 2 & 1 & 0 & 0 & 1 & 0 & | & 4000 \\ \hline -10 & -12 & 0 & 0 & 0 & 1 & | & 0 \end{bmatrix} \quad \begin{array}{c} \rightarrow \\ -2R_1 + R_2 \rightarrow R_2 \\ -R_1 + R_3 \rightarrow R_3 \\ 12R_1 + R_4 \rightarrow R_4 \end{array} \quad \begin{bmatrix} 1 & 1 & 1 & 0 & 0 & 0 & | & 1000 \\ -1 & 0 & -2 & 1 & 0 & 0 & | & 1000 \\ 1 & 0 & -1 & 0 & 1 & 0 & | & 3000 \\ \hline 2 & 0 & 12 & 0 & 0 & 1 & | & 12000 \end{bmatrix}$$

The minimum cost of \$12,000 is obtained by using facility A for 12 weeks, and 0 weeks for B and C.

27. $x =$ days for Factory 1; $y =$ days for Factory 2

Minimize: $C = 10,000x + 20,000y$

Constraints: $80x + 20y \geq 1600 \rightarrow 4x + y \geq 80$

$\qquad\qquad\quad 10x + 10y \geq 500 \rightarrow x + y \geq 50$

$\qquad\qquad\quad 50x + 20y \geq 1900 \rightarrow 5x + 2y \geq 190$

$$\begin{bmatrix} 4 & 1 & | & 80 \\ 1 & 1 & | & 50 \\ 5 & 2 & | & 190 \\ \hline 10000 & 20000 & | & g \end{bmatrix} \quad \text{transpose:} \quad \begin{bmatrix} 4 & 1 & 5 & | & 10000 \\ 1 & 1 & 2 & | & 20000 \\ \hline 80 & 50 & 190 & | & g \end{bmatrix}$$

$$\begin{bmatrix} 4 & 1 & \boxed{5} & 1 & 0 & 0 & | & 10,000 \\ 1 & 1 & 2 & 0 & 1 & 0 & | & 20,000 \\ \hline -80 & -50 & -190 & 0 & 0 & 1 & | & 0 \end{bmatrix} \quad \begin{array}{c} \frac{1}{5}R_1 \rightarrow R_1 \text{ then} \\[4pt] -2R_1 + R_2 \rightarrow R_2 \\ 190R_1 + R_3 \rightarrow R_3 \end{array}$$

$$\begin{bmatrix} \frac{4}{5} & \boxed{\frac{1}{5}} & 1 & \frac{1}{5} & 0 & 0 & | & 2000 \\ -\frac{3}{5} & \frac{3}{5} & 0 & -\frac{2}{5} & 1 & 0 & | & 16,000 \\ \hline 72 & -12 & 0 & 38 & 0 & 1 & | & 380,000 \end{bmatrix} \quad \begin{array}{c} 5R_1 \rightarrow R_1 \text{ then} \quad -\frac{3}{5}R_1 + R_2 \rightarrow R_2 \\[4pt] \rightarrow \\ 12R_1 + R_3 \rightarrow R_3 \end{array}$$

$$\begin{bmatrix} 4 & 1 & 5 & 1 & 0 & 0 & | & 10,000 \\ -3 & 0 & -3 & -1 & 1 & 0 & | & 10,000 \\ \hline 120 & 0 & 60 & 50 & 0 & 1 & | & 500,000 \end{bmatrix}$$

Minimum cost is \$500,000 using 50 days at Factory 1 and 0 days at Factory 2.

29. Minimize: $C = 500T + 100R$

Constraints: $0.9T + 0.6R \geq 63$

$T + R \geq 90$

$$\begin{bmatrix} 0.9 & 0.6 & | & 63 \\ 1 & 1 & | & 90 \\ \hline 500 & 100 & | & g \end{bmatrix} \quad \text{transpose:} \quad \begin{bmatrix} 0.9 & 1 & | & 500 \\ 0.6 & 1 & | & 100 \\ \hline 63 & 90 & | & g \end{bmatrix}$$

$$\begin{bmatrix} 0.9 & 1 & 1 & 0 & 0 & | & 500 \\ 0.6 & \boxed{1} & 0 & 1 & 0 & | & 100 \\ \hline -63 & -90 & 0 & 0 & 1 & | & 0 \end{bmatrix} \quad \begin{array}{l} -R_2 + R_1 \rightarrow R_1 \\ \rightarrow \\ 90R_2 + R_3 \rightarrow R_3 \end{array}$$

$$\begin{bmatrix} 0.3 & 0 & 1 & -1 & 0 & | & 400 \\ \boxed{0.6} & 1 & 0 & 1 & 0 & | & 100 \\ \hline -9 & 0 & 0 & 90 & 1 & | & 9000 \end{bmatrix} \quad \begin{array}{l} -\frac{3}{10}R_2 + R_1 \rightarrow R_1 \\ \tfrac{5}{3}R_2 \rightarrow R_2 \quad \text{then} \quad \rightarrow \\ 9R_2 + R_3 \rightarrow R_3 \end{array} \quad \begin{bmatrix} 0 & -\frac{1}{2} & 1 & -\frac{3}{2} & 0 & | & 350 \\ 1 & \frac{5}{3} & 0 & \frac{5}{3} & 0 & | & \frac{500}{3} \\ \hline 0 & 15 & 0 & 105 & 1 & | & 10{,}500 \end{bmatrix}$$

Minimum cost is \$10,500 using 0 minutes of TV time and 105 minutes of radio time.

31. Minimize: $\text{Cost} = 120x_1 + 140x_2 + 126x_3$

Constraints: $10x_1 + 5x_2 + 6x_3 \geq 230$

$\qquad\qquad 5x_1 + 8x_2 + 6x_3 \geq 240$

$\qquad\qquad 6x_1 + 6x_2 + 6x_3 \geq 210 \quad \rightarrow \quad x_1 + x_2 + x_3 \geq 35$

$$
\left[
\begin{array}{ccccccc|c}
10 & 5 & 1 & 1 & 0 & 0 & 0 & 120 \\
5 & 8 & 1 & 0 & 1 & 0 & 0 & 140 \\
6 & 6 & 1 & 0 & 0 & 1 & 0 & 126 \\
\hline
-230 & -240 & -35 & 0 & 0 & 0 & 1 & 0
\end{array}
\right]
\quad \begin{array}{l} \rightarrow \\[6pt] \frac{1}{8}R_2 \rightarrow R_2 \end{array}
$$

$$
\left[
\begin{array}{ccccccc|c}
10 & 5 & 1 & 1 & 0 & 0 & 0 & 120 \\
0.625 & 1 & 0.125 & 0 & 0.125 & 0 & 0 & 17.5 \\
6 & 6 & 1 & 0 & 0 & 1 & 0 & 126 \\
\hline
-230 & -240 & -35 & 0 & 0 & 0 & 1 & 0
\end{array}
\right]
\quad \begin{array}{l} -5R_2 + R_1 \rightarrow R_1 \\[3pt] \rightarrow \\[3pt] -6R_2 + R_3 \rightarrow R_3 \\[3pt] 240R_2 + R_4 \rightarrow R_4 \end{array}
$$

$$
\left[
\begin{array}{ccccccc|c}
6.875 & 0 & 0.375 & 1 & -0.625 & 0 & 0 & 32.5 \\
0.625 & 1 & 0.125 & 0 & 0.125 & 0 & 0 & 17.5 \\
2.250 & 0 & 0.250 & 0 & -0.750 & 1 & 0 & 21 \\
\hline
-80 & 0 & -5 & 0 & 30 & 0 & 1 & 4200
\end{array}
\right]
\quad \begin{array}{l} \frac{1}{6.875}R_1 \rightarrow R_1 \\[6pt] \rightarrow \end{array}
$$

$$
\left[
\begin{array}{ccccccc|c}
1 & 0 & 0.0545 & 0.1455 & -0.0909 & 0 & 0 & 4.7273 \\
0.625 & 1 & 0.125 & 0 & 0.125 & 0 & 0 & 17.5 \\
2.250 & 0 & 0.250 & 0 & -0.750 & 1 & 0 & 21 \\
\hline
-80 & 0 & -5 & 0 & 30 & 0 & 1 & 4200
\end{array}
\right]
\quad \begin{array}{l} \rightarrow \\[3pt] -0.625R_1 + R_2 \rightarrow R_2 \\[3pt] -2.25R_1 + R_3 \rightarrow R_3 \\[3pt] 80R_1 + R_4 \rightarrow R_4 \end{array}
$$

$$
\left[
\begin{array}{ccccccc|c}
1 & 0 & 0.0545 & 0.1455 & -0.0909 & 0 & 0 & 4.7273 \\
0 & 1 & 0.0909 & -0.0909 & 0.1818 & 0 & 0 & 14.5455 \\
0 & 0 & 0.1273 & -0.3273 & -0.5455 & 1 & 0 & 10.3636 \\
\hline
0 & 0 & -0.6363 & 11.6364 & 22.7273 & 0 & 1 & 4578.1818
\end{array}
\right]
\quad \begin{array}{l} \rightarrow \\[6pt] \frac{1}{0.1273}R_3 \rightarrow R_3 \end{array}
$$

$$
\left[
\begin{array}{ccccccc|c}
1 & 0 & 0.0545 & 0.1455 & -0.0909 & 0 & 0 & 4.7273 \\
0 & 1 & 0.0909 & -0.0909 & 0.1818 & 0 & 0 & 14.5455 \\
0 & 0 & 1 & -2.5709 & -4.2848 & 7.8558 & 0 & 81.4111 \\
\hline
0 & 0 & -0.6363 & 11.6364 & 22.7273 & 0 & 1 & 4578.1818
\end{array}
\right]
\quad \begin{array}{l} -0.0545R_3 + R_1 \rightarrow R_1 \\[3pt] -0.0909R_3 + R_2 \rightarrow R_2 \\[3pt] \rightarrow \\[3pt] 0.6363R_3 + R_4 \rightarrow R_4 \end{array}
$$

$$
\left[
\begin{array}{ccccccc|c}
1 & 0 & 0 & 0.2856 & 0.1426 & -0.4281 & 0 & 0.2904 \\
0 & 1 & 0 & 0.1428 & 0.5713 & -0.7141 & 0 & 7.1452 \\
0 & 0 & 1 & -2.5709 & -4.2848 & 7.8555 & 0 & 81.4111 \\
\hline
0 & 0 & 0 & 10.0005 & 20.0009 & 4.9984 & 1 & 4629.98
\end{array}
\right]
$$

a. DeTurris should buy 10 Georgia packages, 20 Union packages, and 5 Pacific packages.

b. The minimum possible cost is $4630.

33. x = ounces of Food I, y = ounces of Food II, z = ounces of Food III

Minimize: $g = x + 5y + 2z$

Constraints: $2x + 2y + 2z \geq 24$, $x + 5y + z \geq 16$

Equivalent constraints: $x + y + z \geq 12$, $x + 5y + z \geq 16$

$$\begin{bmatrix} 1 & 1 & 1 & | & 12 \\ 1 & 5 & 1 & | & 16 \\ 1 & 5 & 2 & | & g \end{bmatrix} \quad \text{transpose:} \quad \begin{bmatrix} 1 & 1 & | & 1 \\ 1 & 5 & | & 5 \\ 1 & 1 & | & 2 \\ 12 & 16 & | & g \end{bmatrix}$$

$$\begin{bmatrix} 1 & \boxed{1} & 1 & 0 & 0 & 0 & | & 1 \\ 1 & 5 & 0 & 1 & 0 & 0 & | & 5 \\ 1 & 1 & 0 & 0 & 1 & 0 & | & 2 \\ -12 & -16 & 0 & 0 & 0 & 1 & | & 0 \end{bmatrix} \quad \begin{array}{c} \rightarrow \\ -5R_1 + R_2 \rightarrow R_2 \\ -R_1 + R_3 \rightarrow R_3 \\ 16R_1 + R_4 \rightarrow R_4 \end{array} \quad \begin{bmatrix} 1 & 1 & 1 & 0 & 0 & 0 & | & 1 \\ -4 & 0 & -5 & 1 & 0 & 0 & | & 0 \\ 0 & 0 & -1 & 0 & 1 & 0 & | & 1 \\ 4 & 0 & 16 & 0 & 0 & 1 & | & 16 \end{bmatrix}$$

a. Minimum possible cost is \$16.

b. 16 ounces of Food I and 0 ounces of Food II and Food III should be served to give the minimum cost (NOTE: This is one of several possible solutions, another is 11 oz of Food I, 1 oz of Food II, and 0 oz of Food III.)

35. Let y_1 = number of workers who begin on Monday, y_2 = number of workers who begin on Tuesday, etc.

Minimize: $g = y_1 + y_2 + y_3 + y_4 + y_5 + y_6 + y_7$

Constraints: $y_1 + y_2 + y_3 + y_4 + y_5 \geq 22$

$y_2 + y_3 + y_4 + y_5 + y_6 \geq 12$

$y_3 + y_4 + y_5 + y_6 + y_7 \geq 17$

$y_1 + y_4 + y_5 + y_6 + y_7 \geq 20$

$y_1 + y_2 + y_5 + y_6 + y_7 \geq 16$

$y_1 + y_2 + y_3 + y_6 + y_7 \geq 16$

$y_1 + y_2 + y_3 + y_4 + y_7 \geq 20$

$$\begin{bmatrix} 1 & 0 & 0 & 1 & 1 & 1 & 1 & 1 & 0 & 0 & 0 & 0 & 0 & 0 & 0 & | & 1 \\ 1 & 1 & 0 & 0 & 1 & 1 & 1 & 0 & 1 & 0 & 0 & 0 & 0 & 0 & 0 & | & 1 \\ 1 & 1 & 1 & 0 & 0 & 1 & 1 & 0 & 0 & 1 & 0 & 0 & 0 & 0 & 0 & | & 1 \\ 1 & 1 & 1 & 1 & 0 & 0 & 1 & 0 & 0 & 0 & 1 & 0 & 0 & 0 & 0 & | & 1 \\ 1 & 1 & 1 & 1 & 1 & 0 & 0 & 0 & 0 & 0 & 0 & 1 & 0 & 0 & 0 & | & 1 \\ 0 & 1 & 1 & 1 & 1 & 1 & 0 & 0 & 0 & 0 & 0 & 0 & 1 & 0 & 0 & | & 1 \\ 0 & 0 & 1 & 1 & 1 & 1 & 1 & 0 & 0 & 0 & 0 & 0 & 0 & 1 & 0 & | & 1 \\ \hline -22 & -12 & -17 & -20 & -16 & -16 & -20 & 0 & 0 & 0 & 0 & 0 & 0 & 0 & 1 & | & 0 \end{bmatrix}$$

Solve using technology: The minimum number of workers to staff the depot is 25 when 8 employees begin their shift on Monday, none on Tuesday, 5 on Wednesday, 4 on Thursday, 5 on Friday, none on Saturday, and 3 on Sunday.

Exercise 4.5

1. $3x - y \geq 5$

$-3x + y \leq -5$

3. $y \geq 40 - 6x$

$-6x - y \leq -40$

5.   Maximize: $f = 2x + 3y$

Constraints: $7x + 4y \leq 28$

$-3x + y \geq 2$

$x, y \geq 0$

a. $f = 2x + 3y$

$7x + 4y \leq 28$

$3x - y \leq -2$

$x, y \geq 0$

b.
$$\begin{array}{ccccc} x & y & s_1 & s_2 & f \end{array}$$
$$\left[\begin{array}{ccccc|c} 7 & 4 & 1 & 0 & 0 & 28 \\ 3 & \boxed{-1} & 0 & 1 & 0 & -2 \\ \hline -2 & -3 & 0 & 0 & 1 & 0 \end{array} \right]$$

7.   Minimize: $g = 3x + 8y$

Constraints: $4x - 5y \leq 50$

$x + y \leq 80$

$-x + 2y \geq 4$

$x, y \geq 0$

a. $-g = -3x - 8y$ (to maximize)

$4x - 5y \leq 50$

$x + y \leq 80$

$x - 2y \leq -4$

$x, y \geq 0$

b.
$$\begin{array}{ccccccc} x & y & s_1 & s_2 & s_3 & -g \end{array}$$
$$\left[\begin{array}{cccccc|c} 4 & -5 & 1 & 0 & 0 & 0 & 50 \\ 1 & 1 & 0 & 1 & 0 & 0 & 80 \\ 1 & \boxed{-2} & 0 & 0 & 1 & 0 & -4 \\ \hline 3 & 8 & 0 & 0 & 0 & 1 & 0 \end{array} \right]$$

9.   The minimum is $f = 120$ at $x = 6$, $y = 8$, $z = 12$.

11.   Maximize: $f = 4x + y$

Constraints: $5x + 2y \leq 84$, $3x - 2y \leq -4$, $x, y \geq 0$

$$\left[\begin{array}{ccccc|c} 5 & 2 & 1 & 0 & 0 & 84 \\ 3 & \boxed{-2} & 0 & 1 & 0 & -4 \\ \hline -4 & -1 & 0 & 0 & 1 & 0 \end{array} \right] \quad -\frac{1}{2}R_2 \to R_2 \text{ then}$$

$$\begin{array}{c} -2R_2 + R_1 \to R_1 \\ \to \\ R_2 + R_3 \to R_3 \end{array}$$

$$\left[\begin{array}{ccccc|c} \boxed{8} & 0 & 1 & 1 & 0 & 80 \\ -\frac{3}{2} & 1 & 0 & -\frac{1}{2} & 0 & 2 \\ \hline -\frac{11}{2} & 0 & 0 & -\frac{1}{2} & 1 & 2 \end{array} \right] \quad \frac{1}{8}R_1 \to R_1 \text{ then}$$

$$\begin{array}{c} \to \\ \frac{3}{2}R_1 + R_2 \to R_2 \\ \frac{11}{2}R_1 + R_3 \to R_3 \end{array} \quad \left[\begin{array}{ccccc|c} 1 & 0 & \frac{1}{8} & \frac{1}{8} & 0 & 10 \\ 0 & 1 & \frac{3}{16} & -\frac{5}{16} & 0 & 17 \\ 0 & 0 & \frac{11}{16} & \frac{3}{16} & 1 & 57 \end{array} \right]$$

A maximum of 57 occurs when $x = 10$ and $y = 17$.

13.   Maximize: $-f = -2x - 3y$

Constraints: $-x \leq -5$, $y \leq 13$, $x - y \leq -2$, $x \geq 0$, $y \geq 0$

$$\left[\begin{array}{cccccc|c} \boxed{-1} & 0 & 1 & 0 & 0 & 0 & -5 \\ 0 & 1 & 0 & 1 & 0 & 0 & 13 \\ 1 & -1 & 0 & 0 & 1 & 0 & -2 \\ \hline 2 & 3 & 0 & 0 & 0 & 1 & 0 \end{array} \right] \quad -R_1 \to R_1 \text{ then}$$

$$\begin{array}{c} \to \\ \to \\ -R_1 + R_3 \to R_3 \\ -2R_1 + R_4 \to R_4 \end{array}$$

$$\left[\begin{array}{cccccc|c} 1 & 0 & -1 & 0 & 0 & 0 & 5 \\ 0 & 1 & 0 & 1 & 0 & 0 & 13 \\ 0 & \boxed{-1} & 1 & 0 & 1 & 0 & -7 \\ \hline 0 & 3 & 2 & 0 & 0 & 1 & -10 \end{array} \right] \quad -R_3 \to R_3$$

$$\begin{array}{c} \to \\ -R_3 + R_2 \to R_2 \\ \to \\ -3R_3 + R_4 \to R_4 \end{array} \quad \left[\begin{array}{cccccc|c} 1 & 0 & -1 & 0 & 0 & 0 & 5 \\ 0 & 0 & 1 & 1 & 1 & 0 & 6 \\ 0 & 1 & -1 & 0 & -1 & 0 & 7 \\ \hline 0 & 0 & 5 & 0 & 3 & 1 & -31 \end{array} \right]$$

Minimum is 31 at $x = 5$ and $y = 7$.

15. Minimize: $f = 3x + 2y$

Subject to: $-x + y \le 10$

$\qquad x + y \ge 20$

$\qquad x + y \le 35$

$\qquad x, y \ge 0$

Maximize: $-f = -3x - 2y$

Subject to: $-x + y \le 10$

$\qquad -x - y \le -20$

$\qquad x + y \le 35$

$\qquad x, y \ge 0$

$$\left[\begin{array}{cccccc|c} -1 & 1 & 1 & 0 & 0 & 0 & 10 \\ \boxed{-1} & -1 & 0 & 1 & 0 & 0 & -20 \\ 1 & 1 & 0 & 0 & 1 & 0 & 35 \\ \hline -3 & -2 & 0 & 0 & 0 & 1 & 0 \end{array}\right] \begin{array}{l} \\ -R_2 \to R_2 \ \text{ then} \\ \\ \\ \end{array}$$

$$\begin{array}{l} R_2 + R_1 \to R_1 \\ \\ \to \\ \\ -R_2 + R_3 \to R_3 \\ -3R_2 + R_4 \to R_4 \end{array}$$

$$\left[\begin{array}{cccccc|c} 0 & \boxed{2} & 1 & -1 & 0 & 0 & 30 \\ 1 & 1 & 0 & -1 & 0 & 0 & 20 \\ 0 & 0 & 0 & 1 & 1 & 0 & 15 \\ \hline 0 & -1 & 0 & 3 & 0 & 1 & -60 \end{array}\right] \begin{array}{l} \tfrac{1}{2}R_1 \to R_1 \ \text{ then} \\ \\ -R_1 + R_2 \to R_2 \\ \\ \to \\ \\ R_1 + R_4 \to R_4 \end{array}$$

$$\begin{array}{l} \to \end{array} \left[\begin{array}{cccccc|c} 0 & 1 & \tfrac{1}{2} & -\tfrac{1}{2} & 0 & 0 & 15 \\ 1 & 0 & -\tfrac{1}{2} & -\tfrac{1}{2} & 0 & 0 & 5 \\ 0 & 0 & 0 & 1 & 1 & 0 & 15 \\ \hline 0 & 0 & \tfrac{1}{2} & \tfrac{5}{2} & 0 & 1 & -45 \end{array}\right]$$

Maximum of 45 occurs when $x = 5$ and $y = 15$.

17. Maximize: $f = 2x + 5y$

Constraints: $-x + y \le 10,\ \ x + y \le 30,\ \ 2x - y \le 24,\ \ x, y \ge 0$

$$\left[\begin{array}{cccccc|c} -1 & \boxed{1} & 1 & 0 & 0 & 0 & 10 \\ 1 & 1 & 0 & 1 & 0 & 0 & 30 \\ 2 & -1 & 0 & 0 & 1 & 0 & 24 \\ \hline -2 & -5 & 0 & 0 & 0 & 1 & 0 \end{array}\right] \begin{array}{l} \to \\ \\ -R_1 + R_2 \to R_2 \\ R_1 + R_3 \to R_3 \\ 5R_1 + R_4 \to R_4 \end{array}$$

$$\left[\begin{array}{cccccc|c} -1 & 1 & 1 & 0 & 0 & 0 & 10 \\ \boxed{2} & 0 & -1 & 1 & 0 & 0 & 20 \\ 1 & 0 & 1 & 0 & 1 & 0 & 34 \\ \hline -7 & 0 & 5 & 0 & 0 & 1 & 50 \end{array}\right] \begin{array}{l} \tfrac{1}{2}R_2 \to R_2 \ \text{ then} \\ \\ \\ \end{array}$$

$$\begin{array}{l} R_2 + R_1 \to R_1 \\ \\ \to \\ \\ -R_2 + R_3 \to R_3 \\ 7R_2 + R_4 \to R_4 \end{array} \left[\begin{array}{cccccc|c} 0 & 1 & \tfrac{1}{2} & \tfrac{1}{2} & 0 & 0 & 20 \\ 1 & 0 & -\tfrac{1}{2} & \tfrac{1}{2} & 0 & 0 & 10 \\ 0 & 0 & \tfrac{3}{2} & -\tfrac{1}{2} & 1 & 0 & 24 \\ \hline 0 & 0 & \tfrac{3}{2} & \tfrac{7}{2} & 0 & 1 & 120 \end{array}\right]$$

Maximum is 120 at $x = 10, y = 20$.

19. Maximize: $-f = -x - 2y - 3z$

Constraints: $x + z \le 20,\ \ -x - y \le -30,\ \ y + z \le 20,\ \ x \ge 0,\ \ y \ge 0,\ \ z \ge 0$

$$\left[\begin{array}{ccccccc|c} \boxed{1} & 0 & 1 & 1 & 0 & 0 & 0 & 20 \\ -1 & -1 & 0 & 0 & 1 & 0 & 0 & -30 \\ 0 & 1 & 1 & 0 & 0 & 1 & 0 & 20 \\ \hline 1 & 2 & 3 & 0 & 0 & 0 & 1 & 0 \end{array}\right] \begin{array}{l} \to \\ \\ R_1 + R_2 \to R_2 \\ \to \\ -R_1 + R_4 \to R_4 \end{array}$$

$$\left[\begin{array}{ccccccc|c} 1 & 0 & 1 & 1 & 0 & 0 & 0 & 20 \\ 0 & \boxed{-1} & 1 & 1 & 1 & 0 & 0 & -10 \\ 0 & 1 & 1 & 0 & 0 & 1 & 0 & 20 \\ \hline 0 & 2 & 2 & -1 & 0 & 0 & 1 & -20 \end{array}\right] \begin{array}{l} -R_2 \to R_2 \ \text{ then} \\ \\ -R_2 + R_3 \to R_3 \\ -2R_2 + R_4 \to R_4 \end{array}$$

$$\begin{array}{l} \to \end{array} \left[\begin{array}{ccccccc|c} 1 & 0 & 1 & 1 & 0 & 0 & 0 & 20 \\ 0 & 1 & -1 & -1 & -1 & 0 & 0 & 10 \\ 0 & 0 & 2 & 1 & 1 & 1 & 0 & 10 \\ \hline 0 & 0 & 4 & 1 & 2 & 0 & 1 & -40 \end{array}\right]$$

Minimum is 40 at $x = 20, y = 10, z = 0$.

Note: recall basic variables have columns with a single entry of 1 and other entries, 0.

21. Maximize: $f = 2x - y + 4z$

Constraints: $x + y + z \leq 8$, $-x + y - z \leq -4$, $-x - y + z \leq -2$, $x, y, z \geq 0$

$$\begin{bmatrix} 1 & 1 & 1 & 1 & 0 & 0 & 0 & | & 8 \\ -1 & 1 & -1 & 0 & 1 & 0 & 0 & | & -4 \\ \boxed{-1} & -1 & 1 & 0 & 0 & 1 & 0 & | & -2 \\ \hline -2 & 1 & -4 & 0 & 0 & 0 & 1 & | & 0 \end{bmatrix}$$

$-R_3 \to R_3$ then $\to$

$-R_3 + R_1 \to R_1$
$R_3 + R_2 \to R_2$
$\to$
$2R_3 + R_4 \to R_4$

$$\begin{bmatrix} 0 & 0 & 2 & 1 & 0 & 1 & 0 & | & 6 \\ 0 & 2 & \boxed{-2} & 0 & 1 & -1 & 0 & | & -2 \\ 1 & 1 & -1 & 0 & 0 & -1 & 0 & | & 2 \\ \hline 0 & 3 & -6 & 0 & 0 & -2 & 1 & | & 4 \end{bmatrix}$$

$-\frac{1}{2} R_2 \to R_2$ then $\to$

$-2R_2 + R_1 \to R_1$
$\to$
$R_2 + R_3 \to R_3$
$6R_2 + R_4 \to R_4$

$$\begin{bmatrix} 0 & \boxed{2} & 0 & 1 & 1 & 0 & 0 & | & 4 \\ 0 & -1 & 1 & 0 & -\frac{1}{2} & \frac{1}{2} & 0 & | & 1 \\ 1 & 0 & 0 & 0 & -\frac{1}{2} & -\frac{1}{2} & 0 & | & 3 \\ \hline 0 & -3 & 0 & 0 & -3 & 1 & 1 & | & 10 \end{bmatrix}$$

$\frac{1}{2} R_1 \to R_1$ then $\to$

$R_1 + R_2 \to R_2$
$\to$
$3R_1 + R_4 \to R_4$

$$\begin{bmatrix} 0 & 1 & 0 & \frac{1}{2} & \boxed{\frac{1}{2}} & 0 & 0 & | & 2 \\ 0 & 0 & 1 & \frac{1}{2} & 0 & \frac{1}{2} & 0 & | & 3 \\ 1 & 0 & 0 & 0 & -\frac{1}{2} & -\frac{1}{2} & 0 & | & 3 \\ \hline 0 & 0 & 0 & \frac{3}{2} & -\frac{3}{2} & 1 & 1 & | & 16 \end{bmatrix}$$

$2R_1 \to R_1$ then $\to$

$\to$
$\frac{1}{2} R_1 + R_3 \to R_3$
$\frac{3}{2} R_1 + R_4 \to R_4$

$$\begin{bmatrix} 0 & 2 & 0 & 1 & 1 & 0 & 0 & | & 4 \\ 0 & 0 & 1 & \frac{1}{2} & 0 & \frac{1}{2} & 0 & | & 3 \\ 1 & 1 & 0 & \frac{1}{2} & 0 & -\frac{1}{2} & 0 & | & 5 \\ \hline 0 & 3 & 0 & 3 & 0 & 1 & 1 & | & 22 \end{bmatrix}$$

Maximum is 22 at $x = 5$, $y = 0$, $z = 3$.

23. Maximize: $-f = -10x - 30y - 35z$

Constraints: $x + y + z \leq 250$, $-x - y - 2z \leq -150$, $2x + y + z \leq 180$, $x \geq 0$, $y \geq 0$, $z \geq 0$

$$\begin{bmatrix} 1 & 1 & 1 & 1 & 0 & 0 & 0 & | & 250 \\ -1 & -1 & \boxed{-2} & 0 & 1 & 0 & 0 & | & -150 \\ 2 & 1 & 1 & 0 & 0 & 1 & 0 & | & 180 \\ \hline 10 & 30 & 35 & 0 & 0 & 0 & 1 & | & 0 \end{bmatrix}$$

$-\frac{1}{2} R_2 \to R_2$ then $\to$

$-R_2 + R_1 \to R_1$
$\to$
$-R_2 + R_3 \to R_3$
$-35R_2 + R_4 \to R_4$

$$\begin{bmatrix} \frac{1}{2} & \frac{1}{2} & 0 & 1 & \frac{1}{2} & 0 & 0 & | & 175 \\ \frac{1}{2} & \frac{1}{2} & 1 & 0 & -\frac{1}{2} & 0 & 0 & | & 75 \\ \boxed{\frac{3}{2}} & \frac{1}{2} & 0 & 0 & \frac{1}{2} & 1 & 0 & | & 105 \\ \hline -\frac{15}{2} & \frac{25}{2} & 0 & 0 & \frac{35}{2} & 0 & 1 & | & -2625 \end{bmatrix}$$

$\frac{2}{3} R_3 \to R_3$ then $\to$

$-\frac{1}{2} R_3 + R_1 \to R_1$
$-\frac{1}{2} R_3 + R_2 \to R_2$
$\to$
$\frac{15}{2} R_3 + R_4 \to R_4$

$$\begin{bmatrix} 0 & \frac{1}{3} & 0 & 1 & \frac{1}{3} & -\frac{1}{3} & 0 & | & 140 \\ 0 & \frac{1}{3} & 1 & 0 & -\frac{2}{3} & -\frac{1}{3} & 0 & | & 40 \\ 1 & \frac{1}{3} & 0 & 0 & \frac{1}{3} & \frac{2}{3} & 0 & | & 70 \\ \hline 0 & 15 & 0 & 0 & 20 & 5 & 1 & | & -2100 \end{bmatrix}$$

Minimum is 2100 at $x = 70$, $y = 0$, $z = 40$.

25. Maximize: $f = 25x_1 + 25x_2 + 25x_3 + 20x_4 + 20x_5 + 20x_6$

Constraints: $x_1 + x_4 \leq 100, \quad x_2 + x_5 \leq 20, \quad x_3 + x_6 \leq 30$

$$-0.8x_2 + 0.2x_1 + 0.2x_3 \leq 0$$
$$-0.9x_5 + 0.1x_4 + 0.1x_6 \leq 0$$
$$-0.9x_6 + 0.1x_4 + 0.1x_5 \leq 0$$

$$
\begin{bmatrix}
1 & 0 & 0 & 1 & 0 & 0 & 1 & 0 & 0 & 0 & 0 & 0 & 0 & 100 \\
0 & 1 & 0 & 0 & 1 & 0 & 0 & 1 & 0 & 0 & 0 & 0 & 0 & 20 \\
0 & 0 & 1 & 0 & 0 & 1 & 0 & 0 & 1 & 0 & 0 & 0 & 0 & 30 \\
0.2 & -0.8 & 0.2 & 0 & 0 & 0 & 0 & 0 & 0 & 1 & 0 & 0 & 0 & 0 \\
0 & 0 & 0 & 0.1 & -0.9 & 0.1 & 0 & 0 & 0 & 0 & 1 & 0 & 0 & 0 \\
0 & 0 & 0 & 0.1 & 0.1 & -0.9 & 0 & 0 & 0 & 0 & 0 & 1 & 0 & 0 \\
-25 & -25 & -25 & -20 & -20 & -20 & 0 & 0 & 0 & 0 & 0 & 0 & 1 & 0
\end{bmatrix}
$$

Steps: Pivot on 0.2 in R_4C_1. Row Ops: $5R_4 \to R_4$, then $-R + R_1 \to R_1$, $25R_4 + R_7 \to R_7$

Pivot on 1 in R_2C_2. Row Ops: $-4R_2 + R_1 \to R_1$, $4R_2 + R_4 \to R_4$, $125R_2 + R_7 \to R_7$

Pivot on 1 in R_5C_6. Row Ops: $10R_6 \to R_6$, then $-R_6 + R_1 \to R_1$, $-0.1R_6 + R_5 \to R_5$, $20R_6 + R_7 \to R_7$

Pivot on 4 in R_1C_5. Row Ops: $\frac{1}{4}R_1 \to R_1$, then $-R_1 + R_2 \to R_2$, $-R_1 + R_3 \to R_3$, $-4R_1 + R_4 \to R_4$

$$R_1 + R_5 \to R_5, \, 8R_1 + R_6 \to R_6, \, 75R_1 + R_7 \to R_7$$

Pivot on 1.25 in R_3C_3. Row Ops: $\frac{4}{3}R_3 \to R_3$, then $0.25R_3 + R_1 \to R_1$, $-0.25R_3 + R_2 \to R_2$,

$$-2R_3 + R_4 \to R_4, \, 0.25R_3 + R_5 \to R_5, \, 2R_3 + R_6 \to R_6,$$
$$18.75R_3 + R_7 \to R_7$$

$$
\begin{bmatrix}
0 & 0 & 0 & 0 & 1 & 0 & 0.2 & -0.8 & 0.2 & -1 & -2 & 0 & 0 & 10 \\
0 & 1 & 0 & 0 & 0 & 0 & -0.2 & 1.8 & -0.2 & 1 & 2 & 0 & 0 & 10 \\
0 & 0 & 1 & 0 & 0 & 0 & -0.2 & 0.8 & 0.8 & 1 & 1 & 1 & 0 & 20 \\
1 & 0 & 0 & 0 & 0 & 0 & -0.6 & 6.4 & -1.6 & 8 & 7 & -1 & 0 & 20 \\
0 & 0 & 0 & 0 & 0 & 1 & 0.2 & -0.8 & 0.2 & -1 & -1 & -1 & 0 & 10 \\
0 & 0 & 0 & 1 & 0 & 0 & 1.6 & -6 & 1.6 & -8 & -7 & 1 & 0 & 80 \\
0 & 0 & 0 & 0 & 0 & 0 & 15 & 65 & 15 & 50 & 50 & 0 & 1 & 3250
\end{bmatrix}
$$

Maximum is 3250 when $x_1 = 20$, $x_2 = 10$, $x_3 = 20$, $x_4 = 80$, $x_5 = 10$, $x_6 = 10$.

27. Let $x_1 = $ number of filters, $x_2 = $ number of housing units

Minimize: $f = 6.60x_1 + 8.35x_2$ Maximize: $-f = -6.60x_1 - 8.35x_2$

Constraints: $x_1 + x_2 \geq 500$ Constraints: $-x_1 - x_2 \leq -500$

$$20x_1 + 40x_2 \leq 20,000 \qquad\qquad 20x_1 + 40x_2 \leq 20,000$$
$$x_2 \leq 400 \qquad\qquad\qquad\qquad x_2 \leq 400$$

$$
\begin{bmatrix}
\boxed{-1} & -1 & 1 & 0 & 0 & 0 & -500 \\
20 & 40 & 0 & 1 & 0 & 0 & 20000 \\
0 & 1 & 0 & 0 & 1 & 0 & 400 \\
\hline
6.60 & 8.35 & 0 & 0 & 0 & 1 & 0
\end{bmatrix}
$$

$-1R_1 \to R_1$ then

$$\to$$
$$-20R_1 + R_2 \to R_2$$
$$\to$$
$$-6.60R_1 + R_4 \to R_4$$

$$
\begin{bmatrix}
1 & 1 & -1 & 0 & 0 & 0 & 500 \\
0 & 20 & 20 & 1 & 0 & 0 & 10000 \\
0 & 1 & 0 & 0 & 1 & 0 & 400 \\
\hline
0 & 1.75 & 6.6 & 0 & 0 & 1 & -3300
\end{bmatrix}
$$

Nolan Industries should produce 500 filters and no housing units for a weekly cost of \$3300.

29.

	All Beef	Regular	Available
Beef	$0.75x$	$0.18y$	≤ 1020
Spices	$0.2x$	$0.2y$	≥ 500
Pork	--	$0.3y$	≤ 600

Objective function: $f = 0.6x + 0.4y$

$$\begin{bmatrix} \boxed{0.75} & 0.18 & 1 & 0 & 0 & 0 & | & 1020 \\ -0.20 & -0.20 & 0 & 1 & 0 & 0 & | & -500 \\ 0 & 0.3 & 0 & 0 & 1 & 0 & | & 600 \\ \hline -0.60 & -0.40 & 0 & 0 & 0 & 1 & | & 0 \end{bmatrix}$$

$\frac{4}{3}R_1 \to R_1$ then $\quad \to$

$0.2R_1 + R_2 \to R_2$

$\to$

$0.6R_1 + R_4 \to R_4$

$$\begin{bmatrix} 1 & 0.24 & 1.\overline{3} & 0 & 0 & 0 & | & 1360 \\ 0 & \boxed{-0.152} & 0.2\overline{6} & 1 & 0 & 0 & | & -228 \\ 0 & 0.3 & 0 & 0 & 1 & 0 & | & 600 \\ \hline 0 & -0.256 & 0.8 & 0 & 0 & 1 & | & 816 \end{bmatrix}$$

$-0.24R_2 + R_1 \to R_1$

$-\frac{1}{0.152}R_2 \to R_2$ then $\quad \to$

$-0.3R_2 + R_3 \to R_3$

$0.256R_2 + R_4 \to R_4$

$$\approx \begin{bmatrix} 1 & 0 & 1.7544 & 1.5789 & 0 & 0 & | & 1000 \\ 0 & 1 & -1.7544 & -6.5789 & 0 & 0 & | & 1500 \\ 0 & 0 & 0.5263 & \boxed{1.9737} & 1 & 0 & | & 150 \\ \hline 0 & 0 & 0.3509 & -1.6842 & 0 & 1 & | & 1200 \end{bmatrix}$$

$-1.5789R_3 + R_1 \to R_1$

$6.5789R_3 + R_2 \to R_2$

$\frac{1}{1.9737}R_3 \to R_3$ then $\quad \to$

$1.6842R_3 + R_4 \to R_4$

$$\begin{bmatrix} 1 & 0 & 1.\overline{3} & 0 & -0.8 & 0 & | & 880 \\ 0 & 1 & 0 & 0 & 3.\overline{3} & 0 & | & 2000 \\ 0 & 0 & 0.2\overline{6} & 1 & 0.50\overline{6} & 0 & | & 76 \\ \hline 0 & 0 & 0.8 & 0 & 0.85\overline{3} & 1 & | & 1328 \end{bmatrix}$$

Maximum profit is $1328 with 880 lbs of all beef and 2000 lbs of regular.

31.

	Total Needed	Monaca	Hamburg
Heating Components	500	--	--
Produced	--	x	$500 - x$
Profit	--	$400x$	$390(500 - x)$
Domestic Furnaces	750	--	--
Produced	--	y	$750 - y$
Profit	--	$200y$	$215(750 - y)$

Maximize: $f = 400x + 390(500 - x) + 200y + 215(750 - y) = 10x - 15y + 356,250$

Constraints: $x, y \geq 0$

$$x + y \leq 1000 \qquad\qquad \text{Capacity at Monaca}$$

$$(500 - x) + (750 - y) \leq 850 \quad \text{or} \quad x + y \geq 400 \qquad \text{Capacity at Hamburg}$$

$$x \leq 100 + \frac{1}{2}y \quad \text{or} \quad 2x - y \leq 200 \qquad \text{Monaca capacity restriction}$$

$$\begin{bmatrix} 1 & 1 & 1 & 0 & 0 & 0 & | & 1000 \\ -1 & \boxed{-1} & 0 & 1 & 0 & 0 & | & -400 \\ 2 & -1 & 0 & 0 & 1 & 0 & | & 200 \\ \hline -10 & 15 & 0 & 0 & 0 & 1 & | & 356,250 \end{bmatrix} \quad \begin{array}{c} -R_2 + R_1 \to R_1 \\ -R_2 \to R_2 \quad \text{then} \quad \to \\ R_2 + R_3 \to R_3 \\ -15R_2 + R_4 \to R_4 \end{array}$$

$$\begin{bmatrix} 0 & 0 & 1 & 1 & 0 & 0 & | & 600 \\ 1 & 1 & 0 & -1 & 0 & 0 & | & 400 \\ \boxed{3} & 0 & 0 & -1 & 1 & 0 & | & 600 \\ \hline -25 & 0 & 0 & 15 & 0 & 1 & | & 350,250 \end{bmatrix} \quad \tfrac{1}{3}R_3 \to R_3 \quad \text{then}$$

$$\begin{array}{c} \to \\ -R_3 + R_2 \to R_2 \\ \to \\ 25R_3 + R_4 \to R_4 \end{array} \quad \begin{bmatrix} 0 & 0 & 1 & 1 & 0 & 0 & | & 600 \\ 0 & 1 & 0 & -0.\overline{6} & -0.\overline{3} & 0 & | & 200 \\ 1 & 0 & 0 & -0.\overline{3} & 0.\overline{3} & 0 & | & 200 \\ \hline 0 & 0 & 0 & 6.\overline{6} & 8.\overline{3} & 1 & | & 355,250 \end{bmatrix}$$

Maximum profit is \$355,250.

$x = 200$ heating components at Monaca and $500 - 200 = 300$ at Hamburg.

$y = 200$ domestic furnaces at Monaca and $750 - 200 = 550$ at Hamburg.

33.

	Total Needed	Monaca	Hamburg
Heating Components	500	--	--
Produced	--	x	$500 - x$
Cost	--	$380x$	$400(500 - x)$
Domestic Furnaces	750	--	--
Produced	--	y	$750 - y$
Cost	--	$200y$	$185(750 - y)$

$$\text{Cost} = \text{objective function} = C = 380x + 400(500 - x) + 200y + 185(750 - y)$$
$$= -20x + 15y + 338,750$$

Constraints: $x, y \geq 0$

$x + y \leq 1000$	Capacity at Monaca
$(500 - x) + (750y) \leq 850$ or $x + y \geq 400$	Capacity at Hamburg
$x \leq 100 + \dfrac{1}{2}y$ or $2x - y \leq 200$	Monaca capacity restriction

Need to maximize $-C = 20x - 15y - 338,750$

$$\left[\begin{array}{cccccc|c} 1 & 1 & 1 & 0 & 0 & 0 & 1000 \\ -1 & \boxed{-1} & 0 & 1 & 0 & 0 & -400 \\ 2 & -1 & 0 & 0 & 1 & 0 & 200 \\ \hline -20 & 15 & 0 & 0 & 0 & 1 & -338,750 \end{array}\right] \begin{array}{l} \\ -R_2 \to R_2 \text{ then} \\ \\ \end{array}$$

$$\begin{array}{l} -R_2 + R_1 \to R_1 \\ \to \\ R_2 + R_3 \to R_3 \\ -15R_2 + R_4 \to R_4 \end{array}$$

$$\left[\begin{array}{cccccc|c} 0 & 0 & 1 & 1 & 0 & 0 & 600 \\ 1 & 1 & 0 & -1 & 0 & 0 & 400 \\ \boxed{3} & 0 & 0 & -1 & 1 & 0 & 600 \\ \hline -35 & 0 & 0 & 15 & 0 & 1 & -344,750 \end{array}\right] \begin{array}{l} \\ \\ \tfrac{1}{3}R_3 \to R_3 \text{ then} \\ \\ \end{array}$$

$$\begin{array}{l} \to \\ -R_3 + R_2 \to R_2 \\ \to \\ 35R_3 + R_4 \to R_4 \end{array} \left[\begin{array}{cccccc|c} 0 & 0 & 1 & 1 & 0 & 0 & 600 \\ 0 & 1 & 0 & -\tfrac{2}{3} & -\tfrac{1}{3} & 0 & 200 \\ 1 & 0 & 0 & -\tfrac{1}{3} & \tfrac{1}{3} & 0 & 200 \\ \hline 0 & 0 & 0 & \tfrac{10}{3} & \tfrac{35}{3} & 1 & -337,750 \end{array}\right]$$

Minimum cost is \$337,750.
$x = 200$ heating components at Monaca and $500 - 200 = 300$ at Hamburg.
$y = 200$ domestic furnaces at Monaca and $750 - 200 = 550$ at Hamburg.

35.

	Plant 1	Plant 2	Plant 3	Totals
Million gallons	x	y	z	≤ 10
Impurities	$0.20x$	$0.15y$	$0.10z$	$\leq 0.15(x+y+z) \Rightarrow 0.05x - 0.05z \leq 0$
Cost (Thousands)	$20x$	$30y$	$40z$	$= C$
Other constraints	x		z	≥ 6

$$\begin{bmatrix} 1 & 1 & 1 & 1 & 0 & 0 & 0 & 10 \\ 0.05 & 0 & -0.05 & 0 & 1 & 0 & 0 & 0 \\ \boxed{-1} & 0 & -1 & 0 & 0 & 1 & 0 & -6 \\ \hline 20 & 30 & 40 & 0 & 0 & 0 & 1 & 0 \end{bmatrix} \begin{array}{l} \rightarrow \\ 100R_2 \rightarrow R_2 \\ -R_3 \rightarrow R_3 \\ \rightarrow \end{array}$$

$$\begin{bmatrix} 1 & 1 & 1 & 1 & 0 & 0 & 0 & 10 \\ 5 & 0 & -5 & 0 & 100 & 0 & 0 & 0 \\ \boxed{1} & 0 & 1 & 0 & 0 & -1 & 0 & 6 \\ \hline 20 & 30 & 40 & 0 & 0 & 0 & 1 & 0 \end{bmatrix} \begin{array}{l} -R_3 + R_1 \rightarrow R_1 \\ -5R_3 + R_2 \rightarrow R_2 \\ \rightarrow \\ -20R_3 + R_4 \rightarrow R_4 \end{array}$$

$$\begin{bmatrix} 0 & 1 & 0 & 1 & 0 & 1 & 0 & 4 \\ 0 & 0 & \boxed{-10} & 0 & 100 & 5 & 0 & -30 \\ 1 & 0 & 1 & 0 & 0 & -1 & 0 & 6 \\ \hline 0 & 30 & 20 & 0 & 0 & 20 & 1 & -120 \end{bmatrix} \begin{array}{l} -\frac{1}{10}R_2 \rightarrow R_2 \quad \text{then} \\ \\ -R_2 + R_3 \rightarrow R_3 \\ -20R_2 + R_4 \rightarrow R_4 \end{array} \begin{array}{l} \rightarrow \\ \rightarrow \end{array}$$

$$\begin{bmatrix} 0 & 1 & 0 & 1 & 0 & 1 & 0 & 4 \\ 0 & 0 & 1 & 0 & -10 & -\frac{1}{2} & 0 & 3 \\ 1 & 0 & 0 & 0 & 10 & -\frac{1}{2} & 0 & 3 \\ \hline 0 & 30 & 0 & 0 & 200 & 30 & 1 & -180 \end{bmatrix}$$

Minimum cost is $180,000 with plants 1 and 3 each handling 3,000,000 gallons. Plant 2 is zero, not 4, because the y-column still represents a nonbasic variable.

37. $x =$ number of footballs , $y =$ number of soccer balls , $z =$ number of volleyballs

Maximize: $f = 30x + 25y + 20z$

Subject to: $10x + 8y + 8z \geq 10,000 \quad \rightarrow \quad -10x - 8y - 8z \leq -10,000$

$\qquad\qquad 10x + 12y + 8z \leq 20,000$

$$\begin{bmatrix} \boxed{-10} & -8 & -8 & 1 & 0 & 0 & -10,000 \\ 10 & 12 & 8 & 0 & 1 & 0 & 20,000 \\ \hline -30 & -25 & -20 & 0 & 0 & 1 & 0 \end{bmatrix} \begin{array}{l} -\frac{1}{10}R_1 \rightarrow R_1 \quad \text{then} \\ \\ -10R_1 + R_2 \rightarrow R_2 \\ 30R_1 + R_3 \rightarrow R_3 \end{array} \quad \rightarrow$$

$$\begin{bmatrix} 1 & 0.8 & 0.8 & -0.1 & 0 & 0 & 1000 \\ 0 & 4 & 0 & \boxed{1} & 1 & 0 & 10,000 \\ \hline 0 & -1 & 4 & -3 & 0 & 1 & 30,000 \end{bmatrix} \begin{array}{l} 0.1R_2 + R_1 \rightarrow R_1 \\ \rightarrow \\ 3R_2 + R_3 \rightarrow R_3 \end{array} \begin{bmatrix} 1 & 1.2 & 0.8 & 0 & 0.1 & 0 & 2000 \\ 0 & 4 & 0 & 1 & 1 & 0 & 10,000 \\ \hline 0 & 11 & 4 & 0 & 3 & 1 & 60,000 \end{bmatrix}$$

The maximum revenue is $60,000 when 2500 footballs and no soccer balls or volleyballs are produced.

Review Exercises

1.

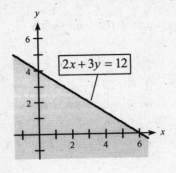

2.

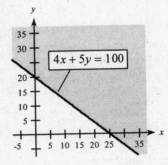

3.

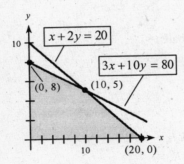

4.

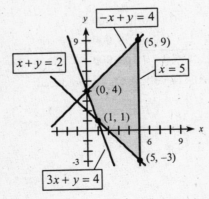

5.

Corners	$f = -x + 3y$
$(0,5)$	15
$(5,10)$	25
$(10,6)$	8
$(12,0)$	-12
$(0,0)$	0

Maximum is 25 at $(5,10)$.
Minimum is -12 at $(12,0)$.

6.

Corners	$f = 6x + 4y$
$(0,30)$	120
$(0,40)$	160
$(17,23)$	194
$(8,14)$	104

Maximum is 194 at $(17,23)$.
Minimum is 104 at $(8,14)$.

7. Solving: $\begin{cases} -3x + 2y = 4 \\ x + 2y = 20 \end{cases}$

$\begin{cases} -3x + 2y = 4 \\ -x - 2y = -20 \end{cases}$

$-4x = -16$

$x = 4$

Corner: $(4,8)$

Solving: $\begin{cases} -3x + 2y = 4 \\ x = 20 \end{cases}$

$-3(20) + 2y = 4$

$2y = 64$

$y = 32$

Corner: $(20,32)$

Corners	$f = 7x - 6y$
$(20,0)$	140
$(4,8)$	-20
$(20,32)$	-52

Maximum is 140 at $(20,0)$.

Minimum is -52 at $(20,32)$.

8. Solving: $\begin{cases} x+2y=19 \\ 3x+2y=29 \end{cases}$

    $\begin{cases} -x-2y=-19 \\ 3x+2y=29 \end{cases}$

    $2x=10$

    $x=10$

    $10+2y=19$

    $2y=9$

    $y=4.5$

    Corner: $(10,4.5)$

Corners	$f=9x+10y$
$(10,4.5)$	135
$(19,0)$	171
$(0,14.5)$	145

    Minimum is 135 at $(10,4.5)$
    There is no maximum because the region is unbounded.

9. Maximize $f=5x+6y$
    Subject to: $x+3y \le 24$

    $4x+3y \le 42$

    $2x+y \le 20$

    Solving $\begin{cases} 4x+3y=42 \\ x+3y=24 \end{cases}$ we get $A(6,6)$.

    Solving $\begin{cases} 2x+y=20 \\ 4x+3y=42 \end{cases}$ we get $B(9,2)$.

Feasible Corners	$f=5x+6y$
$(0,8)$	48
$(6,6)$	66
$(9,2)$	57
$(10,0)$	50

    Maximum is 66 at $(6,6)$.

10. Maximize $f=x+4y$
    Subject to: $7x+3y \le 105$

    $2x+5y \le 59$

    $x+7y \le 70$

    Solving $\begin{cases} 7x+3y=105 \\ 2x+5y=59 \end{cases}$ we get $B(12,7)$.

    Solving $\begin{cases} x+7y=70 \\ 2x+5y=59 \end{cases}$ we get $A(7,9)$.

Feasible Corners	$f=x+4y$
$(0,10)$	40
$(7,9)$	43
$(12,7)$	40
$(15,0)$	15

    Maximum is 43 at $(7,9)$.

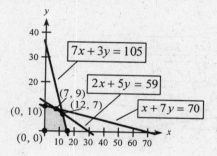

11. Minimize $g = 5x + 3y$

Subject to: $3x + y \geq 12$

$\qquad x + y \geq 6$

$\qquad x + 6y \geq 11$

Solving $\begin{cases} 3x + y = 12 \\ x + y = 6 \end{cases}$ we get $A(3, 3)$.

Solving $\begin{cases} x + 6y = 11 \\ x + y = 6 \end{cases}$ we get $B(5, 1)$.

Feasible Corners	$g = 5x + 3y$
$(0,12)$	36
$(3,3)$	24
$(5,1)$	28
$(11,0)$	55

Minimum is 24 at $(3,3)$.

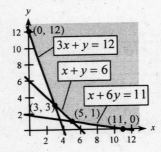

12. Minimize $g = x + 5y$

Subject to: $8x + y \geq 85$

$\qquad x + y \geq 50$

$\qquad x + 4y \geq 80$

$\qquad x + 10y \geq 104$

Solving $\begin{cases} 8x + y = 85 \\ x + y = 50 \end{cases}$ we get $A(5, 45)$.

Solving $\begin{cases} x + y = 50 \\ x + 4y = 80 \end{cases}$ we get $B(40, 10)$.

Solving $\begin{cases} x + 4y = 80 \\ x + 10y = 104 \end{cases}$ we get $C\ (64, 4)$.

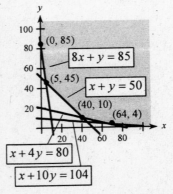

Feasible Corners	$g = x + 5y$
$(0,85)$	425
$(5,45)$	230
$(40,10)$	90
$(64,4)$	84
$(94,0)$	94

Minimum is 84 at $(64,4)$.

13. Maximize $f = 5x + 2y$

Subject to: $\qquad y \leq 20$

$\qquad 2x + y \leq 32$

$\qquad -x + 2y \geq 4$

$\qquad x \geq 0, \ y \geq 0$

Solving: $\begin{cases} 2x + y = 32 \\ -x + 2y = 4 \end{cases}$

$\qquad \begin{cases} -4x - 2y = -64 \\ -x + 2y = 4 \end{cases}$

$\qquad -5x = -60$

$\qquad x = 12, \ y = 8$

Corner: $(12,8)$

Solving: $\begin{cases} y = 20 \\ 2x + y = 32 \end{cases}$

$\qquad 2x + 20 = 32$

$\qquad 2x = 12$

$\qquad x = 6$

Corner: $(6,20)$

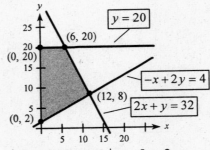

Feasible Corners	$g = 5x + 2y$
$(0,2)$	4
$(12,8)$	76
$(6,20)$	70
$(0,20)$	40

Maximum is 76 at $(12,8)$.

14. Maximize $f = x + 4y$

Subject to:
$$y \leq 30$$
$$3x + 2y \leq 75$$
$$-3x + 5y \geq 30$$
$$x, y \geq 0$$

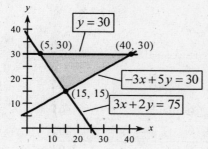

$$
\begin{array}{ll}
y = 30 & y = 30 \\
3x + 2y = 75 & -3x + 5y = 30 \\
\hline
3x = 15 & -3x = -120 \\
x = 5 & x = 40
\end{array}
$$

$(5, 30)$ and $(40, 30)$ are feasible corners.

$$3x + 2y = 75 \qquad (15, 15) \text{ is a feasible corner.}$$
$$\underline{-3x + 5y = 30}$$
$$7y = 105$$
$$y = 15$$

Corner	$f = x + 4y$
$(5, 30)$	125
$(40, 30)$	160
$(15, 15)$	75

Minimum = 75 at $(15, 15)$.

15.
$$
\begin{bmatrix}
7 & 3 & 1 & 0 & 0 & 0 & 105 \\
2 & 5 & 0 & 1 & 0 & 0 & 59 \\
1 & \boxed{7} & 0 & 0 & 1 & 0 & 70 \\
\hline
-7 & -12 & 0 & 0 & 0 & 1 & 0
\end{bmatrix}
\begin{array}{l}
\frac{1}{7}R_3 \to R_3 \text{ then}
\end{array}
\qquad
\begin{array}{l}
-3R_3 + R_1 \to R_1 \\
-5R_3 + R_2 \to R_2 \\
\to \\
12R_3 + R_4 \to R_4
\end{array}
$$

$$
\begin{bmatrix}
\frac{46}{7} & 0 & 1 & 0 & -\frac{3}{7} & 0 & 75 \\
\boxed{\frac{9}{7}} & 0 & 0 & 1 & -\frac{5}{7} & 0 & 9 \\
\frac{1}{7} & 1 & 0 & 0 & \frac{1}{7} & 0 & 10 \\
\hline
-\frac{37}{7} & 0 & 0 & 0 & \frac{12}{7} & 1 & 120
\end{bmatrix}
\begin{array}{l}
\frac{7}{9}R_2 \to R_2 \text{ then}
\end{array}
\qquad
\begin{array}{l}
-\frac{46}{7}R_2 + R_1 \to R_1 \\
\\
-\frac{1}{7}R_2 + R_3 \to R_3 \\
\frac{37}{7}R_2 + R_4 \to R_4
\end{array}
$$

$$
\begin{bmatrix}
0 & 0 & 1 & -\frac{46}{9} & \boxed{\frac{29}{9}} & 0 & 29 \\
1 & 0 & 0 & \frac{7}{9} & -\frac{5}{9} & 0 & 7 \\
0 & 1 & 0 & -\frac{1}{9} & \frac{2}{9} & 0 & 9 \\
\hline
0 & 0 & 0 & \frac{37}{9} & -\frac{11}{9} & 1 & 157
\end{bmatrix}
\begin{array}{l}
\frac{9}{29}R_1 \to R_1 \text{ then}
\end{array}
\qquad
\begin{array}{l}
\to \\
\frac{5}{9}R_1 + R_2 \to R_2 \\
-\frac{2}{9}R_1 + R_3 \to R_3 \\
\frac{11}{9}R_1 + R_4 \to R_4
\end{array}
\begin{bmatrix}
0 & 0 & \frac{9}{29} & -\frac{46}{29} & 1 & 0 & 9 \\
1 & 0 & \frac{5}{29} & -\frac{3}{29} & 0 & 0 & 12 \\
0 & 1 & -\frac{2}{29} & \frac{7}{29} & 0 & 0 & 7 \\
0 & 0 & \frac{11}{29} & \frac{63}{29} & 0 & 1 & 168
\end{bmatrix}
$$

Maximum is 168 at $x = 12$, $y = 7$.

16.
$$\begin{bmatrix} 1 & \boxed{4} & 1 & 0 & 0 & 0 & | & 160 \\ 1 & 2 & 0 & 1 & 0 & 0 & | & 100 \\ 4 & 3 & 0 & 0 & 1 & 0 & | & 300 \\ -3 & -4 & 0 & 0 & 0 & 1 & | & 0 \end{bmatrix}$$
$\frac{1}{4}R_1 \to R_1$ then →
$-2R_1 + R_2 \to R_2$
$-3R_1 + R_3 \to R_3$
$4R_1 + R_4 \to R_4$

$$\begin{bmatrix} \frac{1}{4} & 1 & \frac{1}{4} & 0 & 0 & 0 & | & 40 \\ \boxed{\frac{1}{2}} & 0 & -\frac{1}{2} & 1 & 0 & 0 & | & 20 \\ \frac{13}{4} & 0 & -\frac{3}{4} & 0 & 1 & 0 & | & 180 \\ -2 & 0 & 1 & 0 & 0 & 1 & | & 160 \end{bmatrix}$$
$2R_2 \to R_2$ then
$-\frac{1}{4}R_2 + R_1 \to R_1$
→
$-\frac{13}{4}R_2 + R_3 \to R_3$
$2R_2 + R_4 \to R_4$

$$\begin{bmatrix} 0 & 1 & \frac{1}{2} & -\frac{1}{2} & 0 & 0 & | & 30 \\ 1 & 0 & -1 & 2 & 0 & 0 & | & 40 \\ 0 & 0 & \boxed{\frac{5}{2}} & -\frac{13}{2} & 1 & 0 & | & 50 \\ 0 & 0 & -1 & 4 & 0 & 1 & | & 240 \end{bmatrix}$$
$\frac{2}{5}R_3 \to R_3$ then
$-\frac{1}{2}R_3 + R_1 \to R_1$
$R_3 + R_2 \to R_2$
→
$R_3 + R_4 \to R_4$
$$\begin{bmatrix} 0 & 1 & 0 & \frac{4}{5} & -\frac{1}{5} & 0 & | & 20 \\ 1 & 0 & 0 & -\frac{3}{5} & \frac{2}{5} & 0 & | & 60 \\ 0 & 0 & 1 & -\frac{13}{5} & \frac{2}{5} & 0 & | & 20 \\ 0 & 0 & 0 & \frac{7}{5} & \frac{2}{5} & 1 & | & 260 \end{bmatrix}$$

Solution: $x = 60$, $y = 20$; $f = 260$

17.
$$\begin{bmatrix} 1 & \boxed{4} & 1 & 0 & 0 & 0 & | & 160 \\ 1 & 2 & 0 & 1 & 0 & 0 & | & 100 \\ 4 & 3 & 0 & 0 & 1 & 0 & | & 300 \\ -3 & -8 & 0 & 0 & 0 & 1 & | & 0 \end{bmatrix}$$
$\frac{1}{4}R_1 \to R_1$ then →
$-2R_1 + R_2 \to R_2$
$-3R_1 + R_3 \to R_3$
$8R_1 + R_4 \to R_4$

$$\begin{bmatrix} \frac{1}{4} & 1 & \frac{1}{4} & 0 & 0 & 0 & | & 40 \\ \boxed{\frac{1}{2}} & 0 & -\frac{1}{2} & 1 & 0 & 0 & | & 20 \\ \frac{13}{4} & 0 & -\frac{3}{4} & 0 & 1 & 0 & | & 180 \\ -1 & 0 & 2 & 0 & 0 & 1 & | & 320 \end{bmatrix}$$
$2R_2 \to R_2$ then
$-\frac{1}{4}R_2 + R_1 \to R_1$
→
$-\frac{13}{4}R_2 + R_3 \to R_3$
$R_2 + R_4 \to R_4$
$$\begin{bmatrix} 0 & 1 & \frac{1}{2} & -\frac{1}{2} & 0 & 0 & | & 30 \\ 1 & 0 & -1 & 2 & 0 & 0 & | & 40 \\ 0 & 0 & \frac{5}{2} & -\frac{13}{2} & 1 & 0 & | & 50 \\ 0 & 0 & 1 & 2 & 0 & 1 & | & 360 \end{bmatrix}$$

Maximum is 360 at $x = 40$, $y = 30$.

18.
$$\begin{bmatrix} 1 & 2 & 1 & 0 & 0 & 0 & 0 & | & 48 \\ 1 & 1 & 0 & 1 & 0 & 0 & 0 & | & 30 \\ \boxed{2} & 1 & 0 & 0 & 1 & 0 & 0 & | & 50 \\ 1 & 10 & 0 & 0 & 0 & 1 & 0 & | & 200 \\ -3 & -2 & 0 & 0 & 0 & 0 & 1 & | & 0 \end{bmatrix}$$
$-R_3 + R_1 \to R_1$
$-R_3 + R_2 \to R_2$
$\frac{1}{2}R_3 \to R_3$ then →
$-R_3 + R_4 \to R_4$
$3R_3 + R_5 \to R_5$

$$\begin{bmatrix} 0 & 1.5 & 1 & 0 & -0.5 & 0 & 0 & | & 23 \\ 0 & \boxed{0.5} & 0 & 1 & -0.5 & 0 & 0 & | & 5 \\ 1 & 0.5 & 0 & 0 & 0.5 & 0 & 0 & | & 25 \\ 0 & 9.5 & 0 & 0 & -0.5 & 1 & 0 & | & 175 \\ 0 & -0.5 & 0 & 0 & 1.5 & 0 & 1 & | & 75 \end{bmatrix}$$
$2R_2 \to R_2$ then
$-1.5R_2 + R_1 \to R_1$
→
$-0.5R_2 + R_3 \to R_3$
$-9.5R_2 + R_4 \to R_4$
$0.5R_2 + R_5 \to R_5$
$$\begin{bmatrix} 0 & 0 & 1 & -3 & 1 & 0 & 0 & | & 8 \\ 0 & 1 & 0 & 2 & -1 & 0 & 0 & | & 10 \\ 1 & 0 & 0 & -1 & 1 & 0 & 0 & | & 20 \\ 0 & 0 & 0 & -19 & 9 & 1 & 0 & | & 80 \\ 0 & 0 & 0 & 1 & 1 & 0 & 1 & | & 80 \end{bmatrix}$$

Maximum is 80 at $x = 20$ and $y = 10$.

19.
$$\left[\begin{array}{cccccccc|c} 1 & 0 & 1 & 1 & 0 & 0 & 0 & 7 \\ 3 & 5 & 0 & 0 & 1 & 0 & 0 & 30 \\ \boxed{3} & 1 & 0 & 0 & 0 & 1 & 0 & 18 \\ \hline -39 & -5 & -30 & 0 & 0 & 0 & 1 & 0 \end{array}\right]$$

$\frac{1}{3}R_3 \to R_3$ then

$-R_3 + R_1 \to R_1$
$-3R_3 + R_2 \to R_2$
$\to$
$39R_3 + R_4 \to R_4$

$$\left[\begin{array}{cccccccc|c} 0 & -\frac{1}{3} & \boxed{1} & 1 & 0 & -\frac{1}{3} & 0 & 1 \\ 0 & 4 & 0 & 0 & 1 & -1 & 0 & 12 \\ 1 & \frac{1}{3} & 0 & 0 & 0 & \frac{1}{3} & 0 & 6 \\ \hline 0 & 8 & -30 & 0 & 0 & 13 & 1 & 234 \end{array}\right]$$

$30R_1 + R_4 \to R_4$

$\to$
$\to$
$\to$

$$\left[\begin{array}{cccccccc|c} 0 & -\frac{1}{3} & 1 & 1 & 0 & -\frac{1}{3} & 0 & 1 \\ 0 & \boxed{4} & 0 & 0 & 1 & -1 & 0 & 12 \\ 1 & \frac{1}{3} & 0 & 0 & 0 & \frac{1}{3} & 0 & 6 \\ \hline 0 & -2 & 0 & 30 & 0 & 3 & 1 & 264 \end{array}\right]$$

$\frac{1}{4}R_2 \to R_2$ then

$\frac{1}{3}R_2 + R_1 \to R_1$
$\to$
$-\frac{1}{3}R_2 + R_3 \to R_3$
$2R_2 + R_4 \to R_4$

$$\left[\begin{array}{cccccccc|c} 0 & 0 & 1 & 1 & \frac{1}{12} & -\frac{5}{12} & 0 & 2 \\ 0 & 1 & 0 & 0 & \frac{1}{4} & -\frac{1}{4} & 0 & 3 \\ 1 & 0 & 0 & 0 & -\frac{1}{12} & \frac{5}{12} & 0 & 5 \\ \hline 0 & 0 & 0 & 30 & \frac{1}{2} & \frac{5}{2} & 1 & 270 \end{array}\right]$$

Solution: $x = 5$, $y = 3$, $z = 2$; $f = 270$

20.
$$\left[\begin{array}{ccccccccc|c} 3 & 2 & 2 & 5 & 1 & 0 & 0 & 0 & 0 & 200 \\ 2 & 2 & \boxed{4} & 5 & 0 & 1 & 0 & 0 & 0 & 100 \\ 1 & 1 & 1 & 1 & 0 & 0 & 1 & 0 & 0 & 200 \\ 1 & 0 & 0 & 0 & 0 & 0 & 0 & 1 & 0 & 40 \\ \hline -88 & -86 & -100 & -100 & 0 & 0 & 0 & 0 & 1 & 0 \end{array}\right]$$

$\frac{1}{4}R_2 \to R_2$ then

$-2R_2 + R_1 \to R_1$
$\to$
$-R_2 + R_3 \to R_3$
$\to$
$100R_2 + R_5 \to R_5$

$$\left[\begin{array}{ccccccccc|c} 2 & 1 & 0 & 2.1 & 1 & -0.5 & 0 & 0 & 0 & 150 \\ 0.5 & 0.5 & 1 & 1.25 & 0 & 0.25 & 0 & 0 & 0 & 25 \\ 0.5 & 0.5 & 0 & -0.25 & 0 & -0.25 & 1 & 0 & 0 & 175 \\ \boxed{1} & 0 & 0 & 0 & 0 & 0 & 0 & 1 & 0 & 40 \\ \hline -38 & -36 & 0 & 25 & 0 & 25 & 0 & 0 & 1 & 2500 \end{array}\right]$$

$-2R_4 + R_1 \to R_1$
$-0.5R_4 + R_2 \to R_2$
$-0.5R_4 + R_3 \to R_3$
$\to$
$38R_4 + R_5 \to R_5$

$$\left[\begin{array}{ccccccccc|c} 0 & 1 & 0 & 2.5 & 1 & -0.5 & 0 & -2 & 0 & 70 \\ 0 & \boxed{0.5} & 1 & 1.25 & 0 & 0.25 & 0 & -0.5 & 0 & 5 \\ 0 & 0.5 & 0 & -0.25 & 0 & -0.25 & 1 & -0.5 & 0 & 155 \\ 1 & 0 & 0 & 0 & 0 & 0 & 0 & 1 & 0 & 40 \\ \hline 0 & -36 & 0 & 25 & 0 & 25 & 0 & 38 & 1 & 4020 \end{array}\right]$$

$2R_2 \to R_2$ then

$-R_2 + R_1 \to R_1$
$\to$
$-0.5R_2 + R_3 \to R_3$
$\to$
$36R_2 + R_5 \to R_5$

$$\left[\begin{array}{ccccccccc|c} 0 & 0 & -2 & 0 & 1 & -1 & 0 & -1 & 0 & 60 \\ 0 & 1 & 2 & 2.5 & 0 & 0.5 & 0 & -1 & 0 & 10 \\ 0 & 0 & -1 & -1.5 & 0 & -0.5 & 1 & 0 & 0 & 150 \\ 1 & 0 & 0 & 0 & 0 & 0 & 0 & 1 & 0 & 40 \\ \hline 0 & 0 & 72 & 115 & 0 & 43 & 0 & 2 & 1 & 4380 \end{array}\right]$$

Maximum is 4380 at $x_1 = 40$, $x_2 = 10$, $x_3 = 0$, and $x_4 = 0$.

21.
$$\left[\begin{array}{cccccc|c} 1 & \boxed{5} & 1 & 0 & 0 & 0 & 500 \\ 1 & 2 & 0 & 1 & 0 & 0 & 230 \\ 1 & 1 & 0 & 0 & 1 & 0 & 160 \\ \hline -4 & -4 & 0 & 0 & 0 & 1 & 0 \end{array}\right]$$
$\frac{1}{5}R_1 \to R_1$ then $\quad\to$

$-2R_1 + R_2 \to R_2$
$-R_1 + R_3 \to R_3$
$4R_1 + R_4 \to R_4$

$$\left[\begin{array}{cccccc|c} \frac{1}{5} & 1 & \frac{1}{5} & 0 & 0 & 0 & 100 \\ \boxed{\frac{3}{5}} & 0 & -\frac{2}{5} & 1 & 0 & 0 & 30 \\ \frac{4}{5} & 0 & -\frac{1}{5} & 0 & 1 & 0 & 60 \\ \hline -\frac{16}{5} & 0 & \frac{4}{5} & 0 & 0 & 1 & 400 \end{array}\right]$$
$\frac{5}{3}R_2 \to R_2$ then

$-\frac{1}{5}R_2 + R_1 \to R_1$
$\quad\to$
$-\frac{4}{5}R_2 + R_3 \to R_3$
$\frac{16}{5}R_2 + R_4 \to R_4$

$$\left[\begin{array}{cccccc|c} 0 & 1 & \frac{1}{3} & -\frac{1}{3} & 0 & 0 & 90 \\ 1 & 0 & -\frac{2}{3} & \frac{5}{3} & 0 & 0 & 50 \\ 0 & 0 & \boxed{\frac{1}{3}} & -\frac{4}{3} & 1 & 0 & 20 \\ \hline 0 & 0 & -\frac{4}{3} & \frac{16}{3} & 0 & 1 & 560 \end{array}\right]$$
$3R_3 \to R_3$ then

$-\frac{1}{3}R_3 + R_1 \to R_1$
$\frac{2}{3}R_3 + R_2 \to R_2$
$\quad\to$
$\frac{4}{3}R_3 + R_4 \to R_4$

$$\left[\begin{array}{cccccc|c} 0 & 1 & 0 & 1 & -1 & 0 & 70 \\ 1 & 0 & 0 & -1 & 2 & 0 & 90 \\ 0 & 0 & 1 & -4 & 3 & 0 & 60 \\ \hline 0 & 0 & 0 & 0 & 4 & 1 & 640 \end{array}\right]$$

There are multiple solutions. Maximum is 640 at $x = 90$, $y = 70$.

$$\left[\begin{array}{cccccc|c} 0 & 1 & 0 & \boxed{1} & -1 & 0 & 70 \\ 1 & 0 & 0 & -1 & 2 & 0 & 90 \\ 0 & 0 & 1 & -4 & 3 & 0 & 60 \\ \hline 0 & 0 & 0 & 0 & 4 & 1 & 640 \end{array}\right]$$
$\quad\to$
$R_1 + R_2 \to R_2$
$4R_1 + R_3 \to R_3$
$\quad\to$
$$\left[\begin{array}{cccccc|c} 0 & 1 & 0 & 1 & -1 & 0 & 70 \\ 1 & 1 & 0 & 0 & 1 & 0 & 160 \\ 0 & 4 & 1 & 0 & -1 & 0 & 340 \\ \hline 0 & 0 & 0 & 0 & 4 & 1 & 640 \end{array}\right]$$

Second maximum occurs at $x = 160$, $y = 0$.

$f = 640$ on the line between $(160, 0)$ and $(90, 70)$.

22.
$$\left[\begin{array}{ccccc|c} -4 & \boxed{1} & 1 & 0 & 0 & 40 \\ 1 & -7 & 0 & 1 & 0 & 70 \\ \hline -2 & -5 & 0 & 0 & 1 & 0 \end{array}\right]$$
$\quad\to$
$7R_1 + R_2 \to R_2$
$5R_1 + R_3 \to R_3$
$$\left[\begin{array}{ccccc|c} -4 & 1 & 1 & 0 & 0 & 40 \\ -27 & 0 & 7 & 1 & 0 & 350 \\ \hline -22 & 0 & 5 & 0 & 1 & 200 \end{array}\right]$$
No solution

23.
$$\left[\begin{array}{cc|c} 5 & 2 & 16 \\ 3 & 7 & 27 \\ \hline 7 & 6 & g \end{array}\right]$$
Dual:
$$\left[\begin{array}{cc|c} 5 & 3 & 7 \\ 2 & 7 & 6 \\ \hline 16 & 27 & g \end{array}\right]$$

$$\left[\begin{array}{ccccc|c} 5 & 3 & 1 & 0 & 0 & 7 \\ 2 & \boxed{7} & 0 & 1 & 0 & 6 \\ \hline -16 & -27 & 0 & 0 & 1 & 0 \end{array}\right]$$
$\frac{1}{7}R_2 \to R_2$ then

$-3R_2 + R_1 \to R_1$
$\quad\to$
$27R_2 + R_3 \to R_3$

$$\left[\begin{array}{ccccc|c} \boxed{\frac{29}{7}} & 0 & 1 & -\frac{3}{7} & 0 & \frac{31}{7} \\ \frac{2}{7} & 1 & 0 & \frac{1}{7} & 0 & \frac{6}{7} \\ \hline -\frac{58}{7} & 0 & 0 & \frac{27}{7} & 1 & \frac{162}{7} \end{array}\right]$$
$\frac{7}{29}R_1 \to R_1$ then

$\quad\to$
$-\frac{2}{7}R_1 + R_2 \to R_2$
$\frac{58}{7}R_1 + R_3 \to R_3$

$$\left[\begin{array}{ccccc|c} 1 & 0 & \frac{7}{29} & -\frac{3}{29} & 0 & \frac{31}{29} \\ 0 & 1 & -\frac{2}{29} & \frac{5}{29} & 0 & \frac{16}{29} \\ \hline 0 & 0 & 2 & 3 & 1 & 32 \end{array}\right]$$

Minimum is 32 at $y_1 = 2$, $y_2 = 3$.

24. $\begin{bmatrix} 3 & 1 & 8 \\ 1 & 1 & 6 \\ 2 & 5 & 18 \\ \hline 3 & 4 & g \end{bmatrix}$ $\rightarrow$ $\begin{bmatrix} 3 & 1 & 2 & 3 \\ 1 & 1 & 5 & 4 \\ \hline 8 & 6 & 18 & g \end{bmatrix}$ Maximize $g = 8x_1 + 6x_2 + 18x_3$

Constraints: $3x_1 + x_2 + 2x_3 \le 3$
$x_1 + x_2 + 5x_3 \le 4$
$x_1, x_2, x_3 \ge 0$

$\begin{bmatrix} 3 & 1 & 2 & 1 & 0 & 0 & 3 \\ 1 & 1 & \boxed{5} & 0 & 1 & 0 & 4 \\ \hline -8 & -6 & -18 & 0 & 0 & 1 & 0 \end{bmatrix}$ $\frac{1}{5}R_2 \to R_2$ then

$-2R_2 + R_1 \to R_1$
$\rightarrow$
$18R_2 + R_3 \to R_3$

$\begin{bmatrix} \boxed{\frac{13}{5}} & \frac{3}{5} & 0 & 1 & -\frac{2}{5} & 0 & \frac{7}{5} \\ \frac{1}{5} & \frac{1}{5} & 1 & 0 & \frac{1}{5} & 0 & \frac{4}{5} \\ \hline -\frac{22}{5} & -\frac{12}{5} & 0 & 0 & \frac{18}{5} & 1 & \frac{72}{5} \end{bmatrix}$ $\frac{5}{13}R_1 \to R_1$ then

$\rightarrow$
$-\frac{1}{5}R_1 + R_2 \to R_2$
$\frac{22}{5}R_1 + R_3 \to R_3$

$\begin{bmatrix} 1 & \boxed{\frac{3}{13}} & 0 & \frac{5}{13} & -\frac{2}{13} & 0 & \frac{7}{13} \\ 0 & \frac{2}{13} & 1 & -\frac{1}{13} & \frac{3}{13} & 0 & \frac{9}{13} \\ \hline 0 & -\frac{18}{13} & 0 & \frac{22}{13} & \frac{38}{13} & 1 & \frac{218}{13} \end{bmatrix}$ $\frac{13}{3}R_1 \to R_1$ then

$\rightarrow$
$-\frac{2}{13}R_1 + R_2 \to R_2$
$\frac{18}{13}R_1 + R_3 \to R_3$

$\begin{bmatrix} \frac{13}{3} & 1 & 0 & \frac{5}{3} & -\frac{2}{3} & 0 & \frac{7}{3} \\ -\frac{2}{3} & 0 & 1 & -\frac{1}{3} & \frac{1}{3} & 0 & \frac{1}{3} \\ \hline 6 & 0 & 0 & 4 & 2 & 1 & 20 \end{bmatrix}$

Solution: $y_1 = 4$, $y_2 = 2$; $g = 20$

25. $\begin{bmatrix} 3 & 1 & 8 \\ 1 & 1 & 6 \\ 2 & 5 & 18 \\ \hline 2 & 1 & g \end{bmatrix}$ Dual: $\begin{bmatrix} 3 & 1 & 1 & 2 \\ 1 & 1 & 5 & 1 \\ \hline 8 & 6 & 18 & g \end{bmatrix}$

$\begin{bmatrix} 3 & 1 & 2 & 1 & 0 & 0 & 2 \\ 1 & 1 & \boxed{5} & 0 & 1 & 0 & 1 \\ \hline -8 & -6 & -18 & 0 & 0 & 1 & 0 \end{bmatrix}$ $\frac{1}{5}R_2 \to R_2$ then

$-2R_2 + R_1 \to R_1$
$\rightarrow$
$18R_2 + R_3 \to R_3$

$\begin{bmatrix} \boxed{\frac{13}{5}} & \frac{3}{5} & 0 & 1 & -\frac{2}{5} & 0 & \frac{8}{5} \\ \frac{1}{5} & \frac{1}{5} & 1 & 0 & \frac{1}{5} & 0 & \frac{1}{5} \\ \hline -\frac{22}{5} & -\frac{12}{5} & 0 & 0 & \frac{18}{5} & 1 & \frac{18}{5} \end{bmatrix}$ $\frac{5}{13}R_1 \to R_1$ then

$\rightarrow$
$-\frac{1}{5}R_1 + R_2 \to R_2$
$\frac{22}{5}R_1 + R_3 \to R_3$

$\begin{bmatrix} 1 & \frac{3}{13} & 0 & \frac{5}{13} & -\frac{2}{13} & 0 & \frac{8}{13} \\ 0 & \boxed{\frac{2}{13}} & 1 & -\frac{1}{13} & \frac{3}{13} & 0 & \frac{1}{13} \\ \hline 0 & -\frac{18}{13} & 0 & \frac{22}{13} & \frac{38}{13} & 1 & \frac{82}{13} \end{bmatrix}$ $\frac{13}{2}R_2 \to R_2$ then

$-\frac{3}{13}R_2 + R_1 \to R_1$
$\rightarrow$
$\frac{18}{13}R_2 + R_3 \to R_3$

$\begin{bmatrix} 1 & 0 & -\frac{3}{2} & \frac{1}{2} & -\frac{1}{2} & 0 & \frac{1}{2} \\ 0 & 1 & \frac{13}{2} & -\frac{1}{2} & \frac{3}{2} & 0 & \frac{1}{2} \\ \hline 0 & 0 & 9 & 1 & 5 & 1 & 7 \end{bmatrix}$

Minimum is 7 at $y_1 = 1$, $y_2 = 5$.

26. $\begin{bmatrix} 1 & 1 & | & 100 \\ 2 & 1 & | & 140 \\ 6 & 5 & | & 580 \\ \hline 12 & 11 & | & g \end{bmatrix}$ transpose: $\begin{bmatrix} 1 & 2 & 6 & | & 12 \\ 1 & 1 & 5 & | & 11 \\ \hline 100 & 140 & 580 & | & g \end{bmatrix}$

$\begin{bmatrix} 1 & 2 & \boxed{6} & 1 & 0 & 0 & | & 12 \\ 1 & 1 & 5 & 0 & 1 & 0 & | & 11 \\ \hline -100 & -140 & -580 & 0 & 0 & 1 & | & 0 \end{bmatrix}$ $\frac{1}{6}R_1 \rightarrow R_1$ then $\rightarrow$
$-5R_1 + R_2 \rightarrow R_2$
$580R_1 + R_3 \rightarrow R_3$

$\begin{bmatrix} \frac{1}{6} & \frac{1}{3} & 1 & \frac{1}{6} & 0 & 0 & | & 2 \\ \boxed{\frac{1}{6}} & -\frac{2}{3} & 0 & -\frac{5}{6} & 1 & 0 & | & 1 \\ \hline -\frac{10}{3} & 53\frac{1}{3} & 0 & 96\frac{2}{3} & 0 & 1 & | & 1160 \end{bmatrix}$ $6R_2 \rightarrow R_2$ then

$-\frac{1}{6}R_2 + R_1 \rightarrow R_1$ $\begin{bmatrix} 0 & 1 & 1 & 1 & -1 & 0 & | & 1 \\ 1 & -4 & 0 & -5 & 6 & 0 & | & 6 \\ \hline 0 & 40 & 0 & 80 & 20 & 1 & | & 1180 \end{bmatrix}$
$\frac{10}{3}R_2 + R_3 \rightarrow R_3$

The minimum is 1180 when $y_1 = 80$ and $y_2 = 20$.

27. $\begin{bmatrix} 1 & 2 & 1 & | & 60 \\ 12 & 4 & 3 & | & 120 \\ 2 & 3 & 1 & | & 80 \\ \hline 12 & 5 & 2 & | & g \end{bmatrix}$ transpose: $\begin{bmatrix} 1 & 12 & 2 & | & 12 \\ 2 & 4 & 3 & | & 5 \\ 1 & 3 & 1 & | & 2 \\ \hline 60 & 120 & 80 & | & g \end{bmatrix}$

$\begin{bmatrix} 1 & 12 & 2 & 1 & 0 & 0 & 0 & | & 12 \\ 2 & 4 & 3 & 0 & 1 & 0 & 0 & | & 5 \\ 1 & \boxed{3} & 1 & 0 & 0 & 1 & 0 & | & 2 \\ \hline -60 & -120 & -80 & 0 & 0 & 0 & 1 & | & 0 \end{bmatrix}$ $\frac{1}{3}R_3 \rightarrow R_3$ then

$-12R_3 + R_1 \rightarrow R_1$
$-4R_3 + R_2 \rightarrow R_2$
$\rightarrow$
$120R_3 + R_4 \rightarrow R_4$

$\begin{bmatrix} -3 & 0 & -2 & 1 & 0 & -4 & 0 & | & 4 \\ \frac{2}{3} & 0 & \boxed{\frac{5}{3}} & 0 & 1 & -\frac{4}{3} & 0 & | & \frac{7}{3} \\ \frac{1}{3} & 1 & \frac{1}{3} & 0 & 0 & \frac{1}{3} & 0 & | & \frac{2}{3} \\ \hline -20 & 0 & -40 & 0 & 0 & 40 & 1 & | & 80 \end{bmatrix}$ $\frac{3}{5}R_2 \rightarrow R_2$ then

$2R_2 + R_1 \rightarrow R_1$
$\rightarrow$
$-\frac{1}{3}R_2 + R_3 \rightarrow R_3$
$40R_2 + R_4 \rightarrow R_4$

$\begin{bmatrix} -2.2 & 0 & 0 & 1 & 1.2 & -5.6 & 0 & | & 6.8 \\ 0.4 & 0 & 1 & 0 & 0.6 & -0.8 & 0 & | & 1.4 \\ \boxed{0.2} & 1 & 0 & 0 & -0.2 & 0.6 & 0 & | & 0.2 \\ \hline -4 & 0 & 0 & 0 & 24 & 8 & 1 & | & 136 \end{bmatrix}$ $5R_3 \rightarrow R_3$ then

$2.2R_3 + R_1 \rightarrow R_1$
$-0.4R_3 + R_2 \rightarrow R_2$
$\rightarrow$
$4R_3 + R_4 \rightarrow R_4$

$\begin{bmatrix} 0 & 11 & 0 & 1 & -1 & 1 & 0 & | & 9 \\ 0 & -2 & 1 & 0 & 1 & -2 & 0 & | & 1 \\ 1 & 5 & 0 & 0 & -1 & 3 & 0 & | & 1 \\ \hline 0 & 20 & 0 & 0 & 20 & 20 & 1 & | & 140 \end{bmatrix}$

The minimum is 140 at $y_1 = 0$, $y_2 = 20$, $y_3 = 30$.

28.
$$\begin{bmatrix} 7.5 & 4.5 & 2 & | & 650 \\ 6.5 & 3 & 1.5 & | & 400 \\ 1 & 1.5 & 0.5 & | & 200 \\ \hline 25 & 10 & 4 & | & g \end{bmatrix}$$
transpose:
$$\begin{bmatrix} 7.5 & 6.5 & 1 & | & 25 \\ 4.5 & 3 & 1.5 & | & 10 \\ 2 & 1.5 & 0.5 & | & 4 \\ \hline 650 & 400 & 200 & | & g \end{bmatrix}$$

$$\begin{bmatrix} 7.5 & 6.5 & 1 & 1 & 0 & 0 & 0 & | & 25 \\ 4.5 & 3 & 1.5 & 0 & 1 & 0 & 0 & | & 10 \\ \boxed{2} & 1.5 & 0.5 & 0 & 0 & 1 & 0 & | & 4 \\ \hline -650 & -400 & -200 & 0 & 0 & 0 & 1 & | & 0 \end{bmatrix}$$
$0.5R_3 \to R_3$ then

$-7.5R_3 + R_1 \to R_1$
$-4.5R_3 + R_2 \to R_2$
$\to$
$650R_3 + R_4 \to R_4$

$$\begin{bmatrix} 0 & 0.875 & -0.875 & 1 & 0 & -3.75 & 0 & | & 10 \\ 0 & -0.375 & \boxed{0.375} & 0 & 1 & -2.25 & 0 & | & 1 \\ 1 & 0.75 & 0.25 & 0 & 0 & 0.5 & 0 & | & 2 \\ \hline 0 & 87.5 & -37.5 & 0 & 0 & 325 & 1 & | & 1300 \end{bmatrix}$$
$\frac{8}{3}R_2 \to R_2$ then

$0.875R_2 + R_1 \to R_1$
$\to$
$-0.25R_2 + R_3 \to R_3$
$37.5R_2 + R_4 \to R_4$

$$\begin{bmatrix} 0 & 0 & 0 & 1 & \frac{7}{3} & -9 & 0 & | & 12\frac{1}{3} \\ 0 & -1 & 1 & 0 & \frac{8}{3} & -6 & 0 & | & 2\frac{2}{3} \\ 1 & 1 & 0 & 0 & -\frac{2}{3} & 2 & 0 & | & 1\frac{1}{3} \\ \hline 0 & 50 & 0 & 00 & 100 & 100 & 1 & | & 1400 \end{bmatrix}$$

Minimum is 1400 when $y_1 = 0$, $y_2 = 100$, $y_3 = 100$.

29. Maximize $f = 3x + 5y$

Constraints: $-x - y \le -19$, $x - y \le -1$, $-x + 10y \le 190$, $x, y \ge 0$

$$\begin{bmatrix} -1 & -1 & 1 & 0 & 0 & 0 & | & -19 \\ 1 & \boxed{-1} & 0 & 1 & 0 & 0 & | & -1 \\ -1 & 10 & 0 & 0 & 1 & 0 & | & 190 \\ \hline -3 & -5 & 0 & 0 & 0 & 1 & | & 0 \end{bmatrix}$$
$-R_2 \to R_2$ then

$R_2 + R_1 \to R_1$
$\to$
$-10R_2 + R_3 \to R_3$
$5R_2 + R_4 \to R_4$

$$\begin{bmatrix} \boxed{-2} & -0 & 1 & -1 & 0 & 0 & | & -18 \\ -1 & 1 & 0 & -1 & 0 & 0 & | & 1 \\ 9 & 0 & 0 & 10 & 1 & 0 & | & 180 \\ \hline -8 & 0 & 0 & -5 & 0 & 1 & | & 5 \end{bmatrix}$$
$-\frac{1}{2}R_1 \to R_1$ then

$\to$
$R_1 + R_2 \to R_2$
$-9R_1 + R_3 \to R_3$
$8R_1 + R_4 \to R_4$

$$\begin{bmatrix} 1 & 0 & -\frac{1}{2} & \frac{1}{2} & 0 & 0 & | & 9 \\ 0 & 1 & -\frac{1}{2} & -\frac{1}{2} & 0 & 0 & | & 10 \\ 0 & 0 & \boxed{\frac{9}{2}} & \frac{11}{2} & 1 & 0 & | & 99 \\ \hline 0 & 0 & -4 & -1 & 0 & 1 & | & 77 \end{bmatrix}$$
$\frac{2}{9}R_3 \to R_3$ then

$\frac{1}{2}R_3 + R_1 \to R_1$
$\frac{1}{2}R_3 + R_2 \to R_2$
$\to$
$4R_3 + R_4 \to R_4$

$$\begin{bmatrix} 1 & 0 & 0 & \frac{10}{9} & \frac{1}{9} & 0 & | & 20 \\ 0 & 1 & 0 & \frac{1}{9} & \frac{1}{9} & 0 & | & 21 \\ 0 & 0 & 1 & \frac{11}{9} & \frac{2}{9} & 0 & | & 22 \\ \hline 0 & 0 & 0 & \frac{35}{9} & \frac{8}{9} & 1 & | & 165 \end{bmatrix}$$

Maximum is 165 at $x = 20$, $y = 21$.

30.
$$\begin{bmatrix} 2 & 5 & 1 & 0 & 0 & 0 & | & 37 \\ 5 & -1 & 0 & 1 & 0 & 0 & | & 34 \\ 1 & \boxed{-2} & 0 & 0 & 1 & 0 & | & -4 \\ \hline -4 & -6 & 0 & 0 & 0 & 1 & | & 0 \end{bmatrix}$$

$-5R_3 + R_1 \to R_1$
$R_3 + R_2 \to R_2$
$-\frac{1}{2} R_3 \to R_3$ then $\to$
$6R_3 + R_4 \to R_4$

$$\begin{bmatrix} \boxed{\frac{9}{2}} & 0 & 1 & 0 & \frac{5}{2} & 0 & | & 27 \\ \frac{9}{2} & 0 & 0 & 1 & -\frac{1}{2} & 0 & | & 36 \\ -\frac{1}{2} & 1 & 0 & 0 & -\frac{1}{2} & 0 & | & 2 \\ \hline -7 & 0 & 0 & 0 & -3 & 1 & | & 12 \end{bmatrix}$$

$\frac{2}{9} R_1 \to R_1$ then $\to$
$-\frac{9}{2} R_1 + R_2 \to R_2$
$\frac{1}{2} R_1 + R_3 \to R_3$
$7R_1 + R_4 \to R_4$

$$\begin{bmatrix} 1 & 0 & \frac{2}{9} & 0 & \frac{5}{9} & 0 & | & 6 \\ 0 & 0 & -1 & 1 & -3 & 0 & | & 9 \\ 0 & 1 & \frac{1}{9} & 0 & -\frac{2}{9} & 0 & | & 5 \\ \hline 0 & 0 & \frac{14}{9} & 0 & \frac{8}{9} & 1 & | & 54 \end{bmatrix}$$

Solution: $x = 6$, $y = 5$; $f = 54$

31. Maximize : $-f = -10x - 3y$

Constraints: $x - 10y \le -5$, $-4x - y \le -62$, $x + y \le 50$, $x, y \ge 0$

$$\begin{bmatrix} 1 & \boxed{-10} & 1 & 0 & 0 & 0 & | & -5 \\ -4 & -1 & 0 & 1 & 0 & 0 & | & -62 \\ 1 & 1 & 0 & 0 & 1 & 0 & | & 50 \\ \hline 10 & 3 & 0 & 0 & 0 & 1 & | & 0 \end{bmatrix}$$

$-\frac{1}{10} R_1 \to R_1$ then $\to$
$R_1 + R_2 \to R_2$
$-R_1 + R_3 \to R_3$
$-3R_1 + R_4 \to R_4$

$$\begin{bmatrix} -\frac{1}{10} & 1 & -\frac{1}{10} & 0 & 0 & 0 & | & \frac{1}{2} \\ \boxed{-\frac{41}{10}} & 0 & -\frac{1}{10} & 1 & 0 & 0 & | & -\frac{123}{2} \\ \frac{11}{10} & 0 & \frac{1}{10} & 0 & 1 & 0 & | & \frac{99}{2} \\ \hline \frac{103}{10} & 0 & \frac{3}{10} & 0 & 0 & 1 & | & -\frac{3}{2} \end{bmatrix}$$

$\frac{1}{10} R_2 + R_1 \to R_1$
$-\frac{10}{41} R_2 \to R_2$ then $\to$
$-\frac{11}{10} R_2 + R_3 \to R_3$
$-\frac{103}{10} R_2 + R_4 \to R_4$

$$\begin{bmatrix} 0 & 1 & -\frac{4}{41} & -\frac{1}{41} & 0 & 0 & | & 2 \\ 1 & 0 & \frac{1}{41} & -\frac{10}{41} & 0 & 0 & | & 15 \\ 0 & 0 & \frac{3}{41} & \frac{11}{41} & 1 & 0 & | & 33 \\ \hline 0 & 0 & \frac{2}{41} & \frac{103}{41} & 0 & 1 & | & -156 \end{bmatrix}$$

Minimum is 156 at $x = 15$, $y = 2$.

32. Maximize: $-f = -4x - 3y$

Subject to: $x - y \le -1$, $x + y \le 45$, $-10x - y \le -45$

$$\begin{bmatrix} 1 & \boxed{-1} & 1 & 0 & 0 & 0 & | & -1 \\ 1 & 1 & 0 & 1 & 0 & 0 & | & 45 \\ -10 & -1 & 0 & 0 & 1 & 0 & | & -45 \\ \hline 4 & 3 & 0 & 0 & 0 & 1 & | & 0 \end{bmatrix}$$

$-R_1 \to R_1$ then $\to$
$-R_1 + R_2 \to R_2$
$R_1 + R_3 \to R_3$
$-3R_1 + R_4 \to R_4$

$$\begin{bmatrix} -1 & 1 & -1 & 0 & 0 & 0 & | & 1 \\ 2 & 0 & 1 & 1 & 0 & 0 & | & 44 \\ \boxed{-11} & 0 & -1 & 0 & 1 & 0 & | & -44 \\ \hline 7 & 0 & 3 & 0 & 0 & 1 & | & -3 \end{bmatrix}$$

$R_3 + R_1 \to R_1$
$-2R_3 + R_2 \to R_2$
$-\frac{1}{11} R_3 \to R_3$ then $\to$
$-7R_3 + R_4 \to R_4$

$$\begin{bmatrix} 0 & 1 & -\frac{10}{11} & 0 & -\frac{1}{11} & 0 & | & 5 \\ 0 & 0 & \frac{9}{11} & 1 & \frac{2}{11} & 0 & | & 36 \\ 1 & 0 & \frac{1}{11} & 0 & -\frac{1}{11} & 0 & | & 4 \\ \hline 0 & 0 & \frac{26}{11} & 0 & \frac{7}{11} & 1 & | & -31 \end{bmatrix}$$

Minimum is 31 at $x = 4$, $y = 5$.

33. Maximize: $f = 8x_1 + 10x_2 + 12x_3 + 14x_4$

Constraints: $-6x_1 - 3x_2 - 2x_3 - x_4 \le -350$

$$3x_1 + 2x_2 + 5x_3 + 6x_4 \le 300$$

$$8x_1 + 3x_2 + 2x_3 + x_4 \le 400$$

$$x_1 + x_2 + x_3 + x_4 \le 100$$

$$\begin{bmatrix} -6 & -3 & -2 & -1 & 1 & 0 & 0 & 0 & 0 & -350 \\ 3 & 2 & 5 & 6 & 0 & 1 & 0 & 0 & 0 & 300 \\ 8 & 3 & 2 & 1 & 0 & 0 & 1 & 0 & 0 & 400 \\ 1 & 1 & 1 & 1 & 0 & 0 & 0 & 1 & 0 & 100 \\ \hline -8 & -10 & -12 & -14 & 0 & 0 & 0 & 0 & 1 & 0 \end{bmatrix}$$

Using technology, the maximum is 1000 at $x_1 = 25$, $x_2 = 62.5$, $x_3 = 0$, $x_4 = 12.5$.

34. Maximize: $-g = -10x_1 - 9x_2 - 12x_3 - 8x_4$

Constraints: $-45x_1 - 58.5x_3 \le -4680$

$$-36x_2 - 31.5x_4 \le -4230$$

$$x_1 + x_2 \le 100$$

$$x_3 + x_4 \le 100$$

$$\begin{bmatrix} -45 & 0 & -58.5 & 0 & 1 & 0 & 0 & 0 & 0 & -4680 \\ 0 & -36 & 0 & -31.5 & 0 & 1 & 0 & 0 & 0 & -4230 \\ 1 & 1 & 0 & 0 & 0 & 0 & 1 & 0 & 0 & 100 \\ 0 & 0 & 1 & 1 & 0 & 0 & 0 & 1 & 0 & 100 \\ \hline 10 & 9 & 12 & 8 & 0 & 0 & 0 & 0 & 1 & 0 \end{bmatrix}$$

Using technology, the minimum is 2020 at $x_1 = 0$, $x_2 = 100$, $x_3 = 80$, $x_4 = 20$.

35. Let x = number of large sets and y = number of small sets.

Maximize $f = 100x + 50y$

Constraints: $5x + 2y \le 700$, $x + y \le 185$

Solving the two equations we get $A(110, 75)$.

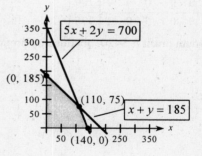

Feasible Corners	$f = 100x + 50y$
$(0, 185)$	9250
$(110, 75)$	14,750
$(140, 0)$	14,000

Maximum profit is $14,750 with 110 large sets and 75 small sets.

36. x_1 = days for factory 1, x_2 = days for factory 2

Minimize: $g = 5000x_1 + 6000x_2$

Subject to: $x_1 + 2x_2 \ge 80$

$$3x_1 + 2x_2 \ge 140$$

$$x_1, x_2 \ge 0$$

$x_1 + 2x_2 = 80$ and $3x_1 + 2x_2 = 140$ intersect at $x_1 = 30, x_2 = 25$.

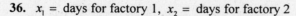

The feasible corners are $(80, 0)$, $(30, 25)$, and $(0, 70)$.

At $(80, 0)$, $g = 400,000$

At $(30, 25)$, $g = 300,000$ ← Minimum

At $(0, 70)$, $g = 420,000$

The minimum cost is $300,000 when factory 1 operates 30 days and factory 2 operates 25 days.

37. Let x = number of Is and y = number of IIs.

Maximize $P = 6x + 4y$

Constraints: $2x + y \leq 100$, $x + y \leq 60$

$$\begin{bmatrix} \boxed{2} & 1 & 1 & 0 & 0 & | & 100 \\ 1 & 1 & 0 & 1 & 0 & | & 60 \\ -6 & -4 & 0 & 0 & 1 & | & 0 \end{bmatrix}$$

$\frac{1}{2}R_1 \rightarrow R_1$ then

$\rightarrow$

$-R_1 + R_2 \rightarrow R_2$

$6R_1 + R_3 \rightarrow R_3$

$$\begin{bmatrix} 1 & \frac{1}{2} & \frac{1}{2} & 0 & 0 & | & 50 \\ 0 & \boxed{\frac{1}{2}} & -\frac{1}{2} & 1 & 0 & | & 10 \\ 0 & -1 & 3 & 0 & 1 & | & 300 \end{bmatrix}$$

$2R_2 \rightarrow R_2$ then

$-\frac{1}{2}R_2 + R_1 \rightarrow R_1$

$\rightarrow$

$R_2 + R_3 \rightarrow R_3$

$$\begin{bmatrix} 1 & 0 & 1 & -1 & 0 & | & 40 \\ 0 & 1 & -1 & 2 & 0 & | & 20 \\ 0 & 0 & 2 & 2 & 1 & | & 320 \end{bmatrix}$$

Maximum profit is $320 with 40 of the Is and 20 of the IIs.

38. Let x = number of Jacob's Ladders, let y = number of locomotive engines.

Maximize $P = 3x + 5y$

Constraints: $x + y \leq 120$ finishing hours

           $\frac{1}{2}x + y \leq 75$ carpentry hours

           $x \leq 100$

           $x, y \geq 0$

$$\begin{bmatrix} 1 & 1 & 1 & 0 & 0 & 0 & | & 120 \\ \frac{1}{2} & \boxed{1} & 0 & 1 & 0 & 0 & | & 75 \\ 1 & 0 & 0 & 0 & 1 & 0 & | & 100 \\ -3 & -5 & 0 & 0 & 0 & 1 & | & 0 \end{bmatrix}$$

$-R_2 + R_1 \rightarrow R_1$

$\rightarrow$

$\rightarrow$

$5R_2 + R_4 \rightarrow R_4$

$$\begin{bmatrix} \frac{1}{2} & 0 & 1 & -1 & 0 & 0 & | & 45 \\ \frac{1}{2} & 1 & 0 & 1 & 0 & 0 & | & 75 \\ 1 & 0 & 0 & 0 & 1 & 0 & | & 100 \\ -\frac{1}{2} & 0 & 0 & 5 & 0 & 1 & | & 375 \end{bmatrix}$$

$2R_1 \rightarrow R_1$ then

$-\frac{1}{2}R_1 + R_2 \rightarrow R_2$

$-R_1 + R_3 \rightarrow R_3$

$\frac{1}{2}R_1 + R_4 \rightarrow R_4$

$$\begin{bmatrix} 1 & 0 & 2 & -2 & 0 & 0 & | & 90 \\ 0 & 1 & -1 & 2 & 0 & 0 & | & 30 \\ 0 & 0 & -2 & 2 & 1 & 0 & | & 10 \\ 0 & 0 & 1 & 4 & 0 & 1 & | & 420 \end{bmatrix}$$

Maximum profit is $420, obtained by producing 90 Jacob's Ladders and 30 locomotive engines.

39. a. $x_1 =$ number of 27-in. HD models, $x_2 =$ number of 32-in. HD models,

$x_3 =$ number of 42-in. HD models, $x_4 =$ number of 42-in. plasma models

b. Maximize: $f = 80x_1 + 120x_2 + 160x_3 + 200x_4$

Constraints: $8x_1 + 10x_2 + 12x_3 + 15x_4 \leq 1870$

$$2x_1 + 4x_2 + 4x_3 + 4x_4 \leq 530$$

$$x_1 + x_2 + x_3 + x_4 \leq 200$$

$$x_3 + x_4 \leq 100$$

$$x_2 \leq 120$$

c.
$$\begin{bmatrix} 8 & 10 & 12 & 15 & 1 & 0 & 0 & 0 & 0 & 0 & 1870 \\ 2 & 4 & 4 & 4 & 0 & 1 & 0 & 0 & 0 & 0 & 530 \\ 1 & 1 & 1 & 1 & 0 & 0 & 1 & 0 & 0 & 0 & 200 \\ 0 & 0 & 1 & 1 & 0 & 0 & 0 & 1 & 0 & 0 & 100 \\ 0 & 1 & 0 & 0 & 0 & 0 & 0 & 0 & 1 & 0 & 120 \\ \hline -80 & -120 & -160 & -200 & 0 & 0 & 0 & 0 & 0 & 1 & 0 \end{bmatrix}$$

Using technology to solve the problem, Nolmaur Electronics can manufacture 15 27-in. HD models, 25 32-in. HD models, and 100 42-in. plasma models for a maximum profit of $24,200.

40. We display the information in a table.

	x	y	
	food I	food II	Requirement
A	2	10	5
B	1	10	30
cost	30 cents/ ounce	20 cents/ ounce	

Minimize: $g = 30x + 20y$

Subject to: $2x + 10y \geq 5$

$$x + 10y \geq 30$$

$$x, y \geq 0$$

$$\begin{bmatrix} 2 & 10 & 5 \\ 1 & 10 & 30 \\ \hline 30 & 20 & g \end{bmatrix} \quad \text{transpose:} \quad \begin{bmatrix} 2 & 1 & 30 \\ 10 & 10 & 20 \\ \hline 5 & 30 & g \end{bmatrix}$$

Maximization dual problem:

$$\begin{bmatrix} 2 & 1 & 1 & 0 & 0 & 30 \\ 10 & \boxed{10} & 0 & 1 & 0 & 20 \\ \hline -5 & -30 & 0 & 0 & 1 & 0 \end{bmatrix} \begin{array}{c} \\ \frac{1}{10}R_2 \to R_2 \text{ then} \\ \\ \end{array} \begin{array}{c} -R_2 + R_1 \to R_1 \\ \to \\ 30R_2 + R_3 \to R_3 \end{array} \begin{bmatrix} 1 & 0 & 1 & -\frac{1}{10} & 0 & 28 \\ 1 & 1 & 0 & \frac{1}{10} & 0 & 2 \\ \hline 25 & 0 & 0 & 3 & 1 & 60 \end{bmatrix}$$

The nutritionist should use none of food I and 3 ounces of food II for a minimum cost of $0.60.

41. Minimize $C = 14A + 16B$

Constraints: $A + 4B \geq 40$

$$2A + B \geq 80$$

$$\begin{bmatrix} 1 & 4 & | & 40 \\ 2 & 1 & | & 80 \\ \hline 14 & 16 & | & C \end{bmatrix} \quad \text{Dual:} \quad \begin{bmatrix} 1 & 2 & | & 14 \\ 4 & 1 & | & 16 \\ \hline 40 & 80 & | & C \end{bmatrix}$$

$$\begin{bmatrix} 1 & \boxed{2} & 1 & 0 & 0 & | & 14 \\ 4 & 1 & 0 & 1 & 0 & | & 16 \\ \hline -40 & -80 & 0 & 0 & 1 & | & 0 \end{bmatrix} \quad \begin{array}{l} \tfrac{1}{2}R_1 \to R_1 \text{ then} \\ \\ 80R_1 + R_3 \to R_3 \end{array} \quad -R_1 + R_2 \to R_2 \quad \begin{bmatrix} \tfrac{1}{2} & 1 & \tfrac{1}{2} & 0 & 0 & | & 7 \\ \tfrac{7}{2} & 0 & -\tfrac{1}{2} & 1 & 0 & | & 9 \\ \hline 0 & 0 & 40 & 0 & 1 & | & 560 \end{bmatrix}$$

The laboratory should purchase 40 pounds of Feed A and none of Feed B for a minimum cost of $5.60.

42. x = days at factory A, y = days of factory B, z = days at factory C

Minimize $g = 200x + 300y + 500z$

Subject to:

$$\begin{array}{lll} 10x + 20z \geq 200 & \to & x + 2z \geq 20 \\ 10x + 20y + 20z \geq 500 & & x + 2y + 2z \geq 50 \\ 10x + 20y + 10z \geq 300 & & x + 2y + z \geq 30 \\ \quad x, y, z \geq 0 \end{array}$$

$$\begin{bmatrix} 1 & 0 & 2 & | & 20 \\ 1 & 2 & 2 & | & 50 \\ 1 & 2 & 1 & | & 30 \\ \hline 200 & 300 & 500 & | & g \end{bmatrix} \quad \text{Dual:} \quad \begin{bmatrix} 1 & 1 & 1 & | & 200 \\ 0 & 2 & 2 & | & 300 \\ 2 & 2 & 1 & | & 500 \\ \hline 20 & 50 & 30 & | & g \end{bmatrix}$$

$$\begin{bmatrix} 1 & 1 & 1 & 1 & 0 & 0 & 0 & | & 200 \\ 0 & \boxed{2} & 2 & 0 & 1 & 0 & 0 & | & 300 \\ 2 & 2 & 1 & 0 & 0 & 1 & 0 & | & 500 \\ \hline -20 & -50 & -30 & 0 & 0 & 0 & 1 & | & 0 \end{bmatrix} \quad \tfrac{1}{2}R_2 \to R_2 \text{ then} \quad \begin{array}{l} -R_2 + R_1 \to R_1 \\ \to \\ -2R_2 + R_3 \to R_3 \\ 50R_2 + R_4 \to R_4 \end{array}$$

$$\begin{bmatrix} 1 & 0 & 0 & 1 & -\tfrac{1}{2} & 0 & 0 & | & 50 \\ 0 & 1 & 1 & 0 & \tfrac{1}{2} & 0 & 0 & | & 150 \\ 2 & 0 & -1 & 0 & -1 & 1 & 0 & | & 200 \\ \hline -20 & 0 & 20 & 0 & 25 & 0 & 1 & | & 7500 \end{bmatrix} \quad \begin{array}{l} \to \\ \\ -2R_1 + R_3 \to R_3 \\ 20R_1 + R_4 \to R_4 \end{array} \quad \begin{bmatrix} 1 & 0 & 0 & 1 & -\tfrac{1}{2} & 0 & 0 & | & 50 \\ 0 & 1 & 1 & 0 & \tfrac{1}{2} & 0 & 0 & | & 150 \\ 0 & 0 & -1 & -2 & 0 & 1 & 0 & | & 100 \\ \hline 0 & 0 & 20 & 20 & 15 & 0 & 1 & | & 8500 \end{bmatrix}$$

The company should operate factory A for 20 days and Factory B for 15 days for a minimum cost of $8500.

43.

	Flour	Shortening	Sugar	Objective Function
Pancake: x	$0.6x$	$0.1x$	$--$	$P = 0.35x + 0.25y$
Cake: y	$0.4y$	$0.1y$	$0.4y$	

Constraints:
$$0.6x + 0.4y \leq 6000 \quad \text{Flour}$$
$$0.1x + 0.1y \geq 500 \quad \text{Shortening}$$
$$0.4y \leq 1200 \quad \text{Sugar}$$
$$x, y \geq 0$$

$$\begin{bmatrix} 0.6 & 0.4 & 1 & 0 & 0 & 0 & | & 6000 \\ \boxed{-0.1} & -0.1 & 0 & 1 & 0 & 0 & | & -500 \\ 0 & 0.4 & 0 & 0 & 1 & 0 & | & 1200 \\ \hline -0.35 & -0.25 & 0 & 0 & 0 & 1 & | & 0 \end{bmatrix}$$

$-0.6R_2 + R_1 \to R_1$
$-10R_2 \to R_2$ then $\to$
$0.35R_2 + R_4 \to R_4$

$$\begin{bmatrix} 0 & -0.2 & 1 & \boxed{6} & 0 & 0 & | & 3000 \\ 1 & 1 & 0 & -10 & 0 & 0 & | & 5000 \\ 0 & 0.4 & 0 & 0 & 1 & 0 & | & 1200 \\ \hline 0 & 0.1 & 0 & -3.5 & 0 & 1 & | & 1750 \end{bmatrix}$$

$\frac{1}{6}R_1 \to R_1$ then $\to$
$10R_1 + R_2 \to R_2$
$\to$
$3.5R_1 + R_4 \to R_4$

$$\begin{bmatrix} 0 & -0.0\overline{3} & 0.1\overline{6} & 1 & 0 & 0 & | & 500 \\ 1 & 0.\overline{6} & 1.\overline{6} & 0 & 0 & 0 & | & 10{,}000 \\ 0 & \boxed{0.4} & 0 & 0 & 1 & 0 & | & 1200 \\ \hline 0 & -0.01\overline{6} & 0.58\overline{3} & 0 & 0 & 1 & | & 3500 \end{bmatrix}$$

$0.0\overline{3}R_3 + R_1 \to R_1$
$-0.\overline{6}R_3 + R_2 \to R_2$
$2.5R_3 \to R_3$ then $\to$
$0.01\overline{6}R_3 + R_4 \to R_4$

$$\begin{bmatrix} 0 & 0 & 0.1\overline{6} & 1 & 0.08\overline{3} & 0 & | & 600 \\ 1 & 0 & 1.\overline{6} & 0 & -1.\overline{6} & 0 & | & 8000 \\ 0 & 1 & 0 & 0 & 2.5 & 0 & | & 3000 \\ \hline 0 & 0 & 0.58\overline{3} & 0 & 0.041\overline{6} & 1 & | & 3550 \end{bmatrix}$$

The company should make 8000 lbs of pancake mix and 3000 lbs of cake mix for a maximum profit of $3550.

44. x_1 = number of desks at Texas, x_2 = number of tables at Texas,
x_3 = number of desks at Louisiana, x_4 = number of tables at Louisiana
Minimize $g = 12x_1 + 20x_2 + 14x_3 + 19x_4$
Subject to: $x_1 + x_2 \le 120,\ x_3 + x_4 \le 150,\ x_1 + x_3 \ge 130,\ x_2 + x_4 \ge 130,\ -x_1 + x_2 \ge 10,\ x_1, x_2, x_3, x_4 \ge 0$

$$\left[\begin{array}{rrrrrrrrrr|r}
1 & 1 & 0 & 0 & 1 & 0 & 0 & 0 & 0 & 0 & 120 \\
0 & 0 & 1 & 1 & 0 & 1 & 0 & 0 & 0 & 0 & 150 \\
\boxed{-1} & 0 & -1 & 0 & 0 & 0 & 1 & 0 & 0 & 0 & -130 \\
0 & -1 & 0 & -1 & 0 & 0 & 0 & 1 & 0 & 0 & -130 \\
1 & -1 & 0 & 0 & 0 & 0 & 0 & 0 & 1 & 0 & -10 \\
12 & 20 & 14 & 19 & 0 & 0 & 0 & 0 & 0 & 1 & 0
\end{array}\right]$$

$-R_3 + R_1 \to R_1$
$\to$
$-R_3 \to R_3$ then
$\to$
$\to$
$\to$
$-R_3 + R_5 \to R_5$
$-12R_3 + R_6 \to R_6$

$$\left[\begin{array}{rrrrrrrrrr|r}
0 & 1 & -1 & 0 & 1 & 0 & 1 & 0 & 0 & 0 & -10 \\
0 & 0 & 1 & 1 & 0 & 1 & 0 & 0 & 0 & 0 & 150 \\
1 & 0 & 1 & 0 & 0 & 0 & -1 & 0 & 0 & 0 & 130 \\
0 & -1 & 0 & -1 & 0 & 0 & 0 & 1 & 0 & 0 & -130 \\
0 & \boxed{-1} & -1 & 0 & 0 & 0 & 1 & 0 & 1 & 0 & -140 \\
0 & 20 & 2 & 19 & 0 & 0 & 12 & 0 & 0 & 1 & -1560
\end{array}\right]$$

$R_5 + R_1 \to R_1$
$\to$
$\to$
$-R_5 + R_4 \to R_4$
$\to$
$20R_5 + R_6 \to R_6$
then $-R_5 \to R_5$

$$\left[\begin{array}{rrrrrrrrrr|r}
0 & 0 & -2 & 0 & 1 & 0 & 2 & 0 & 1 & 0 & -150 \\
0 & 0 & 1 & 1 & 0 & 1 & 0 & 0 & 0 & 0 & 150 \\
1 & 0 & 1 & 0 & 0 & 0 & -1 & 0 & 0 & 0 & 130 \\
0 & 0 & \boxed{1} & -1 & 0 & 0 & -1 & 1 & -1 & 0 & 10 \\
0 & 1 & 1 & 0 & 0 & 0 & -1 & 0 & -1 & 0 & 140 \\
0 & 0 & -18 & 19 & 0 & 0 & 32 & 0 & 20 & 1 & -4360
\end{array}\right]$$

$2R_4 + R_1 \to R_1$
$-R_4 + R_2 \to R_2$
$-R_4 + R_3 \to R_3$
$\to$
$-R_4 + R_5 \to R_5$
$18R_4 + R_6 \to R_6$

$$\left[\begin{array}{rrrrrrrrrr|r}
0 & 0 & 0 & \boxed{-2} & 1 & 0 & 0 & 2 & -1 & 0 & -130 \\
0 & 0 & 0 & 2 & 0 & 1 & 1 & -1 & 1 & 0 & 140 \\
1 & 0 & 0 & 1 & 0 & 0 & 0 & -1 & 1 & 0 & 120 \\
0 & 0 & 1 & -1 & 0 & 0 & -1 & 1 & -1 & 0 & 10 \\
0 & 1 & 0 & 1 & 0 & 0 & 0 & -1 & 0 & 0 & 130 \\
0 & 0 & 0 & 1 & 0 & 0 & 14 & 18 & 2 & 1 & -4180
\end{array}\right]$$

then $-\tfrac{1}{2}R_1 \to R_1$
$R_1 + R_2 \to R_2$
$\tfrac{1}{2}R_1 + R_3 \to R_3$
$-\tfrac{1}{2}R_1 + R_4 \to R_4$
$\tfrac{1}{2}R_1 + R_5 \to R_5$
$\tfrac{1}{2}R_1 + R_6 \to R_6$

$$\left[\begin{array}{rrrrrrrrrr|r}
0 & 0 & 0 & 1 & -\tfrac{1}{2} & 0 & 0 & -1 & \tfrac{1}{2} & 0 & 65 \\
0 & 0 & 0 & 0 & 1 & 1 & 1 & 1 & 0 & 0 & 10 \\
1 & 0 & 0 & 0 & \tfrac{1}{2} & 0 & 0 & 0 & \tfrac{1}{2} & 0 & 55 \\
0 & 0 & 1 & 0 & -\tfrac{1}{2} & 0 & -1 & 0 & -\tfrac{1}{2} & 0 & 75 \\
0 & 1 & 0 & 0 & \tfrac{1}{2} & 0 & 0 & 0 & -\tfrac{1}{2} & 0 & 65 \\
0 & 0 & 0 & 0 & \tfrac{1}{2} & 0 & 14 & 19 & \tfrac{3}{2} & 1 & -4245
\end{array}\right]$$

The company should schedule the plant in Texas to make 55 desks and 65 tables, and the plant in Louisiana to make 75 desks and 65 tables for a minimum cost of $4245.

45. x_1 = number of tons of Grade 1 steel made each week in Midland

x_2 = number of tons of Grade 2 steel made each week in Midland

y_1 = number of tons of Grade 1 steel made each week in Donora

y_2 = number of tons of Grade 2 steel made each week in Donora

Minimize: $f = 100x_1 + 120x_2 + 110y_1 + 90y_2$

Maximize: $-f = -100x_1 - 120x_2 - 110y_1 - 90y_2$

Constraints: $40x_1 + 42x_2 \le 19,460$ Revised Constraints: $40x_1 + 42x_2 \le 19,460$

$44y_1 + 45y_2 \le 21,380$ $44y_1 + 45y_2 \le 21,380$

$x_1 + y_1 \ge 500$ $-x_1 - y_1 \le -500$

$x_2 + y_2 \ge 450$ $-x_2 - y_2 \le -450$

$$\begin{bmatrix} 40 & 42 & 0 & 0 & 1 & 0 & 0 & 0 & 0 & 19,460 \\ 0 & 0 & 44 & 45 & 0 & 1 & 0 & 0 & 0 & 21,380 \\ -1 & 0 & -1 & 0 & 0 & 0 & 1 & 0 & 0 & -500 \\ 0 & -1 & 0 & -1 & 0 & 0 & 0 & 1 & 0 & -450 \\ \hline 100 & 120 & 110 & 90 & 0 & 0 & 0 & 0 & 1 & 0 \end{bmatrix}$$

Using technology to solve the problem, Armstrong Industries should instruct the Midland plant to make 486.5 tons of Grade 1 steel, and instruct the Donora plant to make 13.5 tons of Grade 1 steel and 450 tons of Grade 2 steel for a minimum cost of $90,635.

Chapter Test

1.

Corners	$f = 3x + 5y$
$(0, 24)$	120
$(8, 18)$	114
$(20, 0)$	60
$(0, 0)$	0

Maximum is 120 at $(0, 24)$.

2. a. C, since there is a zero indicator in a nonbasic variable column. Pivot on the 1 in $R_2 C_2$.

Row Ops: $-2R_2 + R_1 \rightarrow R_1; -3R_2 + R_3 \rightarrow R_3$

b. A; Pivot column is column 3, but no pivot element is defined.

c. B　$x_1 = 40, x_2 = 12, x_3 = 0,$
　$s_1 = 0, s_2 = 20, s_3 = 0, f = 170$

3. a.

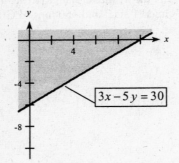

b.

c.

4.
$$\begin{bmatrix} 3 & 5 & 1 & | & 100 \\ 4 & 6 & 3 & | & 120 \\ \hline 2 & 3 & 5 & | & g \end{bmatrix}$$

Take the transpose to obtain:

Maximize: $f = 100x_1 + 120x_2$

Subject to: $3x_1 + 4x_2 \leq 2$
　　　$5x_1 + 6x_2 \leq 3$
　　　　$x_1 + 3x_2 \leq 5$
　　　　$x_1, x_2 \geq 0$

5. Solving: $\begin{cases} 4x + y = 12 \\ x + y = 9 \end{cases}$

$\begin{cases} 4x + y = 12 \\ -x - y = -9 \end{cases}$
　　$3x = 3$
　　　$x = 1, \ y = 9$

Solving: $\begin{cases} x + y = 9 \\ x + 3y = 15 \end{cases}$

$\begin{cases} -x - y = -9 \\ x + 3y = 15 \end{cases}$
　　$2y = 6$
　　　$y = 3, \ x = 6$

Corners	$f = 5x + 2y$
$(0, 12)$	24
$(1, 9)$	23
$(6, 3)$	36
$(15, 0)$	75

Minimum is 23 at $(1, 9)$.

There is no maximum since the region is unbounded.

6.

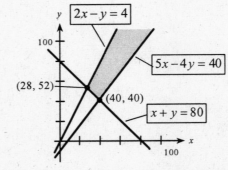

Minimum is 136 at $(28, 52)$. There are only two feasible corners.

7. Maximize (change signs) $-g = -7x - 3y$

Subject to: $x - 4y \leq -4$

$\qquad x - y \leq 5$

$\qquad 2x + 3y \leq 30$

8. Maximum is $f = 658$ at $x_1 = 17, x_2 = 15, x_3 = 0$.

Minimum is $g = 658$ at $y_1 = 4, y_2 = 18, y_3 = 0$.

9. Maximize: $f = 70x + 5y$

Subject to: $x + 1.5y \leq 150$

$\qquad x + 0.5y \leq 90$

$\qquad x, y \geq 0$

Clear the decimals by multiplying each inequality by 2.

$$\begin{bmatrix} 2 & 3 & 1 & 0 & 0 & | & 300 \\ \boxed{2} & 1 & 0 & 1 & 0 & | & 180 \\ \hline -70 & -5 & 0 & 0 & 1 & | & 0 \end{bmatrix} \quad \tfrac{1}{2}R_2 \to R_2 \text{ then}$$

$$\begin{array}{c} -2R_2 + R_1 \to R_1 \\ \to \\ 70R_2 + R_3 \to R_3 \end{array} \begin{bmatrix} 0 & 2 & 1 & -1 & 0 & | & 120 \\ 1 & \tfrac{1}{2} & 0 & \tfrac{1}{2} & 0 & | & 90 \\ 0 & 30 & 0 & 35 & 1 & | & 6300 \end{bmatrix}$$

Maximum is 6300 at $x = 90, y = 0$.

10.

$$\left[\begin{array}{cccccccc|c} 8 & 6 & 1 & 1 & 0 & 0 & 0 & 0 & 108 \\ \boxed{4} & 2 & 1.5 & 0 & 1 & 0 & 0 & 0 & 50 \\ 2 & 1.5 & 0.5 & 0 & 0 & 1 & 0 & 0 & 40 \\ 0 & 0 & 1 & 0 & 0 & 0 & 1 & 0 & 16 \\ \hline -60 & -48 & -36 & 0 & 0 & 0 & 0 & 1 & 0 \end{array}\right]$$

$\frac{1}{4}R_2 \to R_2$ then

$-8R_2 + R_1 \to R_1$

$\to$

$-2R_2 + R_3 \to R_3$

$\to$

$60R_2 + R_5 \to R_5$

$$\left[\begin{array}{cccccccc|c} 0 & \boxed{2} & -2 & 1 & -2 & 0 & 0 & 0 & 8 \\ 1 & 0.5 & 0.375 & 0 & 0.25 & 0 & 0 & 0 & 12.5 \\ 0 & 0.5 & -0.25 & 0 & -0.5 & 1 & 0 & 0 & 15 \\ 0 & 0 & 1 & 0 & 0 & 0 & 1 & 0 & 16 \\ \hline 0 & -18 & -13.5 & 0 & 15 & 0 & 0 & 1 & 750 \end{array}\right]$$

$\frac{1}{2}R_1 \to R_1$ then

$\to$

$-0.5R_1 + R_2 \to R_2$

$-0.5R_1 + R_3 \to R_3$

$\to$

$18R_1 + R_5 \to R_5$

$$\left[\begin{array}{cccccccc|c} 0 & 1 & -1 & 0.5 & -1 & 0 & 0 & 0 & 4 \\ 1 & 0 & \boxed{0.875} & -0.25 & 0.75 & 0 & 0 & 0 & 10.5 \\ 0 & 0 & 0.25 & -0.25 & 0 & 1 & 0 & 0 & 13 \\ 0 & 0 & 1 & 0 & 0 & 0 & 1 & 0 & 16 \\ \hline 0 & 0 & -31.5 & 9 & -3 & 0 & 0 & 1 & 822 \end{array}\right]$$

$R_2 + R_1 \to R_1$

$\to$

$\frac{8}{7}R_2 \to R_2$ then

$-0.25R_2 + R_3 \to R_3$

$-R_2 + R_4 \to R_4$

$31.5R_2 + R_5 \to R_5$

$$\left[\begin{array}{cccccccc|c} \frac{8}{7} & 1 & 0 & \frac{3}{14} & -\frac{2}{7} & 0 & 0 & 0 & 16 \\ \frac{8}{7} & 0 & 1 & -\frac{2}{7} & \frac{7}{8} & 0 & 0 & 0 & 12 \\ -\frac{2}{7} & 0 & 0 & -\frac{5}{28} & -\frac{3}{14} & 1 & 0 & 0 & 10 \\ -\frac{8}{7} & 0 & 0 & \frac{2}{7} & -\frac{7}{8} & 0 & 1 & 0 & 4 \\ \hline 36 & 0 & 0 & 0 & 24 & 0 & 0 & 1 & 1200 \end{array}\right]$$

The maximum is 1200 at $x = 0$, $y = 16$, $z = 12$.

11. Let $x =$ barrels of lager, $y =$ barrels of ale.

Maximize: $P = 35x + 30y$

Subject to: $3x + 2y \leq 1200$

$\qquad\qquad 2x + 2y \leq 1000$

$$\left[\begin{array}{ccccc|c} 3 & 2 & 1 & 0 & 0 & 1200 \\ 2 & 2 & 0 & 1 & 0 & 1000 \\ \hline -35 & -30 & 0 & 0 & 1 & 0 \end{array}\right]$$

$\frac{1}{3}R_1 \to R_1$

$$\left[\begin{array}{ccccc|c} 1 & \frac{2}{3} & \frac{1}{3} & 0 & 0 & 400 \\ 2 & 2 & 0 & 1 & 0 & 1000 \\ \hline -35 & -30 & 0 & 0 & 1 & 0 \end{array}\right]$$

$\to$

$-2R_1 + R_2 \to R_2$

$35R_1 + R_3 \to R_3$

$$\left[\begin{array}{ccccc|c} 1 & \frac{2}{3} & \frac{1}{3} & 0 & 0 & 400 \\ 0 & \frac{2}{3} & -\frac{2}{3} & 1 & 0 & 200 \\ \hline 0 & -\frac{20}{3} & \frac{35}{3} & 0 & 1 & 14{,}000 \end{array}\right]$$

$-R_2 + R_1 \to R_1$

$\to$

$10R_2 + R_3 \to R_3$

$$\left[\begin{array}{ccccc|c} 1 & 0 & 1 & -1 & 0 & 200 \\ 0 & \frac{2}{3} & -\frac{2}{3} & 1 & 0 & 200 \\ \hline 0 & 0 & 5 & 10 & 1 & 16{,}000 \end{array}\right]$$

$\frac{3}{2}R_2 \to R_2$

$$\left[\begin{array}{ccccc|c} 1 & 0 & 1 & -1 & 0 & 200 \\ 0 & 1 & -1 & \frac{3}{2} & 0 & 300 \\ \hline 0 & 0 & 5 & 10 & 1 & 16{,}000 \end{array}\right]$$

River Brewery should produce 200 barrels of lager and 300 barrels of ale for a maximum profit of $16,000.

12. Let x = number of day calls, let y = number of evening calls.

Minimize $C = 3x + 4y$

Subject to: $0.3x + 0.3y \geq 150$

$\qquad\qquad 0.1x + 0.3y \geq 120$

$\qquad\qquad\qquad x, y \geq 0$

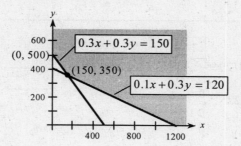

Corners	$C = 3x + 4y$
$(0, 500)$	2000
$(150, 350)$	1850
$(1200, 0)$	3600

The marketing research group should make 150 day calls and 350 evening calls for a minimum cost of $1850.

13. x_1 = tons of Product 1, x_2 = tons of Product 2, x_3 = tons of Product 3, and x_4 = tons of Product 4

Maximize: $f = 40x_1 + 50x_2 + 60x_3 + 70x_4$

Constraints: $0.60x_1 + 0.30x_2 + 0.20x_3 + 0.10x_4 \geq 35$ Revised: $-6x_1 - 3x_2 - 2x_3 - x_4 \geq -350$

$\qquad\qquad 0.20x_1 + 0.20x_2 + 0.20x_3 + 0.20x_4 \leq 20$ $\qquad\qquad x_1 + x_2 + x_3 + x_4 \leq 100$

$\qquad\qquad 0.09x_1 + 0.06x_2 + 0.15x_3 + 0.18x_4 \leq 9$ $\qquad\qquad 9x_1 + 6x_2 + 15x_3 + 18x_4 \leq 900$

$\qquad\qquad 0.08x_1 + 0.03x_2 + 0.02x_3 + 0.01x_4 \leq 4$ $\qquad\qquad 8x_1 + 3x_2 + 2x_3 + x_4 \leq 400$

$$\begin{bmatrix} -6 & -3 & -2 & -1 & 1 & 0 & 0 & 0 & 0 & -350 \\ 1 & 1 & 1 & 1 & 0 & 1 & 0 & 0 & 0 & 100 \\ 9 & 6 & 15 & 18 & 0 & 0 & 1 & 0 & 0 & 900 \\ 8 & 3 & 2 & 1 & 0 & 0 & 0 & 1 & 0 & 400 \\ \hline -40 & -50 & -60 & -70 & 0 & 0 & 0 & 0 & 1 & 0 \end{bmatrix}$$

Using technology to solve the problem, Lawn Rich should use 25 tons of product 1, 62.5 tons of product 2, and 12.5 tons of product 4 for a maximum profit of $5000.

Chapter 5: Exponential and Logarithmic Functions

Exercise 5.1

1. $10^{0.5}$ 10 $\boxed{y^x}$ 0.5 $\boxed{=}$ 3.1623

3. $5^{-2.7}$ 5 $\boxed{y^x}$ 2.7 $\boxed{\pm}$ $\boxed{=}$ 0.012965

5. $3^{1/3}$ 3 $\boxed{y^x}$ 0.3333 $\boxed{=}$ 1.44225

7. e^2 e $\boxed{y^x}$ 2 $\boxed{=}$ 7.3891

9.
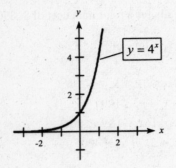

11.
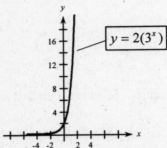

13.
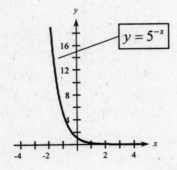

15.
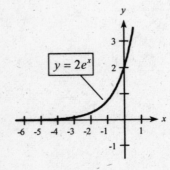

17.
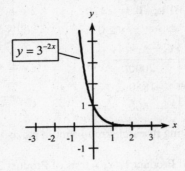

19. a. $\quad y = 3\left(\dfrac{2}{5}\right)^x = 3\left(\dfrac{5}{2}\right)^{-x}$

   b. These functions are decay exponentials. They are algebraically equivalent and an exponential decay function is one of the form $f(x) = cb^{-x}$ where $b > 1$. In this form $b = 5/2$ which is greater than 1.

   c. Graphs are identical and falling.

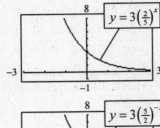

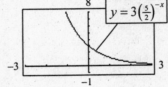

21. a.
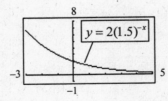

   b.
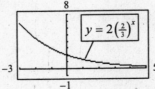

   c. $\quad y = 2(1.5)^{-x} = 2\left(\dfrac{3}{2}\right)^{-x} = 2\left(\dfrac{2}{3}\right)^x$

23. $f(x) = e^{-x}, \quad f(kx) = e^{-kx}$

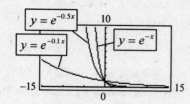

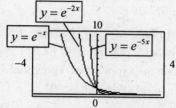

For $k > 1$ the graphs rise and fall more sharply than $y = e^{-x}$. For $k < 1$ the graphs rise and fall more slowly than $y = e^{-x}$.

25. $f(x) = 4^x \qquad f(x) + C = 4^x + C$

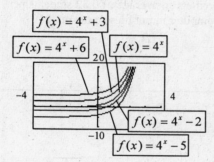

The graphs are the same but $f(x) + C$ is shifted C units on the y-axis.

27. a.

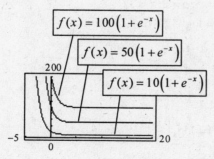

b. As c changes, the y-intercept and the horizontal asymptote change.

29. $S(x) = 1000(1.02)^{4x}$

$S(8) = 1.02 \boxed{y^x} \, 32 \boxed{\times} 1000 \boxed{=} \$1884.54$

31.

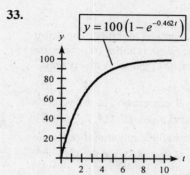

33.

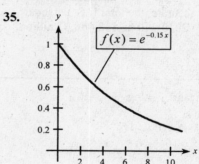

After 10 hours the drug is almost completely in the bloodstream.

35.

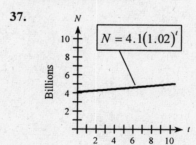

After 10 years about 20% of the television sets are still in service.

37.

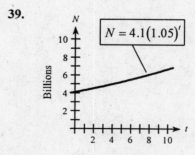

39.

41. a. $N = 12.58e^{0.2096t}$ is an exponential growth model since $e > 1$. Graph rises.

b.

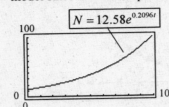

$N = 12.58e^{0.2096t}$

43. Compare the two graphs together. The linear model gives a negative number of processors in 2010. The exponential model is always non-negative.

45. a. Letting $x = 0$ correspond to the year 1960, enter the data values $(0, 411.7)$, $(10, 836.1)$, etc. and obtain the exponential regression equation $y = 434.2404(1.0779)^x$.

b. $x = 50$ in 2010, so
$y = 434.2404(1.0779)^{50} \approx 18466.61$. If model is accurate, the total U.S. personal income in 2010 will be \$18,466.61 billion.

47. From the scatter plot of points, an exponential function of the form $a \cdot b^x$ is the best fit.

a. $y = 2.58(1.043)^x$

b. $f(110) = y = 2.58(1.043)^{110} \approx 264.79$

c. Graph $y_1 = 300$ and $y_2 = y = 2.58(1.043)^x$. The point of intersection is the solution. In 112.55 years the CPI will reach 300. So, the year is 2012-2013.

49. a. $x = 0$ in 1980, $x = 5$ in 1985, etc. Enter the data values $(5, 75)$, $(6, 50)$, $(7, 37)$, etc., and obtain the exponential regression equation $y = 118.18(0.84827)^x$.

b. This is an exponential decay problem. Since $b = 0.84727$ is less than one, the y values will decrease as x increases.

c. In 2010 $x = 30$, so
$y = 118.18(0.84827)^{30} \approx 0.82$. This model predicts an average of 0.82 students per computer in public schools in 2010.

Exercise 5.2

NOTE: $a^y = x$ means $y = \log_a x$.

1. $4 = \log_2 16$
$y = 4, a = 2, x = 16$
$2^4 = 16$

3. $\dfrac{1}{2} = \log_4 2$

$y = \dfrac{1}{2}, a = 4, x = 2$

$4^{1/2} = 2$

5. $\log_2 x = 3$
$2^3 = x$
$x = 8$

7. $\log_8 x = -\dfrac{1}{3}$

$8^{-1/3} = x$

$x = \dfrac{1}{8^{1/3}} = \dfrac{1}{2}$

9. $\log_5(2x+1) = 2$
$$2x+1 = 5^2$$
$$2x = 25 - 1$$
$$x = 12$$

11. $2^5 = 32$
$y = 5, a = 2, x = 32$
$5 = \log_2 32$

13. $4^{-1} = \dfrac{1}{4}$

$y = -1, a = 4, x = \dfrac{1}{4}$

$-1 = \log_4\left(\dfrac{1}{4}\right)$

15.

$y = \log_3 x$

17.

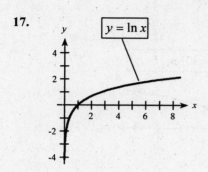

19.

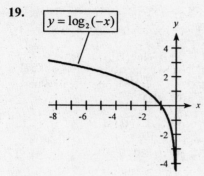

21. a. $\log_3 27 = \log_3 3^3 = 3\log_3 3 = 3 \cdot 1 = 3$

b. $\log_5 \left(\dfrac{1}{5}\right) = \log_5 5^{-1} = -1\log_5 5 = -1 \cdot 1 = -1$

23. $f(x) = \ln(x)$

$f(e^x) = \ln(e^x) = x\ln e = x \cdot 1 = x$

25. $f(x) = e^x$

$f(\ln 3) = e^{\ln 3} = 3 \quad (e^{\ln 3} = e^{\log_e 3})$

27. a. $\log_a(xy) = \log_a x + \log_a y = 3.1 + 1.8 = 4.9$

b. $\log_a \left(\dfrac{x}{z}\right) = \log_a x - \log_a z = 3.1 - 2.7 = 0.4$

c. $\log_a x^4 = 4\log_a x = 4(3.1) = 12.4$

d. $\log_a \sqrt{y} = \dfrac{1}{2}\log_a y = \dfrac{1}{2}(1.8) = 0.9$

29. $\log \left(\dfrac{x}{x+1}\right) = \log x - \log(x+1)$

31. $\log_7 x\sqrt[3]{x+4} = \log_7 x + \log_7 (x+4)^{1/3}$

$= \log_7 x + \dfrac{1}{3}\log_7 (x+4)$

33. $\ln x - \ln y = \ln \left(\dfrac{x}{y}\right)$

35. $\log_5 (x+1) + \dfrac{1}{2}\log_5 x = \log_5 (x+1) + \log_5 x^{1/2}$

$= \log_5 \left[x^{1/2}(x+1)\right]$

37. a. $\ln(4 \cdot 6)^{1/2} = 1.5890$

b. $\dfrac{1}{2}(\ln 4 + \ln 6)$

The expressions are equivalent. Properties V and III are illustrated.

39. a. $\log \left(\dfrac{8}{5}\right)^{1/3} = \log (8 \div 5)^{1/3} = 0.06804$

b. $\dfrac{1}{3}\log 8 - \log 5 = \dfrac{1}{3}(0.9031) - 0.6990$

$\neq 0.06804$

The expressions are not equivalent.

Change (a) to $\log \dfrac{\sqrt[3]{8}}{5}$ to get (a) = (b).

41. a.

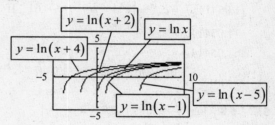

b. For each c, the domain is $x > c$ and the vertical asymptote is at $x = c$.

c. Each x-intercept is at $x = c+1$.

$\ln(x-c) = \ln(c+1-c) = \ln 1 = 0$.

d. The graph of $f(x-c)$ is the graph of $f(x)$ shifted c units on the x-axis.

43. a. $\log_2 17 = \dfrac{\ln 17}{\ln 2} = \dfrac{2.8332}{0.6931} = 4.0875$

b. $\log_5 (0.78) = \dfrac{\ln(0.78)}{\ln 5} = \dfrac{-0.2485}{1.6094} = -0.1544$

45. $y = \log_5 x \qquad y = \dfrac{\ln x}{\ln 5} = \dfrac{1}{1.6094}\ln x$

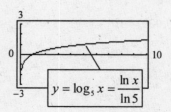

47. $y = \log_{13} x$ $y = \dfrac{\ln x}{\ln 13} = \dfrac{1}{2.5649} \ln x$

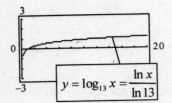

$y = \log_{13} x = \dfrac{\ln x}{\ln 13}$

49. Let $u = \log_a M$ and $v = \log_a N$. Then,
$M = a^u$ and $N = a^v$. So,

$$\log_a\left(\frac{M}{N}\right) = \log_a\left(\frac{a^u}{a^v}\right) = \log_a(a^{u-v})$$
$$= (u - v)\log_a a$$
$$= u - v = \log_a M - \log_a N.$$

51. $y = 3.974(1.05414)^{-x}$

To change to base e we have

$(1.05414)^{-x} = e^{-kx} = \left(e^k\right)^{-x}$ so

$1.05414 = e^k$

$\ln 1.05414 = k$

$k \approx 0.05273$ thus $y = 3.974e^{-0.05273x}$

53. $6.8 = \log\left(\dfrac{I}{I_0}\right)$ $7.7 = \log\left(\dfrac{I}{I_0}\right)$

$10^{6.8} = \dfrac{I}{I_0}$ $10^{7.7} = \dfrac{I}{I_0}$

$\dfrac{10^{7.7}}{10^{6.8}} = 10^{0.9} \approx 7.9$ So the earthquake was
approximately 7.9 times as severe.

55. Intensity of the 1906 San Francisco quake:
$10^{8.25} I_0$
Intensity of the 1989 San Francisco quake:
$10^{7.1} I_0$
The 1906 quake was

$$\dfrac{10^{8.25} I_0}{10^{7.1} I_0} = 10^{8.25 - 7.1} = 10^{1.15} \approx 14$$

times as severe.

57. $L = 10\log(10{,}000) = 10 \cdot 4 = 40$

59. $L = 10\log\left(\dfrac{I}{I_0}\right)$

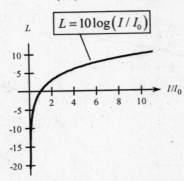

$L = 10\log(I/I_0)$

61. When pH = 1, $1 = -\log[H^+]$
$-1 = \log[H^+]$
$10^{-1} = [H^+]$
$0.1 = \dfrac{1}{10} = [H^+]$
When pH = 14, $[H^+] = 10^{-14}$

63. pH $= \log\left(\dfrac{1}{[H^+]}\right)$

Since $\log\left(\dfrac{1}{[H^+]}\right) = \log 1 - \log[H^+]$ and

$\log 1 = 0$, we have pH $= -\log[H^+]$.

65. $2 = \left(1 + \dfrac{8}{100(4)}\right)^{4t}$

$2 = (1.02)^{4t}$

$\log_{1.02} 2 = 4t$

$\dfrac{\ln 2}{\ln 1.02} = 4t$

$t = \dfrac{\ln 2}{4\ln 1.02} \approx 8.75$ years

67. In 1999,
$L(99) = 11.307 + 14.229\ln(99) \approx 76.691$ and
$\ell(99) = 11.6164 + 14.1442\ln(99) = 76.611$.
In 2010,
$L(110) = 11.307 + 14.229\ln 112 \approx 78.447$
$\ell(112) = 11.6164 + 14.1442\ln 112 \approx 78.356$
The additional data gave predictions that were
very similar to predictions based on the earlier
model.

69. a.

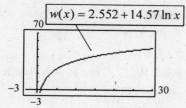

$$w(x) = 2.552 + 14.57 \ln x$$

b. In 2010, $x = 60$, so

$w(60) = 2.552 + 14.57 \ln 60 \approx 62.21$. In 2010, 62.21% of women will be in the workforce.

c. Graph $y_1 = 2.552 + 14.57 \ln x$ and $y_2 = 60$ and find where they intersect. The point of intersection is $(51.57, 60)$. $51.57 \approx 52$, which corresponds to the year 2002. The percent of women in the workforce reached 60% in 2002.

71. a. For x the number of years after 1970, enter the data values $(10, 22.6)$, $(11, 28.3)$, etc., and obtain the logarithmic regression equation $y = 24.926 + 14.504 \ln x$.

b. $y(30) = 24.926 + 14.504 \ln 30 \approx 73.6$

If this model remains valid, the percentage of U.S. households with cable TV in the year 2010 is predicted to be 73.6%.

c. For this model to remain valid the percentage of households cannot exceed 100.

Exercise 5.3

NOTE: Remember $\ln e = 1$.

1.
$$8^{3x} = 32,768$$
$$\ln 8^{3x} = \ln 32768$$
$$3x \ln 8 = \ln 32768$$
$$x = \frac{\ln 32768}{3 \ln 8} \approx 1.667$$

3.
$$0.13P = P(2^{-x})$$
$$\frac{0.13P}{P} = \frac{P(2^{-x})}{P}$$
$$0.13 = 2^{-x}$$
$$\ln 0.13 = \ln 2^{-x}$$
$$\ln 0.13 = -x \ln 2$$
$$x = \frac{\ln 0.13}{-\ln 2} \approx 2.943$$

5.
$$10000 = 1500e^{0.10x}$$
$$\frac{20}{3} = e^{0.10x}$$
$$\ln \frac{20}{3} = \ln e^{0.10x}$$
$$\ln \frac{20}{3} = 0.10x$$
$$x = \frac{\ln \frac{20}{3}}{0.10} \approx 18.972$$

7.
$$78 = 100 - 100e^{-0.01x}$$
$$-22 = -100e^{-0.01x}$$
$$0.22 = e^{-0.01x}$$
$$\ln 0.22 = \ln e^{-0.01x}$$
$$\ln 0.22 = -0.01x$$
$$x = \frac{\ln 0.22}{-0.01} \approx 151.413$$

9.
$$55 = \frac{60}{1 + 5e^{-0.6x}}$$
$$55(1 + 5e^{-0.6x}) = 60$$
$$1 + 5e^{-0.6x} = \frac{12}{11}$$
$$5e^{-0.6x} = \frac{1}{11}$$
$$e^{-0.6x} = \frac{1}{55}$$
$$\ln e^{-0.6x} = \ln \frac{1}{55}$$
$$-0.6x = \ln \frac{1}{55}$$
$$x = \frac{\ln \frac{1}{55}}{-0.6} \approx 6.696$$

11. a.

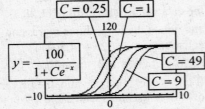

b. Different c values change the y-intercept and how the graph approaches the asymptote.

13. $S = 50,000e^{-0.8x}$

a. $x = 4$

$S = 50,000e^{-3.2}$

$= 50,000(0.0408) = \$2038.11$

b. $S = 1000$

$1000 = 50,000e^{-0.8x}$

$e^{-0.8x} = 0.02$

$-0.8x = \ln 0.02$

$x = \dfrac{\ln 0.02}{-0.8} = \dfrac{-3.912}{-0.8} = 4.89$

In the fifth month sales will drop below $1000.

15. a. $30,000 = 60,000e^{-0.05t}$

$e^{-0.05t} = \dfrac{30,000}{60,000}$

$e^{-0.05t} = 0.5$ take ln of both sides

$-0.05t \approx -0.6931$

$t = \dfrac{-0.6931}{-0.05} \approx 13.9$ years

b. $50,000 = A \cdot e^{-0.05(15)}$

$50,000 = Ae^{-0.75}$

$A = \dfrac{50,000}{e^{-0.75}} = \dfrac{50,000}{0.47237} \approx \$105,849$

17. $Q(t) = 100e^{-0.02828t}$

At $t = 0$ the initial amount is 100 gm. Thus,

$50 = 100e^{-0.02828t}$

$0.5 = e^{-0.02828t}$

$\ln 0.5 = -0.02828t$ take ln of both sides

$t = \dfrac{\ln 0.5}{-0.02828} = \dfrac{-0.6931}{-0.02828} = 24.5$ years

19. $y = P_0 e^{ht}$

$110,517 = 100,000e^{10h}$

$e^{10h} = 1.10517$

$10h = \ln 1.10517$ (Take ln of both sides.)

$10h = 0.099999$

$h = 0.009999 = 0.01$

2015 gives $t = 15$ and $P_0 = 110,517$.

So, $y = 110,517e^{0.01(15)} = 110,517e^{0.15}$

$= 110,517(1.161834) = 128,402$

21. $H = 29.57e^{0.097t}$

$4000 = 29.57e^{0.097t}$

$135.27 = e^{0.097t}$

$\ln 135.27 = \ln e^{0.097t}$

$\ln 135.27 = 0.097t$

$t = \dfrac{\ln 135.27}{0.097} \approx 50.6$

In the year 2011.

23. $p = 100e^{-q/2}$

a. $q = 6$

$p = 100e^{-3} = 100(0.04978) = \$4.98$

b. $p = \$1.83$

$1.83 = 100e^{-q/2}$

$0.0183 = e^{-q/2}$

$-q/2 = \ln(0.0183)$

$q = -2(\ln 0.0183)$

$= -2(-4.000) = 8$ units

25. $p = \dfrac{100e^q}{q+1}$

$q = \dfrac{300}{100} = 3$

$p = \dfrac{100e^3}{4} = 25e^3 = \$502.14$

27. $C(x) = e^{0.1x} + 400$

$C(30) = e^3 + 400 = \$420.09$

29. $p = 200e^{-2} = \$27.07$

100 units will give a revenue of $2707.

31. $S(t) = 8500e^{0.115t}$

   a. $S(1.5) = 8500e^{0.1725} = 8500(1.1883)$

        $= \$10,100.31$

   b. $17,000 = 8500e^{0.115t}$

       $2 = e^{0.115t}$

     $0.115t = \ln 2$

        $t = \dfrac{\ln 2}{0.115} = \dfrac{0.6931}{0.115} = 6.03$

    The account will double in 6.03 years.

33. $S = 5000(1.0075)^t$

   a. $t = 12$

     $S = 5000(1.0075)^{12} = 5000(1.0938)$

      $= \$5469.03$

   b. $S = 10,000$

     $10,000 = 5000(1.0075)^t$

      $1.0075^t = 2$

     $t \ln 1.0075 = \ln 2$

        $t = \dfrac{\ln 2}{\ln 1.0075} = \dfrac{0.6931}{0.00747}$

         $= 92.8 \text{ months}$

    Thus, $t = 7$ years, 9 months.

35. a. $\text{Profit} = R(t) - C(t) = 21.4e^{0.131t} - 18.6e^{0.131t}$

           $= 2.8e^{0.131t}$

    $P(25) = 2.8e^{0.131(25)} \approx \$74.04 \text{ billion}$

   b. $148.08 = 2.8e^{0.131t}$

     $52.89 = e^{0.131t}$

    $\ln 52.89 = \ln e^{0.131t}$

    $\ln 52.89 = 0.131t$

       $t = \dfrac{\ln 52.89}{0.131} \approx 30.29; \text{ the year 2010.}$

    It will only take 5 years after 2005 to double.

   c. September 11, 2001, the "war on terror" and the war in Iraq will all affect these numbers.

37. a. $P(10) = 3.818e^{-0.0506(10)} \approx \$2.30$

    $P(70) = 3.818e^{-0.0506(70)} \approx \$0.11$

    The purchasing power of a 1983 dollar was \$2.30 in 1970 and will only be \$0.11 in the year 2030.

   b. $0.25 = 3.818e^{-0.0506t}$

     $0.0655 = e^{-0.0506t}$

    $\ln 0.0655 = \ln e^{-0.0506t}$

    $\ln 0.0655 = -0.0506t$

       $t = \dfrac{\ln 0.0655}{-0.0506} \approx 53.9; \text{ the year 2020.}$

39. $\quad 50 = 10 + 5\ln(3x + 1)$

     $40 = 5\ln(3x + 1)$

      $8 = \ln(3x + 1)$

     $e^8 = e^{\ln(3x+1)}$

     $e^8 = 3x + 1$

   $e^8 - 1 = 3x$

      $x = \dfrac{e^8 - 1}{3} \approx 993.3$

41. It will take 52.5 years before it will cost \$1 to purchase goods that cost \$0.25 in 1983. The year is 2013.

   $N = 3000(0.2)^{0.6^t}$

   a. $t = 0 \qquad N = 3000(0.2)^1$

               $= 600$

   b. $t = 3 \qquad N = 3000(0.2)^{0.6^t}$

               $= 3000(0.2)^{0.216}$

               $= 3000(0.706354)$

               $= 2119$

   c. The largest value of $(0.2)^{0.6^t}$ occurs when $0.6^t = 0$. Then, $(0.2)^0 = 1$ and the maximum sales are 3000.

   d.

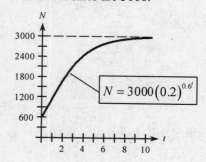

43. $N = 500(0.02)^{0.7^t}$

    a. If $t = 0$, $N = 500(0.02) = 10$ employees when company opens.

    b. $N = 100$

$$100 = 500(0.02)^{0.7^t}$$

$$(0.02)^{0.7^t} = 0.2$$

$$0.7^t \log 0.02 = \log 0.2$$

$$0.7^t = \frac{\log 0.2}{\log 0.02} = \frac{-0.69897}{-1.69897} = 0.41141$$

$$t \log 0.7 = \log 0.41141$$

$$t = \frac{\log 0.41141}{\log 0.7} = \frac{-0.38573}{-0.15490} = 2.49$$

    At least 100 employees are working $2\frac{1}{2}$ years later.

45. $y = 100(1 - e^{-0.462t})$

    a. $t = 1$

$$y = 100(1 - e^{-0.462}) = 100(1 - 0.6300) = 37$$

    b. $y = 50$

$$50 = 100(1 - e^{-0.462t})$$

$$1 - e^{-0.462t} = 0.5$$

$$e^{-0.462t} = 0.5$$

$$-0.462t = \ln 0.5 = -0.6931$$

$$t = \frac{-0.6931}{-0.462} = 1.5 \text{ hours}$$

47. $y = \dfrac{10,000}{1 + 9999e^{-0.99t}}$

    a. $t = 4$

$$y = \frac{10,000}{1 + 9999e^{-3.96}} = \frac{10,000}{1 + 190.61} \approx 52$$

    b. $y = 5000$

$$5000 = \frac{10,000}{1 + 9999e^{-0.99t}}$$

$$1 + 9999e^{-0.99t} = 2$$

$$e^{-0.99t} = \frac{1}{9999}$$

$$-0.99t = \ln 1 - \ln 9999$$

$$t = \frac{-\ln 9999}{-0.99} = \frac{9.2102}{0.99} = 9.30$$

School will close during the tenth day.

49. $y = 40 - 40e^{-0.05t}$

    a. $t = 1$

$$y = 40 - 40e^{-0.05} = 40 - 40(0.95123)$$

$$\approx 1.95\% \approx 2\%$$

    b. $y = 25$

$$25 = 40 - 40e^{-0.05t}$$

$$40e^{-0.05t} = 15$$

$$e^{-0.05t} = \frac{15}{40} = 0.375$$

$$t = \frac{\ln 0.375}{-0.05} = \frac{-0.9808}{-0.05}$$

$$t = 19.6 \text{ or } 20 \text{ months}$$

51. $x = 0.05 + 0.18e^{-0.38t}$

    a. $t = 0$ $x = 0.05 + 0.18(1) = 0.23 \text{km}^3$

    b. 30% of $0.23 = 0.069$.

    So, for $x = 0.069$ we have

$$0.069 = 0.05 + 0.18e^{-0.38t}$$

$$e^{-0.38t} = \frac{0.069 - 0.05}{0.18} = 0.10556$$

$$t = \frac{\ln 0.10556}{-0.38} = \frac{-2.2485}{-0.38} = 5.9 \text{ years}$$

53. $N(t) = \dfrac{8750}{1 + 80.8e^{-0.2286t}}$, t = years past 1975

    a. $N(25) = \dfrac{8750}{1 + 80.8e^{-0.2286(25)}} \approx 6909.7$

    b. $8500 = \dfrac{8750}{1 + 80e^{-0.2286t}}$

$$8500\left(1 + 80e^{-0.2286t}\right) = 8750$$

$$1 + 80e^{-0.2286t} = 1.0294$$

$$80e^{-0.2286t} = 0.0294$$

$$e^{-0.2286t} = 0.0003675$$

$$\ln e^{-0.2286t} = \ln 0.0003675$$

$$-0.2286t = \ln 0.0003675$$

$$t = \frac{\ln 0.0003675}{-0.2286} \approx 35.4$$

$$= \text{the year } 2010$$

55. a. $y(x) = \dfrac{59.57}{1 + 5.22e^{-0.585x}}$

b. $y(8) = \dfrac{59.57}{1 + 5.22e^{-0.585(8)}} \approx 56.8\%$

c. $59.5 = \dfrac{59.57}{1 + 5.22e^{-0.585x}}$

$59.5\left(1 + 5.22e^{-0.585x}\right) = 59.57$

$1 + 5.22e^{-0.585x} = 1.001176$

$5.22e^{-0.585x} = 0.001176$

$e^{-0.585x} = 0.0002254$

$\ln e^{-0.585x} = \ln 0.0002254$

$-0.585x = \ln 0.0002254$

$x = \dfrac{\ln 0.0002254}{-0.585} \approx 14.4$

$= \text{the year 2010}$

Review Exercises

1. **a.** $2^x = y \rightarrow x = \log_2 y$

   b. $3^y = 2x \rightarrow y = \log_3 (2x)$

2. **a.** $\log_7 \left(\dfrac{1}{49}\right) = -2 \rightarrow 7^{-2} = \dfrac{1}{49}$

   b. $\log_4 x = -1 \rightarrow 4^{-1} = x$

3.

$y = e^x$

4.

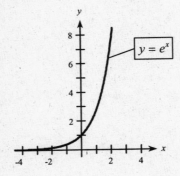

$y = e^{-x}$

5.

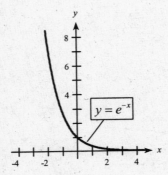

$y = \log_2 x$

6.

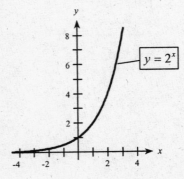

$y = 2^x$

7.

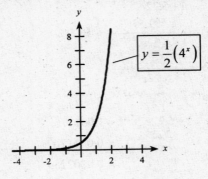

$y = \dfrac{1}{2}(4^x)$

8.

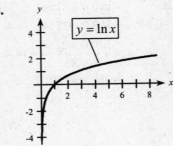

$y = \ln x$

9.

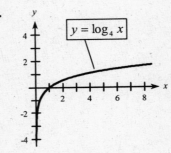

$y = \log_4 x$

10.

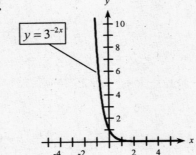

$y = 3^{-2x}$

11.

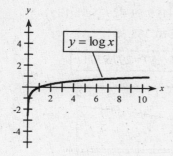

12.

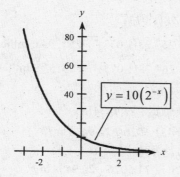

13. $\log_5 1 = 0$

14. $\log_8 64 = \log_8 8^2 = 2$

15. $\log_{25} 5 = \log_{25}(25)^{1/2} = \dfrac{1}{2}$

16. $\log_3\left(\dfrac{1}{3}\right) = \log_3 3^{-1} = -1$

17. $\log_3 3^8 = 8\log_3 3 = 8 \cdot 1 = 8$

18. $\ln e = \ln e^1 = 1$

19. $e^{\ln 5} = 5$

20. $10^{\log 3.15} = 3.15$

21. $\log_a\left(\dfrac{x}{y}\right) = \log_a x - \log_a y = 1.2 - 3.9 = -2.7$

22. $\log_a x = 1.2$

$\log_a \sqrt{x} = \dfrac{1}{2}\log_a x = \dfrac{1}{2}(1.2) = 0.6$

23. $\log_a(xy) = \log_a x + \log_a y = 1.2 + 3.9 = 5.1$

24. $\log_a y = 3.9$

$\log_a(y^4) = 4\log_a y = 4(3.9) = 15.6$

25. $\log(yz) = \log y + \log z$

26. $\ln\sqrt{\dfrac{x+1}{x}} = \dfrac{1}{2}\ln\dfrac{x+1}{x} = \dfrac{1}{2}\big[\ln(x+1) - \ln x\big]$

$\qquad = \dfrac{1}{2}\ln(x+1) - \dfrac{1}{2}\ln x$

27. No, Let $x = y = 1$. Then $\ln 1 + \ln 1 = 0 + 0 = 0$
and $\ln(1+1) = \ln 2 \approx 0.6931$.

28. $f(e^{-2}) = \ln(e^{-2}) = -2$

29. $f(x) = 2^x + \log(7x - 4)$

$f(2) = 2^2 + \log(14 - 4) = 4 + 1 = 5$

30. $f(0) = e^0 + \ln 1 = 1 + 0 = 1$

31. $\quad f(x) = \ln(3e^x - 5)$

$f(\ln 2) = \ln(3e^{\ln 2} - 5)$

$\qquad = \ln(3 \cdot 2 - 5)$

$\qquad = \ln 1 = 0$

32. $\log_9 2158 = \dfrac{\log 2158}{\log 9} \approx 3.494$

33. $\log_{12} 0.0195 = \dfrac{\ln 0.0195}{\ln 12} = \dfrac{-3.9373}{2.4849} = -1.5845$

34.

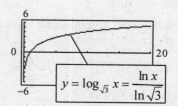

35. $f(x) = \log_{11}(2x - 5)$

$\qquad = \dfrac{\ln(2x-5)}{\ln 11} = \dfrac{\ln(2x-5)}{2.3979}$

36. $\qquad 6^{4x} = 46,656$

$\ln 6^{4x} = \ln 46656$

$4x \ln 6 = \ln 46656$

$x = \dfrac{\ln 46656}{4\ln 6} = 1.5$

37. $8000 = 250(1.07)^x$

$$32 = 1.07^x$$

$$\ln 32 = \ln 1.07^x$$

$$\ln 32 = x \ln 1.07$$

$$x = \frac{\ln 32}{\ln 1.07} \approx 51.224$$

38. $11,000 = 45,000e^{-0.05x}$

$$0.244 = e^{-0.05x}$$

$$\ln 0.244 = \ln e^{-0.05x}$$

$$\ln 0.244 = -0.05x$$

$$x = \frac{\ln 0.244}{-0.05} \approx 28.212$$

39. $312 = 300 + 300e^{-0.08x}$

$$12 = 300e^{-0.08x}$$

$$0.04 = e^{-0.08x}$$

$$\ln 0.04 = \ln e^{-0.08x}$$

$$\ln 0.04 = -0.08x$$

$$x = \frac{\ln 0.04}{-0.08} \approx 40.236$$

40. Growth exponential, because Medicare funding increases faster with time.

41. Exponential because the general shape is similar to the graph of a decay exponential.

42. a. $P(20) = 60,000(0.97)^{20} \approx \$32,627.66$

b.

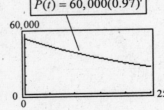

$$P(t) = 60,000(0.97)^t$$

43. a. $y = 267.78(1.0729)^x$

b. $y(27) = 267.78(1.0729)^{27}$

$$\approx \$1790.1 \text{ billion}$$

c. $y(35) = 267.78(1.0729)^{35}$

$$\approx \$3143.1 \text{ billion}$$

d. too high, because the data seem to level off

44. a. $y(35) = -195.59 + 2975.5 \ln(35)$

$$\approx \$10,383.35$$

b.

$$y = -195.59 + 2975.5 \ln x$$

45. a. $y = 267.78(1.073)^x$

b. $y(27) = 267.78(1.073)^{27} \approx \$1794.6 \text{ billion}$

c. $y(35) = 267.78(1.073)^{35} \approx \$3153.3 \text{ billion}$

d. Answers vary.

46. a. $y = -31.5026 + 22.9434 \ln x$

b. $y(70) = -31.5026 + 22.9434 \ln 70$

$$\approx 65.97\%$$

c. $75 = -31.5026 + 22.9434 \ln x$

$$106.5026 = 22.9734 \ln x$$

$$4.642 = \ln x$$

$$e^{4.642} = e^{\ln x}$$

$$e^{4.642} = x$$

$$x \approx 103.75; \text{ the year } 2044$$

47. a. $M = -\frac{5}{2} \log\left(\frac{36.3B_0}{B_0}\right) = -\frac{5}{2}\log 36.3$

$$= -\frac{5}{2}(1.5599) = -3.8998$$

b. $2.1 = -\frac{5}{2}\log\left(\frac{B}{B_0}\right)$

$$\log \frac{B}{B_0} = \frac{-4.2}{5} = -0.84$$

$$\frac{B}{B_0} = 10^{-0.84}$$

$$B = 10^{-0.84} B_0 = 0.14B_0$$

c. $6 = -\frac{5}{2}\log\left(\frac{B}{B_0}\right)$

$$\log \frac{B}{B_0} = \frac{-12}{5} = -2.4$$

$$\frac{B}{B_0} = 10^{-2.4}$$

$$B = 10^{-2.4} B_0 = 0.004B_0$$

d. Yes

48. a. $S = 50,000e^{-0.1(6)} = 50,000(0.5488)$

$= \$27,440.58$

b. $15,000 = 50,000e^{-0.1x}$

$e^{-0.1x} = \dfrac{15,000}{50,000}$ Take ln of each side.

$-0.1x = \ln 15,000 - \ln 50,000$

$x = \dfrac{9.6158 - 10.8198}{-0.1} = 12$ weeks

49. $S = 50,000e^{-0.6x}$

$x = 6$

$S = 50,000e^{-3.6} = 50,000(0.02732) = \$1366.19$

50. $2000 = 1000(1.01)^{12t}$

$(1.01)^{12t} = 2$

$12t = \dfrac{\ln 2}{\ln 1.01} = 69.66$

$t = \dfrac{69.66}{12} = 5.8$ years

51. $S = 5000e^{0.135t}$

a. $t = 0.75$

$S = 5000e^{0.10125} = 5000(1.10655)$

$= \$5532.77$

b. $S = 10,000$

$10,000 = 5000e^{0.135t}$

$2 = e^{0.135t}$

$t = \dfrac{\ln 2}{0.135} = \dfrac{0.6931}{0.135} = 5.13$ years

52. Logistic – begins like an exponential function, then grows at a slower rate.

53. $N = 10,000(0.3)^{0.5^t}$

a. $t = 0$

$N = 10,000(0.3)^1 = \$3000$

b. $t = 3$

$N = 10,000(0.3)^{0.125} = 10,000(0.86028)$

$\approx \$8603$

c. Sales are a maximum when $(0.3)^{0.5^t}$ is a maximum. Since 0.3 is between 0 and 1 we know that 0.3 to a positive power cannot be greater than 1. Thus, maximum sales will be $10,000.

54. a.

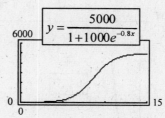

b. $x = 0$ when first discovered.

$y = \dfrac{5000}{1 + 1000e^{-0.8(0)}} = \dfrac{5000}{1001} = 4.995 \approx 5$

5 students had the virus when it was first discovered.

c. When $x = 15$,

$y = \dfrac{5000}{1 + 1000e^{-0.8(15)}} = 4969.4665$.

The total number infected by the virus during the first 15 days is 4969.

d. $3744 = \dfrac{5000}{1 + 1000e^{-0.8x}}$

$3744(1 + 1000e^{-0.8x}) = 5000$

$3744 + 3744000e^{-0.8x} = 5000$

$3744000e^{-0.8x} = 1256$

$e^{-0.8x} = \dfrac{1256}{3744000}$

$e^{-0.8x} \approx 0.0003547$

take ln of both sides

$-0.8x = \ln 0.0003547$

$-0.8x = -7.9442$

$x = \dfrac{-7.9442}{-0.8} \approx 9.9$

$= 10$ to the nearest whole number

In 10 days the number infected will reach 3744.

Chapter Test

1.

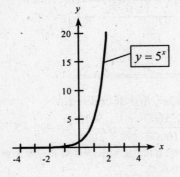

$y = 5^x$

2.

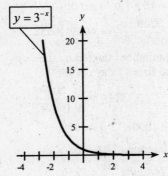

$y = 3^{-x}$

3.

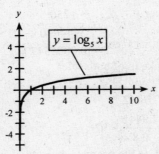

$y = \log_5 x$

4.

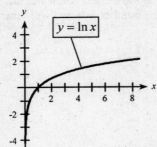

$y = \ln x$

5.

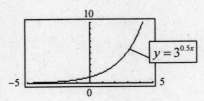

$y = 3^{0.5x}$

6.

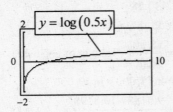

$y = \log(0.5x)$

7.

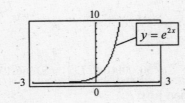

$y = e^{2x}$

8.

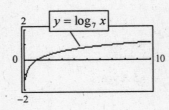

$y = \log_7 x$

9. $\boxed{2^{nd}}$ $\boxed{\ln}$ $\boxed{4}$ $\boxed{=}$ 54.5982

10. $\boxed{2^{nd}}$ $\boxed{\ln}$ $\boxed{(-)}$ 2.1 $\boxed{=}$ 0.1225

11. $\boxed{\ln}$ $\boxed{4}$ $\boxed{=}$ 1.3863

12. $\boxed{\ln}$ $\boxed{21}$ $\boxed{=}$ 3.0445

13. $\log_{17} x = 3.1$
$$x = 17^{3.1}$$
$$x = 6522.163$$

14. $3^{2x} = 27$
$$2x = \log_3 27$$
$$x = \frac{1}{2}\log_3(3^3) = \frac{1}{2} \cdot 3 = \frac{3}{2}$$

15. $\log_2 8 = \log_2 2^3 = 3\log_2 2 = 3 \cdot 1 = 3$

16. $e^{\ln x^4} = x^4$ (Property II)

17. $\log_7 7^3 = 3$ (Property I)

18. $\ln e^{x^2} = x^2$ (Property I)

19. $\ln(M \cdot N) = \ln M + \ln N$

20. $\ln\left(\dfrac{x^3 - 1}{x + 2}\right) = \ln(x^3 - 1) - \ln(x + 2)$

21. $\log_4\left(x^3+1\right)=\dfrac{\ln\left(x^3+1\right)}{\ln 4}$ or $0.721\ln\left(x^3+1\right)$

22.

$$47,500=1500\left(1.015\right)^{6x}$$

$$31.667=\left(1.015\right)^{6x}$$

$$\ln 31.667=\ln 1.015^{6x}$$

$$\ln 31.667=6x\ln 1.015$$

$$x=\frac{\ln 31.667}{6\ln 1.015}\approx 38.679$$

23. A decay exponential is more appropriate.

24. Exponential. If years were placed on horizontal axis and number of subscribers on the vertical axis, you would have a shape approximated by an exponential growth curve.

25. a. $R\left(25\right)=330.42e^{0.0833(25)}\approx \$2653.45$ billion

b.

$$5306.90=330.42e^{0.08333t}$$

$$16.06=e^{0.08333t}$$

$$\ln 16.06=\ln e^{0.08333t}$$

$$\ln 16.06=0.08333t$$

$$t=\frac{\ln 16.06}{0.08333}\approx 33.3$$

$$33.3-25=\text{ about 8.3 years}$$

26.

$$2500=22,000e^{-0.35t}$$

$$0.1136=e^{-0.35t}$$

$$\ln 0.1136=\ln e^{-0.35t}$$

$$\ln 0.1136=-0.35t$$

$$t=\frac{\ln 0.1136}{-0.35}\approx 6.21 \text{ months}$$

27. a.

$$I\left(52\right)=\frac{14892.86}{1+42.1036e^{-0.0993(52)}}$$

$$\approx \$12,001.88 \text{ billion}$$

b.

$$12,500=\frac{14892.86}{1+42.1036e^{-0.0993t}}$$

$$12,500\left(1+42.1036e^{-0.0993t}\right)=14892.86$$

$$1+42.1036e^{-0.0993t}=1.19143$$

$$42.1036e^{-0.0993t}=0.19143$$

$$e^{-0.0993t}=0.0045466$$

$$\ln e^{-0.0993t}=\ln 0.0045466$$

$$-0.0993t=\ln 0.0045466$$

$$t=\frac{\ln 0.0045466}{-0.0993}\approx 54.3$$

In the year 2015.

28. a.

An exponential model is not appropriate; the shape of the scatter plot does not resemble an exponential function (neither growth nor decay).

b. $y_1=-54.846+192.7\ln x$

$$y_2=\frac{495}{1+3.9e^{-0.275x}}$$

c. $y_1\left(25\right)=-54.846+192.7\ln 25$

$$\approx \$565.43 \text{ billion}$$

$$y_2\left(25\right)=\frac{495}{1+3.9e^{-0.275(25)}}\approx \$493.01 \text{ billion}$$

The first model is more optimistic.

Chapter 6: Mathematics of Finance

Exercise 6.1 _____

1. $P = 10,000, t = 6, r = 0.16$
  a. $I = Prt = 10,000(0.16)(6) = \$9600$
  b. $S = P + I = 10,000 + 9600 = \$19,600$

3. $P = 1000, t = \dfrac{1}{4}, r = 0.12$

  a. $I = Prt = 1000(0.12)\left(\dfrac{1}{4}\right) = \$30$
  b. $S = P + I = 1000 + 30 = \$1030$

5. $P = 800, t = \dfrac{1}{2}, r = 0.16$

  $I = Prt = 800(0.16)\left(\dfrac{1}{2}\right) = \$64$

  $S = P + I = \$864$

7. $P = 3500, t = \dfrac{15}{12} = \dfrac{5}{4}, r = 0.08$

  $I = 3500(0.08)\left(\dfrac{5}{4}\right) = \$350$

  $S = P + I = 3500 + 350 = \$3850$

9. $P = 30, t = 1, I = 0.90$
  From $I = Prt$ we have

  $r = \dfrac{I}{Pt} = \dfrac{0.90}{30(1)} = 0.03.$ Thus, the couple earned

  3% from the dividend.
  $P = 30, t = 1, I = 3.90$

  So, $r = \dfrac{3.90}{30(1)} = 0.13.$ The total rate earned on the

  investment is 13%.

11. $S = 10,000, P = 9750, t = \dfrac{1}{2}$

  a. $I = S - P = 250$

  $r = \dfrac{I}{Pt} = \dfrac{250}{(9750)\left(\frac{1}{2}\right)} = 0.051$

  The interest rate is 5.1%.
  b. $P = 9750 + 40 = \$9790$
  $I = S - P = 210$

  $r = \dfrac{210}{(9750)\left(\frac{1}{2}\right)} = 0.0429$

  The interest rate is 4.29%.

13. $S = 12(140) = 1680, t = \dfrac{1}{4}, r = 0.12,$ find P

  $S = P + I = P + Prt = P(1 + rt)$
  Thus,

  $P = \dfrac{S}{1 + rt} = \dfrac{1680}{1 + (0.12)\left(\frac{1}{4}\right)} = \dfrac{1680}{1.03} = \$1631.07$

  The firm must deposit \$1631.07 now in order to
  pay the bill in 90 days.

15. $S = 13,500, P = ?$ $t = \dfrac{10}{12} = \dfrac{5}{6}, r = 0.15$

  $S = P + Prt = P + P(0.15)\left(\dfrac{5}{6}\right) = 1.125P$

  $P = 13500 \div 1.125 = \$12,000$

17. $S = 9000$ $P = \$5000$ $t = ??$ $r = 0.08$
  $I = S - P = \$4000$
  $4000 = 5000\,(0.08)(t)$

  $t = \dfrac{4000}{5000(0.08)} = 10$ years

19. 45 day note: $I = 500,000(0.06)\left(\dfrac{45}{360}\right) = \$3750$

  60 day note:

  $I = 500,000(0.07)\left(\dfrac{60}{360}\right) = \$5833.33$

  Penalty is \$5000. By paying on time the retailer
  will earn \$3750 versus \$833.33 if he waits.

21. a. $P = 2000, t = \dfrac{9}{12}, r = 0.08$

  $S = P(1 + rt) =$

  $2000\left[1 + (0.08)\left(\dfrac{9}{12}\right)\right] = \$2120$

  b. $S = 2120, t = \dfrac{3}{12}, r = 0.10$

  $P = \dfrac{S}{1 + rt} = \dfrac{2120}{1 + \left(\frac{1}{10}\right)\left(\frac{3}{12}\right)} = \$2068.29$

23.

n	1	2	3	4	5	6	7	8	9	10
a_n	3	6	9	12	15	18	21	24	27	30

25. $a_n = \dfrac{(-1)^n}{2n+1}$

n	1	2	3	4	5	6
a_n	$\frac{-1}{3}$	$\frac{1}{5}$	$\frac{-1}{7}$	$\frac{1}{9}$	$-\frac{1}{11}$	$\frac{1}{13}$

27. $a_n = \dfrac{n-4}{n(n+2)}$

$a_1 = \dfrac{1-4}{1(1+2)} = -1 \quad a_2 = \dfrac{2-4}{2(2+2)} = -\dfrac{1}{4}$

$a_3 = \dfrac{3-4}{3(5)} = -\dfrac{1}{15} \quad a_4 = \dfrac{0}{4(6)} = 0$

$a_{10} = \dfrac{10-4}{10(12)} = \dfrac{1}{20}$

29. a. $d = 3, \ a_1 = 2$

    b. 11, 14, 17

31. a. $d = \dfrac{9}{2} - \dfrac{6}{2} = \dfrac{3}{2}, \ a_1 = 3$

    b. $\dfrac{15}{2}, 9, \dfrac{21}{2}$

33. $a_1 = 6, \ d = -\dfrac{1}{2}, \ n = 83$

$a_{83} = 6 + (83-1)\left(-\dfrac{1}{2}\right) = -35$

35. $a_1 = 5, \ a_8 = 19, \ n = 100$

$19 = 5 + (8-1)d \ \text{ or } \ 14 = 7d \ \text{ or } \ d = 2$

$a_{100} = 5 + (100-1)2 = 203$

37. $a_1 = 2, \ d = 3, \ n = 38, \ a_{38} = 113$

$s_{38} = \dfrac{38}{2}(2+113) = 2185$

39. $a_1 = 10, \ d = \dfrac{1}{2}, \ n = 70$

$a_{70} = 10 + (70-1)\left(\dfrac{1}{2}\right) = \dfrac{89}{2}$

$s_{70} = \dfrac{70}{2}\left(10 + \dfrac{89}{2}\right) = \dfrac{3815}{2}$

41. $a_1 = 6, \ d = -\dfrac{3}{2}, \ n = 150$

$a_{150} = 6 + (150-1)\left(-\dfrac{3}{2}\right) = -\dfrac{435}{2}$

$s_{150} = \dfrac{150}{2}\left(6 - \dfrac{435}{2}\right) = -\dfrac{31,725}{2}$

43. Beginning with the third term and following terms: The value of the term is the sum of the values of the preceding two terms. 21, 34, 55

45. $a_1 = -2000, d = 400, n = 12, \ \text{Find } a_{12}$

$a_{12} = -2000 + (12-1)(400) = \$2400$

47. $a_1 = 20,000, d = 1000, n = 10, \ \text{Find } a_{10}$

$a_{10} = 20,000 + (10-1)1000 = \$29,000$

$a_1 = 18,000, d = 1200, n = 10, \ \text{Find } a_{10}$

$a_{10} = 18,000 + (10-1)1200 = \$28,800$

Through 10 years of work, the \$20,000 beginning salary is better.

49. a. The total base salary for 3 years is \$60,000. The total salary received for the 3 years is \$63,000. Thus, the sum of the raises is \$3000.

    b. The total salary received for 3 years is \$64,500. Thus, the sum of the raises is \$4500.

    c. Plan II is better by a total of \$1500.

    d. Total salary for Plan I for 5 years is

$63,000 + 11,500 + 11,500$

$+12,000 + 12,000 = \$110,000$

Total base salary is $5(20,000) = \$100,000$. The sum of raises is \$10,000.

    e. Total salary for Plan II for 5 years is:

$64,500 + 11,800 + 12,100 + 12,400$

$+12,700 = \$113,500$

Sum of the raises is \$13,500.

    f. Plan II is better by \$3500.

    g. Choose Plan II. As the years increase, its pay advantage also increases.

Exercise 6.2

1. a. 8%

    b. 7

    c. $\dfrac{8\%}{4} = 2\% = 0.02$

    d. $4 \cdot 7 = 28$

3. a. 9%

    b. 5

    c. $\dfrac{9\%}{12} = \dfrac{3}{4}\% = 0.0075$

    d. $12 \cdot 5 = 60$

5. $P = 8000$, $\dfrac{r}{m} = \dfrac{0.12}{1} = 0.12$, $mt = 1(10) = 10$

$S = 8000(1 + 0.12)^{10}$

$\quad = 8,000(3.105848) = \$24,846.79$

7. $P = 3200$, $\dfrac{r}{m} = \dfrac{0.08}{4} = 0.02$, $mt = 4(5) = 20$

$S = 3200(1 + 0.02)^{20} = 3200(1.485947)$

$\quad = \$4,755.03$

9. $P = 10,000$, $\dfrac{r}{m} = \dfrac{0.09}{12} = 0.0075$, $mt = (12)3 = 36$

$S = 10,000(1 + 0.0075)^{36}$

$\quad = 10,000(1.308645) = \$13,086.45$

Interest $= \$3086.45$ $(I = S - P)$

11. $S = 40,000$, $P = ??$ $\dfrac{r}{m} = \dfrac{0.10}{12} = 0.008333$

$mt = 12(18) = 216$

$P = \dfrac{40,000}{(1 + 0.008333)^{216}} = \$6661.50$

13. $S = 10,000$ $P = ??$

$\dfrac{r}{m} = \dfrac{0.06}{1} = 0.06$

$mt = 1(10) = 10$

$P = \dfrac{10,000}{(1 + 0.06)^{10}} = \$5583.95$

15. $P = 5100$, $r = 0.09$, $t = 4$, $rt = 0.36$

$S = 5100e^{0.36}$

$\quad = 5100(1.433329) = \$7309.98$

17. $P = 410$, $r = 0.08$, $t = 10$, $rt = 0.8$

$S = 410e^{0.8} = 410(2.225541) = \$912.47$

$I = S - P = \$502.47$

19. $S = 100,000$, $P = ??$, $t = 20$

  a. $r = 0.105$, $rt = 2.1$

$\quad P = \dfrac{S}{e^{rt}} = \dfrac{100,000}{e^{2.1}} = \dfrac{100,000}{8.16617} = \$12,245.64$

  b. $r = 0.11$, $rt = 2.2$

$\quad P = \dfrac{S}{e^{rt}} = \dfrac{100,000}{e^{2.2}} = \dfrac{100,000}{9.02501} = \$11,080.32$

  c. A $\dfrac{1}{2}\%$ increase in the interest rate reduces the investment by $1165.32

21. $P = 1000$, $\dfrac{r}{m} = 0.08$, $mt = 5$

$S = 1000(1 + 0.08)^5$

$\quad = 1000(1.469328) = \$1469.33$

$P = 1000$, $r = 0.07$, $t = 5$, $rt = 0.35$

$S = 1000e^{0.35} = 1000(1.419068) = \$1419.07$

The 8% compounded annually yields
$50.26 more interest.

23. a. $APY = \left(1 + \dfrac{0.073}{12}\right)^{12} - 1$

$\quad = 1.0755 - 1 = 7.55\%$

  b. $S = Pe^{0.06} = P(1.0618)$

$\quad APY = 1.0618 - 1 = 0.0618 = 6.18\%$

25. The highest yield occurs with the most compounding periods. Rank: 8% compounded monthly, 8% compounded quarterly, 8% compounded annually.

27. $S_1 = Pe^r = Pe^{0.082} = P(1.08546)$

$S_2 = P\left(1 + \frac{0.084}{4}\right)^4 = P(1.08668)$

$APY_1 = 1.08546 - 1 = 0.08546 = 8.546\%$

$APY_2 = 1.08668 - 1 = 0.08668 = 8.668\%$

The 8.4% investment is better.

29. The higher graph represents the continuous compounding since it has the higher yield.

31. $S = \$93904.80$, $P = \$10,000$, $n = 18$, $i = ?$

$S = P(1 + i)^n$

$93904.80 = 10000(1 + i)^{18}$

$9.39048 = (1 + i)^{18}$

$9.39048^{1/18} = 1 + i$

$i \approx 1.1325 - 1 \approx 0.1325 \approx 13.25\%$

33. $P = 700$, $S = 700 + 300 = 1000$, $r = 0.119$, $t = ?$

$1000 = 700e^{0.119t}$

$e^{0.119t} = \dfrac{1000}{700}$ Take ln of both sides.

$0.119t = \ln 1000 - \ln 700 = 6.9078 - 6.5511$

$t = 2.997$ or approximately 3 years.

35. $P = 20,000, S = 26,425.82, \dfrac{r}{m} = \dfrac{r}{4}, t = 7$

$$26,425.82 = 20,000\left(1 + \dfrac{r}{4}\right)^{28}$$

$$\left(1 + \dfrac{r}{4}\right) = (1.32129)^{1/28}$$

$$= 1.00999 = 1.01$$

$$\dfrac{r}{4} = 0.01$$

$$r = 0.04 = 4\%$$

37. $P = 1000, \dfrac{r}{m} = \dfrac{0.08}{1} = 0.08, mt = 1(18) = 18$

$$S = 1000(1 + 0.08)^{18} = \$3996.02$$

39. a. $P = 112,860, i = 7.5\%, n = 22, S = ?$

$$S = P(1+i)^n$$

$$S = 112860(1 + 0.075)^{22} = \$554,021.04$$

b. $i = 8.5\%$

$$S = 112860(1 + 0.085)^{22} = \$679,194.70$$

At the higher rate of interest there will be an additional \$125,173.66 available for retirement.

41. $P = 10,000, S = 15,000,$

$$\dfrac{r}{m} = \dfrac{0.08}{4} = 0.02, mt = 4t$$

$$15,000 = 10,000(1 + 0.02)^{4t}$$

$$(1 + 0.02)^{4t} = 1.5000$$

$$4t \cdot \ln(1 + 0.02) = \ln(1.500)$$

$$t = \dfrac{\ln 1.5}{4 \ln 1.02} \approx 5.1 \text{ years}$$

43. $P_1 = 2500, t = 3, r = 0.085$

$$S_1 = 2500\left[1 + 3(0.085)\right] = \$3137.50$$

$$S_1 = P_2 = 3137.50, \ n = 9, \ r = 0.18$$

$$S_2 = 3137.50(1 + 0.18)^9 = \$13,916.24$$

45. This problem is more meaningful if a spreadsheet is used.

a. From a table, the investments will be worth more than \$7500 during the sixth year.

b. $P = 5000 \qquad i = \dfrac{0.063}{4} \qquad\qquad\qquad i = \dfrac{0.063}{12}$

3 years $\quad 5000\left(1 + \dfrac{.063}{4}\right)^{12} = \$6031.31 \quad 5000\left(1 + \dfrac{.063}{12}\right)^{36} = \$6037.22$

7 years $\quad 5000\left(1 + \dfrac{.063}{4}\right)^{28} = \$7744.634 \quad 5000\left(1 + \dfrac{.063}{12}\right)^{84} = 7762.34$

10 years $\quad 5000\left(1 + \dfrac{.063}{4}\right)^{40} = 9342.07 \quad 5000\left(1 + \dfrac{.063}{12}\right)^{120} = \$9372.59$

47. a. 3, 6, 12   Ratio is 2.
Next three terms are 24, 48, 96.

b. 81, 54, 36

$$81 \cdot r = 54 \text{ or } r = \dfrac{54}{81} = \dfrac{2}{3}$$

Next three terms are 24, 16, $\dfrac{32}{3}$.

49. $a_1 = 10, r = 2, n = 13$

$$a_{13} = 10 \cdot 2^{13-1} = 10 \cdot 4096 = 40,960$$

51. $a_1 = 4, r = \dfrac{3}{2}, n = 16$

$$a_{16} = 4\left(\dfrac{3}{2}\right)^{16-1} = 4\left(\dfrac{3}{2}\right)^{15}$$

53. $a_1 = 6, r = 3, n = 17$

$$s_{17} = \frac{6(1-3^{17})}{1-3} = -3(1-3^{17})$$

55. $a_1 = 1,\ r = 3, n = 35$

$$s_{35} = \frac{1-3^{35}}{1-3} = \frac{3^{35}-1}{2}$$

57. $a_1 = 6, r = \dfrac{2}{3}, n = 18$

$$s_{18} = \frac{6\left[1-\left(\frac{2}{3}\right)^{18}\right]}{1-\frac{2}{3}} = 18\left[1-\left(\tfrac{2}{3}\right)^{18}\right]$$

59. $a_1 = 160,000, r = 1.04,$

$n = 21$ (at the end of 20 years)

$$a_{21} = 160,000(1.04)^{20} = \$350,580$$

61. $a_1 = 20(\text{million}), r = 1.02, n = 11$

$$a_{11} = 20(1.02)^{10}$$
$$= 20(1.218994) = 24.38 \text{ million}$$

Note: As in problem 65, n is 11 since 20 million is the value at the beginning of the year.

63. $a_1 = P, a_n = 2P, r = 1.02$

$$2P = P(1.02)^n$$
$$(1.02)^n = 2$$
$$n \ln 1.02 = \ln 2$$
$$n = \frac{\ln 2}{\ln 1.02} = \frac{0.6931}{0.0198} = 35 \text{ years}$$

65. Down: 128 96 72 54

Up: 96 72 54 $\dfrac{81}{2}$

Note that there are two separate sequences.

Ball will bounce up $\dfrac{81}{2}$ or 40.5 ft.

67. Value now: 10,000

Value end of year	1	2	3	4
	8000	6400	5120	4096

69. Number now: 5000

End of Hour	1	2	3	4	5
Number	10,000	20,000	40,000	80,000	160,000

At the end of the 6th hour, 320,000 bacteria are present. $N = 10,000 \cdot 2^{6-1}$

71. Profit now: $8,000,000

End of Year	2003	2004	2005	2006	2007
Profit	$7,840,000	$7,683,200	$7,529,536	$7,378,945	$7,231,366

73. Payment #1 $0.1(12,000) = 1200$ Payment #2 $0.1(10,800) = 1080$

Payment #3 $0.1(9720) = 972$ Payment #4 $0.1(8748) = 874.80$

The sequence of outstanding balances is $10,800, 9720, 8748, \cdots$. The 18th term is

$$a_{18} = 10,800(0.9)^{17} = \$1801.14.$$

75. Five people receive your letter with your name in position 6. They cross out the top name and each sends out 5 letters. Your name is now in place 5 on $25 = 5^2$ letters. Each of these sends out 5 letters and now your name is in position 4 on $25 \cdot 5 = 5^3$ letters. Continuing this pattern we have: position 3 5^4 letters

position 2 5^5 letters

position 1 5^6 letters

You will receive $5^6 = 15,625$ dimes.

77. Total number of letters mailed: $5 + 5^2 + 5^3 + \cdots + 5^{12}$ $s_{12} = \dfrac{5(1-5^{12})}{1-5} = 305,175,780$ letters

Exercise 6.3

1. a. The higher graph is the $1120 per year annuity.

b. $S = 1000 \left[\dfrac{(1+0.08)^{25} - 1}{0.08} \right] = \$73{,}105.94$

$S = 1120 \left[\dfrac{(1+0.08)^{25} - 1}{0.08} \right] = \$81{,}878.65$

The difference is
$\$81{,}878.65 - 73{,}105.94 = \$8772.71$

3. $R = 1300, \quad i = 0.06, \quad n = 5$

$S = 1300 \left[\dfrac{(1+0.06)^{5} - 1}{0.06} \right] = \$7328.22$

5. $R = 80, n = 4 \cdot 3 = 12, i = \dfrac{0.08}{4} = 0.02$

$S = 80 \left[\dfrac{(1+0.02)^{12} - 1}{0.02} \right]$

$= 80(13.4121) = \$1072.97$

7. $R = 500, n = 4 \cdot 2 = 8, i = \dfrac{0.10}{2} = 0.05$

$S = 500 \left[\dfrac{(1+0.05)^{8} - 1}{0.05} \right]$

$= 500(9.5491) = \$4774.55$

9. $S = \$50{,}000, n = 8, i = 0.10$

$R = 50{,}000 \left(\dfrac{0.10}{(1+0.10)^{8} - 1} \right)$

$= 50{,}000(0.0874) = \$4372.20$

11. $S = \$200{,}000, n = 20 \cdot 4 = 80,$

$i = \dfrac{0.076}{4} = 0.019$

$R = 200{,}000 \left(\dfrac{0.019}{(1+0.019)^{80} - 1} \right)$

$= 200{,}000(0.005417) = \$1083.40$

13. $S = \$20{,}000, n = 2 \cdot 12 = 24, i = \dfrac{0.12}{12} = 0.01$

$R = 20{,}000 \left(\dfrac{0.01}{(1+0.01)^{24} - 1} \right)$

$= 20{,}000(0.03707) = \$741.47$

15. $S = \$80{,}000, R = \$2500,$

$i = 0.05/4 = 0.0125, n = ?$

$S = R \left[\dfrac{(1+i)^{n} - 1}{i} \right]$

$80000 = 2500 \left[\dfrac{(1+0.0125)^{n} - 1}{0.0125} \right]$

$32 = \dfrac{(1.0125)^{n} - 1}{0.0125}$

$0.4 = (1.0125)^{n} - 1$

$1.4 = (1.0125)^{n}$

$\ln 1.4 = n \ln 1.0125$

$n = \dfrac{\ln 1.4}{\ln 1.0125} \approx 27.1 \text{ quarters}$

17. The 10% rate is better. A sinking fund is an account that accumulates money.

19. Twin 1: $R = 2000, n = 10, i = 0.08$

$S_1 = 2000 \left[\dfrac{(1+0.08)^{10} - 1}{0.08} \right] = \$28{,}973.12$

$P_2 = 28{,}973.12, i = 0.08, n = 32 \text{ (age 65)}$

$S_2 = 28{,}973.12(1+0.08)^{32} = 340{,}060 \text{ (rounded)}$

Twin 2: $S = 340{,}060, i = 0.08, n = 25 \text{ (age 65)}$

$R = 340{,}060 \left[\dfrac{0.08}{(1+0.08)^{25} - 1} \right] = \$4651.61$

21. $R = 100, n = \dfrac{5}{2} \cdot 4 = 10, i = \dfrac{0.12}{4} = 0.03$

$S_{due} = 100 \left[\dfrac{(1+0.3)^{10} - 1}{0.03} \right] (1+0.03) = \$1180.78$

23. $R = 200, n = 8 \cdot 2 = 16, i = \dfrac{0.06}{2} = 0.03$

$S_{due} = 200 \left[\dfrac{(1+.03)^{16} - 1}{0.03} \right] (1+0.03) = \$4152.32$

25. $S_{due} = 24000, n = 5, i = 0.08$

$R = 24000 \left[\dfrac{0.08}{(1+.08)^{5} - 1} \right] \left(\dfrac{1}{1+0.08} \right) = \$3787.92$

27. a. Ordinary annuity

b. $S = \$300,000$, $n = 25 \cdot 12 = 300$, $i = \dfrac{0.10}{12}$

$$R = 300,000 \left(\frac{0.10/12}{(1+0.10/12)^{300} - 1} \right)$$
$$= 300,000(0.00075) = \$226.10$$

29. a. Ordinary annuity

b. $S = \$50,000$, $R = \$300$, $i = 0.09/12$, $n = ?$

$$50000 = 300 \left[\frac{(1+0.0075)^n - 1}{0.0075} \right]$$

$$166.67 = \frac{(1.0075)^n - 1}{0.0075}$$

$$1.25 = (1.0075)^n - 1$$

$$2.25 = (1.0075)^n$$

$$\ln 2.25 = n \ln 1.0075$$

$$n = \frac{\ln 2.25}{\ln 1.0075} \approx 108.53 \text{ months}$$

31. a. Annuity due

b. $S_{due} = 180,000$, $n = 18 \cdot 12 = 216$,

$$i = \frac{0.12}{12} = 0.01$$

$$R = 180,000 \left[\frac{0.01}{(1+0.1)^{216} - 1} \right] \left(\frac{1}{1+0.01} \right)$$
$$= \$235.16$$

33. a. Annuity due

b. $R = 500$, $n = 9 \cdot 4 = 36$, $i = \dfrac{0.08}{4} = 0.02$

$$S_{due} = 500 \left[\frac{(1+0.02)^{36} - 1}{0.02} \right] (1+0.02)$$
$$= \$26,517.13$$

35. a. Ordinary annuity

b. $S = \$20,000$, $n = 10 \cdot 4 = 40$,

$$i = \frac{0.12}{4} = 0.03$$

$$R = 20,000 \left(\frac{0.03}{(1+0.03)^{40} - 1} \right)$$
$$= 20,000(0.01326) = \$265.25$$

37. $R = 100$, $n = 8 \cdot 12 = 96$, $i = \dfrac{0.09}{12} = 0.0075$

$$S_1 = 100 \left[\frac{(1+0.0075)^{96} - 1}{0.0075} \right]$$
$$= 100(139.8562) = \$13,985.62$$

Now: $P = 13,985.62$, $n = 15 \cdot 12 = 180$,

$i = 0.0075$

$$S_2 = 13,985.62(1+0.0075)^{180} = 53,677.41$$

39. a. $S = 150,000$, $R = 3000$, $i = \dfrac{0.08}{4} = 0.02$, $n = ?$, use the formula $S = R\left[\dfrac{(1+i)^n - 1}{i}\right]$.

$$150,000 = 3000\left[\frac{(1+0.02)^n - 1}{0.02}\right]$$

$$50 = \frac{(1.02)^n - 1}{0.02}$$

$$1 = 1.02^n - 1$$

$$2 = 1.02^n$$

$$\ln 2 = n\ln 1.02$$

$$n = \frac{\ln 2}{\ln 1.02} \approx 35 \text{ quarters}$$

b. Treat this as two problems. The $150,000 will grow at 8% compounded quarterly for 15 years. The second part is the deposits of $5000 each quarter at 8% compounded quarterly.

Part I: $S = P(1+i)^n$, where $P = 150,000$, $i = 0.02$, $n = 15 \cdot 4 = 60$

$$S = 150,000(1+0.02)^{60} = \$492,154.62$$

Part II: $S = R\left[\dfrac{(1+i)^n - 1}{i}\right]$, where $R = 5000$, $i = 0.02$, $n = 60$

$$S = 5000\left[\frac{(1+0.02)^{60} - 1}{0.02}\right] = \$570,257.70$$

So, his total will be $492,154.62 + $570,257.70 = $1,062,412.32

41. Using a calculator: $S = 100\left[\dfrac{(1.006)^{120} - 1}{0.006}\right] = \$17,500.30$ and $S_{due} = S(1.006) = \$17,605.30$

Or, using a spreadsheet:

	A	B	C
1		Future Value	
2	End of Month	Ordinary Ann.	Annuity Due
3	n	S=100(((1+i)^n-1)/i)	S(due)=S(1+i)
4	0	$0.00	$100.00
5	1	$100.00	$100.60
6	2	$200.60	$201.80
7	3	$301.80	$303.61
...	...	...	...
119	115	$16,493.49	$16,592.45
120	116	$16,692.45	$16,792.60
121	117	$16,892.60	$16,993.96
122	118	$17,093.96	$17,196.52
123	119	$17,296.52	$17,400.30
124	120	$17,500.30	$17,605.30

a. $n = 10 \cdot 12 = 120$ for each annuity. Total contributed: $100(120) = \$12,000$

b. The annuity due has more money. Each payment earns one month's interest more than is earned by the payments for the ordinary annuity.

Exercise 6.4

1. $R = 6000, n = 8 \cdot 2 = 16, i = \dfrac{0.08}{2} = 0.04$

$$A_n = 6000 \left[\frac{1 - (1 + .04)^{-16}}{0.04} \right] = \$69,913.77$$

3. $R = 250,000 \ n = 20, i = 0.10$

$$A_n = 250,000 \left[\frac{1 - (1 + .10)^{-20}}{0.10} \right] = \$2,128,390.93$$

5. $A = 135,000, n = 10 \cdot 4 = 40, i = \dfrac{0.064}{4} = 0.016$

$$R = 135,000 \left[\frac{0.016}{1 - (1 + .016)^{-40}} \right] = \$4595.46$$

7. We need to find the present value of $250,000 taken now plus $70,000 payments over 10 years.
$A_n = ?, \ n = 10, \ i = 0.065, \ R = 70,000$

$$A_n = 70000 \left[\frac{1 - (1 + .065)^{-10}}{0.065} \right] = \$503,218.12$$

Combining this with the $250,000, we get $753,218.12, which is a little higher than the $750,000 payment she would receive now.

9. $A_n = 1,500,000, \ i = 0.006, \ n = 20 \cdot 12 = 240$

$$R = 1,500,000 \left[\frac{0.006}{1 - (1 + 0.006)^{-240}} \right]$$
$$= \$11,810.24$$

11. $A_n = \$200,000, \ i = 0.015, \ R = 4500, \ n = ?$

$$200,000 = 4500 \left[\frac{1 - (1 + 0.015)^{-n}}{0.015} \right]$$

$$44.44 = \frac{1 - (1.015)^{-n}}{0.015}$$

$$0.666 = 1 - (1.015)^{-n}$$

$$-0.333 = -(1.015)^{-n}$$

$$0.333 = 1.015^{-n}$$

$$\ln 0.333 = -n \ln 1.015$$

$$n = -\frac{\ln 0.333}{\ln 1.015} = 73.8 \text{ quarters}$$

13. a. The higher graph corresponds to 8%.
 b. $1500 (approximately)
 c. A 10% interest rate has a present value of $9000. An 8% interest rate has a present value of $10,500.

15. The payments are at the end of each period for an ordinary annuity. The payments are at the beginning of each period for an annuity due.

17. $R = 3000, n = 7 \cdot 4 = 28, i = \dfrac{0.058}{4} = 0.0145$

$$A_{(n,due)} = 3000 \left[\frac{1 - (1 + .0145)^{-28}}{0.0145} \right] (1 + .0145)$$
$$= \$69,632.02$$

19. $R = 50,000, n = 12, i = 0.0592$

$$A_{(n,due)} = 50,000 \left[\frac{1 - (1 + .0592)^{-12}}{0.0592} \right] (1 + .0592)$$
$$= \$445,962.23$$

21. $A_{(n,due)} = 25,000, n = 12, i = \dfrac{0.0648}{12} = 0.0054$

$$R = 25,000 \left[\frac{0.0054}{1 - (1 + 0.0054)^{-12}} \right] \left(\frac{1}{1 + 0.0054} \right)$$
$$= \$2145.59$$

23. $A_{(n,due)} = 800,000, \ n = 3 \cdot 2 = 6,$

$$i = \frac{0.077}{2} = 0.0385$$

$$R = 800,000 \left[\frac{0.0385}{1 - (1 + .0385)^{-6}} \right] \left(\frac{1}{1 + .0385} \right)$$
$$= \$146,235.06$$

25. $R = 800, n = \dfrac{5}{2} \cdot 12 = 30, i = \dfrac{0.048}{12} = 0.004$

$$A_{(n,due)} = 800\left[\dfrac{1-(1+.004)^{-30}}{0.004}\right](1+.004)$$

$$= \$22,663.74$$

27. a. ordinary annuity

b. $A_n = 750,000$, $n = 40 \cdot 12 = 480$, $i = 0.007$

$$R = 750,000\left[\dfrac{0.007}{1-(1+0.007)^{-480}}\right] = \$5441.23$$

29. a. annuity due

b. $R = 40,000, n = \dfrac{9}{2} \cdot 2 = 9,$

$$i = \dfrac{0.0668}{2} = 0.0334$$

$$A_{(n,due)} = 40,000\left[\dfrac{1-(1+.0334)^{-9}}{0.0334}\right](1+.0334)$$

$$= \$316,803.61$$

31. a. ordinary annuity

b. $R = 2000$, $n = 16$, $i = \dfrac{0.072}{4} = 0.018$

$$A_n = 2000\left[\dfrac{1-(1+0.018)^{-16}}{0.018}\right] = \$27,590.62$$

33. $P = 2500$, $n = 12 \cdot 12 = 144$, $i = 0.0065$

$S_1 = 2500(1+0.0065)^{144} = \$6355.13$

$$S_2 = 100\left[\dfrac{(1+0.0065)^{144}-1}{0.0065}\right] = \$23,723.86$$

a. After the last deposit, $S_1 + S_2 = \$30,078.99$ is in the account.

b. $\$2500 + \$14,400 = \$16,900$ was deposited.

c. $A_n = 30,078.99$, $n = 5 \cdot 12 = 60$, $i = 0.0065$

$$R = 30,078.99\left[\dfrac{0.0065}{1-(1+0.0065)^{-60}}\right]$$

$$= \$607.02$$

d. The total amount withdrawn is $60(607.02) = \$36,421.20$.

35. The couple wants an annuity with $R = 30,000$ and $n = 8 \cdot 2 = 16$. We assume from part **a** that $i = 0.04$.

$$A_n = 30,000\left[\dfrac{1-(1+0.04)^{-16}}{0.04}\right] = \$349,568.87$$

They need a fund with $\$349,568.87$.

a. $S = 349,568.87$, $i = 0.04$, $n = 18 \cdot 2 = 36$

$$S = R\left[\dfrac{(1+i)^n - 1}{i}\right]$$

$$349,568.87 = R\left[\dfrac{(1+0.04)^{36}-1}{0.04}\right]$$

$$349,568.87 = R(77.598)$$

$$R = \$4504.87$$

b. When the extra $\$38,000$ is invested, there are 10 years left to compound the interest.

$$S = 38,000(1.04)^{20} = \$83,262.68$$

Now the account will have $\$349,568.87 + \$83,262.68 = \$432,831.55$ in it when the withdrawals begin. We need to calculate a new value of n.

$i = 0.04$, $R = 30,000$, $A_n = 432,831.55$

$$432,831.55 = 30,000\left[\dfrac{1-(1+0.04)^{-n}}{0.04}\right]$$

$$14.4277 = \dfrac{1-(1.04)^{-n}}{0.04}$$

$$0.5771 = 1-(1.04)^{-n}$$

$$-0.4229 = -1.04^{-n}$$

$$0.4229 = 1.04^{-n}$$

$$\ln 0.4229 = -n\ln 1.04$$

$$n = -\dfrac{\ln 0.4229}{\ln 1.04}$$

$$= 21.9 \text{ withdrawals}$$

37. $R = 2000$, $n = 8$, $k = 6$, $i = 0.07$

$$= 2000\left[\dfrac{1-(1+0.07)^{-8}}{0.07}\right](1+0.07)^{-6}$$

$$= \$7957.86$$

39. $R = 16,000, n = 7 \text{ (difficult part)},$

$k = 3, i = 0.06$

$$A_{(n,k)} = 16,000\left[\dfrac{1-(1+.06)^{-7}}{0.06}\right](1+.06)^{-3}$$

$$= \$74,993.20$$

41. $R = 10,000, n = 8, k = 36,$

$$i = \frac{0.07}{2} = 0.035$$

$$A_{(8,36)} = 10,000 \left[\frac{1-(1+.035)^{-8}}{0.035} \right] (1+.035)^{-36}$$

$$= \$19,922.97$$

43. $A_{(4,18)} = 1600, \ n = 4, \ k = 18, \ i = 0.06$

$$1600 = R \left[\frac{1-(1+.06)^{-4}}{0.06} \right] (1+.06)^{-18}$$

$$= R(1.21398)$$

$$R = \frac{1600}{1.21398} = \$1317.98 \, (\text{Alternate method})$$

45. $A_{(480,180)} = 2,500,000, \ n = 480, \ k = 180,$

$$i = 0.008$$

$$2,500,000 = R \left[\frac{1-(1+0.008)^{-480}}{0.008} \right] (1+.008)^{-180}$$

$$= R(29.1361)$$

$$R = \frac{2,500,000}{29.1361} = \$85,804.29$$

47. a.

	A	B	C	D
1	End of Month	Acct. Value	Payment	New Balance
2	0	$100,000.00	$0.00	$100,000.00
3	1	$100,650.00	$1,000.00	$99,650.00
4	2	$100,297.73	$1,000.00	$99,297.73
5	3	$99,943.16	$1,000.00	$98,943.16
6	4	$99,586.29	$1,000.00	$98,586.29
7	5	$99,227.10	$1,000.00	$98,227.10
8	6	$98,865.58	$1,000.00	$97,865.58
9	7	$98,501.70	$1,000.00	$97,501.70
10	8	$98,135.47	$1,000.00	$97,135.47
11	9	$97,766.85	$1,000.00	$96,766.85
12	10	$97,395.83	$1,000.00	$96,395.83
⋮	⋮	⋮	⋮	⋮
159	157	$5,938.16	$1,000.00	$4,938.16
160	158	$4,970.26	$1,000.00	$3,970.26
161	159	$3,996.06	$1,000.00	$2,996.06
162	160	$3,015.54	$1,000.00	$2,015.54
163	161	$2,028.64	$1,000.00	$1,028.64
164	162	$1,035.32	$1,000.00	$35.32
164	163	$35.55	$35.55	$0.00

b.

	A	B	C	D
1	End of Month	Acct. Value	Payment	New Balance
2	0	$100,000.00	$0.00	$100,000.00
3	1	$100,650.00	$2,500.00	$98,150.00
4	2	$98,787.98	$2,500.00	$96,287.98
5	3	$96,913.85	$2,500.00	$94,413.85
6	4	$95,027.54	$2,500.00	$92,527.54
7	5	$93,128.97	$2,500.00	$90,628.97
8	6	$91,218.05	$2,500.00	$88,718.05
9	7	$89,294.72	$2,500.00	$86,794.72
10	8	$87,358.89	$2,500.00	$84,858.89
11	9	$85,410.47	$2,500.00	$82,910.47
12	10	$83,449.39	$2,500.00	$80,949.39
⋮	⋮	⋮	⋮	⋮
43	41	$15,902.26	$2,500.00	$13,402.26
44	42	$13,489.37	$2,500.00	$10,989.37
45	43	$11,060.81	$2,500.00	$8,560.81
46	44	$8,616.45	$2,500.00	$6,116.45
47	45	$6,156.21	$2,500.00	$3,656.21
48	46	$3,679.98	$2,500.00	$1,179.98
49	47	$1,187.65	$1,187.65	$0.00

Exercise 6.5

1. a. The 10 year loan requires more payment to principal since the loan must be paid more quickly.

  b. The 25 year loan requires lower payment period since the loan is paid more slowly.

3. $A_n = 8000, \ n = 8, \ i = \dfrac{0.12}{2} = 0.06,$

$$a_{\overline{8}|0.06} = \dfrac{1-(1+0.06)^{-8}}{0.06}$$

$$R = \dfrac{8000}{a_{\overline{8}|0.06}} = 8000 \cdot \dfrac{1}{a_{\overline{8}|0.06}} = \dfrac{8000}{6.209794}$$

$$= 8000(0.161036) = \$1288.29$$

5. $A_n = 18,000, \ n = 10 \cdot 4 = 40,$

$$i = \dfrac{0.042}{4} = 0.0105, \ a_{\overline{40}|0.0105} = ?$$

$$a_{\overline{40}|0.0105} = \dfrac{1-(1+0.0105)^{-40}}{0.0105} = 32.53$$

$$R = \dfrac{18,000}{a_{\overline{40}|0.0105}} = \dfrac{18,000}{32.53} = \$553.34$$

9. $A_n = 100,000, \ n = 3, \ i = 0.09,$

$$a_{\overline{3}|0.09} = \dfrac{1-(1+0.09)^{-3}}{0.09}$$

$$R = \dfrac{100,000}{a_{\overline{3}|0.09}} = \dfrac{100,000}{2.531295} = \$39,505.48$$

Payment − Interest = Balance Reduction

Period	Payment	Interest	Balance Reduction	Unpaid Balance
				100,000.00
1	39,505.48	9000.00	30,505.48	69,494.52
2	39,505.48	6254.51	33,250.98	36,243.55
3	39,505.47	3261.92	36,243.55	0.00

Last payment usually leaves a few cents on balance. Also, method used can affect answer by 2 or 3 cents.

11. $A_n = 20,000, \ n = 4, \ i = \dfrac{0.12}{4} = 0.03, \ a_{\overline{4}|0.03} = \dfrac{1-(1+0.03)^{-4}}{0.03}$

$$R = \dfrac{20,000}{a_{\overline{4}|0.03}} = \dfrac{20,000}{3.717098} = \$5380.54$$

Period	Payment	Interest	Balance Reduction	Unpaid Balance
				20,000.00
1	5380.54	600.00	4780.54	15,219.46
2	5380.54	456.58	4923.96	10,295.50
3	5380.54	308.87	5071.67	5223.83
4	5380.54	156.71	5223.83	0.00

7. $R = 200, \ A_n = ?, \ n = 5 \cdot 12 = 60,$

$$i = \dfrac{0.06}{12} = 0.005, \ a_{\overline{60}|0.005} = ?$$

$$a_{\overline{60}|0.005} = \dfrac{1-(1+0.005)^{-60}}{0.005} = 51.7256$$

$$A_n = R \cdot a_{\overline{60}|0.005} = 200 \cdot 51.7256 = \$10,345.12$$

13. $R = 334.27,\ \ n = 10 \cdot 4,\ \ k = 6,\ \ i = \dfrac{0.06}{4} = 0.015$

Unpaid balance: $334.27 \left[\dfrac{1 - (1 + .015)^{-(40-6)}}{0.015} \right] = \$8{,}852.05$

15. $R = 342.44,\ n = 48,\ k = 30,\ i = \dfrac{0.081}{12} = 0.00675$

Unpaid balance: $342.44 \left[\dfrac{1 - (1 + .00675)^{-(48-30)}}{0.00675} \right] = \$5785.83$

17. $A_n = 100{,}000,\ \ n = 10 \cdot 2 = 20,\ \ i = \dfrac{0.12}{2} = 0.06$

   a. $R = 100{,}000 \left[\dfrac{0.06}{1 - (1 + .06)^{-20}} \right] = \$8718.46$

   b. Total paid = 20(8718.46) = \$174,369.20

   c. Interest paid = 174,869.20 − 100,000 = \$74,369.20

19. $A_n = 8000,\ n = 4 \cdot 2 = 8,\ i = \dfrac{0.10}{2} = 0.05$

   a. $R = 8000 \left[\dfrac{0.05}{1 - (1 + .05)^{-8}} \right] = \$1237.78$

   b. Total paid = 8(1237.78) = \$9902.24

   c. Interest paid = \$9902.24 − 8000 = \$1902.24

21. $R = 300, n = 36, i = 0.01$

$A_n = 300 \left[\dfrac{1 - (1 + .01)^{-36}}{0.01} \right] = \$9032.25$

Down payment is \$12,000 − 9032.25 = \$2967.75.

23. $R = 900, n = 25 \cdot 12 = 300$

   a. $i = \dfrac{0.069}{12} = 0.00575$

$A_n = 900 \left[\dfrac{1 - (1 + 0.00575)^{-300}}{0.00575} \right]$

$= \$128{,}495.66$

Amount available:

\$20,000 + 128,495.66 = \$148,495.66

   b. $i = \dfrac{0.075}{12} = 0.00625$

$A_n = 900 \left[\dfrac{1 - (1 + 0.00625)^{-300}}{0.00625} \right]$

$= \$121{,}787.65$

Amount available:

\$20,000 + \$121,787.65 = \$141,787.65

25. $R = 610.91, n = 360, k = 12, i = 0.006$

   a. Unpaid balance:

$610.91 \left[\dfrac{1 - (1 + .006)^{-(360-12)}}{0.006} \right] = \$89{,}120.53$

   b. Interest paid: 12(610.91) − 879.47

$= \$6451.45$

27. a. $R = ?$, $A_n = 18,000$, $n = 5 \cdot 12 = 60$,

$i = \dfrac{0.084}{12} = 0.007$, $a_{\overline{60}|0.007} = ?$

$a_{\overline{60}|0.007} = \dfrac{1-(1+0.007)^{-60}}{0.007} = 48.8559$

$R = \dfrac{A_n}{a_{\overline{60}|0.007}} = \dfrac{18,000}{48.8559} = \$368.43$

$368.43 is required by the loan, she decides to pay \$383.43 a month.

b. $A_n = 18,000$, $R = 383.43$, $i = 0.007$, $n = ?$

$18,000 = 383.43\left[\dfrac{1-(1+0.007)^{-n}}{0.007}\right]$

$46.94 = \dfrac{1-(1.007)^{-n}}{0.007}$

$0.3286 = 1 - 1.007^{-n}$

$-0.6714 = -1.007^{-n}$

$0.6714 = 1.007^{-n}$

$\ln 0.6714 = -n \ln 1.007$

$n = -\dfrac{\ln 0.6714}{\ln 1.007} = 57.1$

It will take a little over 57 months.

c. Without the extra, she would pay
$368.43 \cdot 60 = \$22,105.80.$
With the extra, she pays
$383.43 \cdot 57.1 = \$21,893.85$ for a savings of
$211.95.

29. a. The line is the total amount paid (amount per month x months). The other curve is the amount paid toward the principal.

b. The length of the vertical line segment from the lower curve to the line at $x = 250$ represents the total interest paid after 250 months.

31. a. $A_n = 15,000, \ n = 12 \cdot 4 = 48$

$$i = \frac{0.08}{12} \ \rightarrow \ R = 15,000 \left[\frac{\left(\frac{.08}{12}\right)}{1 - \left(1 + \frac{.08}{12}\right)^{-48}} \right] = \$366.19$$

$$i = \frac{0.085}{12} \ \rightarrow \ R = 15,000 \left[\frac{\left(\frac{.085}{12}\right)}{1 - \left(1 + \frac{.085}{12}\right)^{-48}} \right] = \$369.72$$

Total interest at:
  8% : $48(366.19) - 15,000 = \$2577.12$
  8.5% : $48(369.72) - 15,000 = \$2746.56$

b. $A_n = 80,000, \ n = 12 \cdot 30 = 360$

$$i = \frac{0.0675}{12} = 0.005625 \ \rightarrow \ R = 80,000 \left[\frac{0.005625}{1 - \left(1 + 0.005625\right)^{-360}} \right] = \$518.88$$

$$i = \frac{0.0725}{12} \ \rightarrow \ R = 80000 \left[\frac{\left(\frac{.0725}{12}\right)}{1 - \left(1 + \frac{.0725}{12}\right)^{-360}} \right] = \$545.74$$

Total interest at:
  6.75%: $360(518.88) - 80,000 = \$106,796.80$
  7.25%: $360(545.74) - 80,000 = \$116.466.40$

c.   The duration of the loan has the greatest effect on the borrower. It affects R and the total interest paid.

33. $A_n = 100,000, \ n = 25 \cdot 12 = 300$

i. $$R = 100,000 \left[\frac{\left(\frac{.075}{12}\right)}{1 - \left(1 + \frac{.075}{12}\right)^{-300}} \right] = \$738.99$$

ii. $$R = 100,000 \left[\frac{\left(\frac{.0725}{12}\right)}{1 - \left(1 + \frac{.0725}{12}\right)^{-300}} \right] = \$722.81$$

iii. $$R = 100,000 \left[\frac{\left(\frac{.07}{12}\right)}{1 - \left(1 + \frac{.07}{12}\right)^{-300}} \right] = \$706.78$$

a.

Payment	Points	Total Paid $\left[(R \cdot 300) + \text{Points}\right]$
$738.99	0	$221,697
722.81	$1000	217,843
706.78	$2000	214,034

b.   The lower total cost is the 7%, 2 points loan.

35. a. To find the loan balance after the grace period, we need to calculate the accrued interest on the four loans.

$i = \dfrac{0.01}{4} = 0.0025$ during the in-school period of four years.

$n_1 = 4 \cdot 4 - 1 = 15, \ S_1 = 4000(1+0.0025)^{15} = \$4152.65, n_2 = 3 \cdot 4 - 1 = 11, \ S_2 = 3500(1+0.0025)^{11} = \$3597.46$

$n_3 = 2 \cdot 4 - 1 = 7, \ S_3 = 4400(1+0.0025)^{7} = \$4477.58, n_4 = 1 \cdot 4 - 1 = 3, \ S_4 = 5000(1+0.0025)^{3} = \$5037.59$

Total at end of 4 years is $17,265.28.

For 6-month grace period, $i = \dfrac{0.03}{4} = 0.0075$. $S = 17,265.28(1+0.0075)^{2} = \$17,525.23$

b. $A_n = 17,525.23, \ i = \dfrac{0.03}{4} = 0.0075, \ n = 10 \cdot 4 = 40, \ R = ?, \quad R = 17525.23\left[\dfrac{0.0075}{1-(1+0.0075)^{-40}}\right] = \$508.76$

c. $R = \$598.76, \ i = 0.0075, \ A_n = 17,525.23, \ n = ?$

$17525.23 = 598.76\left[\dfrac{1-(1+0.0075)^{-n}}{0.0075}\right]$

$29.2692 = \dfrac{1-(1.0075)^{-n}}{0.0075}$

$0.2195 = 1 - 1.0075^{-n}$

$-0.7805 = -1.0075^{-n}$

$0.7805 = 1.0075^{-n}$

$\ln 0.7805 = -n \ln 1.0075$

$n = -\dfrac{\ln 0.7805}{\ln 1.0075} = 33.2$ quarters

d. Without the extra payment, Nolan would've paid $\$508.76 \cdot 40 = \$20,350.40$. With the extra payment, he will pay $\$598.76 \cdot 33.2 = \$19,878.83$, so the savings will be $471.57.

37. $A_n = 16,700, \ i = \dfrac{0.082}{12}, \ n = (4)(12) = 48 \qquad R = 16,700\left[\dfrac{\frac{0.082}{12}}{1-\left(1+\frac{0.082}{12}\right)^{-48}}\right] = \$409.27$

	A	B	C	D	E
1	Period	Payment	Interest	Bal. Reduction	Unpaid Bal.
2	0	$0.00	$0.00	$0.00	$16,700.00
3	1	$409.27	$114.12	$295.15	$16,404.85
4	2	$409.27	$112.10	$297.17	$16,107.68
5	3	$409.27	$110.07	$299.20	$15,808.48
6	4	$409.27	$108.02	$301.25	$15,507.23
7	5	$409.27	$105.97	$303.30	$15,203.93
8	6	$409.27	$103.89	$305.38	$14,898.55
9	7	$409.27	$101.81	$307.46	$14,591.09
10	8	$409.27	$99.71	$309.56	$14,281.52
⋮	⋮	⋮	⋮	⋮	⋮
45	43	$409.27	$16.38	$392.89	$2,004.81
46	44	$409.27	$13.70	$395.57	$1,609.24
47	45	$409.27	$11.00	$398.27	$1,210.97
48	46	$409.27	$8.27	$401.00	$809.98
49	47	$409.27	$5.53	$403.74	$406.24
50	48	$409.02	$2.78	$406.69	$0.00

Review Exercises

1. $a_n = \dfrac{1}{n^2}$ $a_1 = \dfrac{1}{1^2} = 1$ $a_2 = \dfrac{1}{2^2} = \dfrac{1}{4}$

$a_3 = \dfrac{1}{3^2} = \dfrac{1}{9}$ $a_4 = \dfrac{1}{4^2} = \dfrac{1}{16}$

2. $12, 7, 2, -3, \ldots$ arithmetic with $d = -5$.

$\dfrac{1}{6}, \dfrac{2}{6}, \dfrac{3}{6}, \dfrac{4}{6}, \ldots$ arithmetic with $d = \dfrac{1}{6}$.

3. $a_1 = -2, d = 3, n = 80, a_{80} = -2 + (80-1) = 235$

4. $a_3 = 10, a_8 = 25, n = 36$

$\begin{cases} 25 = a_1 + 7d \\ 10 = a_1 + 2d \end{cases} \begin{array}{l} d = 3 \\ a_1 = 4 \end{array}$

$a_{36} = 4 + (36-1)3 = 109$

5. $a_1 = \dfrac{1}{3}, d = \dfrac{1}{6}, n = 60$

$a_{60} = \dfrac{1}{3} + (60-1)\dfrac{1}{6} = \dfrac{61}{6}$

$s_{60} = \dfrac{60}{2}\left(\dfrac{1}{3} + \dfrac{61}{6}\right) = 315$

6. $\dfrac{1}{4}, 2, 16, 128, \ldots$ geometric with $r = 8$.

$16, -12, 9, -\dfrac{27}{4}, \ldots$ geometric with $r = -\dfrac{3}{4}$.

$\left(16r = -12, r = -\dfrac{3}{4}\right)$

7. $a_1 = 64, a_8 = \dfrac{1}{2}, n = 4$

$\dfrac{1}{2} = 64r^7, r^7 = \dfrac{1}{128}$

or $r = \dfrac{1}{2}, a_4 = 64\left(\dfrac{1}{2}\right)^3 = 8$

8. $a_1 = \dfrac{1}{9}, r = 3, n = 16,$

$s_{16} = \left(\dfrac{1}{9}\right)\dfrac{\left(1-3^{16}\right)}{1-3} = 2,391,484\dfrac{4}{9}$

9. $P = 8000, r = 0.12, t = 3$

$S = P + Prt$

$= 8000 + 8000(0.12)(3) = \$10,880$

10. $P = 2000, S = 2100, t = 0.75, r = ?$

$I = S - P = 100;\ 100 = 2000 \cdot r(0.75)$

$r = \dfrac{100}{2000(0.75)} = 6\dfrac{2}{3}\%$

11. $S = 3000, r = 0.06, t = \dfrac{4}{12} = \dfrac{1}{3}$

$3000 = P\left(1 + (.06)\dfrac{1}{3}\right),$

$P = \dfrac{3000}{1.02} = \$2941.18$

12. $a_1 = 10, a_2 = 20, a_3 = 30, \ldots, a_{30} = 300$

$s_{30} = \dfrac{30}{2}(10 + 300) = \$4650$

13. Job #1

$n = 10,\ a_1 = 20,000,$

$d = 1000,\ a_{10} = 29,000$

$s_{10} = \dfrac{10}{2}(20,000 + 29,000) = \$245,000$

Job #2

$n = 10,\ a_1 = 18,000,$

$d = 1200,\ a_{10} = 28,800$

$s_{10} = \dfrac{10}{2}(18,000 + 28,800) = \$234,000$

14. a. $n = 4 \cdot 10 = 40$

b. $i = \dfrac{0.08}{4} = 0.02 = 2\%$

15. a. $S = P\left(1 + \dfrac{r}{m}\right)^{mt} = P(1+i)^n$

b. $S = Pe^{rt}$

16. Monthly compounding will earn more.

17. Interest $= 1000(1.02)^{16} - 1000 = \$372.79$

18. $S = 18000, n = 4 \cdot 12 = 48, i = \dfrac{0.054}{12} = 0.0045$

$P = \dfrac{18000}{(1 + 0.0045)^{48}} = \$14,510.26$

19. $S = 1000e^{(0.08)(6)} = \$1616.07$

20. $S = 100,000, t = 15, r = 0.1031$

$P = \dfrac{100,000}{e^{(.1031)(15)}} = \$21,299.21$

Note: In 21–22 we can use either log or ln.

21. $25,000 = 15,000\left(1+\dfrac{0.06}{4}\right)^{4t}$

$\dfrac{5}{3} = (1.015)^{4t}$

$\ln\left(\dfrac{5}{3}\right) = 4t\ln(1.015)$

$t = \dfrac{\ln(5/3)}{4\ln(1.015)} \approx 8.577$ years

22. a. $\qquad 257,000 = 35,000e^{15r}$

$\ln\left(\dfrac{257,000}{35,000}\right) = 15r$

$\dfrac{\ln 257,000 - \ln 35,000}{15} = r$

$r \approx 0.1329 = 13.29\%$

b. $APY = e^{0.1329} - 1 = 0.1421 = 14.21\%$

23. a. $APY = \left(1+\dfrac{r}{m}\right)^m - 1,\ m = 4,\ r = 0.072$

$APY = \left(1+\dfrac{0.072}{4}\right)^4 - 1$

$= 1.0740 - 1 = 0.0740 = 7.40\%$

b. $APY = e^r - 1 = e^{.072} - 1$

$= 0.07466 = 7.47\%$

24. $a_1 = 1, a_2 = 2, \ldots, r = 2;$

$a_{64} = 1 \cdot 2^{64-1} = 2^{63}$

25. $a_1 = 1, r = 2, a_{32} = 1 \cdot 2^{32-1} = 2^{31}$

$s_{32} = \dfrac{1(1-2^{32})}{1-2} = 2^{32} - 1 \approx 4.295 \times 10^9$

26. $R = 800, n = 20, \dfrac{0.12}{2} = 0.06$

$S = 800 \cdot s_{\overline{20}|0.06} = 800(36.785592)$

$= \$29,428.47$

27. $S = 80,000, n = 10, i = 0.06;$

$R = S \cdot \dfrac{1}{s_{\overline{10}|0.06}} = 80,000(0.075868) = \$6096.44$

28. $R = 800, n = 10 \cdot 2 = 20, i = \dfrac{0.12}{2} = 0.06$

$S_{due} = 800\left[\dfrac{(1+.06)^{20}-1}{0.06}\right](1+.06) = \$31,194.18$

29. $S = 250,000, n = \dfrac{9}{2} \cdot 4 = 18, i = \dfrac{0.102}{4} = 0.0255$

$R = 250,000\left[\dfrac{0.0255}{(1+.0255)^{18}-1}\right]\left(\dfrac{1}{1+.0255}\right)$

$= \$10,841.24$

30. $S = 60,000, n = ?, i = \dfrac{0.072}{4} = 0.018, R = 1200$

$60,000 = 1200\left[\dfrac{(1+0.018)^n-1}{0.018}\right]$

$50 = \dfrac{(1+0.018)^n-1}{0.018}$

$0.9 = (1.018)^n - 1$

$1.9 = 1.018^n$

$\ln 1.9 = n\ln 1.018$

$n = -\dfrac{\ln 1.9}{\ln 1.018} = 36$ quarters

31. $R = 10,000, n = 10 \cdot 2 = 20, i = \dfrac{0.09}{2} = 0.045$

$A_n = 10,000\left[\dfrac{1-(1+.045)^{-20}}{0.045}\right] = \$130,079.36$

32. $R = 3000, n = 6, k = 5 \cdot 12 - 1 = 59,$

$i = \dfrac{0.078}{12} = 0.0065$

$A_{(n,k)} = 3000\left[\dfrac{1-(1+.0065)^{-6}}{0.0065}\right](1+.0065)^{-59}$

$= \$12,007.09$

First payment is before they leave.

33. a. $R = \dfrac{295.7}{25}$ million $= \$11,828,000$

b. $R = 11,828,000, n = 24, i = 0.0591$

$A_n = 11,828,000\left[\dfrac{1-(1+.0591)^{-24}}{0.0591}\right]$

$= \$149,688,218$

Now add first payment at beginning:
$A = \$161,516,218$. This is also an annuity
due with $n = 25$.

34. $A_n = 20,000, n = 12, i = \dfrac{0.066}{12} = 0.0055$

$R = 20,000\left[\dfrac{0.0055}{1-(1+.0055)^{-12}}\right] = \$1,726.85$

35. $A_{due} = 250,000, \; n = 20 \cdot 4 = 80,$

$$i = \frac{0.062}{4} = 0.0155$$

$$A_{due} : R = 250,000 \left[\frac{0.0155}{1-(1+.0155)^{-80}} \right] \left(\frac{1}{1+.0155} \right)$$

$$= \$5390.77$$

36. $R = 40,000, \; i = \frac{0.068}{2} = 0.034, \; A_n = \$488,000, \; n = ?$

$$488,000 = 40,000 \left[\frac{1-(1+0.034)^{-n}}{0.034} \right]$$

$$12.2 = \frac{1-(1.034)^{-n}}{0.034}$$

$$0.4148 = 1-(1.034)^{-n}$$

$$-0.5852 = -1.034^{-n}$$

$$0.5852 = 1.034^{-n}$$

$$\ln 0.5852 = -n \ln 1.034$$

$$n = -\frac{\ln 0.5852}{\ln 1.034} = 16 \text{ half-years, or 8 years}$$

37. $A_n = 1000, \; n = 12, \; i = 0.01$

$$R = 1000 \left[\frac{0.01}{1-(1+.01)^{-12}} \right] = \$88.85$$

38. $R = 1288.29, \; n = 8, \; k = 5, \; i = 0.06$

$$A_{n-k} = 1288.29 \left[\frac{1-(1+.06)^{-(8-5)}}{0.06} \right] = \$3443.61$$

39. Present value of the 18 quarterly payment is

$$A_n = 1500 \cdot a_{\overline{18}|1\%} = 1500 \cdot \frac{1-(1.01)^{-18}}{0.01}$$

or \$24,597.40

Total cost: $24,597.40 + 10,000 = \$34,597.40$

40. $i = 0.00625$

Interest = (Unpaid Balance)(.00625)(1)

Balance Reduction = Payment – Interest

57: Interest = (95042.20)(.00625) = 594.01

Payment	Amount	Interest	Balance Reduction	Unpaid Balance
57	\$699.22	\$594.01	\$105.21	\$94,936.99
58	\$699.22	\$593.36	\$105.86	\$94,831.13

41. $P = 2500, \; S = 38,000, \; n = 18$

$$38,000 = 2500(1+i)^{18}$$

$$(1+i)^{18} = \frac{38,000}{2500}$$

$$18\ln(1+i) = \ln 38,000 - \ln 2500$$

$$18\ln(1+i) = 10.5453 - 7.8240 = 2.7213$$

$$\ln(1+i) = \frac{2.7213}{18} = 0.1512$$

$$1+i = e^{0.1512} = 1.1632$$

$$i = 0.1632 = 16.32\%$$

42. $S = 40,000, \; n = 10 \cdot 12 = 120, \; i = \frac{0.084}{12} = 0.007$

$$R = 40,000 \left[\frac{0.007}{(1+.007)^{120} -1} \right] = \$213.81$$

43. a. $P = 1000$, $r = 8\%$, $t = 6$

$S = P + Prt = 1000 + 1000(0.08)(6) = \$1,480$

b. $P = 1000$, $i = \dfrac{0.08}{2} = 0.04$, $n = 6 \cdot 2 = 12$

$S = P(1+i)^n = 1000(1+0.04)^{12} = \$1601.03$

44. $R = 500$, $i = \dfrac{0.08}{4} = 0.02$, $n = 4 \cdot 4 = 16$

$S = R\left[\dfrac{(1+i)^n - 1}{i}\right] = 500\left[\dfrac{(1+0.02)^{16} - 1}{0.02}\right] = 500(18.6393) = \$9319.64$

45. $P = 8000$, $S = 22,000$, $r = 0.07$

$22,000 = 8000e^{0.07t}$

$e^{0.07t} = \dfrac{22,000}{8000}$

$0.07t = \ln 22,000 - \ln 8000$

$0.07t = 9.9988 - 8.9872 = 1.0116$

$t = 14.45$ years

46. $P = 87.89$, $S = 105.34$, $t = 1/4$

$I = S - P = 105.34 - 87.89 = 17.45$

$17.45 = 87.89 \cdot r \cdot \frac{1}{4}$

$r = \dfrac{4(17.45)}{87.89} = 0.794 = 79.4\%$

47. $R = 400$, $n = 4 \cdot 12 = 48$, $i = \dfrac{0.054}{12} = 0.0045$

$S_{due} = R\left[\dfrac{(1+i)^n - 1}{i}\right](1+i) = 400\left[\dfrac{(1+0.0045)^{48} - 1}{0.0045}\right](1+0.0045) = \$21,474.08$

48. $A_{(n,k)} = 72,000$, $n = 9 \cdot 2 = 18$,

$k = 11 \cdot 2 = 22$, $i = 0.0365$

$A_{(n,k)} : R = 72,000\left[\dfrac{0.0365}{1-(1+.0365)^{-18}}\right](1+.0365)^{22} = \$12,162.06$

49. $APY = \left(1 + \dfrac{r}{m}\right)^m - 1 = \left(1 + \dfrac{0.0652}{4}\right)^4 - 1 = 0.066811$

$APY = e^r - 1 = e^{0.0648} - 1 = 0.066946$

The compounded continuous offer is the higher rate.

50. $r = 1000$, $n = 40$, $i = \dfrac{0.12}{12} = 0.01$

$A_n = 1000\left[\dfrac{1-(1+.01)^{-40}}{0.01}\right] = \$32,834.69$

51. $P = 8000$, $i = 0.12$, $n = 3$

$S = Pe^{rt} = 8000e^{0.36} = 8000(1.433329415) = \$11,466.64$

$S - P = \$3,466.64$

52. $A_n = 92,000$, $n = 25(12) = 300$, $i = \dfrac{6\%}{12} = 0.005$

    a. $R = A_n\left(\dfrac{1}{a_{\overline{n}|i\%}}\right) = 92,000\left(\dfrac{0.005}{1-(1+0.005)^{-300}}\right) = 92,000(0.006443014) = \$592.76$

    b. Total payment: $300(592.76) = \$177,828$

    c. Total interest: $\$177,828 - 92,000 = \$85,828$

    d. Find the present value for $n = 300 - 84 = 216$.

$$A_{n-k} = 592.76\left[\dfrac{1-(1+0.005)^{-216}}{0.005}\right] = \$78,183.79$$

53. Bonus: $P = 12,500$, $n = 150$, $i = \dfrac{10.8\%}{12} = 0.009$

$$S_1 = 12,500(1+0.009)^{150} = \$47,927.52$$

Annuity: $S_2 = 150\left[\dfrac{(1+0.009)^{150}-1}{0.009}\right] = \$47,236.69$

    a. $S = S_1 + S_2 = \$95,164.21$

    b. $R = \$95,164.21\left[\dfrac{0.009}{1-(1+0.009)^{-120}}\right] = \$1300.14$

54. The withdrawals begin one month after the last deposit. Look at the problem this way. Amount needed in account at first withdrawal is $A = 5000 + 5000\left[\dfrac{1-(1+0.0055)^{-3}}{0.0055}\right] = \$19,836.50$. Now we have

$$19,836.50 = R\left[\dfrac{(1+0.0055)^{36}-1}{0.0055}\right](1+0.0055). \text{ Solving we have } R = \$497.04.$$

55. We'll assume that Aruam invests the \$2000 IRA at 8.4% compounded monthly from age 22 to age 67, a total of 45 years. This would give her $S = 2000(1+0.007)^{12\cdot45} = 2000(1.007)^{540} = \$86,485.66$. At age 30, she has 37 years till retirement. She wants to receive \$10,000 a month for 20 years ($n = 12 \cdot 20 = 240$). First we need to figure out what the present value of the annuity will be when she turns 67.

$$R = 10,000, \; i = 0.007, \; n = 240, \text{ so } A_n = 10,000\left[\dfrac{1-(1+0.007)^{-240}}{0.007}\right] \approx \$1,160,760.05$$

She will need to have \$1,160,760.05 when she turns 67. She will need an additional $\$1,160,760.05 - \$86,485.66 = \$1,074,274.39$ from the monthly payments. So now we calculate the necessary monthly payment. $A_n = 1,074,274.39$, $R = ?$, $i = 0.007$, $n = 37\cdot12 = 444$.

$$1,074,274.39 = R\left[\dfrac{(1+0.007)^{444}-1}{0.007}\right]$$

$$1,074,274.39 = R(3019.325)$$

$$R = \$355.80$$

She should make a monthly payment of \$355.80 from age 30 to age 67.

56. First we calculate what the regular payment would be on \$2,600,000 borrowed for 15 years at 5.6% compounded quarterly. $A_n = 2,600,000$, $i = 0.056/4 = 0.014$, $R = ?$, $n = 15 \cdot 4 = 60$

$$R = 2,600,000 \left[\frac{0.014}{1-(1+0.014)^{-60}} \right] = \$64,337.43$$

Next we find the unpaid balance after 2 years, at this time there will be $60 - 8 = 52$ payments remaining.

$$A_{n-k} = \$64,337.43 \left[\frac{1-(1+0.014)^{-52}}{0.014} \right] = \$2,365,237.24$$

Increasing to $R = 70,000$, we solve for n.

$$2,365,237.24 = 70,000 \left[\frac{1-(1+0.014)^{-n}}{0.014} \right]$$

$$33.7891 = \frac{1-(1.014)^{-n}}{0.014}$$

$$0.473 = 1 - 1.014^{-n}$$

$$-0.527 = -1.014^{-n}$$

$$0.527 = 1.014^{-n}$$

$$\ln 0.527 = -n \ln 1.014$$

$$n = -\frac{\ln 0.527}{\ln 1.014} = 46.1 \text{ quarterly payments}$$

Now we calculate the savings. Without altering the monthly payment, the company would have made 60 payments of \$64,337.43 for a total of \$3,860,245.80. Really, they made 8 payments of \$64,337.43 and 46.1 payments of \$70,000 for a total of \$3,741,699.44. Their savings was \$118,546.36.

Chapter Test

1. $S = 47,000, P = 8000, r = 0.07$,

$S = Pe^{rt}$

$47,000 = 8000e^{0.07t}$

$e^{0.07t} = \frac{47,000}{8000} = 5.875$

$0.07t \ln e = \ln 5.875$

$t = \frac{\ln 5.875}{0.07} = 25.3 \text{ years}$

2. $S = 12,000, n = 6 \cdot 2 = 12, i = 0.031$

$R = 12,000 \left[\frac{0.031}{(1+.031)^{12} - 1} \right] = \$840.75$

3. $S = 5000, P = 4756.69, t = 0.75$

$I = S - P = 243.31$

$r = \frac{I}{Pt} = \frac{243.31}{4756.69(.75)} = 0.0682 = 6.82\%$

4. $R = 14,357.78, n = 40, k = 25, i = .041$
Unpaid balance;

$= 14,357.78 \left[\frac{1-(1+.041)^{-(40-25)}}{0.041} \right]$

$= \$158,524.90$

5. $P = 1000, S = 13,500, n = 9$

$13,500 = 1000(1+i)^9$

$1 + i = \left(\frac{13,500}{1000} \right)^{1/9} = 1.3353$

$i = 0.3353 = 33.53\%$

This problem can also be solved by taking ln of both sides.

6. $A_n = 97,000$, $n = 25 \cdot 12 = 300$, $i = 0.006$

a. $R = 97,000 \left[\frac{0.006}{1-(1+.006)^{-300}} \right] = \$698$

b. Interest paid $= 300(698) - 97,000 = \$112,400$

7. $P = 2500, r = 0.04, t = \dfrac{15}{12} = 1.25$

$S = 2500 + 2500(.04)(1.25) = \$2625$

8. $R = 100, n = \dfrac{11}{2} \cdot 12 = 66, i = \dfrac{0.069}{12} = 0.00575$

$S_n = 100\left[\dfrac{(1+.00575)^{66} - 1}{0.00575}\right] = \$7999.41$

9. $i = \dfrac{0.084}{12} = 0.007, APY = (1+.007)^{12} - 1$

$= 0.0873 = 8.73\%$

10. $R = 3000, n = 15 \cdot 4 = 60, i = 0.015$

$A_{due} = 3000\left[\dfrac{1-(1+.015)^{-60}}{0.015}\right](1+.015)$

$= \$119,912.92$

11. $P = 10,000, r = 0.07, t = 20$

$S = 10,000e^{(.07)(20)} = \$40,552$

12. $R = 1500, n = 18, i = 0.02$

$A_n = 1500\left[\dfrac{1-(1+.02)^{-18}}{0.02}\right] = \$22,488 \ \text{(rounded)}$

Cost of car $= \$22,488 + 10,000 = \$32,488$

13. $S = 9,500, n = 5 \cdot 4 = 20, i = 0.017$

$P = \dfrac{9500}{(1+.017)^{20}} = \$6781.17$

14. $S = 500,000, R = 1500, i = 0.041, n = ?$

$500,000 = 1500\left[\dfrac{(1+0.041)^n - 1}{0.041}\right]$

$333.33 = \dfrac{(1.041)^n - 1}{0.041}$

$13.666 = 1.041^n - 1$

$14.666 = 1.041^n$

$\ln 14.666 = n \ln 1.041$

$n = \dfrac{\ln 14.666}{\ln 1.041} = 66.8 \ \text{half years}$

15. $P = 10,000, n = \dfrac{9}{2} \cdot 4 = 18, i = 0.01925$

$S_1 = 10,000(1+.01925)^{18} = 14,094.61$

$S_2 = 50,000 - S_1 = 35,905.39$

$R = 35,905.39\left[\dfrac{0.01925}{(1+.01925)^{18} - 1}\right] = \$1688.02$

16. a. $R = 3000, n = 8 \cdot 2 = 16, i = 0.0375$

$S_1 = 3000\left[\dfrac{(1+.0375)^{16} - 1}{0.0375}\right] = 64,178.22$

$S = 64,178.22(1 + 0.0375)^{40} = \$279,841.35$

b. $A = 279,841.35, n = 20 \cdot 2 = 40, i = 0.0375$

$A_{due} : R = 279,841.35\left[\dfrac{0.0375}{1-(1+.0375)^{-40}}\right](1+.0375)^{-1}$

$= \$13,124.75$

17. $R = 80, n = 15 \cdot 12 = 180, i = 0.005$

$S_{due} = 80\left[\dfrac{(1+.005)^{180} - 1}{0.005}\right](1+.005)$

$= \$23,381.82$

18. Deferred Annuity: $R = 4000, n = 16, k = 40,$ $i = 0.016$

$A = 4000\left[\dfrac{1-(1+.016)^{-16}}{0.016}\right](1+.016)^{-40}$

$= \$29,716.47$

19. a. The difference between successive terms is -5.5.

b. $a_{51} = 298.8 + (51-1)(-5.5) = 23.8$

c. $s_{51} = \dfrac{51}{2}(298.8 + 23.8) = 8226.3$

20. $a_1 = 400, r = 0.6, n = 31$ (Look at sequence.)

$s_{31} = 400\dfrac{(1 - 0.6^{31})}{1 - 0.6} = 1000 \ \text{mg}$

21. a.   After 24 payments there are 336 payments remaining.
$i = 0.084/12 = 0.007$, $R = \$1142.76$, and $n - k = 336$.

$$A_{n-k} = 1142.76 \left[\frac{1 - (1 + 0.007)^{-336}}{0.007} \right] = \$147,585.55$$

b.   The extra \$2000 payment would essentially reduce the unpaid balance to \$145,585.55.
$R = 1142.76$, $A_n = 145,585.55$, $i = 0.007$, $n = ?$

$$145,585.55 = 1142.76 \left[\frac{1 - (1 + 0.007)^{-n}}{0.007} \right]$$

$$127.3982 = \frac{1 - (1.007)^{-n}}{0.007}$$

$$0.8918 = 1 - (1.007)^{-n}$$

$$-0.1082 = -1.007^{-n}$$

$$0.1082 = 1.007^{-n}$$

$$\ln 0.1082 = -n \ln 1.007$$

$$n = -\frac{\ln 0.1082}{\ln 1.007} = 318.8 \text{ monthly payments}$$

Total payment of loan is $\$1142.76(318.8 + 24) + 2000 = 393,738.13$, so \$243,738.13 is interest.

c. $R = 1160$, $A_n = 147,585.55$, $i = 0.007$, $n = ?$

$$147,585.55 = 1160 \left[\frac{1 - (1 + 0.007)^{-n}}{0.007} \right]$$

$$127.2289 = \frac{1 - (1.007)^{-n}}{0.007}$$

$$0.8906 = 1 - (1.007)^{-n}$$

$$-0.1094 = -1.007^{-n}$$

$$0.1094 = 1.007^{-n}$$

$$\ln 0.1094 = -n \ln 1.007$$

$$n = -\frac{\ln 0.1094}{\ln 1.007} = 317.2 \text{ monthly payments}$$

Total payment of loan is $\$1142.75(24) + \$1160(317.2) = \$395,378$, so \$245,378 is interest.

Paying the \$2000 is slightly better; it saves approximately \$1640 in interest.

Chapter 7: Introduction to Probability

Exercise 7.1

1. a. $\Pr(R) = \dfrac{4}{10} = \dfrac{2}{5}$

    b. $\Pr(G) = \dfrac{0}{10} = 0$

    c. $\Pr(R \text{ or } W) = \dfrac{10}{10} = 1$

3. $\Pr(4, 8, 12) = \dfrac{3}{12} = \dfrac{1}{4}$

5. $\Pr(\text{greater than } 0) = \dfrac{6}{6} = 1$

7. a. $\Pr(\text{Red}) = \dfrac{3}{10}$

    b. $\Pr(\text{Odd}) = \dfrac{5}{10} = \dfrac{1}{2}$

    c. $\Pr(\text{Red and Odd}) = \Pr(1 \text{ or } 3) = \dfrac{2}{10} = \dfrac{1}{5}$

    d. $\Pr(\text{Red or Odd})$

    $= \Pr(1, 2, 3, 5, 7, \text{or } 9) = \dfrac{6}{10} = \dfrac{3}{5}$

    e. $\Pr(\text{Not Black})$

    $= 1 - \Pr(\text{Black}) = 1 - \dfrac{3}{10} = \dfrac{7}{10}$

9. a. $\Pr(\text{Queen}) = \dfrac{4}{52} = \dfrac{1}{13}$

    b. $\Pr(\text{Heart or Diamond}) = \dfrac{1}{2}$

    c. $\Pr(\text{Spade}) = \dfrac{1}{4}$

11. Sample space = {HH, HT, TH, TT}

    a. $\Pr(0H) = \dfrac{1}{4}$

    b. $\Pr(1H) = \dfrac{2}{4} = \dfrac{1}{2}$

    c. $\Pr(2H) = \dfrac{1}{4}$

13. a. $\Pr(\text{Sum} = 4) = \dfrac{3}{36} = \dfrac{1}{12}$

    b. $\Pr(\text{Sum} = 10) = \dfrac{3}{36} = \dfrac{1}{12}$

    c. $\Pr(\text{Sum} = 12) = \dfrac{1}{36}$

15. a. $\Pr(4 \le S \le 7) = \dfrac{3+4+5+6}{36} = \dfrac{1}{2}$

    (3 ways to roll a 4, etc.)

    b. $\Pr(8 \le S \le 12) = \dfrac{5+4+3+2+1}{36} = \dfrac{15}{36} = \dfrac{5}{12}$

    (5 ways to roll 8, etc.)

17. a. $\Pr(6) = \dfrac{431}{1200}$

    b. The die is biased since $\Pr(6)$ should be about $1/6$.

19. a. $2 : 3$ or $\dfrac{2}{3}$

    b. $3 : 2$ or $\dfrac{3}{2}$

21. a. $\Pr(\text{Win}) = \dfrac{1}{20+1} = \dfrac{1}{21}$

    b. $\Pr(\text{Lose}) = \dfrac{20}{20+1} = \dfrac{20}{21}$

23. $\Pr(\text{Inner city}) = 0.46$

25. a. $\Pr\left(\text{defective turn signal}\right) = \dfrac{63}{425} = 0.1482$

    b. $\Pr\left(\text{defective tires}\right) = \dfrac{32}{425} = 0.753$

27. a. $\Pr\left(\text{Republican will vote}\right)$

    $= \dfrac{2835}{4500} = \dfrac{63}{100} = 0.63$

    $\Pr\left(\text{Democrat will vote}\right)$

    $= \dfrac{2501}{6100} = \dfrac{41}{100} = 0.41$

    $\Pr\left(\text{Independent will vote}\right)$

    $= \dfrac{1122}{2200} = \dfrac{51}{100} = 0.51$

    b. Probability is highest that a Republican will vote in the next election.

29. a. $\Pr(100\% \text{ discount}) = \dfrac{1}{3601}$

b. $\Pr(50\% \text{ discount}) = \dfrac{100}{3601}$

c. $\Pr(< 50\% \text{ discount}) = \dfrac{3000 + 500}{3601} = \dfrac{3500}{3601}$

d. $\Pr(30\% \text{ discount}) = \dfrac{500}{3601}$;

$\Pr(> 30\% \text{ discount}) = \dfrac{101}{3601}$.

So, more likely to get a 30% discount than a higher discount.

31. a. $\Pr\left(\text{from developed nation}\right)$

$= \dfrac{1188}{6157} = 0.1930$

b. $\Pr\left(\text{age 20-39}\right) = \dfrac{1946}{6157} = 0.3161$

c. $\Pr\left(\text{age 60-79}\right) = \dfrac{545}{6157} = 0.0885$

33. $S = \{A^+, A^-, B^+, B^-, AB^+, AB^-, O^+, O^-\}$

35. a. $\Pr(AB) = 0.04$
b. $\Pr(\text{not } AB) = 0.96$

37. a. $\Pr(A) = 0.10$
b. $\Pr(\text{not } A) = 0.90$

39. $\Pr(\text{Empty}) = \dfrac{60,000}{2,000,000} = \dfrac{3}{100}$

41. $\Pr(\text{has lactose intolerance}) = 0.75$

43. $\Pr(\text{Woman}) = \dfrac{3}{9} = \dfrac{1}{3}$

45. $\Pr(\text{grade}) = \dfrac{2}{6} = \dfrac{1}{3}$

47. $\Pr(\text{No V and DS}) = 0.22$; yes

49. $\Pr\left(\text{Win}\right) = \dfrac{5}{3 + 8} = \dfrac{5}{13}$

51. Sample space $= \{GGG, GGB, GBG, BGG,$ $BBG, BGB, GBB, BBB\}$; $\Pr(2G) = \dfrac{3}{8}$

53. a. No
b. Sample space $= \{GG, GB, BG, BB\}$
c. $\Pr(1B \text{ and } 1G) = \dfrac{2}{4} = \dfrac{1}{2}$

55. $\Pr(Q1 \text{ is correct}) = \dfrac{1}{5}$

57. $\Pr(\text{Defective}) = \dfrac{6}{250} = \dfrac{3}{125}$

59. $\Pr(A) = \dfrac{182}{25,500(365)} \approx 0.0000196$

$\Pr(B) = \dfrac{51}{3890(365)} \approx 0.0000359$

$\Pr(C) = \dfrac{118}{8580(365)} \approx 0.0000377$

Intersection C is the most dangerous.

61. a. $\Pr(A) = \dfrac{557}{1200}$

b. $\Pr(\text{No Vote}) = \dfrac{110}{1200} = \dfrac{11}{120}$

63. a. $\Pr(B) = \dfrac{2}{10} = \dfrac{1}{5}$

$\Pr(G) = \dfrac{8}{10} = \dfrac{4}{5}$

b. $\Pr(B) = \dfrac{1194}{1220 + 1194} \approx 0.4946$

$\Pr(G) = \dfrac{1220}{2414} \approx 0.5054$

c. $\Pr(B) = \Pr(G) = \dfrac{1}{2}$; part b

Exercise 7.2

1. $\Pr(6, 12) = \dfrac{2}{12} = \dfrac{1}{6}$

3. $\Pr(1, 3, 5) = \dfrac{1}{2}$

5. $\Pr(E') = 1 - \Pr(E) = 1 - \dfrac{3}{5} = \dfrac{2}{5}$

7. a. $\Pr(R \cap \text{Even}) = \Pr(2, 4) = \dfrac{2}{14} = \dfrac{1}{7}$

b. $\Pr(R \cup \text{Even})$

$= \Pr(R) + \Pr(\text{Even}) - \Pr(R \cap \text{Even})$

$= \dfrac{5}{14} + \dfrac{7}{14} - \dfrac{2}{14} = \dfrac{10}{14} = \dfrac{5}{7}$

9. Pr(odd or {4, 8, 12})

$$= \frac{6}{12} + \frac{3}{12} - 0 = \frac{9}{12} = \frac{3}{4}$$

11. Pr(odd or {3, 6, 9, 12}) $= \frac{6}{12} + \frac{4}{12} - \frac{2}{12}$

$$= \frac{8}{12} = \frac{2}{3}$$

13. $Pr(W \cup G) = \frac{4}{17} + \frac{6}{17} - 0 = \frac{10}{17}$

15. $Pr(R \cup W) = \frac{2}{6} + \frac{2}{6} - 0 = \frac{4}{6} = \frac{2}{3}$

17. a. $Pr(\text{not } E) = \frac{9}{18} = \frac{1}{2}$

    b. $Pr(R \cap E) = \frac{6}{18} = \frac{1}{3}$

    c. $Pr(R \cup E) = \frac{13}{18} + \frac{9}{18} - \frac{6}{18} = \frac{16}{18} = \frac{8}{9}$

    d. $Pr(R' \cap E') = Pr(15, 17) = \frac{2}{18} = \frac{1}{9}$

27. a. $Pr(\text{female} \cap \text{less than } \$30,000) = 0.35$

    b. $Pr(\text{at least } \$50,000) = 0.08$

    c. $Pr(\text{male} \cup \text{less than } \$30,000)$

$$= Pr(\text{male}) + Pr(\text{less than } \$30,000) - Pr(\text{male} \cap \text{less than } \$30,000) = 0.48 + 0.60 - 0.25 = 0.83$$

29. a. $Pr(\text{developed nation or } 40\text{-}59) = \frac{1188}{6157} + \frac{1200}{6157} - \frac{318}{6157} = \frac{2070}{6157} = 0.336$

    b. $Pr(20\text{-}39 \cap \text{less developed nation}) = \frac{1605}{6157} = 0.261$

    c. $Pr(\text{age 40 or more}) = Pr(40\text{-}59) + Pr(60\text{-}79) + Pr(80+) = \frac{318 + 882}{6157} + \frac{196 + 349}{6157} + \frac{38 + 35}{6157} = \frac{1818}{6157} = 0.295$

31. a. $Pr(\text{Hispanic}) = \frac{\text{\# of Hispanics}}{\text{Total}} = \frac{2674 + 763}{15,311 + 5062} = \frac{3437}{20,373} = 0.1687$

    b. $Pr(\text{female}) = \frac{\text{\# of females}}{\text{Total}} = \frac{5074}{20,373} = 0.2491$

    c. $Pr(\text{female} \cup \text{Hispanic}) = Pr(\text{female}) + Pr(\text{Hispanic}) - Pr(\text{female} \cap \text{Hispanic})$

$$= \frac{5074}{20,373} + \frac{3437}{20,373} - \frac{763}{20,373} = \frac{7748}{20,373} = 0.3803$$

    d. $Pr(\text{male or Black}) = Pr(\text{male}) + Pr(\text{Black}) - Pr(\text{male} \cap \text{Black})$

$$= \frac{15,383}{20,373} + \frac{9945}{20,373} - \frac{6667}{20,373} = \frac{18,661}{20,373} = 0.9160$$

19. $Pr(IC') = 1 - Pr(IC) = 1 - 0.46 = 0.54$

21. a. $Pr(DT') = 1 - Pr(DT) = 1 - \frac{63}{425} = \frac{362}{425}$

    b. $Pr(\text{No Defects}) = 1 - \frac{63 + 32}{425} = \frac{330}{425} = \frac{66}{85}$

23. $Pr(F \cup G) = \frac{24}{100} + \frac{18}{100} - \frac{8}{100} = \frac{34}{100} = \frac{17}{50}$

25. a. $Pr(S \cup R) = \frac{1}{3} + \frac{1}{2} - 0 = \frac{5}{6}$

    b. $Pr(S' \cap R') = 1 - \frac{5}{6} = \frac{1}{6}$

33. a. $\Pr(A \cup B) = \dfrac{7}{12} + \dfrac{7}{12} - \dfrac{3}{12} = \dfrac{11}{12}$

   b. $\Pr(B \cup C) = \dfrac{7}{12} + \dfrac{8}{12} - \dfrac{5}{12} = \dfrac{10}{12} = \dfrac{5}{6}$

35. a. $\Pr(S \cup LA) = \dfrac{12}{32} + \dfrac{4}{32} - 0 = \dfrac{16}{32} = \dfrac{1}{2}$

   b. $\Pr(E \cup S) = \dfrac{16}{32} + \dfrac{12}{32} - 0 = \dfrac{28}{32} = \dfrac{7}{8}$

   c. $\Pr(E \cup F) = \dfrac{16}{32} + \dfrac{13}{32} - \dfrac{5}{32} = \dfrac{24}{32} = \dfrac{3}{4}$

37. $\Pr\left(R \cup {>}S\right) = \dfrac{400+50}{1000} + \dfrac{435}{1000} - \dfrac{300+25}{1000} = \dfrac{560}{1000} = 0.56$

39. $\Pr(W \cup S') = \dfrac{400+350}{1000} + \dfrac{565}{1000} - \dfrac{100+250}{1000} = \dfrac{965}{1000} = 0.965$

41. a. $\Pr(F \cup RA) = \dfrac{390+160}{1000} + \dfrac{560}{1000} - \dfrac{390}{1000} = \dfrac{720}{1000} = 0.72$

   b. $\Pr(M \cup RA) = \dfrac{450}{1000} + \dfrac{560}{1000} - \dfrac{170}{1000} = \dfrac{840}{1000} = 0.84$

   c. $\Pr(M \cup RA') = \dfrac{450}{1000} + \dfrac{440}{1000} - \dfrac{280}{1000} = \dfrac{610}{1000} = 0.61$

43. $\Pr(M \cup B') = \Pr(M) + \Pr(B') - \Pr(M \cap B') = \dfrac{50}{80} + \dfrac{42}{80} - \dfrac{30}{80} = \dfrac{62}{80} = \dfrac{31}{40}$

45. $\Pr(PC') = 1 - 0.87 = 0.13$

Exercise 7.3

1. a. $\Pr(H \mid R) = \dfrac{13 \text{ Hearts}}{26 \text{ Red}} = \dfrac{1}{2}$

   b. $\Pr(K \mid R) = \dfrac{2 \text{ Kings}}{26 \text{ Red}} = \dfrac{1}{13}$

3. a. $\Pr(6 \mid \text{even})$

    $= \dfrac{\Pr(6 \text{ and even})}{\Pr(\text{even})} = \dfrac{\frac{2}{9}}{\frac{6}{9}} = \dfrac{2}{6} = \dfrac{1}{3}$

   b. $\Pr(3 \mid 3 \text{ or } 6)$

    $= \dfrac{\Pr(3 \text{ and } (3 \text{ or } 6))}{\Pr(3 \text{ or } 6)} = \dfrac{\frac{1}{9}}{\frac{3}{9}} = \dfrac{1}{3}$

5. $\Pr(R \mid E) = \dfrac{\Pr(R \text{ and } E)}{\Pr(E)} = \dfrac{\frac{4}{15}}{\frac{7}{15}} = \dfrac{4}{7}$

7. a. $\Pr(W \mid R) = \dfrac{6(W)}{9(\text{Total})} = \dfrac{2}{3}$

   b. $\Pr(R \mid W) = \dfrac{4(R)}{9(\text{Total})} = \dfrac{4}{9}$

   c. First draw has no effect on second draw.

    $\Pr(W) = \dfrac{6}{10} = \dfrac{3}{5}$

9. a. $\Pr(H2 \text{ and } H3 \mid H1)$

    $= \dfrac{\Pr(H1H2H3)}{\Pr(H1)} = \dfrac{\frac{1}{8}}{\frac{1}{2}} = \dfrac{1}{4}$

   b. $\Pr(H3 \mid H1 \text{ and } H2)$

    $= \dfrac{\Pr(H1H2H3)}{\Pr(H1 \text{ and } H2)} = \dfrac{\frac{1}{8}}{\frac{1}{4}} = \dfrac{1}{2}$

11. $\Pr(3 \text{ first} \cap 6 \text{ second}) = \dfrac{1}{6} \cdot \dfrac{1}{6} = \dfrac{1}{36}$

13. a. $\Pr(HHH) = \dfrac{1}{2} \cdot \dfrac{1}{2} \cdot \dfrac{1}{2} = \dfrac{1}{8}$

   b. $\Pr(\text{at least 1 tail})$

      $= 1 - \Pr(\text{no tails}) = 1 - \dfrac{1}{8} = \dfrac{7}{8}$

15. a. $\Pr(R1 \text{ and } W2) = \Pr(R) \cdot \Pr(W) = \dfrac{3}{10} \cdot \dfrac{2}{10} = \dfrac{3}{50}$

   b. $\Pr(R1 \text{ and } W2) = \Pr(R) \cdot \Pr(W \mid R) = \dfrac{3}{10} \cdot \dfrac{2}{9} = \dfrac{1}{15}$

   c. The events in (a) are independent. Since the ball was replaced it did not affect the probability of getting a white ball the second time. The events in (b) were dependent. Not replacing the first ball changed the sample space which affected the probability of getting a white ball the second time.

17. a. $\Pr(RR) = \Pr(R) \cdot \Pr(R) = \dfrac{2}{5} \cdot \dfrac{2}{5} = \dfrac{4}{25}$

   b. $\Pr(WW) = \Pr(W) \cdot \Pr(W) = \dfrac{3}{5} \cdot \dfrac{3}{5} = \dfrac{9}{25}$

   c. $\Pr(R1 \text{ and } W2) = \Pr(R) \cdot \Pr(W) = \dfrac{2}{5} \cdot \dfrac{3}{5} = \dfrac{6}{25}$

   d. $\Pr(Black) = 0$

19. $\Pr(N1Q2N3) = \Pr(N) \cdot \Pr(Q \mid N) \cdot \Pr(N \mid N \cap Q) = \dfrac{9}{18} \cdot \dfrac{5}{17} \cdot \dfrac{8}{16} = \dfrac{5}{68}$

21. a. $\Pr(R1 \text{ and } W2) = \Pr(R) \cdot \Pr(W \mid R) = \dfrac{1}{5} \cdot \dfrac{4}{4} = \dfrac{1}{5}$

   b. $\Pr(W1 \text{ and } W2) = \Pr(W) \cdot \Pr(W \mid W) = \dfrac{4}{5} \cdot \dfrac{3}{4} = \dfrac{3}{5}$

   c. $\Pr(R1 \text{ and } R2) = 0$ (Only one red ball in box)

23. a. $\Pr(S1 \text{ and } S2) = \Pr(S) \cdot \Pr(S \mid S) = \dfrac{13}{52} \cdot \dfrac{12}{51} = \dfrac{1}{17}$

   b. $\Pr(H1 \text{ and } C2) = \Pr(H) \cdot \Pr(C \mid H) = \dfrac{13}{52} \cdot \dfrac{13}{51} = \dfrac{13}{204}$

25. a. $\Pr(R \mid W) = \dfrac{13}{17}$

   b. $\Pr(E1 \text{ and } E2) = \Pr(E) \cdot \Pr(E \mid E) = \dfrac{9}{18} \cdot \dfrac{8}{17} = \dfrac{4}{17}$

   c. $\Pr(ER1 \text{ and } E2) = \Pr(E \text{ and } R) \cdot \Pr(E \mid E \cap R) = \dfrac{6}{18} \cdot \dfrac{8}{17} = \dfrac{8}{51}$

27. $\Pr(F \mid D) = \dfrac{310}{520} = \dfrac{31}{52}; \quad \Pr(F \mid D) = \dfrac{\Pr(F \cap D)}{\Pr(D)}$

29. $\Pr(F \mid R) = \dfrac{125}{480} = \dfrac{25}{96}; \quad \Pr(F \mid R) = \dfrac{\Pr(F \cap R)}{\Pr(R)}$

31. $\Pr(O \mid NW) = \dfrac{25 + 190}{50 + 200} = \dfrac{215}{250} = \dfrac{43}{50}$

33. $\Pr(R \mid F) = \dfrac{300 + 25}{435} = \dfrac{325}{435} = \dfrac{65}{87}$

35. $\Pr(NW \mid F) = \dfrac{25 + 10}{435} = \dfrac{35}{435} = \dfrac{7}{87}$

37. $\Pr(WR \text{ and } O) = \dfrac{100}{1000} = \dfrac{1}{10}$ (No "given that" in this problem)

39. Events are independent. $\Pr(H1 \text{ and } H1) = \dfrac{1}{12,000} \cdot \dfrac{1}{12,000} = \dfrac{1}{144,000,000}$

41. $\Pr(A \text{ and } SD \text{ and } PGM) = (0.337)(0.796)(0.016) = 0.004292032$

43. $\Pr(\text{not identified on test 1} \mid \text{defective}) = 0.3$

    $\Pr(\text{not identified on test 2} \mid \text{defective}) = 0.2$

    $\Pr(\text{not identified on either test} \mid \text{defective}) = (0.3)(0.2) = 0.06$

45. $\Pr(H \text{ and } LI) = \Pr(H) \cdot \Pr(LI \mid H) = (0.09)(0.5) = 0.045$

47. $\Pr(\text{Good}) = 0.95$; $\Pr(\text{GGGGG}) = (0.95)^5$ (Independent events)

49. $\Pr(\text{See and Buy}) = \Pr(\text{See}) \cdot \Pr(\text{Buy} \mid \text{See}) = (0.3)(0.2) = 0.06$

51. Events (each woman) are independent.
    a. $\Pr(0 \text{ pregnancies}) = (0.99)^{100} \approx 0.366$
    b. $\Pr(\text{at least 1 pregnancy}) \approx 1 - 0.366 = 0.634$

53. a. $\Pr(0 \text{ defects}) = (0.77)^3 \approx 0.456$
    b. $\Pr(\text{at least 1 defect}) \approx 1 - 0.456 = 0.544$

55. Events are independent.
    a. $\Pr(\text{all correct}) = \dfrac{1}{3} \cdot \dfrac{1}{3} \cdot \dfrac{1}{3} \cdot \dfrac{1}{5} \cdot \dfrac{1}{5} \cdot \dfrac{1}{5} \cdot \dfrac{1}{5} = \dfrac{1}{16,875}$

    b. $\Pr(\text{no correct}) = \dfrac{2}{3} \cdot \dfrac{2}{3} \cdot \dfrac{2}{3} \cdot \dfrac{4}{5} \cdot \dfrac{4}{5} \cdot \dfrac{4}{5} \cdot \dfrac{4}{5} = \dfrac{2048}{16,875}$

    c. $\Pr(\text{at least 1 correct}) = 1 - \dfrac{2048}{16,875} = \dfrac{14,827}{16,875}$

57. $\Pr(\text{Portia}) = \dfrac{3}{11}$ $\Pr(\text{Trinka}) = \dfrac{1}{11}$

    $\Pr(P \text{ or } T) = \Pr(P) + \Pr(T) = \dfrac{3}{11} + \dfrac{1}{11} = \dfrac{4}{11}$

    Odds for Portia or Trinka $= \dfrac{4}{7}$ or $= 4 : 7$

59. Choose one person. Now choose the other person.
    a. $\Pr(\text{different from birthday \#1}) = \dfrac{364}{365}$

    b. $\Pr(\text{same birthday as \#1}) = \dfrac{1}{365}$

61. a. $\mathrm{Pr(Different\ days)} = 1\ \left(\dfrac{364}{365}\right)\left(\dfrac{363}{365}\right)\ \cdots\ \left(\dfrac{346}{365}\right)$

$$\updownarrow \qquad \updownarrow \qquad \updownarrow \qquad\qquad \updownarrow$$
$$1 \qquad\ 2 \qquad\ 3 \qquad\qquad 20$$

$$\approx 0.588 \approx 0.59$$

b. $\mathrm{Pr(At\ least\ 2\ same)} = 1 - \mathrm{Pr(Different\ days)} \approx 1 - 0.59 = 0.41$

Exercise 7.4

1.

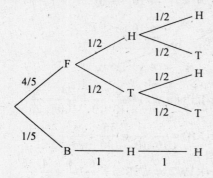

$$\mathrm{Pr(2H)} = \frac{4}{5}\cdot\frac{1}{2}\cdot\frac{1}{2} + \frac{1}{5}\cdot 1\cdot 1 = \frac{2}{5}$$

3.

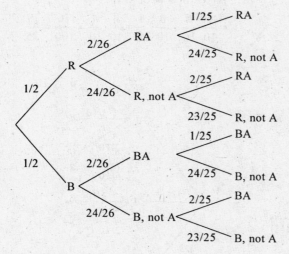

$$\mathrm{Pr(2A\ |\ same\ color)}$$

$$= \frac{1}{2}\cdot\frac{2}{26}\cdot\frac{1}{25} + \frac{1}{2}\cdot\frac{2}{26}\cdot\frac{1}{25} = \frac{1}{325}$$

5.

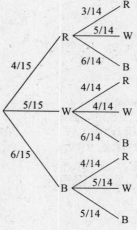

a. $\mathrm{Pr(WW)} = \dfrac{5}{15}\cdot\dfrac{4}{14} = \dfrac{2}{21}$

b. $\mathrm{Pr(RW\ or\ WR)} = \dfrac{4}{15}\cdot\dfrac{5}{14} + \dfrac{5}{15}\cdot\dfrac{4}{14} = \dfrac{4}{21}$

c. $\mathrm{Pr(at\ least\ one\ is\ black)}$

$= \mathrm{Pr(RB\ or\ WB\ or\ BR\ or\ BW\ or\ BB)}$

$= \dfrac{4}{15}\cdot\dfrac{6}{14} + \dfrac{5}{15}\cdot\dfrac{6}{14} + \dfrac{6}{15}\cdot\left(\dfrac{4}{14} + \dfrac{5}{14} + \dfrac{5}{14}\right) = \dfrac{23}{35}$

7.

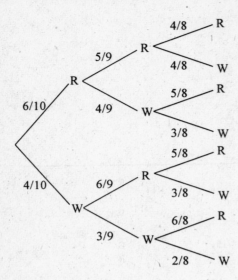

a. $\Pr(WWW) = \dfrac{4}{10} \cdot \dfrac{3}{9} \cdot \dfrac{2}{8} = \dfrac{1}{30}$

b. $\Pr(WRR \text{ or } RWR \text{ or } RRW)$

$= \dfrac{4}{10} \cdot \dfrac{6}{9} \cdot \dfrac{5}{8} + \dfrac{6}{10} \cdot \dfrac{4}{9} \cdot \dfrac{5}{8} + \dfrac{6}{10} \cdot \dfrac{5}{9} \cdot \dfrac{4}{8}$

$= \dfrac{3(4 \cdot 5 \cdot 6)}{8 \cdot 9 \cdot 10} = \dfrac{1}{2}$

c. $\Pr(\text{at least } 1W) = 1 - \Pr(\text{No } W)$

$= 1 - \dfrac{6}{10} \cdot \dfrac{5}{9} \cdot \dfrac{4}{8} = \dfrac{5}{6}$

9.   Referring to the probability tree in exercise 8, we have

$\Pr(HHT \text{ or } HTH \text{ or } THH \text{ or } HHH)$

$= 3\left(\dfrac{4}{5} \cdot \dfrac{1}{2} \cdot \dfrac{1}{2} \cdot \dfrac{1}{2}\right) + \left(\dfrac{4}{5} \cdot \dfrac{1}{2} \cdot \dfrac{1}{2} \cdot \dfrac{1}{2} + \dfrac{1}{5} \cdot 1 \cdot 1 \cdot 1\right) = \dfrac{3}{5}$

11.

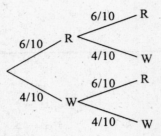

a. $\Pr(R1 \text{ and } W2) = \dfrac{6}{10} \cdot \dfrac{4}{10} = \dfrac{24}{100} = \dfrac{6}{25}$

b. $\Pr(RR) = \dfrac{6}{10} \cdot \dfrac{6}{10} = \dfrac{36}{100} = \dfrac{9}{25}$

c. $\Pr(R1W2 \text{ or } W1R2) = \dfrac{6}{10} \cdot \dfrac{4}{10} + \dfrac{4}{10} \cdot \dfrac{6}{10}$

$= \dfrac{48}{100} = \dfrac{12}{25}$

d. $\Pr(RR \text{ or } RW \text{ or } WW)$

$= \dfrac{6}{10} \cdot \dfrac{6}{10} + \dfrac{6}{10} \cdot \dfrac{4}{10} + \dfrac{4}{10} \cdot \dfrac{4}{10} = \dfrac{76}{100} = \dfrac{19}{25}$

13. a. $\Pr(III \mid G)$

$= \dfrac{\text{product of branch probabilities on III-G path}}{\text{sum of all branch products leading to } G}$

$= \dfrac{\frac{1}{3} \cdot \frac{8}{10}}{\frac{2}{30} + \frac{2}{30} + \frac{8}{30}} = \dfrac{\frac{8}{30}}{\frac{12}{30}} = \dfrac{2}{3}$

a. $\Pr(III)|G)$

$= \dfrac{\Pr(III) \cdot \Pr(G \mid III)}{\Pr(III) \cdot \Pr(G \mid III) + \Pr(II) \cdot \Pr(G \mid II) + \Pr(I) \cdot \Pr(G \mid I)}$

$= \dfrac{\frac{1}{3} \cdot \frac{8}{10}}{\frac{1}{3} \cdot \frac{8}{10} + \frac{1}{3} \cdot \frac{2}{10} + \frac{1}{3} \cdot \frac{2}{10}} = \dfrac{\frac{8}{30}}{\frac{12}{30}} = \dfrac{2}{3}$

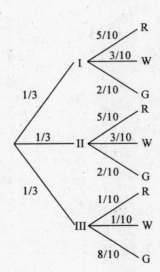

15. a. $\Pr(I \mid G)$

$$= \frac{\text{product of branch probabilities on I-G path}}{\text{sum of all branch probabilities leading to G}}$$

$$= \frac{\frac{1}{3} \cdot 1}{\frac{1}{3} + \frac{1}{6}} = \frac{\frac{1}{3}}{\frac{3}{6}} = \frac{2}{3}$$

b. $\Pr(I \mid G)$

$$= \frac{\Pr(I) \cdot \Pr(G \mid I)}{\Pr(I) \cdot \Pr(G \mid I) + \Pr(II) \cdot \Pr(G \mid II) + \Pr(III) \cdot \Pr(G \mid III)}$$

$$= \frac{\frac{1}{3} \cdot 1}{\frac{1}{3} \cdot 1 + \frac{1}{3} \cdot \frac{1}{2} + \frac{1}{3} \cdot 0} = \frac{\frac{1}{3}}{\frac{3}{6}} = \frac{2}{3}$$

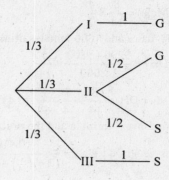

17. $\Pr(W) = 0.76 \qquad \Pr(H) = 0.09 \qquad \Pr(AANA) = 0.15$

$\Pr(I \mid W) = 0.20 \qquad \Pr(I \mid H) = 0.50 \qquad \Pr(I \mid AANA) = 0.75$

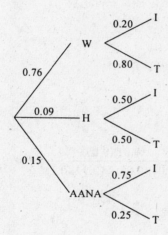

$$\Pr(I) = 0.76(0.20) + 0.09(0.50) + 0.15(0.75) = 0.3095$$

19. a. $\Pr(HHHH) = \dfrac{15}{50} \cdot \dfrac{15}{50} \cdot \dfrac{15}{50} \cdot \dfrac{15}{50} = \dfrac{81}{10,000}$

b. There are 6 successful events, each with the same probability.
Success: HHMM, HMHM, HMMH, MHMH, MMHH, MHHM

$$\Pr(\text{Success}) = 6\left(\frac{15}{50} \cdot \frac{15}{50} \cdot \frac{35}{50} \cdot \frac{35}{50}\right) = \frac{1323}{5000}$$

21. a. $\Pr(G1D2) = \dfrac{12}{15} \cdot \dfrac{3}{14} = \dfrac{6}{35}$

b. $\Pr(D1G2) = \dfrac{3}{15} \cdot \dfrac{12}{14} = \dfrac{6}{35}$

c. $\Pr(G1D2 \text{ or } D1G2) = \dfrac{6}{35} + \dfrac{6}{35} = \dfrac{12}{35}$

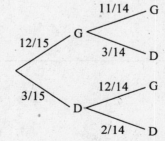

23. a. $\Pr(\text{Male}) = \dfrac{8}{14} = \dfrac{4}{7}$

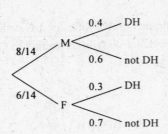

b. 3200 males and 1800 females drink heavily.

$\Pr(\text{DH}) = \dfrac{3200 + 1800}{14,000} = \dfrac{5}{14}$

c. $\Pr(\text{M or DH}) = \dfrac{8}{14} + \dfrac{5}{14} - \dfrac{8}{14} \cdot (0.4) = \dfrac{7}{10}$

d. Using Bayes' formula and the probability tree,

we have $\Pr(\text{M} \mid \text{DH}) = \dfrac{(0.4)\left(\frac{8}{14}\right)}{(0.4)\left(\frac{8}{14}\right) + (0.3)\left(\frac{6}{14}\right)} = \dfrac{16}{25}$

25. $\Pr(\text{F} \mid \text{M}) = \dfrac{170}{170 + 280} = \dfrac{170}{450} = \dfrac{17}{45}$

27. T = Test indicates pregnant (P)
not T = Test shows not pregnant

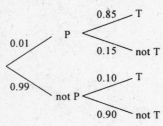

Using Bayes' formula and the probability tree, we have $\Pr(\text{P} \mid \text{T}) = \dfrac{(0.01)(0.85)}{(0.01)(0.85) + (0.99)(0.1)} \approx 0.079$.

Alternate Method:

	T	not T	
P	0.0085		0.01
not P	0.099		0.99
			1

	T	not T	
P	0.0085	0.0015	0.01
not P	0.099	0.891	0.99
	0.1075	0.8925	1

$\Pr(\text{P} \mid \text{T}) = \dfrac{\Pr(\text{P and T})}{\Pr(\text{T})} = \dfrac{0.0085}{0.1075} \approx 0.079$

29. a. $\Pr(\text{Female}) = \dfrac{10 + 12 + 6 + 13 + 8}{30 + 25 + 20 + 15 + 10} = \dfrac{49}{100}$

b. $\Pr(\text{S} \mid \text{F}) = \dfrac{12 \text{ Science Females}}{49 \text{ Females}} = \dfrac{12}{49}$

Exercise 7.5

1. $_6P_4 = \dfrac{6!}{(6-4)!} = \dfrac{6 \cdot 5 \cdot 4 \cdot 3 \cdot 2!}{2!} = 360$

3. $_{10}P_6 = \dfrac{10!}{(10-6)!} = \dfrac{10 \cdot 9 \cdot 8 \cdot 7 \cdot 6 \cdot 5 \cdot 4!}{4!} = 151,200$

5. $_5P_0 = \dfrac{5!}{(5-0)!} = \dfrac{5!}{5!} = 1$

7. a. By Fundamental Counting Principle:
$6 \cdot 5 \cdot 4 \cdot 3 = 360$

b. By Fundamental Counting Principle:
$6 \cdot 6 \cdot 6 \cdot 6 = 1296$

9. $_nP_n = \dfrac{n!}{(n-n)!} = \dfrac{n!}{1} = n!$

11. $\dfrac{(n+1)!}{n!} = \dfrac{(n+1)n!}{n!} = n+1$

13. $(n+1)! = 17n!$

$(n+1)n! = 17n!$

$n+1 = 17$

$n = 16$

15. $_{100}C_{98} = \dfrac{100!}{98!2!} = \dfrac{100 \cdot 99}{2} = 4950$

17. $_4C_4 = \dfrac{4!}{0!4!} = \dfrac{4!}{1 \cdot 4!} = 1$

19. $\dbinom{5}{0} = \dfrac{5!}{0!5!} = 1$

21.

$_nC_6 = {}_nC_4$

$\dfrac{n!}{6!(n-6)!} = \dfrac{n!}{4!(n-4)!}$

$4!(n-4)! = 6!(n-6)!$

$4!(n-4)(n-5)(n-6)! = 6 \cdot 5 \cdot 4!(n-6)!$

$n^2 - 9n + 20 = 30$

$n^2 - 9n - 10 = 0$

$(n-10)(n+1) = 0$

$n = 10$ is the only possible solution.

23. a. By FCP: $2 \cdot 3 \cdot 2 = 12$

b. By FCP: $2 \cdot 3 \cdot 2 \cdot 6 = 72$

25. By FCP: $10 \cdot 9 \cdot 8 \cdot 7 \cdot 6 \cdot 5 \cdot 4 = 604,800$

27. By FCP: $6 \cdot 5 \cdot 4 = 120$

29. By FCP: $4 \cdot 3 \cdot 2 \cdot 1 = 24$

31. By FCP: $4 \cdot 4 \cdot 4 = 64$

33. By FCP: $6 \cdot 5 \cdot 4 \cdot 3 \cdot 2 \cdot 1 = 720$

35. By FCP: $2 \cdot 2 \cdot 2 \cdots 2 = 2^{10} = 1024$

Question | | | |

     1 2 3   10

37. By FCP: $10 \cdot 9 \cdot 8 \cdots 1 = 10! = 3,628,800$

       | | | |

Question 1 2 3   10

39. $_{10}C_5 = \dfrac{10!}{5!5!} = \dfrac{10 \cdot 9 \cdot 8 \cdot 7 \cdot 6 \cdot 5!}{5! \cdot 5 \cdot 4 \cdot 3 \cdot 2 \cdot 1} = 252$

41. $_{30}C_{20} = \dfrac{30!}{20!10!} = 30,045,015$

43. $_{12}C_5 = \dfrac{12!}{5!7!} = \dfrac{12 \cdot 11 \cdot 10 \cdot 9 \cdot 8 \cdot 7!}{5 \cdot 4 \cdot 3 \cdot 2 \cdot 1 \cdot 7!} = 792$

45. $_{10}C_4 = \dfrac{10!}{4!6!} = \dfrac{10 \cdot 9 \cdot 8 \cdot 7 \cdot 6!}{4 \cdot 3 \cdot 2 \cdot 1 \cdot 6!} = 210$

47. "AND" means multiply.

$_{20}C_6 \cdot {}_{22}C_6 = \dfrac{20!}{6!14!} \cdot \dfrac{22!}{6!16!} = 2,891,999,880$

49. By FCP: $10 \cdot 370,000 = 3,700,000$

Exercise 7.6

1. $n(E) = 1$ DOG is the only way for a success.

$n(S) = \underline{6} \cdot \underline{5} \cdot \underline{4} = 120$

So, $\Pr(E) = \dfrac{1}{120}$. Also, $\Pr(E) = \dfrac{1}{_6P_3} = \dfrac{1}{120}$

3. a. By FCP: $5 \cdot 4 \cdot 3 \cdot 2 \cdot 1 = 120$

b. $\Pr(\text{alphabetical order}) = \dfrac{1}{120}$

5. Total License Plates:

$26 \cdot 26 \cdot 26 \cdot 10 \cdot 10 \cdot 10 = 17,576,000$

Letters and numbers different:

$26 \cdot 25 \cdot 24 \cdot 10 \cdot 9 \cdot 8 = 11,232,000$

$\Pr(\text{Success}) = \dfrac{11,232,000}{17,576,000} \approx 0.639$

7. a. $\Pr(\text{Success}) = \dfrac{1}{10 \cdot 10 \cdot 10 \cdot 10} = \dfrac{1}{10,000}$

b. $\Pr(\text{Success}) = \dfrac{1}{10 \cdot 9 \cdot 8 \cdot 7} = \dfrac{1}{5040}$

9. Total Phone Numbers:

$8 \cdot 10 \cdot 10 \cdot 10 \cdot 10 \cdot 10 \cdot 10 = 8,000,000$

There are 8 phone numbers that have the same digits.

$\Pr(\text{Success}) = \dfrac{8}{8,000,000} = \dfrac{1}{1,000,000}$

11. $\Pr\left(\text{John and Jill}\right) = \dfrac{\dbinom{4}{2} \cdot \dbinom{496}{2}}{\dbinom{500}{4}} \approx 2.8 \times 10^{-4}$

13. Total ways to answer test $= \underline{10} \cdot \underline{9} \cdot \underline{8} \cdots \underline{1} = 10!$

    Question: 1 2 3 ··· 10

    Also, order is important, thus $_{10}P_{10} = 10!$ is another way to obtain the total.

    $\text{Pr(All correct)} = \dfrac{1}{10!} = \dfrac{1}{3,628,800}$

15. a. $\text{Pr(DD)} = \dfrac{\binom{3}{2}}{\binom{12}{2}} = \dfrac{3}{66} = \dfrac{1}{22} \approx 0.0455$

    b. $\text{Pr(GG)} = \dfrac{\binom{9}{2}}{\binom{12}{2}} = \dfrac{36}{66} = \dfrac{6}{11} \approx 0.5455$

    c. $\text{Pr(1D)} = 1 - [\text{Pr(DD)} + \text{Pr(GG)}]$
       $\approx 1 - 0.5910 = 0.409$

    Also,

    $\text{Pr(1G1D)} = \dfrac{\binom{9}{1}\binom{3}{1}}{\binom{12}{2}} = \dfrac{9 \cdot 3}{66} = \dfrac{9}{22} \approx 0.409$

17. Method 1:

     Def Good Def Good

    $\text{Pr(E)} = \dfrac{\binom{2}{1} \cdot \binom{98}{4} + \binom{2}{2} \cdot \binom{98}{3}}{\binom{100}{5}}$

    $= \dfrac{7,224,560 + 152,096}{75,287,520} \approx 0.098$

    Method 2:

    $n(\text{1 or 2 Def}) = \text{Total} - \text{All Good} = \binom{100}{5} - \binom{98}{5}$

    $\phantom{n(\text{1 or 2 Def})} = 75,287,520 - 67,910,864$

    $\phantom{n(\text{1 or 2 Def})} = 7,376,656$

    $\text{Pr(E)} = \dfrac{7,376,656}{\binom{100}{5}} \approx 0.098$

19. In the sample of 30, there are 2 minority loans. Thus, $_{10}C_2$ is the numbers of ways to obtain 2 minority loans from the 10 minority loans. Also, $_{90}C_{28}$ is the number of ways of getting 28 non minority loans from the total of non minority loans.

    $\text{Pr(E)} = \dfrac{\binom{90}{28} \cdot \binom{10}{2}}{\binom{100}{30}}$

21. a. $\text{Pr(No minority)} = \dfrac{\binom{6}{4}}{\binom{9}{4}} = \dfrac{15}{126} = \dfrac{5}{42} \approx 0.119$

    b. $\text{Pr(All minority)} = \dfrac{\binom{6}{1} \cdot \binom{3}{3}}{\binom{9}{4}} = \dfrac{6}{126}$
       $= \dfrac{1}{21} \approx 0.0476$

    c. $\text{Pr(1 minority)} = \dfrac{\binom{6}{3} \cdot \binom{3}{1}}{\binom{9}{4}} = \dfrac{20 \cdot 3}{126} = \dfrac{10}{21}$
       ≈ 0.476

23. $\text{Pr(E)} = \dfrac{\binom{6}{5}}{\binom{10}{5}} = \dfrac{6}{252} = \dfrac{1}{42} \approx 0.0238$

25. a. $\text{Pr(2M)} = \dfrac{\binom{23}{2}}{\binom{27}{2}} = \dfrac{253}{351} \approx 0.721$

    b. $\text{Pr(MW)} = \dfrac{\binom{23}{1} \cdot \binom{4}{1}}{\binom{27}{2}} = \dfrac{23 \cdot 4}{351} = \dfrac{92}{351} \approx 0.262$

    c. $\text{Pr(at least 1W)} = 1 - \text{Pr(2M)} \approx 1 - 0.721$
       $= 0.279$

    Also,

    $\text{Pr(at least 1W)} = \dfrac{\binom{23}{1}\binom{4}{1} + \binom{4}{2}}{\binom{27}{2}} = \dfrac{92 + 6}{351}$
    $= \dfrac{98}{351} \approx 0.279$

27. a. $\text{Pr(E)} = \dfrac{1 \text{ most able}}{3 \text{ men}} = \dfrac{1}{3}$

    b. $\text{Pr(E)} = \dfrac{1}{_3P_3} = \dfrac{1}{3 \cdot 2 \cdot 1} = \dfrac{1}{6}$

29. $\text{Pr(E)} = \dfrac{\text{ways to get 10 winning numbers}}{\text{total ways to get 10 numbers}}$

    $= \dfrac{_{20}C_{10}}{_{80}C_{10}} \approx 1.12 \times 10^{-7}$

31. a. $\text{Pr(E)} = \dfrac{\binom{2}{2}\binom{23}{3}}{\binom{25}{5}} = \dfrac{1 \cdot 1771}{53,130} \approx 0.033$

    b. $\text{Pr(E)} = \dfrac{\binom{23}{5}}{\binom{25}{5}} = \dfrac{33,649}{53,130} \approx 0.633$

33. Order is not important.

    a. $\Pr(5 \text{ spades}) = \dfrac{\binom{13}{5}}{\binom{52}{5}} = \dfrac{1287}{2{,}598{,}960} \approx 0.0005$

    b. $\Pr(5 \text{ of same suit})$

      $= \Pr(\text{All S or All C or All D or All H})$

      $\approx 4(0.0005) = 0.002$

35. $\Pr(4S4H4D1C) = \dfrac{\binom{13}{4}\cdot\binom{13}{4}\cdot\binom{13}{4}\cdot\binom{13}{1}}{\binom{52}{13}}$

    $\approx \dfrac{715\cdot715\cdot715\cdot13}{6.35\times10^{11}}$

    ≈ 0.00748

Exercise 7.7

1. Can be since $\dfrac{1}{4} + \dfrac{3}{4} = 1$.

3. Cannot be since $\dfrac{1}{4} + \dfrac{3}{5} + \dfrac{1}{6} \neq 1$.

5. Cannot be since matrix is not square.

7. Can be since entries the sum of entries in each row is 1, and matrix is square.

9. $\begin{bmatrix} 0.2 & 0.8 \end{bmatrix}\begin{bmatrix} 0.1 & 0.9 \\ 0.3 & 0.7 \end{bmatrix} = \begin{bmatrix} 0.26 & 0.74 \end{bmatrix};$

$\begin{bmatrix} 0.26 & 0.74 \end{bmatrix}\begin{bmatrix} 0.1 & 0.9 \\ 0.3 & 0.7 \end{bmatrix} = \begin{bmatrix} 0.248 & 0.752 \end{bmatrix}$

11. $\begin{bmatrix} 0.1 & 0.3 & 0.6 \end{bmatrix}\begin{bmatrix} 0.5 & 0.3 & 0.2 \\ 0.3 & 0.5 & 0.2 \\ 0.1 & 0.1 & 0.8 \end{bmatrix} = \begin{bmatrix} 0.20 & 0.24 & 0.56 \end{bmatrix};$

$\begin{bmatrix} 0.20 & 0.24 & 0.56 \end{bmatrix}\begin{bmatrix} 0.5 & 0.3 & 0.2 \\ 0.3 & 0.5 & 0.2 \\ 0.1 & 0.1 & 0.8 \end{bmatrix} = \begin{bmatrix} 0.228 & 0.236 & 0.536 \end{bmatrix}$

13. $\begin{bmatrix} V_1 & V_2 \end{bmatrix}\begin{bmatrix} 0.1 & 0.9 \\ 0.3 & 0.7 \end{bmatrix} = \begin{bmatrix} V_1 & V_2 \end{bmatrix}$ or $\begin{array}{l} 0.1V_1 + 0.3V_2 = V_1 \\ 0.9V_1 + 0.7V_2 = V_2 \end{array}$ or $\begin{array}{l} -0.9V_1 + 0.3V_2 = 0 \\ 0.9V_1 - 0.3V_2 = 0 \end{array}$

Thus, $V_1 = \dfrac{0.3}{0.9}V_2 = \dfrac{1}{3}V_2$. $V_1 + V_2 = 1$ gives $\dfrac{1}{3}V_2 + V_2 = 1$ or $V_2 = \dfrac{3}{4}$.

Then, $V_1 = \dfrac{1}{4}$. The steady-state vector is $\begin{bmatrix} \dfrac{1}{4} & \dfrac{3}{4} \end{bmatrix}$.

15. $[V_1 \quad V_2 \quad V_3] \begin{bmatrix} 0.5 & 0.3 & 0.2 \\ 0.3 & 0.5 & 0.2 \\ 0.1 & 0.1 & 0.8 \end{bmatrix} = [V_1 \quad V_2 \quad V_3]$

$0.5V_1 + 0.3V_2 + 0.1V_3 = V_1 \qquad -0.5V_1 + 0.3V_2 + 0.1V_3 = 0$

$0.3V_1 + 0.5V_2 + 0.1V_3 = V_2 \quad \text{or} \quad 0.3V_1 - 0.5V_2 + 0.1V_3 = 0$

$0.2V_1 + 0.2V_2 + 0.8V_3 = V_3 \qquad 0.2V_1 + 0.2V_2 - 0.2V_3 = 0$

$$\begin{bmatrix} -0.5 & 0.3 & 0.1 & | & 0 \\ 0.3 & -0.5 & 0.1 & | & 0 \\ 0.2 & 0.2 & -0.2 & | & 0 \end{bmatrix} \rightarrow \begin{bmatrix} 1 & -0.6 & -0.2 & | & 0 \\ 0.3 & -0.5 & 0.1 & | & 0 \\ 0.2 & 0.2 & -0.2 & | & 0 \end{bmatrix} \rightarrow \begin{bmatrix} 1 & -0.6 & -0.2 & | & 0 \\ 0 & -0.32 & 0.16 & | & 0 \\ 0 & 0.32 & -0.16 & | & 0 \end{bmatrix} \rightarrow \begin{bmatrix} 1 & -0.6 & -0.2 & | & 0 \\ 0 & -.32 & 0.16 & | & 0 \\ 0 & 0 & 0 & | & 0 \end{bmatrix}$$

$$\rightarrow \begin{bmatrix} 1 & -0.6 & -0.2 & | & 0 \\ 0 & 1 & -0.5 & | & 0 \\ 0 & 0 & 0 & | & 0 \end{bmatrix} \rightarrow \begin{bmatrix} 1 & 0 & -0.5 & | & 0 \\ 0 & 1 & -0.5 & | & 0 \\ 0 & 0 & 0 & | & 0 \end{bmatrix}$$

So, $V_1 = 0.5V_3$

$V_2 = 0.5V_3$

$V_1 + V_2 + V_3 = 0.5V_3 + 0.5V_3 + V_3 = 1$

Thus, $V_3 = \dfrac{1}{2}$ and the steady-state vector is $\left[\dfrac{1}{4} \quad \dfrac{1}{4} \quad \dfrac{1}{2}\right]$.

17. $[1 \quad 0 \quad 0] \begin{bmatrix} 0.5 & 0.4 & 0.1 \\ 0.4 & 0.5 & 0.1 \\ 0.3 & 0.3 & 0.4 \end{bmatrix} = [0.5 \quad 0.4 \quad 0.1]; \quad [0.5 \quad 0.4 \quad 0.1] \begin{bmatrix} 0.5 & 0.4 & 0.1 \\ 0.4 & 0.5 & 0.1 \\ 0.3 & 0.3 & 0.4 \end{bmatrix} = [0.44 \quad 0.43 \quad 0.13]$

$[0.44 \quad 0.43 \quad 0.13] \begin{bmatrix} 0.5 & 0.4 & 0.1 \\ 0.4 & 0.5 & 0.1 \\ 0.3 & 0.3 & 0.4 \end{bmatrix} = [0.431 \quad 0.430 \quad 0.139]$

$[0.431 \quad 0.430 \quad 0.139] \begin{bmatrix} 0.5 & 0.4 & 0.1 \\ 0.4 & 0.5 & 0.1 \\ 0.3 & 0.3 & 0.4 \end{bmatrix} = [0.4292 \quad 0.4291 \quad 0.1417]$

19.

$$\begin{array}{c} \qquad\quad \text{Daughter} \\ \qquad\quad \text{R} \quad \text{N} \\ \text{Mother} \begin{array}{c} \text{R} \\ \text{N} \end{array} \begin{bmatrix} 0.8 & 0.2 \\ 0.3 & 0.7 \end{bmatrix} \end{array}$$

21. $[0 \quad 1] \begin{bmatrix} 0.8 & 0.2 \\ 0.3 & 0.7 \end{bmatrix} = [0.3 \quad 0.7]$

$[0.3 \quad 0.7] \begin{bmatrix} 0.8 & 0.2 \\ 0.3 & 0.7 \end{bmatrix} = [0.45 \quad 0.55]$

Pr(Granddaughter regular) = 0.45

23.

$$\begin{array}{c} \quad\;\; \text{A} \quad\; \text{F} \quad \text{VW} \\ \begin{array}{c} \text{A} \\ \text{F} \\ \text{VW} \end{array} \begin{bmatrix} 0 & 0.7 & 0.3 \\ 0.6 & 0 & 0.4 \\ 0.8 & 0.2 & 0 \end{bmatrix} \end{array}$$

25. $[0 \quad 1 \quad 0]\begin{bmatrix} 0 & 0.7 & 0.3 \\ 0.6 & 0 & 0.4 \\ 0.8 & 0.2 & 0 \end{bmatrix} = [0.6 \quad 0 \quad 0.4]$

$[0.6 \quad 0 \quad 0.4]\begin{bmatrix} 0 & 0.7 & 0.3 \\ 0.6 & 0 & 0.4 \\ 0.8 & 0.2 & 0 \end{bmatrix} = [0.32 \quad 0.50 \quad 0.18];$

$[0.32 \quad 0.50 \quad 0.18]\begin{bmatrix} 0 & 0.7 & 0.3 \\ 0.6 & 0 & 0.4 \\ 0.8 & 0.2 & 0 \end{bmatrix} = [0.444 \quad 0.260 \quad 0.296]$

$[0.444 \quad 0.260 \quad 0.296]\begin{bmatrix} 0 & 0.7 & 0.3 \\ 0.6 & 0 & 0.4 \\ 0.8 & 0.2 & 0 \end{bmatrix} = [0.3928 \quad 0.37 \quad 0.2372]$

 A $\quad$ F $\quad$ VW

27. $[V_1 \quad V_2 \quad V_3]\begin{bmatrix} 0 & 0.7 & 0.3 \\ 0.6 & 0 & 0.4 \\ 0.8 & 0.2 & 0 \end{bmatrix} = [V_1 \quad V_2 \quad V_3]$

$0.6V_2 + 0.8V_3 = V_1 \qquad -V_1 + 0.6V_2 + 0.8V_3 = 0$

$0.7V_1 \qquad + 0.2V_3 = V_2$ or $0.7V_1 - V_2 + 0.2V_3 = 0$

$0.3V_1 + 0.4V_2 \qquad = V_3 \qquad 0.3V_1 + 0.4V_2 - V_3 = 0$

$\begin{bmatrix} 1 & -0.6 & -0.8 & | & 0 \\ 0.7 & -1 & 0.2 & | & 0 \\ 0.3 & 0.4 & -1 & | & 0 \end{bmatrix} \rightarrow \begin{bmatrix} 1 & -0.60 & -0.80 & | & 0 \\ 0 & -0.58 & 0.76 & | & 0 \\ 0 & 0.58 & -0.76 & | & 0 \end{bmatrix} \rightarrow \begin{bmatrix} 1 & -0.60 & -0.80 & | & 0 \\ 0 & -0.58 & 0.76 & | & 0 \\ 0 & 0 & 0 & | & 0 \end{bmatrix} \rightarrow \begin{bmatrix} 1 & -0.60 & -0.80 & | & 0 \\ 0 & 1 & -1.31 & | & 0 \\ 0 & 0 & 0 & | & 0 \end{bmatrix}$

$\rightarrow \begin{bmatrix} 1 & 0 & -1.586 & | & 0 \\ 0 & 1 & -1.310 & | & 0 \\ 0 & 0 & 0 & | & 0 \end{bmatrix} \quad \begin{array}{l} V_1 = 1.586V_3 \\ V_2 = 1.31V_3 \\ V_1 + V_2 + V_3 = 1 \end{array}$

So, $1.586V_3 + 1.31V_3 + V_3 = 1$ or $V_3 = \dfrac{1}{3.896} \approx 0.257$.

$V_1 \approx 1.586(0.257) \approx 0.407$ and $V_2 \approx 1.31(0.257) \approx 0.336$

Steady-state vector $\approx [0.407 \quad 0.336 \quad 0.257]$.

Actual steady-state vector is $\left[\dfrac{46}{113} \quad \dfrac{38}{113} \quad \dfrac{29}{113}\right]$.

29. R $\quad$ U

$\begin{array}{c} R \\ U \end{array}\begin{bmatrix} 0.7 & 0.3 \\ 0.1 & 0.9 \end{bmatrix}$

$[V_1 \quad V_2]\begin{bmatrix} 0.7 & 0.3 \\ 0.1 & 0.9 \end{bmatrix} = [V_1 \quad V_2]$

$\begin{array}{l} 0.7V_1 + 0.1V_2 = V_1 \\ 0.3V_1 + 0.9V_2 = V_2 \end{array}$ or $\begin{array}{l} -0.3V_1 + 0.1V_2 = 0 \\ 0.3V_1 - 0.1V_2 = 0 \end{array}$;

$V_1 = \dfrac{1}{3}V_2$, $V_1 + V_2 = 1$ gives $V_1 = \dfrac{1}{4}$, $V_2 = \dfrac{3}{4}$.

Steady-state vector is $\left[\dfrac{1}{4} \quad \dfrac{3}{4}\right]$.

31. $\begin{bmatrix} V_1 & V_2 & V_3 \end{bmatrix} \begin{bmatrix} 0.7 & 0.2 & 0.1 \\ 0.1 & 0.6 & 0.3 \\ 0 & 0.1 & 0.9 \end{bmatrix} = \begin{bmatrix} V_1 & V_2 & V_3 \end{bmatrix}$ or $\begin{array}{l} -0.3V_1 + 0.1V_2 \qquad = 0 \\ 0.2V_1 - 0.4V_2 + 0.1V_3 = 0 \\ 0.1V_1 + 0.3V_2 - 0.1V_3 = 0 \end{array}$

$$\begin{bmatrix} 1 & 3 & -1 & | & 0 \\ -0.3 & 0.1 & 0 & | & 0 \\ 0.2 & -0.4 & 0.1 & | & 0 \end{bmatrix} \rightarrow \begin{bmatrix} 1 & 3 & -1 & | & 0 \\ 0 & 1 & -0.3 & | & 0 \\ 0 & -1 & 0.3 & | & 0 \end{bmatrix} \rightarrow \begin{bmatrix} 1 & 0 & -0.1 & | & 0 \\ 0 & 1 & -0.3 & | & 0 \\ 0 & 0 & 0 & | & 0 \end{bmatrix} \begin{array}{l} V_1 = \dfrac{1}{10}V_3 \\[2mm] V_2 = \dfrac{3}{10}V_3 \end{array}$$

$V_1 + V_2 + V_3 = 1$ So, $V_3 = \dfrac{10}{14} = \dfrac{5}{7}$.

This gives $V_1 = \dfrac{1}{14}$ and $V_2 = \dfrac{3}{14}$.

Thus, the steady-state vector is $\begin{bmatrix} \dfrac{1}{14} & \dfrac{3}{14} & \dfrac{5}{7} \end{bmatrix}$.

33. $\begin{bmatrix} V_1 & V_2 & V_3 \end{bmatrix} \begin{bmatrix} 0.7 & 0.2 & 0.1 \\ 0.4 & 0.4 & 0.2 \\ 0.4 & 0.4 & 0.2 \end{bmatrix} = \begin{bmatrix} V_1 & V_2 & V_3 \end{bmatrix}$ or $\begin{array}{l} -0.3V_1 + 0.4V_2 + 0.4V_3 = 0 \\ 0.2V_1 - 0.6V_2 + 0.4V_3 = 0 \\ 0.1V_1 + 0.2V_2 - 0.8V_3 = 0 \end{array}$

$$\begin{bmatrix} 1 & 2 & -8 & | & 0 \\ -0.3 & 0.4 & 0.4 & | & 0 \\ 0.2 & -0.6 & 0.4 & | & 0 \end{bmatrix} \rightarrow \begin{bmatrix} 1 & 2 & -8 & | & 0 \\ 0 & 1 & -2 & | & 0 \\ 0 & -1 & 2 & | & 0 \end{bmatrix} \rightarrow \begin{bmatrix} 1 & 0 & -4 & | & 0 \\ 0 & 1 & -2 & | & 0 \\ 0 & 0 & 0 & | & 0 \end{bmatrix} \begin{array}{l} V_1 = 4V_3 \\ V_2 = 2V_3 \end{array}$$

$4V_3 + 2V_3 + V_3 = 1$ or $V_3 = \dfrac{1}{7}$, $V_1 = \dfrac{4}{7}$, $V_2 = \dfrac{2}{7}$

Steady-state vector is $\begin{bmatrix} \dfrac{4}{7} & \dfrac{2}{7} & \dfrac{1}{7} \end{bmatrix}$.

35. From exercise 34, it is clear that the vector $\begin{bmatrix} 0.49 & 0.42 & 0.09 \end{bmatrix}$ is the steady-state vector for the given transition matrix.

Review Exercises

1. **a.** $\Pr(\text{odd}) = \dfrac{5}{9}$

    b. $\Pr(3, 6, 9) = \dfrac{3}{9} = \dfrac{1}{3}$

    c. $\Pr(3, 9) = \dfrac{2}{9}$

2. **a.** $\Pr(R) = \dfrac{9}{12} = \dfrac{3}{4}$

    b. $\Pr(\text{odd}) = \dfrac{6}{12} = \dfrac{1}{2}$

    c. $\Pr(10, 12) = \dfrac{1}{6}$

    d. $\Pr(W \text{ or } 1, 3, 5, 7, 9) = \dfrac{3+5}{12} = \dfrac{2}{3}$

3. $\Pr(E) = \dfrac{3}{7}$ means Win 3, Lose 4.

    a. Odds for E: 3:4
    b. Odds against E: 4:3

4. Equiprobable sample space:
    {(HH), (HT), (TH), (TT)}

    a. $\Pr(0 \text{ heads}) = \dfrac{1}{4}$

    b. $\Pr(1 \text{ head}) = \dfrac{2}{4} = \dfrac{1}{2}$

    c. $\Pr(2 \text{ heads}) = \dfrac{1}{4}$

5. {HHH, HHT, HTH, HTT, TTT, TTH, THT, THH}

    a. $\Pr(2H) = \dfrac{3}{8}$

    b. $\Pr(3H) = \dfrac{1}{8}$

    c. $\Pr(1H) = \dfrac{3}{8}$

6. $\Pr(\text{Queen or Jack}) = \dfrac{8}{52} = \dfrac{2}{13}$

7. $\Pr(\text{Success}) = \dfrac{16}{52} \cdot \dfrac{16}{52} = \dfrac{16}{169}$

8. $\Pr(L) = 1 - \Pr(W) = \dfrac{3}{4}$

9. $\Pr(A \text{ or } 10) = \dfrac{8}{52} = \dfrac{2}{13}$

10. $\Pr(\text{King or Red}) = \Pr(\text{King}) + \Pr(\text{Red}) - \Pr(\text{King and Red}) = \dfrac{4}{52} + \dfrac{26}{52} - \dfrac{2}{52} = \dfrac{28}{52} = \dfrac{7}{13}$

11. **a.** There are 2 even numbered R. $\Pr(\text{Even and R}) = \dfrac{2}{9}$

    b. $\Pr(R \text{ or } E) = \Pr(R) + \Pr(E) - \Pr(R \text{ and } E) = \dfrac{4}{9} + \dfrac{4}{9} - \dfrac{2}{9} = \dfrac{2}{3}$

    c. $\Pr(W \text{ or } O) = \Pr(W) + \Pr(O) - \Pr(W \text{ and } O) = \dfrac{5}{9} + \dfrac{5}{9} - \dfrac{3}{9} = \dfrac{7}{9}$

12. $\Pr(\text{both balls red}) = \Pr(\text{1st ball red}) \cdot \Pr(\text{2nd ball red} \mid \text{1st ball red}) = \left(\dfrac{4}{7}\right)\left(\dfrac{3}{6}\right) = \dfrac{12}{42} = \dfrac{2}{7}$

13. There are 4 balls in box, given that the first ball drawn is black.

    $S = \{R, R, B, B\}$ Thus, $\Pr(R \mid B) = \dfrac{2}{4} = \dfrac{1}{2}.$

14. $\Pr(R,R,B) = \Pr(R) \cdot \Pr(R) \cdot \Pr(B) = \dfrac{5}{20} \cdot \dfrac{5}{20} \cdot \dfrac{7}{20} = \dfrac{175}{8000} = \dfrac{7}{320}$

15. $\Pr(R,R,B) = \Pr(R) \cdot \Pr(2\text{nd } R \mid 1\text{st } R) \cdot \Pr(3\text{rd } B \mid 1\text{st } R \text{ and } 2\text{nd } R) = \dfrac{5}{20} \cdot \dfrac{4}{19} \cdot \dfrac{7}{18} = \dfrac{140}{6840} = \dfrac{7}{342}$

16. There are $16 \cdot 15 \cdot 14 = 3360$ ways to choose three balls if order matters. There are six orders that yield the desired results: $\{WRB, WBR, BWR, BRW, RWB, RBW\}$. For each of these outcomes, there are 8 ways to choose white, 3 ways to choose red, and 5 ways to choose black. So $8 \cdot 3 \cdot 5 = 120$ ways to get the particular order. Thus, $(6 \text{ orders})(120 \text{ ways to get each order}) = 720$ outcomes that are desired. $\Pr(\text{one of each color}) = \dfrac{720}{3360} = \dfrac{3}{14}$.

17. $\Pr(W \text{ and } R) = \Pr(WR \text{ or } RW) = \Pr(W) \cdot \Pr(R \mid W) + \Pr(R) \cdot \Pr(W \mid R) = \left(\dfrac{6}{10}\right)\left(\dfrac{4}{9}\right) + \left(\dfrac{4}{10}\right)\left(\dfrac{6}{9}\right) = \dfrac{48}{90} = \dfrac{8}{15}$

18. **a.** $\Pr(U1 \text{ and } R) = \Pr(U1) \cdot \Pr(R \mid U1) = \dfrac{1}{2} \cdot \dfrac{3}{7} = \dfrac{3}{14}$

 b. $\Pr(R) = \Pr((U1 \text{ and } R) \text{ or } (U2 \text{ and } R)) = \Pr(U1) \cdot \Pr(R \mid U1) + \Pr(U2) \cdot \Pr(R \mid U2) = \dfrac{1}{2} \cdot \dfrac{3}{7} + \dfrac{1}{2} \cdot \dfrac{5}{7} = \dfrac{8}{14} = \dfrac{4}{7}$

 c. Using Bayes' formula, we have
 $$\Pr(U1 \mid R) = \frac{\Pr(U1) \cdot \Pr(R \mid U1)}{\Pr(U1) \cdot \Pr(R \mid U1) + \Pr(U2) \cdot \Pr(R \mid U2)} = \frac{\frac{1}{2} \cdot \frac{3}{7}}{\frac{1}{2} \cdot \frac{3}{7} + \frac{1}{2} \cdot \frac{5}{7}} = \frac{3}{3+5} = \frac{3}{8}$$

19. $\Pr(\text{Bag B} \mid \text{white ball}) = \dfrac{\Pr(\text{Bag B}) \cdot \Pr(\text{White} \mid \text{Bag B})}{\Pr(\text{Bag B}) \cdot \Pr(\text{White} \mid \text{Bag B}) + \Pr(\text{Bag A}) \cdot \Pr(\text{White} \mid \text{Bag A})} = \dfrac{\frac{1}{2} \cdot \frac{7}{16}}{\frac{1}{2} \cdot \frac{7}{16} + \frac{1}{2} \cdot \frac{5}{14}} = \dfrac{49}{89}$

20. $_6P_2 = \dfrac{6!}{4!} = 30$

21. $_7C_3 = \dfrac{7!}{3!4!} = \dfrac{7 \cdot 6 \cdot 5 \cdot 4!}{3!4!} = \dfrac{7 \cdot 6 \cdot 5}{3 \cdot 2} = 35$

22. $26^3 = 17{,}576$

23. $_8C_5 = \dfrac{8!}{5!3!} = \dfrac{8 \cdot 7 \cdot 6 \cdot 5!}{5!3!} = \dfrac{8 \cdot 7 \cdot 6}{3 \cdot 2} = 56$

24. **a.** It is not a square matrix.
 b. The row sums are not equal to 1.

25. $\begin{bmatrix} 0.1 & 0.9 \end{bmatrix} \begin{bmatrix} 0.4 & 0.6 \\ 0.8 & 0.2 \end{bmatrix} = \begin{bmatrix} 0.76 & 0.24 \end{bmatrix}$

 $\begin{bmatrix} 0.76 & 0.24 \end{bmatrix} \begin{bmatrix} 0.4 & 0.6 \\ 0.8 & 0.2 \end{bmatrix} = \begin{bmatrix} 0.496 & 0.504 \end{bmatrix}$

26. $\begin{bmatrix} V_1 & V_2 \end{bmatrix} \begin{bmatrix} 0.6 & 0.4 \\ 0.1 & 0.9 \end{bmatrix} = \begin{bmatrix} V_1 & V_2 \end{bmatrix}$

 $\begin{array}{l} 0.6V_1 + 0.1V_2 = V_1 \\ 0.4V_1 + 0.9V_2 = V_2 \end{array}$ or $\begin{array}{l} -0.4V_1 + 0.1V_2 = 0 \\ 0.4V_1 - 0.1V_2 = 0 \end{array}$

 Thus $V_1 = \dfrac{-0.1}{-0.4} V_2 = \dfrac{1}{4} V_2$ and $V_1 + V_2 = 1$.

 $\dfrac{1}{4}V_2 + V_2 = 1 \;\rightarrow\; \dfrac{5}{4}V_2 = 1$

 $V_2 = \dfrac{4}{5} = 0.8, \; V_1 = 0.2$

 The steady-state vector is $\begin{bmatrix} 0.2 & 0.8 \end{bmatrix}$.

27. $\Pr(\leq 50) = \dfrac{50{,}000}{80{,}000} = \dfrac{5}{8}$

28. $\Pr(\text{win}) = \dfrac{1}{6+1} = \dfrac{1}{7}$

29. $\Pr(\text{French or German})$
 $= \Pr(\text{Fr}) + \Pr(\text{Ger}) - \Pr(\text{Fr and Ger})$
 $= \dfrac{30}{100} + \dfrac{40}{100} - \dfrac{12}{100} = \dfrac{29}{50} = 0.58$

30. $\Pr(\text{Demo and Med. Pro}) = \dfrac{25}{280} = \dfrac{5}{56}$

31. $\Pr(\text{HP or A}) = \dfrac{75}{280} + \dfrac{120}{280} - \dfrac{30}{280} = \dfrac{165}{280} = \dfrac{33}{56}$

32. No formulas are needed. Read table.

$\Pr(\text{Auth} \mid \text{Med Pro}) = \dfrac{75}{110} = \dfrac{15}{22}$

33.

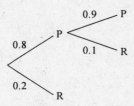

$\Pr\big(\text{pass both}\big) = (0.8)(0.9) = 0.72$

34. a. $\Pr(\text{Color Blind}) = \dfrac{63}{2000}$

   b. $\Pr(\text{M} \mid \text{CB}) = \dfrac{60}{63}$

35.

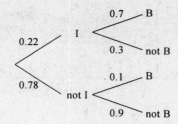

Using Bayes' formula and the probability tree, we have

$\Pr(\text{No intention} \mid \text{Bought})$

$= \dfrac{(0.78)(0.10)}{(0.78)(0.10) + (0.22)(0.7)} = \dfrac{39}{116}$

36. By FCP: $4 \cdot 3 \cdot 2 \cdot 1 = 24$

37. $_8P_4 = \dfrac{8!}{4!} = 1680$

38. $\dbinom{12}{4} = \dfrac{12!}{4!8!} = 495$

39. $_8C_4 = \dfrac{8!}{4!4!} = 70$

40. a. $_{12}C_2 = \dbinom{12}{2} = \dfrac{12!}{2!10!} = 66$

   b. $_{12}C_3 = \dbinom{12}{3} = \dfrac{12!}{3!9!} = 220$

41. There are 36 choices for each position on the plate and repetitions are allowed.
Total plates $= 36 + 36 \cdot 36 + 36 \cdot 36 \cdot 36 + 36^4 + 36^5 = 62,193,780$

42. (# of blood groups) $\cdot$ (pos or neg) $\cdot$ (# blood types) $\cdot$ (#Rh types) $= 4 \cdot 2 \cdot 4 \cdot 8 = 256$
So, the conclusion that there are 288 unique groups is incorrect if the given information is correct.

43. Number of ways to enter courses $= 4! = 24$. $\Pr(\text{alphabetical}) = \dfrac{1}{24}$.

44. Number of three digit possibilities $= 10^3 = 1000$.

Number of permutations of 3 different digits $= 6$. $\Pr(\text{winner}) = \dfrac{6}{1000} = \dfrac{3}{500}$

45. There are 24 ways to win. Assume 0 can be in any position and digits can be repeated.

$\Pr(\text{Win}) = \dfrac{24}{10 \cdot 10 \cdot 10 \cdot 10} = \dfrac{3}{1250}$

46. a. $\Pr(0 \text{ def}) = \dfrac{\binom{180}{10}}{\binom{200}{10}} \approx \dfrac{7.6283 \times 10^{15}}{2.2451 \times 10^{16}} \approx 0.3398$

b. $\Pr(2 \text{ def}) = \dfrac{\binom{20}{2} \cdot \binom{180}{8}}{\binom{200}{10}} \approx \dfrac{(190)(2.3342 \times 10^{13})}{2.2451 \times 10^{16}} \approx 0.1975$

47. $\Pr(3 \text{ best}) = \dfrac{1}{{}_5C_3} = \dfrac{1}{\binom{5}{3}} = \dfrac{1}{10}$

48. a. $\Pr(1 \text{ def}) = \dfrac{\binom{2}{1} \cdot \binom{10}{5}}{\binom{12}{6}} = \dfrac{2(252)}{924} = \dfrac{6}{11} \approx 0.545$

b. $\Pr(\text{At least } 1 \text{ def}) = 1 - \Pr(\text{No def}) = 1 - \dfrac{\binom{10}{6}}{\binom{12}{6}} = 1 - \dfrac{210}{924} = 1 - \dfrac{5}{22} = \dfrac{17}{22} \approx 0.773$

Alternate Method: $\Pr(E) = \dfrac{\binom{2}{1} \cdot \binom{10}{5} + \binom{2}{2} \cdot \binom{10}{4}}{\binom{12}{6}} = \dfrac{504 + 210}{924} \approx 0.773$

49. $\begin{bmatrix} 0.7 & 0.2 & 0.1 \end{bmatrix} \begin{bmatrix} 0.15 & 0.60 & 0.25 \\ 0.15 & 0.35 & 0.50 \\ 0 & 0.20 & 0.80 \end{bmatrix} = \begin{bmatrix} 0.135 & 0.51 & 0.355 \end{bmatrix}$

$\begin{bmatrix} 0.135 & 0.51 & 0.355 \end{bmatrix} \begin{bmatrix} 0.15 & 0.60 & 0.25 \\ 0.15 & 0.35 & 0.50 \\ 0 & 0.20 & 0.80 \end{bmatrix} = \begin{bmatrix} 0.09675 & 0.3305 & 0.57275 \end{bmatrix}$

$\begin{bmatrix} 0.09675 & 0.3305 & 0.57275 \end{bmatrix} \begin{bmatrix} 0.15 & 0.60 & 0.25 \\ 0.15 & 0.35 & 0.50 \\ 0 & 0.20 & 0.80 \end{bmatrix} = \begin{bmatrix} 0.0640875 & 0.288275 & 0.6476375 \end{bmatrix}$

50. $\begin{bmatrix} V_1 & V_2 & V_3 \end{bmatrix} \begin{bmatrix} 0.15 & 0.60 & 0.25 \\ 0.15 & 0.35 & 0.50 \\ 0 & 0.20 & 0.80 \end{bmatrix} = \begin{bmatrix} V_1 & V_2 & V_3 \end{bmatrix}$

$\begin{aligned}
0.15V_1 + 0.15V_2 \qquad\quad &= V_1 \qquad -0.85V_1 + 0.15V_2 \qquad\quad = 0 \\
0.60V_1 + 0.35V_2 + 0.20V_3 &= V_2 \quad \text{or}\ \ 0.60V_1 - 0.65V_2 + 0.20V_3 = 0 \\
0.25V_1 + 0.50V_2 + 0.80V_3 &= V_3 \qquad 0.25V_1 + 0.50V_2 - 0.20V_3 = 0
\end{aligned}$

$\begin{bmatrix} 1 & 2 & -0.80 & | & 0 \\ -0.85 & 0.15 & 0 & | & 0 \\ 0.60 & -0.65 & 0.20 & | & 0 \end{bmatrix} \rightarrow \begin{bmatrix} 1 & 2 & -0.80 & | & 0 \\ 0 & 1.85 & -0.68 & | & 0 \\ 0 & -1.85 & 0.68 & | & 0 \end{bmatrix} \rightarrow \begin{bmatrix} 1 & 2 & -.80 & | & 0 \\ 0 & 1 & -\frac{68}{185} & | & 0 \\ 0 & 0 & 0 & | & 0 \end{bmatrix} \rightarrow \begin{bmatrix} 1 & 0 & -\frac{12}{185} & | & 0 \\ 0 & 1 & -\frac{68}{185} & | & 0 \\ 0 & 0 & 0 & | & 0 \end{bmatrix}$

$V_1 = \dfrac{12}{185}V_3,\ V_2 = \dfrac{68}{185}V_3;\ \text{So}\ \dfrac{12}{185}V_3 + \dfrac{68}{185}V_3 + V_3 = 1\ \text{gives}\ V_3 = \dfrac{37}{53}.$

$V_1 = \dfrac{12}{185} \cdot \dfrac{37}{53} = \dfrac{12}{265} \qquad\qquad V_2 = \dfrac{68}{185} \cdot \dfrac{37}{53} = \dfrac{68}{265}$

The steady-state vector is $\begin{bmatrix} \dfrac{12}{265} & \dfrac{68}{265} & \dfrac{37}{53} \end{bmatrix}$

Chapter Test

1. a. $\Pr(\text{odd}) = \dfrac{4}{7}$

b. $\Pr(\text{white}) = \dfrac{3}{7}$

2. a. $\Pr(\text{black and even}) = \dfrac{2}{7}$

b. $\Pr(\text{white or even}) = \dfrac{5}{7}$

3. a. $\Pr(\text{red}) = 0$
b. $\Pr(< 8) = 1$

4. $\Pr(\text{sum} = 7)$

$= \Pr((1, 6), (2, 5), (3, 4))$

$= \dfrac{3}{\binom{7}{2}} = \dfrac{3}{21} = \dfrac{1}{7}$

5. $\Pr(\text{W and W}) = \dfrac{3}{7} \cdot \dfrac{2}{6} = \dfrac{1}{7}$

6. a. $\Pr(\text{W1 and B2}) = \dfrac{3}{7} \cdot \dfrac{4}{6} = \dfrac{2}{7}$

b. $\Pr(\text{W1B2 or B1W2}) = \dfrac{3}{7} \cdot \dfrac{4}{6} + \dfrac{4}{7} \cdot \dfrac{3}{6} = \dfrac{4}{7}$

7. $\Pr(\text{both odd}) = \dfrac{4}{7} \cdot \dfrac{3}{6} = \dfrac{2}{7}$

8.

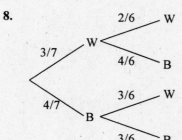

$\Pr(\text{W1W2 or B1W2}) = \dfrac{3}{7} \cdot \dfrac{2}{6} + \dfrac{4}{7} \cdot \dfrac{3}{6} = \dfrac{3}{7}$

9. Using Bayes' formula and the probability tree in exercise 8, we have

$\Pr(\text{B1} \mid \text{W2}) = \dfrac{\frac{4}{7} \cdot \frac{3}{6}}{\frac{4}{7} \cdot \frac{3}{6} + \frac{3}{7} \cdot \frac{2}{6}} = \dfrac{12}{18} = \dfrac{2}{3}$

10. Total outcomes: $26 \cdot 26 \cdot 26$

$\Pr(\text{RAT}) = \dfrac{1}{26^3} = \dfrac{1}{17,576}$

11. Assume a large sample space and events are independent.
$\Pr(\text{2 Am in 3 draws})$

$= \binom{3}{2}(0.35)^2(0.65) = 0.238875$

12. a. $\Pr(\text{left handed}) = \dfrac{4}{20} = \dfrac{1}{5}$

b. $\Pr(\text{ambidextrous}) = \dfrac{1}{20}$

13. a. $\Pr(\text{LL}) = \dfrac{4}{20} \cdot \dfrac{3}{19} = \dfrac{3}{95}$

b. There are other ways to solve this problem.

$\Pr(\text{1R and 1L}) = \dfrac{\binom{15}{1}\binom{4}{1}}{\binom{20}{2}} = \dfrac{6}{19}$

c. $\Pr(\text{2R}) = \dfrac{15}{20} \cdot \dfrac{14}{19} = \dfrac{21}{38}$

d. $\Pr(\text{2 ambidextrous}) = 0$

14. a. $\binom{42}{6} = 5,245,786$

b. $\Pr(\text{Win}) = \dfrac{1}{5,245,786}$

15. a. $\binom{50}{5} = 2,118,760$

b. $\Pr(\text{Win}) = \dfrac{1}{2,118,760}$

16. $\Pr(\text{Def}) = \Pr(\text{\#1 and Def or \#2 and Def})$
$= (0.8)(0.06) + (0.2)(0.08) = 0.064$

17.

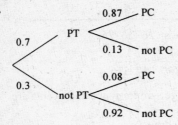

a. Pr(Pass T and Pass C or Fail T and Pass C) $= (0.7)(0.87) + (0.3)(0.08) = 0.633$

b. Pr(Pass T | Pass C) $= \dfrac{(0.7)(0.87)}{(0.7)(0.87) + (0.3)(0.08)} \approx 0.96$

18.

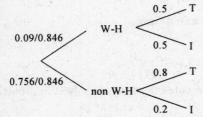

Using Bayes' formula and the probability tree, we have Pr(H | I) $= \dfrac{\left(\frac{0.09}{0.846}\right)(0.5)}{\left(\frac{0.09}{0.846}\right)(0.5) + \left(\frac{0.756}{0.846}\right)(0.2)} \approx 0.229$

19. a. Pr(Def WW) $= \dfrac{70}{350} = \dfrac{1}{5}$

b. Pr(Def TL) $= \dfrac{25}{350} = \dfrac{1}{14}$

c. Pr(No Def TL) $= 1 - \text{Pr(Def TL)} = \dfrac{13}{14}$

20. Pr(Paint) $= 0.14$
Pr(BW and Paint) $= 0.03$
Pr(BW | Paint)

$= \dfrac{\text{Pr(BW and Paint)}}{\text{Pr(Paint)}} = \dfrac{0.03}{0.14} = \dfrac{3}{14}$

21. a. off/on $2 \cdot 2 \cdot 2 \cdot \ldots \cdot 2 = 2^{10} = 1024$
Switch 1 2 3 … 10

b. Pr(open) $= \dfrac{1}{1024}$

c. Pr(open) $= \dfrac{1}{3}$

d. Change the code (sequence).

22. a.

$$A = \begin{bmatrix} 0.80 & 0.20 \\ 0.07 & 0.93 \end{bmatrix} \begin{matrix} \text{Text} \\ \text{Other} \end{matrix}$$

with column labels T   O

b. $[0.25 \quad 0.75] \cdot A = [0.2525 \quad 0.7475]$

$[0.2525 \quad 0.7475] \cdot A = [0.254325 \quad 0.745675]$

$[0.2543 \quad 0.7457] \cdot A = [0.25565725 \quad 0.74434275]$

Percent of market share 3 editions later is $\approx 25.6\%$.

c. $[V_1 \quad V_2] \begin{bmatrix} 0.80 & 0.20 \\ 0.07 & 0.93 \end{bmatrix} = [V_1 \quad V_2]$

$\begin{cases} 0.8V_1 + 0.07V_2 = V_1 \\ -0.2V_1 + 0.07V_2 = 0 \end{cases}$

$\begin{cases} 0.2V_1 + 0.93V_2 = V_2 \\ 0.2V_1 - 0.07V_2 = 0 \end{cases}$

$V_1 = \dfrac{7}{20}V_2$, $V_1 + V_2 = 1$ gives $V_1 = \dfrac{7}{27}$, $V_2 = \dfrac{20}{27}$.

Steady-state vector is $\left[\dfrac{7}{27}, \dfrac{20}{27}\right]$. Hence the percent of market this text will have is $\dfrac{7}{27} \approx 25.9\%$.

Chapter 8: Further Topics in Probability; Statistics

Exercise 8.1

1. $p = 0.3$, $q = 0.7$, $n = 6$, $x = 4$

$$\binom{6}{4}(0.3)^4(0.7)^2 = 15(0.0081)(0.49) \approx 0.0595$$

3. **a.** $p = \dfrac{1}{6}$ **b.** $q = \dfrac{5}{6}$ **c.** $n = 18$

d. $\Pr(6\ 4\text{'s}) = \binom{18}{6}\left(\dfrac{1}{6}\right)^6\left(\dfrac{5}{6}\right)^{12} = 18564\left(\dfrac{1}{46656}\right)\left(\dfrac{244140625}{2176782336}\right) \approx 0.04463$

5. $p = \dfrac{1}{2}$, $q = \dfrac{1}{2}$, $n = 6$

a. $x = 6$ $\binom{6}{6}\left(\dfrac{1}{2}\right)^6\left(\dfrac{1}{2}\right)^0 = 1\left(\dfrac{1}{64}\right)(1) = \dfrac{1}{64}$

b. $x = 3$ $\binom{6}{3}\left(\dfrac{1}{2}\right)^3\left(\dfrac{1}{2}\right)^3 = 20\left(\dfrac{1}{8}\right)\left(\dfrac{1}{8}\right) = \dfrac{5}{16}$

c. $x = 2$ $\binom{6}{2}\left(\dfrac{1}{2}\right)^2\left(\dfrac{1}{2}\right)^4 = 15\left(\dfrac{1}{4}\right)\left(\dfrac{1}{16}\right) = \dfrac{15}{64}$

7. $p = \dfrac{1}{6}$, $q = \dfrac{5}{6}$, $n = 12$, $x = 5$

$$\binom{12}{5}\left(\dfrac{1}{6}\right)^5\left(\dfrac{5}{6}\right)^7 = 792\left(\dfrac{1}{7776}\right)\left(\dfrac{78,125}{279,936}\right) \approx 0.0284$$

9. **a.** $p = \dfrac{3}{5}$, $q = \dfrac{2}{5}$, $n = 5$, $x = 2$

$$\binom{5}{2}\left(\dfrac{3}{5}\right)^2\left(\dfrac{2}{5}\right)^3 = 10\left(\dfrac{9}{25}\right)\left(\dfrac{8}{125}\right) = 0.2304$$

b. $p = \dfrac{2}{5}$, $q = \dfrac{3}{5}$, $n = 5$, $x = 5$

$$\binom{5}{5}\left(\dfrac{2}{5}\right)^5\left(\dfrac{3}{5}\right)^0 = 1\left(\dfrac{32}{3125}\right)(1) = 0.01024$$

c. $\Pr(\text{At least 3B}) = \binom{5}{3}\left(\dfrac{2}{5}\right)^3\left(\dfrac{3}{5}\right)^2 + \binom{5}{4}\left(\dfrac{2}{5}\right)^4\left(\dfrac{3}{5}\right) + \binom{5}{5}\left(\dfrac{2}{5}\right)^5\left(\dfrac{3}{5}\right)^0$

$$= 10\left(\dfrac{8}{125}\right)\left(\dfrac{9}{25}\right) + 5\left(\dfrac{16}{625}\right)\left(\dfrac{3}{5}\right) + 0.01024 = 0.2304 + 0.0768 + 0.01024 = 0.31744$$

11. $p = \dfrac{4}{36} = \dfrac{1}{9}$, $q = \dfrac{8}{9}$, $n = 4$, $x = 2$;

$$\binom{4}{2}\left(\dfrac{1}{9}\right)^2\left(\dfrac{8}{9}\right)^2 = 6\left(\dfrac{1}{81}\right)\left(\dfrac{64}{81}\right) \approx 0.0585$$

13. $p = 0.85$, $q = 0.15$, $n = 10$, $x = 8$

$$\binom{10}{8}(0.85)^8(0.15)^2 \approx 45(0.2725)(0.0225) \approx 0.2759$$

15. $p = \dfrac{1}{2}$, $q = \dfrac{1}{2}$, $n = 4$

    a. $x = 2$ $\binom{4}{2}\left(\dfrac{1}{2}\right)^2\left(\dfrac{1}{2}\right)^2 = 6 \cdot \dfrac{1}{4} \cdot \dfrac{1}{4} = \dfrac{3}{8} = 0.375$

    b. $x = 4$ $\binom{4}{4}\left(\dfrac{1}{2}\right)^4\left(\dfrac{1}{2}\right)^0 = 1 \cdot \dfrac{1}{16} \cdot 1 = 0.0625$

17. $\Pr(\text{Def}) = \dfrac{1}{6}$, $n = 4$

    a. $\Pr(2\text{Def}) = \binom{4}{2}\left(\dfrac{1}{6}\right)^2\left(\dfrac{5}{6}\right)^2 = 6 \cdot \dfrac{1}{36} \cdot \dfrac{25}{36} = \dfrac{25}{216} \approx 0.1157$

    b. $\Pr(0\text{Def}) = \binom{4}{0}\left(\dfrac{1}{6}\right)^0\left(\dfrac{5}{6}\right)^4 = \dfrac{625}{1296} \approx 0.4823$

19. $\Pr(\text{Blue}) = \dfrac{1}{4}$, $n = 4$

    a. $\Pr(1\ \text{Blue}) = \binom{4}{1}\left(\dfrac{1}{4}\right)^1\left(\dfrac{3}{4}\right)^3 = 4 \cdot \dfrac{1}{4} \cdot \dfrac{27}{64} = \dfrac{27}{64}$

    b. $\Pr(2\ \text{Blue}) = \binom{4}{2}\left(\dfrac{1}{4}\right)^2\left(\dfrac{3}{4}\right)^2 = 6 \cdot \dfrac{1}{16} \cdot \dfrac{9}{16} = \dfrac{27}{128}$

    c. $\Pr(0\ \text{Blue}) = \binom{4}{0}\left(\dfrac{1}{4}\right)^0\left(\dfrac{3}{4}\right)^4 = 1 \cdot 1 \cdot \dfrac{81}{256} = \dfrac{81}{256}$

21. $\Pr(\text{Death}) = 0.1$, $n = 5$

    a. $\Pr(2\ \text{Deaths}) = \binom{5}{2}(.1)^2(.9)^3 = 10(.01)(.729) = 0.0729$

    b. $\Pr(0\ \text{Deaths}) = \binom{5}{0}(.1)^0(.9)^5 = 1(1)(.59049) = 0.59049$

    c. $\Pr(0\ \text{or}\ 1\ \text{or}\ 2\ \text{Deaths}) = 0.59049 + \binom{5}{1}(.1)^1(.9)^4 + 0.0729 = .59049 + .32805 + .0729 = 0.9914$

23. $\Pr(\text{Boy}) = \dfrac{105}{205} \approx 0.5122$, $n = 6$

$$\Pr(4\ \text{Boys}) = \binom{6}{4}(.5122)^4(.4878)^2 \approx 15(.06883)(.2379) \approx 0.2457$$

25. $\Pr(\text{Fire}) = 0.004$, $n = 10$

$$\Pr(2\ \text{Fires}) = \binom{10}{2}(.004)^2(.996)^8 \approx 45(.000016)(.96844) \approx 0.0007$$

27. $\Pr(\text{Hit}) = 0.3$, $n = 5$

   a. $\Pr(3 \text{ Hits}) = \binom{5}{3}(.3)^3(.7)^2 = 10(.027)(.49) = 0.1323$

   b. $\Pr(4 \text{ or } 5 \text{ Hits}) = \binom{5}{4}(.3)^4(.7)^1 + \binom{5}{5}(.3)^5(.7)^0 = 5(.0081)(.7) + 1(.00243)(1)$

$$= 0.02835 + 0.00243 = 0.03078$$

29. $\Pr(\text{Def}) = 0.01$, $n = 10$

   a. $\Pr(0 \text{ Def}) = \binom{10}{0}(.01)^0(.99)^{10} = (.99)^{10} \approx 0.9044$

   b. $\Pr(1 \text{ Def}) = \binom{10}{1}(.01)^1(.99)^9 \approx 10(.01)(.9135) = 0.09135$

   c. $\Pr(>1 \text{ Def}) = 1 - \Pr(0 \text{ or } 1 \text{ Def}) \approx 1 - (.9044 + .09135) = 0.00425$

31. $\Pr(\text{correct}) = \dfrac{1}{3}$

   b. To get 60%, student needs to get at least 1 of the last 5 correct.

$$\Pr(\geq 60\%) = 1 - \Pr(0 \text{ correct}) = 1 - \binom{5}{0}\left(\frac{1}{3}\right)^0\left(\frac{2}{3}\right)^5 = 1 - \frac{32}{243} = \frac{211}{243} \approx 0.8683$$

   c. To get 80%, student needs to get at least 3 of the last 5 correct.

$$\Pr(\geq 80\%) = \Pr(3 \text{ or } 4 \text{ or } 5 \text{ correct}) = \binom{5}{3}\left(\frac{1}{3}\right)^3\left(\frac{2}{3}\right)^2 + \binom{5}{4}\left(\frac{1}{3}\right)^4\left(\frac{2}{3}\right) + \binom{5}{5}\left(\frac{1}{3}\right)^5\left(\frac{2}{3}\right)^0$$

$$= 10 \cdot \frac{1}{27} \cdot \frac{4}{9} + 5 \cdot \frac{1}{81} \cdot \frac{2}{3} + \frac{1}{243} \approx 0.1646 + 0.0412 + 0.0041 = 0.2099$$

33. $\Pr(\text{No suicide}) = 0.997$, $n = 100$

$$\Pr(\text{No suicide}) = \binom{100}{100}(.997)^{100}(.003)^0 \approx 0.7405$$

Exercise 8.2

1.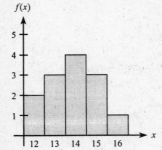

3.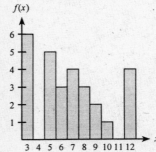

5.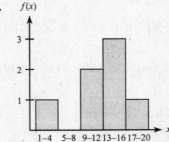

7.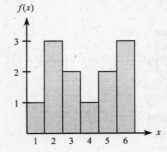

9. The mode is the score that occurs most frequently. 3 (4 times)

11. 13 (3 times)

13. Arrange scores in ascending order.
1, 1, 2, **2**, 2, 3, 4
The median is 2.

15. Arrange scores in ascending order.
0, 0, 1, 1, **1**, 2, 2, 14, 37
The median is 1.

17. Arranged in order: 1, 2, 2, **3**, **6**, 8, 12, 14
The mode is 2. The median is $(3+6) \div 2 = 4.5$.
The mean is $\overline{x} = (1+2+2+3+6+8+12+14) \div 8 = 6$.

19. Arranged in order: 14, 17, **17**, **20**, 31, 42
The mode is 17 since 17 occurs most often.
The median is $(17+20) \div 2 = 18.5$.
The mean is $\overline{x} = (14+17+17+20+31+42) \div 6 = 141 \div 6 = 23.5$.

21. Arranged in order: 2.8, 5.3, **5.3**, 6.4, 6.8
The mode is 5.3. The median is 5.3.
The mean is $\overline{x} = (2.8+5.3+5.3+6.4+6.8) \div 5 = (26.6) \div 5 = 5.32$.

23.

Scores	Class marks	Frequencies
1–4	2.5	1
5–8	6.5	0
9–12	10.5	2
13–16	14.5	3
17–20	18.5	1

mean: $\overline{x} = \dfrac{2.5(1)+6.5(0)+10.5(2)+14.5(3)+18.5(1)}{1+0+2+3+1} \approx 12.21$

mode : 14.5 (most frequent score)
median: 14.5 (middle (4th) score)

25. Range: $11 - 2 = 9$

27. Range: $11 - (-3) = 14$

29. $\overline{x} = (5+7+1+3+0+8+6+2) \div 8 = 4$

$s^2 = \left[1^2 + 3^2 + (-3)^2 + (-1)^2 + (-4)^2 + 4^2 + 2^2 + (-2)^2 \right] \div 7 = \dfrac{60}{7} \approx 8.57$

$s = \sqrt{8.57} \approx 2.93$

31. $\overline{x} = (11+12+13+14+15+16+17) \div 7 = 14$

$s^2 = \left[(-3)^2 + (-2)^2 + (-1)^2 + 0^2 + 1^2 + 2^2 + 3^2 \right] \div 6 = \dfrac{28}{6} \approx 4.67$

$s = \sqrt{4.67} \approx 2.16$

33. $\overline{x} = (3 \cdot 1 + 1 \cdot 2 + 4 \cdot 3 + 2 \cdot 4 + 1 \cdot 5) \div 11 = \dfrac{30}{11} \approx 2.73$

$s^2 = \left[3(-1.73)^2 + 1(-0.73)^2 + 4(0.27)^2 + 2(1.27)^2 + 1(2.27)^2 \right] \div 10 \approx 1.82$

$s = \sqrt{1.82} \approx 1.35$

35. $\bar{x} = (6\cdot3 + 0\cdot4 + 5\cdot5 + 3\cdot6 + 4\cdot7 + 3\cdot8 + 2\cdot9 + 1\cdot10 + 0\cdot11 + 4\cdot12) \div 28 = \dfrac{189}{28} = 6.75$

$s^2 = \dfrac{6(-3.75)^2 + 5(-1.75)^2 + 3(-.75)^2 + 4(.25)^2 + 3(1.25)^2 + 2(2.25)^2 + 1(3.25)^2 + 4(5.25)^2}{27}$

$= \dfrac{237.25}{27} \approx 8.787$

$s = \sqrt{8.787} \approx 2.96$

37. a.

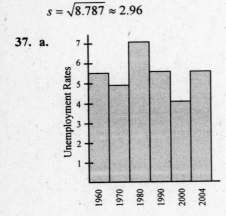

b. $\bar{x} = 5.45$, $s = 1.02$

39. a.

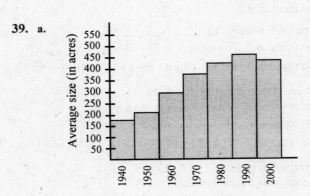

b. $\bar{x} = \dfrac{174 + 213 + 297 + 374 + 460 + 434}{6}$

$= 325.33$ acres

41. Using the mean will give the highest measure.

43. Using the median would give the most representative average.

45. Using technology:
   a. $\bar{x} = 70.09\%$
   b. $s = 8.79\%$

47.

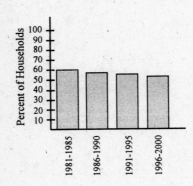

49. $\bar{x} = \dfrac{\Sigma xf}{\Sigma f} = \dfrac{531.8}{160} = 3.32375$ $s^2 = \dfrac{72.970}{159} \approx 0.4589$ $s = 0.6774$

51. **a.** $\bar{x} = \dfrac{160,000(1) + 120,000(1) + 60,000(2) + 40,000(1) + 32,000(5)}{10} = \$60,000$

    b. 32,000, 32,000, 32,000, 32,000, <u>32,000</u>, <u>40,000</u>, 60,000, 60,000, 12,000, 160,000

       Median $= \dfrac{32,000 + 40,000}{2} = \$36,000$

    c. $32,000

53. Using Technology: **a.** $\bar{x} = \$23.33$ **b.** $s = \$5.14$

55. **a.** $\bar{x} = \dfrac{63.03}{8} = 7.88\%$

    b. $\bar{x} = \dfrac{49.03}{8} = 6.13\%$

    c. $\bar{x} = \dfrac{112.06}{16} = 7.00\%$

    d. $s = 1.2480$

57. **a.**

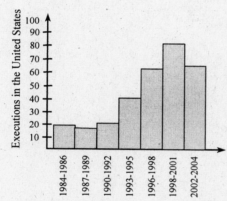

    b. $\bar{x} = \dfrac{19 + 17.3 + 22.7 + 41.7 + 62.3 + 83 + 65}{7} = 44.4$ executions

    c. $s = 26.975$ executions

    d. No

Exercise 8.3

1. No. $\Pr(x)$ must be ≥ 0.

3. Yes. $0 \leq \Pr(x) \leq 1$ and $\sum \Pr(x) = 1$.

5. Yes. $0 \leq \Pr(x) \leq 1$ and $\sum \Pr(x) = 1$.

7. No. $\sum \Pr(x) \neq 1$.

9. $E(x) = \sum x \Pr(x) = 0 \cdot \dfrac{1}{8} + 1 \cdot \dfrac{1}{4} + 2 \cdot \dfrac{1}{4} + 3 \cdot \dfrac{3}{8} = \dfrac{15}{8}$

11. $E(x) = \sum x \Pr(x) = 4 \cdot \dfrac{1}{3} + 5 \cdot \dfrac{1}{3} + 6 \cdot \dfrac{1}{3} + 7 \cdot 0 = 5$

13. $\mu = \sum x \Pr(x) = 0 \cdot \dfrac{1}{4} + 1 \cdot \dfrac{1}{4} + 2 \cdot \dfrac{1}{8} + 3 \cdot \dfrac{3}{8} = \dfrac{13}{8}$

$\sigma^2 = \sum (x-\mu)^2 \Pr(x) = \left(0 - \dfrac{13}{8}\right)^2 \cdot \dfrac{1}{4} + \left(1 - \dfrac{13}{8}\right)^2 \cdot \dfrac{1}{4} + \left(2 - \dfrac{13}{8}\right)^2 \cdot \dfrac{1}{8} + \left(3 - \dfrac{13}{8}\right)^2 \cdot \dfrac{3}{8}$

$= \dfrac{169}{256} + \dfrac{25}{256} + \dfrac{9}{512} + \dfrac{363}{512} = \dfrac{760}{512} = \dfrac{95}{64} \approx 1.48$

$\sigma \approx \sqrt{1.48} \approx 1.22$

15. $\mu = 0 \cdot \dfrac{0}{21} + 1 \cdot \dfrac{1}{21} + 2 \cdot \dfrac{2}{21} + 3 \cdot \dfrac{3}{21} + 4 \cdot \dfrac{4}{21} + 5 \cdot \dfrac{5}{21} + 6 \cdot \dfrac{6}{21} = \dfrac{91}{21} = \dfrac{13}{3}$

$\sigma^2 = \left(-\dfrac{13}{3}\right)^2 \cdot 0 + \left(-\dfrac{10}{3}\right)^2 \cdot \dfrac{1}{21} + \left(-\dfrac{7}{3}\right)^2 \cdot \dfrac{2}{21} + \left(-\dfrac{4}{3}\right)^2 \cdot \dfrac{3}{21} + \left(-\dfrac{1}{3}\right)^2 \cdot \dfrac{4}{21} + \left(\dfrac{2}{3}\right)^2 \cdot \dfrac{5}{21} + \left(\dfrac{5}{3}\right)^2 \cdot \dfrac{6}{21}$

$= \dfrac{1}{21}\left(\dfrac{100}{9} + \dfrac{98}{9} + \dfrac{48}{9} + \dfrac{4}{9} + \dfrac{20}{9} + \dfrac{150}{9}\right) = \dfrac{420}{9 \cdot 21} = \dfrac{20}{9} \approx 2.22$

$\sigma = \sqrt{2.22} \approx 1.49$

17. $\mathrm{E}(x) = \sum x \Pr(x) = 0 \cdot \dfrac{0}{10} + 1 \cdot \dfrac{1}{10} + 2 \cdot \dfrac{2}{10} + 3 \cdot \dfrac{3}{10} + 4 \cdot \dfrac{4}{10} = 3$

19. $\mathrm{E}(x) = 0 \cdot \dfrac{1}{5} + 1 \cdot \dfrac{1}{5} + 2 \cdot \dfrac{1}{5} + 3 \cdot \dfrac{1}{5} + 4 \cdot \dfrac{1}{5} = \dfrac{10}{5} = 2$

21. a. Let $x =$ number of fives

x	$\Pr(x)$
0	$\dbinom{3}{0}\left(\dfrac{1}{6}\right)^0\left(\dfrac{5}{6}\right)^3 = \dfrac{125}{216}$
1	$\dbinom{3}{1}\left(\dfrac{1}{6}\right)^1\left(\dfrac{5}{6}\right)^2 = \dfrac{75}{216} = \dfrac{25}{72}$
2	$\dbinom{3}{2}\left(\dfrac{1}{6}\right)^2\left(\dfrac{5}{6}\right)^1 = \dfrac{15}{216} = \dfrac{5}{72}$
3	$\dbinom{3}{3}\left(\dfrac{1}{6}\right)^3\left(\dfrac{5}{6}\right)^0 = \dfrac{1}{216}$

b. $\mu = 3\left(\dfrac{1}{6}\right) = \dfrac{1}{2}$

c. $\sigma = \sqrt{3\left(\dfrac{1}{6}\right)\left(\dfrac{5}{6}\right)} = \sqrt{\dfrac{15}{36}} = \dfrac{1}{6}\sqrt{15} \approx 0.645$

23. a. $\mu = 60(0.7) = 42$

b. $\sigma = \sqrt{60(0.7)(0.3)} \approx 3.55$

25. a. $\mu = 50\left(\dfrac{3}{5}\right) = 30$

b. $\sigma = \sqrt{50\left(\dfrac{3}{5}\right)\left(\dfrac{2}{5}\right)} = \sqrt{\dfrac{300}{25}} = \dfrac{2\sqrt{3}}{5} \approx 0.693$

27. $\Pr(\text{sum of }6) = \dfrac{5}{36}.$ $E(x) = 900\left(\dfrac{5}{36}\right) = 125.$

29. $(a+b)^6 = \dbinom{6}{0}a^6 + \dbinom{6}{1}a^5b + \dbinom{6}{2}a^4b^2 + \dbinom{6}{3}a^3b^3 + \dbinom{6}{4}a^2b^4 + \dbinom{6}{5}ab^5 + \dbinom{6}{6}b^6$

$\qquad = a^6 + 6a^5b + 15a^4b^2 + 20a^3b^3 + 15a^2b^4 + 6ab^5 + b^6$

31. $(x+h)^4 = \dbinom{4}{0}x^4 + \dbinom{4}{1}x^3h + \dbinom{4}{2}x^2h^2 + \dbinom{4}{3}xh^3 + \dbinom{4}{4}h^4$

$\qquad = x^4 + 4x^3h + 6x^2h^2 + 4xh^3 + h^4$

33. $E(x) = 0(.04) + 1(.35) + 2(.38) + 3(.18) + 4(.05) = 1.85$

35. TV: $E(x) = 100,000(0.01) + 50,000(0.47) + 25,000(0.52) = 37,500$

PA: $E(x) = 80,000(0.02) + 50,000(0.47) + 20,000(0.51) = 35,300$

37. $E(x) = 39,900\left(\dfrac{1}{1500}\right) + 4900\left(\dfrac{1}{1500}\right) + 2400\left(\dfrac{1}{1500}\right) + 1400\left(\dfrac{1}{1500}\right) - 100\left(\dfrac{1496}{1500}\right)$

$\qquad = 26.60 + 3.27 + 1.60 + 0.93 - 99.73 = -67.33$

Expected loss $67.33

39. $E(x) = 15 \cdot \dfrac{1}{13} + 10 \cdot \dfrac{1}{13} + 1 \cdot \dfrac{1}{13} - 4(1) = -\$2.00$

41. Buy 0: There is no profit.

Buy 100: $E(x) = 3(100)(.25) + 3(50)(.20) + 3(10)(.55) - 1(50)(.20) - 1(90)(.55) = \$62$

Buy 200: $E(x) = 3(180)(.25) + 3(50)(.20) + 3(10)(.55) - 1(20)(.25) - 1(150)(.20) - 1(190)(.55) = \$42$

Buy 100 for the best profit.

43. E(cost with policy) = \$100 + 0(0.92) + 100(0.08) = \$108

E(cost without policy) = 0(0.92) + 1000(0.08) = \$80

Save \$28 per year by "taking the chance."

45. No. To be usable the diameter must be within 0.01 inches of 2 inches. Although the average is 2 inches, some of the pipes may fall outside this range.

47. $n = 100$, $p = 0.1$

  a. $\mu = np = 100(0.1) = 10$

  b. $\sigma = \sqrt{npq} = \sqrt{100(0.1)(0.9)} = 3$

49. $n = 100,000$, $p = 0.6$

  a. $\mu = np = 100,000(0.6) = 60,000$

  b. $\sigma = \sqrt{npq} = \sqrt{100,000(0.6)(0.4)} = \sqrt{24,000} \approx 155$

51. In problem 49 we have $\sigma \approx 155$. Thus, 2 standard deviations is $2(155) = 310$ votes. The candidate actually received $60,000 - 310 = 59,690$ votes.

53. $n = 20$, $p = \dfrac{1}{5}$, $q = \dfrac{4}{5}$

  a. $\mu = np = 20 \cdot \dfrac{1}{5} = 4$

  b. $\sigma = \sqrt{npq} = \sqrt{20 \cdot \dfrac{1}{5} \cdot \dfrac{4}{5}} = \sqrt{\dfrac{16}{5}} \approx 1.79$

55. $n = 200$, $\Pr(\text{Def}) = 0.01 = p$, $q = 0.99$

$\mu = np = 200(0.01) = 2$

$\sigma = \sqrt{200(0.01)(0.99)} \approx 1.41$

57. $\Pr(\text{Germinate}) = 0.85$,
$\Pr(\text{Fail to germinate}) = 0.15$
For this question $n = 2000$, $p = 0.15$
$\mu = 2000(0.15) = 300$

Exercise 8.4

1. Problems 1 and 3 are the same from a probability standpoint. $\Pr(0 \le z \le 1.8) = 0.4641$

3. Problems 3 and 1 are the same from a probability standpoint. $\Pr(-1.8 \le z \le 0) = 0.4641$

5. $\Pr(-1.5 \le z \le 0) = 0.4332$ $\Pr(0 \le z \le 2.1) = 0.4821$

Thus, $\Pr(-1.5 \le z \le 2.1) = 0.4332 + 0.4821 = 0.9153$

7. $\Pr(-1.9 \le z \le 0) = 0.4713$ $\Pr(-1.1 \le z \le 0) = 0.3643$

Thus, $\Pr(-1.9 \le z \le -1.1) = 0.4713 - 0.3643 = 0.1070$

9. $\Pr(0 \le z \le 3) = 0.4987$ $\Pr(0 \le z \le 2.1) = 0.4821$

Thus, $\Pr(2.1 \le z \le 3) = 0.4987 - 0.4821 = 0.0166$

11. $\Pr(0 \le z \le 2) = 0.4773$

$\Pr(z > 2) = 0.5 - 0.4773 = 0.0227$

13. $\Pr(0 \le x \le 1.2) = 0.3849$

$\Pr(z < 1.2) = 0.5 + 0.3849 = 0.8849$

15. 20: $z = \dfrac{20 - 20}{5} = 0$ 22.5: $z = \dfrac{22.5 - 20}{5} = 0.5$

$\Pr(20 \le x \le 22.5) = \Pr(0 \le z \le 0.5) = 0.1915$

17. 13.75: $z = \dfrac{13.75 - 20}{5} = -1.25$ 20: $z = \dfrac{20 - 20}{5} = 0$

$\Pr(13.75 \le x \le 20) = \Pr(0 \le z \le 1.25) = 0.3944$

19. 45: $z = \dfrac{45 - 50}{10} = -0.5$ 55: $z = \dfrac{55 - 50}{10} = 0.5$

$\Pr(45 \le x \le 55) = 2\Pr(0 \le z \le 0.5) = 2(0.1915) = 0.3830$

21. 35: $z = \dfrac{35 - 50}{10} = -1.5$ 60: $z = \dfrac{60 - 50}{10} = 1$

$\Pr(35 \le x \le 60) = \Pr(0 \le z \le 1.5) + \Pr(0 \le z \le 1) = 0.4332 + 0.3413 = 0.7745$

23. 134: $z = \dfrac{134 - 110}{12} = 2$

$\Pr(x < 134) = 0.5 + \Pr(0 \le z \le 2) = 0.5 + 0.4773 = 0.9773$

25. $134:\ z = \dfrac{134 - 110}{12} = 2$

$\Pr(x > 134) = 0.5 - \Pr(0 \le z \le 2) = 0.5 - 0.4773 = 0.0227$

27. a. $15:\ z = 0$ $\qquad$ $19:\ z = \dfrac{19 - 15}{4} = 1$ $\qquad$ $\Pr(0 \le z \le 1) = 0.3413$

b. $10:\ z = \dfrac{10 - 15}{4} = -1.25$ $\qquad$ $\Pr(-1.25 \le z \le 0) = 0.3944$

29. $10:\ z = \dfrac{10 - 20}{4} = -2.5$ $\qquad$ $z = \dfrac{30 - 20}{4} = 2.5$

$\Pr(10 \le x \le 30) = \Pr(-2.5 \le z \le 0) + \Pr(0 \le z \le 2.5) = 0.4938 + 0.4938 = 0.9876$

31. a. $160:\ z = 0$ $\qquad$ $181:\ z = \dfrac{181 - 160}{15} = 1.4$ $\quad$ $\Pr(0 \le z \le 1.4) = 0.4192$

b. $190:\ z = \dfrac{190 - 160}{15} = 2$ $\qquad$ $\Pr(z \ge 2) = 0.5000 - 0.4773 = 0.0227$

c. $\Pr(1.4 \le z \le 2) = 0.4773 - 0.4192 = 0.0581$

d. $130:\ z = \dfrac{130 - 160}{15} = -2$

$\Pr(-2 \le z \le 1.4) = \Pr(-2 \le z \le 0) + \Pr(0 \le z \le 1.4) = 0.4773 + 0.4192 = 0.8965$

33. a. $22:\ z = \dfrac{22 - 28}{4} = -1.5$ $\qquad$ $\Pr(z \le -1.5) = 0.5000 - 0.4332 = 0.0668$

b. $30:\ z = \dfrac{30 - 28}{4} = 0.5$ $\qquad$ $\Pr(z \ge 0.5) = 0.5000 - 0.1915 = 0.3085$

c. $26:\ z = \dfrac{26 - 28}{4} = -0.5$

$\Pr(-0.5 \le z \le 0.5) = \Pr(-0.5 \le z \le 0) + \Pr(0 \le z \le 0.5) = 0.1915 + 0.1915 = 0.3830$

35. a. $140:\ z = \dfrac{140 - 120}{12} = 1.67$ $\qquad$ $\Pr(z \ge 1.67) = 0.5000 - 0.4525 = 0.0475$

b. $110:\ z = \dfrac{110 - 120}{12} = -0.83$ $\qquad$ $\Pr(z \le -0.83) = 0.5000 - 0.2967 = 0.2033$

c. $130:\ z = 0.83$ $\qquad$ $\Pr(-0.83 \le z \le 0.83) = 0.2967 + 0.2967 = 0.5934$

37. a. $0.9:\ z = \dfrac{0.9 - 0.7}{0.1} = 2$ $\qquad$ $\Pr(z \ge 2) = 0.5000 - 0.4773 = 0.0227$

b. $0.6:\ z = \dfrac{0.6 - 0.7}{0.1} = -1$ $\qquad$ $\Pr(z \le -1) = 0.5000 - 0.3413 = 0.1587$

c. $\Pr(-1 \le z \le 2) = 0.3413 + 0.4773 = 0.8186$

Review Exercises

1. $\Pr(5 \text{ successes}) = \binom{7}{5}(0.4)^5 (0.6)^2 = 21(0.01024)(0.36) = 0.07741$

2.   **a.** $\Pr(2 \text{ black out of } 4) = \binom{4}{2}\left(\dfrac{5}{12}\right)^2 \left(\dfrac{7}{12}\right)^2 \approx 0.354$

   b. $\Pr(\text{at least } 2) = \Pr(2) + \Pr(3) + \Pr(4) = \binom{4}{2}\left(\dfrac{5}{12}\right)^2 \left(\dfrac{7}{12}\right)^2 + \binom{4}{3}\left(\dfrac{5}{12}\right)^3 \left(\dfrac{7}{12}\right)^1 + \binom{4}{4}\left(\dfrac{5}{12}\right)^4 \left(\dfrac{7}{12}\right)^0$
   $$= 0.3545 + 0.1688 + 0.03014 = 0.5534$$

3.   Pr(At least 2 rolls are greater than 4) = Pr(2 rolls) + Pr(3 rolls) + Pr(4 rolls)
   $$= \binom{4}{2}\left(\frac{1}{3}\right)^2 \left(\frac{2}{3}\right)^2 + \binom{4}{3}\left(\frac{1}{3}\right)^3 \left(\frac{2}{3}\right) + \binom{4}{4}\left(\frac{1}{3}\right)^4 \left(\frac{2}{3}\right)^0 = 6\cdot\frac{1}{9}\cdot\frac{4}{9} + 4\cdot\frac{1}{27}\cdot\frac{2}{3} + 1\cdot\frac{1}{81}\cdot 1 = \frac{11}{27} \approx 0.407$$

4.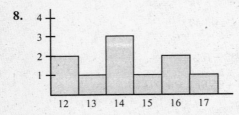

5.   mode = 3

6. $\overline{x} = \dfrac{4\cdot 1 + 6\cdot 2 + 8\cdot 3 + 3\cdot 4 + 5\cdot 5}{4 + 6 + 8 + 3 + 5} \approx 2.96$

7.   Median is between the 13th and 14th score. It is $\dfrac{3+3}{2} = 3$.

8.

9.   Median is between the 5th and 6th score. It is 14.

10.   Mode is 14 (3 times).

11. $\text{mean} = \dfrac{\text{sum of scores}}{10} = \dfrac{143}{10} = 14.3$

12.   Mean: $\overline{x} = \dfrac{4 + 3 + 4 + 6 + 8 + 0 + 2}{7} = \dfrac{27}{7} \approx 3.86$

   Variance: $s^2 = \dfrac{2(0.14)^2 + (-0.86)^2 + (2.14)^2 + (4.14)^2 + (-3.86)^2 + (-1.86)^2}{6} \approx 6.81$

   Standard deviation: $s = \sqrt{6.81} \approx 2.61$

13. $\bar{x} = \dfrac{\Sigma x}{n} = \dfrac{20}{10} = 2$

$s^2 = \dfrac{\Sigma(x-\bar{x})^2}{n-1} = \dfrac{1+0+1+9+1+4+4+0+1+1}{9} = \dfrac{22}{9} = 2.4\overline{4}$

$s \approx 1.56$

14. $E(x) = 1(0.2) + 2(0.3) + 3(0.4) + 4(0.1) = 2.4$

15. $\sum \Pr(x) = \dfrac{1}{15} + \dfrac{2}{15} + \dfrac{3}{15} + \dfrac{4}{15} + \dfrac{5}{15} = 1$

Yes.

16. No, because $\sum \Pr(x) \neq 1$.

17. $\sum \Pr(x) = \dfrac{1}{6} + \dfrac{1}{3} + \dfrac{1}{3} + \dfrac{1}{6} = 1$ Yes.

18. No, because $\Pr(x)$ cannot be negative.

19. $E(x) = 1(0.4) + 2(0.3) + 3(0.2) + 4(0.1) = 2.0$

20. a. $\mu = \sum x \Pr(x) = 1\left(\dfrac{1}{16}\right) + 2\left(\dfrac{2}{16}\right) + 3\left(\dfrac{3}{16}\right) + 4\left(\dfrac{4}{16}\right) + 6\left(\dfrac{6}{16}\right) = \dfrac{66}{16} = 4.125$

b. $\sigma^2 = \sum (x-\mu)^2 \Pr(x)$

$= \left(1-\dfrac{33}{8}\right)^2 \cdot \dfrac{1}{16} + \left(2-\dfrac{33}{8}\right)^2 \cdot \dfrac{2}{16} + \left(3-\dfrac{33}{8}\right)^2 \cdot \dfrac{3}{16} + \left(4-\dfrac{33}{8}\right)^2 \cdot \dfrac{4}{16} + \left(6-\dfrac{33}{8}\right)^2 \cdot \dfrac{6}{16} = \dfrac{1}{16} \cdot \dfrac{175}{4} = 2.734375$

c. $\sigma = \sqrt{2.734375} \approx 1.6536$

21. a. $\mu = \sum x \Pr(x) = 1 \cdot \dfrac{1}{12} + 2 \cdot \dfrac{1}{6} + 3 \cdot \dfrac{1}{3} + 4 \cdot \dfrac{5}{12} = \dfrac{37}{12}$

b. $\sigma^2 = \sum (x-\mu)^2 \cdot \Pr(x) = \left(-\dfrac{25}{12}\right)^2 \cdot \dfrac{1}{12} + \left(-\dfrac{13}{12}\right)^2 \cdot \dfrac{1}{6} + \left(-\dfrac{1}{12}\right)^2 \cdot \dfrac{1}{3} + \left(\dfrac{11}{12}\right)^2 \cdot \dfrac{5}{12} = \dfrac{131}{144} \approx 0.9097$

c. $\sigma = \sqrt{0.9097} \approx 0.9538$

22. $\mu = np = 6\left(\dfrac{2}{3}\right) = 4$ $\qquad \sigma = \sqrt{npq} = \sqrt{6\left(\dfrac{2}{3}\right)\left(\dfrac{1}{3}\right)} = \dfrac{\sqrt{12}}{3} \approx 1.15$

23. $\mu = np = 18\left(\dfrac{1}{6}\right) = 3$

24. $(x+y)^5 = x^5 + 5x^4 y + 10x^3 y^2 + 10x^2 y^3 + 5xy^4 + y^5$

25. $\Pr(-1.6 \leq z \leq 1.9) = 0.4452 + 0.4713 = 0.9165$

26. $\Pr(-1 \leq z \leq -0.5) = 0.3413 - 0.1915 = 0.1498$

27. $\Pr(1.23 \leq z \leq 2.55) = 0.4946 - 0.3907 = 0.1039$

28. $\Pr(25 \leq x \leq 30) = \Pr\left(\dfrac{25-25}{5} \leq \dfrac{x-\mu}{\sigma} \leq \dfrac{30-25}{5}\right) = \Pr(0 \leq z \leq 1) = 0.3413$

29. z scores are -1 and 1 respectively.

$\Pr(-1 \leq z \leq 1) = 0.3413 + 0.3413 = 0.6826$

30. $\Pr(30 \le x \le 35) = \Pr\left(\dfrac{30-25}{5} \le \dfrac{x-\mu}{\sigma} \le \dfrac{35-25}{5}\right) = \Pr(1 \le z \le 2) = 0.4773 - 0.3413 = 0.1360$

31. $\Pr(2 \text{ blond}) = \dbinom{6}{2}\left(\dfrac{1}{4}\right)^2\left(\dfrac{3}{4}\right)^4 = 15 \cdot \dfrac{1}{16} \cdot \dfrac{81}{256} \approx 0.297$

32. $\Pr(x \ge 3) = \dbinom{5}{3}(0.3)^3(0.7)^2 + \dbinom{5}{4}(0.3)^4(0.7)^1 + \dbinom{5}{5}(0.3)^5(0.7)^0 = 0.1323 + 0.02835 + 0.00243 = 0.16308$

33. $\Pr(\text{victim}) = \dfrac{1}{5} \qquad \Pr(\text{not a victim}) = \dfrac{4}{5}$

$\Pr(2 \text{ victims}) = \dbinom{5}{2}\left(\dfrac{1}{5}\right)^2\left(\dfrac{4}{5}\right)^3 = 10\left(\dfrac{1}{25}\right)\left(\dfrac{64}{125}\right) = \dfrac{640}{3125} = 0.2048$

34. a. $\Pr(\text{exactly 1}) = \dbinom{100{,}000}{1}\left(\dfrac{1}{100{,}000}\right)^1\left(\dfrac{99{,}999}{100{,}000}\right)^{99{,}999} \approx 100{,}000\left(\dfrac{1}{100{,}000}\right)(0.36788) \approx 0.37$

b. $\Pr(\text{at least 1}) = 1 - \Pr(0 \text{ get disease})$

$= 1 - \dbinom{100{,}000}{0}\left(\dfrac{1}{100{,}000}\right)^0\left(\dfrac{99{,}999}{100{,}000}\right)^{100{,}000} \approx 1 - 0.3679 \approx 0.63$

35.

36.

Percentages	Class Marks	Frequencies
10–19	14.5	5
20–29	24.5	16
30–39	34.5	25
40–49	44.5	3
50–59	54.5	1

$\bar{x} = \dfrac{5(14.5)+16(24.5)+25(34.5)+3(44.5)+1(54.5)}{50} = \dfrac{1515}{50} = 30.3$

37. $s^2 = \dfrac{5(14.5-30.3)^2+16(24.5-30.3)^2+25(34.5-30.3)^2+3(44.5-30.3)^2+1(54.5-30.3)^2}{49}$

≈ 69.76

$s = \sqrt{69.76} \approx 8.35$

38. Pr(testing positive) = 0.91
Thus, $500(0.91) = 455$ of the 500 are expected to test positively.

39. Each capsule is an independent event.

Expected number of empty capsules $= np = 100\left(\dfrac{60,000}{2,000,000}\right) = 3$.

40. $\Pr(\text{Win}) = \dfrac{3}{3+12} = \dfrac{1}{5}$ Expected value $= 98\left(\dfrac{1}{5}\right) + \dfrac{4}{5}(-2) = \$18.00$

41. $E(x) = (-1)(0.999) + 499(0.001) = -\$0.50$

42.

x	0	1	2	3	4	5
$\Pr(x)$	$\dfrac{1024}{5^5}$	$\dfrac{1280}{5^5}$	$\dfrac{640}{5^5}$	$\dfrac{160}{5^5}$	$\dfrac{20}{5^5}$	$\dfrac{1}{5^5}$

a. $E(x) = 0 \cdot \dfrac{1024}{5^5} + 1 \cdot \dfrac{1280}{5^5} + 2 \cdot \dfrac{640}{5^5} + 3 \cdot \dfrac{160}{5^5} + 4 \cdot \dfrac{20}{5^5} + 5 \cdot \dfrac{1}{5^5} = \dfrac{3125}{5^5} = 1$

b. $\Pr(1 \text{ victim}) = \dbinom{5}{1}\left(\dfrac{1}{5}\right)\left(\dfrac{4}{5}\right)^4 = \left(\dfrac{4}{5}\right)^4 = 0.4096$

43. a. $\Pr(1000 \le x \le 1400) = \Pr\left(\dfrac{1000-1000}{200} \le \dfrac{x-\mu}{\sigma} \le \dfrac{1400-1000}{200}\right) = \Pr(0 \le z \le 2) = 0.4773$

b. $\Pr(1200 \le x \le 1400) = \Pr(1 \le z \le 2) = 0.4773 - 0.3413 = 0.1360$

c. $\Pr(x > 1400) = \Pr(z > 2) = 0.5000 - 0.4773 = 0.0227$

44. $\sigma = 96,000, \ \mu = 611,000$

$z_1 = \dfrac{700,000-611,000}{96,000} \approx 0.93$ $z_2 = \dfrac{800,000-611,000}{96,000} \approx 1.97$

$\Pr(0.93 \le z \le 1.97) = 0.4756 - 0.3238 = 0.1518 \approx 15\%$

Chapter Test

1. a. $\Pr(3H) = \dbinom{5}{3}\left(\dfrac{1}{3}\right)^3\left(\dfrac{2}{3}\right)^2 = \dfrac{40}{243}$

b. $\Pr(3H \text{ or } 4H \text{ or } 5H) = \dfrac{40}{243} + \dbinom{5}{4}\left(\dfrac{1}{3}\right)^4\left(\dfrac{2}{3}\right) + \dbinom{5}{5}\left(\dfrac{1}{3}\right)^5\left(\dfrac{2}{3}\right)^0 = \dfrac{51}{243}$

2. a. For a binomial distribution, $E(x) = np$, so the expected number of heads is $12\left(\dfrac{1}{3}\right) = 4$.

b. $\mu = np = 12\left(\dfrac{1}{3}\right) = 4$

$\sigma = \sqrt{npq} = \sqrt{12\left(\dfrac{1}{3}\right)\left(\dfrac{2}{3}\right)} = \sqrt{\dfrac{8}{3}} = \sqrt{\dfrac{24}{9}} = \dfrac{2}{3}\sqrt{6} \approx 1.6$ The variance $\sigma^2 = \left(\sqrt{\dfrac{8}{3}}\right)^2 = \dfrac{8}{3}$.

3. For each x, $\Pr(x) \ge 0$ and $\sum \Pr(x) = 1$.

4. $E(x) = \sum x \Pr(x) = 0.6 + 1.2 + 0.5 + 0.6 + 1.4 + 0.8 = 5.1$

5. $\mu = E(x) = \sum x \Pr(x) = 1 + 3.6 + 1.5 + 3.6 + 2 + 5 = 16.7$

$\sigma^2 = \sum (x-\mu)^2 \Pr(x) = 26.61$ $\sigma = \sqrt{26.61} \approx 5.16$

6.   mean $\mu = \dfrac{100+147+66+92+48}{5+7+3+4+2} \approx 21.57$

median $= 21$ (11th score), mode $= 21$

7. $z = \dfrac{x-\mu}{\sigma}$

a. $z_1 = \dfrac{14-16}{6} \approx -.33$, $z_2 = \dfrac{22-16}{6} = 1$

$\Pr(14 \le x \le 22) = \Pr(-0.33 \le z \le 1) = 0.1293 + 0.3413 = 0.4706$

b. $\Pr(x \le 22) = \Pr(z \le 1) = 0.5000 + 0.3413 = 0.8413$

c. $\Pr(22 \le x \le 24) = \Pr(1 \le z \le 1.33) = 0.4082 - 0.3413 = 0.0669$

8. $z = \dfrac{x-\mu}{\sigma}$

a. $\Pr(73 \le x \le 97) = \Pr(0.25 \le z \le 2.25) = 0.4878 - 0.0987 = 0.3891$

b. $\Pr(65 \le x \le 84) = \Pr(-0.42 \le z \le 1.17) = 0.1628 + 0.3790 = 0.5418$

c. $\Pr(x \ge 84) = \Pr(z \ge 1.17) = 0.5000 - 0.3790 = 0.1210$

9.

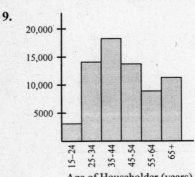

Age of Householder (years)

10.   Class marks: 19.5, 29.5, ..., 59.5, 72.5

$\bar{x} = \dfrac{\Sigma\,(\text{class mark}) \cdot \text{number}}{\Sigma\,\text{number}} = \dfrac{3,221,572}{69,312} \approx 46.48$ years

$s^2 = \dfrac{\Sigma(x-\bar{x})^2 \cdot \text{number}}{n-1} = \dfrac{16,432,190.25}{69,311} \approx 237.079$ $s = \sqrt{237.079} \approx 15.40$

11. a. $\mu = \dfrac{7(60.3)+22(58.7)+37(65.7)+52(51.6)+67(29.2)+82(16.7)}{267} \approx 38.03$

b.   Under 30 would likely be the largest group, which would make the mean age lower.

12. a. $\mu = \dfrac{4(94.5)+11(84.5)+11(74.5)+9(64.5)+5(54.5)+5(44.5)}{45} \approx 71.17$

b.   It would be higher because more people subscribe to use cell phones every year.

13. a. $\Pr(10 \text{ of } 100 \text{ are defective}) = \dbinom{100}{10}(0.02)^{10}(0.98)^{90} \approx 0.00003$

b.   If 1500 chips are shipped, the expected number of defective chips is $(0.02)(1500) = 30$.

14. In this group of 30, $0.06(30) = 1.8$ or 2 could expect to become pregnant.

15. In this group of 30, $0.18(30) = 5.4$ or 5 could expect to become pregnant.

16. In this group of 30, 0.06(0.03)(30) = 0.054 or 0 could expect to become pregnant.

17. Use $z = \dfrac{x - \mu}{\sigma}$, $\mu = 18620$, and $\sigma = 4012$

    a. $z = \dfrac{10000 - 18620}{4012} \approx -2.15$

         $\Pr(x < 10000) = \Pr(z < -2.15) = 0.5000 - 0.4842 = 0.0158$

    b. $z = \dfrac{24000 - 18620}{4012} \approx 1.34$

         $\Pr(x > 24000) = \Pr(z > 1.34) = 0.5000 - 0.4099 = 0.0901$

    c. $\Pr(15000 < x < 21000) = \Pr(-0.90 < z < 0) + \Pr(0 < z < 0.59) = 0.3159 + 0.2224 = 0.5383$

Chapter 9: Derivatives

Exercise 9.1 _____

1. **a.** $\lim\limits_{x \to c} f(x) = -8$

   b. $f(c) = -8$

3. **a.** $\lim\limits_{x \to c} f(x) = 10$

   b. $f(x)$ is not defined at $x = c$.

5. **a.** $\lim\limits_{x \to c} f(x) = 0$

   b. $f(c) = -6$

7. **a.** $\lim\limits_{x \to c^-} f(x) = +\infty$

   b. $\lim\limits_{x \to c^+} f(x) = +\infty$

   c. $\lim\limits_{x \to c} f(x) = +\infty$

   d. $f(c)$ is not defined.

9. **a.** $\lim\limits_{x \to c^-} f(x) = 3$

   b. $\lim\limits_{x \to c^+} f(x) = -6$

   c. $\lim\limits_{x \to c} f(x)$ does not exist.

   d. $f(c) = -6$

11.

x	0.9	0.99	0.999	$\to 1 \leftarrow$	1.001	1.01	1.1
$f(x)$	−2.9	−2.99	−2.999		−3.001	−3.01	−3.1

$\lim\limits_{x \to 1} f(x) = -3$

13.

x	0.9	0.99	0.999	$\to 1 \leftarrow$	1.001	1.01	1.1
$f(x)$	3.5	3.95	3.995		4.995999	4.9599	4.59

$\lim\limits_{x \to 1} f(x)$ does not exist since $4 \neq 5$.

15. $\lim\limits_{x \to -35} (34 + x) = 34 + (-35) = -1$

17. $\lim\limits_{x \to -1} (4x^3 - 2x^2 + 2) = 4(-1) - 2(1) + 2 = -4$

19. $\lim\limits_{x \to -\frac{1}{2}} \dfrac{4x - 2}{4x^2 + 1} = \dfrac{-2 - 2}{1 + 1} = -2$

21. $\lim\limits_{x \to 3} \dfrac{x^2 - 9}{x - 3} = \lim\limits_{x \to 3} \dfrac{(x - 3)(x + 3)}{x - 3} = \lim\limits_{x \to 3}(x + 3) = 6$

23. $\lim\limits_{x \to 7} \dfrac{(x^2 - 8x + 7)}{(x^2 - 6x - 7)} = \lim\limits_{x \to 7} \dfrac{(x - 7)(x - 1)}{(x - 7)(x + 1)} = \dfrac{7 - 1}{7 + 1} = \dfrac{3}{4}$

25. $\lim\limits_{x \to -2} \dfrac{x^2 + 4x + 4}{x^2 + 3x + 2} = \lim\limits_{x \to -2} \dfrac{(x + 2)(x + 2)}{(x + 2)(x + 1)} = \dfrac{0}{-1} = 0$

27. $\lim\limits_{x \to 3^-} f(x) = 4 \quad \lim\limits_{x \to 3^+} f(x) = 6 \quad \lim\limits_{x \to 3} f(x)$ does not exist.

29. $\lim\limits_{x \to (-1)^-} f(x) = -3 \quad \lim\limits_{x \to (-1)^+} f(x) = -3 \quad \lim\limits_{x \to -1} f(x) = -3$

31. $\lim\limits_{x \to 2} \dfrac{x^2 + 6x + 9}{x - 2} = \dfrac{\lim\limits_{x \to 2}(x^2 + 6x + 9)}{\lim\limits_{x \to 2}(x - 2)} = \dfrac{25}{0} =$ does not exist (unbounded near $x = 2$)

33. $\lim\limits_{x \to -1} \dfrac{x^2 + 5x + 6}{x + 1} = \dfrac{\lim\limits_{x \to -1}(x^2 + 5x + 6)}{\lim\limits_{x \to -1}(x + 1)} = \dfrac{2}{0} =$ does not exist (unbounded near $x = -1$)

35. $\lim\limits_{h \to 0} \dfrac{(x + h)^3 - x^3}{h} = \lim\limits_{h \to 0} \dfrac{x^3 + 3x^2 h + 3xh^2 + h^3 - x^3}{h} = \lim\limits_{h \to 0}(3x^2 + 3xh + h^2) = 3x^2$

37. $\lim\limits_{x \to 10} \dfrac{x^2 - 19x + 90}{3x^2 - 30x} = \lim\limits_{x \to 10} \dfrac{(x - 10)(x - 9)}{3x(x - 10)} = \lim\limits_{x \to 10} \dfrac{x - 9}{3x} = \dfrac{1}{30}$

39. $\lim\limits_{x \to -1} \dfrac{x^3 - x}{x^2 + 2x + 1} = \lim\limits_{x \to -1} \dfrac{x(x + 1)(x - 1)}{(x + 1)(x + 1)} = \lim\limits_{x \to -1} \dfrac{x(x - 1)}{x + 1} = \dfrac{2}{0}$ Limit does not exist.

41. $\lim\limits_{x \to -2} \dfrac{x^4 - 4x^2}{x^2 + 8x + 12} = \lim\limits_{x \to -2} \dfrac{x^2(x - 2)(x + 2)}{(x + 6)(x + 2)} = \dfrac{4(-4)}{4} = -4$

43. $\lim\limits_{x \to 4^-} f(x) = 9 \quad \lim\limits_{x \to 4^+} f(x) = 9 \quad \lim\limits_{x \to 4} f(x) = 9$

45.

a	0.1	0.01	0.001	0.0001	0.00001	$\to 0$
$(1 + a)^{1/a}$	2.5937	2.7048	2.7169	2.7181	2.71827	$e \approx 2.71828$

47. a. $\lim\limits_{x \to 3}[f(x) + g(x)] = 4 + (-2) = 2$

    b. $\lim\limits_{x \to 3}[f(x) - g(x)] = 4 - (-2) = 6$

    c. $\lim\limits_{x \to 3}[f(x) \cdot g(x)] = 4(-2) = -8$

    d. $\lim\limits_{x \to 3}\left[\dfrac{g(x)}{f(x)}\right] = -\dfrac{2}{4} = -\dfrac{1}{2}$

49. $\lim\limits_{x \to 100}(1600x - x^2) = 160{,}000 - 10{,}000 = \$150{,}000$

51. a. $\lim\limits_{x \to 4^+}\left(\dfrac{4}{x} + 30 + \dfrac{x}{4}\right) = 1 + 30 + 1 = \$32$ (thousands)

    b. $\lim\limits_{x \to 100^-}\left(\dfrac{4}{x} + 30 + \dfrac{x}{4}\right) = 0.04 + 30 + 25 = \$55.04$ (thousands)

53. a. $S(0) = 400 + \dfrac{2400}{1} = \$2800$

    b. $\lim\limits_{t \to 7}\left(400 + \dfrac{2400}{t + 1}\right) = 400 + \dfrac{2400}{8} = \$700$

    c. $\lim\limits_{t \to 14}\left(400 + \dfrac{2400}{t + 1}\right) = 400 + \dfrac{2400}{15} = \$560$

55. a. $\lim\limits_{t \to 4} \dfrac{128t(t + 6)}{(t^2 + 6t + 18)^2} = \dfrac{128 \cdot 4 \cdot 10}{(16 + 24 + 18)^2} \approx 1.52$ units/hr

    b. $\lim\limits_{t \to 8^-} \dfrac{128t(t + 6)}{(t^2 + 6t + 18)^2} = \dfrac{128 \cdot 8 \cdot 14}{(64 + 48 + 18)^2} \approx 0.85$ units/hr

    c. lunch time

57. a. $\lim\limits_{p \to 100^-} C(p) = 0$ This is basically untreated water.

    b. $\lim\limits_{p \to 0^+} C(p) = \infty$

c.   No. The cost would be extremely large since $C(0)$ is undefined.

59. a. $\lim\limits_{x \to 29,050^-} T(x) = \$4000$

   b. $\lim\limits_{x \to 29,050^+} T(x) = \$4000$

   c. $\lim\limits_{x \to 29,050} T(x) = \$4000$

61. a. $C(x) = \begin{cases} 3.59 + 7.98x & 0 \le x \le 10 \\ 83.39 + 6.78(x - 10) & 10 < x \le 120 \\ 829.19 + 5.43(x - 120) & x > 120 \end{cases}$

   b. $\lim\limits_{x \to 10^-} (3.59 + 7.98x) = 83.39$

   $\lim\limits_{x \to 10^+} (83.39 + 6.78(x - 10)) = 83.39$

   $\lim\limits_{x \to 10} C(x) = 83.39$

63. $\lim\limits_{t \to 9:30AM^+} D(t) \approx 10,216.54$

   This corresponds to the Dow Jones opening average on October 5, 2004.

65. a. $\lim\limits_{t \to 210} f(t) = 1000 \cdot \dfrac{-7.4812(210) + 1560.2}{1.2882(210)^2 - 122.18(210) + 21,483} \approx -0.21$

   b. -0.21% of U.S. Workers will be in farm occupations in 2010

   c.   No, it does not make sense for the percentage of workers to be negative.

Exercise 9.2

1. a.   continuous
   b.   discontinuous; $f(1)$ does not exist.
   c.   discontinuous; $\lim\limits_{x \to 3} f(x)$ does not exist.
   d.   discontinuous; $\lim\limits_{x \to 0} f(x)$ does not exist.
   $f(0)$ does not exist.

3. $f(x)$ is continuous at $x = -2$.

5. $f(x)$ is discontinuous at $x = -3$.
   $f(x)$ is not defined at $x = -3$.

7.   Because $\lim\limits_{x \to 2^-} f(x) = -1$ and $\lim\limits_{x \to 2^+} f(x) = 1$,
   $\lim\limits_{x \to 2} f(x)$ does not exist. The function is not
   continuous at $x = 2$.

9. $f(x)$ is continuous everywhere.

11. $g(x)$ is discontinuous at $x = -2$ since $g(-2)$ is
   not defined and $\lim\limits_{x \to -2} g(x)$ does not exist.

13. $f(x)$ is continuous everywhere.

15. $f(x)$ is continuous everywhere.

17. $y = \dfrac{x^2 - 5x - 6}{x + 1}$. $f(-1)$ is not defined. $f(x)$ is
   discontinuous at $x = -1$.

19. $f(x)$ is discontinuous at $x = 3$ since $\lim\limits_{x \to 3} f(x)$
   does not exist.

21. a.   From the graph, VA: $x = -2$, $\lim\limits_{x \to +\infty} f(x) = 0$,
   $\lim\limits_{x \to -\infty} f(x) = 0$, and HA: $y = 0$.

   b.   The denominator is 0 and the numerator is
   not 0 at $x = -2$, so f has a vertical
   asymptote at $x = -2$.

   $\lim\limits_{x \to +\infty} \dfrac{8}{x + 2} = \lim\limits_{x \to +\infty} \dfrac{\frac{8}{x}}{1 + \frac{2}{x}} = \dfrac{0}{1} = 0$

   $\lim\limits_{x \to -\infty} \dfrac{8}{x + 2} = \lim\limits_{x \to -\infty} \dfrac{\frac{8}{x}}{1 + \frac{2}{x}} = \dfrac{0}{1} = 0$

   Since $\lim\limits_{x \to \pm\infty} f(x) = 0$, there is a horizontal
   asymptote at $y = 0$.

23. a. From the graph, VA: $x = -2$ and $x = 3$,
$\lim\limits_{x \to -\infty} f(x) = 2$, $\lim\limits_{x \to +\infty} f(x) = 2$, and HA:
$y = 2$.

   b. The denominator is 0 and the numerator is not 0 at $x = -2$ and $x = 3$, so f has vertical asymptotes at $x = -2$ and $x = 3$.

$$\lim_{x \to +\infty} \frac{2(x+1)^3(x+5)}{(x-3)^2(x+2)^2} = \lim_{x \to +\infty} \frac{2(x+1)^3(x+5) \cdot \frac{1}{x^4}}{(x-3)^2(x+2)^2 \cdot \frac{1}{x^4}}$$

$$= \lim_{x \to +\infty} \frac{2\left(1+\frac{1}{x}\right)^3\left(1+\frac{5}{x}\right)}{\left(1-\frac{3}{x}\right)^2\left(1+\frac{2}{x}\right)^2} = \frac{2(1)(1)}{(1)(1)} = 2$$

Similarly, $\lim\limits_{x \to -\infty} f(x) = 2$.

Since $\lim\limits_{x \to \pm\infty} f(x) = 2$, there is a horizontal asymptote at $y = 2$.

25. a. $\lim\limits_{x \to +\infty} \dfrac{3}{x+1} = \lim\limits_{x \to +\infty} \dfrac{\frac{3}{x}}{1+\frac{1}{x}} = \dfrac{0}{1+0} = 0$

   b. $y = 0$ is a horizontal asymptote.

27. a. $\lim\limits_{x \to +\infty} \dfrac{x^3-1}{x^3+4} = \lim\limits_{x \to +\infty} \dfrac{1-\frac{1}{x^3}}{1+\frac{4}{x^3}} = \dfrac{1-0}{1+0} = 1$

   b. $y = 1$ is a horizontal asymptote.

29. a. $\lim\limits_{x \to -\infty} \dfrac{5x^3-4x}{3x^3-2} = \lim\limits_{x \to -\infty} \dfrac{5-\frac{4}{x^2}}{3-\frac{2}{x^3}} = \dfrac{5-0}{3-0} = \dfrac{5}{3}$

   b. $y = \dfrac{5}{3}$ is a horizontal asymptote.

31. a. $\lim\limits_{x \to +\infty} \dfrac{3x^2+5x}{6x+1} = \lim\limits_{x \to +\infty} \dfrac{3+\frac{5}{x}}{\frac{6}{x}+\frac{1}{x^2}} \to +\infty$

   Because the limit of the denominator is 0, this limit does not exist.

   b. There is no horizontal asymptote.

33. a.

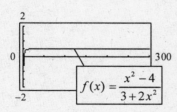

$$\lim_{x \to +\infty} f(x) = 0.5$$

   b. $f(100,000) = 0.49999$

$f(1,000,000) = 0.5$

The table supports the conclusion.

35. $f(x) = \dfrac{1000(x-1)}{x+1000}$

   a. Discontinuous at $x = -1000$

   b. Since $f(x) = \dfrac{1000\left(1-\frac{1}{x}\right)}{1+\frac{1000}{x}}$, $y = 1000$ is a horizontal asymptote.

   c. For large values, an attempt to find the appropriate window is difficult. The asymptote may never be located.

37. $\lim\limits_{x \to \infty} \dfrac{a_n x^n + a_{n-1}x^{n-1} + \ldots + a_1 x + a_0}{b_m x^m + b_{m-1}x^{m-1} + \ldots + b_1 x + b_0} \cdot \dfrac{\left(\frac{1}{x^m}\right)}{\left(\frac{1}{x^m}\right)}$

$= \lim\limits_{x \to \infty} \dfrac{a_n + a_{n-1}\left(\frac{1}{x}\right) + \ldots + a_1\left(\frac{1}{x^{m-1}}\right) + a_0\left(\frac{1}{x^m}\right)}{b_m + b_{m-1}\left(\frac{1}{x}\right) + \ldots + b_1\left(\frac{1}{x^{m-1}}\right) + b_0\left(\frac{1}{x^m}\right)} = \dfrac{a_n}{b_m}$

Similarly, $\lim\limits_{x \to -\infty} f(x) = \dfrac{a_n}{b_n}$. Thus, there is a horizontal asymptote at $y = \dfrac{a_n}{b_n}$.

39. $y = \dfrac{32}{(p+8)^{2/5}}$

   a. If all values are allowed, then y is not continuous at $p = -8$.

   b. Yes

   c. Yes

   d. $p > 0$

41. a. Discontinuous at $q = -1$.

   b. Yes

43. a. $\lim\limits_{n \to \infty} A_n = \lim\limits_{n \to \infty} R\left(\dfrac{1-\frac{1}{(1+i)^n}}{i}\right) = R\left[\dfrac{1-0}{i}\right] = \dfrac{R}{i}$

   b. $i = 0.01$, $R = 100$; $A = \dfrac{100}{0.01} = \$10,000$

45. $p \le 100$ for problem to have meaning. Thus, the function is continuous for the domain of the function.

47. $p = \dfrac{100C}{7300+C} = \dfrac{100}{\frac{7300}{C}+1}$ $\lim\limits_{C \to \infty} p = \dfrac{100}{0+1} = 100\%$

100% of pollution cannot be removed. Cost would be impossible to afford.

49. $R(x)$ is discontinuous at each point that the rate changes. At $x = 14,300$, $58,100$, $117,250$, $178,650$, and $319,100$.

51. **a.** $C(1100) = 49.40 + 0.05(1100 - 500)$
$= \$79.40$

b. $\lim_{x \to 100^-} C(x) = 19.40$

$\lim_{x \to 100^+} C(x) = 19.40 + 0.075(100 - 100)$

$= 19.40$

Thus, $\lim_{x \to 100} C(x) = 19.40$.

By similar procedure, $\lim_{x \to 500} C(x) = 49.40$.

c. Yes, since $\lim_{x \to a} C(x) = C(a)$.

53. **a.** $d(t) = -0.00001719t^4 + 0.004697t^3$
$- 0.4579t^2 + 18.9049t - 268.314$

b. $d(110) = 5.55$

c. $\lim_{t \to +\infty} d(t) = -\infty$

d. No. $d(t) < 0$ for $t > 112$.

e. $t < 30$ and $t > 112$ since $d(t) < 0$.

Exercise 9.3

1. **a.** average rate of change $= \dfrac{f(5) - f(0)}{5 - 0} = \dfrac{18 - (-12)}{5} = \dfrac{30}{5} = 6$

b. average rate of change $= \dfrac{f(10) - f(-3)}{10 - (-3)} = \dfrac{98 - (-6)}{13} = \dfrac{104}{13} = 8$

3. **a.** average rate of change $= \dfrac{f(5) - f(2)}{5 - 2} = \dfrac{30 - 20}{3} = \dfrac{10}{3} = 3.\overline{3}$

b. average rate of change $= \dfrac{f(4) - f(3.8)}{4 - 3.8} = \dfrac{16 - 17}{0.2} = \dfrac{-1}{0.2} = -5$

5. **a.** average rate of change over $[2.9, 3] = \dfrac{f(3) - f(2.9)}{3 - 2.9} = \dfrac{-3 - (-2.61)}{3 - 2.9} = \dfrac{-0.39}{0.1} = -3.9$

average rate of change over $[2.99, 3] = \dfrac{f(3) - f(2.99)}{3 - 2.99} = \dfrac{-3 - (-2.96)}{3 - 2.99} = \dfrac{-0.04}{0.01} = -4$

b. average rate of change over $[3, 3.1] = \dfrac{f(3.1) - f(3)}{3.1 - 3} = \dfrac{-3.41 - (-3)}{3.1 - 3} = \dfrac{-0.41}{0.1} = -4.1$

average rate of change over $[3, 3.01] = \dfrac{f(3.01) - f(3)}{3.01 - 3} = \dfrac{-3.04 - (-3)}{3.01 - 3} = \dfrac{-0.04}{0.01} = 4$

c. The calculations suggest that the instantaneous rate of change of $f(x)$ at $x = 3$ might be -4.

7. **a.** Instantaneous rate of change means $f'(x)$. $f'(4) = 8 \cdot 4 = 32$.

b. Slope of tangent to the graph means $f'(x)$. $f'(4) = 8 \cdot 4 = 32$.

c. $f(4) = 4(4)^2 = 64$ Point on graph: (4, 64)

9. $f(x) = 2x^2 - x$

a. $f(x + h) - f(x) = 2(x + h)^2 - (x + h) - [2x^2 - x] = 4xh + 2h^2 - h$

$\dfrac{f(x + h) - f(x)}{h} = 4x + 2h - 1$

$f'(x) = \lim_{h \to 0}(4x + 2h - 1) = 4x - 1$

b. $f'(-1) = 4(-1) - 1 = -5$

c. $f'(-1) = 4(-1) - 1 = -5$

d. $f(-1) = 2(-1)^2 - (-1) = 3$ Point on graph: (-1, 3)

11. a. $P(x, y) = P(1,1) \quad A(x, y) = A(3,0)$

    b. $m_{\tan} = \dfrac{y_2 - y_1}{x_2 - x_1} = \dfrac{0 - 1}{3 - 1} = -\dfrac{1}{2}$

    c. $f'(1) = -\dfrac{1}{2}$

    d. $f'(1) = -\dfrac{1}{2}$

13. a. $P(x, y) = P(1, 3) \quad A(x, y) = A(0, 3)$

    b. $m_{\tan} = \dfrac{y_2 - y_1}{x_2 - x_1} = \dfrac{3 - 3}{0 - 1} = 0$

    c. $f'(1) = 0$

    d. $f'(1) = 0$

15. $f(x) = 4x^2 - 2x + 1$

Follow the step-by-step procedure of the text.

    a. $f(x + h) - f(x) = [4(x + h)^2 - 2(x + h) + 1] - (4x^2 - 2x + 1) = 8xh + 4h^2 - 2h$

    $\dfrac{f(x + h) - f(x)}{h} = \dfrac{8xh + 4h^2 - 2h}{h} = 8x + 4h - 2$

    $f'(x) = \lim\limits_{h \to 0} \dfrac{f(x + h) - f(x)}{h} = \lim\limits_{h \to 0}(8x + 4h - 2) = 8x - 2$

    b. $f'(x) = 8x - 2, \ f'(-3) = -26$

    c. $f'(-3) = -26$

17. a. $p(q) = q^2 + 4q + 1$

    $p(q + h) - p(q) = [(q + h)^2 + 4(q + h) + 1] - (q^2 + 4q + 1) = 2qh + h^2 + 4h$

    $\dfrac{p(q + h) - p(q)}{h} = \dfrac{2qh + h^2 + 4h}{h} = 2q + h + 4$

    $p'(q) = \lim\limits_{h \to 0}(2q + h + 4) = 2q + 4$

    b. $p'(q) = 2q + 4, \ p'(5) = 14$

    c. $p'(5) = 14$

19. a. $nDeriv(3x^4 - 7x - 5, \ x, \ 2) = 89.000024$

    b. $\dfrac{f(2.0001) - f(2)}{0.0001} \approx \dfrac{29.0089 - 29}{0.0001} \approx 89.0072$

    c. $f'(2) \approx 89$

21. a. $nDeriv((2x - 1)^3, \ x, \ 4) = 294.000008$

    b. $\dfrac{f(4.0001) - f(4)}{0.0001} \approx \dfrac{343.0294 - 343}{0.0001} \approx 294.0084$

    c. $f'(4) \approx 294$

23. $f'(13) = \dfrac{f(12.99) - f(13)}{-0.01} = \dfrac{17.42 - 17.11}{-0.01} = -31$

25.

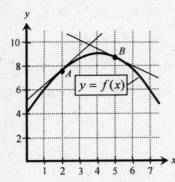

a. $f'(x)$ is greater at point A. The slope of the tangent line is positive and rate of change is greater.

b. Answers may vary. $f'(x)$ at $B \approx -\dfrac{2}{3}$.

27. (a, b) is on the graph of $f(x)$.

If $(a, b) = (-3, -9)$, then $f(a) = -9$.

(a, b) is on the graph of the tangent line at the point of tangency.

$f'(-3)$ is the slope of the tangent line. If

$5x - 2y = 3$, then $y = \dfrac{5}{2}x - \dfrac{3}{2}$ and $f'(-3) = \dfrac{5}{2}$.

29. $f'(1) = 3$

$y - y_1 = m(x - x_1)$

$y - (-1) = 3(x - 1)$

yields $y = 3x - 4$

31. a. $f'(x) > 0$ means $f(x)$ is increasing.

Answer: a, b, d

b. $f'(x) < 0$ means $f(x)$ is decreasing.

Answer: c

c. $f'(x) = 0$ at A, C, and E.

33. a. Function is continuous at A, B, C, and D.

b. Function is differentiable at A and D.

($f'(B)$ and $f'(C)$ are undefined.)

35. a. $f(x+h) - f(x)$

$= [(x+h)^2 + (x+h)] - (x^2 + x)$

$= 2xh + h^2 + h$

$\dfrac{f(x+h) - f(x)}{h} = \dfrac{2xh + h^2 + h}{h} = 2x + h + 1$

$f'(x) = \lim_{h \to 0}(2x + h + 1) = 2x + 1$

b. $f'(2) = 2(2) + 1 = 5$

c. $y - y_1 = m(x - x_1)$

$y - 6 = 5(x - 2)$ yields $y = 5x - 4$

d.

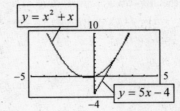

37. a. $f(x+h) - f(x)$

$= [(x+h)^3 + 3] - (x^3 + 3)$

$= 3x^2h + 3xh^2 + h^3$

$\dfrac{f(x+h) - f(x)}{h} = \dfrac{3x^2h + 3xh^2 + h^3}{h}$

$= 3x^2 + 3xh + h^2$

$f'(x) = \lim_{h \to 0}(3x^2 + 3xh + h^2) = 3x^2$

b. $f'(1) = 3(1)^2 = 3$

c. $y - y_1 = m(x - x_1)$

$y - 4 = 3(x - 1)$ yields $y = 3x + 1$

d.

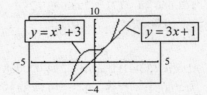

39. a. $\dfrac{f(300) - f(100)}{300 - 100} = \dfrac{14550 - 5950}{200} = \dfrac{8600}{200} = 43$

b. $\dfrac{f(600) - f(300)}{600 - 300} = \dfrac{43200 - 14550}{300} = \dfrac{28650}{300} = 95.50$

c. The average rate of change of total cost of production is greater when 300 to 600 printers are produced.

41. a. $\dfrac{D(25) - D(1)}{25 - 1} = \dfrac{\left(\frac{1000}{\sqrt{25}} - 1\right) - \left(\frac{1000}{\sqrt{1}} - 1\right)}{24} = \dfrac{199 - 999}{24} = -\dfrac{100}{3}$

b. $\dfrac{D(100) - D(25)}{100 - 25} = \dfrac{(100 - 1) - (199)}{75} = -\dfrac{4}{3}$

43. Smallest to largest: A to B, A to C, B to C. The slope of the line from A to B was less than the slope from A to C and the slope of the line from A to C was less than the slope of the line from B to C.

45. $R(x) = 300x - x^2$

    a. $R'(x) = \lim\limits_{h \to 0} \dfrac{[300(x+h)-(x+h)^2]-(300x-x^2)}{h} = 300 - 2x$

    b. $R'(50) = 300 - 2(50) = 200$ A positive value for $R'(x)$ means that $R(x)$ is increasing.

    c. $R'(200) = 300 - 2(200) = -100$ A negative value for $R'(x)$ means that $R(x)$ is decreasing.

    d. $R'(150) = 300 - 2(150) = 0$ A zero value for $R'(x)$ means that $R(x)$ is stationary.

    e. The revenue changes from increasing to decreasing at $x = 150$.

47. $Q(x) = 15,000 + 2x^2$

    $Q'(x) = \lim\limits_{h \to 0} \dfrac{[15,000 + 2(x+h)^2]-(15,000 + 2x^2)}{h} = \lim\limits_{h \to 0}(4x + 2h) = 4x$

    $Q'(50) = 4(50) = 200$

49. $P(x) = 500x - x^2 - 100$

    a. $nDeriv(500x - x^2 - 100, x, 200) = 100$ Profit is increasing.

    b. $nDeriv(500x - x^2 - 100, x, 300) = -100$ Profit is decreasing.

51. $f(x) = 0.009x^2 + 0.139x + 91.875$

    a. $nDeriv(0.009x^2 + 0.139x + 91.875, x, 50) = 1.039$

    b. If humidity changes 1%, the heat index will change by about $1.039°F$.

53. a. The $\overline{MR}$ is greater at 300. The slope of the tangent line to the curve is greater at 300 than it is at 700.

    b. The sale of the 301st cell phone brings in more revenue since the rate of change in revenue is greater at 300 and it indicates the predicted increase in total revenue when one more unit is sold.

Exercise 9.4

1. $y = 4$

    $y'(x) = 0$

    Derivative of a constant is zero.

3. $y = x$

    $\dfrac{dy}{dx} = 1 \cdot x^{1-1} = 1$

5. $f(x) = 2x^3 - x^5$

    $f'(x) = 2 \cdot 3x^2 - 5 \cdot x^4$

    $\quad = 6x^2 - 5x^4$

7. $y = 6x^4 - 5x^2 + x - 2$

    $\dfrac{dy}{dx} = 6 \cdot 4x^3 - 5 \cdot 2x + 1 \cdot x^0 - 0$

    $\quad = 24x^3 - 10x + 1$

9. $g(x) = 10x^9 - 5x^5 + 7x^3 + 5x - 6$

    $g'(x) = 10 \cdot 9x^8 - 5 \cdot 5x^4 + 7 \cdot 3x^2 + 5x^0 - 0$

    $\quad = 90x^8 - 25x^4 + 21x^2 + 5$

11. $y = 4x^2 + 3x$

    $y' = 8x + 3$

    a. At $x = 2$ we have $y' = 8(2) + 3 = 19$.

    b. At $x = 2$ we have $y' = 19$.

13. $P(x) = x^2 - 4x$

    $P'(x) = 2x - 4$

    a. $P'(2) = 2(2) - 4 = 0$

    b. $P'(2) = 0$

15. $y = x^{-5} + x^{-8} - 3$

    $y' = -5x^{-5-1} + (-8)x^{-8-1} - 0$

    $\quad = -5x^{-6} - 8x^{-9}$

17. $y = 3x^{11/3} - 2x^{7/4} - x^{1/2} + 8$

    $y' = 3\left(\dfrac{11}{3}x^{8/3}\right) - 2\left(\dfrac{7}{4}x^{3/4}\right) - \dfrac{1}{2}x^{-1/2} + 0$

    $\quad = 11x^{8/3} - \dfrac{7}{2}x^{3/4} - \dfrac{1}{2}x^{-1/2}$

19. $f(x) = 5x^{-4/5} + 2x^{-4/3}$

$$f'(x) = 5\left(-\frac{4}{5}x^{-9/5}\right) + 2\left(-\frac{4}{3}x^{-7/3}\right)$$

$$= -4x^{-9/5} - \frac{8}{3}x^{-7/3}$$

21. $g(x) = 3x^{-4} + 2x^{-5} + 5x^{1/3}$

$$g'(x) = 3\left(-4x^{-5}\right) + 2\left(-5x^{-6}\right) + 5\left(\frac{1}{3}x^{-2/3}\right)$$

$$= -12x^{-5} - 10x^{-6} + \frac{5}{3}x^{-2/3}$$

23. $y = x^3 - 3x^2 + 5 \qquad y - y_1 = m(x - x_1)$

$\quad y' = 3x^2 - 6x \qquad\qquad y - 3 = -3(x - 1)$

$\qquad\qquad\qquad\qquad\qquad\qquad y = -3x + 6$

At $x = 1$, $y' = 3(1)^2 - 6(1) = -3$

At $x = 1$, $y = 1 - 3 + 5 = 3$

25. $f(x) = 4x^2 - x^{-1}$

$$f'(x) = 8x + x^{-2} = 8x + \frac{1}{x^2}$$

$$f'\left(-\frac{1}{2}\right) = 8\left(-\frac{1}{2}\right) + \frac{1}{\left(-\frac{1}{2}\right)^2} = -4 + 4 = 0$$

$$f\left(-\frac{1}{2}\right) = 4\left(-\frac{1}{2}\right)^2 - \frac{1}{-\frac{1}{2}} = 1 + 2 = 3$$

$$y - y_1 = m(x - x_1)$$

$$y - 3 = 0\left(x - \frac{1}{2}\right)$$

$$y = 3$$

27–29 A horizontal tangent means $f'(x) = 0$

27. $f(x) = -x^3 + 9x^2 - 15x + 6$

$\quad f'(x) = -3x^2 + 18x - 15$

$\qquad\quad = -3(x^2 - 6x + 5)$

$\qquad\quad = -3(x - 5)(x - 1)$

$\quad f'(x) = 0$ if $x = 1$ or 5

Answer: $(1, -1)$, $(5, 31)$

29. $f(x) = x^4 - 4x^3 + 9$

$\quad f'(x) = 4x^3 - 12x^2$

$\qquad\quad = 4x^2(x - 3)$

$\quad f'(x) = 0$ if $x = 0$ or 3

Answer: $(0, 9)$, $(3, -18)$

31. $y = 5 - 2x^{1/2}$

a. $y' = -x^{-1/2} = \dfrac{-1}{\sqrt{x}}$

At $x = 4$, $y' = \dfrac{-1}{\sqrt{4}} = -\dfrac{1}{2}$

b. $\text{nDeriv}\left(5 - 2x^{1/2}, x, 4\right) = -0.500000004$

33. $f(x) = 2x^3 + 5x - \pi^4 + 8$

a. $f'(x) = 6x^2 + 5$

b.

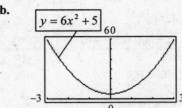

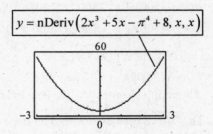

35. $h(x) = 10x^{-3} - 10x^{-2/5} + x^2 + 1$

a. $h'(x) = -30x^{-4} + 4x^{-7/5} + 2x$

$$= \frac{-30}{x^4} + \frac{4}{\sqrt[5]{x^7}} + 2x$$

b.

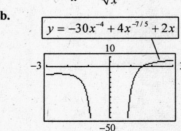

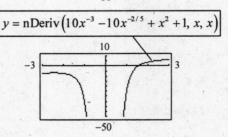

37. a. $f(x) = 3x^2 + 2x$ $y - y_1 = m(x - x_1)$

     $f'(x) = 6x + 2$ $y - 5 = 8(x - 1)$

     $f'(1) = 6 + 2 = 8$ $y = 8x - 3$

b.

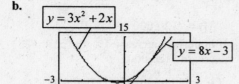

c. Answers may vary.

Starting with the window $[0, 2] \times [0, 10]$ and zooming in four times, we find that, in the window

$[0.99609375, 1.00390625] \times$

$[4.98046875, 5.01953125]$

the function and tangent line cannot be distinguished.

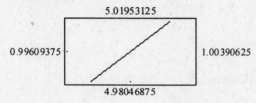

39. $f(x) = 8 - 2x - x^2$

a. $f'(x) = -2 - 2x$

b.

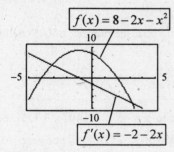

c. $f'(x) = 0$ at $x = -1$.

     $f'(x) > 0$ if $x < -1$.

     $f'(x) < 0$ if $x > -1$.

d. $f(x)$ has a max if $x = -1$.

     $f(x)$ rises if $x < -1$.

     $f(x)$ falls if $x > -1$.

41. $f(x) = x^3 - 12x - 5$

a. $f'(x) = 3x^2 - 12$

b.

c. $f'(x) = 0$ if $x = \pm 2$.

     $f'(x) > 0$ if $x < -2$ and if $x > 2$.

     $f'(x) < 0$ if $-2 < x < 2$.

d. $f(x)$ has a max if $x = -2$.

     $f(x)$ has a min if $x = 2$.

     $f(x)$ rises if $x < -2$ and if $x > 2$.

     $f(x)$ falls if $-2 < x < 2$.

43. $R(x) = 100x - 0.1x^2$

     $R'(x) = 100 - 0.2x$

a. $R'(300) = 100 - 0.2(300) = 40$

The additional revenue from the 301st unit is about \$40.

b. $R'(600) = 100 - 0.2(600) = -20$

The sale of the 601st unit results in a loss of revenue of about \$20.

45. a. $Q'(60) = 200 + 12(60) = 920$

b. $Q(61) - Q(60)$

     $= [200(61) + 6(61)^2] - [200(60) + 6(60)^2]$

     $= 34,526 - 33,600 = 926$

47. $q = 1000p^{-1/2} - 1$

     $q'(p) = -500p^{-3/2} = \dfrac{-500}{\left(\sqrt{p}\right)^3}$

a. $q'(25) = \dfrac{-500}{\left(\sqrt{25}\right)^3} = \dfrac{-500}{125} = -4$

If price increases by \$1, the demand will drop approximately 4 units.

b. $q'(100) = \dfrac{-500}{\left(\sqrt{100}\right)^3} = \dfrac{-500}{1000} = -\dfrac{1}{2}$

If price increases to \$101, the demand will drop approximately $\dfrac{1}{2}$ unit.

49. $\overline{C}(x) = \dfrac{4000}{x} + 55 + 0.1x$

   a. $\overline{C}'(x) = \dfrac{-4000}{x^2} + 0.1$

   b. $\dfrac{-4000}{x^2} + 0.1 = 0$

         $0.1x^2 = 4000$

            $x^2 = 40,000$

             $x = 200$

   c. $C'(x) = 55 + 0.2x$. So $C'(200) = 95$.

      $\overline{C}(200) = \frac{4000}{200} + 55 + 0.1(200) = 95$.

      So $C'(200) = \overline{C}(200)$.

51. $C = \dfrac{120,000}{p} - 1200$

    $C'(p) = -\dfrac{120,000}{p^2}$

   a. $C'(1) = -120,000$

   b. If impurities increase 1%, the cost will decrease $120,000.

53. **a.** $WC = 48.064 + 0.474(15) - 0.020(15s)$

         $-1.85s + 0.304\left(15\sqrt{s}\right) - 27.74\sqrt{s}$

        $= 55.17 - 2.15s - 23.18\sqrt{s}$

   b. $WC'(s) = -2.15 - \dfrac{11.59}{\sqrt{s}}$

      $WC'(25) = -2.15 - \dfrac{11.59}{\sqrt{25}} = -4.468$

   c. If the wind speed increases 1 mph, then the wind-chill will decrease approximately 4.468°.

NOTE: Answers to 55-57 may vary slightly depending on calculator.

55. **a.** $S(x) = 0.0004027x^{4.1674}$

   b. $\dfrac{140.767 - 11.033}{22 - 12} \approx 12.97$

   c. $S'(x) = (4.1674)0.0004027x^{4.1674-1} = 0.001678x^{3.1674}$

   The year 2002 corresponds to $x = 22$, so we find $S'(22) \approx 29.98$. In 2002, the number of U.S. cellular subscribers is increasing at 29.98 million subscribers per year.

57. **a.** $P(t) = -0.0004282x^3 + 0.0567x^2 + 0.8426x + 187.1654$

   b. $P'(t) = -0.0012846x^2 + 0.1134x + 0.8426$

   c. The year 2000 corresponds to $x = 40$ and the year 2025 corresponds to $x = 65$. $P'(40) \approx 3.323$ and $P'(65) \approx 2.786$. In 2000, the U.S. population was increasing at a rate of 3.323 million per year. From this model, predict that the U.S. population will be increasing at a rate of 2.786 million per year.

Exercise 9.5

1. $y = (x+3)(x^2 - 2x)$ $u' = 1$ $v' = 2x - 2$

 $y' = (x+3)(2x-2) + (x^2 - 2x) \cdot 1 = 2x^2 + 4x - 6 + x^2 - 2x = 3x^2 + 2x - 6$

3. $f(x) = (x^{12} + 3x^4 + 4)(4x^3 - 1)$

 $f'(x) = (x^{12} + 3x^4 + 4)(12x^2) + (4x^3 - 1)(12x^{11} + 12x^3)$

 $\quad = 12x^{14} + 36x^6 + 48x^2 + 48x^{14} + 48x^6 - 12x^{11} - 12x^3$

 $\quad = 60x^{14} - 12x^{11} + 84x^6 - 12x^3 + 48x^2$

5. $y = (7x^6 - 5x^4 + 2x^2 - 1)(4x^9 + 3x^7 - 5x^2 + 3x)$

 $\dfrac{dy}{dx} = (7x^6 - 5x^4 + 2x^2 - 1)(36x^8 + 21x^6 - 10x + 3) + (4x^9 + 3x^7 - 5x^2 + 3x)(42x^5 - 20x^3 + 4x)$

7. $y = (x^2 + x + 1)(x^{1/3} - 2x^{1/2} + 5)$

 $\dfrac{dy}{dx} = (x^2 + x + 1)\left(\dfrac{1}{3}x^{-2/3} - x^{-1/2}\right) + (x^{1/3} - 2x^{1/2} + 5)(2x + 1)$

9. $f(x) = (x^2 + 1)(x^3 - 4x)$

 $f'(x) = (x^2 + 1)(3x^2 - 4) + (x^3 - 4x)(2x) = 5x^4 - 9x^2 - 4$

 $f'(-2) = 5(-2)^4 - 9(-2)^2 - 4 = 40$

 a. Slope of tangent line at $(-2, 0)$ is $f'(-2) = 40$.

 b. Instantaneous rate of change at $(-2, 0)$ is $f'(-2) = 40$.

11. $p = \dfrac{q^2 + 1}{q - 2}$, $u' = 2q$, $v' = 1$

 $\dfrac{dp}{dq} = \dfrac{(q-2)(2q) - (q^2 + 1)(1)}{(q-2)^2} = \dfrac{q^2 - 4q - 1}{(q-2)^2}$

13. $y = \dfrac{1 - 2x^2}{x^4 - 2x^2 + 5}$

 $\dfrac{dy}{dx} = \dfrac{(x^4 - 2x^2 + 5)(-4x) - (1 - 2x^2)(4x^3 - 4x)}{(x^4 - 2x^2 + 5)^2} = \dfrac{4x^5 - 4x^3 - 16x}{(x^4 - 2x^2 + 5)^2}$ or $\dfrac{4x(x^4 - x^2 - 4)}{(x^4 - 2x^2 + 5)^2}$

15. $z = x^2 + \dfrac{x^2}{1 - x - 2x^2}$

 $\dfrac{dz}{dx} = 2x + \dfrac{(1 - x - 2x^2)(2x) - x^2(-1 - 4x)}{(1 - x - 2x^2)^2} = 2x + \dfrac{-x^2 + 2x}{(1 - x - 2x^2)^2}$

17. $p = \dfrac{3\sqrt[3]{q}}{1 - q}$, $u' = \dfrac{1}{q^{2/3}}$, $v' = -1$

 $\dfrac{dp}{dq} = \dfrac{(1-q)\left(\dfrac{1}{q^{2/3}}\right) - 3q^{1/3}(-1)}{(1-q)^2} = \dfrac{(1-q) - 3q(-1)}{q^{2/3}(1-q)^2} = \dfrac{1 + 2q}{q^{2/3}(1-q)^2}$

19. $y = \dfrac{x(x^2 + 4)}{x - 2} = \dfrac{x^3 + 4x}{x - 2}$

$y' = \dfrac{(x-2)(3x^2 + 4) - (x^3 + 4x)(1)}{(x-2)^2} = \dfrac{2x^3 - 6x^2 - 8}{(x-2)^2}$

21. $f(x) = \dfrac{x^2 + 1}{x + 3}$ $f'(x) = \dfrac{(x+3)(2x) - (x^2 + 1)(1)}{(x+3)^2} = \dfrac{x^2 + 6x - 1}{(x+3)^2}$ $f'(2) = \dfrac{4 + 12 - 1}{25} = \dfrac{15}{25} = \dfrac{3}{5}$

    a.   The slope of the tangent line at $(2, 1)$ is $\dfrac{3}{5}$.

    b.   The instantaneous rate of change is also $\dfrac{3}{5}$ at this point.

23. $y = (9x^2 - 6x + 1)(1 + 2x)$ $\dfrac{dy}{dx} = (9x^2 - 6x + 1)(2) + (1 + 2x)(18x - 6)$

    At $x = 1$, $\dfrac{dy}{dx} = 4(2) + 3(12) = 44$. At $x = 1, y = 4(3) = 12$.

    $y - y_1 = m(x - x_1)$

    $y - 12 = 44(x - 1)$

        $y = 44x - 32$

25. $f(x) = \dfrac{3x^4 - 2x - 1}{4 - x^2}$ $f'(x) = \dfrac{(4-x^2)(12x^3 - 2) - (3x^4 - 2x - 1)(-2x)}{(4-x^2)^2} = \dfrac{-6x^5 + 48x^3 - 2x^2 - 2x - 8}{(4-x^2)^2}$

    At $x = 1$ we have $f(x) = \dfrac{3 - 2 - 1}{3} = 0$.   At $x = 1$ we have $f'(x) = \dfrac{-6 + 48 - 2 - 2 - 8}{9} = \dfrac{30}{9} = \dfrac{10}{3}$.

    $y - y_1 = m(x - x_1)$ $\rightarrow$ $y - 0 = \dfrac{10}{3}(x - 1)$ $\rightarrow$ $y = \dfrac{10}{3}x - \dfrac{10}{3}$

27. $nDeriv\left(\left(4\sqrt{x} + 3x^{-1}\right) \cdot \left(3x^{1/3} - 5x^{-2} - 25\right), x, 1\right) = 104.0002584$

29. $nDeriv\left((4x - 4)/\left(3x^{2/3}\right), x, 1\right) = 1.333334074$

31. $f(x) = (x^2 + 4x + 4)(x - 7)$

  a. $f'(x) = (x^2 + 4x + 4)(1) + (x - 7)(2x + 4)$

      $= 3x^2 - 6x - 24$

      $= 3(x^2 - 2x - 8)$

      $= 3(x - 4)(x + 2)$

  b.

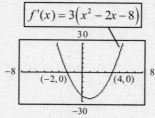

The slope of the tangent is 0 at $x = -2$ and at $x = 4$.

  c.

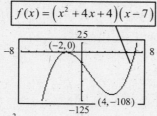

33. $y = \dfrac{x^2}{x - 2}$

  a. $\dfrac{dy}{dx} = \dfrac{(x - 2)(2x) - x^2(1)}{(x - 2)^2}$

      $= \dfrac{x^2 - 4x}{(x - 2)^2} = \dfrac{x(x - 4)}{(x - 2)^2}$

  b.

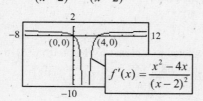

37. $f'(x) = \lim\limits_{h \to 0} \dfrac{\frac{u(x+h)}{v(x+h)} - \frac{u(x)}{v(x)}}{h} = \lim\limits_{h \to 0} \dfrac{u(x+h)v(x) - u(x)v(x+h)}{h \cdot v(x)v(x+h)}$

    $= \lim\limits_{h \to 0} \dfrac{u(x+h)v(x) - u(x)v(x) + u(x)v(x) - u(x)v(x+h)}{h \cdot v(x)v(x+h)}$

    $= \lim\limits_{h \to 0} \dfrac{v(x)\left[\frac{u(x+h)-u(x)}{h}\right] - u(x)\left[\frac{v(x+h)-v(x)}{h}\right]}{v(x)v(x+h)} = \dfrac{v(x)u'(x) - u(x)v'(x)}{[v(x)]^2}$

39. $C(p) = \dfrac{8100 p}{100 - p} \qquad C'(p) = \dfrac{(100 - p)(8100) - 8100 p(-1)}{(100 - p)^2} = \dfrac{810{,}000}{(100 - p)^2}$

The slope of the tangent is 0 at $x = 0$ and at $x = 4$.

  c.

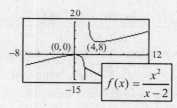

35. $f(x) = \dfrac{10x^2}{x^2 + 1}$

  a. $f'(x) = \dfrac{(x^2 + 1)(20x) - 10x^2(2x)}{(x^2 + 1)^2} = \dfrac{20x}{(x^2 + 1)^2}$

  b.

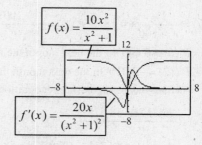

  c. $f'(x) = 0$ if $x = 0$.

    $f'(x) > 0$ if $x > 0$.

    $f'(x) < 0$ if $x < 0$.

    Read graph of $f'(x)$.

  d. f has a min at $x = 0$.

    f is increasing for $x > 0$.

    f is decreasing for $x < 0$.

41. $R(x) = \dfrac{60x^2 + 74x}{2x + 2}$

$R'(x) = \dfrac{(2x+2)(120x+74) - (60x^2 + 74x)2}{(2x+2)^2} = \dfrac{120x^2 + 240x + 148}{(2x+2)^2}$ $R'(49) = \dfrac{300,028}{(100)^2} \approx \$30$

The revenue from the sale of the next unit is approximately \$30.

43. $R(x) = (25+x)(300-10x) = 7500 + 50x - 10x^2$ $R'(x) = 50 - 20x$ $R'(5) = -50$

The revenue will decrease approximately \$50 if the group adds one person.

45. $R(x) = 500x^2 - \dfrac{1}{3}x^3$ $R'(x) = 1000x - x^2$

47. $R(n) = \dfrac{nr}{1 + nr - r}$, r is a constant. $R'(n) = \dfrac{(1 + nr - r)r - nr(r)}{(1 + nr - r)^2} = \dfrac{r(1-r)}{[1 + (n-1)r]^2}$

49. $P(t) = \dfrac{13t}{t^2 + 100} + 0.18$ $P'(t) = \dfrac{(t^2 + 100)13 - 13t(2t)}{(t^2 + 100)^2} = \dfrac{-13t^2 + 1300}{(t^2 + 100)^2}$

  a. $P'(6) \approx 0.045$ In the next month the recognition will increase about 4.5%.

  b. $P'(12) \approx -0.010$ In the next month the recognition will decrease about 1%.

  c. Positive means increasing recognition.

51. $f(x) = \dfrac{289.173 - 58.5731x}{x + 1}$

$f'(x) = \dfrac{(x+1)(-58.5731) - (289.173 - 58.5731x)(1)}{(x+1)^2} = \dfrac{-347.7461}{(x+1)^2}$

  a. $f'(20) \approx -0.79$

  b. At 0°F, if the windspeed increases 1 mph, the wind-chill will decrease about 0.79°F.

53. a. $B'(t) = (0.01t + 3)(0.04766t - 9.79) + (0.01)(0.02383t^2 - 9.79t + 3097.19)$

$= 0.0004766t^2 + 0.04508t - 29.37 + 0.0002383t^2 - 0.0979t + 30.9719$

$= 0.0007149t^2 - 0.05282t + 1.6019$

  b. Instantaneous rate of change in 2010 $= B'(60) = 1.006$

  This indicates the number of beneficiaries (in millions) is predicted to increase by 1.006 from 2010 to 2011.

  c. From 2000 to 2010 $= \dfrac{53.3 - 44.8}{10} = 0.85$ From 2010 to 2020 $= \dfrac{68.8 - 53.3}{10} = 1.55$

  From 2000 to 2020 $= \dfrac{68.8 - 44.8}{20} = 1.2$

  The average rate of change from 2000 to 2020 best approximates the instantaneous rate of change in 2010.

  d. The instantaneous rate of change appears to be the smallest around 1987. The graph of the derivative is the smallest when t is around 37.

55. a. $f'(t) = 1000 \left[\dfrac{\left(1.2882t^2 - 122.18t + 21.483\right)\left(-7.4812\right) - \left(-7.4812t + 1560.2\right)\left(2.5764t - 122.18\right)}{\left(1.2882t^2 - 122.18t + 21.483\right)^2} \right]$

$= 1000 \left[\dfrac{9.6373t^2 - 4019.7t + 29906.6}{\left(1.2882t^2 - 122.18t + 21.483\right)^2} \right]$

b. The year 1870 corresponds to $t = 70$ and the year 2000 corresponds to $t = 200$. So we find $f'(70) \approx -0.55$ and $f'(200) \approx -0.16$.

c. $f'(70)$ means that from 1870 to 1871, the model predicts a change of about -0.55% in U.S. workers in farm occupations. $f'(200)$ means that from 2000 to 2001, the model predicts a change of about -0.16% in U.S. workers in farm occupations.

Exercise 9.6

Unless otherwise specified the Power Rule will be used in calculating the derivatives.

$$y = u^n \qquad \frac{dy}{dx} = nu^{n-1} \cdot \frac{du}{dx}$$

1. $y = u^3$ and $u = x^2 + 1$

$\dfrac{dy}{du} = 3u^2; \quad \dfrac{du}{dx} = 2x;$

$\dfrac{dy}{dx} = \dfrac{dy}{du} \cdot \dfrac{du}{dx} = 3u^2 \cdot 2x = 6x\left(x^2 + 1\right)^2$

3. $y = u^4$ and $u = 4x^2 - x + 8$

$\dfrac{dy}{du} = 4x^3; \quad \dfrac{du}{dx} = 8x - 1;$

$\dfrac{dy}{dx} = \dfrac{dy}{du} \cdot \dfrac{du}{dx} = 4u^3(8x - 1)$

$= 4(8x - 1)(4x^2 - x + 8)^3$

5. $f'(x) = 25\left(2x^4 - 5\right)^{24}\left(8x^3\right)$

$= 200x^3\left(2x^4 - 5\right)^{24}$

7. $h'(x) = \dfrac{2}{3} \cdot 8\left(x^6 + 3x^2 - 11\right)^7\left(6x^5 + 6x\right)$

$= \dfrac{16\left(6x^5 + 6x\right)\left(x^6 + 3x^2 - 11\right)^7}{3}$

$= \dfrac{96x\left(x^4 + 1\right)\left(x^6 + 3x^2 - 11\right)^7}{3}$

$= 32x\left(x^4 + 1\right)\left(x^6 + 3x^2 - 11\right)^7$

9. $g(x) = (x^2 + 4x)^{-2}$

$g'(x) = -2(x^2 + 4x)^{-3}(2x + 4)$

$= -4(x + 2)(x^2 + 4x)^{-3}$

11. $f(x) = \dfrac{1}{(x^2 + 2)^3} = (x^2 + 2)^{-3}$

Preferred: $f'(x) = -3(x^2 + 2)^{-4}(2x) = \dfrac{-6x}{(x^2 + 2)^4}$

Acceptable:

$f'(x) = \dfrac{(x^2 + 2)^3(0) - 1 \cdot 3(x^2 + 2)^2(2x)}{(x^2 + 2)^6}$

$= \dfrac{-6x(x^2 + 2)^2}{(x^2 + 2)^6} = \dfrac{-6x}{(x^2 + 2)^4}$

13. $g(x) = \left(2x^3 + 3x + 5\right)^{-3/4}$

$g'(x) = -\dfrac{3}{4}\left(2x^3 + 3x + 5\right)^{-7/4}\left(6x^2 + 3\right)$

$= \dfrac{-18x^2 - 9}{4(2x^3 + 3x + 5)^{7/4}}$

15. $y = \left(x^2 + 4x + 5\right)^{1/2}$

$y' = \dfrac{1}{2}\left(x^2 + 4x + 5\right)^{-1/2}(2x + 4) = \dfrac{x + 2}{\sqrt{x^2 + 4x + 5}}$

17. $y = \dfrac{8}{5}\left(x^2 - 3\right)^5$

$y' = \dfrac{8}{5} \cdot 5\left(x^2 - 3\right)^4(2x) = 16x\left(x^2 - 3\right)^4$

19. $y = \dfrac{1}{7}\left[(3x+1)^5 - 3x\right]$

$y' = \dfrac{1}{7}\left[5(3x+1)^4(3) - 3\right] = \dfrac{3}{7}\left[5(3x+1)^4 - 1\right]$

21. $y = f(x) = (x^3 + 2x)^4$

$f'(x) = 4(x^3 + 2x)^3(3x^2 + 2)$

$f'(2) = 4(8+4)^3(12+2) = 96{,}768$

a. The slope of the tangent line at $x = 2$ is 96,768.

b. The instantaneous rate of change of the function at $x = 2$ is also 96,768.

$nDeriv\left((x^3 + 2x)^4, x, 2\right) = 96{,}768.28378$

23. $f(x) = (x^3 + 1)^{1/2}$

$f'(x) = \dfrac{1}{2}(x^3 + 1)^{-1/2}(3x^2) = \dfrac{3x^2}{2\sqrt{x^3 + 1}}$

$f'(2) = \dfrac{12}{2\sqrt{8+1}} = \dfrac{6}{\sqrt{9}} = 2$

a. The slope of the tangent line at $(2, 3)$ is 2.

b. The instantaneous rate of change of the function at $(2, 3)$ is also 2.

$nDeriv\left((x^3 + 1)^{1/2}, x, 2\right) = 2$

25. $f(x) = (x^2 - 3x + 3)^3$

$f'(x) = 3(x^2 - 3x + 3)^2(2x - 3)$

$f'(1) = 3(1)(-1) = -3$

$y - y_1 = m(x - x_1)$

$y - 1 = -3(x - 1)$

$y = -3x + 4$

27. $f(x) = (3x^2 - 2)^{1/2}$

$f'(x) = \dfrac{1}{2}(3x^2 - 2)^{-1/2}(6x) = \dfrac{3x}{\sqrt{3x^2 - 2}}$

$f(3) = \sqrt{27 - 2} = 5$

$f'(3) = \dfrac{9}{\sqrt{27 - 2}} = \dfrac{9}{5}$

$y - y_1 = m(x - x_1)$

$y - 5 = \dfrac{9}{5}(x - 3)$

$9x - 5y = 2$

29. $f(x) = (x^2 - 4)^3 + 12$

a. $f'(x) = 3(x^2 - 4)^2(2x) = 6x(x-2)^2(x+2)^2$

b.

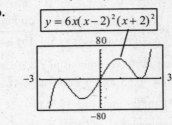

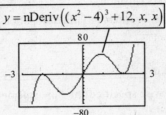

c. $f'(x) = 0$ at $x = 0, 2, -2$

d. At $\begin{cases} x = & 0 \quad 2 \quad -2 \\ y = & -52 \quad 12 \quad 12 \end{cases}$

Points: $(0, -52), (2, 12), (-2, 12)$

e.

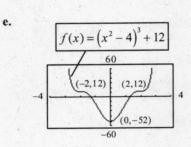

31. $f(x) = 12 - 3(1 - x^2)^{4/3}$

　a. $f'(x) = -4(1 - x^2)^{1/3}(-2x) = 8x(1 - x^2)^{1/3}$

　b.

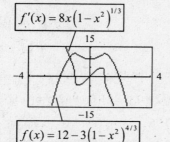

$$\boxed{f'(x) = 8x(1 - x^2)^{1/3}}$$

$$\boxed{f(x) = 12 - 3(1 - x^2)^{4/3}}$$

　c.

x		-1		0		1	
$f'(x)$	$+$	0	$-$	0	$+$	0	$-$

　d.　max at $x = -1$ and $x = 1$.　min at $x = 0$.
　　$f(x)$ increasing for $x < -1$ and $0 < x < 1$.
　　$f(x)$ decreasing for $x > 1$ and $-1 < x < 0$.

33. a.　$y = \dfrac{2}{3}x^3$　　$\dfrac{dy}{dx} = \dfrac{2}{3} \cdot 3x^2 = 2x^2$

　b.　$y = \dfrac{2}{3}x^{-3}$　　$\dfrac{dy}{dx} = \dfrac{2}{3}(-3x^{-4}) = -2x^{-4} = -\dfrac{2}{x^4}$

　c.　$y = \dfrac{1}{3}(2x)^3$　　$\dfrac{dy}{dx} = \dfrac{1}{3} \cdot 3(2x)^2 \cdot 2 = 2(2x)^2$

　d.　$y = 2(3x)^{-3}$

　　　$\dfrac{dy}{dx} = 2(-3)(3x)^{-4}(3) = -\dfrac{18}{(3x)^4}$

35.　$s = 27 - (3 - 10t)^3$

　　$s'(t) = 0 - 3(3 - 10t)^2(0 - 10) = 30(3 - 10t)^2$

　　$s'\left(\dfrac{1}{10}\right) = 30(3 - 1)^2 = 120\,\text{in/sec} = 10\,\text{ft/sec}$

37.　$R = 1500x + 3000(2x + 3)^{-1} - 1000$

　　$R'(x) = 1500 - 3000(2x + 3)^{-2}(2)$

　　$= 1500 - \dfrac{6000}{(2x + 3)^2}$

　　$R'(100) = 1500 - \dfrac{6000}{203^2} = \$1499.85$

So if the sales go from 100 units sold to 101 units sold, the revenue will increase by about $1499.85.

39.　$y = 32(3p + 1)^{-2/5}$

　　$y'(p) = 32\left(-\dfrac{2}{5}\right)(3p + 1)^{-7/5}(3)$

　　$= \dfrac{-192}{5(3p + 1)^{7/5}}$

　a.　$y'(21) = \dfrac{-192}{5(64)^{7/5}} \approx \dfrac{-192}{5(337.79)} \approx -0.114$

　b.　If the price increases $1, the sales volume will decrease by 114 units.

41.　$p = 200,000(q + 1)^{-2}$

　　$p'(q) = -400,000(q + 1)^{-3}\,(1)$

　b.　$p'(49) = \dfrac{-400,000}{50^3} = -\$3.20$

　c.　If the quantity demanded increases one unit, the price will decrease about $3.20.

43.　$y = k(x - x_0)^{8/5}$　k and x_0 are constants.

　　$\dfrac{dy}{dx} = k \cdot \dfrac{8}{5}(x - x_0)^{3/5} = \dfrac{8k}{5}(x - x_0)^{3/5}$

45.　$p = 100(2q + 1)^{-1/2}$

　　$p'(q) = -50(2q + 1)^{-3/2}(2)$

　　$= \dfrac{-100}{(2q + 1)^{3/2}}$

47.　$K_c = 4\sqrt{4v + 1}$

　　$K_c' = 4 \cdot \dfrac{1}{2}(4v + 1)^{-1/2} \cdot 4 = \dfrac{8}{\sqrt{4v + 1}}$

49. $S = 1000\left[1 + \dfrac{0.01r}{12}\right]^{240}$

$S'(r) = 240,000\left[1 + \dfrac{0.01r}{12}\right]^{239}\left(\dfrac{0.01}{12}\right) = 200\left[1 + \dfrac{0.01r}{12}\right]^{239}$

   a. $S'(6) = 200[1 + 0.005]^{239} = 200(3.2937) = \$658.75$

   b. $S'(12) = 200[1 + 0.01]^{239} = 200(10.7847) = \$2156.94$

In each case the future value of the investment would have an increase of the above values.

51. a. $d'(t) = -0.197(4)(0.1t + 3)^3(0.1) + 5.32(3)(0.1t + 3)^2(0.1) - 51.3(2)(0.1t + 3)(0.1) + 209.6(0.1)$

        $= -0.0788(0.1t + 3)^3 + 1.596(0.1t + 3)^2 - 10.26(0.1t + 3) + 20.96$

    For the year 1980, $t = 50$, thus $d'(50) \approx 0.68$. For the year 2000, $t = 70$, so $d'(70) \approx -0.84$.

    This model predicts that the percentage of federal expenditures devoted to the payment of interest on the public debt will change by about 0.60% from 1980 to 1981 and about −0.84% from 2000 to 2001.

   b. Take the average rate between 1975 to 1980: $\dfrac{12.7 - 9.8}{5} = 0.58\%$

53. a. $p'(t) = -0.001905(3)(t + 60)^2 + 0.4783(2)(t + 60) - 39.37$

        $= -0.005715(t + 60)^2 + 0.9566(t + 60) - 39.37$

    The year 1970 corresponds to $t = 10$, thus $p'(10) \approx -0.41$. The year 1998 corresponds to $t = 38$, thus $p'(38) \approx -0.51$. This model predicts the number of people in the U.S. who live below the poverty line decreased by 0.41 million between 1970 and 1971, and decreased by 0.51 million between 1998 and 1999.

   b. $\dfrac{33.6 - 25.9}{15} = 0.513$

   c. $p'(15) \approx 0.228$, $p'(20) \approx 0.582$, $p'(26) \approx 0.629$, $p'(30) \approx 0.433$, so 1980 has the closest average rate of change.

Exercise 9.7

1. $f(x) = \pi^4$ $f'(x) = 0$
Derivative of a constant.

9. $f(x) = \dfrac{x^3 + 1}{x^2} = \dfrac{x^3}{x^2} + \dfrac{1}{x^2} = x + x^{-2}$

    Preferred: $f'(x) = 1 - 2x^{-3} = 1 - \dfrac{2}{x^3} = \dfrac{x^3 - 2}{x^3}$

3. $g(x) = 4x^{-4}$ $g'(x) = 4(-4)x^{-5} = \dfrac{-16}{x^5}$

5. $g(x) = 5x^3 + 4x^{-1}$

    $g'(x) = 15x^2 - 4x^{-2} = 15x^2 - \dfrac{4}{x^2}$

    Acceptable: $f'(x) = \dfrac{x^2(3x^2) - (x^3 + 1)(2x)}{x^4}$

                   $= \dfrac{x^4 - 2x}{x^4} = \dfrac{x^3 - 2}{x^3}$

7. $y = (x^2 - 2)(x + 4)$

    $y' = (x^2 - 2)(1) + (x + 4)(2x) = 3x^2 + 8x - 2$

11. $y = \dfrac{1}{10}(x^3 - 4x)^{10}$

    $y' = \dfrac{1}{10} \cdot 10(x^3 - 4x)^9(3x^2 - 4)$

    $= (3x^2 - 4)(x^3 - 4x)^9$

13. $y = \dfrac{5}{3}x^3\left(4x^5 - 5\right)^3$

$y' = \dfrac{5}{3}x^3 \cdot 3\left(4x^5 - 5\right)^2\left(20x^4\right) + \left(4x^5 - 5\right)^3 \cdot \dfrac{5}{3} \cdot 3x^2$

$= \left(4x^5 - 5\right)^2\left[\left(5x^3\right)\left(20x^4\right) + 5x^2\left(4x^5 - 5\right)\right]$

$= 5x^2\left(4x^5 - 5\right)^2\left[20x^5 + \left(4x^5 - 5\right)\right]$

$= 5x^2\left(4x^5 - 5\right)^2\left(24x^5 - 5\right)$

15. $f(x) = (x-1)^2\left(x^2 + 1\right)$

$f'(x) = (x-1)^2(2x) + \left(x^2 + 1\right)(2)(x-1)(1)$

$= (x-1)\left[(x-1)2x + 2\left(x^2 + 1\right)\right]$

$= (x-1)\left(4x^2 - 2x + 2\right)$

$= 2(x-1)\left(2x^2 - x + 1\right)$

17. $y = \dfrac{\left(x^2 - 4\right)^3}{x^2 + 1}$

$y' = \dfrac{\left(x^2 + 1\right) \cdot 3\left(x^2 - 4\right)^2(2x) - \left(x^2 - 4\right)^3(2x)}{\left(x^2 + 1\right)^2} = \dfrac{2x\left(x^2 - 4\right)^2\left[3\left(x^2 + 1\right) - \left(x^2 - 4\right)\right]}{\left(x^2 + 1\right)^2} = \dfrac{2x\left(x^2 - 4\right)^2\left(2x^2 + 7\right)}{\left(x^2 + 1\right)^2}$

19. $p = (q+1)^3\left(q^3 - 3\right)^3$

$p' = (q+1)^3 \cdot 3\left(q^3 - 3\right)^2\left(3q^2\right) + \left(q^3 - 3\right)^3 \cdot 3(q+1)^2(1) = 3\left(q^3 - 3\right)^2\left[(q+1)^3\left(3q^2\right) + \left(q^3 - 3\right)(q+1)^2\right]$

$= 3\left(q^3 - 3\right)^2(q+1)^2\left[(q+1)\left(3q^2\right) + \left(q^3 - 3\right)\right] = 3\left(q^3 - 3\right)^2(q+1)^2\left(4q^3 + 3q^2 - 3\right)$

21. $R(x) = x^8\left(x^2 + 3x\right)^4$

$R'(x) = x^8 \cdot 4\left(x^2 + 3x\right)^3(2x+3) + \left(x^2 + 3x\right)^4 \cdot 8x^7 = 4\left(x^2 + 3x\right)^3\left[x^8(2x+3) + \left(x^2 + 3x\right)2x^7\right]$

$= 4x^7\left(x^2 + 3x\right)^3\left[x(2x+3) + 2\left(x^2 + 3x\right)\right] = 4x^7\left(x^2 + 3x\right)^3\left[4x^2 + 9x\right] = 4x^8\left(x^2 + 3x\right)^3(4x+9)$

23. $y = \dfrac{(2x-1)^4}{\left(x^2 + x\right)^4}$

$y' = \dfrac{\left(x^2 + x\right)^4 \cdot 4(2x-1)^3(2) - (2x-1)^4 \cdot 4\left(x^2 + x\right)^3(2x+1)}{\left(x^2 + x\right)^8}$

$= \dfrac{4\left(x^2 + x\right)^3\left[\left(x^2 + x\right)(2x-1)^3(2) - (2x-1)^4(2x+1)\right]}{\left(x^2 + x\right)^8} = \dfrac{4(2x-1)^3\left(-2x^2 + 2x + 1\right)}{\left(x^2 + x\right)^5}$

25. $g(x) = (8x^4 + 3)^2(x^3 - 4x)^3$

$g'(x) = (8x^4 + 3)^2 \cdot 3(x^3 - 4x)^2(3x^2 - 4) + (x^3 - 4x)^3 \cdot 2(8x^4 + 3)(32x^3)$

$= (x^3 - 4x)^2[3(8x^4 + 3)^2(3x^2 - 4) + 2(x^3 - 4x)(8x^4 + 3)(32x^3)]$

$= (x^3 - 4x)^2(8x^4 + 3)[3(8x^4 + 3)(3x^2 - 4) + 64x^3(x^3 - 4x)]$

$= (x^3 - 4x)^2(8x^4 + 3)(136x^6 - 352x^4 + 27x^2 - 36)$

27. $f(x) = \dfrac{(x^2 + 5)^{1/3}}{4 - x^2}$

$f'(x) = \dfrac{(4 - x^2) \cdot \frac{1}{3}(x^2 + 5)^{-2/3}(2x) - (x^2 + 5)^{1/3}(-2x)}{(4 - x^2)^2}$

$= \dfrac{2[x(4 - x^2) - (x^2 + 5)(-3x)]}{3(x^2 + 5)^{2/3}(4 - x^2)^2} = \dfrac{2(2x^3 + 19x)}{3(x^2 + 5)^{2/3}(4 - x^2)^2} = \dfrac{2x(2x^2 + 19)}{3(x^2 + 5)^{2/3}(4 - x^2)^2}$

29. $y = x^2(4x-3)^{1/4}$

$$y' = x^2 \cdot \frac{1}{4}(4x-3)^{-3/4}(4) + (4x-3)^{1/4}(2x) = \frac{x^2}{(4x-3)^{3/4}} + \frac{2x(4x-3)^{1/4}}{1} = \frac{x^2 + 2x(4x-3)}{(4x-3)^{3/4}} = \frac{9x^2 - 6x}{(4x-3)^{3/4}}$$

31. $c(x) = 2x(x^3+1)^{1/2}$

$$c'(x) = (2x)\cdot\frac{1}{2}(x^3+1)^{-1/2}(3x^2) + (x^3+1)^{1/2}\cdot(2)$$

$$= \frac{3x^3}{(x^3+1)^{1/2}} + \frac{2(x^3+1)^{1/2}}{1} = \frac{3x^3 + 2(x^3+1)}{(x^3+1)^{1/2}} = \frac{5x^3+2}{(x^3+1)^{1/2}}$$

33. a. $F_1(x) = \frac{3}{5}(x^4+1)^5$ $\quad F_1'(x) = \frac{3}{5}\cdot 5(x^4+1)^4(4x^3) = 12x^3(x^4+1)^4$

b. $F_2(x) = \frac{3}{5}(x^4+1)^{-5}$ $\quad F_2'(x) = \frac{3}{5}\cdot(-5)(x^4+1)^{-6}(4x^3) = \frac{-12x^3}{(x^4+1)^6}$

c. $F_3(x) = \frac{1}{5}(3x^4+1)^5$ $\quad F_3'(x) = \frac{1}{5}\cdot 5(3x^4+1)^4(12x^3) = 12x^3(3x^4+1)^4$

d. $F_4(x) = 3\left(5x^4+1\right)^{-5}$ $\quad F_4'(x) = 3\cdot(-5)\left(5x^4+1\right)^{-6}\left(20x^3\right) = \frac{-300x^3}{\left(5x^4+1\right)^6}$

35. $P = 10(3x+1)^3 - 10$ $\quad \frac{dP}{dx} = 30(3x+1)^2(3) = 90(3x+1)^2$

37. $R(x) = 60,000x + 40,000(10+x)^{-1} - 4000$ $\quad R'(x) = 60,000 - 40,000(10+x)^{-2}(1)$

a. $R'(10) = 60,000 - \frac{40,000}{20^2} = \$59,900$

b. The revenue is increasing.

39. $C(y) = 2(y+1)^{1/2} + 0.4y + 4$ $\quad C'(y) = 2\cdot\frac{1}{2}(y+1)^{-1/2}(1) + 0.4 = \frac{1}{(y+1)^{1/2}} + 0.4$

41. $V = x(12-2x)^2$

$$V'(x) = x\cdot 2(12-2x)(0-2) + (12-2x)^2\cdot 1 = (12-2x)[-4x + (12-2x)]$$

$$= (12-2x)(12-6x) = 144 - 96x + 12x^2$$

43. $S(t) = \dfrac{200t}{(t+1)^2}$

$$S'(t) = \frac{(t+1)^2(200) - 200t\cdot 2(t+1)\cdot 1}{(t+1)^4} = \frac{200(t+1)[(t+1)-2t]}{(t+1)^4} = \frac{200(1-t)}{(t+1)^3} \quad S'(9) = \frac{200(-8)}{10^3} = -1.6$$

From the 9th to the 10th week, sales will decrease approximately $1.60 thousand or $1,600.

45. a. $f(t) = \dfrac{1000\left[-7.4812(t+20)+1560.2\right]}{1.2882(t+20)^2 -122.18(t+20)-21,483}$

Let $d(t)$ be the denominator of $f(t)$.

$f'(t) = 1000\left[\dfrac{d(t)(-7.4812)-\left(-7.4812(t+20)+1560.2\right)d'(t)}{\left[d(t)\right]^2}\right]$

$= 1000\left[\dfrac{\begin{array}{l}-9.6373(t+20)^2 +914.053(t+20)-160718\\ \quad +\left(7.4812(t+20)-1560.2\right)\left(2.5764(t+20)-122.18\right)\end{array}}{\left[d(t)\right]^2}\right]$

$= 1000\left[\dfrac{\begin{array}{l}-9.6373(t+20)^2 +914.053(t+20)-160718+19.275(t+20)^2\\ \quad -914.053(t+20)-4019.699(t+20)+190625\end{array}}{\left[d(t)\right]^2}\right]$

$= 1000\left[\dfrac{9.6377(t+20)^2 -4019.699(t+20)+29,907}{\left(1.2882(t+20)^2 -122.18(t+20)-21,483\right)^2}\right]$

b. The year 1850 corresponds to $t=30$, thus $f'(30)\approx -0.425$. The year 1950 corresponds to $t=130$, thus $f'(130)\approx -0.345$.

c. $f'(30)$ means that from 1850 to 1851, the model predicts a change of –0.425% in U.S. workers in farm occupations. $f'(130)$ means that from 1950 to 1951, the model predicts a change of –0.345% in U.S. workers in farm occupations.

Exercise 9.8

1. $y = 10x^3 -x^2 +14x+3$
$y' = 30x^2 -2x+14$
$y'' = 60x-2$

3. $g(x) = x^3 -x^{-1}$
$g'(x) = 3x^2 +x^{-2}$
$g''(x) = 6x-2x^{-3} = 6x-\dfrac{2}{x^3}$

5. $y = x^3 -x^{1/2}$
$y' = 3x^2 -\dfrac{1}{2}x^{-1/2}$
$y'' = 6x+\dfrac{1}{4}x^{-3/2} = 6x+\dfrac{1}{4x^{3/2}}$

7. $y = x^5 -16x^3 +12$
$y' = 5x^4 -48x^2$
$y'' = 20x^3 -96x$
$y''' = 60x^2 -96$

9. $f(x) = 2x^9 -6x^6$
$f'(x) = 18x^8 -36x^5$
$f''(x) = 144x^7 -180x^4$
$f'''(x) = 1008x^6 -720x^3$

11. $y = x^{-1}$
$y' = -1x^{-2}$
$y'' = 2x^{-3}$
$y''' = -6x^{-4}$

13. $y = x^5 - x^{1/2}$

$$\frac{dy}{dx} = 5x^4 - \frac{1}{2}x^{-1/2}$$

$$\frac{d^2y}{dx^2} = 20x^3 + \frac{1}{4}x^{-3/2}$$

15. $f(x) = (x+1)^{1/2}$

$$f'(x) = \frac{1}{2}(x+1)^{-1/2}(1)$$

$$f''(x) = -\frac{1}{4}(x+1)^{-3/2}(1)$$

$$f'''(x) = \frac{3}{8}(x+1)^{-5/2}(1) = \frac{3}{8}(x+1)^{-5/2}$$

17. $y = 4x^3 - 16x$

$y' = 12x^2 - 16$

$y'' = 24x$

$y''' = 24$

$y^{(4)} = 0$

19. $f(x) = x^{1/2}$

From problem 15 we have $f^{(3)}(x) = \frac{3}{8}x^{-5/2}$.

$$f^{(4)}(x) = -\frac{15}{16}x^{-7/2}$$

21. $y' = (4x-1)^{1/2}$

$$y'' = \frac{1}{2}(4x-1)^{-1/2}(4) = 2(4x-1)^{-1/2}$$

$$y''' = -1(x-1)^{-3/2}(4) = -4(4x-1)^{-3/2}$$

$$y^{(4)} = \frac{12}{2}(4x-1)^{-5/2}(4) = 24(4x-1)^{-5/2}$$

23. $f^{(4)}(x) = \frac{x}{x+1}$

$$f^{(5)}(x) = \frac{(x+1)(1) - x(1)}{(x+1)^2} = (x+1)^{-2}$$

$$f^{(6)}(x) = -2(x+1)^{-3}(1) = -\frac{2}{(x+1)^3}$$

25. $f(x) = 16x^2 - x^3$

$f'(x) = 32x - 3x^2$

$f''(x) = 32 - 6x$

At $x = 1$, $f''(1) = 32 - 6 = 26$.

27. $f''(3) \approx nDeriv(nDeriv(x^3 - 27/x, x, x,), x, 3) = 15.99999925$

29. $f''(21) \approx nDeriv(nDeriv(\sqrt{(x^2+4)}, x, x,), x, 21) = 0.000426$

31. $f(x) = x^3 - 3x^2 + 5$

   a. $f'(x) = 3x^2 - 6x$

       $f''(x) = 6x - 6$

   b.

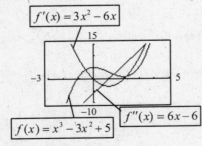

   c. $f''(x) = 0$ at $x = 1$.

       $f''(x) > 0$ if $x > 1$;

       $f''(x) < 0$ if $x < 1$.

   d. $f'(x)$ [NOT $f(x)$] min at $x = 1$.

       $f'(x)$ increasing if $x > 1$. ($f''(x) > 0$)

       $f'(x)$ decreasing if $x < 1$. ($f''(x) < 0$)

   e. $f''(x) < 0$

   f. $f''(x) > 0$ [For now look at graphs.]

33. $f(x) = -\frac{1}{3}x^3 - x^2 + 3x + 7$

   a. $f'(x) = -x^2 - 2x + 3$

      $f''(x) = -2x - 2$

   b.

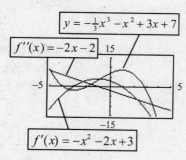

$y = -\frac{1}{3}x^3 - x^2 + 3x + 7$

$f''(x) = -2x - 2$

$f'(x) = -x^2 - 2x + 3$

   c. $f''(x) = 0$ at $x = -1$.

      $f''(x) > 0$ if $x < -1$;

      $f''(x) < 0$ if $x > -1$.

   d. $f'(x)$ [NOT $f(x)$] max at $x = -1$.

      $f'(x)$ increasing if $x < -1$. $(f''(x) > 0)$

      $f'(x)$ decreasing if $x > -1$. $(f''(x) < 0)$

   e. $f''(x) < 0$

   f. $f''(x) > 0$

      [For now look at graphs.]

35. $s(t) = 100 + 10t + 0.01t^3$ distance

     $s'(t) = 10 + 0.03t^2$ velocity

     $s''(t) = 0.06t$ acceleration

     $s''(2) = 0.12$

37. $R(x) = 100x - 0.01x^2$

     $R'(x) = 100 - 0.02x$

     $R''(x) = -0.02$

39. $R = m^2\left(\frac{c}{2} - \frac{m}{3}\right) = \frac{c}{2}m^2 - \frac{1}{3}m^3$

   a. $R' = cm - m^2$

   b. $R'' = c - 2m$

   c. Second derivative

41. $R(x) = 15x + 30(4x + 1)^{-1} - 30$

   $\overline{MR} = R'(x) = 15 - 30(4x + 1)^{-2}(4)$

              $= 15 - 120(4x + 1)^{-2}$

   $\overline{MR}' = R''(x) = 0 + 240(4x + 1)^{-3}(4) = \dfrac{960}{(4x + 1)^3}$

   a. $R''(25) = \dfrac{960}{101^3} \approx 0.0009$

   b. When the next unit is sold the marginal revenue will increase about \$0.90 per unit.

43. $S(t) = 1 + 3(t + 3)^{-1} - 18(t + 3)^{-2}$

   a. $S'(t) = -3(t + 3)^{-2} + 36(t + 3)^{-3}$

         $= \dfrac{-3}{(t + 3)^2} + \dfrac{36}{(t + 3)^3}$

   b. $S''(t) = 6(t + 3)^{-3} - 108(t + 3)^{-4}$

         $= \dfrac{6}{(t + 3)^3} - \dfrac{108}{(t + 3)^4}$

      $S''(15) = \dfrac{6}{18^3} - \dfrac{108}{18^4} = 0$

   c. The rate of change of the rate of change in sales (after 15 weeks) is zero.

45. $p(t) = -0.0022605t^3 + 0.154805t^2 - 2.78785t + 41.038$

   a. $p'(t) = -0.0067815t^2 + 0.309610t - 2.78785$

   b. $p''(t) = -0.0135630t + 0.309610$

      1980: $p''(20) = 0.0384$ 1998: $p''(38) = -0.2058$

   c. $p'(20) = 0.6918$ means that in the next year (1981), the number of people who lived below the poverty level was expected to increase by about 691,800 people.

      $p''(20) = 0.0384$ means that in the next year (1981), the rate of change of the number of people who lived below the poverty level was expected to increase by about 38,400 people. From 1980 to 1981 the number of people who lived below the poverty level was increasing at an increasing rate.

47. a. $p(t) = 2586t^{0.4077}$

b. $p'(t) = 2586(0.4077)t^{-0.5923} = 1054.3t^{-0.5923}$

c. $p''(t) = 1054.3(-0.5923)t^{-1.5923} = -624.46t^{-1.5923}$

The year 1995 corresponds to $t = 15$, thus $p''(15) \approx -8.377$. The year 2002 corresponds to $t = 22$, thus $p''(22) \approx -4.550$.

d. $p'(22) \approx 168.98$. The model predicts that in 2002 the poverty threshold income for a single person was changing at the rate of \$168.98 per year (the level was increasing). Also, in 2002, the rate of change of the poverty threshold income was changing at the rate of -4.5500 dollars per year per year (the rate was decreasing). Thus, in 2002, $p(t)$ was increasing at a decreasing rate.

Exercise 9.9

1. a. $\overline{MR} = R'(x) = 4$

b.   Each unit sold brings in \$4 revenue.

3. $R(x) = 36x - 0.01x^2$

a. $R(100) = 3600 - 0.01(10,000)$

$R(100) = \$3500$

100 units produce \$3500 of revenue.

b. $\overline{MR} = R'(x) = 36 - 0.02x$

c. $R'(100) = 36 - 2 = 34$

The sale of the next unit will increase the revenue by about \$34. The sale of the next 3 units will increase the revenue by about \$102.

d. $R(101) - R(100) = 3533.99 - 3500$

$\qquad\qquad\qquad = \$33.99$

(Actual revenue from 101st unit.)

5. a. $R(x) = px = (80 - 0.4x)x = 80x - 0.4x^2$

(in hundreds of dollars)

b. $50 = 80 - 0.4x$

$0.4x = 80 - 50 = 30$

$x = \dfrac{30}{0.4} = 75$ hundreds

(or 7500 subscribers)

$R = 50 \cdot 7500 = \$375,000$

c.   More subscribers are attracted by lower prices.

d. $R'(x) = 80 - 0.8x$

$p = 50 \;\rightarrow\; x = 75$

$R'(75) = 80 - 0.8(75) = \$20$

If the number of subscribers increases from 75 to 76 (hundred) the revenue increases $\$20(100) = \$2,000$. Increasing the number of subscribers will occur by lowering the monthly charges.

7. a. $\overline{MR} = 36 - 0.02x$

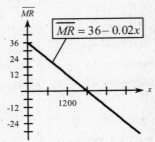

b.   Maximum revenue at $x = 1800$ $(\overline{MR} = 0)$

c. $R(1800) = \$32,400$

9. $C(x) = 40 + 8x;\; \overline{MC} = C'(x) = 8$

11. $C(x) = 500 + 13x + x^2;\; \overline{MC} = C'(x) = 13 + 2x$

13. $C = x^3 - 6x^2 + 24x + 10$

$\overline{MC} = C' = 3x^2 - 12x + 24$

15. $C = 400 + 27x + x^3;\; \overline{MC} = C' = 27 + 3x^2$

17. a. $C(x) = 40 + x^2$

$\overline{MC} = C'(x) = 2x$

$C'(5) = 2(5) = 10$

The cost to produce the 6th unit is predicted to be \$10.

b. $C(6) - C(5) = 76 - 65 = \$11$

19. $C(x) = x^3 - 4x^2 + 30x + 20$

$\overline{MC} = C'(x) = 3x^2 - 8x + 30$

$C'(4) = 48 - 32 + 30 = 46$

The cost of producing 1 additional unit will increase by \$46. The cost of producing 3 additional units will increase by $3 \times \$46 = \$138$.

21. $C(x) = 300 + 4x + x^2$ $\overline{MC} = C'(x) = 4 + 2x$

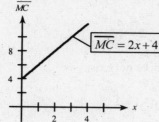

$\boxed{\overline{MC} = 2x + 4}$

23. a. Remember that the slope gives the cost of the next unit. $C'(100)$ is greater than $C'(500)$. Thus the 101st unit will cost more than the 501st unit.

   b. Since the slope is decreasing, the cost of each additional unit is less than the previous unit. The manufacturing process is more efficient.

25. $P(x) = 5x - 25$, $\overline{MP} = P'(x) = 5$

The next unit sold earns a $5 profit.

27. $P(x) = R(x) - C(x) = 30x - x^2 - 200$

   a. $P(20) = 600 - 400 - 200 = 0$

   b. $\overline{MP} = P'(x) = 30 - 2x$

   c. $P'(20) = 30 - 40 = -10$

   The total profit will decrease by approximately $10 on the sale of the next (21st) unit.

   d. $P(21) - P(20) = (-11) - (0) = -\$11$ (The actual loss on the sale of the 21st unit.)

29. a. From $P = R - C$ we obtain the following profits in order from smallest to largest: 100, 700, and 400 units. There is a loss at 100 units since the cost curve is above the revenue curve.

   b. $MP = MR - MC$ Look at the slopes of each curve at $x = 100$, 400, and 700. Ranking from high to low:

   $\overline{MP}(100) > \overline{MP}(400) > \overline{MP}(700)$.

   $\overline{MP}(700) < 0$ since $R' - C' < 0$.

31. a. From low to high: $A < B < C$.
   The graph shows a loss at A.

   b. $P' > 0$ at each point.
   Evaluate the slope at each point to obtain $P'(C) < P'(B) < P'(A)$.

33. $\overline{MP} = 30 - 2x$

   a.

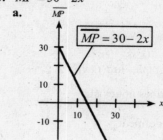

$\boxed{\overline{MP} = 30 - 2x}$

   b. $\overline{MP} = 0$ if $30 - 2x = 0$. Thus, $x = 15$.

   c. Profit is a maximum at the vertex of the parabola $P(x) = 30x - x^2 - 200$.

   Vertex is at $x = \dfrac{-b}{2a} = \dfrac{-30}{2(-1)} = 15$.

   d. $P(15) = 450 - 225 - 200 = \$25$

35. $P(x) = R(x) - C(x) = 300x - (160 + x)x$
$$= 140x - x^2$$

$P'(x) = 140 - 2x$

$P'(x) = 0$ at $140 - 2x = 0$ or $x = 70$.

Note that cost is per unit. Total cost is the cost per unit times the number of units.
Production of 70 units will maximize profit.

37. $P(x) = R(x) - C(x) = 50x - (10 + 2x)x$
$$= 40x - 2x^2$$

$P'(x) = 40 - 4x$

$P'(x) = 0$ at $x = 10$.

Maximum profit $= P(10) = 400 - 200 = \$200$

Review Exercises

1. **a.** $f(-2) = 2$
   b. $\lim\limits_{x \to -2} f(x) = 2$

2. **a.** From the graph, $f(-1) = 0$.
   b. $\lim\limits_{x \to -1} f(x) = 0$

3. **a.** $f(4) = 2$
   b. $\lim\limits_{x \to 4^-} f(x) = 1$

4. **a.** From the graph, $\lim\limits_{x \to 4^+} f(x) = 2$.
   b. $\lim\limits_{x \to 4} f(x)$ does not exist.

5. **a.** $f(1)$ is not defined
   b. $\lim\limits_{x \to 1} f(x) = 2$

6. **a.** From the graph, $f(2)$ does not exist.
   b. $\lim\limits_{x \to 2} f(x)$ does not exist.

7. $\lim\limits_{x \to 4}(3x^2 + x + 3) = 48 + 4 + 3 = 55$

8. $\lim\limits_{x \to 4} \dfrac{x^2 - 16}{x + 4} = \dfrac{4^2 - 16}{4 + 4} = 0$

9. $\lim\limits_{x \to -1} \dfrac{x^2 - 1}{x + 1} = \lim\limits_{x \to -1} \dfrac{(x+1)(x-1)}{x+1}$
$= \lim\limits_{x \to -1}(x - 1) = -2$

10. $\lim\limits_{x \to 3} \dfrac{x^2 - 9}{x - 3} = \lim\limits_{x \to 3} \dfrac{(x-3)(x+3)}{x-3}$
$= \lim\limits_{x \to 3}(x + 3) = 3 + 3 = 6$

11. $\lim\limits_{x \to 2} \dfrac{4x^3 - 8x^2}{4x^3 - 16x} = \lim\limits_{x \to 2} \dfrac{4x^2(x-2)}{4x(x-2)(x+2)}$
$= \lim\limits_{x \to 2} \dfrac{x}{x+2} = \dfrac{2}{4} = \dfrac{1}{2}$

12. $\lim\limits_{x \to -\frac{1}{2}} \dfrac{x^2 - \frac{1}{4}}{6x^2 + x - 1} = \lim\limits_{x \to -\frac{1}{2}} \dfrac{\left(x + \frac{1}{2}\right)\left(x - \frac{1}{2}\right)}{(3x-1)(2)\left(x + \frac{1}{2}\right)}$
$= \lim\limits_{x \to -\frac{1}{2}} \dfrac{x - \frac{1}{2}}{2(3x-1)} = \dfrac{-1}{-5} = \dfrac{1}{5}$

13. $\lim\limits_{x \to 3} \dfrac{x^2 - 16}{x - 3} = \dfrac{-7}{0}$
Thus, the limit does not exist.

14. $\lim\limits_{x \to -3} \dfrac{x^2 - 9}{x - 3} = \dfrac{(-3)^2 - 9}{-3 - 3} = \dfrac{0}{-6} = 0$

15. $\lim\limits_{x \to 1} \dfrac{x^2 - 9}{x - 3} = \dfrac{-8}{-2} = 4$

16. $\lim\limits_{x \to 2} \dfrac{x^2 - 8}{x - 2}$ does not exist.

17. $\lim\limits_{x \to 1^-} f(x) = 4 - 1 = 3$
$\lim\limits_{x \to 1^+} f(x) = 2(1) + 1 = 3$
Thus, $\lim\limits_{x \to 1} f(x) = 3$

18. $\lim\limits_{x \to -2^-} f(x) = -8 + 2 = -6$
$\lim\limits_{x \to -2^+} f(x) = 2 - 4 = -2$
Since these two limits are not equal, $\lim\limits_{x \to -2} f(x)$
does not exist.

19. $\lim\limits_{h \to 0} \dfrac{3(x+h)^2 - 3x^2}{h}$
$= \lim\limits_{h \to 0} \dfrac{3x^2 + 6xh + 3h^2 - 3x^2}{h}$
$= \lim\limits_{h \to 0} \dfrac{6xh + 3h^2}{h} = \lim\limits_{h \to 0}(6x + 3h) = 6x$

20. $\lim\limits_{h \to 0} \dfrac{[(x+h) - 2(x+h)^2] - (x - 2x^2)}{h}$
$= \lim\limits_{h \to 0} \dfrac{x + h - 2x^2 - 4xh - 2h^2 - x + 2x^2}{h}$
$= \lim\limits_{h \to 0} \dfrac{h - 4xh - 2h^2}{h}$
$= \lim\limits_{h \to 0}(1 - 4x - 2h) = 1 - 4x$

21. $\lim\limits_{x \to 2} \dfrac{(x+12)(x-2)}{(x-3)(x-2)}$
$= \lim\limits_{x \to 2} \dfrac{x + 12}{x - 3} = \dfrac{14}{-1} = -14$

22. $\lim\limits_{x \to -\frac{1}{2}} \dfrac{\left(x + \frac{1}{2}\right)\left(x - \frac{1}{3}\right)}{\left(x + \frac{1}{2}\right)\left(x + \frac{1}{3}\right)}$
$= \lim\limits_{x \to -\frac{1}{2}} \dfrac{x - \frac{1}{3}}{x + \frac{1}{3}} = \dfrac{-\frac{1}{2} - \frac{1}{3}}{-\frac{1}{2} + \frac{1}{3}} = 5$

23. **a.** From the graph, $f(-1) = 0 = \lim\limits_{x \to -1} f(x)$, so
     $f(x)$ is continuous at $x = -1$.
   b. From the graph, $f(1)$ does not exist, so
     $f(x)$ is not continuous at $x = 1$.

24. a. From the graph, $f(-2) = 2 = \lim_{x \to -2} f(x)$, so

$f(x)$ is continuous at $x = -2$.

b. From the graph, $f(2)$ and $\lim_{x \to 2} f(x)$ do not

exist, so $f(x)$ is not continuous at $x = 2$.

25. $\lim_{x \to -1^+} f(x) = (-1)^2 + 1 = 2$

$\lim_{x \to -1^-} f(x) = (-1)^2 + 1 = 2$

Thus, $\lim_{x \to -1} f(x) = 2$

26. $\lim_{x \to 0^-} f(x) = 0^2 + 1 = 1$

$\lim_{x \to 0^+} f(x) = 0$

Thus, $\lim_{x \to 0} f(x)$ does not exist.

27. $\lim_{x \to 1^+} f(x) = 2(1)^2 - 1 = 1$

$\lim_{x \to 1^-} f(x) = 1$

Thus, $\lim_{x \to 1} f(x) = 1$

28. $f(x)$ is not continuous at $x = 0$ since the $\lim_{x \to 0} f(x)$

does not exist.

29. $f(1) = 2(1)^2 - 1 = 1 = \lim_{x \to 1} f(x)$

Yes, $f(x)$ is continuous at $x = 1$.

30. Since $f(-1) = 2 = \lim_{x \to -1} f(x)$, $f(x)$ is continuous at

$x = -1$.

31. Function is discontinuous at $x = 5$.
(Function is not defined.)

32. $f(x)$ is not continuous at $x = 2$ since $f(2)$ does

not exist.

33. $f(2) = 4$

$\lim_{x \to 2} f(x) = 4$

Function is continuous everywhere.

34. $f(x)$ is not continuous at $x = 1$ since $\lim_{x \to 1} f(x)$

does not exist.

35. a. $f(x)$ is discontinuous at $x = 0$ and $x = 1$.

b. $\lim_{x \to +\infty} f(x) = 0$

c. $\lim_{x \to -\infty} f(x) = 0$

36. a. From the graph, $f(x)$ is discontinuous at $x = 0$ and $x = -1$.

b. $\lim_{x \to +\infty} f(x) = \frac{1}{2}$

c. $\lim_{x \to -\infty} f(x) = \frac{1}{2}$

37. $\lim_{x \to -\infty} \frac{2x^2}{1 - x^2} = \lim_{x \to -\infty} \frac{2}{\frac{1}{x^2} - 1} = \frac{2}{0 - 1} = -2$

38. $\lim_{x \to +\infty} \frac{3x^{2/3}}{x + 1} = \lim_{x \to +\infty} \frac{\frac{3}{x^{1/3}}}{1 + \frac{1}{x}} = \frac{0}{1} = 0$

39. $\text{Avg} = \frac{f(2) - f(-1)}{2 - (-1)} = \frac{33 - 12}{3} = 7$

40. True.

41. False. The given expression gives the slope of the tangent line at $x = c$.

42. $f(x) = 3x^2 + 2x - 1$

$f'(x) = \lim_{h \to 0} \frac{[3(x + h)^2 + 2(x + h) - 1] - (3x^2 + 2x - 1)}{h} = \lim_{h \to 0} \frac{6xh + 3h^2 + 2h}{h} = \lim_{h \to 0} (6x + 3h + 2) = 6x + 2$

43. $f'(x) = \lim_{h \to 0} \frac{x + h - (x + h)^2 - (x - x^2)}{h} = \lim_{h \to 0} \frac{h - 2xh - h^2}{h} = \lim_{h \to 0} (1 - 2x - h) = 1 - 2x$

44. $[-3, 0]$: $\text{Avg} = \frac{f(0) - f(-3)}{0 - (-3)} = \frac{1 - 0}{3} = \frac{1}{3}$

$[-1, 0]$: $\text{Avg} = \frac{f(0) - f(-1)}{0 - (-1)} = \frac{1 - 0}{1} = 1$

From $[-1, 0]$ the change is greater.

45. a. No   **b.** No

46. a. From the graph, $f(x)$ is differentiable at $x = -2$.

b. $f(x)$ is not differentiable at $x = 2$.

47. a. $nDeriv\left((4x)^{1/3}/(3x^2-10)^2, x, 2\right)$

$= -5.917062766$

b. $\dfrac{f(2+.0001)-f(2)}{0.0001}$

$= \dfrac{0.49941-0.5}{0.0001} = -5.91$

48. a. $Avg = \dfrac{g(5)-g(2)}{5-2} = \dfrac{18.1-13.2}{3} = 1.633$

b. $g'(4) = \dfrac{g(4.3)-g(4)}{4.3-4} = \dfrac{2.1}{0.3} = 7$

49. Approximate the tangent through $(0,2)$ and

$(8,0)$. $f'(4) \approx \dfrac{0-2}{8-0} = -\dfrac{1}{4}$

50. $A: f'(2) \approx 0$

$B: f'(6) \approx \dfrac{-1-1}{8-4} = -\dfrac{1}{2}$

$C: Avg = \dfrac{-1-1.2}{10-2} = -0.275$

Rank: $B < C < A$

51. $c = 4x^5 - 6x^3$

$c' = 4(5x^4) - 6(3x^2) = 20x^4 - 18x^2$

52. $f(x) = 4x^2 - 1$ $f'(x) = 8x$

53. $p = 3q + \sqrt{7}$ $\dfrac{dp}{dq} = 3 + 0 = 3$

54. $y = \sqrt{x} = x^{1/2}$ $y' = \dfrac{1}{2}x^{-1/2} = \dfrac{1}{2\sqrt{x}}$

55. $f(z) = 2^{4/3}$ $f'(z) = 0$

56. $v(x) = \dfrac{4}{\sqrt[3]{x}} = 4x^{-1/3}$

$v'(x) = -\dfrac{4}{3}x^{-4/3} = -\dfrac{4}{3\sqrt[3]{x^4}}$

57. $y = x^{-1} - x^{-1/2}$

$y' = -1(x^{-2}) - \left(-\dfrac{1}{2}x^{-3/2}\right) = -\dfrac{1}{x^2} + \dfrac{1}{2x^{3/2}}$

58. $f(x) = \dfrac{3}{2x^2} - \sqrt[3]{x} + 4^5 = \dfrac{3}{2}x^{-2} - x^{1/3} + 4^5$

$f'(x) = -3x^{-3} - \dfrac{1}{3}x^{-2/3} = -\dfrac{3}{x^3} - \dfrac{1}{3\sqrt[3]{x^2}}$

59. $f(x) = 3x^5 - 6$ $f'(x) = 15x^4$

$f(1) = 3-6 = -3$ $f'(1) = 15$

So, $y - (-3) = 15(x-1)$ or $y = 15x - 18$.

60. $f(x) = 3x^3 - 2x$ $f(2) = 20$

$f'(x) = 9x^2 - 2$ $f'(2) = 34$

So, $y - 20 = 34(x-2)$ or $y = 34x - 48$.

61. $f'(x) = 3x^2 - 6x = 3x(x-2)$

a. $f'(x) = 0$ if $x = 0, 2$.

b. $f(0) = 0-0+1 = 1$ Point: $(0, 1)$

$f(2) = 8-12+1 = -3$ Point: $(2, -3)$

c.

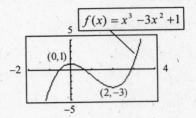

62. $f(x) = x^6 - 6x^4 + 8$

$f'(x) = 6x^5 - 24x^3 = 6x^3(x^2-4)$

$= 6x^3(x+2)(x-2)$

a. $f'(x) = 0$ when $x = 0$, $x = -2$, and when $x = 2$.

b. The associated points where the slope equals zero are $(0, 8)$, $(-2, -24)$ and $(2, -24)$.

c.

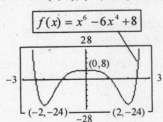

63. $f(x) = (3x-1)(x^2-4x)$

$f'(x) = (3x-1)(2x-4) + (x^2-4x)(3)$

$= 9x^2 - 26x + 4$

64. $y = (x^2+1)(3x^3+1)$

$y' = (x^2+1)(9x^2) + (3x^3+1)(2x)$

$= 15x^4 + 9x^2 + 2x$

65. $p = \dfrac{2q-1}{q^2}$

$\dfrac{dp}{dq} = \dfrac{q^2(2)-(2q-1)2q}{q^4} = \dfrac{-2q^2+2q}{q^4} = \dfrac{2-2q}{q^3}$

66. $s = \dfrac{\sqrt{t}}{3t+1} = \dfrac{t^{1/2}}{3t+1}$ $\dfrac{ds}{dt} = \dfrac{(3t+1)\cdot\frac{1}{2}t^{-1/2} - t^{1/2}\cdot 3}{(3t+1)^2} = \dfrac{\frac{1}{2}t^{-1/2}(3t+1-6t)}{(3t+1)^2} = \dfrac{1-3t}{2\sqrt{t}(3t+1)^2}$

67. $y = \sqrt{x}(3x+2) = 3x^{3/2} + 2x^{1/2}$ $\dfrac{dy}{dx} = \dfrac{9}{2}x^{1/2} + x^{-1/2} = \dfrac{9x+2}{2\sqrt{x}}$

68. $C = \dfrac{5x^4 - 2x^2 + 1}{x^3 + 1}$

$\dfrac{dC}{dx} = \dfrac{(x^3+1)(20x^3 - 4x) - (5x^4 - 2x^2 + 1)(3x^2)}{(x^3+1)^2}$

$= \dfrac{20x^6 - 4x^4 + 20x^3 - 4x - 15x^6 + 6x^4 - 3x^2}{(x^3+1)^2} = \dfrac{5x^6 + 2x^4 + 20x^3 - 3x^2 - 4x}{(x^3+1)^2}$

69. $y = \left(x^3 - 4x^2\right)^3$ $y' = 3\left(x^3 - 4x^2\right)^2\left(3x^2 - 8x\right) = \left(x^3 - 4x^2\right)^2\left(9x^2 - 24x\right)$

70. $y = \left(5x^6 + 6x^4 + 5\right)^6$ $y' = 6\left(5x^6 + 6x^4 + 5\right)^5\left(30x^5 + 24x^3\right)$

71. $y = \left(2x^4 - 9\right)^9$ $\dfrac{dy}{dx} = 9\left(2x^4 - 9\right)^8 \cdot 8x^3 = 72x^3\left(2x^4 - 9\right)^8$

72. $g(x) = \dfrac{1}{\sqrt{x^3 - 4x}} = \left(x^3 - 4x\right)^{-1/2}$ $g'(x) = -\dfrac{1}{2}\left(x^3 - 4x\right)^{-3/2}\left(3x^2 - 4\right) = -\dfrac{3x^2 - 4}{2\sqrt{\left(x^3 - 4x\right)^3}}$

73. $f(x) = x^2\left(2x^4 + 5\right)^8$

$f'(x) = x^2 \cdot 8\left(2x^4 + 5\right)^7\left(8x^3\right) + \left(2x^4 + 5\right)^8(2x) = 2x\left(2x^4 + 5\right)^7\left[32x^4 + \left(2x^4 + 5\right)\right] = 2x\left(2x^4 + 5\right)^7\left(34x^4 + 5\right)$

74. $S = \dfrac{(3x+1)^2}{x^2 - 4}$

$S' = \dfrac{(x^2 - 4)(2)(3x+1)(3) - (3x+1)^2(2x)}{(x^2 - 4)^2} = \dfrac{(3x+1)[(6x^2 - 24) - (6x^2 + 2x)]}{(x^2 - 4)^2} = -\dfrac{2(3x+1)(x+12)}{(x^2 - 4)^2}$

75. $y = (3x+1)^{12}(2x^3 - 1)^{12}$

$\dfrac{dy}{dx} = (3x+1)^{12} \cdot 12\left(2x^3 - 1\right)^{11}\left(6x^2\right) + \left(2x^3 - 1\right)^{12} \cdot 12(3x+1)^{11}(3)$

$= 12\left(2x^3 - 1\right)^{11}\left[6x^2(3x+1)^{12} + 3\left(2x^3 - 1\right)(3x+1)^{11}\right]$

$= 36\left(2x^3 - 1\right)^{11}(3x+1)^{11}\left[2x^2(3x+1) + \left(2x^3 - 1\right)\right]$

$= 36\left(2x^3 - 1\right)^{11}(3x+1)^{11}\left(8x^3 + 2x^2 - 1\right)$

76. $y = \left(\dfrac{x+1}{1-x^2}\right)^3 = \left(\dfrac{1}{1-x}\right)^3 = (1-x)^{-3}$ $y' = -3(1-x)^{-4}(-1) = \dfrac{3}{(1-x)^4}$

77. $y = x(x^2 - 4)^{1/2}$

$y' = x \cdot \dfrac{1}{2}(x^2 - 4)^{-1/2}(2x) + (x^2 - 4)^{1/2} \cdot 1 = \dfrac{x^2}{\sqrt{x^2 - 4}} + \dfrac{\sqrt{x^2 - 4}}{1} = \dfrac{x^2 + (x^2 - 4)}{\sqrt{x^2 - 4}} = \dfrac{2x^2 - 4}{\sqrt{x^2 - 4}}$

78. $y = \dfrac{x}{\sqrt[3]{3x-1}} = \dfrac{x}{(3x-1)^{1/3}}$ $\dfrac{dy}{dx} = \dfrac{(3x-1)^{1/3}\cdot 1 - x\cdot\frac{1}{3}(3x-1)^{-2/3}\cdot 3}{(3x-1)^{2/3}} = \dfrac{(3x-1)^{-2/3}\left[(3x-1)-x\right]}{(3x-1)^{2/3}} = \dfrac{2x-1}{(3x-1)^{4/3}}$

79. $y = x^{1/2} - x^2$ $y' = \dfrac{1}{2}x^{-1/2} - 2x$ $y'' = -\dfrac{1}{4}x^{-3/2} - 2$

80. $y = x^4 - x^{-1}$ $y' = 4x^3 + x^{-2}$ $y'' = 12x^2 - 2x^{-3} = 12x^2 - \dfrac{2}{x^3}$

81. $y = (2x+1)^4$ Since the largest power in any term is 4, the fifth derivative equals 0.

82. $y = \dfrac{1}{24}(1-x)^6$ $y' = -\dfrac{1}{4}(1-x)^5$ $y'' = \dfrac{5}{4}(1-x)^4$

$y''' = -5(1-x)^3$ $y^{(4)} = 15(1-x)^2$ $y^{(5)} = -30(1-x)$

83. $\dfrac{dy}{dx} = (x^2-4)^{1/2}$

$\dfrac{d^2y}{dx^2} = \dfrac{1}{2}(x^2-4)^{-1/2}(2x) = \dfrac{x}{(x^2-4)^{1/2}}$

$\dfrac{d^3y}{dx^3} = \dfrac{(x^2-4)^{1/2}(1) - x\left[\frac{x}{(x^2-4)^{1/2}}\right]}{(x^2-4)} \cdot \dfrac{(x^2-4)^{1/2}}{(x^2-4)^{1/2}} = \dfrac{x^2-4-x^2}{(x^2-4)^{3/2}} = \dfrac{-4}{(x^2-4)^{3/2}}$

84. $\dfrac{d^2y}{dx^2} = \dfrac{x}{x^2+1}$

$\dfrac{d^3y}{dx^3} = \dfrac{x^2+1-x(2x)}{(x^2+1)^2} = \dfrac{-x^2+1}{(x^2+1)^2}$

$\dfrac{d^4y}{dx^4} = \dfrac{(x^2+1)^2(-2x) - (-x^2+1)(2)(x^2+1)(2x)}{(x^2+1)^4} = \dfrac{(x^2+1)[-2x(x^2+1) - 4x(-x^2+1)]}{(x^2+1)^4} = \dfrac{2x^3-6x}{(x^2+1)^3} = \dfrac{2x(x^2-3)}{(x^2+1)^3}$

85. a. $\lim\limits_{x\to 4000} R(x) = 140(4000) - 0.01(4000)^2 = 400{,}000$

b. $\lim\limits_{x\to 4000} C(x) = 60(4000) + 70{,}000 = 310{,}000$

c. $\lim\limits_{x\to 4000} P(x) = \lim\limits_{x\to 4000} R(x) - \lim\limits_{x\to 4000} C(x) = 90{,}000$

86. a. $\lim\limits_{x\to 0^+} C(x) = 60(0) + 70{,}000 = 70{,}000$ The fixed costs of the company (at zero units) are \$70,000.

b. $\lim\limits_{x\to 1000} P(x) = \lim\limits_{x\to 1000}\left[R(x) - C(x)\right] = \lim\limits_{x\to 1000}\left[(140x - 0.01x^2) - (60x + 70000)\right]$

$= \left(140(1000) - 0.01(1000)^2\right) - \left(60(1000) + 70000\right) = 0$

1000 units is the break-even point for the company.

87. a. $\lim\limits_{x\to 0^+} \bar{R}(x) = \lim\limits_{x\to 0^+} \dfrac{140x - 0.01x^2}{x} = \lim\limits_{x\to 0^+}(140 - 0.01x) = 140 - 0.01(0) = 140$

b. $\lim\limits_{x\to 0^+} \bar{C}(x) = \lim\limits_{x\to 0^+} \dfrac{60x + 70000}{x} = \lim\limits_{x\to 0^+}\left(60 + \dfrac{70000}{x}\right) = \infty$

88. a. $\lim\limits_{x\to\infty} C(x) = \lim\limits_{x\to\infty}(60x + 70000) = \infty$ The more units made, the greater the overall cost.

b. $\lim\limits_{x\to\infty} \bar{C}(x) = \lim\limits_{x\to\infty} \dfrac{60x + 70000}{x} = \lim\limits_{x\to\infty}\left(60 + \dfrac{70000}{x}\right) = 60$ The average cost per unit approaches \$60 as the

number of units increases.

89. a. Men: $\text{Avg} = \dfrac{18.6 - 63.1}{100} = -0.445\%$

b. Women: $\text{Avg} = \dfrac{10 - 8.3}{100} = 0.017\%$

90. a. 1990-2000: $\text{Avg} = \dfrac{18.6 - 17.6}{10} = 0.1\%$

Increased possibly because men chose to work past age 65.

b. $1950 - 1960$: $\text{Avg} = \dfrac{10.3 - 7.8}{10} = 0.25\%$

Increased possibly because women were needed because of so many men killed in WW II.

91. $x(p) = \dfrac{100}{p} - 1 \quad x'(p) = \dfrac{-100}{p^2}$

a. $x'(10) = \dfrac{-100}{10^2} = -1$

As the price per unit increases from \$10 to \$11, the demand is expected to drop by 1 unit.

b. $x'(20) = \dfrac{-100}{20^2} = -\dfrac{1}{4}$

As the price per unit increases from \$20 to \$21, the demand is expected to drop by $\dfrac{1}{4}$ units.

92. $P(65) = 6.61(65)^{0.559} = 68.17$; in 2005 the model predicts 68.17% of U.S. roads will be paved.

$P'(t) = 6.61(0.559)t^{-0.441} = 3.695t^{-0.441}$, thus $P'(65) \approx 0.586$; in 2005, the number of paved roads will increase at a rate of 0.586% per year.

93. The $(A+1)$st item will produce more revenue. Reason: $R'(A) > R'(B)$.

94. $R(x) = (830 + 30x)(100 - x)$

$\begin{aligned} R'(x) &= (830 + 30x)(-1) + (100 - x)(30) \\ &= -830 - 30x + 3000 - 30x = -60x + 2170 \end{aligned}$

$R'(10) = -60(10) + 2170 = 1570$

This tells us that rent should be raised. An 11th rent increase of \$30 (and hence an 11th vacancy) would change revenue by about \$1570.

95. $P(x) = 3 + \dfrac{70x^2}{x^2 + 1000}$

$P(20) = 3 + \dfrac{70(20)^2}{(20)^2 + 1000} = 23$ means productivity is 23 units per hour after 20 hours of training and experience.

$P'(x) = 0 + \dfrac{(x^2 + 1000)(140x) - 70x^2(2x)}{(x^2 + 1000)^2} = \dfrac{140x^3 + 140000x - 140x^3}{(x^2 + 1000)^2} = \dfrac{140000x}{(x^2 + 1000)^2}$

$P'(20) = \dfrac{140000(20)}{(20^2 + 1000)^2} \approx 1.4$ means that the 21st hour of training or experience will change productivity by about 1.4 units per hour.

96. $q = 10,000 - 50(0.02p^2 + 500)^{1/2} \qquad q' = -25(0.02p^2 + 500)^{-1/2}(0.04p) = -\dfrac{p}{\sqrt{0.02p^2 + 500}}$

97. $x(p) = \sqrt{p-1}$ $x'(p) = \dfrac{1}{2\sqrt{p-1}}$ $x'(10) = \dfrac{1}{2\sqrt{9}} = \dfrac{1}{6}$

If the price increases \$1, the number of units supplied will increase about 1/6.

98. $s(t) = 16 + 140t + 8t^{1/2}$

$v(t) = s'(t) = 140 + 4t^{-1/2}$

$a(t) = v'(t) = -2t^{-3/2}; \quad a(4) = -2(4)^{-3/2} = -\dfrac{2}{8} = -\dfrac{1}{4}$

The acceleration at 4 seconds is -0.25 ft/s^2.

99. $P(x) = 70x - 0.1x^2 - 5500$

$P'(x) = 70 - 0.2x; \quad P'(300) = 70 - 0.2(300) = 10$

$P''(x) = -0.2; \quad P''(300) = -0.2$

The marginal profit is \$10 per unit. The profit on the 301st unit is \$10. The marginal profit decreases at a constant rate of \$0.20 per unit per unit.

100. $C(x) = 3x^2 + 6x + 600$

   a. $C'(x) = 6x + 6 = \overline{MC}$

   b. When $x = 30$, $C'(30) = 186$.

   c. If a 31st unit is produced, costs will increase by about \$186.

101. $C(x) = 400 + 5x + x^3$ $\overline{MC} = C'(x) = 5 + 3x^2$ $C'(4) = 5 + 48 = 53$

It will cost about \$53 for the 5th unit. Also, the 5th unit produced will increase total costs by \$53.

102. $R = 40x - 0.02x^2$

   a. $R' = 40 - 0.04x$

   b. $\overline{MR} = 0$ when $40 - 0.04x = 0$ or when $x = 1000$.

103. $P(x) = 60x - (200 + 10x + 0.1x^2)$ $P'(x) = 50 - 0.2x$ $P'(10) = 50 - 2 = 48$

The next unit sold will have a profit of about \$48.

104. $R = 80x - 0.04x^2$

   a. $R' = 80 - 0.08x$

   b. At $x = 100$, $R' = 80 - 0.08(100) = 72$.

   c. Selling the 101st unit is expected to result in \$72 extra revenue.

105. $R(x) = \dfrac{60x^2}{2x+1}$ $R'(x) = \dfrac{(2x+1)(120x) - 60x^2(2)}{(2x+1)^2} = \dfrac{120x^2 + 120x}{(2x+1)^2}$

106. $C(x) = 45,000 + 100x + x^3$ $R(x) = 4600x$

$P(x) = R(x) - C(x) = 4600x - (45,000 + 100x + x^3) = -x^3 + 4500x - 45,000$

$P'(x) = \overline{MP} = -3x^2 + 4500$

107. $P(x) = 46x - \left(100 + 30x + \dfrac{x^2}{10}\right) = 16x - 100 - \dfrac{x^2}{10}$ $P'(x) = 16 - \dfrac{x}{5} = 16 - 0.2x$

108. Answers are decided from C', R', and $R' - C' = P'$

   a. A, $C'(A) < C'(B) < C'(C)$

   b. Profit is greatest at B since $R - C$ is a maximum distance at B.

   c. $R'(A) > R'(B) > R'(C)$ $C'(A) < C'(B) < C'(C)$

      The marginal profit is greatest at A since $R'(A) - C'(A)$ has its largest value.

   d. Profit is reduced at C since $R'(C) - C'(C) < 0$.

Chapter Test

1. **a.** $\lim\limits_{x\to-2}\dfrac{4x-x^2}{4x-8}=\lim\limits_{x\to-2}\dfrac{x(4-x)}{4(x-2)}=\dfrac{(-2)(6)}{4(-4)}=\dfrac{3}{4}$

   b. $\lim\limits_{x\to\infty}\dfrac{8x^2-4x+1}{2+x-5x^2}=\lim\limits_{x\to\infty}\dfrac{8-\frac{4}{x}+\frac{1}{x^2}}{-5+\frac{1}{x}+\frac{2}{x^2}}=-\dfrac{8}{5}$

   c. $\lim\limits_{x\to7}\dfrac{x^2-5x-14}{x^2-6x-7}=\lim\limits_{x\to7}\dfrac{(x-7)(x+2)}{(x-7)(x+1)}=\dfrac{7+2}{7+1}=\dfrac{9}{8}$

   d. $\lim\limits_{x\to-5}\dfrac{5x-25}{x+5}$ does not exist. (division by zero)

2. **a.** $f'(x)=\lim\limits_{h\to0}\dfrac{f(x+h)-f(x)}{h}$

   b. $f'(x)=\lim\limits_{h\to0}\dfrac{3(x+h)^2-(x+h)+9-(3x^2-x+9)}{h}=\lim\limits_{h\to0}\dfrac{6xh+3h^2-h}{h}=\lim\limits_{h\to0}(6x+3h-1)=6x-1$

3. $f(x)=\dfrac{4x}{x(x-8)}$ $f(x)$ is not continuous at $x=0$ and $x=8$ as there is division by zero.

4. **a.** $B=0.523W-5176;\ B'=0.523$

   b. $p=9t^{10}-6t^7-17t+23;\ p'=90t^9-42t^6-17$

   c. $y=\dfrac{3x^3}{2x^7+11};\ \dfrac{dy}{dx}=\dfrac{(2x^7+11)(9x^2)-3x^3(14x^6)}{(2x^7+11)^2}=\dfrac{-24x^9+99x^2}{(2x^7+11)^2}$

   d. $f(x)=(3x^5-2x+3)(4x^{10}+10x^4-17)$
       $f'(x)=(3x^5-2x+3)(40x^9+40x^3)+(4x^{10}+10x^4-17)(15x^4-2)$

   e. $g(x)=\dfrac{3}{4}(2x^5+7x^3-5)^{12};\ g'(x)=9(2x^5+7x^3-5)^{11}(10x^4+21x^2)$

   f. $y=(x^2+3)(2x+5)^6$
       $y'=(x^2+3)\cdot6(2x+5)^5(2)+(2x+5)^6\cdot2x$
       $=2(2x+5)^5[6(x^2+3)+x(2x+5)]$
       $=2(2x+5)^5(8x^2+5x+18)$

   g. $f(x)=12x^{1/2}-10x^{-2}+17;\ f'(x)=6x^{-1/2}+20x^{-3}=\dfrac{6}{\sqrt{x}}+\dfrac{20}{x^3}$

5. $y=x^3-x^{-3};\ y'=3x^2+3x^{-4};\ y''=6x-12x^{-5};\ y'''=\dfrac{d^3y}{dx^3}=6+60x^{-6}$

6. $f(x)=x^3-3x^2-24x-10$

   a. $f'(x)=3x^2-6x-24=3(x-4)(x+2)$
       $f(-1)=(-1)^3-3(-1)^2-24(-1)-10=10$
       $f'(-1)=3(-1)^2-6(-1)-24=-15$
       $y-y_1=m(x-x_1)$
       $y-10=-15(x+1)$ or $y=-15x-5$

   b. $f'(x)=0$ at $x=4$ and $x=-2$ (from **a**).
       $f(4)=-90,\ f(-2)=18$
       Points: $(4,-90),(-2,18)$

7. $f(x) = 4 - x - 2x^2$ over $[1,6]$. Avg rate $= \dfrac{f(6) - f(1)}{6 - 1} = \dfrac{-74 - 1}{5} = -15$

8. a. $\lim\limits_{x \to 5} f(x) = 2$

    b. $\lim\limits_{x \to 5} g(x) = \text{DNE}$

    c. $\lim\limits_{x \to 5^-} g(x) = -4$

9. $g(x) = \begin{cases} 6 - x \text{ if } x \le -2 \\ x^3 \text{ if } x > -2 \end{cases}$

    $g(-2) = 6 - (-2) = 8$

    $\lim\limits_{x \to -2^-} g(x) = 8 \qquad \lim\limits_{x \to -2^+} g(x) = -8$

    $g(x)$ is not continuous at $x = -2$ since the left hand limit does not equal the right hand limit.

10. a. $R'(x) = 250 - 0.02x$

    b. $R(72) = 17948.16$; the revenue for 72 units is $17,948.16. $R'(72) = 248.56$; when 72 units have been produced, the 73rd unit will increase the revenue $248.56.

11. $C(x) = 200x + 10000 \qquad R(x) = 250x - 0.01x^2$

    a. $P(x) = R(x) - C(x) = -0.01x^2 + 50x - 10000$

    b. $\overline{MP} = P'(x) = -0.02x + 50$

    c. $P'(1000) = 30$. The company will make approximately $30 on the sale of the next unit.

12. $f'(3) \approx \dfrac{f(3) - f(2.999)}{3 - 2.999} = \dfrac{0.104}{0.001} = 104$

13. a. $f(1) = -5$ b. $\lim\limits_{x \to 6} f(x) = -1$ c. $\lim\limits_{x \to 3^-} f(x) = 4$

    d. $\lim\limits_{x \to -4} f(x) = \text{DNE}$ e. $\lim\limits_{x \to -\infty} f(x) = 2$ f. $f'(x) = \text{DNE}$ at $x = -4, 1, 3,$ and 6.

    g. $f(x)$ is not continuous at $x = -4, 3,$ and 6.

    h. $f'(4) \approx \dfrac{3}{2}$ i. $f'(-2) < \text{average rate} < f'(2)$

14. $y = \dfrac{2}{3}x - 8$ is tangent to $f(x)$ at $x = 6$.

    a. $f'(6) = \dfrac{2}{3}$ b. $f(6) = \dfrac{2}{3}(6) - 8 = -4$

    c. At $x = 6$, the instantaneous rate of change is $\dfrac{2}{3}$.

14. a. B $R(x) - C(x)$ is greatest at this point.

    b. A $C(x) - R(x) > 0$

    c. A and B $R'(x) - C'(x) > 0$ at each of these points.

    d. C $R'(x) - C'(x) < 0$ at this point.

Chapter 10: Applications of Derivatives

Exercise 10.1

1. **a.** $(1, 5)$
  b. $(4, 1)$
  c. $(-1, 2)$

3. **a.** $(1, 5)$
  b. $(4, 1)$
  c. $(-1, 2)$

5. **a.** Critical values: $x = 3$ and $x = 7$
  b. Increasing: $3 < x < 7$
  c. Decreasing: $x < 3$, $x > 7$
  d. Max at $x = 7$
  e. Min at $x = 3$

7. $y = 2x^3 - 12x^2 + 6$
  $y' = 6x^2 - 24x = 6x(x - 4)$
  $y' = 0$ if $x = 0$ or 4

9. $y = 2x^3 - 12x^2 + 6$
  $y' = 6x^2 - 24x = 6x(x - 4)$

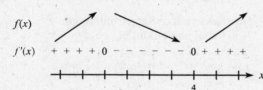

11. $y = x^3 - 3x + 4$
  a. Rel. max: $(-1, 6)$
    Rel. min: $(1, 2)$
  b. $f'(x) = 3x^2 - 3 = 3(x^2 - 1)$
    $f'(x) = 0$ if $3(x + 1)(x - 1) = 0$ or
    $x = -1, 1$.
  c. At $x = -1$ we have
    $y = (-1)^3 - 3(-1) + 4 = 6$.
    At $x = 1$ we have $y = 1^3 - 3 \cdot 1 + 4 = 2$
    Critical points $(-1, 6)$, $(1, 2)$
  d. Yes.

13. **a.** HPI: $(-1, -3)$
  b. $y = x^3 + 3x^2 + 3x - 2$
    $f'(x) = 3x^2 + 6x + 3$
    Critical values: $3(x^2 + 2x + 1) = 0$ or $x = -1$
  c. Critical point: $(-1, -3)$
  d. Yes. From the graph, $(-1, -3)$ is a
    horizontal point of inflection.

15. $y = \frac{1}{2}x^2 - x$
  a. $f'(x) = x - 1$
  b. $f'(x) = 0$ if $x = 1$
  c. At $x = 1$ we have $y = \frac{1}{2}(1)^2 - 1 = -\frac{1}{2}$.
    Critical point: $\left(1, -\frac{1}{2}\right)$
  d.

x	0	1	2
$f'(x)$	$-$	0	$+$

  Decreasing if $x < 1$
  Increasing if $x > 1$
  e.

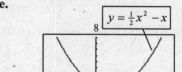

17. $y = \frac{1}{3}x^3 + \frac{1}{2}x^2 - 2x + 1$
  a. $f'(x) = x^2 + x - 2$
  b. $x^2 + x - 2 = 0$
    $(x + 2)(x - 1) = 0$
    Critical values are $x = -2, 1$.
  c. $f(-2) = -\frac{8}{3} + 2 + 4 + 1 = \frac{13}{3}$
    $f(1) = \frac{1}{3} + \frac{1}{2} - 2 + 1 = -\frac{1}{6}$
    Critical points: $\left(-2, \frac{13}{3}\right), \left(1, -\frac{1}{6}\right)$
  d.

x		-2		1	
$f'(x)$	$+$	0	$-$	0	$+$

  Increasing: $x < -2$ and $x > 1$
  Decreasing: $-2 < x < 1$
  e.

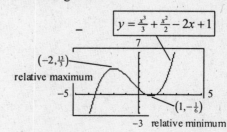

19. $y = x^{2/3}$

   a. $f'(x) = \dfrac{2}{3x^{1/3}}$

   b. Critical value is $x = 0$ since $f'(x)$ does not exist at $x = 0$.

   c. Critical point: $(0, 0)$

   d. Decreasing: $x < 0$ ($f'(x) < 0$)

      Increasing: $x > 0$

   e.

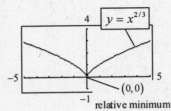

21. **a.** $y' = 0$ if $x = -\dfrac{1}{2}$; $y'(x) > 0$ if $x < -\dfrac{1}{2}$;

      $y'(x) < 0$ if $x > -\dfrac{1}{2}$

   b. $y'(x) = -1 - 2x$

      $-1 - 2x = 0$ if $x = -\dfrac{1}{2}$

      Substitute -1 and $+1$ to verify conclusion.

23. **a.** $y'(x) = 0$ at $x = 0, 3, -3$

      $y'(x) > 0$ for $-3 < x < 3$, $x \ne 0$

      $y'(x) < 0$ for $x < -3$ and for $x > 3$

   b. $y'(x) = 3x^2 - \dfrac{1}{3}x^4 = 3x^2\left(1 - \dfrac{1}{9}x^2\right)$

      Substitute $-4, -3, -1, 0, 1, 3,$ and 4 to verify conclusion.

25.

$$\boxed{y = \tfrac{1}{3}x^3 - x^2 + x + 1}$$

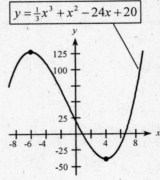

$y = \dfrac{1}{3}x^3 - x^2 + x + 1$

$\dfrac{dy}{dx} = x^2 - 2x + 1 = (x - 1)^2$

Critical value is $x = 1$.

$f(1) = \dfrac{1}{3} - 1 + 1 + 1 = \dfrac{4}{3}$

Critical point: $\left(1, \dfrac{4}{3}\right)$

x		1	
$f'(x)$	+	0	+

No relative maximum or minimum.
Horizontal point of inflection at $x = 1$.

27.

$$\boxed{y = \tfrac{1}{3}x^3 + x^2 - 24x + 20}$$

$y = \dfrac{1}{3}x^3 + x^2 - 24x + 20$

$\dfrac{dy}{dx} = x^2 + 2x - 24 = (x + 6)(x - 4)$

Critical values are $x = -6, 4$.

$f(-6) = 128$; $f(4) = -\dfrac{116}{3}$

Critical points: $(-6, 128)$, $\left(4, -\tfrac{116}{3}\right)$

x		-6		4	
$f'(x)$	+	0	−	0	+

Relative maximum at $x = -6$.
Relative minimum at $x = 4$.

29.

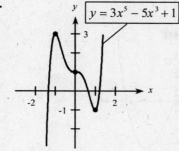

$y = 3x^5 - 5x^3 + 1$

$y' = 15x^4 - 15x^2 = 15x^2(x^2 - 1)$

Critical values are $x = 0, 1, -1$.

Critical points are $(0, 1)$, $(1, -1)$, $(-1, 3)$

x		-1		0		1	
$f'(x)$	$+$	0	$-$	0	$-$	0	$+$

HPI at $x = 0$.

Relative maximum at $x = -1$.

Relative minimum at $x = 1$.

31.

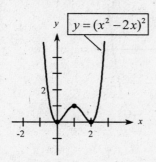

$y = (x^2 - 2x)^2$

$y' = 4x(x - 1)(x - 2)$

Critical values at $x = 0, 1, 2$.

$f(0) = 0;\ f(1) = 1;\ f(2) = 0$

Critical points: $(0, 0)$, $(1, 1)$, $(2, 0)$

x		0		1		2	
$f'(x)$	$-$	0	$+$	0	$-$	0	$+$

Relative minimum at $x = 0, 2$.

Relative maximum at $x = 1$.

33.

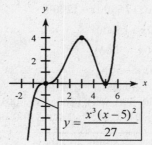

$y = \dfrac{1}{27}x^3(x - 5)^2$

$y' = \dfrac{5x^2(x - 3)(x - 5)}{27}$

Critical values are $x = 0, 5, 3$.

Critical points are $(0, 0)$, $(5, 0)$, $(3, 4)$.

x		0		3		5	
$f'(x)$	$+$	0	$+$	0	$-$	0	$+$

Relative maximum at $x = 3$.

Relative minimum at $x = 5$.

HPI at $x = 0$.

35.

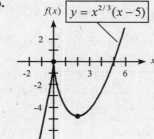

$f(x) = x^{2/3}(x - 5)$

$f'(x) = \dfrac{5(x - 2)}{3x^{1/3}}$

Critical values are $x = 0, 2$.

Critical points are $(0, 0)$ and $\left(2, -3\sqrt[3]{4}\right)$.

Note that $-3\sqrt[3]{4} \approx -4.8$.

x	-1	0	1	2	3
$f'(x)$	$+$	$*$	$-$	0	$+$

* means $f'(0)$ is undefined.

Relative maximum at $x = 0$.

Relative minimum at $x = 2$.

37. $f(x) = x^3 - 225x^2 + 15000x - 12000$

$f'(x) = 3x^2 - 450x + 15000$

$\qquad = 3(x^2 - 150x + 5000)$

$\qquad = 3(x - 100)(x - 50)$

Critical values: $x = 50$ and $x = 100$

Critical points: (50, 300,500), (100, 238,000)

Viewing window: $0 \le x \le 150$, $0 \le y \le 400,000$

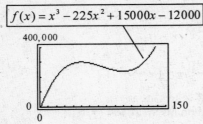

$\boxed{f(x) = x^3 - 225x^2 + 15000x - 12000}$

400,000

0   150
0

39. $f(x) = x^4 - 160x^3 + 7200x^2 - 40000$

$f'(x) = 4x^3 - 480x^2 + 14400x$

$\qquad = 4x(x^2 - 120x + 3600)$

$\qquad = 4x(x - 60)(x - 60)$

Critical values: $x = 0$ and $x = 60$

Critical points: (0, –40,000) and (60, 4,280,000)

Viewing window:

$-30 \le x \le 90$, $-500,000 \le y \le 5,000,000$

(There are other windows.)

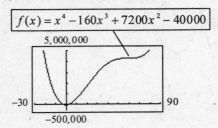

$\boxed{f(x) = x^4 - 160x^3 + 7200x^2 - 40000}$

5,000,000

–30   90
–500,000

41. $y = 7.5x^4 - x^3 + 2$

$y' = 30x^3 - 3x^2 = 3x^2(10x - 1)$

Critical values: $x = 0, 0.1$

Critical points: (0, 2) and (0.1, 1.99975)

Viewing window:

$-0.1 \le x \le 0.2$, $1.9997 \le y \le 2.0007$

Viewing windows can vary.

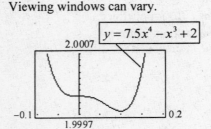

$\boxed{y = 7.5x^4 - x^3 + 2}$

2.0007

–0.1   0.2
1.9997

43. Possible graph for $f(x)$.

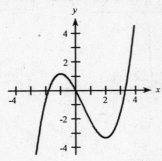

$f'(x) = x^2 - x - 2 = (x - 2)(x + 1)$

Note: The given graph is $f'(x)$ and NOT $f(x)$.

Critical values: $x = -1$, $x = 2$

$f(x)$ is increasing if $x < -1$ and $x > 2$.

$f(x)$ is decreasing if $-1 < x < 2$.

Thus, rel max at $x = -1$; rel min at $x = 2$.

45. Possible graph of $f(x)$.

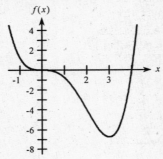

$f'(x) = x^3 - 3x^2 = x^2(x - 3)$

Note: The given graph is $f'(x)$ and NOT $f(x)$.

Critical values: $x = 0$, $x = 3$

$f(x)$ is increasing if $x > 3$.

$f(x)$ is decreasing if $x < 3$, $x \ne 0$.

Thus, relative min at $x = 3$; HPI at $x = 0$.

47. Graph on the left is $f(x)$.

Graph on the right is $f'(x)$.

If $f'(x) > 0$, then $f(x)$ is increasing (rising).

If $f'(x) < 0$, then $f(x)$ is decreasing (falling).

49. $S(t) = 1000 + 400(t + 1)^{-1}$ $\qquad S'(t) = \dfrac{-400}{(t+1)^2}$

$S'(t) < 0$ for all t. Thus, $S(t)$ is always

decreasing for $t \ge 0$.

51. $P(t) = 27t + 6t^2 - t^3$

$P'(t) = 27 + 12t - 3t^2 = 3(9 + 4t - t^2)$

a. $t = \dfrac{-4 \pm \sqrt{16+36}}{-2} = \dfrac{4 \pm 2\sqrt{13}}{2} = 2 \pm \sqrt{13}$

b. $t = 2 + \sqrt{13}$ is the only value in the domain of the function.

c.

t		$2+\sqrt{13}$	
$P'(t)$	+	0	−

$P(t)$ increasing if $0 \le t < 2 + \sqrt{13}$.

d.

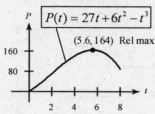

53. $\bar{C}(x) = 5000x + 125{,}000x^{-1}$

$\bar{C}'(x) = 5000 - 125{,}000x^{-2}$

a. $\bar{C}'(x) = 0$ if $x^2 - 25 = 0$

Critical value: $x = 5$

b. $\bar{C}'(x) < 0$ if $x < 5$.

c. $\bar{C}'(x) > 0$ if $x > 5$.

55. a. Increasing at $x = 150$, decreasing at $x = 350$, changing from increasing to decreasing at $x = 250$.

b. $R'(x) > 0$ if $x < 250$

c. $x = 250$ units

57. $R(t) = \dfrac{50t}{t^2 + 36}$

$R'(t) = \dfrac{(t^2+36)(50) - 50t(2t)}{(t^2+36)^2} = \dfrac{1800 - 50t^2}{(t^2+36)^2}$

a. $R'(t) = 0$ if $50(36 - t^2) = 0$ or $t = 6$. (-6 is not in the domain.)

b.

t		6	
$R'(t)$	+	0	−

Revenue will increase for 6 weeks.

59. $P(t) = \dfrac{13t}{t^2 + 100} + 0.18$

$P'(t) = \dfrac{(t^2+100)(13) - 13t(2t)}{(t^2+100)^2} = \dfrac{13(100 - t^2)}{(t^2+100)^2}$

Domain = $\{t : t \ge 0\}$. Critical value is $t = 10$.

t		10	
$P'(t)$	+	0	−

a. Ten months after the campaign starts the recognition is a maximum.

b. Begins on Jan. 1.

61. a. $f(x) = -2.0x^3 + 35.3x^2 - 106.0x - 205.4$

$f'(x) = -6.0x^2 + 70.6x - 106.0$

Using the quadratic formula,

$f'(x) = 0$ if $x \approx 1.76$ or $x = 10$.

$f'(x)$ changes from − to + at $x \approx 1.76$.

$f'(x)$ changes from + to − at $x = 10$.

Maximum surplus from model is in 2000. Maximum surplus from the data is in 2000.

b. September 11, 2001 and wars in Iraq and Afghanistan caused huge budget deficits.

63. a. cubic

b. $y = -0.000316x^3 + 0.0174x^2$
$\qquad\qquad - 0.0688x + 4.313$

c. The graph indicates a maximum at $x = 34.61$ or in 1984.

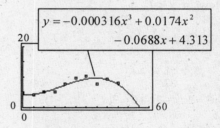

65. a. Use a cubic model.

b. $y = 6.1636x^3 - 90.5303x^2$
$\qquad\qquad + 373.5975x + 202.867$

c. $(2.97, 675)$

d. In the early 90's the maximum grant was approximately $675 million.

e. Absolute max of $1040 million in 2000.

Exercise 10.2

1. $f(x) = x^3 - 3x^2 + 1$

    $f'(x) = 3x^2 - 6x$

    $f''(x) = 6x - 6 = 6(x-1)$

    a. $f''(-2) < 0$ concave downward

    b. $f''(3) > 0$ concave upward

3. $f(x) = 2x^3 + 4x - 8$

    $f'(x) = 6x^2 + 4$

    $f''(x) = 12x$

    a. $f''(-1) < 0$ concave downward

    b. $f''(4) > 0$ concave upward

5. (a, c) and (d, e)

7. (c, d) and (e, f)

9. c, d and e

11. $f(x) = x^3 - 6x^2 + 5x + 6$

    $f'(x) = 3x^2 - 12x + 5$

    $f''(x) = 6x - 12$

    $f''(x) = 0$ if $x = 2$. We have $f(2) = 0$.

x		2	
$f''(x)$	$-$	0	$+$

    The point $(2, 0)$ is a point of inflection.

    $f(x)$ is concave up if $x > 2$. $f(x)$ is concave down if $x < 2$.

13. $y = f(x) = \dfrac{1}{4}x^4 + \dfrac{1}{2}x^3 - 3x^2 + 3$

    $y' = x^3 + \dfrac{3}{2}x^2 - 6x$

    $y'' = 3x^2 + 3x - 6 = 3(x+2)(x-1)$

    $f(-2) = -9$ $f(1) = \dfrac{3}{4}$

x		-2		1	
y''	$+$	0	$-$	0	$+$

    The points $(-2, -9)$ and $\left(1, \dfrac{3}{4}\right)$ are points of

    inflection.

    $f(x)$ is concave up if $x < -2$ and if $x > 1$.

    $f(x)$ is concave down if $-2 < x < 1$.

15. $y = x^2 - 4x + 2$

    $y' = 2x - 4$

    $y'' = 2$

    Critical value: $x = 2$ $y'' > 0$ at $x = 2$.

    There is no point of inflection.

    There is a relative minimum at $(2, -2)$.

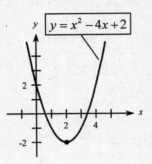

17. $y = \dfrac{1}{3}x^3 - 2x^2 + 3x + 2$

    $y' = x^2 - 4x + 3 = (x-3)(x-1)$

    $y'' = 2x - 4$

    Point of inflection at $x = 2$. (concavity changes)

    Relative minimum at $x = 3$ ($y'' > 0$).

    Relative maximum at $x = 1$ ($y'' < 0$).

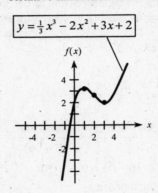

19. $y = x^4 - 16x^2$

$y' = 4x^3 - 32x = 4x(x^2 - 8)$

$y'' = 12x^2 - 32 = 4(3x^2 - 8)$

Critical values are $x = 0, \pm 2\sqrt{2}$.

Possible inflection points at $x = \pm \frac{2\sqrt{6}}{3}$

x		$-\frac{2\sqrt{6}}{3}$		$\frac{2\sqrt{6}}{3}$	
y''	+	0	−	0	+

$\left(\dfrac{-2\sqrt{6}}{3}, -\dfrac{320}{9}\right)$ and $\left(\dfrac{2\sqrt{6}}{3}, -\dfrac{320}{9}\right)$ are inflection points.

x		$-2\sqrt{2}$		0		$2\sqrt{2}$	
y'	−	0	+	0	−	0	+

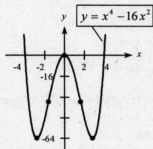

Relative maximum at (0, 0). Relative minima at $(-2\sqrt{2}, -64)$ and $(2\sqrt{2}, -64)$.

21.

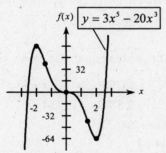

Critical values are $x = -2, 0, 2$.

Possible points of inflection at $x = -\sqrt{2}$, $x = 0$, and $x = \sqrt{2}$.

x		$-\sqrt{2}$		0		$\sqrt{2}$	
$f''(x)$	−	0	+	0	−	0	+

$(-\sqrt{2}, 39.6)$, $(0, 0)$ and $(\sqrt{2}, -39.6)$ are points of inflection.

$f''(0) = 0$ means second derivative test fails.

$f'(x) < 0$ for values near 0, both to the left and right of 0 means HPI at (0, 0).

$f''(-2) < 0$ means relative maximum at $(-2, 64)$.

$f''(2) > 0$ means relative minimum at $(2, -64)$.

23.

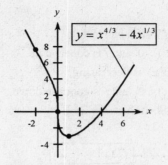

Critical values are $x = 0, 1$.

Possible points of inflection are at $x = -2, 0$,

x		-2		0	
$f''(x)$	+	0	−	*	+

* means $f''(0)$ is undefined.

$(0,0)$ and $(-2, 7.6)$ are points of inflection.

x		0		1	
$f'(x)$	−	*	−	0	+

* means $f'(0)$ is undefined

$(1, -3)$ is a relative minimum.

25. a. $f''(x) > 0$ if $x < 1$

$f''(x) < 0$ if $x > 1$

$f''(x) = 0$ if $x = 1$

b. $f'(x)$ [NOT $f(x)$] has rel max at $x = 1$.

c. $f(x) = -\dfrac{1}{3}x^3 + x^2 + 8x - 12$

$f'(x) = -x^2 + 2x + 8$

$f''(x) = -2x + 2$

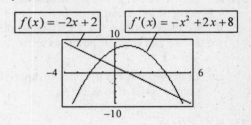

27. $f'(x) = 4x - x^2$

   a. If $x < 2$, $f'(x)$ is increasing.

      Thus, $f''(x) > 0$ and f is concave up.

      If $x > 2$, $f'(x)$ is decreasing.

      Thus, $f''(x) < 0$ and f is concave down.

   b. Point of inflection at $x = 2$.

   c. $f''(x) = 4 - 2x$

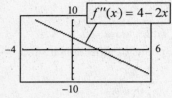

   d. All graphs have the same basic form.

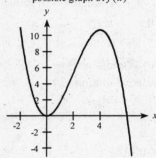

possible graph of $f(x)$

29. a. G

   b. C

   c. F

   d. H

   e. I

31. a. Concave up if $x < 0$; concave down if $x > 0$; point of inflection at $x = 0$.

   b. Concave up if $-1 < x < 1$; concave down if $x < -1$ and if $x > 1$; points of inflection at $x = \pm 1$.

   c. Concave up if $x > 0$; concave down if $x < 0$; point of inflection at $x = 0$.

33. a. $P'(t)$ or $\dfrac{dP}{dt}$

   b. $P''(t) = 0$ at B.

   c. C

35. a. Concavity changes at C.

   b. $S''(t) > 0$ means concave up or to the right of C.

   c. Yes. $S'(t)$ is the rate of change of sales. $S''(t)$ changes from $-$ to $+$ at C.

37. $P(t) = 27t + 12t^2 - t^3$

   $P'(t) = 27 + 24t - 3t^2 = 3(9 - t)(1 + t)$

   $P''(t) = 24 - 6t = 6(4 - t)$

   a.

t		9	
$P'(t)$	+	0	−

      Production is maximized at $t = 9$. But the domain is $0 \le t \le 8$. Thus, production is maximized when $t = 8$.

   b. $P''(t)$ changes signs at $t = 4$. Point of diminishing returns begins at $t = 4$.

39. $S(t) = 3(t+3)^{-1} - 18(t+3)^{-2} + 1$

   a. $S'(t) = \dfrac{-3}{(t+3)^2} + \dfrac{36}{(t+3)^3}$

      $= \dfrac{-3(t+3) + 36}{(t+3)^3} = \dfrac{27 - 3t}{(t+3)^3}$

      Critical value is $t = 9$.

t		9	
$S'(t)$	+	0	−

      Sales volume is maximized after 9 days.

   b. $S''(t) = \dfrac{6}{(t+3)^3} - \dfrac{108}{(t+3)^4}$

      $= \dfrac{6(t+3) - 108}{(t+3)^4} = \dfrac{6t - 90}{(t+3)^4}$

      Critical value is $t = 15$.

t		15	
$S''(t)$	−	0	+

      Point of diminishing returns is $t = 15$ days.

41. $y = -0.03354^3 + 1.4436x^2 - 10.134x + 53.554$

   $y' = -0.10062x^2 + 2.8872x - 10.134$

   $y'' = -0.20124x + 2.8872$

   $y'' = 0$ if $x \approx 14.35$

   The graph of y'' changes from $+$ to $-$ at $x \approx 14.35$. Thus, the rate of change of international phone traffic began to decrease in the year 2000.

43. a. $y = 0.0000591x^3 - 0.00675x^2$
                  $+ 0.0523x + 14.147$

   b. $y' = 0.0001773x^2 - 0.0135x + 0.0523$

      y' changes from $-$ to $+$ at $x \approx 72$.

      Critical point is $(72, 5)$.

   c. Percentage of foreign born is a minimum of approximately 5% in 1972.

45. a. $y = -0.4052x^3 + 6.0161x^2 - 17.9605x + 52$

b. $y' = -1.2156x^2 + 12.0322x - 17.9605$

The graphing utility shows that a minimum occurs at $x \approx 1.83$ or in 1913,

c. A maximum occurs at $x \approx 8.07$ or in 1920.

Exercise 10.3

1. $f(x) = x^3 - 2x^2 - 4x + 2, \; [-1, 3]$

$f'(x) = 3x^2 - 4x - 4 = (3x + 2)(x - 2)$

$f'(x) = 0$ if $x = 2$ or $x = -\dfrac{2}{3}$

x	-1	$-\frac{2}{3}$	2	3
$f(x)$	3	$\frac{94}{27}$	-6	-1

Maximum is $\dfrac{94}{27}$ at $x = -\dfrac{2}{3}$.

Minimum is -6 at $x = 2$.

3. $f(x) = x^3 + x^2 - x + 1, \; [-2, 0]$

$f'(x) = 3x^2 + 2x - 1 = (3x - 1)(x + 1)$

$f'(x) = 0$ if $x = -1$ or $x = \dfrac{1}{3}$

x	-2	-1	0	$\left(\frac{1}{3} \text{ not in domain}\right)$
$f(x)$	-1	2	1	

Maximum is 2 at $x = -1$.

Minimum is -1 at $x = -2$.

5. a. $R(x) = 36x - 0.01x^2 \quad R'(x) = 36 - 0.02x$

$R'(x) = 0$ if $36 - 0.02x = 0$ or $x = 1800$ units

$R'(1000) > 0; \quad R'(1800) = 0; \quad R'(2000) < 0$

Total revenue is maximized at 1800 units. $R(1800) = \$32,400$ is the maximum revenue.

b. $R(1500) = \$31,500$

7. $R(x) = 2000x - 20x^2 - x^3$

$R'(x) = 2000 - 40x - 3x^2$

$\quad\;\; = (100 + 3x)(20 - x)$

$R'(x) = 0$ if $x = 20$ units $\quad$ (Domain ≥ 0)

$R'(10) > 0; \quad R'(20) = 0; \quad R'(30) < 0$

Total revenue is maximized at 20 units.

$R(20) = 40,000 - 8,000 - 8,000 = \$24,000$

9. Let $x =$ number of people above 30.

$R =$ (number of people)(price per person)

$\quad = (30 + x)(10 - 0.20x)$

$\quad = 300 + 4x - 0.20x^2$

$R'(x) = 4 - 0.40x$

$R'(x) = 0$ if $x = 10$

x	0	10	20
$R(x)$	300	320	300

Therefore, $30 + 10 = 40$ people will maximize revenue.

11. Let $x =$ new customers

$R(x) = (1000 + x)(20 - 0.01x)$

$\quad = 20,000 + 10x - 0.01x^2$

Since the price must be positive, we have $0 \leq x \leq 2000$.

$R'(x) = 10 - 0.02x$

$R'(x) = 0$ if $x = 500$.

x	0	500	2000
$R(x)$	$20,000$	$22,500$	0

We maximize revenue with 500 new customers and a price of \$15. The maximum revenue is $R(500) = 1500(15) = \$22,500$.

13. $R(x) = 2000x + 20x^2 - x^3$

$\bar{R}(x) = \dfrac{R(x)}{x} = 2000 + 20x - x^2$

a. $\bar{R}'(x) = 20 - 2x$

$\bar{R}'(x) = 0$ if $x = 10$.

$\bar{R}'(0) > 0; \ \bar{R}'(10) = 0; \ \bar{R}'(20) < 0$

$\bar{R}(10) = 2000 + 200 - 100 = \$2100$ is the maximum average revenue.

b. $\bar{R} = \dfrac{R(x)}{(x)} \quad \bar{R}'(x) = \dfrac{x \cdot R'(x) - R(x) \cdot 1}{x^2}$

$\bar{R}'(x) = 0$ if $x R'(x) - R(x) = 0$ or

$R'(x) = \dfrac{R(x)}{x}$.

Note: $\dfrac{R(x)}{x} = \bar{R}(x)$ and $R'(x) = \overline{MR}$.

$\bar{R}(x)$ is a maximum where $\bar{R}(x) = \overline{MR}$.

$2000 + 20x - x^2 = 2000 + 40x - 3x^2$

$2x^2 - 20x = 0$

$2x(x - 10) = 0$

$x = 10$

15. $\bar{C}(x) = \dfrac{25}{x} + 13 + x$

$\bar{C}'(x) = -\dfrac{25}{x^2} + 1$

$\bar{C}'(x) = 0$ if $x = 5$.

$\bar{C}'(4) < 0; \ \bar{C}'(5) = 0; \ \bar{C}'(6) > 0$

$\bar{C}(5) = \dfrac{25}{5} + 13 + 5 = \$23$

Minimum average cost is $23 at 5 units of production.

17. $\bar{C}(x) = \dfrac{100}{x} + x$

$\bar{C}'(x) = -\dfrac{100}{x^2} + 1$

$\bar{C}'(x) = 0$ if $x = 10$.

$\bar{C}'(9) < 0; \ \bar{C}'(10) = 0; \ \bar{C}'(11) > 0$

$\bar{C}(10) = \dfrac{100}{10} + 10 = \$20$

Minimum average cost is $20 at 10 units of production.

19. $\bar{C}(x) = \dfrac{(x+4)^3}{x}$

$\bar{C}'(x) = \dfrac{x(3)(x+4)^2 - (x+4)^3 \cdot 1}{x^2}$

$= \dfrac{(x+4)^2(2x-4)}{x^2}$

$\bar{C}'(x) = 0$ if $2x - 4 = 0$ or $x = 2$.

Total number of units is $2(100) = 200$.

$\bar{C}'(1) < 0; \ \bar{C}'(2) = 0; \ \bar{C}'(3) > 0$

$\bar{C}(2) = \dfrac{6^3}{2} = \$108$ per 100 units.

Minimum average cost per unit is $\dfrac{108}{100} = \$1.08$.

21. $\bar{C}(x) = \dfrac{C(x)}{x} \qquad \bar{C}'(x) = \dfrac{x \cdot C'(x) - C(x) \cdot 1}{x^2}$

$\bar{C}'(x) = 0$ if $x \cdot C'(x) - C(x) = 0$ or $C'(x) = \dfrac{C(x)}{x}$

But, $\dfrac{C(x)}{x} = \bar{C}(x)$ and $C'(x) = \overline{MC}$. Thus, the minimum average cost occurs where $\bar{C}(x) = \overline{MC}$. Note that this is true for all cost functions. Using the $C(x)$ of problem 15:

$\overline{MC} = 13 + 2x = \dfrac{25}{x} + 13 + x = \bar{C}(x)$

$x^2 = 25$ or $x = 5$

The same answer as for problem 15.

23. a. A line from $(0, 0)$ to $(x, C(x))$ has slope $\dfrac{C(x)}{x} = \bar{C}(x)$. This is minimized when the line has the least rise. This occurs when the line is tangent to $C(x)$.

b. The level is approximately 600 units.

25. $P(x) = 5600x + 85x^2 - x^3 - 200,000$

$P'(x) = 5600 + 170x - 3x^2 = (80 - x)(70 + 3x)$

$P'(x) = 0$ if $x = 80$ units, (domain: $x \geq 0$)

$P'(70) > 0; \ P'(80) = 0; \ P'(90) < 0$

$P(80) = \$280,000$ is the maximum profit.

27. $P(x) = 4600x - (45,000 + 100x + x^3)$

$\qquad = 4500x - x^3 - 45,000$

$P'(x) = 4500 - 3x^2$

$P'(x) = 0$ if $3x^2 = 4500$ or if $x^2 = 1500$ or

$x = 10\sqrt{15} \approx 39$ units.

$P'(30) > 0;\ P'(39) \approx 0;\ P'(40) < 0$

$P(39) = \$71,181$ is the maximum profit.

29. $P(x) = 250x - 0.01x^2 - (300 + 200x)$

$\qquad = -0.01x^2 + 50x - 300$

$P'(x) = -0.02x + 50 \qquad P'(x) = 0$ if $x = 2500$.

Since $P'(x) > 0$ for all x in the domain

($0 \le x \le 1000$), produce the maximum of 1000

units and obtain the maximum profit. That

maximum profit is $P(1000) = \$39,700$.

31. a. B

    b. B

    c. B ($P' = 0$ if $R' - C' = 0$)

    d. P is a max when $P' = 0$ or $\overline{MR} = \overline{MC}$

33. Let $x =$ number of vacant units.

$P(x) = (720 + 20x)(50 - x) - 12(50 - x)$

$\qquad = 35,400 + 292x - 20x^2$

$P'(x) = 292 - 40x$

$P'(x) = 0$ if $x = 7.3$ units

x	0	7	8	50
$P(x)$	35,400	36,464	36,456	0

Using $x = 7$ units , maximum profit is obtained

if rent is $720 + 20(7) = \$860$.

35. $C(x) = 45,000 + 100x + x^2$

$R(x) = 1600x$

$P(x) = R(x) - C(x) = -x^2 + 1500x - 45,000$

$P'(x) = -2x + 1500$

$P'(x) = 0$ if $x = 750$, but this is not in the

domain.

x	0	600
$P(x)$	−45,000	495,000

Maximum profit is at 600 units due to limited

production. $P(600) = \$495,000$

37. $R(x) = p \cdot x = 600x - \dfrac{1}{2}x^2$

$C(x) = \overline{C}(x) \cdot x = 300x + 2x^2$

$P(x) = \left(600x - \dfrac{1}{2}x^2\right) - (300x + 2x^2)$

$\qquad = 300x - \dfrac{5}{2}x^2$

$P'(x) = 300 - 5x$

$P''(x) = -5 < 0$ for all x.

$P'(x) = 0$ if $x = 60$.

    a. Maximum profit is at $x = 60$.

    b. Selling price is $p = 600 - \dfrac{1}{2} \cdot 60 = \$570$.

    c. Maximum profit is $P(60) = \$9000$.

39. $R(x) = p \cdot x = 1960x - \dfrac{1}{3}x^3$

$C(x) = \overline{C}(x) \cdot x = 1000x + 2x^2 + x^3$

$P(x) = 960x - 2x^2 - \dfrac{4}{3}x^3,\ 0 \le x \le 10$

1000 units = 10 hundreds

$P'(x) = 960 - 4x - 4x^2$

$\qquad = 4(16 + x)(15 - x)$

$P'(x) = 0$ if $x = 15$, but this is not in the domain.

    a. Maximum profit at 1000 units ($x = 10$) due

        to limited production.

    b. $P(10) = \$8066.67$

41. $R(x) = p \cdot x = 120x - 0.015x^2$

$C(x) = \overline{C}(x) \cdot x$

$\qquad = 10,000 + 60x - 0.03x^2 + 0.00001x^3$

$P(x) = -10,000 + 60x + 0.015x^2 - 0.00001x^3$

$P'(x) = 60 + 0.03x - 0.00003x^2$

$P'(x) = 0$ if

$x = \dfrac{-0.03 \pm \sqrt{0.0009 + 0.0072}}{-0.00006}$

$\quad = \dfrac{-0.03 \pm 0.09}{-0.00006}$

$\quad = 2000$ (positive value only)

$P'(1090) > 0;\ P'(2000) = 0;\ P'(2010) < 0$

Maximum profit is at $x = 2000$ units.

Price is $p = 120 - 0.015(2000) = \$90$ per unit.

Maximum profit is $P(2000) = \$90,000$.

43. a. $y = 0.0002524x^3 - 0.02785x^2$
$+ 1.6300x + 2.1550$

b. $y' = 0.0007572x^2 - 0.05570x + 1.6300$

$y'' = 0.0015144x - 0.05570$

y'' changes signs (equals zero) at

$x = \dfrac{0.05570}{0.0015144} = 36.78$.

Point of inflection is $(36.78, 36.99)$.

c.

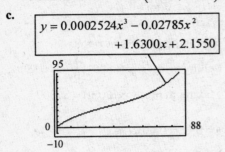

$y = 0.0002524x^3 - 0.02785x^2$
$+ 1.6300x + 2.1550$

95

0   88

−10

d. The rate of change (y') was decreasing until 1987 (from b). After 1987, the rate of change has been increasing ($y'' > 0$). Since 1987 the number of beneficiaries has been increasing at an increasing rate.

45. a. Absolute max was reached in May.

b. Absolute min was reached in September. This was triggered by the 9-11 terrorist attack.

47. a. From the graph, the absolute maximum for $f(t)$ is 16.5.

b. From the graph, the absolute minimum for $f(t)$ is 1.9.

c. The graph suggests that Social Security taxes might rise in the early twenty-first century in order that fewer people can pay in enough to cover those already retired.

Exercise 10.4

1. $S_1 = 30 + 20x_1 - 0.4x_1^2$

$S_2 = 20 + 36x_2 - 1.3x_2^2$

a. $S_1' = 20 - 0.8x_1$

$S_1' = 0$ if $x_1 = \$25$ million

$S_2' = 36 - 2.6x_2$

$S_2' = 0$ if $x_2 \approx \$13.846$ million

b. Total money needed is
$25 + 13.846 = \$38.846$ million .

3. $P(x) = 200x - x^2$

$P'(x) = 200 - 2x$

$P'(x) = 0$ if $x = 100$

$P'(90) > 0;\ P'(100) = 0;\ P'(110) < 0$

Thus, 100 trees per acre will maximize the profit.

5. $y = 70t + \frac{1}{2}t^2 - t^3$

$y' = 70 + t - 3t^2 = (5 - t)(14 + 3t)$

a. $y' = 0$ if $t = 5$.

$y'(4) > 0$

$y'(5) = 0$

$y'(6) < 0$

Production is maximized after 5 hours.

b. $y(5) = 350 + 12.5 - 125 = 237.5$

7. $E(p) = 10,000p - 100p^2$

$E'(p) = 10,000 - 200p$

$E'(p) = 0$ if $p = 50$

$E'(40) > 0,\ E'(50) = 0,\ E'(60) < 0$

Expenditure is greatest at $p = \$50$.

9. $R = \frac{c}{2}m^2 - \frac{m^3}{3}$

$R' = cm - m^2$

$R' = 0$ if $m = 0$ or $m = c$.

$R'(0) = 0;\ R'\left(\frac{c}{2}\right) > 0;\ R'(c) = 0;\ R'(2c) < 0$

Reaction is a max when $m = c$.

11. $S(t) = \frac{200t}{(t+1)^2}$

$S'(t) = \frac{(t+1)^2(200) - 200t \cdot 2 \cdot (t+1)}{(t+1)^4}$

$= \frac{200(1-t)}{(t+1)^3}$

$S'(t) = 0$ if $t = 1$

$S'(0) > 0,\ S'(1) = 0,\ S'(2) < 0$

Sales are maximized after one week.

13. $p(t) = \frac{6.4t}{t^2 + 64} + 0.05$

$p'(t) = \frac{(t^2+64)(6.4) - (6.4t)(2t)}{(t^2+64)^2}$

$= \frac{6.4(64 - t^2)}{(t^2+64)^2}$

$p'(t) = 0$ if $t = 8$.

$p'(7) > 0;\ p'(8) = 0;\ p'(9) < 0$

Awareness is maximized after 8 days.
Maximum percentage that identify 2 defendants
is $p(8) = 45\%$.

15. Quantity being minimized: $F = 4x + 3y$

Another equation: $A = xy = 1200$ or $y = \frac{1200}{x}$

Substituting: $F(x) = 4x + \frac{3600}{x}$

$F'(x) = 4 - \frac{3600}{x^2}$

$F'(x) = 0$ if $x = 30$.

$F'(29) < 0;\ F'(30) = 0;\ F'(31) > 0$

Fence needed is $F(30) = 120 + 120 = 240$ feet .

17. Quantity being minimized: $C = 20x + 5(2y)$

Another equation: $A = xy = 45,000$ or

$$y = \frac{45,000}{x}$$

Substituting: $C(x) = 20x + \frac{450,000}{x}$

$$C'(x) = 20 - \frac{450,000}{x^2}$$

$C'(x) = 0$ if $x^2 = 22,500$ or $x = 150$

$C'(140) < 0; C'(150) = 0; C'(160) > 0$

Then $y = \frac{45,000}{150} = 300$.

Dimensions are 150 ft for fence parallel to the river and 300 ft for each of the other sides.

19. Quantity being maximized: $A = xy$

Another equation:

$800 = 10(2x) + 10(2y) + 20(2y)$

$800 = 20x + 60y$ or $x = 40 - 3y$

Substituting: $A(y) = y(40 - 3y) = 40y - 3y^2$

$$A'(y) = 40 - 6y = 0 \text{ if } y = \frac{20}{3}$$

$A'(6) > 0; A'\left(\frac{20}{3}\right) = 0; A'(7) < 0$

Then $x = 40 - 3\left(\frac{20}{3}\right) = 20$.

Dimensions are $\frac{20}{3}$ ft for the side parallel to the dividers and 20 ft for the other side.

21. $256 = x(2x)h$

$$h = \frac{128}{x^2}$$

$C = 10(x)(2x) + 5(x)(2x) + 5(2)(h)2x + 5(2x)(h)$

$$= 30x^2 + 30xh$$

$$C(x) = 30x^2 + 30x\left(\frac{128}{x^2}\right) = 30x^2 + \frac{3840}{x}$$

$$C'(x) = 60x - \frac{3840}{x^2}$$

$C'(x) = 0$ if $60x^3 = 3840$ or

$x^3 = 64$ or $x = 4$

$C'(3) < 0; C'(4) = 0; C'(5) > 0$

Minimum cost is at $x = 4$. Then $h = \frac{128}{16} = 8$.

Dimensions are 8" by 4" by 8" (high).

23. $C(x) = \frac{1,500,000}{x}(600) + 1,500,000(15) + \frac{x}{2} \cdot 2$

$$C'(x) = \frac{-900 \times 10^6}{x^2} + 1$$

$C'(x) = 0$ if $x^2 = 900 \times 10^6$ or $x = 30,000$.

$C'(1) < 0$, $C'(30,000) = 0$, $C'(40,000) > 0$

30,000 items should be produced in each run.

25. $C(x) = \frac{150,000}{x}(360) + 150,000(7) + \frac{x}{2} \cdot \frac{3}{4}$

$$C'(x) = \frac{-54,000,000}{x^2} + \frac{3}{8}$$

$C'(x) = 0$ if $x^2 = \frac{8}{3}(54,000,000)$ or $x = 12,000$

$C'(11,000) < 0; C'(12,000) = 0; C'(13,000) > 0$

12,000 items should be produced in each run.

27. $V(x) = x(12 - 2x)^2$

$V'(x) = x(2)(12 - 2x)(-2) + (12 - 2x)^2(1)$

$\quad = (12 - 2x)(12 - 6x)$

$V'(x) = 0$ if $x = 2$

Domain: $0 < x < 6$

$V'(1) > 0; V'(2) = 0; V'(3) < 0$

Volume is a maximum at $x = 2$.

29. Let x = time in weeks to pick oranges.

Quantity being maximized:

$R(x) = (8 - 0.5x)(5 + 0.5x), 0 \le x \le 5$

$R'(x) = (8 - 0.5x)(0.5) + (5 + 0.5x)(-0.5)$

$\quad = 1.5 - 0.50x$

$R'(x) = 0$ if $1.5 - 0.5x = 0$ or $x = \frac{1.5}{0.5} = 3$

x	0	3	5
$R(x)$	40	42.25	41.25

Thus, 3 weeks from now they should be picked.

31. Let x = number of plates and y = number of impressions.

Quantity being minimized:

$$C = 2x + 12.50\left(\frac{y}{1000}\right) = 2x + 0.0125y$$

Another equation: $xy = 100,000$ or $y = \frac{100,000}{x}$

Substituting: $C(x) = 2x + \frac{1250}{x}$

$$C'(x) = 2 - \frac{1250}{x^2}$$

$C'(x) = 0$ if $x = 25$.

$C'(24) < 0; C'(25) = 0; C'(26) > 0$

Thus, 25 plates will minimize the cost.

33. The graphing calculator shows a maximum at
$t = 5.3$ or in 2005.

Exercise 10.5

1. **a.** $x = 2$

    b. $\lim_{x \to \infty} f(x) = 1$

    c. $y = 1$

    d. $\lim_{x \to -\infty} f(x) = 1$

    The denominator of f is 0 when $x = 2$, but the numerator is not 0 when $x = 2$, so $x = 2$ is a vertical asymptote.

$$\lim_{x \to \infty} \frac{x-4}{x-2} = \lim_{x \to \infty} \frac{1 - \frac{4}{x}}{1 - \frac{2}{x}} = \frac{1-0}{1-0} = 1$$

$$\lim_{x \to -\infty} \frac{x-4}{x-2} = \lim_{x \to -\infty} \frac{1 - \frac{4}{x}}{1 - \frac{2}{x}} = \frac{1-0}{1-0} = 1$$

3. $f(x) = \dfrac{3(x^4 + 2x^3 + 6x^2 + 2x + 5)}{(x^2 - 4)^2}$

    a. $x = \pm 2$

    b. $\lim_{x \to \infty} f(x) = 3$

    c. $y = 3$

    d. $\lim_{x \to -\infty} f(x) = 3$

    When $x = \pm 2$, the denominator of f is 0, but the numerator is not, so $x = \pm 2$ are equations of vertical asymptotes.

$$\lim_{x \to \infty} \frac{3(x^4 + 2x^3 + 6x^2 + 2x + 5)}{(x^2 - 4)^2}$$

$$= 3 \lim_{x \to \infty} \frac{1 + \frac{2}{x} + \frac{6}{x^2} + \frac{2}{x^3} + \frac{5}{x^4}}{1 - \frac{8}{x} + \frac{16}{x^2}} = 3$$

Likewise, $\lim_{x \to -\infty} f(x) = 3$. An equation of the horizontal asymptote is $y = 3$.

5. $y = \dfrac{2x}{x - 3}$

$$\lim_{x \to \infty} \frac{2x}{x - 3} = \lim_{x \to \infty} \frac{2}{1 - \frac{3}{x}} = 2$$

Horizontal asymptote: $y = 2$
Vertical asymptote: $x = 3$

7. $y = \dfrac{x+1}{(x+2)(x-2)}$

$$\lim_{x \to \infty} \frac{x+1}{x^2 - 4} = \lim_{x \to \infty} \frac{\frac{1}{x} - \frac{1}{x^2}}{1 - \frac{4}{x^2}} = \frac{0}{1} = 0$$

Horizontal asymptote: $y = 0$
Vertical asymptotes: $x = \pm 2$

9. $y = \dfrac{3x^3 - 6}{x^2 + 4}$

Vertical asymptote: None since $x^2 + 4 \neq 0$
Horizontal asymptote: None since degree of numerator > degree of denominator.

11. $f(x) = \dfrac{2x + 2}{x - 3}$

$$f'(x) = \frac{(x-3)2 - (2x+2)}{(x-3)^2} = \frac{-8}{(x-3)^2}$$

VA; $x = 3$

$$\lim_{x \to \infty} f(x) = \lim_{x \to \infty} \frac{2 + \frac{2}{x}}{1 - \frac{3}{x}} = 2$$

HA: $y = 2$

No critical values. Thus, there can be no maximum or minimum points.

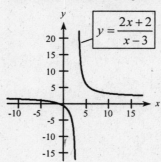

$$y = \frac{2x + 2}{x - 3}$$

13. $y = \dfrac{x^2 + 4}{x}$

$\dfrac{dy}{dx} = \dfrac{x(2x) - (x^2 + 4)}{x^2}$

$\qquad = \dfrac{(x+2)(x-2)}{x^2}$

Critical values: $x = -2, 2$

x		-2		0		2	
y'	$+$	0	$-$	$*$	$-$	0	$+$

* means y' is undefined at $x = 0$. In fact, 0 is not in the domain.

Relative maximum point: $(-2, -4)$

Relative minimum point: $(2, 4)$

$y > 0$ if $x > 0$ and $y < 0$ if $x < 0$.

VA: $x = 0$

HA: None (Deg $x^2 + 4 >$ Deg x)

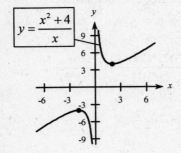

15. $y = \dfrac{27x^2}{(x+1)^3}$

$y' = \dfrac{(x+1)^3 \cdot 54x - 27x^2 \cdot 3(x+1)^2}{(x+1)^6}$

$\quad = \dfrac{27x(x+1)^2[2(x+1) - 3x]}{(x+1)^6} = \dfrac{27x(2-x)}{(x+1)^4}$

VA: $x = -1$

$\displaystyle \lim_{x \to \infty} y = \lim_{x \to \infty} \dfrac{27x^2}{x^3 + 3x^2 + 3x + 1}$

$\qquad = \displaystyle \lim_{x \to \infty} \dfrac{\frac{27}{x}}{1 + \frac{3}{x} + \frac{3}{x^2} + \frac{1}{x^3}} = \dfrac{0}{1} = 0$

HA: $y = 0$

Critical values: $x = 0, 2$

x		-1		0		2	
y'	$-$	$*$	$-$	0	$+$	0	$-$

*means y' is undefined at $x = -1$. In fact, -1 is not in the domain.

Relative minimum point: $(0, 0)$

Relative maximum point: $(2, 4)$

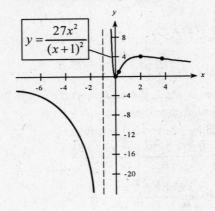

17. $f(x) = \dfrac{16x}{x^2 + 1}$

$f'(x) = \dfrac{(x^2+1)16 - 16x(2x)}{(x^2+1)^2} = \dfrac{16 - 16x^2}{(x^2+1)^2}$

Critical values: $x = -1, 1$

x		-1		1	
$f'(x)$	$-$	0	$+$	0	$-$

Relative maximum point: (1, 8)
Relative minimum point: (−1, −8)
VA: None since $x^2 + 1 \neq 0$. HA: $y = 0$ since
degree $(x^2 + 1) >$ degree $(16x)$.

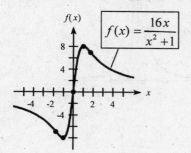

19. Critical value is $x = -1$.
($x = 1$ is not in the domain of the function.)

x		-1		1	
y'	$-$	0	$+$	$*$	$-$

* means y' is undefined at $x = 1$. In fact, 1 is
not in the domain.

Relative minimum point: $\left(-1, -\dfrac{1}{4}\right)$

$y > 0$ if $x > 0$ and $y < 0$ if $x < 0$.
(0, 0) is an x-intercept.

$\lim\limits_{x \to \infty} \dfrac{x}{(x-1)^2} = 0$

HA: $y = 0$, VA: $x = 1$

Point of inflection at $x = -2$ (y'' changes signs)

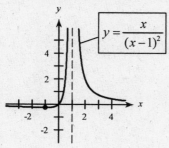

21. VA: $x = 3$, Critical values: $x = 2, 4$

x		2		3		4	
$f'(x)$	$+$	0	$-$	$*$	$-$	0	$+$

* means $f'(3)$ is undefined. In fact, 3 is not in
the domain.
Relative maximum at $x = 2$.
Relative minimum at $x = 4$.
$y'' < 0$ when $x < 3$, and
$y'' > 0$ when $x > 3$ but $x = 3$ is not in the
domain of the function, so there are no inflection
points.

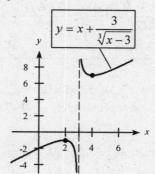

23. VA: $x = 0$

$$\lim_{x \to \infty} f(x) = \lim_{x \to \infty} \frac{9(x-2)^{2/3}}{(x^3)^{2/3}} = \lim_{x \to \infty} 9\left(\frac{x-2}{x^3}\right)^{2/3} = \lim_{x \to \infty} 9\left(\frac{\frac{1}{x^2} - \frac{2}{x^3}}{1}\right)^{2/3} = 9 \cdot \left(\frac{0}{1}\right)^{2/3} = 0$$

HA: $y = 0$, Critical values: $x = 3$ and $x = 2$

x	0		2		3	
$f'(x)$	+	*	−	*	+	0	−

* means $f'(0)$ and $f'(2)$ are undefined. In fact, 0 is not in the domain.

Relative minimum: (2, 0)
Relative maximum: (3, 1)
Possible inflection points: $x = 2$ and

$$x = \frac{42 \pm \sqrt{42^2 - 4(7)(54)}}{2(7)} = \frac{42 \pm 6\sqrt{7}}{14} = 3 \pm \frac{3\sqrt{7}}{7} \approx 1.87 \text{ and } 4.13$$

x	0		$3 - \frac{3\sqrt{7}}{7}$		2		$3 + \frac{3\sqrt{7}}{7}$	
$f''(x)$	+	*	+	0	−	*	−	0	+

* means $f''(0)$ and $f''(2)$ are undefined.

Points of inflection at approximately (1.87, 0.68) and (4.13, 0.87).

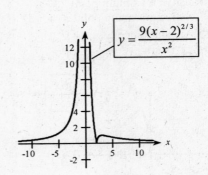

25. a. HA: approximately $y = -2$; VA: approximately $x = 4$

b. $f(x) = \frac{9x}{17 - 4x} = \frac{9}{\frac{17}{x} - 4}$, HA: $y = -\frac{9}{4}$; VA: $x = \frac{17}{4}$

27. a. HA: approximately $y = 2$; VA: approximately $x = \pm 2.5$

b. $f(x) = \frac{20x^2 + 98}{9x^2 - 49} = \frac{20 + \frac{98}{x^2}}{9 - \frac{49}{x^2}}$, HA: $y = \frac{20}{9}$; VA: $x = \pm \frac{7}{3}$

29. a. Standard Viewing Window

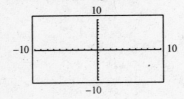

c.

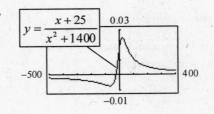

b. $f(x) = \frac{x + 25}{x^2 + 1400}$

HA: $y = 0$

degree $(x^2 + 1400) > $ degree $(x + 25)$

$$f'(x) = \frac{(x^2 + 1400) \cdot 1 - (x + 25)2x}{(x^2 + 1400)^2}$$

$$= \frac{(70 + x)(20 - x)}{(x^2 + 1400)^2}$$

x		−70		20	
$f'(x)$	−	0	+	0	−

Relative minimum at $x = -70$. Relative
maximum at $x = 20$.

31. a.   Standard Viewing Window

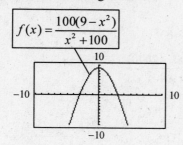

b. $f(x) = \dfrac{100(9 - x^2)}{x^2 + 100}$

$\lim\limits_{x \to \infty} f(x) = \lim\limits_{x \to \infty} \dfrac{100\left(\frac{9}{x^2} - 1\right)}{1 + \frac{100}{x^2}} = -100$

HA: $y = -100$

$f'(x) = 100\left[\dfrac{(x^2 + 100)(-2x) - (9 - x^2)2x}{(x^2 + 100)^2}\right]$

$ = \dfrac{100(-218x)}{(x^2 + 100)^2}$

x		0	
$f'(x)$	+	0	−

Relative maximum: $(0, 9)$
For viewing window try: x: -75 to 75
y: -120 to 20.

c.

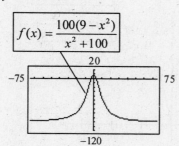

33. a. Standard viewing window: See graphing section below problem.

b. $f(x) = \dfrac{1000(x-4)}{(x-50)(x+40)}$ $\displaystyle\lim_{x\to\infty} f(x) = \lim_{x\to\infty} \dfrac{\frac{1000}{x} - \frac{4000}{x^2}}{1 - \frac{10}{x} - \frac{2000}{x^2}} = \dfrac{0}{1} = 0$

Horizontal asymptote: $y = 0$, Vertical asymptotes: $x = -40$, $x = 50$

$f'(x) = 1000 \cdot \dfrac{(x^2 - 10x - 2000) \cdot 1 - (x-4)(2x-10)}{[(x-50)(x+40)]^2} = \dfrac{1000(-x^2 + 8x - 2040)}{[(x-50)(x+40)]^2}$

Numerator is never zero. No relative maximum or relative minimum.
For viewing window, try x: -200 to 200, y: -200 to 200

a.

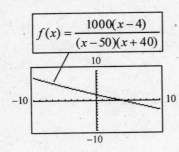

c.

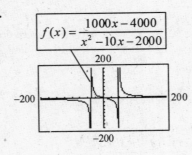

35. $p(C) = \dfrac{100C}{7300 + C}$

$p'(C) = \dfrac{(7300 + C)100 - 100C(1)}{(7300 + C)^2}$

$= \dfrac{730000}{(7300 + C)^2}$

a. Domain $= \{C : C \ge 0\}$. Thus $p(C)$ exists for all C in the domain.

b. $p'(C)$ is always positive. Thus, $p(C)$ is increasing if $C \ge 0$.

c. $\displaystyle\lim_{C\to\infty} \dfrac{100C}{7300 + C} = 100$

Horizontal asymptote: $p = 100$

d. No

37. $R(t) = \dfrac{50t}{t^2 + 36}$

$R'(t) = \dfrac{(t^2 + 36)50 - 50t(2t)}{(t^2 + 36)^2} = \dfrac{50(36 - t^2)}{(t^2 + 36)^2}$

Critical value: $t = 6$

$R'(5) > 0$, $R'(6) = 0$, $R'(7) < 0$

Horizontal asymptote: $y = 0$ (for $t > 0$)

a.

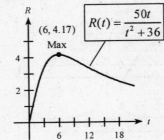

b. Revenue is maximized at $t = 6$.

c. Revenue is decreasing for $t > 6$. The video will be released in $6 + 4 + 12 = 22$ weeks from date of original release.

39. $f(x) = \dfrac{289.173 - 58.5731x}{x + 1}$

a. Yes; $x = -1$

b. No; Domain: $x \ge 5$

c. Yes; $y = -58.5731$

d. At high wind speeds any additional wind will have little effect on the wind-chill. Interpretation is meaningful.

41. a. $P = C$
   b. C
   c. 0
   d. 0

43. a. $\lim\limits_{t \to \infty} f(t) = 0$ (degree denominator > degree numerator)
   b. The percentage of workers in farm occupations approaches 0.
   c. No. Denominator > 0 $(b^2 - 4ac < 0)$.

45. a. No. Barometric pressure is always greater than zero. It can drop off the scale as shown.
   b. A blizzard struck the East Coast, causing $3-6 billion in damages and approximately 270 deaths.

Review Exercises

1. $y = -x^2$

$y' = -2x$

Critical value is at $x = 0$. Critical point is (0, 0).

$y'(-1) > 0$, $y'(0) = 0$, $y'(1) < 0$

(0, 0) is a maximum point.

2. $p = q^2 - 4q - 5$

$p' = 2q - 4$

Since $p' = 0$ for $q = 2$, the point (2, –9) is a critical point. To the left of (2, –9) we have $p' < 0$ and to the right of (2, –9) we have $p' > 0$. The point (2, –9) is a minimum point.

3. $f(x) = 1 - 3x + 3x^2 - x^3$

$f'(x) = -3 + 6x - 3x^2$

$\qquad = -3(x^2 - 2x + 1)$

$\qquad = -3(x - 1)^2$

$f'(x) = 0$ if $x = 1$

$f(1) = 1 - 3 + 3 - 1 = 0$

Critical point is (1, 0).

x		1	
$f'(x)$	–	0	–

(1, 0) is a horizontal point of inflection.

4. $f(x) = \dfrac{3x}{x^2 + 1}$

$f'(x) = \dfrac{(x^2 + 1)(3) - (3x)(2x)}{(x^2 + 1)^2}$

$\qquad = \dfrac{-3x^2 + 3}{(x^2 + 1)^2}$

$\qquad = \dfrac{-3(x + 1)(x - 1)}{(x^2 + 1)^2}$

$f'(x) = 0$ if $x = -1$ or $x = 1$. Critical points are

$\left(-1, -\dfrac{3}{2}\right)$ and $\left(1, \dfrac{3}{2}\right)$.

$f'(-2) < 0$; $f'(-1) = 0$; $f'(0) > 0$;

$f'(1) = 0$; $f'(2) < 0$ means that we have a

relative minimum at $\left(-1, -\dfrac{3}{2}\right)$ and a relative

maximum at $\left(1, \dfrac{3}{2}\right)$.

5. $f(x) = x^3 + x^2 - x - 1$

$f'(x) = 3x^2 + 2x - 1 = (3x - 1)(x + 1)$

a. Critical points are (–1, 0) and $\left(\dfrac{1}{3}, -\dfrac{32}{27}\right)$.

x		–1		$\frac{1}{3}$	
$f'(x)$	+	0	–	0	+

b. Relative maximum: (–1, 0)

Relative minimum: $\left(\dfrac{1}{3}, -\dfrac{32}{27}\right)$

c. No horizontal points of inflection.

d.

6. $f(x) = 4x^3 - x^4$

a. $f'(x) = 12x^2 - 4x^3 = 4x^2(3 - x)$

Critical values: 0, 3

b. $f'(x) > 0$ when $x < 3$, $x \neq 0$.

$f'(x) < 0$ when $x > 3$.

Relative maximum at (3, 27).

c. Horizontal point of inflection: (0, 0)

d.

7. $f(x) = x^3 - \dfrac{15}{2}x^2 - 18x + \dfrac{3}{2}$

$f'(x) = 3x^2 - 15x - 18 = 3(x-6)(x+1)$

$f'(x) = 0$ at $x = -1, 6$.

a. Critical points: $(-1, 11)$ and $(6, -160.5)$.

x	-1		6	
$f'(x)$	$+$	0	$-$	0	$+$

b. Relative maximum: $(-1, 11)$
Relative minimum: $(6, -160.5)$

c. No horizontal points of inflection.

d.

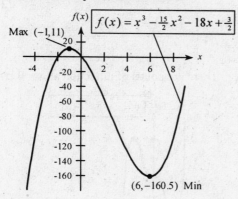

8. $y = f(x) = 5x^7 - 7x^5 - 1$

a. $y' = f'(x) = 35x^6 - 35x^4$

$= 35x^4(x+1)(x-1)$

Critical values: $0, -1, 1$

b. $f'(x) > 0$ when $x < -1$ and when $x > 1$.

$f'(x) < 0$ when $-1 < x < 1$, $x \neq 0$.

Relative maximum: $(-1, 1)$
Relative minimum: $(1, -3)$

c. Horizontal point of inflection: $(0, -1)$

d.

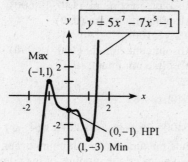

9. $y = x^{2/3} - 1$

$y'(x) = \dfrac{2}{3x^{1/3}}$

a. Critical value is at $x = 0$.

$y'(-1) < 0$, $y'(0)$ is undefined, $y'(1) > 0$

b. $(0, -1)$ is a minimum point.

c. No horizontal points of inflection.

d.

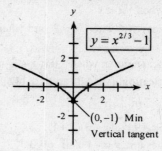

10. $y = f(x) = x^{2/3}(x-4)^2$

a. $f'(x) = x^{2/3} \cdot 2(x-4) + (x-4)^2 \cdot \dfrac{2}{3}x^{-1/3}$

$= \dfrac{2}{3}(x-4)x^{-1/3}(4x-4)$

Critical values: $0, 1, 4$

b. $f'(x) > 0$ when $0 < x < 1$ and when $x > 4$.

$f'(x) < 0$ when $x < 0$ and when $1 < x < 4$.

Relative minimum: $(0, 0)$; Relative minimum: $(4, 0)$; Relative maximum: $(1, 9)$

c. No horizontal points of inflection.

d.

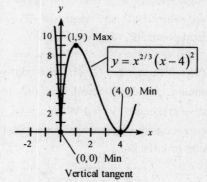

11. $y = x^4 - 3x^3 + 2x - 1$

$y' = 4x^3 - 9x^2 + 2$

$y'' = 12x^2 - 18x$

At $x = 2$ we have $y'' = 12$. Thus, the curve is concave up.

12. $y = f(x) = x^4 - 2x^3 - 12x^2 + 6$

$f'(x) = 4x^3 - 6x^2 - 24x$

$f''(x) = 12x^2 - 12x - 24$

$\qquad = 12(x^2 - x - 2)$

$\qquad = 12(x-2)(x+1)$

$f''(x) > 0$ when $x > 2$ and when $x < -1$.

$f''(x) < 0$ when $-1 < x < 2$.

$f(x)$ is concave up when $x > 2$ and when $x < -1$;

$f(x)$ is concave down when $-1 < x < 2$.

Points of inflection: $(-1, -3)$ and $(2, -42)$.

13. $y = x^3 - 3x^2 - 9x + 10$

$y' = 3x^2 - 6x - 9 = 3(x-3)(x+1)$

Critical values are $x = -1$ and $x = 3$.

$y'' = 6x - 6$. So, possible point of inflection is at $x = 1$.

x	-1	1	3
y''	$-$	0	$+$

There is a point of inflection at $(1, -1)$.

$f''(-1) < 0$ means there is a relative maximum at $(-1, 15)$.

$f''(3) > 0$ means there is a relative minimum at $(3, -17)$.

14. $y = x^3 - 12x$

$y' = 3x^2 - 12$

$y' = 0$ when $x = 2$ and when $x = -2$.

Critical points: $(2, -16)$ and $(-2, 16)$

$y'' = f''(x) = 6x$

$f''(2) > 0$ means that $(2, -16)$ is a relative minimum. $f''(-2) < 0$ means that $(-2, 16)$ is a relative maximum. $y'' = 0$ when $x = 0$. $y'' < 0$ when $x < 0$ and $y'' > 0$ when $x > 0$. So, $(0, 0)$ is a point of inflection.

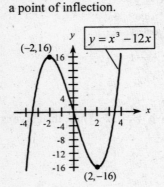

15. $y = 2 + 5x^3 - 3x^5$

$y' = 15x^2 - 15x^4 = 15x^2(1 - x^2)$

Critical values: $x = 0, 1, -1$

x		-1		0		1	
y'	$-$	0	$+$	0	$+$	0	$-$

Relative minimum: $(-1, 0)$

Relative maximum: $(1, 4)$

$y'' = 30x - 60x^3 = 30x(1 - 2x^2)$

Possible points of inflection at $x = 0, \dfrac{1}{\sqrt{2}}, -\dfrac{1}{\sqrt{2}}$.

x		$-\frac{1}{\sqrt{2}}$		0		$\frac{1}{\sqrt{2}}$	
y''	$+$	0	$-$	0	$+$	0	$-$

There are points of inflection at $x = 0, \pm\dfrac{1}{\sqrt{2}}$.

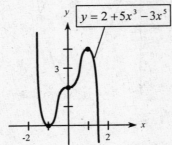

16. $R(x) = 280x - x^2$

a. $0 \le x \le 200$

$R'(x) = 280 - 2x$

$R'(x) = 0$ when $x = 140$

$R(0) = 0$, $R(140) = 19,600$,

$R(200) = 16,000$

Absolute maximum: $(140, 19,600)$

Absolute minimum: $(0, 0)$

b. $0 \le x \le 100$

Absolute maximum: $(100, 18,000)$

Absolute minimum: $(0, 0)$

17. $y = 6400x - 18x^2 - \dfrac{x^3}{3}$

$y' = 6400 - 36x - x^2 = (100 + x)(64 - x)$

Absolute maximum and minimum occur at critical values or at endpoints of the domain.

a. No critical values.

$y(50) \approx 233,333$ absolute maximum

$y(0) = 0$ absolute minimum

b. Critical value is at $x = 64$.

$y(64) \approx 248,491$ absolute maximum

$y(0) = 0$ absolute minimum

$y(100) \approx 126,667$

18. a. Vertical asymptote: $x = 1$
  b. Horizontal asymptote: $y = 0$
  c. $\lim\limits_{x \to \infty} f(x) = 0$
  d. $\lim\limits_{x \to -\infty} f(x) = 0$

19. a. $x = -1$
  b. $y = \dfrac{1}{2}$
  c. $\dfrac{1}{2}$
  d. $\dfrac{1}{2}$

20. $y = \dfrac{3x + 2}{2x - 4}$

Vertical asymptote: $x = 2$
Horizontal asymptote:

$$y = \lim_{x \to \infty} \frac{3x + 2}{2x - 4} = \lim_{x \to \infty} \frac{3 + \frac{2}{x}}{2 - \frac{4}{x}} = \frac{3}{2}$$

21. $y = \dfrac{x^2}{1 - x^2}$

$$\lim_{x \to \infty} \frac{x^2}{1 - x^2} = \lim_{x \to \infty} \frac{1}{\frac{1}{x^2} - 1} = -1$$

Horizontal asymptote: $y = -1$
Vertical asymptotes: $x = \pm 1$

22. $y = \dfrac{3x}{x + 2}$
  a. Vertical asymptote: $x = -2$

  Horizontal asymptote: $y = \lim\limits_{x \to \infty} \dfrac{3x}{x + 2} = 3$

  b. $y' = \dfrac{(x + 2)(3) - (3x)}{(x + 2)^2} = \dfrac{6}{(x + 2)^2}$

  y' is never zero, no maximum nor minimum.

  c.

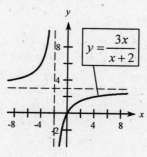

23. $y = \dfrac{8x - 16}{x^2}$

$$\frac{dy}{dx} = \frac{x^2 \cdot 8 - (8x - 16)2x}{x^4} = \frac{8(4 - x)}{x^3}$$

Critical value at $x = 4$.
$y'(3) > 0$, $y'(4) = 0$, $y'(5) < 0$
  a. Horizontal asymptote: $y = 0$
  Vertical asymptote: $x = 0$
  b. Relative maximum: $(4, 1)$
  c.

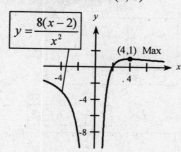

24. $y = \dfrac{x^2}{x - 1}$
  a. Vertical asymptote: $x = 1$
  Horizontal asymptote: none
  b. $y' = \dfrac{(x - 1)(2x) - x^2}{(x - 1)^2} = \dfrac{x^2 - 2x}{(x - 1)^2}$

  $y' = 0$ when $x = 0$ and when $x = 2$.
  Critical points: $(0, 0)$, $(2, 4)$
  $y' = f'(-1) > 0$, $f'(0.5) < 0$, $f'(1.5) < 0$, $f'(3) > 0$
  Relative maximum: $(0, 0)$
  Relative minimum: $(2, 4)$
  c.

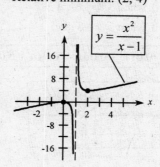

25. a. $f'(x) > 0$ if $x < \frac{2}{3}$ and if $x > 2$.

$f'(x) < 0$ if $\frac{2}{3} < x < 2$.

$f'(x) = 0$ if $x = \frac{2}{3}$ and if $x = 2$.

b. $f''(x) > 0$ if $x > \frac{4}{3}$.

$f''(x) < 0$ if $x < \frac{4}{3}$.

$f''(x) = 0$ if $x = \frac{4}{3}$.

c.

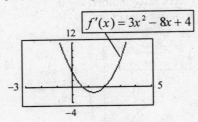

$f'(x) = 3x^2 - 8x + 4$

d.

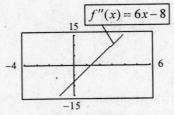

$f''(x) = 6x - 8$

26. $f(x) = 0.0025x^4 + 0.02x^3 - 0.48x^2 + 0.08x + 4$

a. From the graph, the estimated values are

$f'(x) > 0$ when $-13 < x < 0$ and $x > 7$.

$f'(x) < 0$ when $x < -13$ and $0 < x < 7$.

$f'(x) = 0$ when $x = -13$, $x = 0$ and $x = 7$.

b. Estimates from the graph:

$f''(x) > 0$ when $x < -8$ and $x > 4$.

$f''(x) < 0$ when $-8 < x < 4$.

$f''(x) = 0$ when $x = -8$ and $x = 4$.

c.

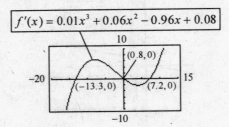

$f'(x) = 0.01x^3 + 0.06x^2 - 0.96x + 0.08$

d.

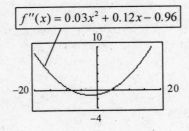

$f''(x) = 0.03x^2 + 0.12x - 0.96$

27. a. $f(x)$ is increasing when $f'(x) > 0$.

$f(x)$ is decreasing when $f'(x) < 0$.

Increasing if $x < -5$ and if $x > 1$.

Decreasing if $-5 < x < 1$.

Relative maximum at $x = -5$ (+ to −)

Relative minimum at $x = 1$ (− to +)

b. If $f'(x)$ is increasing, then $f''(x) > 0$.

If $f'(x)$ is decreasing, then $f''(x) < 0$.

$f''(x) > 0$ if $x > -2$, $f''(x) < 0$ if $x < -2$

$f''(x) = 0$ if $x = -2$.

c. $f(x) = \frac{x^3}{3} + 2x^2 - 5x$ $f'(x) = x^2 + 4x - 5$

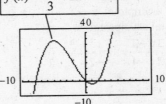

$f(x) = \frac{x^3}{3} + 2x^2 - 5x$

d.

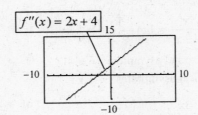

$f''(x) = 2x + 4$

28. $f'(x) = 6x^2 - x^3$

   a. From the graph,

       $f(x)$ is increasing when $x < 6$, $x \neq 0$.

       $f(x)$ is decreasing when $x > 6$.

       $f(x)$ has a relative maximum at $x = 6$.

       $f(x)$ has a horizontal point of inflection at $x = 0$.

   b. From the graph,

       $f''(x) > 0$ when $0 < x < 4$.

       $f''(x) < 0$ when $x < 0$ and when $x > 4$.

       $f''(x) = 0$ when $x = 0$ and $x = 4$.

   c. $f(x) = 2x^3 - \dfrac{x^4}{4}$, $f'(x) = 6x^2 - x^3$

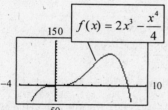

   d. $f''(x) = 12x - 3x^2 = 3x(4 - x)$

       Graph of f''

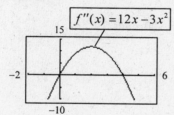

29. a. $f(x)$ is concave up if $f''(x) > 0$.

       $f(x)$ is concave down if $f''(x) < 0$.

       Concave up if $x < 4$.

       Concave down if $x > 4$.

       There is a point of inflection at $x = 4$.

   b. $f(x) = 2x^2 - \dfrac{x^3}{6}$, $f'(x) = 4x - \dfrac{1}{2}x^2$,

       $f''(x) = 4 - x$

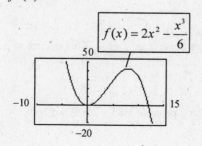

30. $f''(x) = 6 - x - x^2$

   a. From the given graph,

       $f(x)$ is concave up when $-3 < x < 2$.

       $f(x)$ is concave down when $x < -3$, and when $x > 2$. $f(x)$ has a point of inflection at $x = -3$ and at $x = 2$.

   b. $f(x) = 3x^2 - \dfrac{x^3}{6} - \dfrac{x^4}{12}$

       $f'(x) = 6x - \dfrac{1}{2}x^2 - \dfrac{1}{3}x^3$

       $f''(x) = 6 - x - x^2$

       Graph of f:

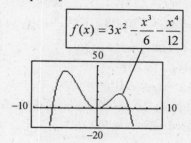

31. $\overline{C}(x) = 3x + 15 + \dfrac{75}{x}$

   $\overline{C}'(x) = 3 - \dfrac{75}{x^2}$

   $\overline{C}'(x) = 0$ if $x^2 = 25$ or $x = 5$.

x		5	
$\overline{C}'(x)$	$-$	0	$+$

   $\overline{C}(5) = 15 + 15 + 15 = 45$

   Minimum average cost is \$45 at 5 units.

32. $R(x) = 32x - 0.01x^2$

   a. $R'(x) = 32 - 0.02x$

       $R'(x) = 0$ when $x = 1600$.

x		1600	
$R'(x)$	$+$	0	$-$

       $R(1600) = \$25,600$

       The maximum revenue is \$25,600 when $x = 1600$ units are produced.

   b. If production is limited to 1500, then the maximum revenue occurs when $x = 1500$. So, the maximum revenue would then be \$25,500.

33. $P(x) = 1080x + 9.6x^2 - 0.1x^3 - 50,000$

$P'(x) = 1080 + 19.2x - 0.3x^2$

$P'(x) = 0$ if $-0.3(x^2 - 64x - 3600)$ or if

$(x - 100)(x + 36) = 0$.

x		100	
$P'(x)$	+	0	−

Maximum profit is at $x = 100$.

$P(100) = \$54,000$ is the maximum profit.

34. $R(x) = 46x - 0.01x^2$

$C(x) = 0.05x^2 + 10x + 1100$

$P(x) = R(x) - C(x)$

$\quad = (46x - 0.01x^2) - (0.05x^2 + 10x + 1100)$

$\quad = -0.06x^2 + 36x - 1100$

$P'(x) = -0.12x + 36$

$P'(x) = 0$ when $x = 300$

Since $P''(x) = -0.12 < 0$ for all x, the profit is maximized when $x = 300$.

35. $P(x) = 80x - \frac{1}{4}x^2 - (800 + 4x)$

$\quad = -\frac{1}{4}x^2 + 76x - 800$

$P'(x) = -\frac{1}{2}x + 76$

$P'(x) = 0$ if $x = 152$.

x		152	
$P'(x)$	+	0	−

Since 152 is not in the domain and P is increasing for $0 \le x \le 150$, the maximum profit is at 150 units.

36. $C = 2x^2 + 54x + 98$

$\bar{C} = 2x + 54 + \frac{98}{x}$

$\bar{C}' = 2 - \frac{98}{x^2} = \frac{2x^2 - 98}{x^2}$

$\bar{C}' = 0$ when $x = 7$.

Since $\bar{C}'' = \frac{196}{x^3} > 0$ for all $x > 0$, the average cost is minimized when $x = 7$.

37. Profit is maximized when $\overline{MP}$ changes from + to −. This occurs at $x = 500$ units.

38. $P'(t)$ is to be maximized (not $P(t)$).

$P(t) = \frac{95t^2}{t^2 + 2700} + 5$

$P'(t) = \frac{(t^2 + 2700) \cdot 190t - 95t^2 \cdot 2t}{(t^2 + 2700)^2} = \frac{513,000t}{(t^2 + 2700)^2}$

$P''(t) = \frac{(t^2 + 2700)^2 \cdot 513,000 - 513,000t \cdot 2(t^2 + 2700) \cdot 2t}{(t^2 + 2700)^4}$

$\quad = \frac{513,000(t^2 + 2700)[(t^2 + 2700) - 4t^2]}{(t^2 + 2700)^4} = \frac{513,000(2700 - 3t^2)}{(t^2 + 2700)^3}$

$P''(t)$ changes from + to − at $t = 30$. Point of diminishing returns is at $t = 30$ hours.

39. a. Diminishing returns or point of inflection occurs at $x = 60$.

b. $m = \frac{f(I) - 0}{I - 0} = \frac{f(I)}{I} = $ Average output

c. Maximum average output occurs when slope of average output is closest to $f'(I)$. This occurs at $x = 70$.

40. $R = (54 + 10x)(385 - 25x) = 20.790 + 2500x - 250x^2$

$R' = 2500 - 500x$

R' changes from + to − at $x = 5$. Revenue is maximized with selling price of $385 - 25(5) = \$260$.

41. $P = 20,790 + 2500x - 250x^2 - (200 \cdot 10x) = 20,790 + 500x - 250x^2$

$P' = 500 - 500x$

Profit is maximized when $P' = 0$ or $x = 1$. The selling price will be $385 - 25(1) = \$360$.

42. Equilibrium price means supply = demand.

$1200 - 2x = 200 + 2x$ gives equilibrium quantity $x = 250$. Equilibrium price is $p = 200 + 2(250) = 700$.

Revenue $= 700x$. Cost $= 12,000 + 50x + x^2$.

$P(x) = 700x - (12,000 + 50x + x^2) = -12,000 + 650x - x^2$

$P'(x) = 650 - 2x$

$P'(x) = 0$ if $x = 325$.

x	325	
$P'(x)$	$+$ $\quad$ 0	$-$

$P(325) = -12,000 + 211,250 - 105,625 = \$93,625$

43. $\overline{C} = 200 + x,\ C = 200x + x^2,\ p = 800 - x,\ R = 800x - x^2$

   a. $P(x) = 800x - x^2 - 200x - x^2 = 600x - 2x^2$

       $P'(x) = 600 - 4x$

       $P'(x) = 0$ when $x = 150$.

       $P''(x) = -4 < 0$ for all x means that $x = 150$ maximizes profit.

   b. The selling price is $p = 800 - 150 = \$650$.

44. $R(x) = 7000x - 10x^2 - \dfrac{x^3}{3} \qquad C(x) = 40,000 + 600x + 8x^2$

$P(x) = -\dfrac{x^3}{3} - 18x^2 + 6400x - 40,000$

$P'(x) = -x^2 - 36x + 6400 = (64 - x)(100 + x)$

Profit is a maximum at $x = 64$. $\left(P'(64) = 0\right)$

x	64	
$P'(x)$	$+$ $\quad$ 0	$-$

$P(64) = -87,381.33 - 73,728 + 409,600 - 40,000 = \$208,490.67$

45. $R(x) = x^2\left(500 - \dfrac{x}{3}\right) = 500x^2 - \dfrac{x^3}{3} \qquad R'(x) = 1000x - x^2$

$R'(x) = 0$ when $x = 0$ and when $x = 1000$.

$R''(x) = 1000 - 2x$ and $R''(1000) < 0$. So, the maximum reaction occurs at $x = 1000$.

46. $N(t) = 4 + 3t^2 - t^3 \qquad N'(t) = 6t - 3t^2 = 3t(2 - t)$

Critical values: $t = 0, 2$

$N''(t) = 6 - 6t,\ N''(2) = -6 < 0$

Thus, maximum at $t = 2$. Maximum production occurs at $8:00 + 2$ hrs $= 10:00$ a.m.

47. $P = 300 + 10t - t^2,\ t = 0$ is 2000 $\ 0 \le t \le 10$

$P' = 10 - 2t = 0$ when $t = 5$.

t	5	
$P'(t)$	$+$ $\quad$ 0	$-$

When $t = 5$, $P = 325$. The largest graduating class is the class of 2005 with 325 graduates.

48. $b(x) = \dfrac{8k}{x^2} + \dfrac{k}{(30-x)^2}$

$b'(x) = \dfrac{-16k}{x^3} - \dfrac{2k}{(30-x)^3}(-1)$

Set $b'(x) = 0$ and simplify.

$x^3 = 8(30-x)^3$

$\quad x = 2(30-x)$

$3x = 60$ or $x = 20$

x		20	
$b'(x)$	$-$	0	$+$

Build the observatory 20 miles from A.

49. Maximize $A = xy$ where $2x + 2y = 16$.

$A = x(8-x) = 8x - x^2 \qquad A' = 8 - 2x$

$A' = 0$ when $x = 4$.

Since $A'' < 0$ for all x, a 4×4 playpen gives a maximum area.

50. Quantity to be minimized: $A = (x+2)\left(y + \dfrac{7}{4}\right)$

Another equation: $xy = 56$ or $y = \dfrac{56}{x}$

Substituting: $A = (x+2)\left(\dfrac{56}{x} + \dfrac{7}{4}\right)$

$\qquad A = 56 + \dfrac{7}{4}x + \dfrac{112}{x} + \dfrac{7}{2}$

$\qquad A' = \dfrac{7}{4} - \dfrac{112}{x^2}$

Set $A' = 0$ and simplify. $7x^2 = 448$

$\qquad\qquad\qquad x^2 = 64$ or $x = 8$

x		8	
A'	$-$	0	$+$

So, $x = 8$ and $y = 7$ minimizes A.

Page dimensions are $(8+2)$ by $\left(7 + 1\dfrac{3}{4}\right)$ or $10''$

by $8\dfrac{3}{4}''$.

51. $R(x) = x^2\left(500 - \dfrac{x}{3}\right) = 500x^2 - \dfrac{x^3}{3}$

$R'(x) = 1000x - x^2$

$R''(x) = 1000 - 2x$

$R''(x) = 0$ when $x = 500$.

x		500	
$R''(x)$	$+$	0	$-$

So, the dosage $x = 500$ maximizes sensitivity.

52. a. $T(x) = -0.848x^3 + 33.17x^2 - 151.4x + 2920$

$T'(x) = -2.544x^2 + 66.34x - 151.4$

$T''(x) = -5.088x + 66.34$

$T''(x) = 0$ when $x \approx 13.04$ (year 1994).

b. $T''(x)$ changes from $+$ to $-$ at $x \approx 13.04$, so this is an inflection point for $T(x)$.

53. Total production cost: $\left(\dfrac{288,000}{x}\right)(1500)+(288,000)(30)$ Total storage cost: $\left(\dfrac{x}{2}\right)(1.5)$

$$C(x)=\left(\frac{288,000}{x}\right)(1500)+(288,000)(30)+\frac{1.5x}{2}=\frac{432,000,000}{x}+8,640,000+0.75x$$

$$C'(x)=-\frac{432,000,000}{x^2}+0.75 \quad C'(x)=0 \text{ when } x^2=576,000,000 \text{ or when } x=24,000.$$

$$C''(x)=\frac{864,000,000}{x^3}>0 \text{ for } x>0.$$

So, the minimum cost occurs for a run of $x=24,000$.

54. a. $\overline{C}(x)=\dfrac{4500+120x+0.05x^2}{x}$

Vertical asymptote at $x=0$.

Horizontal asymptote: $\displaystyle\lim_{x\to\infty}\frac{4500+120x+0.05x^2}{x}=\lim_{x\to\infty}\left(\frac{4500}{x}+120+0.05x\right)=\infty$ (so none).

b.

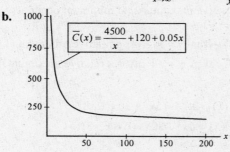

55. a. $M(0)=\dfrac{3.8(0)^2+3}{0.1(0)^2+1}=3\%$

b. Horizontal asymptote: $\displaystyle\lim_{t\to\infty}\frac{3.8t^2+3}{0.1t^2+1}=\lim_{t\to\infty}\frac{3.8+\dfrac{3}{t^2}}{0.1+\dfrac{1}{t^2}}=\frac{3.8}{0.1}=38\%$

The limit for the company's percent share of the market is 38%.

Chapter Test

1. $f(x)=x^3+6x^2+9x+3$

$f'(x)=3x^2+12x+9=3(x+3)(x+1)$

x		-3		-1	
$f'(x)$	$+$	0	$-$	0	$+$

Relative maximum at $x=-3$ $(-3,3)$
Relative minimum at $x=-1$ $(-1,-1)$
Polynomials have no asymptotes.

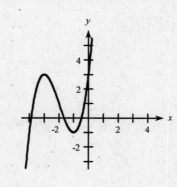

2. $y = 4x^3 - x^4 - 10$

$y' = 12x^2 - 4x^3 = 4x^2(3-x)$

x		0		3	
$f'(x)$	+	0	+	0	−

Maximum at $(3, 17)$.

$y'' = 24x - 12x^2 = 12x(2-x)$

x		0		2	
$f''(x)$	−	0	+	0	−

Inflection points at $(0, -2)$ and $(2, 6)$.

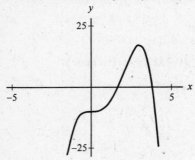

3. $y = \dfrac{x^2 - 3x + 6}{x - 2}$

$y' = \dfrac{(x-2)(2x-3) - (x^2 - 3x + 6)(1)}{(x-2)^2} = \dfrac{x(x-4)}{(x-2)^2}$

x		0		2		4	
y'	+	0	−	*	−	0	+

* means y' is undefined when $x = 2$.

In fact, $x = 2$ is not in the domain.

Relative maximum at $x = 0$ $(0, -3)$

Relative minimum at $x = 4$ $(4, 5)$

Vertical asymptote: $x = 2$

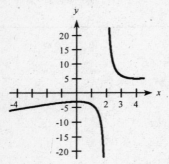

4–6. $y = 3x^5 - 5x^3 + 2$

$y' = 15x^4 - 15x^2 = 15x^2(x+1)(x-1)$

$y'' = 60x^3 - 30x$

$\quad = 30x(2x^2 - 1)$

$\quad = 30x\left[2\left(x + \dfrac{1}{\sqrt{2}}\right)\left(x - \dfrac{1}{\sqrt{2}}\right)\right]$

4. Concave up means $y'' > 0$.

$\left(\dfrac{1}{\sqrt{2}} \approx 0.707\right)$ $\quad -\dfrac{1}{\sqrt{2}} < x < 0, \quad x > \dfrac{1}{\sqrt{2}}$

5. y'' changes signs for inflection points.

$x = 0, \pm\dfrac{1}{\sqrt{2}} \approx \pm 0.707$

6. Relative maximum and minimum occur at $x = \pm 1$.

$y'' > 0$ at $x = 1$ gives a relative minimum $(1, 0)$.

$y'' < 0$ at $x = -1$ gives a relative maximum $(-1, 4)$.

7. $f(x) = 2x^3 - 15x^2 + 3$; $[-2, 8]$

$f'(x) = 6x^2 - 30x = 6x(x-5)$

x	−2	0	5	8
$f(x)$	−73	3	−122	67

Absolute maximum of 67 at $x = 8$.

Absolute minimum of -122 at $x = 5$.

8. $f(x) = \dfrac{200x - 500}{x + 300} = \dfrac{200 - \frac{500}{x}}{1 + \frac{300}{x}}$

Horizontal asymptote: $y = 200$

Vertical asymptote: $x = -300$

9.

Point	f	f'	f''
A	−	+	−
B	+	−	0
C	+	0	+

$f(x) < 0$ if below x-axis

$f'(x)$ is positive if $f(x)$ is increasing.

$f''(x)$ is positive if $f(x)$ is concave up.

10. a. $\lim\limits_{x \to -\infty} f(x) = -2$ (Horizontal asymptote)

b. $x = -1$ is the vertical asymptote

11. $f(6) = 10, f'(6) = 0, f''(6) = -3$

By the 2nd Derivative Test there is a local maximum at $(6, 10)$.

12. $y = 0.19x^2 - 16.59x + 1038.29$

$y' = 0.38x - 16.59$

There is a critical value at $x = \dfrac{16.59}{0.38} \approx 43.66$.

x		43.66	
y'	$-$	0	$+$

a. The number is a minimum during 1943 (1900 + 43.7).

b. This is the inverse of (a). Thus the ratio of priests to Catholics was highest in 1943.

13. $R(x) = 164x \qquad C(x) = 0.01x^2 + 20x + 300$

$P(x) = 164x - (0.01x^2 + 20x + 300)$

$\qquad = 144x - 0.01x^2 - 300$

a. $P'(x) = 144 - 0.02x$

Critical value at $x = \dfrac{144}{0.02} = 7200$.

x		7200	
$P'(x)$	$+$	0	$-$

$x = 7200$ gives maximum profit

b. $P(7200) = \$518,100$

14. $C(x) = 100 + 20x + 0.01x^2$

$\bar{C}(x) = \dfrac{C(x)}{x} = \dfrac{100}{x} + 20 + 0.01x$

$\bar{C}'(x) = -\dfrac{100}{x^2} + 0.01$ Critical value at $x = 100$.

x		100	
$\bar{C}'(x)$	$-$	0	$+$

Minimum occurs when $\bar{C}'(x) = 0$, or $x = 100$.

15. Let x = number of additional units sold.

$R(x) = (100 + x)(300 - 2x) = 30,000 + 100x - 2x^2$

$R'(x) = 100 - 4x$

$R'(x) = 0$ if $x = 25$

x		25	
$R'(x)$	$+$	0	$-$

Price to get maximum revenue
$= 300 - 2(25) = \$250$.

16.

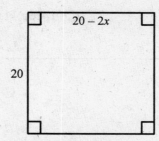

x = length of removed square.

$V = x(20 - 2x)^2$

$V'(x) = x(2)(20 - 2x)(-2) + (20 - 2x)^2(1)$

$\qquad = (20 - 2x)[-4x + (20 - 2x)]$

$\qquad = (20 - 2x)(20 - 6x)$

Critical value occurs if

$V'(x) = 0$ and $x = \dfrac{20}{6} = \dfrac{10}{3}$ cm.

($x = 10$ is not in domain)

x		$\frac{10}{3}$	
$V'(x)$	$+$	0	$-$

$x = \dfrac{10}{3}$ gives maximum volume.

17. Inventory problem

x = number in each production run

$C = \dfrac{784000}{x}(2500) + 784000(420) + \dfrac{x}{2}(5)$

Production cost $\qquad$ Storage cost

$C' = \dfrac{-784000(2500)}{x^2} + \dfrac{5}{2}$

$C' = 0$ if $x^2 = \dfrac{2(2500)(784000)}{5}$

Critical value occurs when $x = 28,000$ units.

x		28000	
C'	$-$	0	$+$

$x = 28,000$ gives minimum costs.

18. a. $y = 0.00253x^3 - 0.05708x^2$
$\qquad\qquad\qquad + 0.5842x + 3.110$

b. $y' = 0.00759x^2 - 0.11416x + 0.5842$

$y'' = 0.0158x - 0.11416$

$y'' = 0$ when $x \approx 7.5$.

y'' changes from $-$ to $+$ at $x \approx 7.5$ so the rate of change of the national debt reaches its minimum in 1998.

c. This represents a point of inflection on the original function.

Chapter 11: Derivatives Continued

Exercise 11.1

1. $f(x) = 4 \ln x$

$$f'(x) = 4 \cdot \frac{1}{x} = \frac{4}{x}$$

3. $y = \ln 8x$

$$\frac{dy}{dx} = \frac{1}{8x}(8) = \frac{1}{x}$$

5. $y = \ln x^4$

$$\frac{dy}{dx} = \frac{1}{x^4} \cdot 4x^3 = \frac{4}{x}$$

7. $f(x) = \ln(4x + 9)$

$$f'(x) = \frac{1}{4x+9}(4) = \frac{4}{4x+9}$$

9. $y = \ln(2x^2 - x) + 3x$

$$\frac{dy}{dx} = \frac{1}{2x^2 - x}(4x - 1) + 3$$

$$= \frac{4x - 1}{2x^2 - x} + 3$$

11. $p = \ln(q^2 + 1)$

$$\frac{dp}{dq} = \frac{1}{q^2 + 1} \cdot 2q = \frac{2q}{q^2 + 1}$$

13. a. $y = \ln x - \ln(x - 1)$

$$\frac{dy}{dx} = \frac{1}{x} - \frac{1}{x-1} = \frac{x - 1 - x}{x(x-1)} = \frac{-1}{x(x-1)}$$

b. $y = \ln \dfrac{x}{x-1}$

$$\frac{dy}{dx} = \frac{1}{\frac{x}{x-1}}\left[\frac{(x-1)(1) - x(1)}{(x-1)^2}\right]$$

$$= \frac{x-1}{x}\left[\frac{-1}{(x-1)^2}\right] = \frac{-1}{x(x-1)}$$

15. a $y = \dfrac{1}{3}\ln(x^2 - 1)$

$$\frac{dy}{dx} = \frac{1}{3}\left[\frac{1}{x^2-1} \cdot 2x\right] = \frac{2x}{3(x^2-1)}$$

b. Using properties of logs.

$$y = \ln \sqrt[3]{x^2 - 1} = \ln(x^2 - 1)^{1/3} = \frac{1}{3}\ln(x^2 - 1)$$

Thus, $\dfrac{dy}{dx} = \dfrac{2x}{3(x^2 - 1)}$.

17. a. $y = \ln(4x - 1) - 3\ln x$

$$\frac{dy}{dx} = \frac{1}{4x-1} \cdot 4 - 3 \cdot \frac{1}{x}$$

$$= \frac{4x - 3(4x-1)}{x(4x-1)} = \frac{3 - 8x}{x(4x-1)}$$

b. $y = \ln\left(\dfrac{4x-1}{x^3}\right) = \ln(4x - 1) - 3\ln x$

$$\frac{dy}{dx} = \frac{1}{4x-1} \cdot 4 - 3 \cdot \frac{1}{x}$$

$$= \frac{4}{4x-1} - \frac{3}{x} = \frac{3 - 8x}{x(4x-1)}$$

19. $p = \ln\left(\dfrac{q^2 - 1}{q}\right)$

$$= \ln(q^2 - 1) - \ln q$$

$$\frac{dp}{dq} = \frac{2q}{q^2-1} - \frac{1}{q} = \frac{2q^2 - q^2 + 1}{q(q^2-1)} = \frac{q^2 + 1}{q(q^2-1)}$$

21. $y = \ln\left(\dfrac{t^2+3}{\sqrt{1-t}}\right) = \ln(t^2 + 3) - \dfrac{1}{2}\ln(1-t)$

$$\frac{dy}{dt} = \frac{2t}{t^2+3} - \frac{-1}{2(1-t)} = \frac{4t - 4t^2 + t^2 + 3}{2(1-t)(t^2+3)}$$

$$= \frac{3 + 4t - 3t^2}{2(1-t)(t^2+3)} = \frac{-(3t^2 - 4t - 3)}{2(1-t)(t^2+3)}$$

23. $y = \ln(x^3 \sqrt{x+1}) = 3\ln x + \dfrac{1}{2}\ln(x+1)$

$$\frac{dy}{dx} = 3 \cdot \frac{1}{x} + \frac{1}{2} \cdot \frac{1}{x+1} \cdot 1 = \frac{3}{x} + \frac{1}{2(x+1)} \quad \text{or}$$

$$= \frac{3(2)(x+1) + x \cdot 1}{2x(x+1)} = \frac{7x + 6}{2x(x+1)}$$

25. $y = x - \ln x$

$$\frac{dy}{dx} = 1 - \frac{1}{x}$$

27. $y = \dfrac{\ln x}{x}$

$$\frac{dy}{dx} = \frac{x \cdot \frac{1}{x} - (\ln x)(1)}{x^2} = \frac{1 - \ln x}{x^2}$$

29. $y = \ln(x^4 + 3)^2 = 2\ln(x^4 + 3)$

$$\frac{dy}{dx} = 2 \cdot \frac{1}{x^4 + 3} \cdot 4x^3 = \frac{8x^3}{x^4 + 3}$$

31. $y = (\ln x)^4$

$$\frac{dy}{dx} = 4(\ln x)^3 \cdot \frac{1}{x} = \frac{4(\ln x)^3}{x}$$

33. $y = \left[\ln(x^4 + 3) \right]^2$

$$\frac{dy}{dx} = 2\left[\ln(x^4 + 3) \right] \cdot \frac{1}{x^4 + 3} \cdot 4x^3$$

$$= \frac{8x^3 \ln(x^4 + 3)}{x^4 + 3}$$

35. $y = \log_4 x = \frac{\ln x}{\ln 4} = \frac{1}{\ln 4}(\ln x)$

$$\frac{dy}{dx} = \frac{1}{\ln 4} \cdot \frac{1}{x} = \frac{1}{x \ln 4}$$

37. $y = \log_6 (x^4 - 4x^3 + 1) = \frac{1}{\ln 6} \ln(x^4 - 4x^3 + 1)$

$$\frac{dy}{dx} = \frac{1}{\ln 6} \cdot \frac{1}{x^4 - 4x^3 + 1} \cdot (4x^3 - 12x^2)$$

$$= \frac{4x^3 - 12x^2}{(x^4 - 4x^3 + 1) \ln 6}$$

39. $y = x \ln x$

$$y' = x \cdot \frac{1}{x} + (\ln x)(1) = 1 + \ln x$$

Set $y' = 0$.

$1 + \ln x = 0$ gives $\ln x = -1$ or $x = e^{-1}$

$y = e^{-1} \ln e^{-1} = e^{-1}(-1) = -e^{-1}$

Rel min at $(e^{-1}, -e^{-1})$.

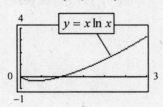

41. $y = x^2 - 8\ln x$

$$y' = 2x - 8 \cdot \frac{1}{x}$$

$2x^2 - 8 = 0$ or $x^2 - 4 = 0$ gives $x = \pm 2$.

Note: $\ln(-2)$ is not defined.

$x = 2$ gives $y = 2^2 - 8\ln 2 = 4 - 8\ln 2$

Rel min at $\left(2, \ 4 - 8\ln 2\right)$.

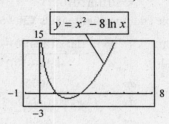

43. $C(x) = 1500 + 200\ln(2x + 1)$

a. $C'(x) = 200 \cdot \dfrac{1}{2x + 1} \cdot 2 = \dfrac{400}{2x + 1}$

b. $C(200) = \dfrac{400}{401} \approx \$1.00$

It will cost approximately $1.00 to make the next unit.

c. $\overline{MC} > 0$. Yes.

45. $R(x) = \dfrac{2500x}{\ln(10x+10)}$

a. $R'(x) = \dfrac{2500\ln(10x+10) - 2500x\left(\frac{1}{10x+10}\cdot 10\right)}{(\ln(10x+10))^2} = \dfrac{2500\ln(10x+10) - \frac{2500x}{x+1}}{(\ln(10x+10))^2}$

$= \dfrac{2500(x+1)\ln(10x+10) - 2500x}{(x+1)\left[\ln(10x+10)\right]^2} = \dfrac{2500\left[(x+1)\ln(10x+10) - x\right]}{(x+1)\left[\ln(10x+10)\right]^2}$

b. $R'(100) = \dfrac{2500(101)\ln 1010 - 250,000}{101(\ln 1010)^2} = 309.67$

Selling one additional unit yields \$309.67.

47. $p(x) = \dfrac{4000}{\ln(x+10)}$ $\dfrac{dp}{dx} = \dfrac{(\ln(x+10))(0) - \frac{4000}{x+10}}{(\ln(x+10))^2} = -\dfrac{4000}{(x+10)(\ln(x+10))^2}$

a. When $x = 40$, $\dfrac{dp}{dx} = -\dfrac{4000}{(40+10)(\ln(40+10))^2} = -5.23$

b. When $x = 90$, $\dfrac{dp}{dx} = -\dfrac{4000}{(90+10)(\ln(90+10))^2} = -1.89$

c. $p''(x) = \dfrac{0 + 4000\left[(x+10)2\ln(x+10)\cdot\frac{1}{x+10} + (\ln(x+10))^2(1)\right]}{\left[(x+10)(\ln(x+10))^2\right]^2} = \dfrac{4000\left[2\ln(x+10) + (\ln(x+10))^2\right]}{\left[(x+10)(\ln(x+10))^2\right]^2}$

When $x = 40$, $p''(x) > 0$. Thus, $p'(x)$ is increasing at 40 units.

49. $y = A\ln t - Bt + C$

$y' = \dfrac{A}{t} - B$

Solving $0 = \dfrac{A}{t} - B$ for t gives $t = \dfrac{A}{B}$.

$y'' = -\dfrac{A}{t^2}$, which is < 0 for all t.

Thus, $t = \dfrac{A}{B}$ is a maximum.

51. $R = \dfrac{1}{\ln 10}(\ln I - \ln I_0)$

$\dfrac{dR}{dI} = \dfrac{1}{\ln 10}\left(\dfrac{1}{I} - 0\right) = \dfrac{1}{I\ln 10}$

53. a. $y = -195.59 + 2975.50\ln x$

b. $y' = 2975.50\cdot\dfrac{1}{x}$

The year 2010 corresponds to $x = 30$, thus we find $y'(30) \approx 99.18$. In 2010, the average poverty threshold for individuals will be increasing by \$99.18 per year.

Exercise 11.2

1. $y = 5e^x - x$

$y' = 5e^x - 1$

3. $f(x) = e^x - x^e$

$f'(x) = e^x - ex^{e-1}$

5. $y = e^{x^3}$

$\dfrac{dy}{dx} = e^{x^3}\cdot 3x^2 = 3x^2 e^{x^3}$

7. $y = 6e^{3x^2}$

$\dfrac{dy}{dx} = 6e^{3x^2}\cdot 6x = 36xe^{3x^2}$

9. $y = 2e^{(x^2+1)^3}$

$$\frac{dy}{dx} = 2e^{(x^2+1)^3} \cdot 3(x^2+1)^2(2x)$$

$$= 12x(x^2+1)^2 e^{(x^2+1)^3}$$

11. $y = e^{\ln x^3} = x^3$

$y' = 3x^2$

13. $y = e^{-1/x} = e^{-(x^{-1})}$

$$y' = e^{-(x^{-1})} \cdot (1x^{-2}) = \frac{e^{-1/x}}{x^2}$$

15. $y = e^{-1/x^2} + e^{-x^2} = e^{-(x^{-2})} + e^{-x^2}$

$$y' = e^{-(x^{-2})}(2x^{-3}) + e^{-x^2}(-2x) = \frac{2e^{-1/x^2}}{x^3} - 2xe^{-x^2}$$

17. $s = t^2 e^t$

$s' = t^2(e^t) + e^t(2t) = te^t(t+2)$

19. $y = e^{x^4} - (e^x)^4 = e^{x^4} - e^{4x}$

$y' = e^{x^4} \cdot 4x^3 - e^{4x}(4) = 4x^3 e^{x^4} - 4e^{4x}$

21. $y = \ln(e^{4x} + 2)$

$$y' = \frac{1}{e^{4x}+2} \cdot e^{4x}(4) = \frac{4e^{4x}}{e^{4x}+2}$$

23. $y = e^{-3x} \ln(2x)$

$$y' = e^{-3x} \cdot \frac{1}{2x} \cdot 2 + \ln(2x) \cdot e^{-3x}(-3)$$

$$= \frac{e^{-3x}}{x} - 3e^{-3x} \ln(2x)$$

25. $y = \dfrac{1+e^{5x}}{e^{3x}} = \dfrac{1}{e^{3x}} + e^{5x-3x} = e^{-3x} + e^{2x}$

$y' = e^{-3x}(-3) + e^{2x}(2) = 2e^{2x} - 3e^{-3x}$

27. $y = (e^{3x}+4)^{10}$

$y' = 10(e^{3x}+4)^9(e^{3x} \cdot 3) = 30e^{3x}(e^{3x}+4)^9$

29. $y = 6^x$

$y' = 6^x \cdot \ln 6$

31. $y = 4^{x^2}$

$y' = 4^{x^2}(2x \ln 4)$

33. a. $y = xe^{-x}$

$y' = xe^{-x}(-1) + e^{-x}(1) = e^{-x} - xe^{-x}$

$y'(1) = e^{-1} - 1e^{-1} = 0$

b. If $x = 1$, then $y = 1e^{-1} = e^{-1}$.

$y - e^{-1} = 0(x-1)$

$y = e^{-1}$

35. a. $y = \dfrac{1}{\sqrt{2\pi}} e^{-z^2/2}$

$$y' = \frac{1}{\sqrt{2\pi}} e^{-z^2/2}(-z)$$

$y' = 0$ at $z = 0$.

Maximum occurs at $z = 0$.

b.

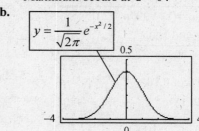

$$y = \frac{1}{\sqrt{2\pi}} e^{-x^2/2}$$

37. $y = \dfrac{e^x}{x}$

$$y' = \frac{xe^x - e^x(1)}{x^2} = \frac{e^x(x-1)}{x^2}$$

$y' = 0$ if $e^x(x-1) = 0$.

Since $e^x \neq 0$ for any x, the critical value is $x = 1$. Relative minimum at $(1, e)$.

39. $y = x - e^x$

$y' = 1 - e^x$

$1 - e^x = 0$ gives $x = 0$ as the critical value.

Relative maximum at $(0, -1)$.

41. $S = Pe^{0.1n}$

a. $\dfrac{dS}{dn} = Pe^{0.1n}(0.1)$

b. When $n = 1$, $\dfrac{dS}{dn} = 0.1Pe^{0.1}$

c. Yes, since $e^{0.1} > 1$.

43. $S(t) = 100,000e^{-0.5t}$

a. $S'(t) = 100,000e^{-0.5t}(-0.5) = -50,000e^{-0.5t}$

b. The function is decay exponential. The derivative is always negative.

45. $C(x) = 10,000 + 20xe^{x/600}$

$$C'(x) = 20x \cdot e^{x/600} \cdot \frac{1}{600} + e^{x/600}(20)$$

$$= e^{x/600}\left(\frac{x}{30} + 20\right)$$

$C'(600) = e(20 + 20) = 40e \approx \$108.73/\text{unit}$

47. $y = 100(1 - e^{-0.462t}) = 100 - 100e^{-0.462t}$

a. $y' = -100e^{-0.462t}(-0.462) = 46.2e^{-0.462t}$

b. $y'(1) = 46.2e^{-0.462(1)} = 29.107$

49. $y = \dfrac{10,000}{1 + 9999e^{-0.99t}}$

$$y' = \frac{0 - 10,000(9999e^{-0.99t})(-0.99)}{(1 + 9999e^{-0.99t})^2}$$

$$= \frac{98,990,100e^{-0.99t}}{(1 + 9999e^{-0.99t})^2}$$

51. $x(t) = 0.05 + 0.18e^{-0.38t}$

$x'(t) = (0.18)e^{-0.38t}(-0.38) = -0.068e^{-0.38t}$

53. a. $P(t) = \dfrac{10.94}{1 + 3.97e^{-0.029t}}$

$$P'(t) = \frac{0 - \left(-0.11513e^{-0.029t}\right)(10.94)}{\left(1 + 3.97e^{-0.029t}\right)^2}$$

$$= \frac{1.25952e^{-0.029t}}{\left(1 + 3.97e^{-0.029t}\right)^2}$$

b. $P'(100) \approx -0.0462$ The model predicts that in the year 2045, the population is expected to increase at the rate of 0.0467 billion people per year.

c. $P''(95) < 0$ means the rate is decreasing.

55. $\dfrac{I}{I_0} = 10^R$

Multiplying both sides by I_0 to be in the form $y = a^u$, we obtain $I = I_0 10^R$, so

$I' = I_0 10^R \ln 10$.

If $I_0 = 1$, $\dfrac{dI}{dR} = 10^R \ln 10$.

57. $H(t) = 29.57e^{0.097t}$

$H'(t) = 29.57e^{0.097t} \cdot (0.097) = 2.868e^{0.097t}$

$H'(48) \approx \$301.8$ billion per year

59. a. $d'(t) = 1.1391e^{0.0828t}$

b. The year 1940 corresponds to $t = 40$, thus $d'(40) \approx \$3.82$ billion per year.

The year 2000 corresponds to $t = 100$, thus $d'(100) \approx \$548.64$ billion per year.

c. 1999-2000: Avg. $= \dfrac{5674.2 - 5656.3}{1}$

$= \$17.9$ billion per year

2000-2001: Avg. $= \dfrac{5807.5 - 5674.2}{1}$

$= \$133.3$ billion per year

Change the model since average rates are smaller than predicted rates.

d. The terrorist attacks of 9-11-01 changed many programs.

61. $P(x) = 3.655e^{-0.049x}$

a. $P'(x) = 3.655e^{-0.049x}(-0.049)$

$= -0.179e^{-0.049x}$

The year 2002 corresponds to $t = 42$, thus $P'(42) \approx 0.029$.

b. The purchasing power of a dollar is decreasing at the rate of \$0.029 per year in the year 2002.

c. 2001-2002: $\dfrac{0.556 - 0.565}{2002 - 2001} = -\$0.009$ per yr

63. a. $y(x) = 257.8(1.0693)^x$

b. $y'(x) = 257.8(1.0693)^x \ln(1.0693)$

$= 17.27(1.043)^x$

The year 2010 corresponds to $t = 60$, thus $y'(60) \approx \$962.41$, which is an increase of \$962.41 per capita per year.

65. a. $P(x) = \dfrac{1,817,670}{1 + 5.0135e^{-0.1375x}}$

b. $P(20) = 1,376,770$ corresponds to the year 2000. It means that the model predicts the prison population is 1,376,770 in 2000.

$$P'(x) = \frac{0 - 1817670\left(5.0135e^{-0.1375x}(-0.1375)\right)}{\left(1 + 5.0135e^{-0.1375x}\right)^2}$$

$$= \frac{1253022e^{-0.1375x}}{\left(1 + 5.0135e^{-0.1375x}\right)^2}$$

$P'(20) \approx 45,932$

In 2002, the prison population will be increasing by 45,932 prisoners per year.

Exercise 11.3

1. $x^2 - 4y - 17 = 0$
   $2x - 4y' = 0$
   At $(1, -4)$, $y' = \dfrac{1}{2}$.

3. $xy^2 = 8$
   $x \cdot 2yy' + y^2 = 0$
   At $(2, 2)$, $8y' + 4 = 0$.
   $8y' = -4$
   $y' = -\dfrac{1}{2}$

5. $x^2 + 3xy - 4 = 0$
   $2x + 3[xy' + y(1)] = 0$
   At $(1, 1)$, $2(1) + 3[(1)y' + (1)(1)] = 0$
   $2 + 3y' + 3 = 0$
   $3y' = -5$
   $y' = -\dfrac{5}{3}$

7. $x^2 + 2y^2 - 4 = 0$
   $2x + 4yy' - 0 = 0$
   $y' = -\dfrac{2x}{4y} = -\dfrac{x}{2y}$

9. $x^2 + 4x + y^2 - 3y + 1 = 0$
   $2x + 4 + 2yy' - 3y' + 0 = 0$
   $y' = -\dfrac{2x + 4}{2y - 3}$

11. $x^2 + y^2 = 4$
    $2x + 2yy' = 0$
    $y' = -\dfrac{x}{y}$

13. $xy^2 - y^2 = 1$
    $x \cdot 2yy' + y^2(1) - 2yy' = 0$
    $y'(2xy - 2y) = -y^2$
    $y' = \dfrac{-y^2}{2xy - 2y}$
    $= \dfrac{-y}{2(x - 1)}$

15. $p^2q = 4p - 2$
    $p^2(1) + q \cdot 2p\dfrac{dp}{dq} = 4\dfrac{dp}{dq}$
    $(2qp - 4)\dfrac{dp}{dq} = -p^2$
    $\dfrac{dp}{dq} = \dfrac{p^2}{4 - 2qp}$

17. $3x^5 - 5y^3 = 5x^2 + 3y^5$
    $15x^4 - 15y^2\dfrac{dy}{dx} = 10x + 15y^4\dfrac{dy}{dx}$
    $\dfrac{dy}{dx}(-15y^2 - 15y^4) = 10x - 15x^4$
    $y' = -\dfrac{5(2x - 3x^4)}{15(y^2 + y^4)} = -\dfrac{2x - 3x^4}{3(y^2 + y^4)}$

19. $x^4 + 2x^3y^2 = x - y^3$
    $4x^3 + 2\left[x^3 \cdot 2y\dfrac{dy}{dx} + y^2 \cdot 3x^2\right] = 1 - 3y^2\dfrac{dy}{dx}$
    $3y^2\dfrac{dy}{dx} + 2x^3 \cdot 2y\dfrac{dy}{dx} = 1 - 4x^3 - 6x^2y^2$
    $\dfrac{dy}{dx} = \dfrac{1 - 4x^3 - 6x^2y^2}{3y^2 + 4x^3y}$

21. $x^4 + 3x^3y^2 - 2y^5 = (2x + 3y)^2$
    $4x^3 + 3x^3 \cdot 2y\dfrac{dy}{dx} + y^2 \cdot 9x^2 - 10y^4\dfrac{dy}{dx} = 2(2x + 3y)\left(2 + 3\dfrac{dy}{dx}\right)$
    $(6x^3y - 10y^4)\dfrac{dy}{dx} = -4x^3 - 9x^2y^2 + 8x + 12x\dfrac{dy}{dx} + 12y + 18y\dfrac{dy}{dx}$
    $(6x^3y - 10y^4 - 12x - 18y)\dfrac{dy}{dx} = -4x^3 - 9x^2y^2 + 8x + 12y$
    $\dfrac{dy}{dx} = \dfrac{-4x^3 - 9x^2y^2 + 8x + 12y}{6x^3y - 10y^4 - 12x - 18y} = \dfrac{4x^3 + 9x^2y^2 - 8x - 12y}{10y^4 + 12x + 18y - 6x^3y}$

23. $x^2 + 4x + y^2 + 2y - 4 = 0$

$2x + 4 + 2yy' + 2y' = 0$

$2(y+1)y' = -2(x+2)$

$y' = -\dfrac{x+2}{y+1}$

At $(1, -1)$, y' is undefined.

25. $x^2 + 2xy + 3 = 0$

$2x + 2xy' + y(2) = 0$

$2xy' = -2x - 2y$

$y' = \dfrac{-2x - 2y}{2x} = -\dfrac{x+y}{x}$

At $(-1, 2)$, $y' = 1$.

27. $x^2 - 2y^2 + 4 = 0$

$2x - 4yy' = 0$

$y' = \dfrac{x}{2y}$

At $(2, 2)$ we have $y' = \dfrac{2}{4} = \dfrac{1}{2}$.

The equation of the tangent line is

$y - 2 = \dfrac{1}{2}(x-2)$ or $y = \dfrac{1}{2}x + 1$.

29. $4x^2 + 3y^2 - 4y - 3 = 0$

$8x + 6yy' - 4y' = 0$

At $(-1, 1)$, we have $8(-1) + 6(1)y' - 4y' = 0$

or $2y' = 8$ or $y' = 4$. So, $m = 4$.

The equation of the tangent line

$y - 1 = 4(x - (-1))$

$y - 1 = 4x + 4$

$y = 4x + 5$.

31. $\ln x = y^2$

$\dfrac{1}{x} = 2y \dfrac{dy}{dx}$

$\dfrac{dy}{dx} = \dfrac{1}{2xy}$

33. $y^2 \ln x = 4$

$y^2 \left(\dfrac{1}{x} \right) + \ln x (2yy') = 0$

$y' = \dfrac{-y^2}{2xy \ln x} = \dfrac{-y}{2x \ln x}$

35. $x^2 + \ln y = 4$

$2x + \dfrac{1}{y} \cdot \dfrac{dy}{dx} = 0$

$\dfrac{dy}{dx} = -2xy$

At $(2, 1)$, $\dfrac{dy}{dx} = -2(2)(1) = -4$.

37. $xe^y = 6$

$xe^y \cdot \dfrac{dy}{dx} + e^y (1) = 0$

$\dfrac{dy}{dx} = \dfrac{-e^y}{xe^y} = -\dfrac{1}{x}$

39. $e^{xy} = 4$

$e^{xy} \left(x \cdot \dfrac{dy}{dx} + y \cdot 1 \right) = 0$

$\dfrac{dy}{dx} = \dfrac{-ye^{xy}}{xe^{xy}} = -\dfrac{y}{x}$

41. $ye^x - y = 3$

$y \cdot e^x + e^x \cdot \dfrac{dy}{dx} - \dfrac{dy}{dx} = 0$

$\dfrac{dy}{dx} = \dfrac{ye^x}{1 - e^x}$

43. $ye^x = y^2 + x - 2$

$ye^x + e^x \dfrac{dy}{dx} = 2y \dfrac{dy}{dx} + 1$

$ye^x - 1 = 2y \dfrac{dy}{dx} - e^x \dfrac{dy}{dx}$

$ye^x - 1 = \dfrac{dy}{dx} \left(2y - e^x \right)$

$\dfrac{dy}{dx} = \dfrac{ye^x - 1}{2y - e^x}$

$\dfrac{dy}{dx} \bigg|_{(0,2)} = \dfrac{2e^0 - 1}{2(2) - e^0} = \dfrac{2 - 1}{4 - 1} = \dfrac{1}{3}$

45. $xe^y = 2y + 3$

$$xe^y \frac{dy}{dx} + e^y = 2\frac{dy}{dx}$$

$$e^y = 2\frac{dy}{dx} - xe^y \frac{dy}{dx}$$

$$e^y = \frac{dy}{dx}\left(2 - xe^y\right)$$

$$\frac{dy}{dx} = \frac{e^y}{2 - xe^y}$$

$$\left.\frac{dy}{dx}\right|_{(3,0)} = \frac{e^0}{2 - 3e^0} = \frac{1}{2-3} = -1$$

Line: $y - 0 = -1(x - 3)$

$$y = -x + 3$$

47. $x^2 + 4y^2 - 4x - 4 = 0$

$$2x + 8yy' - 4 = 0$$

$$y' = \frac{4 - 2x}{8y} = \frac{2 - x}{4y}$$

a. There is a horizontal tangent if
$y' = 0$ or $x = 2$. Then,

$$4 + 4y^2 - 8 - 4 = 0$$

$$4y^2 = 8$$

$$y^2 = 2 \text{ or } y = \pm\sqrt{2}.$$

Horizontal tangents at $(2, \sqrt{2})$ and at $(2, -\sqrt{2})$.

b. There is a vertical tangent when $y = 0$.
Then, $x^2 - 4x - 4 = 0$ gives $x = 2 \pm 2\sqrt{2}$.
Vertical tangents at
$(2 + 2\sqrt{2}, 0)$ and at $(2 - 2\sqrt{2}, 0)$.

49. $y' = -\dfrac{x}{y} = \dfrac{-x}{y}$

a. $y'' = \dfrac{y(-1) - (-x)y'}{y^2} = \dfrac{-y + xy'}{y^2}$

b. $y'' = \dfrac{-y + x\left(-\frac{x}{y}\right)}{y^2} \cdot \dfrac{y}{y}$

$$= \frac{-y^2 - x^2}{y^3} = \frac{-\left(x^2 + y^2\right)}{y^3}$$

c. Yes. $x^2 + y^2 = 4$

51. $\sqrt{x} + \sqrt{y} = 1$

$$\frac{1}{2}x^{-1/2} + \frac{1}{2}y^{-1/2}y' = 0$$

$$y' = -\frac{x^{-1/2}}{y^{-1/2}} = -\frac{y^{1/2}}{x^{1/2}}$$

$$y'' = \frac{-x^{1/2} \cdot \frac{1}{2}y^{-1/2}y' - (-y^{1/2}) \cdot \frac{1}{2}x^{-1/2}}{x}$$

$$= \frac{\frac{y^{1/2}}{x^{1/2}} - \frac{x^{1/2}}{y^{1/2}}y'}{2x} = \frac{\frac{y^{1/2}}{x^{1/2}} - \frac{x^{1/2}}{y^{1/2}} \cdot \frac{-y^{1/2}}{x^{1/2}}}{2x}$$

$$= \frac{\frac{y^{1/2}}{x^{1/2}} + 1}{2x} \cdot \frac{x^{1/2}}{x^{1/2}} = \frac{y^{1/2} + x^{1/2}}{2x^{3/2}}$$

Since $y^{1/2} + x^{1/2} = 1$ we have

$$y'' = \frac{1}{2x^{3/2}} = \frac{1}{2x\sqrt{x}}.$$

53. $x^2 + y^2 - 9 = 0$

$$2x + 2yy' = 0$$

$$y' = \frac{-2x}{2y} = -\frac{x}{y}$$

$$-\frac{x}{y} = 0 \text{ gives } x = 0.$$

Max at $x = 0$, $y = 3$, and min at $x = 0$, $y = -3$.

55. $xy - 20x + 10y = 0$

$$xy' + y(1) - 20 + 10y' = 0$$

$$y' = \frac{20 - y}{x + 10}$$

At $x = 10$, we have $y = 10$.

Thus $y' = \dfrac{20 - 10}{10 + 10} = \dfrac{1}{2} = 0.5$.

57. $(x+1)^{3/4}(y+2)^{1/3} = 384$

$(x+1)^{3/4} \cdot \frac{1}{3}(y+2)^{-2/3} \cdot y' + (y+2)^{1/3} \cdot \frac{3}{4}(x+1)^{-1/4} \cdot 1 = 0$

Multiplying both sides by $\left(12(x+1)^{1/4}(y+2)^{2/3}\right)$ and simplifying and solving for y' we have $y' = \dfrac{-9(y+2)}{4(x+1)}$.

At $(255, 214)$ we have $y' = \dfrac{-9(216)}{4(256)} = -\dfrac{243}{128} \approx -1.898$.

59. $p(q+1)^2 = 200{,}000$

$p \cdot 2(q+1)\dfrac{dq}{dp} + (q+1)^2 \cdot 1 = 0$

$\dfrac{dq}{dp} = -\dfrac{q+1}{2p}$

At $p = 80$, we have $q = 49$.

Thus, $\dfrac{dq}{dp} = \dfrac{-50}{160} = \dfrac{-5}{16}$.

If the price is increased by \$1.00, the demand will decrease by $\dfrac{5}{16}$ unit.

61. $-0.000436t = \ln y - \ln(100)$

$-0.000436 = \dfrac{1}{y} \cdot \dfrac{dy}{dt}$

So, $\dfrac{dy}{dt} = -0.000436 y$

63. $\text{THI} = t - 0.55(1-h)(t-58)$

$0 = 1 - 0.55\left[(1-h)(1) + (t-58)\left(-\dfrac{dh}{dt}\right) \right]$

$0.55(t-58)\dfrac{dh}{dt} = -1 + 0.55(1-h)$

$\dfrac{dh}{dt} = \dfrac{-0.55h - 0.45}{0.55(t-58)}$

At $t = 70$, $\dfrac{dh}{dt} = \dfrac{-0.55h - 0.45}{0.55(12)} = \dfrac{-h}{12} - \dfrac{3}{44}$

Exercise 11.4

1. $y = x^3 - 3x$

$\dfrac{dy}{dt} = 3x^2 \dfrac{dx}{dt} - 3\dfrac{dx}{dt}$

If $x = 2$ and $\dfrac{dx}{dt} = 4$,

$\dfrac{dy}{dt} = 3(2)^2(4) - 3(4) = 36$

3. $xy = 4$ gives $y = \dfrac{4}{x}$

$\dfrac{dy}{dt} = -\dfrac{4}{x^2} \cdot \dfrac{dx}{dt}$

If $x = 8$ and $\dfrac{dx}{dt} = -2$,

$\dfrac{dy}{dt} = -\dfrac{4}{8^2}(-2) = \dfrac{1}{8}$

5. $x^2 + y^2 = 169$

$\dfrac{d}{dt}(x^2 + y^2) = \dfrac{d}{dt}(169)$

$2x \cdot \dfrac{dx}{dt} + 2y \cdot \dfrac{dy}{dt} = 0$

$\dfrac{dx}{dt} = -\dfrac{y}{x} \cdot \dfrac{dy}{dt}$

$\dfrac{dx}{dt} = -\dfrac{12}{5} \cdot 2 = -\dfrac{24}{5}$

7. $y^2 = 2xy + 24$

$$\frac{d}{dt}(y^2) = \frac{d}{dt}(2xy + 24)$$

$$2y \cdot \frac{dy}{dt} = 2x \cdot \frac{dy}{dt} + 2y \cdot \frac{dx}{dt} + 0$$

$$\frac{dx}{dt} = \frac{y - x}{y} \cdot \frac{dy}{dt}$$

$$\frac{dx}{dt} = \frac{12 - 5}{12} \cdot 2 = \frac{7}{6}$$

9. $x^2 + y^2 = z^2$

$$\frac{d}{dt}(x^2 + y^2) = \frac{d}{dt}(z^2)$$

$$2x \cdot \frac{dx}{dt} + 2y \cdot \frac{dy}{dt} = 2z \cdot \frac{dz}{dt}$$

$$\frac{dy}{dt} = \frac{z \cdot \frac{dz}{dt} - x \cdot \frac{dx}{dt}}{y}$$

$x^2 + y^2 = z^2$ yields $3^2 + 4^2 = z^2$.

So, $z^2 = 25$ or $z = \pm 5$.

When $z = 5$, $\dfrac{dy}{dt} = \dfrac{5 \cdot 2 - 3 \cdot 10}{4} = -5$.

When $z = -5$, $\dfrac{dy}{dt} = \dfrac{-5 \cdot 2 - 3 \cdot 10}{4} = -10$.

11. $y = -4x^2$

$$\frac{dy}{dt} = -8x \cdot \frac{dx}{dt} = -8(5)(2) = -80 \text{ units/sec.}$$

13. $A = \pi r^2$

$$\frac{dA}{dt} = A'(t) = 2\pi r \cdot \frac{dr}{dt}$$

$$A'(t) = 2\pi \cdot 3 \cdot 2 = 12\pi \text{ sq ft/min}$$

15. $V = x^3$

$$V'(t) = 3x^2 \cdot \frac{dx}{dt}$$

$$64 = 3 \cdot 36 \cdot \frac{dx}{dt}$$

$$\frac{dx}{dt} = \frac{64}{108} = \frac{16}{27} \text{ in./sec}$$

17. $P = 180x - \dfrac{1}{1000}x^2 - 2000$

$$P'(t) = 180 \cdot \frac{dx}{dt} - \frac{1}{500}x \cdot \frac{dx}{dt}$$

If $x = 100$ and $\dfrac{dx}{dt} = 10$, then

$$P'(t) = 180(10) - \frac{1}{500}(100)(10)$$

$$= \$1798 / \text{day}$$

19. $p = \dfrac{1000 - 10x}{400 - x}$

$$\frac{dp}{dt} = \frac{(400 - x)(-10) - (1000 - 10x)(-1)}{(400 - x)^2} \cdot \frac{dx}{dt}$$

If $\dfrac{dx}{dt} = -20$ and $x = 20$,

then $\dfrac{dp}{dt} = \dfrac{380(-10) - 800(-1)}{(380)^2}(-20)$

$$= \frac{(-3000)(-20)}{380 \cdot 380}$$

$$= 0.42 \text{ dollars/day.}$$

21. $x = 30y + 20y^2$

$$\frac{dx}{dt} = 30\frac{dy}{dt} + 40y\frac{dy}{dt}$$

If $y = 10$ and $\dfrac{dy}{dt} = 1$, then

$$\frac{dx}{dt} = 30(1) + 40(10)(1) = 430 \text{ units/month}$$

Note: y is the number of thousands.

23. $V = \dfrac{4}{3}\pi r^3$

$$\frac{dV}{dt} = 3 \cdot \frac{4}{3}\pi r^2 \cdot \frac{dr}{dt}$$

Substituting $\dfrac{dr}{dt} = -1$ and $r = 3$, we have

$$\frac{dV}{dt} = 4\pi \cdot 3^2(-1)$$

$$= -36\pi \text{ mm}^3 / \text{month.}$$

V is decreasing at the rate of 36π mm^3/month.

25. $W = kL^3$

$$\frac{dW}{dt} = 3kL^2 \frac{dL}{dt}$$

Percentage rate of change:

$$\frac{\frac{dW}{dt}}{W} = \frac{3kL^2 \frac{dL}{dt}}{W}$$

$$\frac{\frac{dW}{dt}}{W} = \frac{3kL^2 \frac{dL}{dt}}{kL^3} = \frac{3\frac{dL}{dt}}{L} = 3\left(\frac{\frac{dL}{dt}}{L}\right)$$

27. $C = 0.11W^{1.54}$

$$\frac{dC}{dt} = 1.54\left(0.11W^{0.54}\right) \cdot \frac{dW}{dt}$$

$$= 0.1694W^{0.54} \cdot \frac{dW}{dt}$$

$$\frac{\frac{dC}{dt}}{C} = \frac{0.1694W^{0.54} \cdot \frac{dW}{dt}}{C}$$

$$= \frac{0.1694W^{0.54}\frac{dW}{dt}}{0.11W^{1.54}} = 1.54\left(\frac{\frac{dW}{dt}}{W}\right)$$

29. $V = \frac{4}{3}\pi r^3$

$$\frac{dV}{dt} = \frac{4}{3}\pi \cdot 3r^2 \cdot \frac{dr}{dt}$$

Substituting $\frac{dV}{dt} = 4$, $r = 2$, we have

$$4 = 4\pi \cdot 2^2 \cdot \frac{dr}{dt} \text{ or } \frac{dr}{dt} = \frac{1}{4\pi} \text{ micrometers/day.}$$

31. $V = \frac{4}{3}\pi r^3$

$$\frac{dV}{dt} = \frac{4}{3}\pi \cdot 3r^2 \cdot \frac{dr}{dt}$$

Substituting $\frac{dV}{dt} = 5 \text{ in}^3 / \text{min}$, $r = 5$ gives

$$5 = 4\pi \cdot 5^2 \cdot \frac{dr}{dt} \text{ or } \frac{dr}{dt} = \frac{5}{100\pi} = \frac{1}{20\pi} \text{ in./min.}$$

33. Let y be the height of the ladder on the wall and let x be the distance from the wall to the bottom of the ladder.

$$\frac{dx}{dt} = 1 \text{ ft/ sec}$$

$$30^2 = x^2 + y^2$$

$$2x\frac{dx}{dt} + 2y\frac{dy}{dt} = 0$$

When $x = 18$, $y = \sqrt{576} = 24$.

$$2(18)(1) + 2(24)\frac{dy}{dt} = 0$$

$$48\frac{dy}{dt} = -36 \text{ or } \frac{dy}{dt} = -\frac{36}{48} = -0.75 \text{ ft/sec}$$

The ladder is sliding down at the rate of $\frac{3}{4}$ ft/sec.

35. Let x be the horizontal distance from the plane to the observer and let y be the direct distance.

$$\frac{dx}{dt} = -300 \text{ mph}$$

$$y^2 = x^2 + 1^2$$

$$2y\frac{dy}{dt} = 2x\frac{dx}{dt}$$

When $y = 5$, $x = \sqrt{24}$.

$$2(5)\frac{dy}{dt} = 2\sqrt{24}(-300) \text{ or}$$

$$\frac{dy}{dt} = -\sqrt{24}(60) \approx -293.94 \text{ mph}.$$

The plane is approaching the observer at the rate of 293.94 mph.

37. Let x be the distance of car A from the intersection, let y be the distance of car B from the intersection, and let z be the distance between them.

$$\frac{dx}{dt} = -40, \frac{dy}{dt} = -55$$

$$z^2 = x^2 + y^2$$

$$2z\frac{dz}{dt} = 2x\frac{dx}{dt} + 2y\frac{dy}{dt}$$

When $x = 15$ and $y = 8$, $z = 17$.

$$2(17)\frac{dz}{dt} = 2(15)(-40) + 2(8)(-55) \text{ or}$$

$$\frac{dz}{dt} \approx -61.18 \text{ mph.}$$

The distance between the cars is decreasing at the rate of 61.18 mph.

39. $V = 10(25)(r) = 250r$, $\dfrac{dV}{dt} = 10$

Find $\dfrac{dr}{dt}$ when $r = 4$.

$\dfrac{dV}{dt} = 250\dfrac{dr}{dt}$; $10 = 250\dfrac{dr}{dt}$ or $\dfrac{dr}{dt} = \dfrac{1}{25}$ ft/hr.

Exercise 11.5

1. $p + 4q = 80$

$1 + 4\dfrac{dq}{dp} = 0$ gives $\dfrac{dq}{dp} = -\dfrac{1}{4}$

   a. At (10, 40) we have $\eta = -\dfrac{40}{10}\left(-\dfrac{1}{4}\right) = 1$.

   b. Since the demand is unitary, there will be no change in the revenue with a price increase.

3. $p^2 + 2p + q = 49$

$2p + 2 + \dfrac{dq}{dp} = 0$ gives $\dfrac{dq}{dp} = -2p - 2$

   a. At (1, 6) we have $\eta = -\dfrac{6}{1}(-14) = 84$.

   b. Since the demand is elastic, an increase in price will decrease the total revenue.

5. $pq + p = 5000$

   a. $p \cdot \dfrac{dq}{dp} + q \cdot 1 + 1 = 0$ gives $\dfrac{dq}{dp} = \dfrac{-(q+1)}{p}$.

   At (99, 50) the elasticity is

   $\eta = -\dfrac{50}{99}\left(-\dfrac{100}{50}\right) = \dfrac{100}{99}$.

   b. $\eta > 1$, so demand is elastic.

   c. A price increase will decrease the total revenue.

7. $pq + p + 100q = 50{,}000$

   a. $p \cdot \dfrac{dq}{dp} + q \cdot 1 + 1 + 100\dfrac{dq}{dp} = 0$ gives

   $\dfrac{dq}{dp} = \dfrac{-(q+1)}{p+100}$. At $p = 401$ we have $q = 99$.

   Thus, elasticity is $\eta = -\dfrac{401}{99}\left(-\dfrac{100}{501}\right) \approx 0.81$

   at $(99, 401)$.

   b. $\eta < 1$, so demand is inelastic.

   c. An increase in price will result in an increase in total revenue.

9. $p = \dfrac{1}{2}[\ln(5000 - q) - \ln(q + 1)]$

   a. $1 = \dfrac{1}{2}\left[\dfrac{1}{5000 - q}(-1) \cdot \dfrac{dq}{dp} - \dfrac{1}{q+1} \cdot \dfrac{dq}{dp}\right]$

   $2 = -\left[\dfrac{q + 1 + 5000 - q}{(5000 - q)(q + 1)}\right] \cdot \dfrac{dq}{dp}$

   $\dfrac{dq}{dp} = \dfrac{-2(5000 - q)(q + 1)}{5001}$

   At $p = 3.71$, we were given that $q = 2$.

   Elasticity is $\eta = -\dfrac{3.71}{2}\left(\dfrac{-29{,}988}{5001}\right) \approx 11.1$.

   b. Demand is elastic.

11. $p = 120\sqrt[3]{125 - q}$

   a. $1 = 40(125 - q)^{-2/3}(-1) \cdot \dfrac{dq}{dp}$ gives

   $\dfrac{dq}{dp} = \dfrac{-(125 - q)^{2/3}}{40}$

   Elasticity is

   $\eta = -\dfrac{p}{q} \cdot \dfrac{dq}{dp}$

   $= -\dfrac{120(125 - q)^{1/3}}{q} \cdot \dfrac{-(125 - q)^{2/3}}{40}$

   $= \dfrac{3(125 - q)}{q}$

   b. $\eta = 1$ gives $1 = \dfrac{375 - 3q}{q}$

   $4q = 375$ or $q = 93.75$ for unitary elasticity.

   Inelastic means $\dfrac{375 - 3q}{q} < 1$ or $q > 93.75$.

   Elastic means $\dfrac{375 - 3q}{q} > 1$ or $0 < q < 93.75$.

   c. Revenue increases when $\eta > 1$ or $0 < q < 93.75$.
   Revenue decreases when $\eta < 1$ or $q > 93.75$.
   Revenue is maximized at $q = 93.75$.

d. $R = pq = 120q \sqrt[3]{125 - q}$

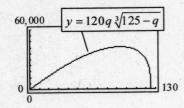

Yes.

13. After taxation the supply function is
$p = 6 + 2q + t$.
Then, $6 + 2q + t = 30 - q$ yields $t = 24 - 3q$.
Total tax revenue: $T = t \cdot q = 24q - 3q^2$.
$T'(q) = 24 - 6q = 0$ if $q = 4$.
Since $T''(q) < 0$, the tax revenue is maximized at
$q = 4$ and $t = 24 - 12 = \$12$.

15. After taxation the supply function is
$p = 100 + 0.5q + t$.
Then, $100 + 0.5q + t = 800 - 2q$ yields
$t = 700 - 2.5q$.
Total tax revenue: $T = t \cdot q = 700q - 2.5q^2$.
$T'(q) = 700 - 5q = 0$ if $q = 140$.
Since $T''(q) < 0$, the tax revenue is maximized at
$q = 140$ and $t = 700 - 2.5(140) = \$350$.

17. After taxation the supply function is $p = 20 + 3q + t$. Then, $20 + 3q + t = 200 - 2q^2$ yields $t = 180 - 3q - 2q^2$.
Total tax revenue: $T = t \cdot q = 180q - 3q^2 - 2q^3$.
$T'(q) = 180 - 6q - 6q^2$.
$T'(q) = 0$ if $6(6 + q)(5 - q) = 0$ and $T''(q) < 0$.
Total tax revenue is maximized at $q = 5$ and $t = 180 - 15 - 50 = \$115 \,/\, \text{item}$.

19. After taxation the supply function is $p = 0.02q^2 + 0.55q + 7.4 + t$.
Then, $0.02q^2 + 0.55q + 7.4 + t = 840 - 2q$ yields $t = 832.6 - 2.55q - 0.02q^2$.
$T(q) = 832.6q - 2.55q^2 - 0.02q^3$ and $T'(q) = 832.6 - 5.1q - 0.06q^2$.
$T'(q) = 0$ if $q = \dfrac{5.1 \pm \sqrt{26.01 + 199.824}}{-0.12} = \dfrac{5.1 \pm 15.03}{-0.12} \approx 83$.
Tax per item is $t = 832.6 - 2.55(83) - 0.02(83)^2 = \$483.17 \,/\, \text{item}$.
Total tax revenue (to nearest dollar) = \$40,100.

21. After taxation the supply function is $p = 300 + 5q + 0.5q^2 + t$.
Then, $300 + 5q + 0.5q^2 + t = 2100 - 10q - 0.5q^2$ gives $t = 1800 - 15q - q^2$.
Total tax revenue $T = t \cdot q = 1800q - 15q^2 - q^3$.
$T'(q) = 1800 - 30q - 3q^2 = 3(30 + q)(20 - q) = 0$ if $q = 20$. Note $T''(q) < 0$.
Tax per unit that will maximize total tax revenue is $t = 1800 - 300 - 400 = \$1100$.

Review Exercises

1. $y = e^{3x^2 - x}$

$$\frac{dy}{dx} = e^{3x^2 - x}(6x - 1)$$

2. $y = \ln e^{x^2} = x^2 \ln e = x^2$

$y' = 2x$

3. $p = \ln q - \ln(q^2 - 1)$

$$\frac{dp}{dq} = \frac{1}{q} - \frac{2q}{q^2 - 1} = \frac{q^2 - 1 - 2q^2}{q(q^2 - 1)} = \frac{-1 - q^2}{q(q^2 - 1)}$$

4. $y = xe^{x^2}$

$$\frac{dy}{dx} = x\left(e^{x^2} \cdot 2x\right) + e^{x^2}(1) = e^{x^2}(2x^2 + 1)$$

5. $f(x) = 5e^{2x} - 40e^{-0.1x} + 11$

$f'(x) = 5e^{2x} \cdot 2 - 40\left(e^{-0.1x}\right)(-0.1)$

$\qquad = 10e^{2x} + 4e^{-0.1x}$

6. $g(x) = \left(2e^{3x+1} - 5\right)^3$

$g'(x) = 3\left(2e^{3x+1} - 5\right)^2 \left(2e^{3x+1} \cdot 3\right)$

$\qquad = 18e^{3x+1}\left(2e^{3x+1} - 5\right)^2$

7. $y = \ln\left(3x^4 + 7x^2 - 12\right)$

$$\frac{dy}{dx} = \frac{1}{3x^4 + 7x^2 - 12}\left(12x^3 + 14x\right)$$

$$\qquad = \frac{12x^3 + 14x}{3x^4 + 7x^2 - 12}$$

8. $s = \dfrac{3}{4}\ln\left(x^{12} - 2x^4 + 5\right)$

$$\frac{ds}{dx} = \frac{3}{4} \cdot \frac{1}{x^{12} - 2x^4 + 5}\left(12x^{11} - 8x^3\right)$$

$$\qquad = \frac{3\left(3x^{11} - 2x^3\right)}{x^{12} - 2x^4 + 5}$$

9. $y = 3^{3x-4}$

$$\frac{dy}{dx} = 3^{3x-4}(3)\ln 3 = 3^{3x-3}(\ln 3)$$

10. $y = 1 + \log_8\left(x^{10}\right) = 1 + \dfrac{10}{\ln 8}\ln x$

$$\frac{dy}{dx} = \frac{10}{\ln 8} \cdot \frac{1}{x} = \frac{10}{x\ln 8}$$

11. $y = \dfrac{\ln x}{x}$

$$\frac{dy}{dx} = \frac{x \cdot \frac{1}{x} - \ln x \cdot 1}{x^2} = \frac{1 - \ln x}{x^2}$$

12. $y = \dfrac{1 + e^{-x}}{1 - e^{-x}}$

$$\frac{dy}{dx} = \frac{(1 - e^{-x})(-e^{-x}) - (1 + e^{-x})(e^{-x})}{(1 - e^{-x})^2}$$

$$\qquad = \frac{-2e^{-x}}{(1 - e^{-x})^2}$$

13. $y = 4e^{x^3}$

$y' = 4e^{x^3}\left(3x^2\right)$

At $x = 1$, $y' = 12e$

Tangent line: $y - 4e = 12e(x - 1)$ or

$y = 12ex - 8e$.

14. $y = x\ln x$ (If $x = 1$, $y = 0$)

$$\frac{dy}{dx} = x \cdot \frac{1}{x} + \ln x(1) = 1 + \ln x$$

At $x = 1$, $\dfrac{dy}{dx} = 1$.

Thus, $y - 0 = 1(x - 1)$ or $y = x - 1$.

15. $y\ln x = 5y$

$$y \cdot \frac{1}{x} + \ln x \cdot \frac{dy}{dx} = 5\frac{dy}{dx} \text{ gives}$$

$$\frac{dy}{dx} = \frac{y}{x} \div (5 - \ln x) = \frac{y}{x(5 - \ln x)}.$$

16. $\qquad e^{xy} = y$

$$e^{xy}\left(x\frac{dy}{dx} + y\right) = \frac{dy}{dx}$$

$$e^{xy}y = \frac{dy}{dx}\left(1 - xe^{xy}\right)$$

$$\frac{dy}{dx} = \frac{ye^{xy}}{1 - xe^{xy}}$$

17. $y^2 = 4x - 1$

$$2y\frac{dy}{dx} = 4 \text{ gives } \frac{dy}{dx} = \frac{2}{y}$$

18. $x^2 + 3y^2 + 2x - 3y + 2 = 0$

$2x + 6y\dfrac{dy}{dx} + 2 - 3\dfrac{dy}{dx} = 0$

$\dfrac{dy}{dx}(6y - 3) = -(2x + 2)$

$\dfrac{dy}{dx} = -\dfrac{2x + 2}{6y - 3} = \dfrac{2(x + 1)}{3(1 - 2y)}$

19. $3x^2 + 2x^3 y^2 - y^5 = 7$

$6x + 2x^3 \cdot 2y \cdot \dfrac{dy}{dx} + y^2(6x^2) - 5y^4 \cdot \dfrac{dy}{dx} = 0$

gives $\dfrac{dy}{dx} = \dfrac{6x + 6x^2 y^2}{5y^4 - 4x^3 y} = \dfrac{6x(1 + xy^2)}{y(5y^3 - 4x^3)}$

20. $x^2 + y^2 = 1$

$2x + 2y\dfrac{dy}{dx} = 0$

$\dfrac{dy}{dx} = \dfrac{-x}{y}$

$y'' = \dfrac{y(-1) - (-x)y'}{y^2}$

$= \dfrac{-y - \frac{x^2}{y}}{y^2} = -\dfrac{y^2 + x^2}{y^3} = -\dfrac{1}{y^3}$

21. $x^2 + 4x - 3y^2 + 6 = 0$

$2x + 4 - 6y\dfrac{dy}{dx} = 0$ gives $\dfrac{dy}{dx} = \dfrac{x + 2}{3y}$. At $(3, 3)$,

we have $\dfrac{dy}{dx} = \dfrac{5}{9}$.

22. $x^2 + 4x - 3y^2 + 6 = 0$

$2x + 4 - 6y\dfrac{dy}{dx} = 0$

$\dfrac{dy}{dx} = \dfrac{-2x - 4}{-6y} = \dfrac{x + 2}{3y}$

$\dfrac{dy}{dx} = 0$ when $x = -2$.

When $x = -2$, $y = \pm\sqrt{\dfrac{2}{3}}$.

So, horizontal tangents at $\left(-2, \pm\sqrt{\dfrac{2}{3}}\right)$.

23. $3x^2 - 2y^3 = 10y$

$6x\dfrac{dx}{dt} - 6y^2\dfrac{dy}{dt} = 10\dfrac{dy}{dt}$

$\dfrac{dy}{dt} = \dfrac{6x \cdot \frac{dx}{dt}}{10 + 6y^2}$

At $(10, 5)$, we have $\dfrac{dy}{dt} = \dfrac{6(10)2}{10 + 6(25)} = \dfrac{120}{160} = \dfrac{3}{4}$.

24. $A = \dfrac{1}{2}xy$

$\dfrac{dA}{dt} = \dfrac{1}{2}x\dfrac{dy}{dt} + y\left(\dfrac{1}{2}\dfrac{dx}{dt}\right)$

Use $\dfrac{dx}{dt} = 2$, $\dfrac{dy}{dt} = 5$, $x = 4$, and $y = 1$.

$\dfrac{dA}{dt} = \dfrac{1}{2} \cdot 4 \cdot 5 + 1 \cdot \dfrac{1}{2} \cdot 2 = 11$ units^2 / min

25. $y = -3.91435 + 2.62196 \ln t$

a. $y'(t) = \dfrac{2.62196}{t}$

b. $y(50) \approx 6.343$ is the predicted number of hectares of deforestation in 2000.
$y'(50) \approx 0.05244$ is the predicted increase in the number of hectares of deforestation in 2001.

26. $S(n) = 1000e^{0.12n}$

$S'(n) = 120e^{0.12n}$

$S'(1) = 120(1.1275) \approx \$135.30 / \text{yr}$

27. $S = 1000e^{0.12n}$

a. $\dfrac{dS}{dn} = 1000e^{0.12n}(0.12)$

$= 1000e^{0.12(2)}(0.12)$

$\approx \$152.55 / \text{yr}$

b. At $n = 1$,

$\dfrac{dS}{dn} = e^{0.12(1)}(0.12) \approx \$135.30 / \text{yr}$

At the end of 2 years the value is growing

$\dfrac{152.55}{135.50} \approx 1.13$ times as fast.

28. $A(t) = A_0 e^{-0.00002876t}$

  a. $A'(t) = A_0 e^{-0.00002876t}(-0.00002876)$

     $A'(0) = -0.00002876 A_0$

  b. $A'(1) = -0.00002876 A_0$

     Note that $e^{-0.00002876} \approx e^0 = 1$.

  c. The rate of decay when $t = 24101$ is approximately -1.44×10^{-5} and the rate when $t = 1$ is approximately -2.88×10^{-5}.

29. $\overline{C} = 600 e^{x/600}$

  $C = 600 x e^{x/600}$

  $\overline{MC} = C' = 600 x e^{x/600} \cdot \dfrac{1}{600} + 600 e^{x/600} \cdot$

    $= e^{x/600}(x + 600)$

  When $x = 600$, $\overline{MC} = 1200e$.

30. $P(t) = 20{,}000 e^{-0.0495t}$

  $P'(t) = -990 e^{-0.0495t}$

  $P'(10) = -990 e^{-0.495}$

    $= -990(0.60957)$

    $= -\$603.48/\text{yr}$

31. $V = \dfrac{4}{3}\pi r^3$

  $\dfrac{dV}{dt} = 4\pi r^2 \dfrac{dr}{dt}$

  Using $r = 2.5$ and $\dfrac{dV}{dt} = -1$ gives

  $-1 = 4\pi(2.5)^2 \dfrac{dr}{dt}$ $\dfrac{dr}{dt} = -\dfrac{1}{25\pi}\,\text{mm/min}$

32. Let y be the height of the sign above the workers' hands and let z be the length of the guide line.

  $\dfrac{dy}{dt} = 2\,\text{ft/min}$

  $z^2 = y^2 + 7^2$

  $2z\dfrac{dz}{dt} = \dfrac{dy}{dt}(2y) + 0$

  If $z = 25$, $y = 24$.

  $2(25)\dfrac{dz}{dt} = 2(24)(2)$ gives $\dfrac{dz}{dt} = \dfrac{48}{25}\,\text{ft/min}$.

33. $S = kA^{1/3}$

  $\dfrac{dS}{dt} = \dfrac{1}{3} \cdot k \cdot \dfrac{1}{A^{2/3}} \cdot \dfrac{dA}{dt}$

  $\dfrac{\frac{dS}{dt}}{S} = \dfrac{1}{S} \cdot \dfrac{k}{3} \cdot \dfrac{1}{A^{2/3}} \cdot \dfrac{dA}{dt}$

  $= \dfrac{1}{kA^{1/3}} \cdot \dfrac{k}{3} \cdot \dfrac{1}{A^{2/3}} \cdot \dfrac{dA}{dt} = \dfrac{1}{3}\left(\dfrac{\frac{dA}{dt}}{A}\right)$

34. Yes.

35. After taxation, the supply function is $p = 400 + 2q + t$.

  $400 + 2q + t = 2800 - 8q - \dfrac{1}{3}q^2$ gives

  $t = 2400 - 10q - \dfrac{1}{3}q^2$.

  $T(q) = tq = 2400q - 10q^2 - \dfrac{1}{3}q^3$

  $T'(q) = 2400 - 20q - q^2 = (60 + q)(40 - q)$

  $T(q)$ is maximized at $q = 40$ and

  $t = 2400 - 400 - \dfrac{1600}{3} = \$1466.67/\text{unit}$. The maximum tax revenue is $T(40) \approx 58{,}666.80$.

36. After taxation, the supply function is $p = 40 + 20q + t$.

  $40 + 20q + t = \dfrac{5000}{q + 1}$ gives $t = \dfrac{5000}{q + 1} - 40 - 20q$

  $T = tq = \dfrac{5000q}{q + 1} - 40q - 20q^2$ and

  $T'(q) = \dfrac{(q + 1)5000 - 5000q(1)}{(q + 1)^2} - 40 - 40q$

  Set $T'(q) = 0$. Then, $5000 - 40(q + 1)^3 = 0$ or $q + 1 = 5$ or $q = 4$.

  So, $t = \dfrac{5000}{5} - 40 - 80 = \$880/\text{unit}$. The maximum tax revenue is $T(4) \approx \$3520$.

37. a. $pq = 27$

  $p \cdot \dfrac{dq}{dp} + q \cdot 1 = 0$ or $\dfrac{dq}{dp} = -\dfrac{q}{p}$

  $\eta = -\dfrac{p}{q} \cdot \dfrac{dq}{dp} = -\dfrac{p}{q} \cdot \left(-\dfrac{q}{p}\right) = 1$

  So at $(9, 3)$ we have $\eta = 1$.

  b. Since $\eta = 1$, there is no change in total revenue with a price increase.

38. $p^2(2q+1) = 10,000$

$$p^2 \cdot 2 \cdot \frac{dq}{dp} + (2q+1)2p = 0$$

$$\frac{dq}{dp} = \frac{-(2q+1)}{p}$$

$$\eta = -\frac{p}{q} \cdot \frac{dq}{dp} = -\frac{20}{12}\left(-\frac{25}{20}\right) = \frac{25}{12}$$

39. $p = 100e^{-0.1q}$

$$1 = 100e^{-0.1q}(-0.1)\frac{dq}{dp} \text{ or } \frac{dq}{dp} = \frac{-e^{0.1q}}{10}$$

At $p = 36.79$, $q = 10$ we have

$$\eta = -\frac{36.79}{10}\left(-\frac{e}{10}\right) \approx 1.$$

40. $p = 100 - 0.5q$

a. $1 = -0.5\frac{dq}{dp}$ or $\frac{dq}{dp} = -2$

$$\eta = -\frac{p}{q}(-2) = \frac{2(100-0.5q)}{q}$$

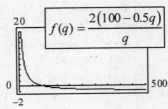

b. $\dfrac{2(100-0.5q)}{q} - 1 = 0$ if $q = 100$.

c. $R(q) = 100q - 0.5q^2$

Max revenue occurs at vertex.

$$q = \frac{-b}{2a} = \frac{-100}{-1} = 100$$

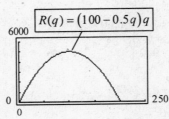

d. Maximum revenue occurs when $\eta = 1$.

Chapter Test

1. $y = 5e^{x^3} + x^2$

$y' = 5 \cdot e^{x^3}(3x^2) + 2x = 15x^2 e^{x^3} + 2x$

2. $y = 4\ln(x^3+1)$

$y' = 4 \cdot \dfrac{1}{x^3+1} \cdot 3x^2 = \dfrac{12x^2}{x^3+1}$

3. $y = \ln(x^4+1)^3 = 3\ln(x^4+1)$

$y' = 3 \cdot \dfrac{1}{x^4+1} \cdot 4x^3 = \dfrac{12x^3}{x^4+1}$

4. $f(x) = 10(3^{2x})$

$f'(x) = 10(3^{2x} \cdot 2\ln 3) = 20(3^{2x})\ln 3$

5. $S = te^{t^4}$

$S'(t) = t\left(4t^3 e^{t^4}\right) + e^{t^4}(1) = e^{t^4}(4t^4+1)$

6. $y = \dfrac{e^{x^3+1}}{x}$

$y' = \dfrac{x(3x^2 e^{x^3+1}) - e^{x^3+1}(1)}{x^2} = \dfrac{e^{x^3+1}(3x^3-1)}{x^2}$

7. $y = \dfrac{\ln x}{x}$

$y' = \dfrac{x \cdot \frac{1}{x} - \ln x(1)}{x^2} = \dfrac{1 - \ln x}{x^2}$

8. $g(x) = 2\log_5(4x+7)$

$g'(x) = 2 \cdot \dfrac{1}{(4x+7)} \cdot \dfrac{4}{\ln 5} = \dfrac{8}{(4x+7)\ln 5}$

9. $3x^4 + 2y^2 + 10 = 0$

$12x^3 + 4yy' = 0$

$$y' = \frac{-3x^3}{y}$$

10. $x^2 + y^2 = 100,\ \dfrac{dx}{dt} = 2,\ x = 6,\ y = 8$

$$2x \cdot \dfrac{dx}{dt} + 2y\dfrac{dy}{dt} = 0$$

$$2(6) \cdot 2 + 2(8)\dfrac{dy}{dt} = 0$$

$$\dfrac{dy}{dt} = -\dfrac{3}{2}$$

11. $\qquad xe^y = 10y$

$$x(e^y y') + e^y(1) = 10y'$$

$$y' = \dfrac{e^y}{10 - xe^y}$$

12. $R(x) = 300x - 0.001x^2$

$C(x) = 4000 + 30x$

$P(x) = -0.001x^2 + 270x - 4000$

$\dfrac{dP}{dt} = -0.002x\dfrac{dx}{dt} + 270\dfrac{dx}{dt}$

If $x = 50,\ \dfrac{dx}{dt} = 5$ then

$\dfrac{dP}{dt} = (-0.002)(50)(5) + 270(5) = \$1349.50\,/\text{week}$

13. $p^2 + 3p + q = 1500$

$p = 30$ gives $q = 510$.

$q = 1500 - p^2 - 3p$

$\dfrac{dq}{dp} = -2p - 3$ If $p = 30,\ \dfrac{dq}{dp} = -63$.

$\eta = -\dfrac{p}{q} \cdot \dfrac{dq}{dp} = \dfrac{-30}{510}(-63) \approx 3.71$

Revenue decreases.

14. $(p+1)q^2 = 10,000$

Find $\dfrac{dq}{dp}$ if $p = 99$ $(q = 10)$.

$(p+1)\left(2q\dfrac{dq}{dt}\right) + q^2(1) = 0$

$\dfrac{dq}{dt} = \dfrac{-q}{2(p+1)}$

$\dfrac{dq}{dt} = \dfrac{-10}{2(99+1)} = -\dfrac{1}{20}$ units/dollar

15. $S = 80,000e^{-0.4t}$

$S'(t) = 80,000(e^{-0.4t} \cdot (-0.4)) = -32,000e^{-0.4t}$

$S'(10) = -32,000e^{-4} = -586$ sales/day

16. $y(x) = 0.4843e^{0.1513(x-50)}$

$y'(x) = 0.4843e^{0.1513(x-50)}(0.1513)$

$\qquad = 0.0733e^{0.1513(x-50)}$

a. The year 1990 corresponds to $x = 90$, thus $y'(90) \approx 31.14$, which means that the average daily shares are increasing at a rate of 31.14 million shares per year.

b. The year 2010 corresponds to $x = 110$, thus $y'(110) \approx 641.96$, which means that the average daily shares are increasing at a rate of 641.96 million shares per year.

17. $D: p = 1100 - 5q \quad S: p = 20 + 0.4q$

New supply: $p = 20 + 0.4q + t$ $(t = \text{tax/unit})$

Break-even: $1100 - 5q = 20 + 0.4q + t$

$\qquad\qquad t = 1080 - 5.4q$

Total tax revenue: $T = t\,q = 1080q - 5.4q^2$

$T'(q) = 1080 - 10.8q$

Tax revenue is maximized when $T' = 0$ or $q = 100$.

Then $t = 1080 - 5.4(100) = \$540\,/\,\text{unit}$.

18. a. $y(x) = -31.503 + 22.943\ln x$

b. $y'(x) = 22.943 \cdot \dfrac{1}{x}$

c. $y(70) \approx 65.97$ corresponds to the year 2010. The model predicts that in 2010 the percent of paved streets will be 65.97%. $y'(70) \approx 0.328$; the model predicts that in 2010 the percent of paved streets would be increasing at a rate of 0.328% per year.

19. $P(t) = 3.655(0.9522)^t$

$P'(t) = 3.655(0.9522)^t \cdot \ln 0.9522$

$\qquad = -0.179(0.9522)^t$

The year 2010 corresponds to $t = 50$, thus $P'(50) \approx -0.0155$ tells us the model predicts that the purchasing power of the 1983 dollar will be decreasing at the rate of \$0.0155 per year in the year 2010.

Chapter 12: Indefinite Integrals

Exercise 12.1

1. $f'(x) = 4x^3$

$f(x) = \dfrac{4x^{3+1}}{4} + C = x^4 + C$

3. $f'(x) = x^6$

$f(x) = \dfrac{x^{6+1}}{7} + C = \dfrac{x^7}{7} + C$

5. $\displaystyle\int x^7 dx = \dfrac{x^8}{8} + C$

7. $8\displaystyle\int x^5 dx = 8\left(\dfrac{x^6}{6}\right) + C = \dfrac{4}{3}x^6 + C$

9. $\displaystyle\int (3^3 + x^{13})dx = 27x + \dfrac{x^{14}}{14} + C$

11. $\displaystyle\int (3 - x^{3/2})dx = 3x - \dfrac{x^{5/2}}{\frac{5}{2}} = 3x - \dfrac{2}{5}x^{5/2} + C$

13. $\displaystyle\int (x^4 - 9x^2 + 3)dx = \dfrac{x^5}{5} - \dfrac{9x^3}{3} + 3x + C$

$\qquad = \dfrac{1}{5}x^5 - 3x^3 + 3x + C$

15. $\displaystyle\int (2 + 2\sqrt{x})dx = \displaystyle\int (2 + 2x^{1/2})dx$

$\qquad = 2x + 2 \cdot \dfrac{x^{3/2}}{\frac{3}{2}} + C$

$\qquad = 2x + \dfrac{4}{3}x^{3/2} + C$

$\qquad = 2x + \dfrac{4}{3}x\sqrt{x} + C$

17. $\displaystyle\int 6\sqrt[4]{x}\,dx = \displaystyle\int 6x^{1/4}dx = 6 \cdot \dfrac{x^{5/4}}{\frac{5}{4}} + C$

$\qquad = \dfrac{24}{5}x^{5/4} + C$

$\qquad = \dfrac{24}{5}x\sqrt[4]{x} + C$

19. $\displaystyle\int \dfrac{5}{x^4}dx = \displaystyle\int 5x^{-4}dx = 5 \cdot \dfrac{x^{-3}}{-3} + C$

$\qquad = -\dfrac{5}{3}x^{-3} + C$

$\qquad = -\dfrac{5}{3x^3} + C$

21. $\displaystyle\int \dfrac{dx}{2\sqrt[3]{x^2}} = \dfrac{1}{2}\displaystyle\int x^{-2/3}dx$

$\qquad = \dfrac{1}{2} \cdot \dfrac{x^{1/3}}{\frac{1}{3}} + C$

$\qquad = \dfrac{3}{2}x^{1/3} + C$

$\qquad = \dfrac{3}{2}\sqrt[3]{x} + C$

23. $\displaystyle\int\left(x^3 - 4 + \dfrac{5}{x^6}\right)dx = \dfrac{x^4}{4} - 4x + 5 \cdot \dfrac{x^{-5}}{-5} + C$

$\qquad = \dfrac{1}{4}x^4 - 4x - \dfrac{1}{x^5} + C$

25. $\displaystyle\int\left(x^9 - \dfrac{1}{x^3} + \dfrac{2}{\sqrt[3]{x}}\right)dx = \dfrac{x^{10}}{10} - \dfrac{x^{-2}}{-2} + 2 \cdot \dfrac{x^{2/3}}{\frac{2}{3}} + C$

$\qquad = \dfrac{1}{10}x^{10} + \dfrac{1}{2x^2} + 3x^{2/3} + C$

27. $\displaystyle\int (4x^2 - 1)^2 x^3 dx = \displaystyle\int (16x^4 - 8x^2 + 1)(x^3)dx$

$\qquad = \displaystyle\int (16x^7 - 8x^5 + x^3)dx$

$\qquad = 16 \cdot \dfrac{x^8}{8} - 8 \cdot \dfrac{x^6}{6} + \dfrac{x^4}{4} + C$

$\qquad = 2x^8 - \dfrac{4}{3}x^6 + \dfrac{1}{4}x^4 + C$

29. $\displaystyle\int \dfrac{x+1}{x^3}dx = \displaystyle\int\left(\dfrac{1}{x^2} + \dfrac{1}{x^3}\right)dx$

$\qquad = \displaystyle\int (x^{-2} + x^{-3})dx$

$\qquad = \dfrac{x^{-1}}{-1} + \dfrac{x^{-2}}{-2} + C$

$\qquad = -\dfrac{1}{x} - \dfrac{1}{2x^2} + C$

31. $\displaystyle\int (2x + 3)dx = 2 \cdot \dfrac{x^2}{2} + 3x + C = x^2 + 3x + C$

$\boxed{\begin{array}{c} f(x) = x^2 + 3x + C \\ (C = -8, -4, 0, 4, \text{ and } 8) \end{array}}$

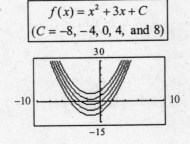

33. $\int f(x)dx = 2x^9 - 7x^5 + C$

$$f(x) = \frac{d}{dx}\left(2x^9 - 7x^5 + C\right)$$
$$= 18x^8 - 35x^4$$

35. $F(x) = 5x - \dfrac{x^2}{4} + C$

$$F'(x) = 5 - \frac{1}{4} \cdot 2x = 5 - \frac{1}{2}x$$
$$\int\left(5 - \frac{1}{2}x\right)dx$$

37. $F(x) = x^3 - 3x^2 + C$

$$F'(x) = 3x^2 - 6x$$
$$\int(3x^2 - 6x)dx$$

39. $\overline{MR} = -0.4x + 30$

$$R(x) = \int(-0.4x + 30)\,dx$$
$$= -0.4 \cdot \frac{x^2}{2} + 30x + C$$

$C = 0$ since $R(0) = 0$.

$$R(x) = -0.2x^2 + 30x$$

41. $\overline{MR} = -0.3x + 450$

$$R(x) = \int(-0.3x + 450)\,dx$$
$$= -0.3 \cdot \frac{x^2}{2} + 450x + C$$

$C = 0$ since $R(0) = 0$.

$$R(50) = \frac{-0.3(50)^2}{2} + 450(50) = \$22{,}125$$

43. $\int(t^3 + 4t^2 + 6)dt = \dfrac{t^4}{4} + 4 \cdot \dfrac{t^3}{3} + 6t + C$

When $t = 0$, $C = 0$. So, $P(t) = \dfrac{1}{4}t^4 + \dfrac{4}{3}t^3 + 6t$.

45. $\dfrac{dx}{dt} = \dfrac{1}{600}t^{3/4}$

a. $x = \dfrac{1}{600} \cdot \dfrac{t^{7/4}}{\frac{7}{4}} + C$

$t = 0$ means $x = 0$. Thus $C = 0$ and

$$x(t) = \frac{1}{1050}t^{7/4}.$$

b. $x(52) = \dfrac{1}{1050}(52)^{7/4}$

$$\approx \frac{1}{1050}(1006.95)$$
$$\approx 0.96 \text{ tons}$$

47. $\overline{C}'(x) = \dfrac{1}{4} - \dfrac{100}{x^2} = \dfrac{1}{4} - 100x^{-2}$

a. $\overline{C}(x) = \dfrac{1}{4}x - \dfrac{100x^{-1}}{-1} + K = \dfrac{1}{4}x + \dfrac{100}{x} + K$

Since $\overline{C}(20) = 40$, we have

$40 = \dfrac{1}{4}(20) + \dfrac{100}{20} + K$. So, $K = 30$. Thus,

$$\overline{C}(x) = \frac{1}{4}x + \frac{100}{x} + 30.$$

b. $\overline{C}(100) = \dfrac{1}{4}(100) + \dfrac{100}{100} + 30 = \$56$

49. $\dfrac{dB}{dt} = -2.1136t + 8.2593$

a. $\dfrac{dB}{dt} = 0$ if $-2.1136t + 8.2593 = 0$

$t \approx 3.91$

Balance begins to decrease 3.9 years after 2000.

b. $B = \int(-2.1136t + 8.2593)\,dt$

$$= \frac{-2.1136}{2}t^2 + 8.2593t + C$$

$B(0) = 74.07$ gives $C = 74.07$

$B = -1.0568t^2 + 8.2593t + 74.07$

c. Graphing $B(t)$ shows that $B = 0$ about 13 years after 2000.

51. $\dfrac{dA}{dt} = 160.869t^{0.5307}$

   a. Yes. $\dfrac{dA}{dt} > 0$ also means A is increasing.

   b. $A = \int 160.869t^{0.5307}\, dt$

$$= 160.869\left(\dfrac{t^{1.5307}}{1.5307}\right) + C$$

$$1234.5 \text{ (billion)} = 105.095 \cdot \left(5^{1.5307}\right) + C$$

gives $C = 0$.

$$A = 105.095t^{1.5307}$$

53. $\dfrac{dH}{dt} = -0.00126t^2 + 1.532t - 5.08$

   a. $H(t) = \int \left(0.00126t^2 + 1.532t - 5.08\right) dt$

$$= 0.0042t^3 + 0.766t^2 - 5.08t + C$$

Using the initial condition, $H(0) = 26.7$,

we solve for C to find $C = 26.7$.

$$H(t) = 0.0042t^3 + 0.766t^2 - 5.08t + 26.7$$

   b. The year 2010 corresponds to $t = 50$:
$H(50) \approx 2212.7$ billion. The model
predicts the national health care
expenditures for 2010 is \$2212.7 billion.

Exercise 12.2

1. $u = 2x^5 + 9;\;\; du = 10x^4 dx$

3. Let $u = x^2 + 3$. Then, $du = 2x dx$

$$\int (x^2 + 3)^3 (2x\, dx) = \int u^3\, du$$

$$= \dfrac{u^4}{4} + C$$

$$= \dfrac{1}{4}(x^2 + 3)^4 + C$$

5. Let $u = 15x^2 + 10$. Then, $du = 30x dx$

$$\int (15x^2 + 10)^4 (30x\, dx) = \int u^4\, du$$

$$= \dfrac{u^5}{5} + C$$

$$= \dfrac{1}{5}(15x^2 + 10)^5 + C$$

7. Let $u = 3x - x^3$. Then, $du = (3 - 3x^2)dx$.

$$\int (3x - x^3)^2 (3 - 3x^2)dx = \int u^2\, du$$

$$= \dfrac{u^3}{3} + C$$

$$= \dfrac{1}{3}(3x - x^3)^3 + C$$

9. Let $u = x^2 + 5$. Then, $du = 2x\, dx$.

$$\int (x^2 + 5)^3 x\, dx = \int (x^2 + 5)^3 \cdot \dfrac{1}{2}(2x\, dx)$$

$$= \dfrac{1}{2}\int u^3\, du$$

$$= \dfrac{1}{2} \cdot \dfrac{u^4}{4} + C$$

$$= \dfrac{1}{8}(x^2 + 5)^4 + C$$

11. $\int 7(4x - 1)^6\, dx = 7 \cdot \dfrac{1}{4}\int (4x - 1)^6 (4\, dx)$

$$= \dfrac{7}{4} \cdot \dfrac{(4x - 1)^7}{7} + C$$

$$= \dfrac{1}{4}(4x - 1)^7 + C$$

13. $\int (x^2 + 1)^{-3} x\, dx = \dfrac{1}{2}\int (x^2 + 1)^{-3}(2x\, dx)$

$$= \dfrac{1}{2} \cdot \dfrac{(x^2 + 1)^{-2}}{-2} + C$$

$$= -\dfrac{1}{4}(x^2 + 1)^{-2} + C$$

15. $\int (x - 1)(x^2 - 2x + 5)^4\, dx$

$$= \dfrac{1}{2}\int (2x - 2)(x^2 - 2x + 5)^4\, dx$$

$$= \dfrac{(x^2 - 2x + 5)^5}{10} + C$$

17. $\int 2(x^3 - 1)(x^4 - 4x + 3)^{-5}\, dx$

$$= \dfrac{1}{2}\int (4x^3 - 4)(x^4 - 4x + 3)^{-5}\, dx$$

$$= \dfrac{1}{2} \cdot \dfrac{(x^4 - 4x + 3)^{-4}}{-4} + C$$

$$= -\dfrac{1}{8}(x^4 - 4x + 3)^{-4} + C$$

19. $\int 7x^3 \sqrt{x^4 + 6}\, dx = 7 \cdot \dfrac{1}{4}\int 4x^3 (x^4 + 6)^{1/2}\, dx$

$$= \dfrac{7}{4} \cdot \dfrac{(x^4 + 6)^{3/2}}{\frac{3}{2}} + C$$

$$= \dfrac{7}{6}(x^4 + 6)^{3/2} + C$$

21. $\int (x^3+1)^2(3x)dx = \int (x^6+2x^3+1)(3x)dx$

$\qquad = \int (3x^7+6x^4+3x)dx$

$\qquad = \dfrac{3}{8}x^8 + \dfrac{6}{5}x^5 + \dfrac{3}{2}x^2 + C$

23. $\int (3x^2-1)^2(8x^2)dx = \int (9x^4-6x^2+1)(8x^2)dx$

$\qquad = \int (72x^6-48x^4+8x^2)dx$

$\qquad = \dfrac{72}{7}x^7 - \dfrac{48}{5}x^5 + \dfrac{8}{3}x^3 + C$

25. $\int \sqrt{x^3-3x}(x^2-1)dx$

$\qquad = \dfrac{1}{3}\int (x^3-3x)^{1/2}(3x^2-3)dx$

$\qquad = \dfrac{1}{3}\cdot\dfrac{(x^3-3x)^{3/2}}{\frac{3}{2}} + C$

$\qquad = \dfrac{2}{9}(x^3-3x)^{3/2} + C$

27. $\int \dfrac{3x^4\,dx}{(2x^5-5)^4} = 3\cdot\dfrac{1}{10}\int (2x^5-5)^{-4}(10x^4)dx$

$\qquad = \dfrac{3}{10}\cdot\dfrac{(2x^5-5)^{-3}}{-3} + C$

$\qquad = -\dfrac{1}{10}(2x^5-5)^{-3} + C$

$\qquad = \dfrac{-1}{10(2x^5-5)^3} + C$

29. $\int \dfrac{x^3-1}{(x^4-4x)^3}dx = \int (x^4-4x)^{-3}(x^3-1)dx$

$\qquad = \dfrac{1}{4}\int (x^4-4x)^{-3}(4x^3-4)dx$

$\qquad = \dfrac{1}{4}\cdot\dfrac{(x^4-4x)^{-2}}{-2} + C$

$\qquad = \dfrac{-1}{8(x^4-4x)^2} + C$

31. $\int \dfrac{x^2-4x}{\sqrt{x^3-6x^2+2}}dx$

$\qquad = \dfrac{1}{3}\int (x^3-6x^2+2)^{-1/2}\cdot 3(x^2-4x)dx$

$\qquad = \dfrac{1}{3}\cdot\dfrac{(x^3-6x^2+2)^{1/2}}{\frac{1}{2}} + C$

$\qquad = \dfrac{2}{3}(x^3-6x^2+2)^{1/2} + C$

33. $\int f(x)dx = (7x-13)^{10} + C$

$\qquad f(x) = \dfrac{d}{dx}\Big[(7x-13)^{10}+C\Big]$

$\qquad = 10(7x-13)^9(7)$

$\qquad = 70(7x-13)^9$

35. a. $\int x(x^2-1)^3 dx = \dfrac{1}{2}\int 2x(x^2-1)^3 dx$

$\qquad = \dfrac{1}{2}\cdot\dfrac{(x^2-1)^4}{4} + C = \dfrac{1}{8}(x^2-1)^4 + C$

b.

$\boxed{\begin{array}{c} f(x)=\frac{1}{8}\left(x^2-1\right)^4+C \\ (C=-5,0,5) \end{array}}$

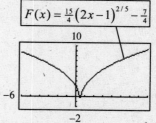

37. a. $\int \dfrac{3\,dx}{(2x-1)^{3/5}} = 3\cdot\dfrac{1}{2}\int (2x-1)^{-3/5}\cdot 2\,dx$

$\qquad = \dfrac{3}{2}\cdot\dfrac{(2x-1)^{2/5}}{\frac{2}{5}} + C$

$\qquad = \dfrac{15}{4}(2x-1)^{2/5} + C$

b. When $x=0$, $y=2$, so we have

$\qquad 2 = \dfrac{15}{4}(-1)^{2/5} + C$

$\qquad 2 = \dfrac{15}{4} + C$

$\qquad -\dfrac{7}{4} = C$

$\qquad F(x) = \dfrac{15}{4}(2x-1)^{2/5} - \dfrac{7}{4}$

$\boxed{F(x)=\frac{15}{4}\left(2x-1\right)^{2/5}-\frac{7}{4}}$

c. $f(x)$ is not defined at $x=\dfrac{1}{2}$.

d. The tangent line to $F(x)$ at $x=\dfrac{1}{2}$ is vertical.

39. $F(x) = (x^2 - 1)^{4/3} + C$

$F'(x) = \dfrac{4}{3}(x^2 - 1)^{1/3}(2x) = \dfrac{8}{3}x(x^2 - 1)^{1/3}$

The indefinite integral that gives the family is

$\displaystyle\int \dfrac{8x(x^2 - 1)^{1/3}}{3}\, dx.$

41. Only (b) can be integrated using methods we have studied:

b. $\displaystyle\int 7x^2 \left(x^3 + 4\right)^{-2} dx = \dfrac{7}{3}\dfrac{(x^3 + 4)^{-1}}{-1} + C$

$\qquad\qquad = \dfrac{-7}{3\left(x^3 + 4\right)} + C$

d. One form is $\displaystyle\int x\left(x^3 + 4\right)^{-p} dx$. ($p > 0$)

43. $\overline{MR} = \dfrac{-30}{(2x+1)^2} + 30$

$\displaystyle\int (-30(2x+1)^{-2} + 30)\,dx = \dfrac{-30}{2}\cdot\dfrac{(2x+1)^{-1}}{-1} + 30x + C$

$\qquad\qquad = \dfrac{15}{2x+1} + 30x + C$

$R(0) = 0$ gives $C = -15$. Thus,

$R(x) = \dfrac{15}{2x+1} + 30x - 15.$

45. $P(x) = \displaystyle\int 90(x+1)^2\, dx$

$\qquad = \dfrac{90(x+1)^3}{3} + C$

$\qquad = 30(x+1)^3 + C$

$P(0) = 0$ or $0 = 30(1)^3 + C$ or $C = -30$ and

$P(x) = 30(x+1)^3 - 30.$

$P(4) = 30(5)^3 - 30 = 3720$

47. a. $\displaystyle\int 5(x+1)^{-1/2}\, dx = 5\cdot\dfrac{(x+1)^{1/2}}{\frac{1}{2}} + C$

$\qquad\qquad = 10(x+1)^{1/2} + C$

$10 = 10(0+1)^{1/2} + C$ gives $C = 0$.

Thus, $s(x) = 10\sqrt{x+1}$.

b. $s(24) = 10\sqrt{24+1} = 50$

49. $A(t) = \displaystyle\int [-100(t+10)^{-2} + 2000(t+10)^{-3}]\, dt$

$\qquad = \dfrac{-100(t+10)^{-1}}{-1} + \dfrac{2000(t+10)^{-2}}{-2} + C$

$\qquad = \dfrac{100}{t+10} - \dfrac{1000}{(t+10)^2} + C$

$A(0) = 0$ or $0 = 10 - 10 + C$ or $C = 0$.

a. $A(t) = \dfrac{100}{t+10} - \dfrac{1000}{(t+10)^2}$

b. $A(10) = \dfrac{100}{20} - \dfrac{1000}{400} = 2.5$ (million)

51. $N(x) = \displaystyle\int -300(x+9)^{-1/2}\, dx$

$\qquad = -300\cdot\dfrac{(x+9)^{1/2}}{\frac{1}{2}} + C$

$\qquad = -600(x+9)^{1/2} + C$

$8000 = -600(0+9)^{1/2} + C$ or $C = 9800$

$N(x) = -600(x+9)^{1/2} + 9800$ and

$N(7) = -600(16)^{1/2} + 9800 = 7400.$

53. $\dfrac{dp}{dt} = -0.005715(t+60)^2 + 0.9566(t+60) - 39.37$

   a. $p(t) = \int\!\left(-0.005715(t+60)^2 + 0.9566(t+60) - 39.37\right) dt$

     Let $u = t+60$; $du = dt$.

     $p(t) = \int\!\left(-0.005715u^2 + 0.9566u - 39.37\right) du$

         $= -0.001905u^3 + 0.4783u^2 - 39.37u + C$

         $= -0.001905(t+60)^3 + 0.4783(t+60)^2 - 39.37(t+60) + C$

     Using $p(30) = 33.6$:

     $33.6 = -0.001905(30+60)^3 + 0.4783(30+60)^2 - 39.37(30+60) + C$

     And solving for C gives $C = 1091.415$.

     Thus, $p(t) = -0.001905(t+60)^3 + 0.4783(t+60)^2 - 39.37(t+60) + 1091.415$.

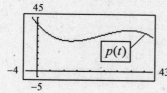

   b.

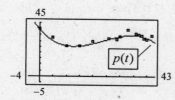

   c.   There is a good fit except for the mid 90's.

55. $D'(t) = -592.5(0.1t-1)^2 + 671.48(0.1t-1) - 106.47$

   a. $D(t) = \int\!\left(-592.5(0.1t-1)^2 + 671.48(0.1t-1) - 106.47\right) dt$

     Let $u = 0.1t-1$; $du = 0.1dt$.

     $D(t) = 10\int\!\left(-592.5u^2 + 671.48u - 106.47\right) du$

         $= -1975u^3 + 3357.4u^2 - 106.47u + C$

         $= -1975(0.1t-1)^3 + 3357.4(0.1t-1)^2 - 1064.7(0.1t-1) + C$

     Using $U(10) = -221.2$:

     $-221.2 = -1975(0.1(10)-1)^3 + 3357.4(0.1(10)-1)^2 - 1064.7(0.1(10)-1) + C$

     And solving for C gives $C = -221.2$.

     $D(t) = -1975(0.1t-1)^3 + 3357.4(0.1t-1)^2 - 1064.7(0.1t-1) - 221.2$

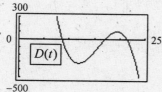

   b.

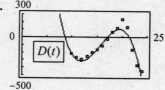

   c.   There is a good fit, except for the maximum.

Exercise 12.3

1. Let $u = 3x$. Then $du = 3dx$.

$\int e^{3x}(3dx) = \int e^u\, du = e^u + C = e^{3x} + C$

3. Let $u = -x$. Then $du = (-1)dx$.

$\int e^{-x}dx = (-1)\int e^{-x}(-1dx)$

$\qquad = -1\int e^u\, du$

$\qquad = (-1)e^u + C$

$\qquad = -e^{-x} + C$

5. $\int 1000e^{0.1x}\, dx = \dfrac{1000}{0.1}\int e^{0.1x}(0.1)dx$

$\qquad\qquad = 10{,}000e^{0.1x} + C$

7. $\int 840e^{-0.7x}\, dx = \dfrac{840}{-0.7}\int e^{-0.7x}(-0.7)dx$

$\qquad\qquad = -1200e^{-0.7x} + C$

9. Let $u = 3x^4$. Then, $du = 12x^3 dx$.

$\int x^3 e^{3x^4}\, dx = \dfrac{1}{12}\int e^{3x^4}(12x^3 dx) = \dfrac{1}{12}e^{3x^4} + C$

11. $\int \dfrac{3}{e^{2x}}\, dx = \dfrac{3}{-2}\int e^{-2x}(-2)dx = -\dfrac{3}{2}e^{-2x} + C$

13. $\int \dfrac{x^5}{e^{2-3x^6}}\, dx = \dfrac{1}{18}\int e^{3x^6-2}(18x^5 dx)$

$\qquad\qquad = \dfrac{1}{18}e^{3x^6-2} + C$

15. $\int\left(e^{4x} - \dfrac{3}{e^{x/2}}\right)dx$

$= \dfrac{1}{4}\int e^{4x}(4)dx - 3(-2)\int e^{-x/2}\left(-\dfrac{1}{2}\right)dx$

$= \dfrac{1}{4}e^{4x} + 6e^{-x/2} + C$

17. Let $u = x^3 + 4$. Then, $du = 3x^2 dx$.

$\int \dfrac{3x^2}{x^3+4}\, dx = \int\dfrac{du}{u} = \ln|u| + C = \ln|x^3+4| + C$

19. $\int \dfrac{dz}{4z+1} = \dfrac{1}{4}\int(4z+1)^{-1}(4\,dz) = \dfrac{1}{4}\ln|4z+1| + C$

21. Let $u = x^4 + 1$. Then, $du = 4x^3 dx$.

$\int \dfrac{x^3}{x^4+1}\, dx = \dfrac{1}{4}\int\dfrac{4x^3}{x^4+1}\, dx$

$\qquad\qquad = \dfrac{1}{4}\int\dfrac{du}{u}$

$\qquad\qquad = \dfrac{1}{4}\ln|u| + C$

$\qquad\qquad = \dfrac{1}{4}\ln|x^4+1| + C$

23. Let $u = x^2 - 4$. Then, $du = 2x\,dx$.

$\int \dfrac{4x}{x^2-4}\, dx = 2\int\dfrac{2x}{x^2-4}\, dx$

$\qquad\qquad = 2\int\dfrac{du}{u}$

$\qquad\qquad = 2\ln|u| + C$

$\qquad\qquad = 2\ln|x^2-4| + C$

25. Let $u = x^3 - 2x$. Then, $du = (3x^2 - 2)dx$.

$\int \dfrac{3x^2-2}{x^3-2x}\, dx = \int\dfrac{du}{u} = \ln|u| + C = \ln|x^3-2x| + C$

27. $\int \dfrac{z^2+1}{z^3+3z+17}\, dz = \dfrac{1}{3}\int\dfrac{3z^2+3}{z^3+3z+17}\, dz$

$\qquad\qquad = \dfrac{1}{3}\ln|z^3+3z+17| + C$

29.

$\begin{array}{r} x^2 \\ x-1\overline{)x^3-x^2+1} \\ \underline{x^3-x^2} \\ \text{Rem } 1 \end{array}$

$\int \dfrac{x^3-x^2+1}{x-1}\, dx = \int\left(x^2 + \dfrac{1}{x-1}\right)dx$

$\qquad\qquad = \dfrac{1}{3}x^3 + \ln|x-1| + C$

31. $\int \dfrac{x^2+x+3}{x^2+3}\, dx = \int\left(\dfrac{x^2+3}{x^2+3} + \dfrac{x}{x^2+3}\right)dx$

$\qquad\qquad = \int\left(1 + \dfrac{x}{x^2+3}\right)dx$

$\qquad\qquad = x + \dfrac{1}{2}\ln|x^2+3| + C$

33. $h(x) = f(x) = -e^{-(1/2)x}$

$$\int f(x)dx = \int (-e^{-(1/2)x})dx$$

$$= 2\int e^{-(1/2)x}\left(-\frac{1}{2}\right)dx$$

$$= 2e^{-(1/2)x} + C$$

$$= g(x) + C$$

35. $\int \dfrac{1}{3-x}dx = -\ln|3-x| + C = F(x)$

$0 = -\ln|3-0| + C$ gives $C = \ln 3$.

$F(x) = \ln 3 - \ln|3 - x|$

$$\boxed{F(x) = \ln 3 - \ln|3 - x|}$$

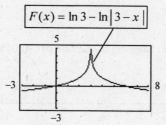

37. $F'(x) = f(x)$;

$f(x) = 1 + \dfrac{1}{x}$;

$F(x) = \int\left(1 + \dfrac{1}{x}\right)dx$

39. $F'(x) = f(x)$;

$f(x) = 5x(-e^{-x}) + e^{-x}(5)$;

$F(x) = \int (5e^{-x} - 5xe^{-x})dx$

41. Only (c) and (d) can be integrated using methods we have studied:

c. $\dfrac{1}{3}\int \dfrac{3(x^2 + 2x)}{x^3 + 3x^2 + 7}dx = \dfrac{1}{3}\ln\left(x^3 + 3x^2 + 7\right) + C$

d. $\dfrac{5}{8}\int e^{2x^4}\left(8x^3\right)dx = \dfrac{5}{8}e^{2x^4} + C$

43. $R(x) = 6\int e^{0.01x}dx = \dfrac{6}{0.01}\int e^{0.01x}(0.01dx)$

$$= 600e^{0.01x} + K$$

$R(0) = 0$ gives $K = -600$. Thus,

$R(x) = 600e^{0.01x} - 600$.

$R(100) = 600e^{0.01(100)} - 600 = 600e - 600$

$$= \$1030.97$$

45. $n = \int n_0(-K)e^{-Kt}dt = n_0\int e^{-Kt}(-K\,dt) = n_0e^{-Kt}$

(Given $C = 0$.)

47. $v(t) = \int \dfrac{40}{t+1}dt = 40\ln(t+1) + C$

$v(0) = 0$ gives $C = 0$. Recall $\ln 1 = 0$.

$v(3) = 40\ln 4 \approx 55.45$ or 55 words.

49. a. $S(n) = P\int e^{0.1n}(0.1dn) = Pe^{0.1n}$ $(C = 0)$

b. $S = 2P$

$2P = Pe^{0.1n}$

$0.1n = \ln 2$

$n = \dfrac{\ln 2}{0.1} \approx 6.9$ years

51. a. $p = 46.645\int e^{-0.491t}(-1dt)$

$$= \dfrac{46.645}{0.491}\int e^{-0.491t}(-0.491dt)$$

$$= 95e^{-0.491t} + C$$

$p = 95$ when $t = 0$ gives $C = 0$.

b. $p(0.1) = 95e^{-0.0491} \approx 90.45$

53. $\dfrac{dl}{dt} = \dfrac{14.202}{t+20}$

a. $l(t) = 14.202\ln|t+20| + C$

Solving $72.6 = 14.202\ln|55 + 20| + C$

gives $C = 11.283$.

$$\boxed{l(t) = 14.202\ln|t+20| + 11.283}$$

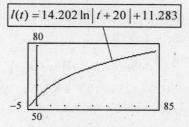

b. $\boxed{l(t) = 14.202\ln|t+20| + 11.283}$

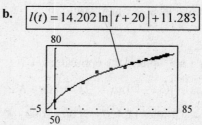

c. The model is a really good fit.

55. $\dfrac{dP}{dt} = -0.0548e^{-0.03443t}$

a. Yes. The exponential factor is always positive, hence, multiplying by -0.0548 makes dP/dt always negative. $dP/dt < 0$ means $P(t)$ is decreasing.

b. $P(t) = \int\left(-0.0548e^{-0.0343t}\right)dt$

Let $u = -0.0343t$; $du = -0.0343dt$.

$P(t) = \dfrac{1}{-0.0343}\int\left(-0.0548e^{u}\right)du = \dfrac{1}{-0.0343} - 0.0548e^{u} + C = 1.598e^{-0.0343t} + C$

$P(t) = 1.598e^{-0.0343t} + C$

Using $P(32) = 0.556$: $0.556 = 1.598e^{-0.0343(32)} + C$

And solving for C gives $C = 0.0228$.

$P(t) = 1.598e^{-0.0343} + 0.0228$

c. $P(45) \approx 0.36418$ and $P'(45) \approx -0.0117$

The model predicts that, in 2015, the CPI will be $0.36418 and it will be changing at the rate of -0.0117 dollars per year.

Exercise 12.4

1. $C(x) = \int(2x+100)dx = x^2 + 100x + K$

From the fact that $C(0) = 200$ we have

$200 = 0 + 0 + K$ or $K = 200$.

$C(x) = x^2 + 100x + 200$

3. $C(x) = \int(4x+2)dx = 2x^2 + 2x + K$

$C(10) = 300 = 200 + 20 + K$ or $K = 80$.

$C(x) = 2x^2 + 2x + 80$

5. $C(x) = \int(4x+40)dx = 2x^2 + 40x + K$

$C(25) = 3000 = 1250 + 1000 + K$ gives $K = 750$.

$C(30) = 1800 + 1200 + 750 = \$3750$

7. a. Optimal level is when $\overline{MR} = \overline{MC}$.

Thus, $44 - 5x = 3x + 20$

$24 = 8x$

So, $x = 3$ units is the optimal level.

b. $C(x) = \int(3x+20)dx = \dfrac{3}{2}x^2 + 20x + K$

Now, $C(80) = 11,400 = 9600 + 1600 + K$

gives $K = 200$.

$R(x) = \int(44-5x)dx = 44x - \dfrac{5}{2}x^2$

$P(x) = R(x) - C(x)$

$= 44x - \dfrac{5}{2}x^2 - \left(\dfrac{3}{2}x^2 + 20x + 200\right)$

$= 24x - 4x^2 - 200$

c. $P(3) = 72 - 36 - 200 = -\$164$ (a loss)

9. $C(x) = \int 30(x+4)^{1/2}\,dx = \dfrac{30(x+4)^{3/2}}{\frac{3}{2}} + K$

$= 20(x+4)^{3/2} + K$

$C(0) = 1000$ gives $1000 = 20(8) + K$ or

$K = 840$.

$R(x) = \int 900\,dx = 900x$

$P(x) = 900x - 20(x+4)^{3/2} - 840$

a. $P(5) = 4500 - 540 - 840 = \$3120$

b. $30\sqrt{x+4} = 900$

$\sqrt{x+4} = 30$

$x+4 = (30)^2 = 900$

Thus, $x = 896$ units will yield maximum profit.

11. $\overline{C}(x) = \int\left(-6x^{-2} + \dfrac{1}{6}\right)dx$

$= \dfrac{-6x^{-1}}{-1} + \dfrac{1}{6}x + K$

$= \dfrac{6}{x} + \dfrac{x}{6} + K$

$\overline{C}(6) = 10 = \dfrac{6}{6} + \dfrac{6}{6} + K$ gives $K = 8$.

a. $\overline{C}(x) = \dfrac{6}{x} + \dfrac{x}{6} + 8$

b. $\overline{C}(12) = \dfrac{6}{12} + \dfrac{12}{6} + 8 = \$10.50$

13. a. $C(x) = \int 1.05(x+180)^{0.05}\,dx$

$$= 1.05 \cdot \frac{(x+180)^{1.05}}{1.05} + K$$

$$= (x+180)^{1.05} + K$$

$C(0) = 180^{1.05} + K = 200$ or $K \approx -33.365$.

$C(x) = (x+180)^{1.05} - 33.365$

$$R(x) = \int\left(\frac{1}{\sqrt{0.5x+4}} + 2.8\right)dx$$

$$= \int((0.5x+4)^{-1/2} + 2.8)\,dx$$

$$= 2 \cdot \frac{(0.5x+4)^{1/2}}{\frac{1}{2}} + 2.8x + C$$

$$= 4(0.5x+4)^{1/2} + 2.8x + C$$

$R(0) = 0 = 4(0+4)^{1/2} + 0 + C$ or $C = -8$

$R(x) = 4(0.5x+4)^{1/2} + 2.8x - 8$

b.

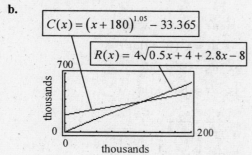

$$C(x) = (x+180)^{1.05} - 33.365$$

$$R(x) = 4\sqrt{0.5x+4} + 2.8x - 8$$

c. Using the graph, the maximum profit is at $x = 200$ thousand units.

$P(x) = 4\sqrt{0.5x+4} + 2.8x - 8$

$\qquad -(x+180)^{1.05} + 33.365$

$P(200) \approx \$114.743$ thousand

15. $\dfrac{dC}{dy} = 0.80$ $C(y) = \int 0.80\,dy = 0.80y + K$

$C(0) = 7 = 0.80(0) + K$ gives $K = 7$.

So, $C(y) = 0.80y + 7$.

17. $C(y) = \int[0.3 + 0.2(y^{-1/2})]\,dy$

$$= 0.3y + 0.2 \cdot \frac{y^{1/2}}{\frac{1}{2}} + K$$

$C(0) = 8$ gives $K = 8$. Thus,

$C(y) = 0.3y + 0.4\sqrt{y} + 8$.

19. $C(y) = \int[(y+1)^{-1/2} + 0.4]\,dy$

$$= 2(y+1)^{1/2} + 0.4y + K$$

$C(0) = 6$ gives $K = 4$.

Thus, $C(y) = 2(y+1)^{1/2} + 0.4y + 4$.

21. $C(y) = \int(0.7 - e^{-2y})\,dy$

$$= 0.7y + \frac{1}{2}e^{-2y} + K$$

$C(0) = 5.65$ gives $5.65 = 0 + 0.5 + K$ or

$K = 5.15$.

Thus, $C(y) = 0.7y + \dfrac{1}{2}e^{-2y} + 5.15$.

23. $\dfrac{dS}{dy} = 1 - \dfrac{dC}{dy}$

So, $0.15 = 1 - \dfrac{dC}{dy}$ or $C'(y) = 0.85$.

Then, $C(y) = \int 0.85\,dy = 0.85y + K$

$C(0) = 5.15$ means $5.15 = 0 + K$ or $K = 5.15$.

Thus, $C(y) = 0.85y + 5.15$.

25. $0.2 - \dfrac{1}{\sqrt{3y+7}} = 1 - C'(y)$ or

$C'(y) = 0.8 + (3y+7)^{-1/2}$

$C(y) = \int[0.8 + (3y+7)^{-1/2}]\,dy$

$$= 0.8y + \frac{2}{3}(3y+7)^{1/2} + K$$

$C(0) = 6 = 0 + \dfrac{2}{3}(0+7)^{1/2} + K$ gives $K \approx 4.24$.

Thus, $C(y) = 0.8y + \dfrac{2}{3}(3y+7)^{1/2} + 4.24$.

Exercise 12.5

1. $4y - 2xy' = 0 \qquad y = x^2$ gives $y' = 2x$
$4(x^2) - 2x(2x) = 0$
$4x^2 - 4x^2 = 0 \quad \checkmark$

3. $y = 3x^2 + 1 \qquad y' = 6x$
$2y\,dx - x\,dy = 2\,dx$ or $(2y - 2)\,dx = x\,dy$
Then, $\dfrac{dy}{dx} = \dfrac{2y-2}{x}$
$6x = \dfrac{2(3x^2+1)-2}{x} = \dfrac{6x^2+2-2}{x} = 6x \quad \checkmark$

5. $dy = xe^{x^2+1}dx$
$y = \int xe^{x^2+1}dx = \dfrac{1}{2}\int e^{x^2+1}(2x)dx = \dfrac{1}{2}e^{x^2+1} + C$

7. $2y\,dy = 4x\,dx$
$\int 2y\,dy = \int 4x\,dx$
$2\cdot\dfrac{y^2}{2} = 4\cdot\dfrac{x^2}{2} + C$
$y^2 = 2x^2 + C$

9. $3y^2\,dy = (2x-1)dx$
$\int 3y^2\,dy = \int(2x-1)dx$
$3\cdot\dfrac{y^3}{3} = 2\cdot\dfrac{x^2}{2} - x + C$
$y^3 = x^2 - x + C$

11. $y' = e^{x-3}$
$y = \int e^{x-3}dx$
$y = e^{x-3} + C$
$y(0) = e^{0-3} + C = 2$ gives $C = 2 - e^{-3}$.
So, $y = e^{x-3} + 2 - e^{-3}$

13. $dy = \left(\dfrac{1}{x} - x\right)dx$
$y = \int\left(\dfrac{1}{x} - x\right)dx = \int(x^{-1} - x)dx$
$y = \ln|x| - \dfrac{x^2}{2} + C$
$y(1) = \ln 1 - \dfrac{(1)^2}{2} + C = 0$ or $0 - \dfrac{1}{2} + C = 0$ or
$C = \dfrac{1}{2}$. So, $y = \ln|x| - \dfrac{x^2}{2} + \dfrac{1}{2}$.

15. $\dfrac{dy}{dx} = \dfrac{x^2}{y}$
$y\,dy = x^2\,dx$
$\int y\,dy = \int x^2\,dx$
$\dfrac{y^2}{2} = \dfrac{x^3}{3} + C$

17. $dx = x^3 y\,dy$
$x^{-3}dx = y\,dy$
$\int x^{-3}dx = \int y\,dy$
$\dfrac{x^{-2}}{-2} = \dfrac{y^2}{2} + \overline{C}$
or $\dfrac{1}{2x^2} + \dfrac{y^2}{2} = C \quad (C = -\overline{C})$

19. $dx = (x^2 y^2 + x^2)dy$
$dx = x^2(y^2+1)dy$
$x^{-2}dx = (y^2+1)dy$
$\int x^{-2}dx = \int(y^2+1)dy$
$\dfrac{x^{-1}}{-1} = \dfrac{y^3}{3} + y + \overline{C}$
or $C = \dfrac{1}{x} + y + \dfrac{1}{3}y^3$

21. $y^2\,dx = x\,dy$
$x^{-1}dx = y^{-2}dy$
$\int x^{-1}dx = \int y^{-2}dy$
$\ln|x| = \dfrac{y^{-1}}{-1} + C$
$C = \dfrac{1}{y} + \ln|x|$

23. $\dfrac{dy}{dx} = \dfrac{x}{y}$
$x\,dx = y\,dy$
$\int x\,dx = \int y\,dy$
$\dfrac{x^2}{2} = \dfrac{y^2}{2} + \overline{C}$
$x^2 - y^2 = C$

25. $(x+1)\dfrac{dy}{dx} = y$

$$y^{-1}dy = (x+1)^{-1}dx$$

$$\int y^{-1}dy = \int (x+1)^{-1}dx$$

$$\ln|y| = \ln|x+1| + C$$

$$\ln|y| = \ln|x+1| + \ln\bar{C}$$

$$\ln|y| = \ln|\bar{C}(x+1)|$$

$$y = \bar{C}(x+1)$$

$$(\ln\bar{C} = C)$$

27. $e^{2x}y\,dy = (y+1)dx$

$$e^{-2x}dx = \dfrac{y}{y+1}dy$$

$$-\dfrac{1}{2}\int e^{-2x}(-2)dx = \int \dfrac{y}{y+1}dy$$

$$-\dfrac{1}{2}e^{-2x} = \int\left(1 - \dfrac{1}{y+1}\right)dy$$

$$-\dfrac{1}{2}e^{-2x} = y - \ln|y+1| + C$$

(C can be on either side.)

29. $y^3dy = x^2dx$ gives $\dfrac{1}{4}y^4 = \dfrac{1}{3}x^3 + C$.

At $x = 1$, $y = 1$, we have

$$\dfrac{1}{4} = \dfrac{1}{3} + C \text{ or } C = -\dfrac{1}{12}.$$

Using this value for C and clearing fractions, we have $3y^4 - 4x^3 + 1 = 0$.

31. $\dfrac{2\,dx}{x^2} = \dfrac{3\,dy}{y^2}$ gives $\dfrac{2x^{-1}}{-1} = \dfrac{3y^{-1}}{-1} + C$ or

$-\dfrac{2}{x} + \dfrac{3}{y} = C$. At $x = 2$, $y = -1$, we have

$$-1 + (-3) = C \text{ or } C = -4.$$

Thus, $\dfrac{3}{y} - \dfrac{2}{x} + 4 = 0$ or $y = \dfrac{3x}{2-4x}$.

33. $e^{2y}dy = \dfrac{x^3+1}{x^2}dx$

$$\dfrac{1}{2}(e^{2y}\,2\,dy) = (x + x^{-2})dx$$

$$\dfrac{1}{2}e^{2y} = \dfrac{x^2}{2} + \dfrac{x^{-1}}{-1} + C$$

At $(1,0)$ we have $\dfrac{1}{2}e^0 = \dfrac{1}{2} - \dfrac{1}{1} + C$ which gives

$C = 1$. $(e^0 = 1)$

Thus, $\dfrac{1}{2}e^{2y} = \dfrac{1}{2}x^2 - \dfrac{1}{x} + 1$ or $xe^{2y} = x^3 - 2 + 2x$.

35. $\dfrac{2y}{y^2+1}dy = \dfrac{dx}{x}$ gives

$$\ln(y^2+1) = \ln|x| + \ln C = \ln(C|x|) \text{ or}$$

$$y^2 + 1 = C|x|$$

At $(1,2)$, we have $4 + 1 = C\cdot 1$ or $C = 5$.

Thus, $y^2 + 1 = 5|x|$.

37. $\dfrac{dy}{y} = k\dfrac{dx}{x}$ gives $\ln|y| = k\ln|x| + \ln|C| = \ln|Cx^k|$

or $y = |Cx^k|$.

39. a. $\dfrac{dx}{dt} = rx$; $r = 0.06$

$$\dfrac{dx}{x} = r\,dt$$

$$\ln x = rt + C$$

$$x = e^{rt+C} = e^{rt}\cdot e^C = Ke^{rt}$$

At $t = 0$, $x = 10{,}000$ and therefore,

$x = 10{,}000e^{rt}$.

b. $x = 10{,}000e^{0.06} = \$10{,}618.37$

$x = 10{,}000e^{0.30} = \$13{,}498.59$

c. $20{,}000 = 10{,}000e^{0.06t}$

$$2 = e^{0.06t}$$

$$\ln 2 = 0.06t$$

$$t = \dfrac{\ln 2}{0.06} \approx 11.55 \text{ years}$$

41. Rearrange $\dfrac{dP}{dt} = kP$ to get $\dfrac{dP}{P} = kdt$.

Then integrate: $\displaystyle\int \dfrac{dP}{P} = \int kdt$

$\ln P = kt + C$

$P = e^{kt+C} = e^{kt} \cdot e^C = Ke^{kt}$

At $t = 0$, $P = 100{,}000$:

$\quad 100{,}000 = Ke^{k(0)} \Rightarrow K = 100{,}000$

At $t = 15$, $P = 211{,}700$:

$\quad 211{,}700 = 100{,}000e^{15k}$

$\quad 2.117 = e^{15k}$

$\quad \ln 2.117 = 15k$

$\quad k = \dfrac{\ln 2.117}{15} \approx 0.05$

So, $P = 100{,}000e^{0.05t}$.

43. $\dfrac{dy}{y} = k\, dt$ gives $\ln|y| = kt + C$ or $y = Ae^{kt}$

At $t = 0$, we have $y = 10{,}000$ and, therefore, $A = 10{,}000$. Thus, $y = 10{,}000e^{kt}$. At $t = 2$, we have $y = 30{,}000$ and, therefore, $3 = e^{2k}$ or $2k = \ln 3$ or $k \approx 0.55$. Then, if $y = 1{,}000{,}000$, we have $100 = e^{0.55t}$ or $\ln 100 = 0.55t$ or $t \approx 8.4$ hr.

45. $\dfrac{dy}{dp} = -\dfrac{2}{5}\left(\dfrac{y}{p+8}\right)$

$\dfrac{dy}{y} = -\dfrac{2}{5}\dfrac{dp}{p+8}$

$\ln y = -\dfrac{2}{5}\ln|p+8| + \ln C = \ln[(p+8)^{-2/5}C]$

$\quad y = C(p+8)^{-2/5}$

At $p = 24$, $y = 8$, we have $8 = C(24+8)^{-2/5}$ or $C = 2^3(32)^{2/5} = 2^5 = 32$. Thus, $y = 32(p+8)^{-2/5}$

47. $\dfrac{dy}{dt} = -kt$ gives, after several steps, $y = Ae^{-kt}$.

At $t = 0$, $y = 100$ and, thus, $A = 100$. So, $y = 100e^{-kt}$. At $t = 10$, $0.9997(100) = 99.97$ pounds remain. Solving $99.97 = 100e^{-10k}$ for k gives $k = 0.00003$. With $y = 50$ (half the original 100), we have $50 = 100e^{-0.00003t}$. Solving, we have $-0.00003t = \ln(0.5)$ or $t \approx 23{,}100$ years.

49. $\dfrac{dx}{dt} = 5(0.06) - 5 \cdot \dfrac{x}{100} = \dfrac{3}{10} - \dfrac{x}{20} = \dfrac{6-x}{20}$

Thus, $\dfrac{dx}{6-x} = \dfrac{dt}{20}$ gives $-\ln|6-x| = \dfrac{t}{20} + C$ or

$\ln|6-x| = -\dfrac{t}{20} - C$. So, $6 - x = ke^{-t/20}$ or

$x = 6 - ke^{-t/20}$. At $t = 0$, we have $x = 0$ and, thus, $k = 6$. Then, $x = 6 - 6e^{-t/20}$.

51. $\dfrac{dx}{dt} = 5(0.1) - 5 \cdot \dfrac{x}{200} = \dfrac{20-x}{40}$

Thus, $\dfrac{dx}{20-x} = \dfrac{dt}{40}$ gives $-\ln|20-x| = \dfrac{t}{40} + C$ or

$\ln|20-x| = -\dfrac{t}{40} - C$ So, $20 - x = ke^{-t/40}$.

At $t = 0$, we have $x = 10$ and, thus, $k = 10$.

Then, $x = 20 - 10e^{-t/40}$. $\left(\dfrac{t}{40} = 0.025t\right)$

53. $\dfrac{dV}{V} = 0.2e^{-0.1t}dt$ gives $\ln|V| = -2e^{-0.1t} + C$.

At $t = 0$, $V = 1.86$ and $\ln 1.86 = -2(1) + C$ or

$C = 2 + \ln 1.86$. So, $\ln|V| = -2e^{-0.1t} + 2 + \ln 1.86$

or $\ln\left(\dfrac{|V|}{1.86}\right) = 2 - 2e^{-0.1t}$. Thus, $V = 1.86e^{2-2e^{-0.1t}}$.

55. $\dfrac{dV}{dt} = kV^{2/3}$ or $V^{-2/3}dV = k\, dt$ gives

$\dfrac{V^{1/3}}{\frac{1}{3}} = kt + C$ or $3V^{1/3} = kt + C$. At $t = 0$,

$V = 0$ and $3(0)^{1/3} = k(0) + C$ gives $C = 0$. So,

$3V^{1/3} = kt$ or $V^{1/3} = \dfrac{kt}{3}$ or $V = \dfrac{k^3t^3}{27}$.

57. $\dfrac{du}{u-T} = k\, dt$ gives $\ln|u-T| = kt + C$ and after

several steps $u - T = Ae^{kt}$.

At $t = 0$, $u - T = 0 - 20 = A \cdot 1$ gives $A = -20$.

Thus, $u = 20 - 20e^{kt}$ ($T = 20$)

At $t = 1$, $u = 8$, we have $\quad 8 = 20 - 20e^{k(1)}$

$\quad\quad\quad\quad\quad\quad\quad\quad\quad\quad -12 = -20e^k$

$\quad\quad\quad\quad\quad\quad\quad\quad\quad\quad e^k = 0.6$

$\ln 0.6 = k \ln e$ or $k \approx -0.51$

Thus, $u = 20 - 20e^{-0.51t}$. If $u = 18$, then

$18 = 20 - 20e^{-0.51t}$ gives $t \approx 4.5$ hours.

59. a. $\dfrac{dE}{dt} = 0.0315E$ or $\dfrac{dE}{E} = 0.043\,dt$ gives

$\ln E = 0.0315t + C$ or $E = ke^{0.0315t}$

Then, $0.13 = ke^0$ gives $k = 0.13$.

Thus, $E(t) = 0.13e^{0.0315t}$.

b.

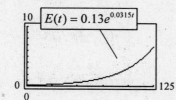

The two graphs have a good comparison.

61. a. $\dfrac{dP}{dt} = -0.05P$ or $\dfrac{dP}{P} = -0.05dt$ gives

$\ln P = -0.05t + C$ or $P = ke^{-0.05t}$.

Using $P(0) = 50,000$:

$50,000 = ke^0 \Rightarrow k = 50,000$

$P(t) = 50,000e^{-0.05t}$

b. $P(30) \approx \$11,156$ The model predicts that the purchasing power of a \$50,000 pension after 30 years will be \$11,156.

Review Exercises

1. $\int x^6 \, dx = \dfrac{1}{7} x^7 + C$

2. $\int x^{1/2} \, dx = \dfrac{2}{3} x^{3/2} + C$

3. $\int \left(12x^3 - 3x^2 + 4x + 5 \right) dx$
$= 3x^4 - x^3 + 2x^2 + 5x + C$

4. $\int 7 \left(x^2 - 1 \right)^2 dx = 7 \int \left(x^4 - 2x^2 + 1 \right) dx$
$\qquad = 7 \left(\dfrac{1}{5} x^5 - \dfrac{2}{3} x^3 + x \right) + C$
$\qquad = \dfrac{7}{5} x^5 - \dfrac{14}{3} x^3 + 7x + C$

5. $\int 7x \left(x^2 - 1 \right)^2 dx$
Let $u = x^2 - 1;\ du = 2x \, dx.$
$\int \dfrac{7}{2} u^2 \, du = \dfrac{7}{2} \cdot \dfrac{1}{3} u^3 + C = \dfrac{7}{6} \left(x^2 - 1 \right)^3 + C$

6. $\int (x^3 - 3x^2)^5 (x^2 - 2x) \, dx$
$= \dfrac{1}{3} \int (x^3 - 3x^2)^5 (3x^2 - 6x) \, dx$
$= \dfrac{1}{18} (x^3 - 3x^2)^6 + C$

7. $\int \left(x^3 + 4 \right)^2 3x \, dx = \int \left(x^6 + 8x^3 + 16 \right) 3x \, dx$
$\qquad = \int \left(3x^7 + 24x^4 + 48x \right) dx$
$\qquad = \dfrac{3}{8} x^8 + \dfrac{24}{5} x^5 + 24x^2 + C$

8. $\int 5x^2 \left(3x^3 + 7 \right)^6 dx$
Let $u = 3x^3 + 7;\ du = 9x^2 \, dx.$
$= \dfrac{5}{9} \int u^6 \, du = \dfrac{5}{9} \cdot \dfrac{u^7}{7} + C = \dfrac{5}{63} \left(3x^3 + 7 \right)^7 + C$

9. $\int \dfrac{x^2}{x^3 + 1} \, dx = \dfrac{1}{3} \int \dfrac{3x^2}{x^3 + 1} \, dx$
$\qquad = \dfrac{1}{3} \ln \left| x^3 + 1 \right| + C$

10. $\int \dfrac{x^2}{(x^3 + 1)^2} \, dx = \dfrac{1}{3} \int \dfrac{3x^2}{(x^3 + 1)^2} \, dx$
$\qquad = \dfrac{-1}{3(x^3 + 1)} + C$

11. $\int (x^3 - 4)^{-1/3} x^2 \, dx = \dfrac{1}{3} \int (x^3 - 4)^{-1/3} (3x^2 \, dx)$
$\qquad = \dfrac{1}{3} \cdot \dfrac{(x^3 - 4)^{2/3}}{\frac{2}{3}} + C$
$\qquad = \dfrac{1}{2} (x^3 - 4)^{2/3} + C$

12. $\int \dfrac{x^2}{x^3 - 4} \, dx = \dfrac{1}{3} \int \dfrac{3x^2 \, dx}{x^3 - 4} = \dfrac{1}{3} \ln \left| x^3 - 4 \right| + C$

13. $\int \left(\dfrac{x^3 + 1}{x^2} \right) dx = \int (x + x^{-2}) \, dx$
$\qquad = \dfrac{x^2}{2} + \dfrac{x^{-1}}{-1} + C$
$\qquad = \dfrac{1}{2} x^2 - \dfrac{1}{x} + C$

14. $\int \dfrac{x^3 - 3x + 1}{x - 1} \, dx = \int \left(x^2 + x - 2 - \dfrac{1}{x - 1} \right) dx$
$\qquad = \dfrac{1}{3} x^3 + \dfrac{1}{2} x^2 - 2x - \ln \left| x - 1 \right| + C$

15. $\int y^2 e^{y^3} \, dy = \dfrac{1}{3} \int e^{y^3} (3y^2 \, dy) = \dfrac{1}{3} e^{y^3} + C$

16. $\int (x - 1)^2 \, dx = \dfrac{1}{3} (x - 1)^3 + C$ or $\dfrac{1}{3} x^3 - x^2 + x + \overline{C}$
$\left(\overline{C} = C - \dfrac{1}{3} \right)$

17. $\int \dfrac{3x^2}{2x^3 - 7} \, dx = \dfrac{1}{2} \int 2(3x^2)(2x^3 - 7)^{-1} \, dx$
$\qquad = \dfrac{1}{2} \ln \left| 2x^3 - 7 \right| + C$

18. $\int \dfrac{5 \, dx}{e^{4x}} = -\dfrac{5}{4} \int e^{-4x} (-4 \, dx)$
$\qquad = -\dfrac{5}{4} e^{-4x} + C$
$\qquad = -\dfrac{5}{4e^{4x}} + C$

19. $\int (x^3 - e^{3x}) \, dx = \int x^3 \, dx - \dfrac{1}{3} \int e^{3x} (3 \, dx)$
$\qquad = \dfrac{1}{4} x^4 - \dfrac{1}{3} e^{3x} + C$

20. $\displaystyle\int xe^{1+x^2}\,dx = \frac{1}{2}\int e^{1+x^2}(2x\,dx)$

$\displaystyle = \frac{1}{2}e^{1+x^2} + C$

21. $\displaystyle\int \frac{6x^7}{(5x^8+7)^3}\,dx = 6\cdot\frac{1}{40}\int (5x^8+7)^{-3}(40x^7\,dx)$

$\displaystyle = \frac{6}{40}\cdot\frac{(5x^8+7)^{-2}}{-2}$

$\displaystyle = \frac{-3}{40(5x^8+7)^2} + C$

22. $\displaystyle\int \frac{7x^3}{(1-x^4)^{1/2}}\,dx = -\frac{7}{4}\int \frac{-4x^3}{(1-x^4)^{1/2}}\,dx$

$\displaystyle = -\frac{7}{4}[2(1-x^4)^{1/2}] + C$

$\displaystyle = -\frac{7}{2}\sqrt{1-x^4} + C$

23. $\displaystyle\int \left(\frac{e^{2x}}{2} + \frac{2}{e^{2x}}\right)dx$

$\displaystyle = \frac{1}{2}\cdot\frac{1}{2}\int e^{2x}(2\,dx) + 2\left(-\frac{1}{2}\right)\int e^{-2x}(-2\,dx)$

$\displaystyle = \frac{1}{4}e^{2x} - e^{-2x} + C$

24. $\displaystyle\int \left(x - \frac{1}{(x+1)^2}\right)dx = \frac{1}{2}x^2 + \frac{1}{x+1} + C$

25. a. $\displaystyle\int (x^2-1)^4(x\,dx) = \frac{1}{2}\int (x^2-1)^4(2x\,dx)$

$\displaystyle = \frac{1}{10}(x^2-1)^5 + C$

b. $\displaystyle\int (x^2-1)^{10}(x\,dx) = \frac{1}{2}\int (x^2-1)^{10}(2x\,dx)$

$\displaystyle = \frac{1}{22}(x^2-1)^{11} + C$

c. $\displaystyle\int (x^2-1)^7(3x\,dx) = \frac{3}{2}\int (x^2-1)^7(2x\,dx)$

$\displaystyle = \frac{3}{16}(x^2-1)^8 + C$

d. $\displaystyle\int (x^2-1)^{-2/3}(x\,dx) = \frac{1}{2}\int (x^2-1)^{-2/3}(2x\,dx)$

$\displaystyle = \frac{3}{2}(x^2-1)^{1/3} + C$

26. a. $\displaystyle\int \frac{2x}{x^2-1}\,dx = \ln\left|x^2-1\right| + C$

b. $\displaystyle\int \frac{2x}{(x^2-1)^2}\,dx = -\frac{1}{x^2-1} + C$

c. $\displaystyle\int \frac{3x\,dx}{(x^2-1)^{1/2}} = \frac{3}{2}\int \frac{2x\,dx}{(x^2-1)^{1/2}}$

$\displaystyle = 3(x^2-1)^{1/2} + C$

$\displaystyle = 3\sqrt{x^2-1} + C$

d. $\displaystyle\int \frac{3x\,dx}{x^2-1} = \frac{3}{2}\int \frac{2x\,dx}{x^2-1} = \frac{3}{2}\ln\left|x^2-1\right| + C$

27. $\displaystyle\frac{dy}{dt} = 4.6e^{-0.05t}$

$dy = 4.6e^{-0.05t}\,dt$

$\displaystyle\int dy = -\frac{1}{0.05}\int 4.6e^{-0.05t}(-0.05\,dt)$

$y = -20(4.6)e^{-0.05t} + C$

$y = C - 92e^{-0.05t}$

28. $dy = (64 + 76x - 36x^2)\,dx$

$\displaystyle\int dy = \int (64 + 76x - 36x^2)\,dx$

$y = 64x + 38x^2 - 12x^3 + C$

29. $\displaystyle\frac{dy}{dx} = \frac{4x}{y-3}$ or $(y-3)\,dy = 4x\,dx$ gives

$\displaystyle\frac{y^2}{2} - 3y = 2x^2 + C.$

Also: $\displaystyle\frac{(y-3)^2}{2} = 2x^2 + C_1$ or $(y-3)^2 = 4x^2 + C$

30. $\displaystyle t\,dy = \frac{dt}{y+1}$

$\displaystyle\int (y+1)\,dy = \int \frac{dt}{t}$

$\displaystyle\frac{(y+1)^2}{2} = \ln|t| + C_1$

$(y+1)^2 = 2\ln|t| + C$

31. $\displaystyle\frac{dy}{dx} = \frac{x}{e^y}$ or $e^y\,dy = x\,dx$ gives $e^y = \frac{x^2}{2} + C.$

32. $\dfrac{dy}{dt} = \dfrac{4y}{t}$

Write in separated form and integrate.

$\displaystyle\int \dfrac{dy}{y} = 4 \int \dfrac{dt}{t}$

$\ln|y| = 4\ln|t| + C$

Assume $y > 0$, $\ln y = \ln t^4 + \ln C_1$ where

$C = \ln C_1$.

$\ln y = \ln C_1 t^4$

$y = C_1 t^4$

33. $y' = \dfrac{x^2}{y+1}$ or $x^2 dx = (y+1)dy$ gives

$\dfrac{x^3}{3} = \dfrac{(y+1)^2}{2} + C.$ $y(0) = 4$ or $\dfrac{0}{3} = \dfrac{(4+1)^2}{2} + C$

gives $C = -\dfrac{25}{2}$. Then, $3(y+1)^2 = 2x^3 + 75$.

34. $(1+2y)dy = 2x\,dx$ gives $y + y^2 = x^2 + C.$

At $(2,0)$ we have $0 + 0 = 4 + C$ or $C = -4$.

So, $x^2 = y + y^2 + 4$.

35. $\overline{MR} = 6x + 12$

$\displaystyle\int (6x + 12)dx = 6 \cdot \dfrac{x^2}{2} + 12x = 3x^2 + 12x$

When $x = 4$, revenue $= 3(4)^2 + 12(4) = 96$.

36. $\dfrac{dp}{dt} = 27 + 24t - 3t^2$

$p(t) = 27t + 12t^2 - t^3 + C$

At $t = 0$, $p = 0$ and this yields $C = 0$.

Then $p(8) = 216 + 768 - 512 = 472$.

37. $P'(t) = 400\left[\dfrac{5}{(t+5)^2} - \dfrac{50}{(t+5)^3}\right]$

$P(t) = 400 \displaystyle\int [5(t+5)^{-2} - 50(t+5)^{-3}]dt$

$= 400\left[\dfrac{5(t+5)^{-1}}{-1} - \dfrac{50(t+5)^{-2}}{-2}\right] + C$

$= 400\left[\dfrac{-5}{t+5} + \dfrac{25}{(t+5)^2}\right] + C$

Since $P(0) = 400$, $400 = 400(-1+1) + C$ gives

$C = 400$.

$P(t) = 400\left[1 - \dfrac{5}{t+5} + \dfrac{25}{(t+5)^2}\right]$

38. $\dfrac{dp}{dt} = \dfrac{100,000}{(t+100)^2}$

$p(t) = 100,000 \cdot \dfrac{(t+100)^{-1}}{-1} + C$

$p(1) = 1000$ gives $1000 = \dfrac{-100,000}{101} + C$ or

$C \approx 1990.099$. Thus,

$p(t) = 1990.099 - \dfrac{100,000}{t+100}$

39. a. $\dfrac{dy}{dt} = 2.4e^{-0.04t}$ or $dy = 2.4e^{-0.04t}dt$ gives

$y = -60e^{-0.04t} + C$

$0 = -60(1) + C$ or $C = 60$

$y = 60 - 60e^{-0.04t}$

b. $y = 60 - 60e^{-0.04(12)} \approx 23\%$

40. $R'(x) = \dfrac{800}{x+1}$

$R(x) = 800\ln(x+1)$ (Note $C = 0$.)

41. $\overline{MC} = 6x + 4$

a. $C(x) = \displaystyle\int (6x + 4)dx = 3x^2 + 4x + K$

$3(100)^2 + 4(100) + K = 31,400$ gives

$K = 1000$ fixed costs.

b. $C(x) = 3x^2 + 4x + 1000$

42. $\overline{MR} = \overline{MC}$

$46 = 30 + \dfrac{1}{5}x$

$16 = \dfrac{1}{5}x$

$x = 80$

So, 80 units are needed to maximize profit.

$C(x) = 30x + \dfrac{1}{10}x^2 + K.$ $C(0) = 200$ gives

$K = 200$. Since $R(x) = 46x$ we have

$P(x) = 16x - \dfrac{1}{10}x^2 - 200.$

$P(80) = 1280 - 640 - 200 = \$440$

43. $\dfrac{dC}{dy} = (2y+16)^{-1/2} + 0.6$ or

$dC = \dfrac{1}{2}(2y+16)^{-1/2}(2\,dy) + 0.6\,dy$ gives

$C = \dfrac{1}{2} \cdot \dfrac{(2y+16)^{1/2}}{\frac{1}{2}} + 0.6y + K = \sqrt{2y+16} + 0.6y + K.$

$8.5 = \sqrt{0+16} + 0 + K$ gives $K = 4.5$. So,

$C = \sqrt{2y+16} + 0.6y + 4.5.$

44. $\dfrac{dS}{dy} = 0.2 - 0.1e^{-2y}$

$\dfrac{dC}{dy} = 1 - \dfrac{dS}{dy}$

$\dfrac{dC}{dy} = 0.80 + 0.1e^{-2y}$

$C(y) = \int (0.80 + 0.1e^{-2y})dy$

$C(y) = 0.8y - \dfrac{0.1}{2}e^{-2y} + K$

Using $C(0) = 7.8$ to find K gives $K = 7.85$.

So, $C(y) = 0.8y - 0.05e^{-2y} + 7.85.$

45. $\dfrac{dW}{dL} = \dfrac{3W}{L}$ or $\dfrac{dW}{W} = 3\dfrac{dL}{L}$ gives

$\ln W = 3\ln L + \ln C = \ln(L^3 \cdot C).$

Thus, $W = CL^3$

46. a. $\dfrac{dP}{dt} = kP \;\Rightarrow\; \dfrac{dP}{P} = kdt$

Integrating: $\ln|P| = kt + C$

b. $P = e^{kt+C} = e^{kt} \cdot e^{C} = Ke^{kt}$

c. Using $P(0) = 50,000$:

$50000 = Ke^{k(0)} \;\Rightarrow\; K = 50,000$

$P(t) = 50,000e^{kt}$

Using $P(10) = 135,914$:

$135914 = 50000e^{k(10)}$

$2.7183 = e^{10k}$

$\ln 2.7183 = 10k$

$k = \dfrac{\ln 2.7183}{10} \approx 0.1$

$P(t) = 50,000e^{0.1t}$

d. k represents the annual interest rate, which is 10%.

47. $\dfrac{dx}{dt} = kx$ As before $x = Ce^{kt}$.

When $t = 0$, $x = 10$ and from this information we have $C = 10$. So, $x = 10e^{kt}$. Use the half-life, $t = 4.6$ million years and $x = 5$, to determine k. The result is $k \approx -0.15$. So, $x = 10e^{-0.15t}$.

With 20% left, $x = 2$. Now determine the corresponding t.

$2 = 10e^{-0.15t}$

$\ln 0.2 = -0.15t$

$t \approx 10.73$ million years

48. $\dfrac{dx}{dt} = 4 \cdot 3 - 4 \cdot \dfrac{x}{120} = \dfrac{360 - x}{30}$

Thus, $\dfrac{dx}{360 - x} = \dfrac{dt}{30}$ gives

$-\ln(360 - x) = \dfrac{1}{30}t + C_1$ or

$\ln(360 - x) = -\dfrac{t}{30} + C$. After algebraic

manipulation, we have $360 - x = Ae^{-t/30}$. Then,

$360 - 0 = A \cdot 1$ or $A = 360$ and

$x = 360 - 360e^{-t/30}.$

49. $\dfrac{dx}{dt} = 3 \cdot 2 - 3 \cdot \dfrac{x}{300}$

$\dfrac{dx}{dt} = 6 - \dfrac{x}{100}$

$\dfrac{dx}{dt} = \dfrac{600 - x}{100}$

$\int \dfrac{dx}{600 - x} = \dfrac{1}{100}\int dt$

$-\ln(600 - x) = \dfrac{1}{100}t + C$

$x = 100$ when $t = 0$.

Using $x = 100$ when $t = 0$ gives $C = -\ln 500$.

So, $-\ln(600 - x) = 0.01t - \ln 500$

$\ln\left(\dfrac{600 - x}{500}\right) = \ln e^{-0.01t}$

$600 - x = 500e^{-0.01t}$

$x = 600 - 500e^{-0.01t}$

When $x = 500$ we have $500 = 600 - 500e^{-0.01t}$

$0.2 = e^{-0.01t}$

$\ln 0.2 = -0.01t$

$t \approx 161$ minutes

Chapter Test _____

1. $\int(6x^2+8x-7)dx = \dfrac{6x^3}{3}+\dfrac{8x^2}{2}-7x+C$

$\qquad = 2x^3+4x^2-7x+C$

2. $\int(4+x^{1/2}-x^{-2})dx = 4x+\dfrac{x^{3/2}}{\frac{3}{2}}-\dfrac{x^{-1}}{-1}+C$

$\qquad = 4x+\dfrac{2}{3}x^{3/2}+\dfrac{1}{x}+C$

3. $5\int(4x^3-7)^9(x^2\,dx) = \dfrac{5}{12}\int(4x^3-7)^9(12x^2dx)$

$\qquad = \dfrac{5}{12}\dfrac{(4x^3-7)^{10}}{10}+C$

$\qquad = \dfrac{1}{24}(4x^3-7)^{10}+C$

4. $\int(3x^2-6x+1)^{-3}(2x-2)\,dx$

Let $u = 3x^2-6x$;

$\qquad du = (6x-6)\,dx = 3(2x-2)\,dx.$

$\qquad = \dfrac{1}{3}\int u^{-3}\,du = \dfrac{1}{3}\cdot\dfrac{u^{-2}}{-2}+C$

$\qquad = -\dfrac{1}{6}(3x^2-6x+1)^{-2}+C$

5. $\int\dfrac{s^3}{2s^4-5}\,ds = \dfrac{1}{8}\int\dfrac{8s^3}{2s^4-5}\,ds = \dfrac{1}{8}\ln\left|2s^4-5\right|+C$

6. $100\int e^{-0.01x}\,dx = \dfrac{100}{-0.01}\int e^{-0.01x}(-0.01\,dx)$

$\qquad = -10,000e^{-0.01x}+C$

7. $5\int e^{2y^4-1}(y^3\,dy) = \dfrac{5}{8}\int e^{2y^4-1}(8y^3\,dy)$

$\qquad = \dfrac{5}{8}e^{2y^4-1}+C$

8. $\int\left(e^x+\dfrac{5}{x}-1\right)dx = e^x+5\ln|x|-x+C$

9.

$$\begin{array}{r}x-1\\x+1\overline{)x^2}\\\underline{x^2+x}\\-x\\\underline{-x-1}\\\text{Remainder: }1\end{array}$$

$\int\dfrac{x^2}{x+1}\,dx = \int\left(x-1+\dfrac{1}{x+1}\right)dx$

$\qquad = \dfrac{x^2}{2}-x+\ln|x+1|+C$

10. $\int f(x)dx = 2x^3-x+5e^x+C.$ $f(x)$ is the

derivative of the right side. $f(x) = 6x^2-1+5e^x$

11. $y' = 4x^3+3x^2;\ y(0) = 4$

$\qquad y = x^4+x^3+C$

$\qquad 4 = 0^4+0^3+C$

$\qquad y = x^4+x^3+4$

12. $\dfrac{dy}{dx} = e^{4x};\ y(0) = 2$

$\qquad y = \dfrac{1}{4}e^{4x}+C$

$\qquad 2 = \dfrac{1}{4}(1)+C$ or $C = \dfrac{7}{4}$

$\qquad y = \dfrac{1}{4}e^{4x}+\dfrac{7}{4}$

13. $\dfrac{dy}{dx} = x^3y^2$

$\qquad \dfrac{dy}{y^2} = x^3\,dx$

$\qquad \dfrac{y^{-1}}{-1} = \dfrac{x^4}{4}+\bar{C}$ or $\dfrac{x^4}{4}+\dfrac{1}{y} = K$

$\qquad \dfrac{1}{y} = \dfrac{4K-x^4}{4}$ or $y = \dfrac{4}{C-x^4}$

14. $p'(t) = 2000t^{1.04};\ p(0) = 50,000;\ p(10) = ?$

$\qquad p(t) = 2000\dfrac{t^{2.04}}{2.04}+C$

$\qquad 50,000 = 0+C = C$

$\qquad p(10) = \dfrac{2000}{2.04}10^{2.04}+50,000 = 157,498$

15. $\overline{MC} = 4x + 50$; $\overline{MR} = 500$; $C(10) = 1000$

$C = 2x^2 + 50x + K$; $R = 500x$

$1000 = 2(10)^2 + 50(10) + K$ gives $K = 300$

$P(x) = 500x - (2x^2 + 50x + 300)$

$\qquad = 450x - 2x^2 - 300$

16. $\dfrac{dS}{dy} = 0.22 - \dfrac{0.25}{\sqrt{0.5y+1}}$ Since $\dfrac{dC}{dy} = 1 - \dfrac{dS}{dy}$, we

have $\dfrac{dC}{dy} = 0.78 + \dfrac{0.25}{\sqrt{0.5y+1}}$.

$C = 0.78y + \sqrt{0.5y+1} + K$

$6.6 = 0 + \sqrt{0+1} + K$ gives $K = 5.6$

$C(y) = 0.78y + \sqrt{0.5y+1} + 5.6$

17. $\dfrac{dx}{dt} = kx$; $\dfrac{dx}{x} = k\,dt$; $\ln x = kt + C$ gives

$x = e^{kt+C} = e^{kt} \cdot e^{C} = Ae^{kt}$. At $t = 0$, $x = A$. Then

at $t = 100$, $x = \dfrac{1}{2}A$.

$\dfrac{1}{2}A = Ae^{100k}$ gives $k = \dfrac{\ln\left(\frac{1}{2}\right)}{100} \approx -0.00693$

$x = Ae^{-0.00693t}$; $x = 0.1A$ gives $0.1A = Ae^{-0.00693t}$

gives $t = \dfrac{\ln(0.1)}{-.00693} \approx 332.3$ days

Chapter 13: Definite Integrals; Techniques of Integration

Exercise 13.1

1. $f(x) = 4x - x^2$

Width of rectangles $= \dfrac{2-0}{2} = 1$.

Height of each rectangle: $f(1) = 4(1) - 1^2 = 3$

$\qquad\qquad\qquad\qquad f(2) = 4(2) - 2^2 = 4$

$A \approx 1(3+4) = 7$

3. $f(x) = 9 - x^2$

Width of rectangles $= \dfrac{3-1}{4} = \dfrac{1}{2}$

Height of rectangles:

$f(1.5) = 9 - 2.25 = 6.75 \qquad f(2) = 9 - 4 = 5$

$f(2.5) = 9 - 6.25 = 2.75 \qquad f(3) = 9 - 9 = 0$

$A = 0.5(6.75 + 5 + 2.75 + 0) = 7.25$

5. $f(x) = 4x - x^2$

Width of rectangles $= \dfrac{2-0}{2} = 1$

Height of each rectangle:

$f(0) = 4(0) - 0^2 = 0$

$f(1) = 4(1) - 1^2 = 3$

$A \approx 1(0+3) = 3$

7. $f(x) = 9 - x^2$

Width of rectangles $= \dfrac{3-1}{4} = \dfrac{1}{2}$

Height of rectangles:

$f(1) = 9 - 1 = 8 \qquad f\left(\frac{3}{2}\right) = 9 - 2.25 = 6.75$

$f(2) = 9 - 4 = 5 \qquad f\left(\frac{5}{2}\right) = 9 - 6.25 = 2.75$

$\frac{1}{2}(8 + 6.75 + 5 + 2.75) = 11.25$

9. $S_L(10) = \dfrac{14}{3} - \dfrac{6}{10} + \dfrac{4}{300} = \dfrac{1224}{300} = 4.08$

$S_R(10) = \dfrac{14(100) + 180 + 4}{300} = \dfrac{1584}{300} = 5.28$

11. $\displaystyle\lim_{n\to\infty} S_L = \lim_{n\to\infty}\left(\dfrac{14}{3} - \dfrac{6}{n} + \dfrac{4}{3n^2}\right)$

$\qquad\qquad = \dfrac{14}{3} - 0 + 0 = \dfrac{14}{3}$

$\displaystyle\lim_{n\to\infty} S_R = \lim_{n\to\infty}\left(\dfrac{14}{3} + \dfrac{6}{n} + \dfrac{4}{3n^2}\right)$

$\qquad\qquad = \dfrac{14}{3} + 0 + 0 = \dfrac{14}{3}$

13. If a point within each subinterval is selected and the area of each rectangle is determined, then the total area A would satisfy $S_L \le A \le S_R$.

Since $\displaystyle\lim_{n\to\infty} S_L = \lim_{n\to\infty} S_R = \dfrac{14}{3}$, it follows that

$\displaystyle\lim_{n\to\infty} A = \dfrac{14}{3}$.

15. $\displaystyle\sum_{i=1}^{4} x_i = 3 + (-1) + 3 + (-2) = 3$

17. $\displaystyle\sum_{j=2}^{5} (j^2 - 3)$

$\qquad = (4-3) + (9-3) + (16-3) + (25-3)$

$\qquad = 42$

No formulas are used because j begins at 2 and the formulas require that j begin at 1.

19. $\displaystyle\sum_{j=0}^{4} (j^2 - 4j + 1)$

$\qquad = (1) + (-2) + (-3) + (-2) + (1) = -5$

21. $\displaystyle\sum_{j=1}^{60} 3 = 3(60) = 180$

23. $\displaystyle\sum_{k=1}^{30} (k^2 + 4k) = \dfrac{30(31)(61)}{6} + \dfrac{4(30)(31)}{2}$

$\qquad\qquad = 11,315$

25. $\displaystyle\sum_{i=1}^{n}\left(1 - \dfrac{2i}{n} + \dfrac{i^2}{n^2}\right)\left(\dfrac{3}{n}\right)$

$\qquad = \dfrac{3}{n}\left[n - \dfrac{2}{n}\cdot\dfrac{n(n+1)}{2} + \dfrac{1}{n^2}\cdot\dfrac{n(n+1)(2n+1)}{6}\right]$

$\qquad = 3 - \dfrac{3}{n}(n+1) + \dfrac{(n+1)(2n+1)}{2n^2}$

$\qquad = \dfrac{2n^2 - 3n + 1}{2n^2}$

27. a. $f(x) = 2x$

Area 1st rectangle: $\dfrac{1}{n} \cdot f\left(\dfrac{0}{n}\right) = \dfrac{1}{n} \cdot 0$

Area 2nd rectangle: $\dfrac{1}{n} \cdot f\left(\dfrac{1}{n}\right) = \dfrac{1}{n} \cdot \dfrac{2}{n}$

$\vdots$

Area i th rectangle: $\dfrac{1}{n} \cdot f\left(\dfrac{i-1}{n}\right) = \dfrac{1}{n} \cdot \dfrac{2i-2}{n}$

$$S(n) = \dfrac{1}{n}\left[0 + \dfrac{2}{n} + \dfrac{4}{n} + \cdots + \dfrac{2n-2}{n}\right]$$

$$= \dfrac{1}{n} \cdot \dfrac{2}{n}(0 + 1 + 2 + \cdots + (n-1)]$$

$$= \dfrac{2}{n^2} \cdot \dfrac{n}{2}(0 + n - 1) = \dfrac{(n-1)}{n}$$

b. $S(10) = \dfrac{10-1}{10} = \dfrac{9}{10}$

c. $S(100) = \dfrac{100-1}{100} = \dfrac{99}{100}$

d. $S(1000) = \dfrac{1000-1}{1000} = \dfrac{999}{1000}$

e. $\displaystyle\lim_{n\to\infty} S(n) = \lim_{n\to\infty}\left(1 - \dfrac{1}{n}\right) = 1$

29. $f(x) = x^2$ (See the partition in problem 27.)

a. Area of 1st rectangle: $\dfrac{1}{n} \cdot f(1/n) = \dfrac{1}{n} \cdot \dfrac{1}{n^2}$

Area of 2nd rectangle: $\dfrac{1}{n} \cdot f(2/n) = \dfrac{1}{n} \cdot \dfrac{4}{n^2}$

$\vdots$

Area of i th rectangle: $\dfrac{1}{n} \cdot f(i/n) = \dfrac{1}{n} \cdot \dfrac{i^2}{n^2}$

$$S(n) = \dfrac{1}{n}\left[\dfrac{1}{n^2} + \dfrac{4}{n^2} + \dfrac{9}{n^2} + \cdots + \dfrac{n^2}{n^2}\right]$$

$$= \dfrac{1}{n^3}\left[\dfrac{n(n+1)(2n+1)}{6}\right]$$

$$= \dfrac{(n+1)(2n+1)}{6n^2}$$

b. $S(10) = \dfrac{11\cdot 21}{600} = \dfrac{77}{200}$

c. $S(100) = \dfrac{101\cdot 201}{60,000} = \dfrac{6767}{20,000}$

d. $S(1000) = \dfrac{1001\cdot 2001}{6,000,000} = \dfrac{667,667}{2,000,000}$

e. $\displaystyle\lim_{n\to\infty} S(n) = \lim_{n\to\infty}\dfrac{2n^2 + 3n + 1}{6n^2}$

$$= \lim_{n\to\infty}\left(\dfrac{2}{6} + \dfrac{1}{2n} + \dfrac{1}{6n^2}\right)$$

$$= \dfrac{2}{6} + 0 + 0 = \dfrac{1}{3}$$

31. $f(x) = x^2 - 6x + 8$

Use right hand endpoints.

Area of 1st rectangle: $\dfrac{2}{n} \cdot f(2/n)$

Area of 2nd rectangle: $\dfrac{2}{n} \cdot f(4/n)$

$\vdots$

Area of ith rectangle: $\dfrac{2}{n} \cdot f(2i/n)$

$$S(n) = \frac{2}{n}\left[\left\{\left(\frac{2}{n}\right)^2 - 6\left(\frac{2}{n}\right) + 8\right\} + \left\{\left(\frac{4}{n}\right)^2 - 6\left(\frac{4}{n}\right) + 8\right\} + \cdots + \left\{\left(\frac{2n}{n}\right)^2 - 6\left(\frac{2n}{n}\right) + 8\right\}\right]$$

We are going to group like terms.

$$S(n) = \frac{2}{n}\left[\left\{\left(\frac{2}{n}\right)^2 + \left(\frac{4}{n}\right)^2 + \cdots + \left(\frac{2n}{n}\right)^2\right\} - 6\left\{\frac{2}{n} + \frac{4}{n} + \cdots + \frac{2n}{n}\right\} + \{8n\}\right]$$

$$= \frac{2}{n}\left[\left\{4\left(\frac{1}{n}\right)^2 + 4\left(\frac{2}{n}\right)^2 + \cdots + 4\left(\frac{n}{n}\right)^2\right\} - 12\left\{\frac{1}{n} + \frac{2}{n} + \cdots + \frac{n}{n}\right\} + 8n\right]$$

$$= \frac{2}{n}\left[\frac{4}{n^2}(1 + 2^2 + \cdots + n^2) - \frac{12}{n}(1 + 2 + \cdots + n) + 8n\right]$$

$$= \frac{2}{n}\left[\frac{4}{n^2} \cdot \frac{n(n+1)(2n+1)}{6} - \frac{12}{n} \cdot \frac{n(n+1)}{2} + 8n\right] = \frac{20n^2 - 24n + 4}{3n^2}$$

$$A = \lim_{n \to \infty}\left(\frac{20}{3} - \frac{24}{3n} + \frac{4}{3n^2}\right) = \frac{20}{3}$$

33. $A \approx 65 \cdot 1 + 82 \cdot 1 + 87 \cdot 1 \approx \$234$ billion

35. There are approximately 90 squares under the curve.

Each square represents $10 \times \dfrac{1 \text{ hr}}{3600} \times \dfrac{1 \text{ miles}}{\text{hr}} = \dfrac{1}{360}$ miles.

Area represents $\approx 90 \cdot \dfrac{1}{360} = \dfrac{1}{4}$ miles.

37. $A \approx 10(0 + 15 + 18 + 18 + 30 + 27 + 24 + 23) = 1550$ sq ft

39. Width of rectangles: $\dfrac{10 - 5}{10} = \dfrac{5}{10} = \dfrac{1}{2}$

Area $\approx 0.5(S(5.5) + S(6.0) + S(6.5) + \ldots + S(9.5) + S(10.0))$

$\approx 0.5(67.97 + 73.585 + 79.2 + 84.815 + 90.43 + 96.045 + 101.66 + 107.275 + 112.89 + 118.505)$

≈ 466.1875 square units.

This represents the total spent for wireless services between 2000 and 2005: \$466.1875 billion

Exercise 13.2

1. $\int_0^3 4x\,dx = 4 \cdot \frac{x^2}{2}\Big|_0^3 = 2\left[3^2 - 0^2\right] = 18$

3. $\int_2^4 dx = x\Big|_2^4 = 4 - 2 = 2$

5. $\int_2^4 x^3\,dx = \frac{1}{4}x^4\Big|_2^4 = \frac{256}{4} - \frac{16}{4} = 60$

7. $\int_0^5 4x^{2/3}\,dx = \frac{12}{5}x^{5/3}\Big|_0^5$

$= \frac{12}{5}\left(5^{5/3} - 0\right)$

$= \frac{12}{5}\left[5(5)^{2/3}\right]$

$= 12\sqrt[3]{25}$

9. $\int_2^4 \left(4x^3 - 6x^2 - 5x\right)dx$

$= \left(x^4 - 2x^3 - \frac{5}{2}x^2\right)\Big|_2^4$

$= (256 - 128 - 40) - (16 - 16 - 10)$

$= 98$

11. $\int_3^4 (x-4)^9\,dx = \frac{1}{10}(x-4)^{10}\Big|_3^4$

$= \frac{1}{10}\left[0^{10} - (-1)^{10}\right] = -\frac{1}{10}$

13. $\int_2^4 \left(x^2+2\right)^3 x\,dx = \frac{1}{2}\int_2^4 \left(x^2+2\right)^3(2x)\,dx$

$= \frac{1}{2}\cdot\frac{1}{4}\left(x^2+2\right)^4\Big|_2^4$

$= \frac{1}{8}(104{,}976 - 1296)$

$= 12{,}960$

15. $\int_{-1}^2 \left(x^3 - 3x^2\right)^3\left(x^2 - 2x\right)dx$

$= \frac{1}{3}\int_{-1}^2 \left(x^3 - 3x^2\right)^3\left(3x^2 - 6x\right)dx$

$= \frac{1}{3}\cdot\frac{1}{4}\left(x^3 - 3x^2\right)^4\Big|_{-1}^2 = \frac{1}{12}[256 - 256] = 0$

17. $\int_0^4 \sqrt{4x+9}\,dx$

$= \frac{1}{4}\int_0^4 (4x+9)^{1/2}(4x\,dx)$

$= \frac{1}{4}\cdot\frac{2}{3}(4x+9)^{3/2}\Big|_0^4 = \frac{1}{6}\left[25^{3/2} - 9^{3/2}\right]$

$= \frac{1}{6}[125 - 27] = \frac{1}{6}(98) = \frac{49}{3} = 16\frac{1}{3}$

19. $\int_1^3 \frac{3}{y^2}\,dy = \int_1^3 3y^{-2}\,dy$

$= 3\cdot\frac{y^{-1}}{-1}\Big|_1^3 = -3\left[3^{-1} - 1^{-1}\right]$

$= -3\left[\frac{1}{3} - 1\right] = 2$

21. $\int_0^1 e^{3x}\,dx = \frac{1}{3}\int_0^1 e^{3x}(3\,dx)$

$= \frac{1}{3}e^{3x}\Big|_0^1 = \frac{1}{3}\left[e^3 - e^0\right] = \frac{1}{3}\left(e^3 - 1\right)$

23. $\int_1^e \frac{4}{z}\,dz = 4\int_1^e z^{-1}\,dz$

$= 4\ln|z|\Big|_1^e$

$= 4[\ln e - \ln 1] = 4$

25. $\int_0^2 8x^2 e^{-x^3}\,dx$

$= 8\left(-\frac{1}{3}\right)\int_0^2 e^{-x^3}\left(-3x^2\,dx\right) = -\frac{8}{3}e^{-x^3}\Big|_0^2$

$= -\frac{8}{3}\left(e^{-8} - e^0\right) = -\frac{8}{3}\left(\frac{1}{e^8} - 1\right) = \frac{8}{3}\left(1 - \frac{1}{e^8}\right)$

27. a. $\int_3^6 \frac{x}{3x^2+4}\,dx = \frac{1}{6}\int_3^6 \frac{6x}{3x^2+4}\,dx$

$= \frac{1}{6}\ln\left(3x^2+4\right)\Big|_3^6 = \frac{1}{6}[\ln 112 - \ln 31]$

$= \frac{1}{6}\ln\left(\frac{112}{31}\right)$

b. Graphing utility gives 0.2140853.

29. a. $\displaystyle\int_1^2 \frac{x^2+3}{x}\,dx = \int_1^2 \left(x + 3x^{-1}\right)dx$

$\displaystyle = \left(\frac{x^2}{2} + 3\ln|x|\right)\Bigg|_1^2 = 2 + 3\ln 2 - \left(\frac{1}{2} + 3\ln 1\right)$

$\displaystyle = \frac{3}{2} + 3\ln 2 = 3.5794415$

b. Graphing utility gives 3.5794415.

31. a. A, C
b. B

33. a. $\displaystyle\int_0^4 \left(2x - \frac{1}{2}x^2\right)dx$

b. $\displaystyle A = \left[2\cdot\frac{x^2}{2} - \frac{1}{2}\cdot\frac{x^3}{3}\right]\Bigg|_0^4$

$\displaystyle = \left(x^2 - \frac{1}{6}x^3\right)\Bigg|_0^4 = 16 - \frac{1}{6}(64) - 0$

$\displaystyle = \frac{96 - 64}{6} = \frac{16}{3}$

35. a. $\displaystyle A = \int_{-1}^0 (x^3 + 1)\,dx$

b. $\displaystyle A = \left(\frac{x^4}{4} + x\right)\Bigg|_{-1}^0 = 0 - \left(\frac{1}{4} - 1\right) = \frac{3}{4}$

37. $\displaystyle\int_1^2 \left(-x^2 + 3x - 2\right)dx$

$\displaystyle = \left(-\frac{1}{3}x^3 + \frac{3}{2}x^2 - 2x\right)\Bigg|_1^2$

$\displaystyle = \left(-\frac{8}{3} + 6 - 4\right) - \left(-\frac{1}{3} + \frac{3}{2} - 2\right) = \frac{1}{6}$

39. $\displaystyle\int_1^3 e^{x^2} x\,dx = \frac{1}{2}\int_1^3 e^{x^2}\left(2x\,dx\right)$

$\displaystyle = \frac{1}{2}e^{x^2}\Bigg|_1^3 = \frac{1}{2}\left(e^9 - e\right)$

41. $\displaystyle\int_0^a g(x)\,dx > \int_0^a f(x)\,dx$ because there is more positive area under $g(x)$ than under $f(x)$ on this interval.

43. The integrals differ only in sign.

45. $\displaystyle\int_1^2 \left(2x - x^2\right)dx + \int_2^4 \left(2x - x^2\right)dx = (-1)\int_1^2 \left(x^2 - 2x\right)dx + (-1)\int_2^4 \left(x^2 - 2x\right)dx = (-1)\int_1^4 \left(x^2 - 2x\right)dx$

Thus, we have $\displaystyle\int_1^2 \left(2x - x^2\right)dx + \int_2^4 \left(2x - x^2\right)dx = \frac{2}{3} + \left(-\frac{20}{3}\right) = -6.$ So, $\displaystyle\int_1^4 \left(x^2 - 2x\right)dx = (-1)(-6) = 6.$

47. $\displaystyle\int_4^4 \sqrt{x^2 - 2}\,dx = 0$

49. a. $D = 50{,}000 \cdot 5 + 30{,}000 \cdot 5 = \$400{,}000$ (RHP)

or $60{,}000 \cdot 5 + 50{,}000 \cdot 5 = \$550{,}000$ (LHP)

or $\frac{1}{2}(10)(60{,}000 + 30{,}000) = \$450{,}000$ (trapezoid)

b. $\displaystyle D = \int_0^{10} 3000(20 - t)\,dt = 3000\left(20t - \frac{t^2}{2}\right)\Bigg|_0^{10} = 3000(200 - 50) - 0 = \$450{,}000$

51. a. $\displaystyle\int_0^7 \left(300t - 3t^2\right)dt = \left(150t^2 - t^3\right)\Big|_0^7 = 7350 - 343 = \$7007$

b. $\displaystyle\int_7^{14} \left(300t - 3t^2\right)dt = \left(150t^2 - t^3\right)\Big|_7^{14} = (29{,}400 - 2744) - 7007 = \$19{,}649$

53. $\displaystyle\int_0^3 120e^{0.01t}\,dt = \frac{120}{0.01}\int_0^3 e^{0.01t}\left(0.01\,dt\right) = 12{,}000e^{0.01t}\Big|_0^3 = 12{,}000\left(e^{0.03} - e^0\right)$

$= 12{,}000(1.03045 - 1) = 365.40$ thousands $= \$365{,}400$

55. $\displaystyle\int_{15}^{25} \left(72.11t + 81.16\right)dt = \left(36.055t^2 + 81.16t\right)\Big|_{15}^{25}$

$= \left(36.055(25)^2 + 81.16(25)\right) - \left(36.055(15)^2 + 81.16(15)\right)$

$\approx \$15{,}233.6$ billion predicted in telecommunications revenue between 2000 and 2010

57. $\int_0^{0.3}\left(0.3-3.33r^2\right)2\pi r\,dr = \int_0^{0.3}\left(0.6\pi r - 6.66\pi\,r^3\right)dr = \left(0.6\pi\cdot\dfrac{r^2}{2}-6.66\pi\cdot\dfrac{r^4}{4}\right)\Big|_0^{0.3}$

$$= \left(0.3\pi r^2 - 1.665\pi r^4\right)\Big|_0^{0.3} = \left[0.3\pi(0.3)^2 - 1.665\pi(0.3)^4\right]-(0-0)$$

$$= \pi\left(0.027 - 0.0134865\right) = 0.04\text{ cm}^3$$

59. $x = \int_0^5 200\left(1+\dfrac{400}{(t+40)^2}\right)dt = 200\left(t+\dfrac{400(t+40)^{-1}}{-1}\right)\Big|_0^5 = 200\left[\left(5-\dfrac{400}{45}\right)-(0-10)\right] = 1222.22$

61. $\int_8^{10}\left(0.012t^2 - 0.0012t^3\right)dt = \left(0.004t^3 - 0.003t^4\right)\Big|_8^{10}$

$$= \left(0.004(10)^3 - 0.003(10)^4\right)-\left(0.004(8)^3 - 0.003(8)^4\right) = 0.1808$$

63. a. $\int_0^3 0.3e^{-0.3t}\,dt = -\int_0^3 e^{-0.3t}\left(-0.3\,dt\right) = -e^{-0.3t}\Big|_0^3 = -e^{-0.3(3)}-\left(-e^{-0.3(0)}\right) = -e^{-0.9}+1 \approx 0.5934$

    b. $\int_5^{10} 0.3e^{-0.3t}\,dt = -\int_5^{10} e^{-0.3t}\left(-0.3\,dt\right) = -e^{-0.3t}\Big|_5^{10} = -e^{-0.3(10)}-\left(-e^{-0.3(5)}\right) = -e^{-3}+e^{-1.5} \approx 0.1733$

65. a. Using technology, $E(t) = 0.001t^4 - 0.058t^3 + 0.978t^2 - 6.76t + 204.1$

    b. $\int_{20}^{30} E(t)\,dt = \int_{20}^{30}\left(0.001t^4 - 0.058t^3 + 0.978t^2 - 6.76t + 204.1\right)dt$

$$= \left(0.0002t^5 - 0.0145t^4 + 0.326t^3 - 3.38t^2 + 204.1t\right)\Big|_{20}^{30}$$

$$\approx 1340\text{ billion short tons of carbon monoxide emissions during the 1990's.}$$

Exercise 13.3

1. a. $\int_0^2\left(4-x^2\right)dx$

    b. $\int_0^2\left(4-x^2\right)dx = \left(4x-\dfrac{x^3}{3}\right)\Big|_0^2 = 4(2)-\dfrac{2^3}{3} = 8-\dfrac{8}{3} = \dfrac{24-8}{3} = \dfrac{16}{3}$

3. a. $A = \int_1^8\left(x^{1/3}-2+x\right)dx$

    b. $A = \left|\left(\dfrac{3}{4}x^{4/3}-2x+\dfrac{x^2}{2}\right)\Big|_1^8\right| = 28-\left(-\dfrac{3}{4}\right) = 28\dfrac{3}{4}$

5. a. $\int_1^2\left[\left(4-x^2\right)-\left(\dfrac{1}{4}x^3-2\right)\right]dx$

    b. $\int_1^2\left(6-x^2-\dfrac{1}{4}x^3\right)dx = \left(6x-\dfrac{1}{3}x^3-\dfrac{1}{16}x^4\right)\Big|_1^2 = \left(12-\dfrac{8}{3}-1\right)-\left(6-\dfrac{1}{3}-\dfrac{1}{16}\right) = \dfrac{131}{48}$

7. a.
$$x + 2 = x^2$$
$$x^2 - x - 2 = 0$$
$$(x - 2)(x + 1) = 0$$
$$x = 2 \text{ or } x = -1$$
If $x = 2$, $y = 4$.
If $x = -1$, $y = 1$.

b., c. $\displaystyle \int_{-1}^{2} \left[(x + 2) - x^2 \right] dx = \left(\frac{x^2}{2} + 2x - \frac{x^3}{3} \right)\Big|_{-1}^{2} = \left(2 + 4 - \frac{8}{3} \right) - \left(\frac{1}{2} - 2 + \frac{1}{3} \right) = \frac{18 - 8}{3} - \frac{3 - 12 + 2}{6} = \frac{20}{6} + \frac{7}{6} = \frac{9}{2}$

9. a.
$$x^2 - 4x = x - x^2$$
$$2x^2 - 5x = 0$$
$$x(2x - 5) = 0$$
$$x = 0 \text{ or } x = \frac{5}{2} \quad \text{If } x = 0,\ y = 0. \text{ If } x = \frac{5}{2},\ y = -\frac{15}{4}$$

b., c. $\displaystyle \int_{0}^{5/2} \left[(x - x^2) - (x^2 - 4x) \right] dx = \int_{0}^{5/2} (5x - 2x^2) \, dx = \left(\frac{5}{2}x^2 - \frac{2}{3}x^3 \right)\Big|_{0}^{5/2} = \left(\frac{125}{8} - \frac{250}{24} \right) - 0 = \frac{125}{24}$

11. a.
$$x^3 - 2x = 2x$$
$$x^3 - 4x = 0$$
$$x(x^2 - 4) = 0$$
$$x(x + 2)(x - 2) = 0$$
$$x = 0,\ x = -2,\ x = 2$$
If $x = 0$, then $y = 0$. If $x = -2$, then $y = -4$. If $x = 2$, then $y = 4$.

b., c. $\displaystyle \int_{-2}^{0} (x^3 - 2x - 2x) \, dx + \int_{0}^{2} \left[2x - (x^3 - 2x) \right] dx = \int_{-2}^{0} (x^3 - 4x) \, dx + \int_{0}^{2} (-x^3 + 4x) \, dx$

$$= \left[\frac{x^4}{4} - 4 \cdot \frac{x^2}{2} \right]\Big|_{-2}^{0} + \left[-\frac{x^4}{4} + 4 \cdot \frac{x^2}{2} \right]\Big|_{0}^{2} = 0 - \left(\frac{16}{4} - 2(-2)^2 \right) + \left(-\frac{16}{4} + 2(2)^2 \right) - 0 = 8$$

13. $f(1) = 3$, $g(1) = -1$, $f(x) \ge g(x)$

$$\int_{0}^{2} \left[(x^2 + 2) - (-x^2) \right] dx$$

$$= \left(\frac{2}{3}x^3 + 2x \right)\Big|_{0}^{2} = \frac{16}{3} + 4 = \frac{28}{3}$$

15. $x^3 - 1 = x - 1$

$x^3 - x = x(x-1)(x+1) = 0$

$x = 0, \; x = 1$

$f\left(\dfrac{1}{2}\right) = \dfrac{1}{8} - 1 = -\dfrac{7}{8}$

$g\left(\dfrac{1}{2}\right) = \dfrac{1}{2} - 1 = -\dfrac{1}{2}$

$g(x) \ge f(x)$

$\displaystyle\int_0^1 \left[(x-1) - (x^3 - 1)\right] dx$

$= \displaystyle\int_0^1 \left(x - x^3\right) dx$

$= \left(\dfrac{x^2}{2} - \dfrac{x^4}{4}\right)\Bigg|_0^1 = \dfrac{1}{2} - \dfrac{1}{4} = \dfrac{1}{4}$

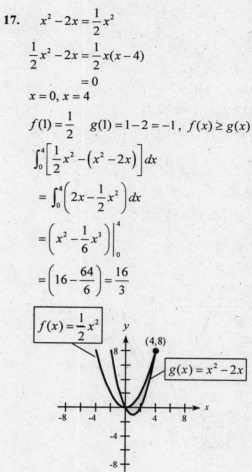

17. $x^2 - 2x = \dfrac{1}{2}x^2$

$\dfrac{1}{2}x^2 - 2x = \dfrac{1}{2}x(x-4)$

$\qquad\qquad = 0$

$x = 0, \; x = 4$

$f(1) = \dfrac{1}{2} \quad g(1) = 1 - 2 = -1, \; f(x) \ge g(x)$

$\displaystyle\int_0^4 \left[\dfrac{1}{2}x^2 - \left(x^2 - 2x\right)\right] dx$

$= \displaystyle\int_0^4 \left(2x - \dfrac{1}{2}x^2\right) dx$

$= \left(x^2 - \dfrac{1}{6}x^3\right)\Bigg|_0^4$

$= \left(16 - \dfrac{64}{6}\right) = \dfrac{16}{3}$

19. $x^2 = \sqrt{x}$

$x^2 - \sqrt{x} = x^{1/2}\left(x^{3/2} - 1\right)$

$\qquad\qquad = 0$

$x = 0, \; x = 1$

$h\left(\dfrac{1}{4}\right) = \dfrac{1}{16}, \; k\left(\dfrac{1}{4}\right) = \dfrac{1}{2}, \; k(x) \ge h(x)$

$\displaystyle\int_0^1 \left(\sqrt{x} - x^2\right) dx = \left(\dfrac{2}{3}x^{3/2} - \dfrac{1}{3}x^3\right)\Bigg|_0^1$

$= \left(\dfrac{2}{3} - \dfrac{1}{3}\right) = \dfrac{1}{3}$

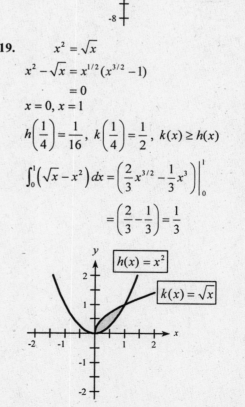

21.
$$x^3 = x^2 + 2x$$
$$x^3 - x^2 - 2x = x(x-2)(x+1)$$
$$= 0$$
$$x = -1, \ x = 0, \ x = 2$$
$$f(1) = 1, \ g(1) = 3,$$
$$g(x) \geq f(x) \text{ over } [0,2]$$
$$\int_{-1}^{0} \left[x^3 - \left(x^2 + 2x \right) \right] dx + \int_{0}^{2} \left[\left(x^2 + 2x \right) - x^3 \right] dx$$
$$= \left(\frac{x^4}{4} - \frac{x^3}{3} - x^2 \right) \Big|_{-1}^{0} + \left(\frac{x^3}{3} + x^2 - \frac{x^4}{4} \right) \Big|_{0}^{2}$$
$$= \left[0 - \left(\frac{1}{4} + \frac{1}{3} - 1 \right) \right] + \left[\left(\frac{8}{3} + 4 - 4 \right) - 0 \right]$$
$$= \frac{5}{12} + \frac{8}{3} = \frac{37}{12}$$

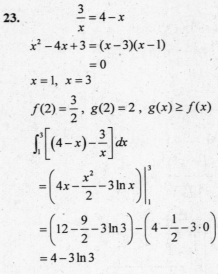

23.
$$\frac{3}{x} = 4 - x$$
$$x^2 - 4x + 3 = (x-3)(x-1)$$
$$= 0$$
$$x = 1, \ x = 3$$
$$f(2) = \frac{3}{2}, \ g(2) = 2, \ g(x) \geq f(x)$$
$$\int_{1}^{3} \left[(4-x) - \frac{3}{x} \right] dx$$
$$= \left(4x - \frac{x^2}{2} - 3 \ln x \right) \Big|_{1}^{3}$$
$$= \left(12 - \frac{9}{2} - 3 \ln 3 \right) - \left(4 - \frac{1}{2} - 3 \cdot 0 \right)$$
$$= 4 - 3 \ln 3$$

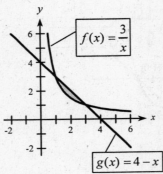

25. $\sqrt{x+3} = 2$
$$x + 3 = 4 \text{ or } x = 1, \ x = -3$$
$$f(0) = \sqrt{3}, \ g(0) = 2, \ g(x) \geq f(x)$$
$$\int_{-3}^{1} (2 - \sqrt{x+3}) dx = \left(2x - \frac{2}{3}(x+3)^{3/2} \right) \Big|_{-3}^{1}$$
$$= \left(2 - \frac{2}{3} \cdot 8 \right) - (-6 - 0) = \frac{8}{3}$$

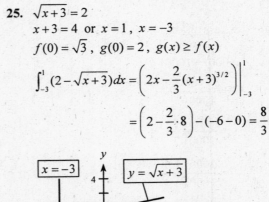

27. Avg value $= \dfrac{1}{b-a} \displaystyle\int_a^b f(x)\,dx$

$$= \frac{1}{3-0} \int_0^3 (9-x^2)\,dx$$

$$= \frac{1}{3}\left(9x - \frac{x^3}{3}\right)\Bigg|_0^3 = \frac{1}{3}(27-9) = 6$$

29. Avg value $= \dfrac{1}{b-a} \displaystyle\int_a^b f(x)\,dx$

$$= \frac{1}{1-(-1)} \int_{-1}^1 (x^3 - x)\,dx$$

$$= \frac{1}{2}\left(\frac{x^4}{4} - \frac{x^2}{2}\right)\Bigg|_{-1}^1 = \frac{1}{2}\left(\frac{1}{4} - \frac{1}{4}\right) = 0$$

31. Avg value $= \dfrac{1}{4-1} \displaystyle\int_1^4 \left(\sqrt{x} - 2\right)dx$

$$= \frac{1}{3}\left(\frac{2}{3}x^{3/2} - 2x\right)\Bigg|_1^4$$

$$= \frac{1}{3}\left[-\frac{8}{3} - \left(-\frac{4}{3}\right)\right] = -\frac{4}{9}$$

33. Use appropriate technology. Answer: 11.83

35. $AP = \dfrac{1}{x_1 - x_0} \displaystyle\int_{x_0}^{x_1} \left(R(x) - C(x)\right)dx$

37. $C(x) = x^2 + 400x + 2000$

    a. $\overline{C}(x) = \dfrac{C(x)}{x} = x + 400 + \dfrac{2000}{x}$

          $\overline{C}(1000) = 1000 + 400 + 2 = \$1402$

    b. $AV = \dfrac{1}{1000-0} \displaystyle\int_0^{1000} (x^2 + 400x + 2000)\,dx$

$$= \frac{1}{1000}\left(\frac{1}{3}x^3 + 200x^2 + 2000x\right)\Bigg|_0^{1000}$$

$$= \frac{1000}{1000}\left(\frac{1,000,000}{3} + 200(1000) + 2000\right)$$

$$= \$535,333\frac{1}{3}$$

39. **a.**   Average $= \dfrac{1}{20-0} \displaystyle\int_0^{20}\left[100e^{-x^2}\cdot x + 100\right]dx$

$$= \frac{1}{20}\left(-50e^{-x^2} + 100x\right)\Bigg|_0^{20}$$

$$= \frac{1}{20}\left[(0+2000) - (-50+0)\right]$$

$$= 102.5$$

    b. $\dfrac{1}{30-20} \displaystyle\int_{20}^{30}\left(100xe^{-x^2} + 100\right)dx$

$$= \frac{1}{10}\left(-50e^{-x^2} + 100x\right)\Bigg|_{20}^{30}$$

$$\approx \frac{1}{10}(3000 - 2000) = 100$$

41. $AV = \displaystyle\int_{50}^{60} 257.8e^{0.067t}\,dt$

$$= 257.8\left(\frac{1}{0.067}\right)\int_{50}^{60} e^{0.067t}\,(0.067\,dt)$$

$$= \frac{257.8}{0.067}e^{0.067t}\Bigg|_{50}^{60} = \frac{257.8}{0.067}\left(e^{0.067(60)} - e^{0.067(50)}\right)$$

$$\approx \$10,466.02 \text{ average federal tax per capita}$$

43.   Average $= \dfrac{1}{4-0} \displaystyle\int_0^4 \left(30x^{18/7} - 240x^{11/7} + 480x^{4/7}\right)dx$

$$= \frac{1}{4}\left(\frac{7(30)}{25}\cdot x^{25/7} - \frac{7(240)}{18}\cdot x^{18/7} + \frac{7(480)}{11}\cdot x^{11/7}\right)\Bigg|_0^4$$

$$= \frac{1}{4}(1187.12 - 3297.55 + 2697.99) = 146.89 = 147 \text{ mg}$$

45. The Gini coefficient of income for 1988 is $2\int_0^1 \left(x - x^{2.3521}\right)dx = 2\left(\dfrac{x^2}{2} - \dfrac{x^{3.3521}}{3.3521}\right)\Big|_0^1 \approx 2(0.2017 - 0) \approx 0.4034$.

The Gini coefficient of income for 2000 is $2\int_0^1 \left(x - x^{2.4870}\right)dx = 2\left(\dfrac{x^2}{2} - \dfrac{x^{3.4870}}{3.4870}\right)\Big|_0^1 \approx 2(0.2132 - 0) \approx 0.4264$.

Income distribution was more nearly equal in 1988. This does not support the convention about Republicans and Democrats, but many other factors may have been to blame.

47. The Gini coefficient of income for blacks is $2\int_0^1 \left(x - x^{2.5938}\right)dx = 2\left(\dfrac{x^2}{2} - \dfrac{x^{3.5938}}{3.5938}\right)\Big|_0^1 \approx 2(0.2217 - 0) \approx 0.4434$.

The Gini coefficient of income for Asians is $2\int_0^1 \left(x - x^{2.5070}\right)dx = 2\left(\dfrac{x^2}{2} - \dfrac{x^{3.5070}}{3.5070}\right)\Big|_0^1 \approx 2(0.2149 - 0) \approx 0.4298$.

Income was more evenly distributed among Asians than blacks in 2003.

Exercise 13.4

1. Total income $= \int_0^{10} 12,000\,dt$

$= 12,000t\,\Big|_0^{10} = \$120,000$

3. Total income $= \int_0^{12} 24,000e^{0.03t}\,dt$

$= 800,000e^{0.03t}\,\Big|_0^{12}$

$= 800,000(1.4333 - 1)$

$= \$346,664$

5. Total income $= \int_0^{10} 80e^{-0.1t}\,dt$

$= -800e^{-0.1t}\,\Big|_0^{10}$

$= -800(0.3678 - 1)$

$= \$505.70$ thousand

$= \$505,700$

7. Total Income $= \int_6^{12} 3000e^{0.004t}\,dt$

$= 750,000e^{0.004t}\,\Big|_6^{12}$

$= 750,000(0.0249)$

$= \$18,675$

9. Present Value $= \int_0^8 12,000e^{0.04t}\cdot e^{-0.08t}\,dt$

$= \int_0^8 12,000e^{-0.04t}\,dt$

$= -300,000e^{-0.04t}\,\Big|_0^8$

$= -300,000(0.72615 - 1)$

$= \$82,155$

11. Present Value $= \int_0^5 63,000e^{-0.07t}\,dt$

$= -900,000e^{-0.07t}\,\Big|_0^5$

$= -900,000(0.70469 - 1)$

$= \$265,781$

$FV = e^{35}\int_0^5 63,000e^{-0.07t}\,dt$

$= e^{35}(-900,000e^{-0.07t})\,\Big|_0^5$

$= \$377,161$

13. Present Value $= \int_0^{10} 97.5e^{-0.2(t+3)}e^{-0.06t}\,dt$

$= 97.5\int_0^{10} e^{-0.26t-0.6}\,dt$

$= -\dfrac{97.5}{0.26}e^{-0.26t-0.6}\,\Big|_0^{10}$

$= -\dfrac{97.5}{0.26}\left[e^{-3.2} - e^{-0.6}\right]$

$= 190.519$ thousand dollars

$= \$190,519$

By pattern in 11, $FV = e^6 PV$

$= \$347,148$

15. $PV = \int_0^7 30,000e^{-0.1t}$

$= -300,000e^{-0.1t}\Big|_0^7$

$= -300,000(0.4966 - 1)$

$= \$151,024$

Present value of gift shop is $151,024

$PV = \int_0^7 21,600e^{0.08t}\left(e^{-0.1t}\right)dt$

$= \int_0^7 21,600e^{-0.02t}dt$

$= -1,080,000e^{-0.02t}\Big|_0^7$

$= -1,080,000(0.8694 - 1)$

$= \$141,048$

Present value of video store is $141,048. The gift shop is the better buy.

17. $9 = 34 - x^2$ gives $x^2 = 25$ or $x = 5$.

Equilibrium point is $(5, 9)$.

$CS = \int_0^5 \left(34 - x^2\right)dx - 5 \cdot 9$

$= \left(34x - \frac{1}{3}x^3\right)\Big|_0^5 - 45$

$= 170 - \frac{125}{3} - 45 = \frac{250}{3} = \$83.33$

19. $p = \dfrac{200}{8 + 2} = 20$.

Equilibrium point is $(8, 20)$.

$CS = \int_0^8 \frac{200}{x + 2}dx - 8 \cdot 20$

$= 200\ln(x + 2)\Big|_0^8 - 160$

$= 200(\ln 10 - \ln 2) - 160$

$= \$161.89$

21. $x^2 + 4x + 11 = 81 - x^2$

$2x^2 + 4x - 70 = 0$

$2(x + 7)(x - 5) = 0$

Equilibrium point is $(5, 56)$, since

$p(5) = 81 - 25 = 56$.

$CS = \int_0^5 \left(81 - x^2\right)dx - 5 \cdot 56$

$= \left(81x - \frac{1}{3}x^3\right)\Big|_0^5 - 280$

$= 405 - \frac{125}{3} - 280 = \$83.33$

23. $\dfrac{12}{x + 1} = 1 + 0.2x$

$12 = 1 + 1.2x + 0.2x^2$

$2x^2 + 12x - 110 = 0$

$2(x + 11)(x - 5) = 0$

Equilibrium point is $(5, 2)$.

$CS = \int_0^5 \frac{12}{x + 1}dx - 5 \cdot 2$

$= 12\ln(x + 1)\Big|_0^5 - 10$

$= 12(\ln 6 - \ln 1) - 10 = \$11.50$

25. $R = px = 360x - 3x^2 - 2x^3$

$P = R - C = -2x^3 - 9x^2 + 240x - 1000$

$P'(x) = -6x^2 - 18x + 240$

$= -6(x^2 + 3x - 40)$

$= -6(x + 8)(x - 5)$

Maximum profit is at $x = 5$ with

$p(5) = 360 - 15 - 50 = 295$

$CS = \int_0^5 \left(360 - 3x - 2x^2\right)dx - 5(295)$

$= \left(360x - \frac{3}{2}x^2 - \frac{2}{3}x^3\right)\Big|_0^5 - 1475$

$= 1800 - 37.50 - 83.33 - 1475$

$= \$204.17$

27. $422 = 4x^2 + 2x + 2$

$4x^2 + 2x - 420 = 0$

$2(2x + 21)(x - 10) = 0$

Equilibrium point is $(10, 422)$.

$PS = 10 \cdot 422 - \int_0^{10} \left(4x^2 + 2x + 2\right)dx$

$= 4220 - \left(\frac{4}{3}x^3 + x^2 + 2x\right)\Big|_0^{10}$

$= 4220 - (1333.33 + 100 + 20)$

$= \$2766.67$

29. $p(x) = 10e^{x/3}$

$p(15) = 10e^5 = 1484.13$

$PS = 15(1484.13) - \int_0^{15} 10e^{x/3}dx$

$= 22,261.95 - \left(30e^{x/3}\right)\Big|_0^{15}$

$= \$17,839.58$

31. $x^2 + 4x + 11 = 81 - x^2$

$2x^2 + 4x - 70 = 2(x+7)(x-5) = 0$

At $x = 5$, we have $p = 81 - 25 = 56$.

$PS = 5(56) - \int_0^5 \left(x^2 + 4x + 11\right)dx$

$= 280 - \left(\dfrac{1}{3}x^3 + 2x^2 + 11x\right)\Big|_0^5$

$= 280 - (41.67 + 50 + 55)$

$= \$133.33$

33. See problem 23 to find equilibrium point $(5,2)$.

$PS = 5 \cdot 2 - \int_0^5 (1 + 0.2x)\,dx$

$= 10 - \left(x + 0.1x^2\right)\Big|_0^5$

$= 10 - (5 + 2.5)$

$= \$2.50$

35. $x^2 + 33x + 48 = 144 - 2x^2$

$3x^2 + 33x - 96 = 3(x^2 + 11x - 32) = 0$

Using the quadratic formula to solve for x and the supply function to find the equilibrium price, we have that the equilibrium point is $(2.39, 132.58)$.

$PS = 2.39(132.58) - \int_0^{2.39} \left(x^2 + 33x + 48\right)dx$

$= 316.87 - \left(\dfrac{1}{3}x^3 + \dfrac{33}{2}x^2 + 48x\right)\Big|_0^{2.39}$

$= 316.87 - 213.52 = \$103.35$

Exercise 13.5

1. #5: $\displaystyle\int \dfrac{dx}{16 - x^2} = \int \dfrac{dx}{4^2 - x^2} = \dfrac{1}{2(4)} \ln \left|\dfrac{4 + x}{4 - x}\right| + C$

3. #11: $\displaystyle\int_1^4 \dfrac{dx}{x\sqrt{9 + x^2}} = \int_1^4 \dfrac{dx}{x\sqrt{3^2 + x^2}} = -\dfrac{1}{3}\ln\left(\dfrac{3 + \sqrt{9 + x^2}}{x}\right)\Big|_1^4 = -\dfrac{1}{3}\left[\ln 2 - \ln\left(3 + \sqrt{10}\right)\right]$

$= \dfrac{1}{3}\left[\ln\left(3 + \sqrt{10}\right) - \ln 2\right] = \dfrac{1}{3}\ln\left[\dfrac{3 + \sqrt{10}}{2}\right]$

5. #14: $\displaystyle\int \ln w\,dw = w(\ln w - 1) + C$

7. #12: $\displaystyle\int_0^2 \dfrac{q\,dq}{6q + 9} = \left[\dfrac{q}{6} - \dfrac{9}{36}\ln(6q + 9)\right]\Big|_0^2 = \left(\dfrac{2}{6} - \dfrac{1}{4}\ln 21\right) - \left(0 - \dfrac{1}{4}\ln 9\right) = \dfrac{1}{3} + \dfrac{1}{4}(\ln 9 - \ln 21) = \dfrac{1}{3} + \dfrac{1}{4}\ln\dfrac{3}{7}$

9. #13: $\displaystyle\int \dfrac{dv}{v(3v + 8)} = \dfrac{1}{8}\ln\left|\dfrac{v}{3v + 8}\right| + C$

11. #7: $\displaystyle\int_5^7 \sqrt{x^2 - 25}\,dx = \int_5^7 \sqrt{x^2 - 5^2}\,dx = \dfrac{1}{2}\left[x\sqrt{x^2 - 25} - 25\ln\left(x + \sqrt{x^2 - 25}\right)\right]\Big|_5^7$

$= \dfrac{1}{2}\left[7\sqrt{24} - 25\ln\left(7 + \sqrt{24}\right) - (0 - 25\ln 5)\right] = \dfrac{1}{2}\left[7\sqrt{24} - 25\ln\left(7 + \sqrt{24}\right) + 25\ln 5\right]$

13. #16: $\displaystyle\int w\sqrt{4w + 5}\,dw = \dfrac{2(12w - 10)(4w + 5)^{3/2}}{15(16)} + C = \dfrac{(6w - 5)(4w + 5)^{3/2}}{60} + C$

15. #3: $\displaystyle\int x5^{x^2}\,dx = \dfrac{1}{2}\int 5^{x^2}(2x\,dx) = \dfrac{1}{2} \cdot \dfrac{5^{x^2}}{\ln 5} + C$ or $\dfrac{1}{2}\left(5^{x^2}\right)\log_5 e + C$

17. #1: $\int_0^3 x\sqrt{x^2+4}\,dx = \frac{1}{2}\int_0^3 (x^2+4)^{1/2}(2x\,dx) = \frac{1}{2}\cdot\frac{(x^2+4)^{3/2}}{\frac{3}{2}}\Big|_0^3 = \frac{1}{3}(x^2+4)^{3/2}\Big|_0^3 = \frac{1}{3}(13^{3/2}-8)$

19. #9: $5\int\frac{dx}{x\sqrt{4-9x^2}} = 5\int\frac{3\,dx}{3x\sqrt{2^2-(3x)^2}} = 5\cdot\left(-\frac{1}{2}\right)\ln\left|\frac{2+\sqrt{4-9x^2}}{3x}\right|+C = -\frac{5}{2}\ln\left|\frac{2+\sqrt{4-9x^2}}{3x}\right|+C$

21. #10: $\int\frac{dx}{\sqrt{9x^2-4}} = \frac{1}{3}\int\frac{3\,dx}{\sqrt{(3x)^2-2^2}} = \frac{1}{3}\ln\left|3x+\sqrt{9x^2-4}\right|+C$

23. #15: $\int\frac{3x\,dx}{(2x-5)^2} = 3\cdot\frac{1}{2^2}\left(\ln|2x-5|+\frac{-5}{2x-5}\right)+C = \frac{3}{4}\left(\ln|2x-5|-\frac{5}{2x-5}\right)+C$

25. #8: $\int\frac{dx}{\sqrt{(3x+1)^2+1}} = \frac{1}{3}\int\frac{(3\,dx)}{\sqrt{(3x+1)^2+1}} = \frac{1}{3}\ln\left|3x+1+\sqrt{(3x+1)^2+1}\right|+C$

27. #6: $\int_0^3 x\sqrt{(x^2+1)^2+9}\,dx = \frac{1}{2}\int_0^3\sqrt{(x^2+1)^2+3^2}(2x\,dx)$

$$= \frac{1}{4}\left[(x^2+1)\sqrt{(x^2+1)^2+9}+9\ln\left((x^2+1)+\sqrt{(x^2+1)^2+9}\right)\right]\Big|_0^3$$

$$= \frac{1}{4}\left[10\sqrt{109}+9\ln\left(10+\sqrt{109}\right)-\sqrt{10}-9\ln\left(1+\sqrt{10}\right)\right]$$

29. #2: $\int\frac{x\,dx}{7-3x^2} = -\frac{1}{6}\int\frac{1}{7-3x^2}(-6x\,dx) = -\frac{1}{6}\ln\left|7-3x^2\right|+C$

31. #8: $\int\frac{dx}{\sqrt{4x^2+7}} = \frac{1}{2}\int\frac{(2\,dx)}{\sqrt{(2x)^2+7}} = \frac{1}{2}\ln\left|2x+\sqrt{4x^2+7}\right|+C$

33. Using technology: $\int_2^3\frac{e^{\sqrt{x-1}}}{\sqrt{x-1}}\,dx \approx 2.7899$

35. Using technology: $\int_0^1\frac{x^3\,dx}{(4x^2+5)^2} \approx 0.004479$

37. At $x = 20$, we have $p = 40+200\ln 21 = 648.90$

$PS = 20(648.90) - \int_0^{20}\left[40+200\ln(x+1)\right]dx = 12{,}978 - \left[40x+200(x+1)(\ln(x+1)-1)\right]\Big|_0^{20}$

$= 12{,}978 - \left[800+4200(\ln 21-1)-(0+200(-1))\right] = 12{,}978-800-12{,}787+4200-200 = \$3391$

39. a. $C(x) = \int\sqrt{x^2+9}\,dx = \frac{1}{2}\left(x\sqrt{x^2+9}+9\ln\left(x+\sqrt{x^2+9}\right)\right)+K$

(By formula 6)

With the interpretation that $C(0) = 300$, we get a K value of 295.06.

b. $C(4) = 19.888+295.06 = 314.95$

41. $TI = \int_0^{120}\left(10\ln(t+1)-0.1t\right)dt = \left[10(t+1)\left(\ln(t+1)-1\right)-\dfrac{.1t^2}{2}\right]\Big|_0^{120}$

$\qquad = 10\cdot121(\ln 121-1)-\dfrac{1440}{2}-\left[10(-1)-0\right]$ (by formula 14)

$\qquad \approx \$3882.9$ thousands

Exercise 13.6

1. $\int xe^{2x}dx = uv - \int v\,du$

$\quad u = x \qquad dv = e^{2x}dx$

$\quad du = dx \qquad v = \dfrac{1}{2}e^{2x}$

$\quad \int xe^{2x}dx = \dfrac{1}{2}xe^{2x}-\dfrac{1}{2}\int e^{2x}dx$

$\qquad\qquad = \dfrac{1}{2}xe^{2x}-\dfrac{1}{4}e^{2x}+C$

3. $\int x^2\ln x\,dx = uv - \int v\,du$

$\quad u = \ln x \qquad dv = x^2dx$

$\quad du = \dfrac{1}{x}dx \qquad v = \dfrac{1}{3}x^3$

$\quad \int x^2\ln x\,dx = \dfrac{1}{3}x^3\ln x-\dfrac{1}{3}\int x^3\cdot\dfrac{1}{x}dx$

$\qquad\qquad = \dfrac{1}{3}x^3\ln x-\dfrac{1}{9}x^3+C$

5. $\int_4^6 q\sqrt{q-4}\,dq = \left[uv - \int v\,du\right]\Big|_4^6$

$\quad u = q \qquad dv = (q-4)^{1/2}dq$

$\quad du = dq \qquad v = \dfrac{2}{3}(q-4)^{3/2}$

$\quad \int_4^6 q\sqrt{q-4}\,dq$

$\quad = \left[\dfrac{2q}{3}(q-4)^{3/2}-\dfrac{2}{3}\int(q-4)^{3/2}\,dq\right]\Big|_4^6$

$\quad = \left[\dfrac{2q}{3}(q-4)^{3/2}-\dfrac{4}{15}(q-4)^{5/2}\right]\Big|_4^6$

$\quad = 4\left(2^{3/2}\right)-\dfrac{4}{15}\left(2^{5/2}\right)-0$

$\quad = 4\left(2^{3/2}\right)-\dfrac{8}{15}\left(2^{3/2}\right)$

$\quad = \dfrac{52}{15}\left(2^{3/2}\right) = \dfrac{104}{15}\sqrt{2}$

7. $\int\dfrac{\ln x}{x^2}dx = uv - \int v\,du$

$\quad u = \ln x \qquad dv = x^{-2}dx$

$\quad du = \dfrac{dx}{x} \qquad v = -\dfrac{1}{x}$

$\quad \int\dfrac{\ln x}{x^2}dx = -\dfrac{1}{x}\ln x+\int\dfrac{dx}{x^2} = -\dfrac{1}{x}\ln x-\dfrac{1}{x}+C$

9. $\int_1^e \ln x\,dx = \left[uv - \int v\,du\right]\Big|_1^e$

$\quad u = \ln x \qquad dv = dx$

$\quad du = \dfrac{dx}{x} \qquad v = x$

$\quad \int_1^e \ln x\,dx = \left[x\ln x-\int dx\right]\Big|_1^e$

$\qquad\qquad = \left[x\ln x-x\right]\Big|_1^e$

$\qquad\qquad = (e\ln e-e)-(0-1)$

$\qquad\qquad = e-e+1 = 1$

11. $\int x\ln(2x-3)\,dx = uv - \int v\,du$

$\quad u = \ln(2x-3) \qquad dv = xdx$

$\quad du = \dfrac{2\,dx}{2x-3} \qquad v = \dfrac{x^2}{2}$

$\quad \int x\ln(2x-3)\,dx$

$\quad = \dfrac{1}{2}x^2\ln(2x-3)-\int\dfrac{x^2}{2x-3}dx$

$\quad = \dfrac{1}{2}x^2\ln(2x-3)-\int\left(\dfrac{1}{2}x+\dfrac{3}{4}+\dfrac{9/4}{2x-3}\right)dx$

$\quad = \dfrac{1}{2}x^2\ln(2x-3)-\dfrac{1}{4}x^2-\dfrac{3}{4}x-\dfrac{9}{8}\ln(2x-3)+C$

13. $\frac{1}{2}\int q^2\left(\sqrt{q^2-3}\cdot 2q\,dq\right)=uv-\int v\,du$

$\quad u=q^2 \qquad dv=\sqrt{q^2-3}\cdot 2q\,dq$

$\quad du=2q\,dq \qquad v=\frac{2}{3}\left(q^2-3\right)^{3/2}$

$\frac{1}{2}\int q^2\left(\sqrt{q^2-3}\cdot 2q\,dq\right)$

$\quad =\frac{1}{2}\left[\frac{2}{3}q^2\left(q^2-3\right)^{3/2}-\frac{2}{3}\int\left(q^2-3\right)^{3/2}\left(2q\,dq\right)\right]$

$\quad =\frac{1}{3}q^2\left(q^2-3\right)^{3/2}-\frac{1}{3}\cdot\frac{\left(q^2-3\right)^{5/2}}{5/2}+C$

$\quad =\left(q^2-3\right)^{3/2}\left[\frac{1}{3}q^2-\frac{2}{15}\left(q^2-3\right)\right]+C$

$\quad =\left(q^2-3\right)^{3/2}\cdot\frac{3q^2+6}{15}+C$

$\quad =\frac{\left(q^2-3\right)^{3/2}\left(q^2+2\right)}{5}+C$

15. $\frac{1}{2}\int_0^4 x^2\left(x^2+9\right)^{1/2}2x\,dx=\frac{1}{2}\left[uv-\int v\,du\right]\Big|_0^4$

$\quad u=x^2 \qquad dv=\left(x^2+9\right)^{1/2}\left(2x\,dx\right)$

$\quad du=2x\,dx \qquad v=\frac{2}{3}\left(x^2+9\right)^{3/2}$

$\frac{1}{2}\int_0^4 x^2\left(x^2+9\right)^{1/2}2x\,dx$

$\quad =\frac{1}{2}\left[\frac{2}{3}x^2\left(x^2+9\right)^{3/2}-\frac{2}{3}\int\left(x^2+9\right)^{3/2}\left(2x\,dx\right)\right]\Big|_0^4$

$\quad =\left[\frac{1}{3}x^2\left(x^2+9\right)^{3/2}-\frac{1}{3}\cdot\frac{\left(x^2+9\right)^{5/2}}{5/2}\right]\Big|_0^4$

$\quad =\left[\frac{16}{3}(125)-\frac{2}{15}(3125)\right]-\left[0-\frac{2}{15}(243)\right]$

$\quad =666.67-416.67+32.4=282.4$

17. $\int x^2 e^{-x}\,dx$

$\quad u=x^2 \qquad dv=e^{-x}\,dx$

$\quad du=2x\,dx \qquad v=-e^{-x}$

$\int x^2 e^{-x}\,dx=-e^{-x}x^2-\int\left(-e^{-x}\right)2x\,dx$

$\qquad\qquad =-x^2 e^{-x}+2\int xe^{-x}\,dx$

$\quad u=x \qquad dv=e^{-x}\,dx$

$\quad du=dx \qquad v=-e^{-x}$

$\int x^2 e^{-x}\,dx$

$\quad =-x^2 e^{-x}+2\left(-xe^{-x}+\int e^{-x}\,dx\right)$

$\quad =-x^2 e^{-x}-2xe^{-x}-2e^{-x}+C$

$\quad =-e^{-x}\left(x^2+2x+2\right)+C$

19. $\int_0^2 x^3 e^{x^2}\,dx=\int_0^2 x^2 e^{x^2}x\,dx$

$\quad u=x^2 \qquad dv=e^{x^2}x\,dx$

$\quad du=2x\,dx \qquad v=\frac{1}{2}e^{x^2}$

$\int_0^2 x^3 e^{x^2}\,dx=\frac{1}{2}x^2 e^{x^2}\Big|_0^2-\frac{1}{2}\int_0^2 e^{x^2}2x\,dx$

$\qquad\qquad =\left(\frac{1}{2}x^2 e^{x^2}-\frac{1}{2}e^{x^2}\right)\Big|_0^2$

$\qquad\qquad =\frac{1}{2}(4)e^4-\frac{1}{2}e^4-\left(0-\frac{1}{2}\right)$

$\qquad\qquad =2e^4-\frac{1}{2}e^4+\frac{1}{2}=\frac{3e^4+1}{2}$

21. $\int x^3\left(\ln x\right)^2\,dx$

$\quad u=\left(\ln x\right)^2 \qquad dv=x^3\,dx$

$\quad du=\frac{2\ln x}{x}\,dx \qquad v=\frac{1}{4}x^4$

$\int x^3\left(\ln x\right)^2\,dx=\frac{1}{4}x^4\left(\ln x\right)^2-\frac{1}{2}\int x^3\ln x\,dx$

$\quad u=\ln x \qquad dv=x^3\,dx$

$\quad du=\frac{dx}{x} \qquad v=\frac{1}{4}x^4$

$\int x^3\left(\ln x\right)^2\,dx$

$\quad =\frac{1}{4}x^4\left(\ln x\right)^2-\frac{1}{2}\left[\frac{1}{4}x^4\ln x-\int\frac{1}{4}x^3\,dx\right]$

$\quad =\frac{1}{4}x^4\left(\ln x\right)^2-\frac{1}{8}x^4\ln x+\frac{1}{32}x^4+C$

23. $\int e^{2x}\sqrt{e^x+1}\,dx = \int\left(e^x+1\right)^{1/2}e^x e^x\,dx$

$u = e^x \qquad dv = \left(e^x+1\right)^{1/2}e^x\,dx$

$du = e^x\,dx \qquad v = \dfrac{2}{3}\left(e^x+1\right)^{3/2}$

$\int e^{2x}\sqrt{e^x+1}\,dx$

$= \dfrac{2}{3}e^x\left(e^x+1\right)^{3/2} - \dfrac{2}{3}\int\left(e^x+1\right)^{3/2}e^x\,dx$

$= \dfrac{2}{3}e^x\left(e^x+1\right)^{3/2} - \dfrac{2}{3}\cdot\dfrac{2}{5}\left(e^x+1\right)^{5/2} + C$

$= \left(e^x+1\right)^{3/2}\left[\dfrac{2e^x}{3} - \dfrac{4}{15}\left(e^x+1\right)\right] + C$

$= (e^x+1)^{3/2}\left[\dfrac{6e^x}{15} - \dfrac{4}{15}\right] + C$

$= \dfrac{2}{15}\left(e^x+1\right)^{3/2}\left(3e^x-2\right) + C$

25. II. $\int e^{x^2}x\,dx = \dfrac{1}{2}\int e^u\,du$ where $u = x^2$ and

$du = 2x\,dx$.

$\int e^{x^2}x\,dx = \dfrac{1}{2}e^{x^2} + C$

27. IV. $\int\sqrt{e^x+1}\,e^x\,dx = \int\left(e^x+1\right)^{1/2}e^x\,dx = \int u^{1/2}\,du$

where $u = e^x+1$ and $du = e^x\,dx$.

$\int\sqrt{e^x+1}\,e^x\,dx = \dfrac{2}{3}\left(e^x+1\right)^{3/2} + C$

29. I. $\displaystyle\int_0^4\dfrac{t}{e^t}\,dt = \int_0^4 e^{-t}t\,dt$

$u = t \qquad dv = e^{-t}\,dt$

$du = dt \qquad v = -e^{-t}$

$\displaystyle\int_0^4\dfrac{t}{e^t}\,dt = -te^{-t}\Big|_0^4 + \int_0^4 e^{-t}\,dt$

$= -te^{-t}\Big|_0^4 + \left(-e^{-t}\right)\Big|_0^4$

$= -4e^{-4} + \left(-e^{-4} + e^0\right)$

$= -5e^{-4} + 1$

31. At $x = 30$, we have $p = 30 + 100\ln 61 = 441.09$.

$PS = 30(441.09) - \displaystyle\int_0^{30}\left[30 + 100\ln(2x+1)\right]dx = 13{,}232.62 - \left(30x + \dfrac{100}{2}(2x+1)\left[\ln(2x+1)-1\right]\right)\Bigg|_0^{30}$

$= 13{,}232.62 - \left\{900 + 50(61)\left[\ln 61 - 1\right] - \left[0 + 50(\ln 1 - 1)\right]\right\}$

$= 13{,}232.62 - 900 - 3050(3.11087) - 50 = \$2794.46$

(by formula 14)

33. Present Value

$= \displaystyle\int_0^5\left(10{,}000 - 500t\right)e^{-0.1t}\,dt = \left[-100{,}000e^{-0.1t} - \int 500te^{-0.1t}\,dt\right]\Bigg|_0^5$

$u = t \qquad dv = e^{-0.1t}\,dt$

$du = dt \qquad v = -10e^{-0.1t}$

(So, $\int te^{-0.1t}\,dt = -10te^{-0.1t} - 100e^{-0.1t}$.)

$= \left[-100{,}000e^{-0.1t} + 5000te^{-0.1t} + 50{,}000e^{-0.1t}\right]\Bigg|_0^5 = \left[5000te^{-0.1t} - 50{,}000e^{-0.1t}\right]\Bigg|_0^5$

$= \left[e^{-0.1t}\left(5000t - 50{,}000\right)\right]\Bigg|_0^5 = 0.6065\left[-25{,}000\right] - 1\left[0 - 50{,}000\right] = \$34{,}837$

35. Gini coeff $= \dfrac{\int_0^1 \left(x - xe^{x-1}\right) dx}{1/2}$

For xe^{x-1}, let $u = x \qquad dv = e^{x-1}dx$

$\qquad\qquad\qquad du = dx \qquad v = e^{x-1}$

$\int xe^{x-1}dx = xe^{x-1} - \int e^{x-1}dx = xe^{x-1} - e^{x-1}$

Gini coeff $= 2\left(\dfrac{x^2}{2} - xe^{x-1} + e^{x-1} \right)\Big|_0^1 = 2\left(\dfrac{1}{2} - 1 + 1 - \left(e^{-1}\right) \right) = 0.264$

37. Total receipts $= \int_{20}^{30} \left(-54.846 + 192.7 \ln t\right) dt = -54.846 \int_{20}^{30} dt + 192.7 \int_{20}^{30} \ln t \, dt$

First part: $-54.846 \int_{20}^{30} dt = -54.846\, t \,\big|_{20}^{30} = -54.846(30 - 20) = -548.46$

Second part uses integration by parts: $u = \ln t, \; du = \dfrac{1}{t}dt, \; dv = dt, \; v = t$

$192.7 \int_{20}^{30} \ln t \, dt = 192.7\left(t \ln t \,\big|_{20}^{30} - \int_{20}^{30} t \cdot \dfrac{1}{t} dt \right) = 192.7\left[30 \ln 30 - 20 \ln 20 - \int_{20}^{30} dt \right]$

$\qquad\qquad\qquad = 192.7\left[42.1213 - \left(t \,\big|_{20}^{30}\right) \right] = 192.7\left[42.1213 - (30 - 20) \right] \approx 6189.77$

Together: $-548.46 + 6189.77 \approx \$5641.3$ billion per year

Exercise 13.7

1. $\displaystyle\int_1^\infty \dfrac{dx}{x^6} = \lim_{a \to \infty} \int_1^a x^{-6}dx$

$\qquad = \dfrac{1}{5}\lim_{a \to \infty}\left(-x^{-5}\right)\Big|_1^a$

$\qquad = \dfrac{1}{5}\lim_{a \to \infty}\left(-\dfrac{1}{a^5} - (-1)\right) = \dfrac{1}{5}$

3. $\displaystyle\int_1^\infty \dfrac{dt}{t^{3/2}} = \lim_{a \to \infty} \int_1^a t^{-3/2}dt$

$\qquad = \lim_{a \to \infty} -2t^{-1/2}\Big|_1^a$

$\qquad = \lim_{a \to \infty}\left(-2\dfrac{1}{a^{1/2}} - (-2(1))\right) = 2$

5. $\displaystyle\int_1^\infty e^{-x}dx = (-1)\lim_{b \to \infty} \int_1^b e^{-x}(-1)dx$

$\qquad = (-1)\lim_{b \to \infty} e^{-x}\Big|_1^b$

$\qquad = (-1)\lim_{b \to \infty}\left(\dfrac{1}{e^b} - \dfrac{1}{e}\right)$

$\qquad = (-1)\left(-\dfrac{1}{e}\right) = \dfrac{1}{e}$

7. $\displaystyle\int_1^\infty \dfrac{dt}{t^{1/3}} = \lim_{a \to \infty} \int_1^a t^{-1/3}dt$

$\qquad = \lim_{a \to \infty} \dfrac{3}{2} t^{2/3}\Big|_1^a$

$\qquad = \lim_{a \to \infty}\left(\dfrac{3}{2}a^{2/3} - \dfrac{3}{2}\right)$

Thus, the integral diverges.

9. $\displaystyle\int_0^\infty e^{3x}dx = \lim_{a \to \infty} \dfrac{1}{3}\int_0^a e^{3x} \cdot 3\, dx$

$\qquad = \lim_{a \to \infty} \dfrac{1}{3}e^{3x}\Big|_0^a$

$\qquad = \lim_{a \to \infty} \dfrac{1}{3}\left(e^{3a} - 1\right)$

Thus, the integral diverges.

11. $\displaystyle\int_{-\infty}^{-1} 10x^{-2}dx = \lim_{a \to \infty}\int_{-a}^{-1} 10x^{-2}dx$

$\qquad = \lim_{a \to \infty} \dfrac{10x^{-1}}{-1}\Big|_{-a}^{-1}$

$\qquad = \lim_{a \to \infty}\left(\dfrac{-10}{-1} - \dfrac{-10}{-a}\right) = 10$

13. $\displaystyle\int_{-\infty}^{0} x^2 e^{-x^3}\,dx = -\frac{1}{3}\lim_{a\to\infty}\int_{-a}^{0} e^{-x^3}\left(-3x^2\,dx\right)$

$\displaystyle = -\frac{1}{3}\lim_{a\to\infty} e^{-x^3}\Big|_{-a}^{0}$

$\displaystyle = -\frac{1}{3}\lim_{a\to\infty}\left(1-e^{a^3}\right) = \infty$

Thus, the integral diverges.

$\left[-(-a)^3 = -(-a^3) = a^3\right]$

15. $\displaystyle\int_{-\infty}^{-1}\frac{6}{x}\,dx = \lim_{a\to\infty}\int_{-a}^{-1}\frac{6}{x}\,dx$

$\displaystyle = \lim_{a\to\infty} 6\ln|x|\,\Big|_{-a}^{-1}$

$\displaystyle = \lim_{a\to\infty}\left(6\ln 1 - 6\ln|a|\right) = -\infty$

Thus, the integral diverges.

17. $\displaystyle\int_{-\infty}^{\infty}\frac{2x}{x^2+1}\,dx = \lim_{a\to\infty}\int_{-a}^{0}\frac{2x}{x^2+1}\,dx + \lim_{b\to\infty}\int_{0}^{b}\frac{2x}{x^2+1}\,dx$

$\displaystyle = \lim_{-a\to\infty}\ln\left(x^2+1\right)\Big|_{-a}^{0} + \lim_{b\to\infty}\ln\left(x^2+1\right)\Big|_{0}^{b} = \lim_{-a\to\infty}\left[\ln 1 - \ln\left(a^2+1\right)\right] + \lim_{b\to\infty}\left[\ln\left(b^2+1\right) - \ln 1\right]$

Since each of these integrals diverges, the given integral diverges. (Ignore $\infty - \infty$).

19. $\displaystyle\int_{-\infty}^{\infty} x^3 e^{-x^4}\,dx = -\frac{1}{4}\left[\lim_{a\to\infty}\int_{-a}^{0} e^{-x^4}\left(-4x^3\,dx\right) + \lim_{b\to\infty}\int_{0}^{b} e^{-x^4}\left(-4x^3\,dx\right)\right]$

$\displaystyle = -\frac{1}{4}\left[\lim_{a\to\infty} e^{-x^4}\Big|_{-a}^{0} + \lim_{b\to\infty} e^{-x^4}\Big|_{0}^{b}\right] = -\frac{1}{4}\left[\lim_{a\to\infty}\left(e^0 - 1/e^{a^4}\right) + \lim_{b\to\infty}\left(1/e^{b^4} - e^0\right)\right] = -\frac{1}{4}(1-0+0-1) = 0$

21. $\displaystyle\int_{0}^{\infty}\frac{c}{e^{0.5t}}\,dt = 1$

$\displaystyle\lim_{a\to\infty}\int_{0}^{a} ce^{-0.5t}\,dt = c\lim_{a\to\infty}-2\int_{0}^{a} e^{-0.5t}\left(-0.5\right)\,dt = c\lim_{a\to\infty}(-2)e^{-0.5t}\Big|_{0}^{a} = -2c\lim_{a\to\infty}\left(e^{-0.5a}-1\right) = 2c$

So, $2c = 1$ gives $c = \dfrac{1}{2}$.

23. $\displaystyle\int_{1}^{\infty}\frac{x}{e^{x^2}}\,dx = \int_{1}^{\infty} e^{-x^2} x\,dx$

$\displaystyle = \lim_{a\to\infty}-\frac{1}{2}\int_{1}^{a} e^{-x^2}\left(-2\right)x\,dx$

$\displaystyle = \lim_{a\to\infty}-\frac{1}{2}\left(e^{-x^2}\right)\Big|_{1}^{a}$

$\displaystyle = \lim_{a\to\infty}-\frac{1}{2}\left[e^{-a^2} - e^{-1}\right] = \frac{1}{2e}$

25. $\displaystyle\int_{1}^{\infty}\frac{1}{\sqrt[3]{x^5}}\,dx = \lim_{a\to\infty}\int_{1}^{a} x^{-5/3}\,dx$

$\displaystyle = \lim_{a\to\infty}-\frac{3}{2}x^{-2/3}\Big|_{1}^{a}$

$\displaystyle = \lim_{a\to\infty}-\frac{3}{2}\left[a^{-2/3} - 1\right] = \frac{3}{2}$

27. $\displaystyle\int_{-\infty}^{\infty} f(x)\,dx = \int_{-\infty}^{10} f(x)\,dx + \int_{10}^{\infty} f(x)\,dx$

$\displaystyle = \int_{-\infty}^{10} 0\cdot dx + \int_{10}^{\infty} 200x^{-3}\,dx$

$\displaystyle = 0 + \lim_{b\to\infty}\frac{200}{-2}x^{-2}\Big|_{10}^{b}$

$\displaystyle = \lim_{b\to\infty}\left(-\frac{100}{b^2} - \frac{-100}{100}\right) = 1$

29. $\displaystyle\int_{-\infty}^{\infty} f(x)\,dx = \int_{-\infty}^{1} f(x)\,dx + \int_{1}^{\infty} f(x)\,dx$

$\displaystyle = \int_{-\infty}^{1} 0\cdot dx + \int_{1}^{\infty} cx^{-2}\,dx$

$\displaystyle = 0 + \lim_{b\to\infty} c\cdot\frac{x^{-1}}{-1}\Big|_{1}^{b}$

$\displaystyle = (-c)\lim_{b\to\infty}\left(\frac{1}{b} - \frac{1}{1}\right) = c$

Thus, $f(x)$ is a probability density function if $c = 1$.

31. $\displaystyle\int_{-\infty}^{\infty} f(x)\,dx = \int_{-\infty}^{0} f(x)\,dx + \int_{0}^{\infty} f(x)\,dx$

$\displaystyle = \int_{-\infty}^{0} 0 \cdot dx + \int_{0}^{\infty} ce^{-x/4}\,dx$

Therefore, $\displaystyle\int_{-\infty}^{\infty} f(x)\,dx = 0 + \lim_{b\to\infty}\frac{c}{-.25}e^{-x/4}\Big|_{0}^{b}$

$\displaystyle = -4c\lim_{b\to\infty}\left(\frac{1}{e^{b/4}} - e^{0}\right)$

Thus, $-4c(0-1) = 1$ or $c = \dfrac{1}{4}$.

33. Mean $\displaystyle= \int_{-\infty}^{\infty} x\,f(x)\,dx$

$\displaystyle = \int_{-\infty}^{10} x\,f(x)\,dx + \int_{10}^{\infty} x\,f(x)\,dx$

$\displaystyle = 0 + \int_{10}^{\infty} x\left(200x^{-3}\right)dx$

$\displaystyle = \lim_{b\to\infty} 200\cdot\frac{x^{-1}}{-1}\Big|_{10}^{b}$

$\displaystyle = -200\lim_{b\to\infty}\left(\frac{1}{b} - \frac{1}{10}\right) = 20$

35. $A = 8\displaystyle\int_{0}^{\infty} x\,e^{-3x}\,(3\,dx)$ (Alternate form used)

$u = x \qquad dv = e^{-3x}\cdot 3\,dx$

$du = dx \qquad v = -e^{-3x}$

$A = 8\left[-xe^{-3x}\Big|_{0}^{\infty} + \displaystyle\int_{0}^{\infty} e^{-3x}\,dx\right]$

$\displaystyle = 8\left[\lim_{b\to\infty}\left(-\frac{b}{e^{3b}} + 0\right) + \lim_{b\to\infty}\left(-\frac{1}{3e^{3b}} + \frac{1}{3}\right)\right]$

$= \dfrac{8}{3}$

If necessary ask your instructor why $\displaystyle\lim_{b\to\infty}\frac{b}{e^{3b}} = 0$.

37. $\displaystyle\int_{0}^{\infty} Ae^{-rt}\,dt = \lim_{b\to\infty}\int_{0}^{b} Ae^{-rt}\,dt$

$\displaystyle = \lim_{b\to\infty} -\frac{A}{r}e^{-rt}\Big|_{0}^{b} = -\frac{A}{r}\lim_{b\to\infty}\left(\frac{1}{e^{br}} - e^{0}\right)$

$\displaystyle = -\frac{A}{r}(0-1) = \frac{A}{r}$

39. $CV = \displaystyle\int_{0}^{\infty} 120e^{0.04t}\,e^{-0.09t}\,dt$

$\displaystyle = \lim_{b\to\infty}\int_{0}^{b} 120e^{-0.05t}\,dt$

$\displaystyle = \lim_{b\to\infty}\left(-2400e^{-0.05t}\right)\Big|_{0}^{b}$

$\displaystyle = \lim_{b\to\infty}\left(-\frac{2400}{e^{0.05b}} - \left(-2400e^{0}\right)\right)$

$= \$2400$ thousands

$= \$2,400,000$

41. $\displaystyle\int_{0}^{\infty} 56{,}000e^{0.02t}\,e^{-0.1t}\,dt = \lim_{a\to\infty}\int_{0}^{a} 56{,}000e^{-0.08t}\,dt = \lim_{a\to\infty} 56{,}000\cdot\frac{1}{-0.08}\int_{0}^{a} e^{-0.08t}\,(-0.08)\,dt$

$\displaystyle = \lim_{a\to\infty}\left(-700{,}000e^{-0.08t}\right)\Big|_{0}^{a} = -700{,}000\lim_{a\to\infty}\left[e^{+0.08a} - 1\right] = \$700{,}000$

43. a. $\displaystyle\int_{2}^{\infty} 0.5e^{-0.5t}\,dt = -\int_{2}^{\infty} e^{-0.5t}\,(-0.5t\,dt) = -\lim_{b\to\infty}\int_{2}^{b} e^{-0.5t}\,(-0.5t\,dt) = -\lim_{b\to\infty} e^{-0.5t}\Big|_{2}^{b} = -\lim_{b\to\infty}\left(e^{-0.5b} - e^{-1}\right) = \frac{1}{e} \approx 0.368$

b. Similarly, $\displaystyle\int_{8}^{\infty} 0.5e^{-0.5t}\,dt = -\lim_{b\to\infty}\int_{8}^{b} e^{-0.5t}\,(-0.5t\,dt) = -\lim_{b\to\infty} e^{-0.5t}\Big|_{8}^{b} = -\lim_{b\to\infty}\left(e^{-0.5b} - e^{-4}\right) = \frac{1}{e^{4}} \approx 0.0183$

45. $P(>24) = \int_{24}^{\infty} 0.08e^{-0.08t}\,dt = \lim_{b\to\infty} \int_{24}^{b} e^{-0.08t}(-0.08)(-1)\,dt$

$= (-1)\lim_{b\to\infty} e^{-0.08t}\Big|_{24}^{b} = (-1)\lim_{b\to\infty}\left(\dfrac{1}{e^{0.08b}} - \dfrac{1}{e^{1.92}}\right) = (-1)\left(0 - \dfrac{1}{6.821}\right) = 0.1466$

47. a. $500\int_{0}^{b} te^{-0.03(b-t)}\,dt$

Using integration by parts: $\quad u = t \qquad dv = e^{-0.03(b-t)}\,dt$

$$du = dt \qquad v = \dfrac{1}{0.03}e^{-0.03(b-t)}$$

$500\int_{0}^{b} te^{-0.03(b-t)}\,dt = \left[\dfrac{500te^{-0.03(b-t)}}{0.03} - \dfrac{500}{0.03}\int e^{-0.03(b-t)}\,dt\right]\Bigg|_{0}^{b} = \left[\dfrac{500te^{-0.03(b-t)}}{0.03} - \dfrac{500e^{-0.03(b-t)}}{(0.03)(0.03)}\right]\Bigg|_{0}^{b}$

$= \dfrac{500e^{-0.03(b-t)}}{0.03}\left(t - \dfrac{1}{0.03}\right)\Bigg|_{0}^{b} = \dfrac{500e^{0}}{0.03}\left(\dfrac{0.03b-1}{0.03}\right) - \dfrac{500e^{-0.03b}}{0.03}\left(0 - \dfrac{1}{0.03}\right)$

$= \dfrac{500}{0.0009}\left(0.03b - 1 + e^{-0.03b}\right)$

b. $\lim_{b\to\infty} \int_{0}^{b} f(t)\,dt = \lim_{b\to\infty}\dfrac{500}{0.0009}\left(0.03b - 1 + \dfrac{1}{e^{0.03b}}\right) = \infty$

Waste is produced more rapidly than existing waste decays.

Exercise 13.8

1. $[0,2]\quad n = 4 \quad h = \dfrac{2-0}{4} = \dfrac{1}{2}$

$x_0 = 0,\ x_1 = \dfrac{1}{2},\ x_2 = 1,\ x_3 = \dfrac{3}{2},\ x_4 = 2$

3. $[1,4]\quad n = 6 \quad h = \dfrac{4-1}{6} = \dfrac{1}{2}$

$x_0 = 1,\ x_1 = \dfrac{3}{2},\ x_2 = 2,\ x_3 = \dfrac{5}{2},\ x_4 = 3,\ x_5 = \dfrac{7}{2},\ x_6 = 4$

5. $[-1,4]\quad n = 5 \quad h = \dfrac{4-(-1)}{5} = 1$

$x_0 = -1,\ x_1 = 0,\ x_2 = 1,\ x_3 = 2,\ x_4 = 3,\ x_5 = 4$

7. $f(x) = x^2$ $[0,3]$ $n = 6$ $h = \dfrac{1}{2}$

   a. $\displaystyle\int_0^3 x^2\,dx \approx \dfrac{h}{2}\Big[f(0)+2f\left(\tfrac{1}{2}\right)+2f(1)+2f\left(\tfrac{3}{2}\right)+2f(2)+2f\left(\tfrac{5}{2}\right)+f(3)\Big]$

$$= \dfrac{1}{4}\left[0+\dfrac{1}{2}+2+\dfrac{9}{2}+8+\dfrac{25}{2}+9\right] = \dfrac{1}{4}\left[\dfrac{73}{2}\right] = 9.13$$

   b. $\displaystyle\int_0^3 x^2\,dx \approx \dfrac{h}{3}\Big[f(0)+4f\left(\tfrac{1}{2}\right)+2f(1)+4f\left(\tfrac{3}{2}\right)+2f(2)+4f\left(\tfrac{5}{2}\right)+f(3)\Big]$

$$= \dfrac{1}{6}\big[0+1+2+9+8+25+9\big] = 9$$

   c. $\displaystyle\int_0^3 x^2\,dx = \dfrac{x^3}{3}\bigg|_0^3 = \dfrac{3^3}{3}-\dfrac{0^3}{3} = 9$

   d. Simpson's Rule is more accurate.

9. $f(x) = \dfrac{1}{x^2}$ $[1,2]$ $n = 4$ $h = \dfrac{1}{4}$

   a. $\displaystyle\int_1^2 \dfrac{1}{x^2}\,dx \approx \dfrac{h}{2}\Big[f(1)+2f\left(\tfrac{5}{4}\right)+2f\left(\tfrac{3}{2}\right)+2f\left(\tfrac{7}{4}\right)+f(2)\Big] = \dfrac{1}{8}\left[1+\dfrac{32}{25}+\dfrac{8}{9}+\dfrac{32}{49}+\dfrac{1}{4}\right] \approx \dfrac{1}{8}[4.072] = 0.51$

   b. $\displaystyle\int_1^2 \dfrac{1}{x^2}\,dx \approx \dfrac{h}{3}\Big[f(1)+4f\left(\tfrac{5}{4}\right)+2f\left(\tfrac{3}{2}\right)+4f\left(\tfrac{7}{4}\right)+f(2)\Big] = \dfrac{1}{12}\left[1+\dfrac{64}{25}+\dfrac{8}{9}+\dfrac{64}{49}+\dfrac{1}{4}\right] \approx \dfrac{1}{12}[6.01] = 0.50$

   c. $\displaystyle\int_1^2 x^{-2}\,dx = -x^{-1}\Big|_1^2 = -\left(\tfrac{1}{2}-1\right) = \dfrac{1}{2}$

   d. Simpson's Rule is more accurate.

11. $f(x) = x^{1/2}$ $[0,4]$ $n = 8$ $h = \dfrac{1}{2}$

   a. $\displaystyle\int_0^4 x^{1/2}\,dx \approx \dfrac{h}{2}\Big[f(0)+2f\left(\tfrac{1}{2}\right)+2f(1)+2f\left(\tfrac{3}{2}\right)+2f(2)+2f\left(\tfrac{5}{2}\right)+2f(3)+2f\left(\tfrac{7}{2}\right)+f(4)\Big]$

$$= \dfrac{1}{4}\big[0+1.4142+2+2.4495+2.8284+3.1623+3.4641+3.7417+2\big] \approx \dfrac{1}{4}[21.0602] = 5.27$$

   b. $\displaystyle\int_0^4 x^{1/2}\,dx \approx \dfrac{h}{3}\Big[f(0)+4f\left(\tfrac{1}{2}\right)+2f(1)+4f\left(\tfrac{3}{2}\right)+2f(2)+4f\left(\tfrac{5}{2}\right)+2f(3)+4f\left(\tfrac{7}{2}\right)+f(4)\Big]$

$$= \dfrac{1}{6}\big[0+2.8284+2+4.8990+2.8284+6.3246+3.4641+7.4833+2\big] \approx \dfrac{1}{6}[31.8278] = 5.30$$

   c. $\displaystyle\int_0^4 x^{1/2}\,dx = \dfrac{2}{3}x^{3/2}\bigg|_0^4 = \dfrac{2}{3}\left(4^{3/2}-0\right) = 5.33$

   d. Simpson's Rule is more accurate.

13. $f(x) = \sqrt{x^3 + 1}$ $[0,2]$ $n = 4$ $h = \dfrac{1}{2}$

    a. $\displaystyle\int_0^2 \sqrt{x^3 + 1}\, dx \approx \dfrac{h}{2}\left[f(0) + 2f\left(\tfrac{1}{2}\right) + 2f(1) + 2f\left(\tfrac{3}{2}\right) + f(2) \right]$

$$= \dfrac{1}{4}\left[1 + 2.121 + 2.828 + 4.183 + 3\right] = \dfrac{1}{4}\left[13.132\right] = 3.283$$

    b. $\displaystyle\int_0^2 \sqrt{x^3 + 1}\, dx \approx \dfrac{h}{3}\left[f(0) + 4f\left(\tfrac{1}{2}\right) + 2f(1) + 4f\left(\tfrac{3}{2}\right) + f(2) \right]$

$$= \dfrac{1}{6}\left[1 + 4.243 + 2.828 + 8.367 + 3\right] = \dfrac{1}{6}\left[19.438\right] = 3.240$$

15. $f(x) = e^{-x^2}$ $[0,1]$ $n = 4$ $h = \dfrac{1}{4}$

    a. $\displaystyle\int_0^1 e^{-x^2}\, dx \approx \dfrac{h}{2}\left[f(0) + 2f\left(\tfrac{1}{4}\right) + 2f\left(\tfrac{1}{2}\right) + 2f\left(\tfrac{3}{4}\right) + f(1) \right]$

$$= \dfrac{h}{2}\left[1 + 1.879 + 1.558 + 1.140 + 0.368\right] = \dfrac{1}{8}\left[5.945\right] = 0.743$$

    b. $\displaystyle\int_0^1 e^{-x^2}\, dx \approx \dfrac{h}{3}\left[f(0) + 4f\left(\tfrac{1}{4}\right) + 2f\left(\tfrac{1}{2}\right) + 4f\left(\tfrac{3}{4}\right) + f(1) \right]$

$$= \dfrac{h}{3}\left[1 + 3.758 + 1.558 + 2.279 + 0.368\right] = \dfrac{1}{12}\left[8.963\right] = 0.747$$

17. $f(x) = \ln\left(x^2 - x + 1\right)$ $[1,5]$ $n = 4$ $h = 1$

    a. $\displaystyle\int_1^5 \ln\left(x^2 - x + 1\right) dx \approx \dfrac{h}{2}\left[f(1) + 2f(2) + 2f(3) + 2f(4) + f(5) \right]$

$$= \dfrac{h}{2}\left[0 + 2.197 + 3.892 + 5.130 + 3.045\right] = \dfrac{1}{2}\left[14.264\right] = 7.132$$

    b. $\displaystyle\int_1^5 \ln\left(x^2 - x + 1\right) dx \approx \dfrac{h}{3}\left[f(1) + 4f(2) + 2f(3) + 4f(4) + f(5) \right]$

$$= \dfrac{h}{3}\left[0 + 4.394 + 3.892 + 10.260 + 3.045\right] = \dfrac{1}{3}\left[21.591\right] = 7.197$$

19. $f(x) = f(x)$ $[1,4]$ $n = 5$ $h = \dfrac{3}{5}$

$$\int_1^4 f(x)\, dx \approx \dfrac{h}{2}\left[f(1) + 2f(1.6) + 2f(2.2) + 2f(2.8) + 2f(3.4) + f(4) \right]$$

$$= \dfrac{3}{10}\left[1 + 4.4 + 3.6 + 5.8 + 9.2 + 2.1\right] = \dfrac{3}{10}\left[26.1\right] = 7.8$$

21. $f(x) = f(x)$ $[1.2,\ 3.6]$ $n = 6$ $h = 0.4$

$$\int_{1.2}^{3.6} f(x)\, dx \approx \dfrac{h}{3}\left[f(1.2) + 4f(1.6) + 2f(2) + 4f(2.4) + 2f(2.8) + 4f(3.2) + f(3.6) \right]$$

$$= \dfrac{2}{15}\left[6.1 + 19.2 + 6.2 + 8.0 + 5.6 + 22.4 + 9.7\right] = \dfrac{2}{15}\left[77.2\right] = 10.3$$

23. $f(t) = 100 \dfrac{e^{0.1t}}{t+1}$ $[0,2]$ $n=4$ $h=\dfrac{1}{2}$

$$\int_0^2 f(t)\,dt \approx \dfrac{h}{3}\Big[f(0) + 4f\big(\tfrac{1}{2}\big) + 2f(1) + 4f\big(\tfrac{3}{2}\big) + f(2)\Big]$$

$$= \dfrac{1}{6}\big[100 + 280.34 + 110.52 + 185.89 + 40.71\big] = \dfrac{1}{6}\big[717.46\big] = 119.58$$

25. $C(x) = \left(x^2 + 1\right)^{3/2} + 1000$ $[30,33]$ $n=3$ $h=1$

$$\dfrac{1}{33-30}\int_{30}^{33} C(x)\,dx \approx \dfrac{h}{6}\big[f(30) + 2f(31) + 2f(32) + f(33)\big]$$

$$= \dfrac{1}{6}\big[28,045 + 61,675 + 67,632 + 36,987\big] = \dfrac{1}{6}\big[194,339\big] = \$32,390$$

Note: This is the average total cost for that number of units.

27. $PS = p_1 x_1 - \text{area under supply curve}$

$$\text{Area} \approx \dfrac{h}{3}\big[f(0) + 4f(10) + 2f(20) + 4f(30) + 2f(40) + 4f(50) + f(60)\big]$$

$$= \dfrac{10}{3}\big[120 + 1040 + 760 + 1800 + 1080 + 2520 + 680\big] = \dfrac{10}{3}\big[8000\big] = 26,666.67$$

$PS = 680(60) - 26,666.67 = \$14,133.33$

29. $N \text{ of units} \approx \dfrac{h}{2}\big[f(0) + 2f(1) + 2f(2) + 2f(3) + 2f(4) + f(5)\big]$

$$= \dfrac{1}{2}\big[250 + 495.2 + 490.8 + 486.6 + 482.6 + 239.5\big] = \dfrac{1}{2}\big[2444.70\big] = 1222.35$$

31. Simplify first: $2\int_0^1 (x - L(x))\,dx = 2\int_0^1 x\,dx - 2\int_0^1 L(x)\,dx = x^2\Big|_0^1 - 2\int_0^1 L(x)\,dx = 1 - 2\int_0^1 L(x)\,dx$

Approximate $\int_0^1 L(x)\,dx$ using a numerical method; we'll use the Trapezoid rule with $h = 0.2$:

(keep in mind that the percentages in the table must be converted to decimal form)

In 1990: $\int_0^1 L(x)\,dx = \dfrac{0.2}{2}\big[L(0) + 2f(0.2) + 2f(0.4) + 2f(0.6) + 2f(0.8) + f(1.0)\big]$

$$= 0.1\big[0 + 2(0.031) + 2(0.111) + 2(0.260) + 2(0.511) + 1.00\big] = 0.2826$$

Gini coefficient of income for blacks in 1990: $1 - 2(0.2826) = 0.4348$

In 2003: $\int_0^1 L(x)\,dx = \dfrac{0.2}{2}\big[L(0) + 2f(0.2) + 2f(0.4) + 2f(0.6) + 2f(0.8) + f(1.0)\big]$

$$= 0.1\big[0 + 2(0.029) + 2(0.111) + 2(0.258) + 2(0.498) + 1.00\big] = 0.2792$$

Gini coefficient of income for blacks in 2003: $1 - 2(0.2792) = 0.4416$

Income was more equally distributed in 1990, but only slightly.

33. $\text{Area} \approx \dfrac{h}{3}\big[f(0) + 4f(10) + 2f(20) + 4f(30) + 2f(40) + 4f(50) + 2f(60) + 4f(70) + f(80)\big]$

$$= \dfrac{10}{3}\big[0 + 60 + 36 + 72 + 60 + 108 + 48 + 92 + 0\big] = \dfrac{10}{3}\big[476\big] = 1586.67 \text{ sq ft}$$

Review Exercises

1. $\displaystyle\sum_{k=1}^{8}(k^2+1)=\sum_{k=1}^{8}k^2+\sum_{k=1}^{8}1$

$\qquad=\dfrac{8(9)(17)}{6}+8\cdot1$

$\qquad=204+8=212$

2. $\displaystyle\sum_{i=1}^{n}\dfrac{3i}{n^3}=\dfrac{3}{n^3}\sum_{i=1}^{n}i=\dfrac{3}{n^3}\cdot\dfrac{n(n+1)}{2}=\dfrac{3(n+1)}{2n^2}$

3. $f(x)=3x^2$

$\text{Area}=\dfrac{1}{6}\cdot f\!\left(\dfrac{1}{6}\right)+\dfrac{1}{6}\cdot f\!\left(\dfrac{2}{6}\right)+\cdots+\dfrac{1}{6}\cdot f\!\left(\dfrac{6}{6}\right)$

$\qquad=\dfrac{1}{6}\cdot\dfrac{3}{36}+\dfrac{1}{6}\cdot3\cdot\dfrac{4}{36}+\dfrac{1}{6}\cdot3\cdot\dfrac{9}{36}+\cdots+\dfrac{1}{6}\cdot3\cdot\dfrac{36}{36}$

$\qquad=\dfrac{3}{216}(1+4+9+16+25+36)$

$\qquad=\dfrac{1}{72}(91)=\dfrac{91}{72}$

4. $A=\displaystyle\lim_{n\to\infty}3\!\left(\dfrac{i}{n}\right)^2\!\left(\dfrac{1}{n}\right)$

$\qquad=\displaystyle\lim_{n\to\infty}\dfrac{3}{n^3}\sum_{i=1}^{n}i^2=\lim_{n\to\infty}\dfrac{3}{n^3}\!\left[\dfrac{n(n+1)(2n+1)}{6}\right]$

$\qquad=\displaystyle\lim_{n\to\infty}\dfrac{(n+1)(2n+1)}{2n^2}=1$

5. $\text{Area}=\displaystyle\int_{0}^{1}3x^2\,dx=x^3\Big|_{0}^{1}=1-0=1$

6. $\displaystyle\int_{1}^{3}(x^3-4x+5)\,dx=\left(\dfrac{x^4}{4}-\dfrac{4x^2}{2}+5x\right)\Big|_{1}^{3}$

$\qquad=\left(\dfrac{81}{4}-18+15\right)-\left(\dfrac{1}{4}-2+5\right)$

$\qquad=14$

7. $\displaystyle\int_{1}^{4}4x^{3/2}\,dx=\dfrac{8}{5}x^{5/2}\Big|_{1}^{4}=\dfrac{8}{5}(32-1)=\dfrac{8}{5}(31)=\dfrac{248}{5}$

8. $\displaystyle\int_{-3}^{2}(x^3-3x^2+4x+2)\,dx$

$\qquad=\left(\dfrac{1}{4}x^4-x^3+2x^2+2x\right)\Big|_{-3}^{2}$

$\qquad=4-8+8+4-\left(\dfrac{81}{4}+27+18-6\right)$

$\qquad=8-\dfrac{237}{4}=-\dfrac{205}{4}$

9. $\displaystyle\int_{0}^{5}(x^3+4x)\,dx=\left(\dfrac{1}{4}x^4+2x^2\right)\Big|_{0}^{5}$

$\qquad=\dfrac{625}{4}+50=\dfrac{825}{4}$

10. $\displaystyle\int_{-2}^{3}(x+2)^2\,dx=\dfrac{1}{3}(x+2)^3\Big|_{-2}^{3}=\dfrac{125}{3}-0=\dfrac{125}{3}$

11. $\displaystyle\int_{-3}^{-1}(x+1)\,dx=\left(\dfrac{1}{2}x^2+x\right)\Big|_{-3}^{-1}$

$\qquad=\left(\dfrac{1}{2}-1\right)-\left(\dfrac{9}{2}-3\right)=-2$

12. $\displaystyle\int_{2}^{3}\dfrac{x^2}{2x^3-7}\,dx=\dfrac{1}{6}\int_{2}^{3}\dfrac{6x^2}{2x^3-7}\,dx$

$\qquad=\dfrac{1}{6}\ln\!\left(2x^3-7\right)\Big|_{2}^{3}$

$\qquad=\dfrac{1}{6}\ln47-\dfrac{1}{6}\ln9$

13. $\displaystyle\int_{-1}^{2}(x^2+x)\,dx=\left(\dfrac{1}{3}x^3+\dfrac{1}{2}x^2\right)\Big|_{-1}^{2}$

$\qquad=\left(\dfrac{8}{3}+2\right)-\left(-\dfrac{1}{3}+\dfrac{1}{2}\right)$

$\qquad=4\dfrac{1}{2}=\dfrac{9}{2}$

14. $\displaystyle\int_{1}^{4}(x^{-1}+x^{1/2})\,dx=\left(\ln x+\dfrac{2}{3}x^{3/2}\right)\Big|_{1}^{4}=\ln4+\dfrac{14}{3}$

15. $\displaystyle\int_{0}^{4}(2x+1)^{1/2}\,dx=\dfrac{1}{2}\cdot\dfrac{(2x+1)^{3/2}}{3/2}\Big|_{0}^{4}$

$\qquad=\dfrac{1}{3}(27-1)=\dfrac{26}{3}$

16. $\displaystyle\int_{0}^{1}\dfrac{x}{x^2+1}\,dx=\dfrac{1}{2}\int_{0}^{1}\dfrac{2x}{x^2+1}\,dx$

$\qquad=\dfrac{1}{2}\ln\!\left(x^2+1\right)\Big|_{0}^{1}=\dfrac{1}{2}\ln2-0=\dfrac{1}{2}\ln2$

17. $\displaystyle\int_{0}^{1}e^{-2x}\,dx=-\dfrac{1}{2}e^{-2x}\Big|_{0}^{1}$

$\qquad=-\dfrac{1}{2}\left(\dfrac{1}{e^2}-e^0\right)=\dfrac{1}{2}\left(1-e^{-2}\right)$

18. $\int_0^1 xe^{x^2}\,dx = \frac{1}{2}\int_0^1 2xe^{x^2}\,dx = \frac{1}{2}e^{x^2}\Big|_0^1 = \frac{1}{2}e - \frac{1}{2}$

19. $x = 1:\ f(x) = x^2 - 3x + 2\quad g(x) = x^2 + 4$

   $\qquad\ f(1) = 0\qquad\qquad g(1) = 5$

   $A = \int_0^5\left[\left(x^2 + 4\right) - \left(x^2 - 3x + 2\right)\right]dx$

   $\quad = \int_0^5 \left(2 + 3x\right)dx$

   $\quad = \left(2x + \frac{3}{2}x^2\right)\Big|_0^5 = \left(10 + \frac{75}{2}\right) - 0 = \frac{95}{2}$

20. $\qquad\qquad x^2 = 4x + 5$

   $\qquad x^2 - 4x - 5 = 0$

   $\qquad (x - 5)(x + 1) = 0$

   $\qquad\qquad\quad x = 5, -1$

   $\int_{-1}^5 \left(4x + 5 - x^2\right)dx = \left(2x^2 + 5x - \frac{1}{3}x^3\right)\Big|_{-1}^5$

   $\qquad\qquad = 75 - \frac{125}{3} - \left(-3 + \frac{1}{3}\right)$

   $\qquad\qquad = \frac{108}{3} = 36$

21. $A = \int_{-1}^0 \left(x^3 - x\right)dx$

   $\quad = \left(\frac{1}{4}x^4 - \frac{1}{2}x^2\right)\Big|_{-1}^0 = 0 - \left(\frac{1}{4} - \frac{1}{2}\right) = \frac{1}{4}$

22. $\qquad\qquad x^3 - 1 = x - 1$

   $\qquad\qquad\quad x^3 - x = 0$

   $\qquad x(x + 1)(x - 1) = 0$

   $\qquad\qquad\qquad x = 0, -1,\ 1$

   $\int_{-1}^0\left[x^3 - 1 - (x - 1)\right]dx + \int_0^1\left[x - 1 - (x^3 - 1)\right]dx$

   $= \int_{-1}^0\left(x^3 - x\right)dx + \int_0^1\left(x - x^3\right)dx$

   $= \left(\frac{1}{4}x^4 - \frac{1}{2}x^2\right)\Big|_{-1}^0 + \left(\frac{1}{2}x^2 - \frac{1}{4}x^4\right)\Big|_0^1$

   $= -\frac{1}{4} + \frac{1}{2} + \frac{1}{2} - \frac{1}{4} = \frac{1}{2}$

23. $\int\sqrt{x^2 - 4}\,dx$

   $= \frac{1}{2}\left(x\sqrt{x^2 - 4} - 4\ln\left|x + \sqrt{x^2 - 4}\right|\right) + C$

   (By formula 7)

24. $\int_0^1 3^x\,dx = 3^x\log_3 e\Big|_0^1 = 3\log_3 e - 1\log_3 e = 2\log_3 e$

25. $\frac{1}{2}\int\ln x^2\,(2x\,dx) = \frac{1}{2}x^2\left(\ln x^2 - 1\right) + C$

   (By formula 14)

26. $\int\frac{dx}{x(3x + 2)} = \frac{1}{2}\ln\left|\frac{x}{3x + 2}\right| + C$

27. $\int x^5\ln x\,dx$

   $\quad u = \ln x\qquad dv = x^5\,dx$

   $\quad du = \frac{1}{x}\,dx\qquad v = \frac{1}{6}x^6$

   $\int x^5\ln x\,dx = \frac{1}{6}x^6\ln|x| - \frac{1}{6}\int x^5\,dx$

   $\qquad\qquad = \frac{1}{6}x^6\ln|x| - \frac{1}{6}\cdot\frac{x^6}{6} + C$

   $\qquad\qquad = \frac{1}{6}x^6\ln|x| - \frac{1}{36}x^6 + C$

28. $\int xe^{-2x}\,dx = uv - \int v\,du$

   $\quad u = x\qquad dv = e^{-2x}\,dx$

   $\quad du = dx\qquad v = -\frac{1}{2}e^{-2x}$

   $\int xe^{-2x}\,dx = -\frac{1}{2}xe^{-2x} + \frac{1}{2}\int e^{-2x}\,dx$

   $\qquad\qquad = -\frac{1}{2}xe^{-2x} - \frac{1}{4}e^{-2x} + C$

29. $\int x(x + 5)^{-1/2}\,dx$

   $\quad u = x\qquad dv = (x + 5)^{-1/2}\,dx$

   $\quad du = dx\qquad v = 2(x + 5)^{1/2}$

   $\int x(x + 5)^{-1/2}\,dx = 2x\sqrt{x + 5} - 2\int(x + 5)^{1/2}\,dx$

   $\qquad\qquad = 2x\sqrt{x + 5} - \frac{4}{3}(x + 5)^{3/2} + C$

30. $\int_1^e \ln x\,dx = uv - \int v\,du$

   $\quad u = \ln x\qquad dv = dx$

   $\quad du = \frac{1}{x}\,dx\qquad v = x$

   $\int_1^e \ln x\,dx = x\ln|x|\Big|_1^e - \int_1^e dx$

   $\qquad\qquad = (x\ln x - x)\Big|_1^e = e - e - (0 - 1) = 1$

31. $\displaystyle\int_1^\infty \frac{1}{x}\,dx = \lim_{b\to\infty}\int_1^b \frac{1}{x}\,dx = \lim_{b\to\infty}\ln x\Big|_1^b = \lim_{b\to\infty}\left(\ln b - \ln 1\right) = \infty$

Thus, the integral diverges.

32. $\displaystyle\int_{-\infty}^{-1} \frac{200}{x^3}\,dx = \lim_{a\to\infty}\int_{-a}^{-1} \frac{200}{x^3}\,dx = \lim_{a\to\infty} -100x^{-2}\Big|_{-a}^{-1} = -100$

33. $\displaystyle\int_0^\infty 5e^{-3x}\,dx = \lim_{b\to\infty}\int_0^b 5e^{-3x}\,dx = -\frac{5}{3}\lim_{b\to\infty} e^{-3x}\Big|_0^b = -\frac{5}{3}\lim_{b\to\infty}\left(\frac{1}{e^{3b}} - e^0\right) = \frac{5}{3}$

34. $\displaystyle\int_{-\infty}^0 \frac{x}{\left(x^2+1\right)^2}\,dx = \lim_{a\to\infty}\int_{-a}^0 x\left(x^2+1\right)^{-2}\,dx = \lim_{a\to\infty}\frac{1}{2}\int_{-a}^0 2x\left(x^2+1\right)^{-2}\,dx = \lim_{a\to\infty} -\frac{1}{2}\left(x^2+1\right)^{-1}\Big|_{-a}^0 = -\frac{1}{2}$

35. a. $\displaystyle\int_1^3 2x^{-3}\,dx = -\frac{1}{x^2}\Big|_1^3 = -\left(\frac{1}{9}-1\right) = \frac{8}{9}$ or 0.889

b. $\displaystyle\int_1^3 \frac{2}{x^3}\,dx \approx \frac{h}{2}\left[f(1)+2f(1.5)+2f(2)+2f(2.5)+f(3)\right]$

$= \frac{1}{4}\left[2+1.185+0.500+0.256+0.074\right] = \frac{1}{4}\left[4.015\right] = 1.004$

c. $\displaystyle\int_1^3 \frac{2}{x^3}\,dx \approx \frac{h}{3}\left[f(1)+4f(1.5)+2f(2)+4f(2.5)+f(3)\right]$

$= \frac{1}{6}\left[2+2.370+0.500+0.512+0.074\right] = \frac{1}{6}\left[5.456\right] = 0.909$

36. $\displaystyle\int_0^1 \frac{4}{x^2+1}\,dx \approx \frac{h}{2}\left[f(0)+2f(0.2)+2f(0.4)+2f(0.6)+2f(0.8)+f(1)\right]$

$= \frac{1}{10}\left[4+7.692+6.897+5.882+4.878+2\right] = \frac{1}{10}\left[31.349\right] = 3.135$

37. $\displaystyle\int_1^{2.2} f(x)\,dx \approx \frac{h}{3}\left[f(1)+4f(1.3)+2f(1.6)+4f(1.9)+f(2.2)\right] = \frac{1}{10}\left[0+11.2+10.2+16.8+0.6\right] = \frac{1}{10}\left[38.8\right] = 3.9$

38. a. $n = 5$
b. $n = 6$

39. $\displaystyle M = \int_0^9 14{,}000\left(t+16\right)^{-1/2}\,dt = 14{,}000\cdot 2\left(t+16\right)^{1/2}\Big|_0^9 = 28{,}000\left(5-4\right) = \$28{,}000$

40. $\displaystyle\int_0^2 1.4e^{-1.4x}\,dx = -e^{-1.4x}\Big|_0^2 = -e^{-2.8} + e^0 = 1 - e^{-2.8} = 0.9392$

So the probability that it lasts 2 years is $1 - 0.9392 \approx 0.061$.

41. $\displaystyle \text{Average} = \frac{1}{5-0}\int_0^5 1000e^{0.1t}\,dt = \frac{1}{5}\cdot\frac{1000}{0.1}e^{0.1t}\Big|_0^5 = 2000\left(1.6487-1\right) = \$1297.44$

42. $\displaystyle AI = \frac{1}{4}\int_0^4 50e^{0.2t}\,dt = \frac{250}{4}\int_0^4 0.2e^{0.2t}\,dt = \frac{250}{4}e^{0.2t}\Big|_0^4 \approx \frac{250}{4}\left[2.2255-1\right] \approx \$76.60$

43. In 1969, the Gini coefficient was $2\int_0^1\left(x-x^{2.1936}\right)dx=2\left(\dfrac{1}{2}x^2-\dfrac{x^{3.1936}}{3.1936}\right)\Bigg|_0^1\approx0.3737$.

In 2000, the Gini coefficient was $2\int_0^1\left(x-x^{2.4870}\right)dx=2\left(\dfrac{1}{2}x^2-\dfrac{x^{3.4870}}{3.4870}\right)\Bigg|_0^1\approx0.4264$.

Income was more equally distributed in 1969.

44. a. $\sqrt{64-4x}=x-1$

$\qquad 64-4x=x^2-2x+1$

$\qquad\quad 0=(x+9)(x-7)$

So, equilibrium is at $(7,6)$.

b. $CS=\int_0^7\left(64-4x\right)^{1/2}dx-7\cdot6=-\dfrac{1}{4}\cdot\dfrac{\left(64-4x\right)^{3/2}}{3/2}\Bigg|_0^7-42=-\dfrac{1}{6}\left(36^{3/2}-64^{3/2}\right)-42$

$\qquad=-\dfrac{1}{6}(-296)-42=49.33-42=\$7.33$

45. $\qquad x-1=\sqrt{64-4x}$

$\qquad x^2-2x+1=64-4x$

$\qquad x^2+2x-63=0$

$\qquad(x+9)(x-7)=0$

$\qquad\qquad x=7$ (only positive values)

$PS=(7)(6)-\int_0^7(x-1)\,dx=42-\left[\left(\dfrac{1}{2}x^2-x\right)\Bigg|_0^7\right]=42-17.5=24.50$

46. $TI=\int_0^{10}125e^{0.05t}\,dt=2500e^{0.05t}\Big|_0^{10}=2500(1.64872-1)=\$1621.803$ thousands $=\$1,621,803$

47. a. $PV=\int_0^5150e^{-0.2t}e^{-0.08t}\,dt=\int_0^5150e^{-0.28t}\,dt=\dfrac{150}{-0.28}\int_0^5-0.28e^{-0.28t}\,dt$

$\qquad=-535.71\left[e^{-0.28t}\Big|_0^5\right]=-535.71(0.2466-1)\approx403.604$ thousands or $\$403,604$.

b. $FV=e^{0.40}(PV)=1.4918(PV)=\$602,106$

48. Average Cost $=\dfrac{1}{150-0}\int_0^{150}\left(x^2+40,000\right)^{1/2}dx$

$\qquad=\dfrac{1}{150}\cdot\dfrac{1}{2}\left[x\sqrt{x^2+40,000}+40,000\ln\left(x+\sqrt{x^2+40,000}\right)\right]\Bigg|_0^{150}$

$\qquad=\dfrac{1}{300}\left[150\sqrt{62,500}+40,000\ln\left(150+\sqrt{62,500}\right)-\left(0+40,000\ln\sqrt{40,000}\right)\right]$

$\qquad=\dfrac{1}{300}\left[150(250)+40,000(5.9915)-40,000(5.2983)\right]=\dfrac{1}{300}(65,228)=\$217.43$

49. $PS=70(1000)-\int_0^{1000}\left(0.02x+50-\dfrac{10}{\sqrt{x^2+1}}\right)dx$

$\qquad=70,000-\left(0.01x^2+50x-10\ln\left|x+\sqrt{x^2+1}\right|\right)\Bigg|_0^{1000}=70,000-\left[(10000+50000-76.01)-(0)\right]\approx\$10,076$

50. $PV = 9000 \int_0^5 te^{-0.08t} \, dt$ Use integration by parts.

$$PV = 9000\left[\frac{-t}{0.08}e^{-0.08t}\Big|_0^5 + \frac{1}{0.08}\int_0^5 e^{-0.08t}\,dt\right] = 9000\left[\frac{-t}{0.08}e^{-0.08t}\Big|_0^5 + \frac{1}{-(0.08)^2}\int_0^5 -0.08e^{-0.08t}\,dt\right]$$

$$= 9000[-41.895] - 1,406,250\left[e^{-0.08t}\Big|_0^5\right] = 9000[-41.895] - 1,406,250[-0.32968] = \$86,557.44$$

51. $C(x) = \int\left[3 + 60(x+1)\ln(x+1)\right]dx$

Using integration by parts: $u = \ln(x+1)$ $dv = 60(x+1)\,dx$

$$du = \frac{dx}{x+1} \qquad v = 30(x+1)^2$$

$$\int\left[3 + 60(x+1)\ln(x+1)\right]dx = 3x + \ln(x+1)\cdot 30(x+1)^2 - \int 30(x+1)^2 \cdot \frac{dx}{x+1}$$

$$= 3x + 30(x+1)^2\ln(x+1) - 30\int(x+1)\,dx$$

$$= 3x + 30(x+1)^2\ln(x+1) - 30\left(\tfrac{1}{2}x^2 + x\right) + K$$

$$= 3x + 30(x+1)^2\ln(x+1) - 15x^2 - 30x + K$$

$$C(x) = 30(x+1)^2\ln(x+1) - 15x^2 - 27x + K$$

Using the concept that fixed cost is K, we have $C(x) = 30(x+1)^2\ln(x+1) - 15x^2 - 27x + 2000$.

52. $\int_1^\infty 1.4e^{-1.4x}\,dx = \lim_{b\to\infty}\left[-\int_1^b -1.4e^{-1.4x}\,dx\right] = -\lim_{b\to\infty}e^{-1.4x}\Big|_1^b = -\lim_{b\to\infty}\left(e^{-1.4b} - e^{-1.4}\right) = e^{-1.4} \approx 0.247$

53. $CV = \int_0^\infty 120e^{.03t}\cdot e^{-0.06t}\,dt = \int_0^\infty 120e^{-0.03t}\,dt = \lim_{a\to\infty}\int_0^a 120e^{-0.03t}\,dt$

$$= \frac{120}{-0.03}\lim_{a\to\infty}e^{-0.03t}\Big|_0^a = -4000\lim_{a\to\infty}\left(\frac{1}{e^{0.03a}} - 1\right) = \$4000 \text{ thousands}$$

54. $\int_0^2 100e^{-0.01t^2}\,dt \approx \frac{h}{3}\left[f(0) + 4f(0.5) + 2f(1) + 4f(1.5) + f(2)\right]$

$$= \frac{1}{6}[100 + 399.00 + 198.01 + 390.10 + 96.08] = \frac{1}{6}[1183.19] = \$197.198 \text{ (thousands)}$$

55. $\int_0^{10}\overline{MR}\,dx \approx \frac{h}{2}\left[f(0) + 2f(2) + 2f(4) + 2f(6) + 2f(8) + f(10)\right]$

$$= 1[0 + 960 + 1440 + 1440 + 960 + 0] = \$4,800 \text{ (hundreds) or } \$480,000$$

Chapter Test

1.

x_i	0	.5	1	1.5
$f(x_1)$	2	1.936	1.732	1.323

$$A \approx (0.5)\sum f(x_i)$$
$$= (0.5)(6.991)$$
$$= 3.496$$

2. $f(x) = 5 - 2x \quad [0, 1]$

$$\Delta x = \frac{1-0}{n} = \frac{1}{n}$$

Rectangle 1: $\Delta x(5 - 2\Delta x)$

Rectangle 2: $\Delta x(5 - 4\Delta x)$

$\cdots$

Rectangle i: $\Delta x(5 - 2i\Delta x)$

a. $S = \sum_{i=1}^{n}\left(5 - 2i \cdot \frac{1}{n}\right)\frac{1}{n}$

$$= \frac{1}{n}\left(5n - \frac{2}{n} \cdot \frac{n(n+1)}{2}\right)$$

$$= 5 - \frac{n+1}{n} \text{ or } 4 - \frac{1}{n}$$

b. $\lim_{n\to\infty} S_n = \lim_{n\to\infty}\left(4 - \frac{1}{n}\right) = 4$

3. $\int_0^6 \left(12 + 4x - x^2\right) dx = \left(12x + 2x^2 - \frac{x^3}{3}\right)\Big|_0^6 = 72$

4. a $\int_0^4 (9 - 4x)\, dx = \left(9x - 2x^2\right)\Big|_0^4 = 4 - 0 = 4$

b. $\int_0^3 \left(8x^2 + 9\right)^{-1/2} (x\, dx) = \frac{1}{16}\left(\frac{\left(8x^2 + 9\right)^{1/2}}{1/2}\right)\Big|_0^3$

$$= \frac{1}{8}(9 - 3) = \frac{3}{4}$$

c. $\int_1^4 \frac{5}{4x - 1}\, dx = \frac{5}{4}\ln(4x - 1)\Big|_1^4 = \frac{5}{4}\ln 5$

d. $\int_1^\infty \frac{7}{x^2}\, dx = \lim_{a\to\infty}\left(-\frac{7}{x}\right)\Big|_1^a = \lim_{a\to\infty}\left(-\frac{7}{a} + 7\right) = 7$

e. $\int_4^4 \sqrt{x^3 + 10}\, dx = 0$

f. $\int_0^1 5x^2 e^{2x^3}\, dx = 5 \cdot \frac{1}{6}\int_0^1 e^{2x^3}\left(6x^2\, dx\right)$

$$= \frac{5}{6}e^{2x^3}\Big|_0^1 = \frac{5}{6}\left(e^2 - e^0\right)$$

$$= \frac{5}{6}\left(e^2 - 1\right)$$

5. a. $\int 3xe^x\, dx = 3xe^x - \int 3e^x\, dx = 3xe^x - 3e^x + C$

$$u = 3x \qquad dv = e^x\, dx$$
$$du = 3\, dx \qquad v = e^x$$

b. $\int x\ln(2x)\, dx = \frac{x^2 \ln 2x}{2} - \int \frac{x}{2}\, dx$

$$= \frac{x^2 \ln 2x}{2} - \frac{x^2}{4} + C$$

6. $\int_1^3 f(x)\, dx + \int_3^4 f(x)\, dx = \int_1^4 f(x)\, dx$, so

$$2\int_1^3 f(x)\, dx = 2\left[\int_1^4 f(x)\, dx - \int_3^4 f(x)\, dx\right]$$

$$= 2[3 - 7] = -8$$

7. a. $\int \ln(2x)\, dx = \frac{1}{2}\int \ln(2x)(2\, dx)$

$$= \frac{1}{2} \cdot 2x\left(\ln(2x) - 1\right)$$

$$= x\left(\ln(2x) - 1\right) + C$$

(Formula 14)

b. $\int x\sqrt{3x - 7}\,(dx)$

$$= \left[\frac{2(3 \cdot 3x + 2 \cdot 7)(3x - 7)^{3/2}}{15 \cdot 9}\right] + C$$

(Formula 16)

8. $\int_1^4 \sqrt{x^3 + 10}\, dx \approx 16.089$

9. $80 = 120 - 0.2x$ gives an equilibrium quantity $x = 200$.

$$CS = \int_0^{200} (120 - 0.2x)\, dx - 80 \cdot 200$$

$$= \left(120x - 0.1x^2\right)\Big|_0^{200} - 16000 = \$4000$$

$$PS = 80 \cdot 200 - \int_0^{200} \left(40 + 0.001x^2\right) dx$$

$$= 16000 - \left(40x + \frac{.001}{3}x^3\right)\Big|_0^{200} = \frac{\$16000}{3}$$

10. a. $TI = \int_0^{12} 85 e^{-0.01t}\, dt$

$$= \frac{85}{-0.01} e^{-0.01t}\Big|_0^{12} = -8500(0.8869 - 1)$$

$$= \$961.18 \text{ thousands}$$

b. $PV = \int_0^{12} 85 e^{-0.01t}\, e^{-0.07t}\, dt$

$$= 85 \int_0^{12} e^{-0.08t}\, dt = \frac{85}{-0.08} e^{-0.08t}\Big|_0^{12}$$

$$= \$655.68 \text{ thousands}$$

c. $CV = \int_0^{\infty} 85 e^{-0.01t}\, e^{-0.07t}\, dt$

$$= \lim_{a \to \infty} \int_0^a 85 e^{-0.08t}\, dt = \lim_{a \to \infty} \left[\frac{85}{-0.08} e^{-0.08t}\Big|_0^a\right]$$

$$= \frac{85}{-0.08} \lim_{a \to \infty}\left(\frac{1}{e^{0.08a}} - 1\right) = \$1062.5 \text{ thousand}$$

11. First, find the points of intersection.

$$x^2 - x = 2x + 4$$
$$0 = x^2 - 3x - 4 = (x - 4)(x + 1)$$

Points are −1 and 4.

$$A = \int_{-1}^{4} \left[(2x + 4) - (x^2 - x)\right] dx$$

$$= \int_{-1}^{4} \left(3x + 4 - x^2\right) dx = \left(\frac{3}{2}x^2 + 4x - \frac{x^3}{3}\right)\Big|_{-1}^{4}$$

$$= \frac{56}{3} - \left(-\frac{13}{6}\right) = \frac{125}{6}$$

12.

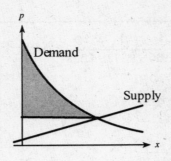

13. Gini coeff $= \dfrac{\int_0^1 \left(x - 0.998x^{2.6}\right) dx}{1/2}$

$$= 2\left(\frac{x^2}{2} - \frac{0.998}{3.6}x^{3.6}\right)\Big|_0^1$$

$$= 2(0.223) = 0.446$$

Gini coeff $= \dfrac{\int_0^1 \left[x - \left(0.57x^2 + 0.43x\right)\right] dx}{1/2}$

$$= 2\left(\frac{x^2}{2} - \frac{0.57}{3}x^3 - \frac{0.43}{2}x^2\right)\Big|_0^1$$

$$= 2(0.095) = 0.19$$

The change decreases the difference in income.

14. This problem is more easily solved by using the graphing calculator.

  a. $\text{Avg.} = \dfrac{1}{20-5} \displaystyle\int_5^{20} \left(3.722x^3 - 145.3x^2 + 1836.8x - 5735.2\right) dx$

  $\qquad = \dfrac{1}{15}\left(0.9305x^4 - 48.43x^3 + 918.4x^2 - 5735.2x\right)\Big|_5^{20}$

  $\qquad = \dfrac{1}{15}\left[14,096 - (-11,188.19)\right] = 1,685.61 \; \dfrac{\text{thousand barrels}}{\text{day}}$

  b. $\text{Avg.} = \dfrac{1}{20-10} \displaystyle\int_{10}^{20} f(x)\,dx = \dfrac{1}{10}\left[14096 - (-4637)\right] = 1,873.3 \; \dfrac{\text{thousand barrels}}{\text{day}}$

15. $\displaystyle\int_1^3 x \ln x\,dx \approx \dfrac{h}{2}\left[f(1) + 2f(1.5) + 2f(2) + 2f(2.5) + f(3)\right]$

  $\qquad = \dfrac{1}{4}\left[0 + 1.216 + 2.773 + 4.581 + 3.296\right] = \dfrac{1}{4}\left[11.866\right] = 2.967$

16. $\text{Area} \approx \dfrac{h}{3}\left[f(0) + 4f(40) + 2f(80) + 4f(120) + 2f(160) + 4f(200) + f(240)\right]$

  $\qquad = \dfrac{40}{3}\left[0 + 88 + 70 + 192 + 64 + 96 + 0\right] = \dfrac{40}{3}\left[510\right] = 6800 \text{ sq ft}$

Chapter 14: Functions of Two or More Variables

Exercise 14.1

1. $z = x^2 + y^2$

Domain $= \{(x,y): x \text{ and } y \text{ are real}\}$

3. $z = \dfrac{4x-3}{y}$

Domain $= \{(x,y): x \text{ and } y \text{ are real and } y \neq 0\}$

5. $z = \dfrac{4x^3 y - x}{2x - y}$

Domain $=$
$\{(x,y): x \text{ and } y \text{ are real and } 2x - y \neq 0\}$

7. $q = \sqrt{p_1} + 3p_2$

Domain $=$
$\{(p_1, p_2): p_1 \text{ and } p_2 \text{ are real and } p_1 \geq 0\}$

9. $z = x^3 + 4xy + y^2$

$x = 1,\ y = -1$

$z = 1 - 4 + 1 = -2$

11. $z = \dfrac{x-y}{x+y}$

$x = 4,\ y = -1$

$z = \dfrac{4+1}{4-1} = \dfrac{5}{3}$

13. $C(x_1, x_2) = 600 + 4x_1 + 6x_2$

$x_1 = 400,\ x_2 = 50$

$C(400, 50) = 600 + 1600 + 300 = 2500$

15. $q_1(p_1, p_2) = \dfrac{p_1 + 4p_2}{p_1 - p_2}$

$q_1(40, 35) = \dfrac{40+140}{40-35} = 36$

17. $z(x,y) = xe^{x+y}$

$z(3,-3) = 3e^0 = 3$

19. $f(x,y) = \dfrac{\ln(xy)}{x^2 + y^2}$

$f(-3,-4) = \dfrac{\ln 12}{9+16} = \dfrac{\ln 12}{25}$

21. $w = \dfrac{x^2 + 4yz}{xyz}$

At $(1,3,1)$ we have $w = \dfrac{1+12}{3} = \dfrac{13}{3}$.

23. $S = f(P,t) = Pe^{0.06t}$

$f(2000, 20) = 2000e^{1.2} = \$6640.23$

When \$2000 is invested for 20 years the compound amount is \$6640.23.

25. $Q = f(K,M,h) = \sqrt{\dfrac{2KM}{h}}$

$f(200,625,1) = \sqrt{\dfrac{2(200)(625)}{1}} = 500$

For the given numbers, the most economical order size is 500 units.

27. $S = 1.98T - 1.09(1-H)(T-58) - 56.9$

$A = 2.70 + 0.885T - 78.7H + 1.20TH$

Max: 97.8°F $H = 44\%$

Min: 74.7°F $H = 80\%$

Max:

$S = 1.98(97.8) - 1.09(1-0.44)(97.8-58) - 56.9 = 112.5°F$

$A = 2.70 + 0.885(97.8) - 78.7(0.44) + 1.2(97.8)(0.44) = 106.3°F$

Min:

$S = 1.98(74.7) - 1.09(1-0.8)(74.7-58) - 56.9 = 87.4°F$

$A = 2.70 + 0.885(74.7) - 78.7(0.8) + 1.2(74.7)(0.8) = 77.6°F$

29. a. $f(90,20,8) = \$752.80$ (Table 1, Row 5, Col 4)

When \$90,000 is borrowed for 20 years at 8%, the monthly payment needed to pay off the loan is   \$752.80.

b. $f(160,15,9) = \$1622.82$ (Table 2, Row 9, Col 3) When \$160,000 is borrowed for 15 years at 9%, the monthly payment needed to pay off the loan is \$1622.82.

31. $U = xy^2$

If $x = 9$, $y = 6$, then $U = 9 \cdot 6^2 = 324$.

a. $324 = x \cdot 9^2$ or $x = 4$ units

b. $324 = 81 \cdot y^2$ or $y = 2$ units

c.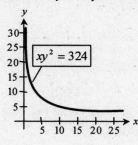

$xy^2 = 324$

33. $Q = 30K^{1/4}L^{3/4}$

a. $Q = 30(10,000)^{1/4}(625)^{3/4}$
$= 30(10)(125) = 37,500$

b. $Q = 30(2K)^{1/4}(2L)^{3/4}$
$= 30(2^{1/4})K^{1/4}(2^{3/4})L^{3/4}$
$= 2\left[30K^{1/4}L^{3/4}\right]$

This is 2 times the original quantity.

c.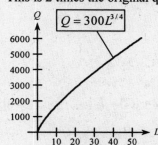

$Q = 300L^{3/4}$

35. $z = 20xy$

a. $z = 20(12)(30) = 7200$

b. $z = 20(10)(25) = 5000$

37. $C(x, y) = 20x + 200y$

$C(14,000, 20) = 20(14,000) + 200(20) = \$284,000$

Exercise 14.2

1. $z = x^4 - 5x^2 + 6x + 3y^3 - 5y + 7$

$\dfrac{\partial z}{\partial x} = 4x^3 - 10x + 6$

$\dfrac{\partial z}{\partial y} = 9y^2 - 5$

3. $z = x^3 + 4x^2y + 6y^2$

$z_x = 3x^2 + 8yx$

$z_y = 4x^2 + 12y$

5. $f(x, y) = \left(x^3 + 2y^2\right)^3$

$\dfrac{\partial f}{\partial x} = 3\left(x^3 + 2y^2\right)^2 3x^2 = 9x^2\left(x^3 + 2y^2\right)^2$

$\dfrac{\partial f}{\partial y} = 3\left(x^3 + 2y^2\right)^2 4y = 12y\left(x^3 + 2y^2\right)^2$

7. $f(x, y) = \sqrt{2x^2 - 5y^2}$

$f_x = \dfrac{1}{2}\left(2x^2 - 5y^2\right)^{-1/2} 4x = \dfrac{2x}{\sqrt{2x^2 - 5y^2}}$

$f_y = \dfrac{1}{2}\left(2x^2 - 5y^2\right)^{-1/2}(-10y) = \dfrac{-5y}{\sqrt{2x^2 - 5y^2}}$

9. $C(x, y) = 600 - 4xy + 10x^2y$

$\dfrac{\partial C}{\partial x} = -4y + 20xy$

$\dfrac{\partial C}{\partial y} = -4x + 10x^2$

11. $Q(s, t) = \dfrac{2s - 3t}{s^2 + t^2}$

$\dfrac{\partial Q}{\partial s} = \dfrac{\left(s^2 + t^2\right)(2) - (2s - 3t)(2s)}{\left(s^2 + t^2\right)^2}$

$= \dfrac{2s^2 + 2t^2 - 4s^2 + 6st}{\left(s^2 + t^2\right)^2} = \dfrac{2\left(t^2 + 3st - s^2\right)}{\left(s^2 + t^2\right)^2}$

$\dfrac{\partial Q}{\partial t} = \dfrac{\left(s^2 + t^2\right)(-3) - (2s - 3t)(2t)}{\left(s^2 + t^2\right)^2}$

$= \dfrac{-3s^2 - 3t^2 - 4st + 6t^2}{\left(s^2 + t^2\right)^2} = \dfrac{3t^2 - 4st - 3s^2}{\left(s^2 + t^2\right)^2}$

13. $z = e^{2x} + y \ln x$

$z_x = e^{2x}(2) + y \cdot \dfrac{1}{x} = 2e^{2x} + \dfrac{y}{x}$

$z_y = \ln x$

15. $f(x,y) = \ln(xy+1)$

$\dfrac{\partial f}{\partial x} = \dfrac{y}{xy+1}$

$\dfrac{\partial f}{\partial y} = \dfrac{x}{xy+1}$

17. $f(x,y) = 4x^3 - 5xy + y^2$

$f_x(x,y) = 12x^2 - 5y$

$f_x(1,2) = 12 - 5(2) = 2$

19. $z = 5x^3 - 4xy$

$z_x = 15x^2 - 4y$

$z_x(1,2) = 15 - 4(2) = 7$

21. $z = 3x + 2y - 7e^{xy}$

$z_y = 2 - 7xe^{xy}$

$z_y(3,0,2) = 2 - 7(3)e^{2(0)} = 2 - 21 = -19$

23. $u = y^2 - x^2 z + 4x$

  a. $\dfrac{\partial u}{\partial w} = 0$

  b. $\dfrac{\partial u}{\partial x} = 4 - 2xz$

  c. $\dfrac{\partial u}{\partial y} = 2y$

  d. $\dfrac{\partial u}{\partial z} = -x^2$

25. $C = 4x_1^2 + 5x_1 x_2 + 6x_2^2 + x_3$

  a. $\dfrac{\partial C}{\partial x_1} = 8x_1 + 5x_2$

  b. $\dfrac{\partial C}{\partial x_2} = 5x_1 + 12x_2$

  c. $\dfrac{\partial C}{\partial x_3} = 1$

27. $z = x^2 + 4x - 5y^3$

$z_x = 2x + 4, \quad z_y = -15y^2$

  a. $z_{xx} = 2$

  b. $z_{xy} = 0$

  c. $z_{yx} = 0$

  d. $z_{yy} = -30y$

29. $z = x^2 y - 4xy^2$

$z_x = 2xy - 4y^2, \quad z_y = x^2 - 8xy$

  a. $z_{xx} = 2y$

  b. $z_{xy} = 2x - 8y$

  c. $z_{yx} = 2x - 8y$

  d. $z_{yy} = -8x$

31. $f(x,y) = x^2 + e^{xy}$

$\dfrac{\partial f}{\partial x} = 2x + ye^{xy}, \quad \dfrac{\partial f}{\partial y} = xe^{xy}$

  a. $\dfrac{\partial^2 f}{\partial x^2} = 2 + y^2 e^{xy}$

  b. $\dfrac{\partial^2 f}{\partial y \, \partial x} = yxe^{xy} + e^{xy} \cdot 1 = e^{xy}(xy+1)$

  c. $\dfrac{\partial^2 f}{\partial x \, \partial y} = xye^{xy} + e^{xy} \cdot 1 = e^{xy}(xy+1)$

  d. $\dfrac{\partial^2 f}{\partial y^2} = x^2 e^{xy}$

33. $f(x,y) = y^2 - \ln xy = y^2 - \ln x - \ln y$

$\dfrac{\partial f}{\partial x} = -\dfrac{1}{x}, \quad \dfrac{\partial f}{\partial y} = 2y - \dfrac{1}{y}$

  a. $\dfrac{\partial^2 f}{\partial x^2} = \dfrac{1}{x^2}$

  b. $\dfrac{\partial^2 f}{\partial y \, \partial x} = 0$

  c. $\dfrac{\partial^2 f}{\partial x \, \partial y} = 0$

  d. $\dfrac{\partial^2 f}{\partial y^2} = 2 + \dfrac{1}{y^2}$

35. $f(x, y) = x^3 y + 4xy^4$

$$\frac{\partial f}{\partial x} = 3x^2 y + 4y^4$$

$$\frac{\partial^2 f}{\partial x^2} = 6xy$$

$$\frac{\partial^2}{\partial x^2} f(x, y) \Big|_{(1,-1)} = 6(1)(-1) = -6$$

37. $f(x, y) = \dfrac{2x}{x^2 + y^2}$

a. $\dfrac{\partial f}{\partial x} = \dfrac{\left(x^2 + y^2\right)(2) - 2x(2x)}{\left(x^2 + y^2\right)^2} = \dfrac{2y^2 - 2x^2}{\left(x^2 + y^2\right)^2}$

$\dfrac{\partial^2 f}{\partial x^2} = \dfrac{\left(x^2 + y^2\right)^2(-4x) - \left(2y^2 - 2x^2\right)2\left(x^2 + y^2\right)2x}{\left(x^2 + y^2\right)^4}$

$= \dfrac{-4x\left(x^2 + y^2\right)\left[\left(x^2 + y^2\right) + \left(2y^2 - 2x^2\right)\right]}{\left(x^2 + y^2\right)^4} = \dfrac{-4x\left(3y^2 - x^2\right)}{\left(x^2 + y^2\right)^3}$

$\dfrac{\partial^2 f}{\partial x^2}\Big|_{(-1,4)} = \dfrac{4(47)}{17^3} = \dfrac{188}{4913}$

b. $\dfrac{\partial f}{\partial y} = 2x(-1)\left(x^2 + y^2\right)^{-2}(2y) = \dfrac{-4xy}{\left(x^2 + y^2\right)^2}$

$\dfrac{\partial^2 f}{\partial y^2} = \dfrac{\left(x^2 + y^2\right)^2(-4x) + (4xy)2\left(x^2 + y^2\right)2y}{\left(x^2 + y^2\right)^4} = \dfrac{4x\left(x^2 + y^2\right)\left[\left(-x^2 - y^2\right) + 4y^2\right]}{\left(x^2 + y^2\right)^4} = \dfrac{4x\left(3y^2 - x^2\right)}{\left(x^2 + y^2\right)^3}$

$\dfrac{\partial^2 f}{\partial y^2}\Big|_{(-1,4)} = \dfrac{-4(47)}{17^3} = -\dfrac{188}{4913}$

39. $z = x^2 y + y e^{x^2}$

$z_y = x^2 + e^{x^2}$

$z_{yx} = 2x + e^{x^2}(2x)$

$z_{yx}\Big|_{(1,2)} = 2 + e^1(2) = 2 + 2e$

41. $z = x^2 - xy + 4y^3$

$z_x = 2x - y$

$z_{xy} = -1$

$z_{xyx} = 0$

43. $w = 4x^3 y + y^2 z + z^3$

$w_x = 12x^2 y$

a. $w_{xx} = 24xy, \quad w_{xxy} = 24x$

b. $w_{xy} = 12x^2, \quad w_{xyx} = 24x$

c. $w_{xyz} = 0$

45. $R = f(A, i)$

a. For a loan of $100,000 at 8% interest, the monthly payment is $1289.

b. If the interest rate changes from 8% to 9% (on this $100,000 loan), the monthly payment increases approximately $62.51.

47. Explanation is based on business interpretation and not the mathematics of exponents.

  a. $\dfrac{\partial Q}{\partial M}$ is the change in quantity ordered per unit change in sales. If the sales increase by one unit, then the order quantity should also increase.

  b. $\dfrac{\partial Q}{\partial h}$ is the change in quantity ordered per unit change in holding costs. If holding costs increase, then the order quantity should decrease.

49. $f(x,y) = 10{,}000 - 6500e^{-0.01x} - 3500e^{-0.02y}$

$f_x = 0 - 6500e^{-0.01x}(-0.01) - 0 = 65e^{-0.01x}$

$f_x(100,100) = 65e^{-0.01(100)} \approx 23.912$

If brand 2 is held constant and brand 1 is increased from 100 to 101 liters, approximately 24,000 additional insects are killed.

51. $U = x^2 y^2$

  a. $\dfrac{\partial U}{\partial x} = 2xy^2$

  b. $\dfrac{\partial U}{\partial y} = 2x^2 y$

53. $Q = 75K^{1/3}L^{2/3}$

$\dfrac{\partial Q}{\partial K} = 25K^{-2/3}L^{2/3}, \quad \dfrac{\partial Q}{\partial L} = 50K^{1/3}L^{-1/3}$

At $K = 729$, $L = 5832$ we have

$\dfrac{\partial Q}{\partial K} = 25 \cdot \dfrac{1}{81} \cdot 324 = 100$ and

$\dfrac{\partial Q}{\partial L} = 50 \cdot 9 \cdot \dfrac{1}{18} = 25.$

If labor hours remain at 5832 and K increases by $1000, Q will increase about 100 thousand units. If capital expenditures remain at $729,000 and L increases by one hour, Q will increase about 25 thousand units.

55. $WC = 48.064 + 0.474t - 0.020ts - 1.85s$

$\qquad + 0.304t\sqrt{s} - 27.74\sqrt{s}$

  a. $\dfrac{\partial WC}{\partial s} = -0.020t - 1.85 + \dfrac{0.152t}{\sqrt{s}} - \dfrac{13.87}{\sqrt{s}}$

  b. At $s = 25$, $t = 10$,

  $\dfrac{\partial WC}{\partial s} = -0.2 - 1.85 + \dfrac{1.52}{5} - \dfrac{13.87}{5} = -4.52$

  At this temperature and wind speed, the wind chill temperature will decrease 4.52°F if the wind speed increases 1 mph.

Exercise 14.3

1. $C(x,y) = 30 + 3x + 5y$

$C(4,3) = 30 + 12 + 15 = \$57$

3. $C(x,y) = 30 + 2x + 4y + \dfrac{xy}{50}$

  a. $C_x = 2 + \dfrac{y}{50}$

  b. $C_y = 4 + \dfrac{x}{50}$

5. $C(x,y) = 20x + 70y + \dfrac{x^2}{1000} + \dfrac{xy^2}{100}$

  a. $C_x = 20 + \dfrac{x}{500} + \dfrac{y^2}{100}$

  $C_x(10,12) = 20 + \dfrac{1}{50} + \dfrac{144}{100} = \$21.46$

  The total cost will increase approximately $21.46 if raw material costs increase to $11 per pound.

  b. $C_y = 70 + \dfrac{xy}{50}$

  $C_y(10,12) = 70 + \dfrac{120}{50} = \$72.40$

  The total cost will increase $72.40 if labor costs increase to $13 per hour.

7. $C(x,y) = 30 + x^2 + 3y + 2xy$

   a. $C_x = 2x + 2y$

     $C_x(8,10) = 16 + 20 = \$36$

     If y remains at 10, the expected change in cost for a 9th unit of x is \$36.

   b. $C_y = 3 + 2x$

     $C_y = (8,10) = 3 + 16 = \$19$

     If x remains at 8, the expected change in cost for an 11th unit of y is \$19.

9. $C(x,y) = x\sqrt{y^2 + 1}$

   a. $C_x = \sqrt{y^2 + 1}$

   b. $C_y = x \cdot \dfrac{1}{2}\left(y^2 + 1\right)^{-1/2}(2y) = \dfrac{xy}{\sqrt{y^2 + 1}}$

11. $C(x,y) = 1200 \ln(xy + 1) + 10{,}000$

   a. $C_x = 1200 \cdot \dfrac{1}{xy + 1} \cdot y = \dfrac{1200y}{xy + 1}$

   b. $C_y = 1200 \cdot \dfrac{1}{xy + 1} \cdot x = \dfrac{1200x}{xy + 1}$

13. $z = \sqrt{4xy} = 2x^{1/2}y^{1/2}$

   a. $z_x = 2y^{1/2} \cdot \dfrac{1}{2}x^{-1/2} = \sqrt{\dfrac{y}{x}}$

   b. $z_y = 2x^{1/2} \cdot \dfrac{1}{2}y^{-1/2} = \sqrt{\dfrac{x}{y}}$

15. $z = x^{1/2}\ln(y + 1)$

   a. $z_x = \dfrac{1}{2}x^{-1/2}\ln(y + 1) = \dfrac{\ln(y + 1)}{2\sqrt{x}}$

   b. $z_y = x^{1/2} \cdot \dfrac{1}{y + 1} \cdot 1 = \dfrac{\sqrt{x}}{y + 1}$

For 17-19, $z = \dfrac{11xy - 0.0002x^2 - 5y}{0.03x + 3y}$.

17. At $x = 300$ and $y = 500$ we have $z = \dfrac{11(150{,}000) - 0.0002(90{,}000) - 2500}{9 + 1500} \approx 1092$

19. $z_x = \dfrac{(0.03x + 3y)(11y - 0.0004x) - \left(11xy - 0.0002x^2 - 5y\right)(0.03)}{(0.03x + 3y)^2}$

   $z_x(300,500) = \dfrac{1509(5499.88) - \left(1{,}650{,}000 - 18 - 2500\right)(0.03)}{(1509)^2} \approx 3.62$

If 500 acres are planted, the expected change in the productivity for the 301st hour of labor is 3.62 crates.

21. $z = 400x^{3/5}y^{2/5}$

   a. $\dfrac{\partial z}{\partial x} = \dfrac{240y^{2/5}}{x^{2/5}}$

   b.

     [graph: 5000 vertical axis, 50 horizontal axis, curve with box $z_x = \dfrac{3840}{x^{2/5}}$]

   a. $\dfrac{\partial z}{\partial y} = \dfrac{160x^{3/5}}{y^{3/5}}$

   b.

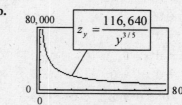

   c. An increase in capital investment or work hours results in an increase in productivity since both partials are positive. As capital investment or work hours continue to increase, the increase in productivity is diminishing.

23. $q_1 = 300 - 80 - 32 = 188$

$q_2 = 400 - 50 - 80 = 270$

25. Setting $q_1 = q_2$, we have

$300 - 8p_1 - 4p_2 = 400 - 5p_1 - 10p_2$

$-8p_1 + 5p_1 = 100 - 10p_2 + 4p_2$

$p_1 = -\dfrac{100}{3} + 2p_2 \quad (1)$

Any values p_1 and p_2 satisfying (1) and that make q_1 and q_2 nonnegative will be a solution. Two solutions are $(6.67, 20)$ and $(20, 26.67)$.

27. $q_A = 400 - 3p_A - 2p_B$, $q_B = 250 - 5p_A - 6p_B$

a. $\dfrac{\partial q_A}{\partial p_A} = -3$

b. $\dfrac{\partial q_A}{\partial p_B} = -2$

c. $\dfrac{\partial q_B}{\partial p_B} = -6$

d. $\dfrac{\partial q_B}{\partial p_A} = -5$

e. Complementary since (b) and (d) are negative.

29. $q_A = 5000 - 50p_A - \dfrac{600}{p_B + 1}$

$q_B = 10,000 - \dfrac{400}{p_A + 4} + \dfrac{400}{p_B + 4}$

a. $\dfrac{\partial q_A}{\partial p_A} = -50$

b. $\dfrac{\partial q_A}{\partial p_B} = -600(-1) \cdot \dfrac{1}{(p_B + 1)^2} = \dfrac{600}{(p_B + 1)^2}$

c. $\dfrac{\partial q_B}{\partial p_B} = \dfrac{-400}{(p_B + 4)^2}$

d. $\dfrac{\partial q_B}{\partial p_A} = \dfrac{400}{(p_A + 4)^2}$

e. Competitive since (b) and (d) are positive.

Exercise 14.4

1. $z = 9 - x^2 - y^2$

$z_x = -2x,\ z_y = -2y$

Critical point is at $x = 0$, $y = 0$.

$z_{xx} = -2,\ z_{yy} = -2,\ z_{xy} = 0$

$D = (-2)(-1) - (0)^2 > 0$ with $z_{xx} < 0$

A relative maximum at $(0, 0, 9)$.

3. $z = x^2 + y^2 + 4$

$z_x = 2x,\ z_y = 2y$

Critical point at $x = 0$, $y = 0$.

$z_{xx} = 2,\ z_{yy} = 2,\ z_{xy} = 0$

$D = (2)(2) - 0^2 > 0$ with $z_{xx} > 0$

A relative minimum at $(0, 0, 4)$.

5. $z = x^2 + y^2 - 2x + 4y + 5$

$z_x = 2x - 2,\ z_y = 2y + 4$

$z_x = 0$ at $x = 1$ and $z_y = 0$ at $y = -2$. Critical point is at $(1, -2)$.

$z_{xx} = 2,\ z_{yy} = 2,\ z_{xy} = 0$

$D = (2)(2) - 0^2 > 0$ with $z_{xx} > 0$

A relative minimum at $(1, -2, 0)$.

7. $z = x^2 + 6xy + y^2 + 16x$

$z_x = 2x + 6y + 16,\ z_y = 6x + 2y$

$z_y = 0$ gives $y = -3x$. $z_x = 0$ and $y = -3x$

gives $2x + 6(-3x) + 16 = 0$.

Thus, $x = 1$ and $y = -3(1) = -3$. So, $(1, -3)$ is a critical point.

$z_{xx} = 2,\ z_{yy} = 2,\ z_{xy} = 6$

$D = (2)(2) - 6^2 < 0$ and thus, there is neither a relative maximum nor a relative minimum at $(1, -3, 8)$. It is a saddle point.

9. $z = \dfrac{1}{9}(x^2 - y^2)$

$z_x = \dfrac{2}{9}x, \ z_y = -\dfrac{2}{9}y$

Critical point is at $(0,0)$.

$z_{xx} = \dfrac{2}{9}, \ z_{yy} = -\dfrac{2}{9}, \ z_{xy} = 0$

$D = \left(\dfrac{2}{9}\right)\left(-\dfrac{2}{9}\right) - 0^2 < 0$ and thus, there is no

relative maximum nor minimum at $(0,0,0)$. It is a saddle point.

11. $z = 24 - x^2 + xy - y^2 + 36y$

$z_x = -2x + y, \ z_y = x - 2y + 36$

$z_x = 0$ yields $y = 2x$. $z_y = 0$ and $y = 2x$ yields

$x - 2(2x) + 36 = 0$.

Solving, we have $x = 12$ and $y = 24$ are the critical values.

$z_{xx} = -2, \ z_{yy} = -2, \ z_{xy} = 1$

$D = (-2)(-2) - 1^2 > 0$ with $z_{xx} < 0$

There is a relative maximum at $(12, 24, 456)$.

13. $z = x^2 + xy + y^2 - 4y + 10x$

$z_x = 2x + y + 10, \ z_y = x + 2y - 4$

$z_x = 0$ or $2x + y = -10$ and $z_y = 0$ or

$x + 2y = 4$.

Solving these two equations, we have a critical point at $(-8, 6)$.

$z_{xx} = 2, \ z_{yy} = 2, \ z_{xy} = 1$

$D = (2)(2) - 1^2 > 0$ with $z_{xx} > 0$

There is a relative minimum at $(-8, 6, -52)$.

15. $z = x^3 + y^3 - 6xy$

$z_x = 3x^2 - 6y, \ z_y = 3y^2 - 6x$

$z_x = 0$ or $y = \dfrac{1}{2}x^2$. $z_y = 0$ and $y = \dfrac{1}{2}x^2$ gives

$\dfrac{3}{4}x^4 - 6x = 0$. $3x^4 - 24x = 0$ yields

$3x(x^3 - 8) = 0$. At $x = 0$ we have $y = 0$ and at

$x = 2$ we have $y = 2$.

Thus, the critical points are $(0,0)$ and $(2,2)$.

$z_{xx} = 6x, \ z_{yy} = 6y, \ z_{xy} = -6$

At $(0,0)$: $D = (0)(0) - (-6)^2 < 0$

There is a saddle point at $(0,0,0)$.

At $(2,2)$: $D = (12)(12) - (-6)^2 > 0$ with $z_{xx} > 0$.

There is a relative minimum at $(2, 2, -8)$.

17. $\sum x = 18, \ \sum y = 97, \ \sum xy = 465, \ \sum x^2 = 86$

$b = \dfrac{18(97) - 4(465)}{324 - 4(86)} = 5.7$

$a = \dfrac{97 - 5.7(18)}{4} = -1.4$

$\hat{y} = a + bx = 5.7x - 1.4$

19. $P(x,y) = 10x + 6.4y - 0.001x^2 - 0.025y^2$

$P_x = 10 - 0.002x, \ P_y = 6.4 - 0.05y$

$P_x = 0$ if $x = 5000$ and $P_y = 0$ if $y = 128$.

Critical point is $(5000, 128)$.

$P_{xx} = -0.002, \ P_{yy} = -0.05, \ P_{xy} = 0$

$D = (-0.002)(-0.05) - 0^2 > 0$

$D > 0$ with $P_{xx} < 0$ means that there is a maximum profit when 5000 pounds of Kisses and 128 pounds of Kreams are sold. The maximum profit is $P(5000, 128) = \$25409.60$.

21. $W = 13xy(20 - x - 2y)$

$\qquad = 260xy - 13x^2 y - 26xy^2$

$W_x = 260y - 26xy - 26y^2 = 26y(10 - x - y^2)$

$W_y = 260x - 13x^2 - 52xy = 13x(20 - x - 4y)$

If $W_x = 0$, then $26y(10 - x - y) = 0$. Thus,

$y = 0$ or $y = 10 - x$.

If $W_y = 0$, then $13x(20 - x - 4y) = 0$. Thus,

$x = 0$ or $x = 20 - 4y$.

Critical points are $(0,0)$, $(0,10)$, and $(20,0)$

using $x = 0$ and $y = 0$.

Using $x = 20 - 4(10 - x)$, we have $x = \dfrac{20}{3}$. For

$x = \dfrac{20}{3}$, we have $y = 10 - \dfrac{20}{3} = \dfrac{10}{3}$. So,

$\left(\dfrac{20}{3}, \dfrac{10}{3}\right)$ is also a critical point.

$W_{xx} = -26y$, $W_{yy} = -52x$, $W_{xy} = 260 - 26x - 52y$

At $(0,0)$, $(0,10)$, and $(20,0)$, $D < 0$ and thus,

there is no maximum or minimum there. At

$\left(\dfrac{20}{3}, \dfrac{10}{3}\right)$, $D > 0$ and since $W_{xx} < 0$, there is a

maximum weight gain with $x = \dfrac{20}{3}$ and $y = \dfrac{10}{3}$.

The maximum weight is

$W = 13\left(\dfrac{20}{3}\right)\left(\dfrac{10}{3}\right)\left(20 - \dfrac{20}{3} - 2\left(\dfrac{10}{3}\right)\right) \approx 1926$ lb.

23. $P = 3.78x^2 + 1.5y^2 - 0.09x^3 - 0.01y^3$

$P_x = 7.56x - 0.27x^2$, $P_y = 3y - 0.03y^2$

$P_x = 0$ if $0.27x(28 - x) = 0$ or $x = 0, 28$.

$P_y = 0$ if $3y(1 - 0.01y) = 0$ or $y = 0, 100$.

Critical points are $(0,0)$, $(0,100)$, $(28,0)$, and

$(28,100)$.

$P_{xx} = 7.56 - 0.54x$, $P_{yy} = 3 - 0.06y$, $P_{xy} = 0$

At $(0,0)$ $D = (7.56)(3) - 0^2 > 0$. But $P_{xx} > 0$.

At $(0,100)$ $D = (7.56)(-3) - 0^2 < 0$.

At $(28,0)$ $D = (-7.56)(3) - 0^2 < 0$.

At $(28,100)$ $D = (-7.56)(-3) - 0^2 > 0$.

Since $P_{xx} < 0$, there is a maximum at $x = 28$ and

$y = 100$. The maximum production is

$P(28,100) = 5987.84 \approx 5988$.

25. $C = xy$

$R = p_1 x + p_2 y$

$\qquad = (70 - x)x + (80 - y)y$

$\qquad = 70x - x^2 + 80y - y^2$

$P = R - C = 70x - x^2 + 80y - y^2 - xy$

$P_x = 70 - 2x - y$

$P_y = 80 - 2y - x$

$P_x = 0$ if $y = 70 - 2x$. $P_y = 0$ if $x = 80 - 2y$.

Solving this system, we get $x = 20$ and $y = 30$.

Thus, the critical point is $(20,30)$.

$P_{xx} = -2$, $P_{yy} = -4$, $P_{xy} = -1$

$D = (-2)(-4) - (-1)^2 > 0$.

Since $P_{xx} < 0$, there is a maximum at $(20,30)$.

The maximum is at $P(20,30) = \$1900$ thousand.

27. $A = xy + 2y \cdot \dfrac{500,000}{xy} + 2x \cdot \dfrac{500,000}{xy}$

$\qquad = xy + \dfrac{1,000,000}{x} + \dfrac{1,000,000}{y}$

$A_x = y - \dfrac{10^6}{x^2}$, $A_y = x - \dfrac{10^6}{y^2}$

Combining $A_x = 0$ and $A_y = 0$, we have

$x = \dfrac{10^6}{\frac{10^{12}}{x^4}} = \dfrac{x^4}{10^6}$, $x \neq 0$.

Thus, $x^3 = 10^6$ or $x = 100$.

The critical point is $(100,100)$.

$A_{xx} = \dfrac{2 \cdot 10^6}{x^3}$, $A_{yy} = \dfrac{2 \cdot 10^6}{y^3}$, $A_{xy} = 1$

$D = (2)(2) - 1^2 > 0$ with $A_{xx} > 0$.

There is a minimum for length and width equal

100 and height equal 50.

29. $P = R - C = 12x - 18y - 2x^2 + 2xy - y^2 - 11$

$P_x = 12 - 4x + 2y$, $P_y = 18 + 2x - 2y$

$P_x = 0$ if $y = 2x - 6$ and $P_y = 0$ if $y = x + 9$.

Solving the system yields $(15,24)$ as the critical

point.

$P_{xx} = -4$, $P_{yy} = -2$, $P_{xy} = 2$

$D = (-4)(-2) - 2^2 > 0$

Since $P_{xx} < 0$, maximum profit occurs with 15

A's and 24 B's. The maximum profit is $\$295$

thousand.

For 31-33 use the linear regression feature of the graphing calculator.

31. a. $y = 0.113x + 4.860$

y is the median age.

b. $y = 0.113(35) + 4.860 = 8.82$

c. Higher prices and better quality are two reasons for the median age to increase.

d. Each year the median age for automobiles decreases by 0.113 years.

33. a. $y = 0.06254x + 6.1914$ (where x is the number of years after 2000 and y is in billions)

b. $y(12) = 6.94188$ billion

c. The world population is changing at the rate of 0.06254 billion persons per year after the year 2000.

Exercise 14.5

1. The objective function is
$$F(x, y, \lambda) = x^2 + y^2 + \lambda(x + y - 6).$$
$$F_x = 2x + \lambda, \ F_y = 2y + \lambda, \ F_\lambda = x + y - 6$$
Setting each of these equal to zero and combining, we have $x - y = 0$ and $x + y = 6$. This yields $x = 3$ and $y = 3$. The minimum is $z = 3^2 + 3^2 = 18$.

3. The objective function is
$$F(x, y, \lambda) = 3x^2 + 5y^2 - 2xy + \lambda(x + y - 5).$$
$$F_x = 6x - 2y + \lambda, \ F_y = 10y - 2x + \lambda$$
$$F_\lambda = x + y - 5$$
Setting $F_x = 0$ and $F_y = 0$, we have
$$6x - 2y = 10y - 2x \text{ or } x = \frac{3}{2}y.$$
Using $F_\lambda = 0$, we have $\frac{3}{2}y + y - 5 = 0$ or $y = 2$.
Now $x = \frac{3}{2}(2) = 3$.
Thus, the minimum occurs at $(3, 2)$.
The minimum is $z = 27 + 20 - 12 = 35$.

5. The objective function is
$$F(x, y, \lambda) = x^2 y + \lambda(x + y - 6).$$
$$F_x = 2xy + \lambda, \ F_y = x^2 + \lambda, \ F_\lambda = x + y - 6$$
Using $F_x = 0$ and $F_y = 0$, we have $x^2 = 2xy$ or $x(x - 2y) = 0$. Thus, $x = 0$ or $x = 2y$.
Using $F_\lambda = 0$, we have $x = 0$ and $y = 6$.
Also, $2y + y - 6 = 0$ yields $y = 2$ and so $x = 4$.
At $(0, 6)$, $z = 0$ and at $(4, 2)$, $z = 16 \cdot 2 = 32$.
The maximum is at $(4, 2)$.

7. The objective function is
$$F(x, y, \lambda) = 2xy - 2x^2 - 4y^2 + \lambda(x + 2y - 8).$$
$$F_x = 2y - 4x + \lambda, \ F_y = 2x - 8y + 2\lambda$$
$$F_\lambda = x + 2y - 8$$
$F_x = 0$ means $\lambda = 4x - 2y$. $F_y = 0$ means $\lambda = -x + 4y$. Combining we have $5x - 6y = 0$.
$F_\lambda = 0$ gives $x + 2y = 8$.
These two equations yield $x = 3$, $y = \frac{5}{2}$.
The maximum is $F = 15 - 18 - 25 = -28$.

9. The objective function is
$$F(x, y, \lambda) = x^2 + y^2 + \lambda(2x + y + 1).$$
$$F_x = 2x + 2\lambda, \ F_y = 2y + \lambda, \ F_\lambda = 2x + y + 1$$
$F_x = 0$ and $F_y = 0$ yields $x - 2y = 0$. $F_\lambda = 0$ gives $2x + y = -1$.
These two equations yield $x = -\frac{2}{5}$, $y = -\frac{1}{5}$.
The minimum is $z = \frac{4}{25} + \frac{1}{25} = \frac{1}{5}$.

11. The objective function is
$$F(x, y, z, \lambda) = x^2 + y^2 + z^2 + \lambda(x + y + z - 3).$$
$$F_x = 2x + \lambda, \ F_y = 2y + \lambda,$$
$$F_z = 2z + \lambda, \ F_\lambda = x + y + z - 3$$
$$F_x = F_y = F_z = 0 \text{ or } x = y = z = -\frac{\lambda}{2}.$$
$F_\lambda = 0$ yields $-\frac{3\lambda}{2} = 3$ or $\lambda = -2$.
Then, $x = y = z = 1$.
The minimum is $1 + 1 + 1 = 3$.

13. The objective function is
$$F(x,y,z,\lambda) = xz + y + \lambda\left(x^2 + y^2 + z^2 - 1\right).$$
$$F_x = z + 2\lambda x, \quad F_y = 1 + 2\lambda y,$$
$$F_z = x + 2\lambda z, \quad F_\lambda = x^2 + y^2 + z^2 - 1$$
$F_x = F_z = 0$ gives $x - 2\lambda x = z - 2\lambda z$ or
$x(1 - 2\lambda) - z(1 - 2\lambda) = 0.$ So, $x = z$ or
$1 - 2\lambda = 0.$ If $x = z$, then $1 + 2\lambda = 0.$ The
solutions are $\lambda = \pm\dfrac{1}{2}.$ If $\lambda = \dfrac{1}{2}$, then $F_y = 0$
gives $y = -1.$ Also, $z + x = 0$ and $F_\lambda = 0$
means that $x^2 + 1 + z^2 - 1 = 0$ or $x = z = 0.$ The
critical point is $(0, -1, 0).$ If $\lambda = -\dfrac{1}{2}$, then
$F_y = 0$ gives $y = 1.$ From $F_x = F_z$ we get
$z = x.$ From $F_\lambda = 0$ we get $x = z = 0.$ The
critical point is $(0, 1, 0).$ At $(0, -1, 0)$, we have
$w = 0 - 1 = -1.$ At $(0, 1, 0)$, we have
$w = 0 + 1 = 1.$ The maximum is $w = 1$ at
$(0, 1, 0).$

15. The objective function is
$$F(x,y,\lambda) = x^2 y + \lambda(x + 2y - 6).$$
$$F_x = 2xy + \lambda, \quad F_y = x^2 + 2\lambda, \quad F_\lambda = x + 2y - 6$$
$F_x = F_y = 0$ or $-\dfrac{x^2}{2} = -2xy$ or $x^2 - 4xy = 0$ or
$x(x - 4y) = 0$ thus $x = 0$ or $x = 4y.$
$F_\lambda = 0$: If $x = 0$, then $y = 3.$ If $x = 4y$, then
$6y - 6 = 0$ or $y = 1.$ If $y = 1$, then $x = 4$. At
$(0, 3)$, we have $U = 0 \cdot 3 = 0$ and at $(4, 1)$ we
have $U = 16 \cdot 1 = 16.$ The maximum occurs at
$(4, 1).$

17. The objective function is
$$F(x,y,\lambda) = x^2 y + \lambda(2x + 3y - 120).$$
$$F_x = 2xy + 2\lambda, \quad F_y = x^2 + 3\lambda, \quad F_\lambda = 2x + 3y - 120$$
$F_x = 0$ gives $\lambda = -xy.$ $F_y = 0$ and $F_x = 0$ give
$x^2 - 3xy = 0$ so $x = 0$ or $x = 3y$.
$F_\lambda = 0$: If $x = 0$, then $y = 40.$
If $x = 3y$, then $y = \dfrac{40}{3}$ and $x = 40.$
At $(0, 40)$, we have $U = 0 \cdot 40 = 0.$
At $\left(40, \dfrac{40}{3}\right)$, we have $U = 1600 \cdot \dfrac{40}{3} = \dfrac{64,000}{3}.$
The maximum is at $\left(40, \dfrac{40}{3}\right).$

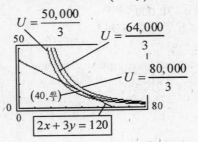

19. a. The objective function is
$$F(x,y,\lambda) = 400x^{0.6} y^{0.4}$$
$$+ \lambda(150x + 100y - 100,000)$$
$$F_x = \frac{240y^{0.4}}{x^{0.4}} + 150\lambda, \quad F_y = \frac{160x^{0.6}}{y^{0.6}} + 100\lambda$$
$$F_\lambda = 150x + 100y - 100,000$$
$F_x = 0$ and $F_y = 0$ give $\dfrac{240y^{0.4}}{150x^{0.4}} = \dfrac{160x^{0.6}}{100y^{0.6}}$
or $x = y.$
$F_\lambda = 0$: $250x - 100,000 = 0$ or $x = 400.$
$250y - 100,000 = 0$ or $y = 400.$

b. $-\lambda = \dfrac{240(400)^{0.4}}{150(400)^{0.4}} = \dfrac{8}{5} = 1.6$

An additional 1.6 units are produced for
each additional dollar spent on production.

c.
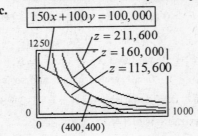

21. The objective function is
$$F(x, y, \lambda) = x^2 + 1200 + 3y^2 + 800$$
$$+ \lambda(x + y - 1200).$$
$F_x = 2x + \lambda, \; F_y = 6y + \lambda, \; F_\lambda = x + y - 1200$
$F_x = F_y = 0$ yields $x - 3y = 0$.
$F_\lambda = 0$ yields $x + y = 1200$.
Solving these 2 equations yields $(900, 300)$.
To minimize cost let $x = 900$ and $y = 300$.

23. The objective function is
$$F(x, y, \lambda) = 20x + y^2 + 4xy + \lambda(x + y - 30,000).$$
$F_x = 20 + 4y + \lambda, \; F_y = 2y + 4x + \lambda,$
$F_\lambda = x + y - 30,000$
$F_x = F_y = 0$ yields $2x - y = 10$.
$F_\lambda = 0$ yields $x + y = 30,000$.
Solving these 2 equations gives maximum
revenue if $x = \$10,003.33$ and $y = \$19,996.67$.

25. The objective function is
$$F(x, h, \lambda) = x^2 + 4xh + \lambda(x^2h - 500,000).$$
$F_x = 2x + 4h + 2xh\lambda, \; F_h = 4x + x^2\lambda$
$F_\lambda = x^2h - 500,000$
$F_h = 0$ means $x = 0$ or $x\lambda = -4$. x cannot be 0.
$F_x = 0$ and $x\lambda = -4$ gives $2x + 4h - 8h = 0$ or
$x = 2h$. $F_\lambda = 0$ gives $4h^3 = 500,000$ or
$h = 50$. With $x = 2h = 100$, the dimensions are
100 by 100 by 50.

Review Exercises

1. $z = \dfrac{3}{2x - y}$

Domain $= \{(x, y): x, y \in \text{Reals and } y \neq 2x\}$.

2. $z = \dfrac{3x + 2\sqrt{y}}{x^2 + y^2}$

Domain: $\{(x, y): x \text{ and } y \text{ are real}$

$\text{with } y \geq 0 \text{ and } (x, y) \neq (0, 0)\}$

3. $w(x, y, z) = x^2 - 3yz$

$w(2, 3, 1) = 4 - 3(3)(1) = -5$

4. $Q(K, L) = 70K^{2/3}L^{1/3}$

$Q(64,000, 512) = 70(1600)(8) = 896,000$

5. $z = 5x^3 + 6xy + y^2$

$z_x = 15x^2 + 6y$

6. $z = 12x^5 - 14x^3y^3 + 6y^4 - 1$

$\dfrac{\partial z}{\partial y} = -42x^3y^2 + 24y^3$

7. $z = 4x^2y^3 + \dfrac{x}{y}$

$z_x = 8xy^3 + \dfrac{1}{y}, \quad z_y = 12x^2y^2 - \dfrac{x}{y^2}$

8. $z = \left(x^2 + 2y^2\right)^{1/2}$

$z_x = \dfrac{1}{2}\left(x^2 + 2y^2\right)^{-1/2}(2x) = \dfrac{x}{\sqrt{x^2 + 2y^2}}$

$z_y = \dfrac{1}{2}\left(x^2 + 2y^2\right)^{-1/2}(4y) = \dfrac{2y}{\sqrt{x^2 + 2y^2}}$

9. $z = (xy + 1)^{-2}$

$z_x = -2(xy + 1)^{-3}(y) = -\dfrac{2y}{(xy + 1)^3}$

$z_y = -2(xy + 1)^{-3}(x) = -\dfrac{2x}{(xy + 1)^3}$

10. $z = e^{x^2y^3}$

$z_x = e^{x^2y^3} \cdot 2xy^3 = 2xy^3 e^{x^2y^3}$

$z_y = e^{x^2y^3} \cdot 3x^2y^2 = 3x^2y^2 e^{x^2y^3}$

11. $z = e^{xy} + y \ln x$

$z_x = e^{xy}(y) + y \cdot \dfrac{1}{x} = ye^{xy} + \dfrac{y}{x}$

$z_y = e^{xy}(x) + \ln x = xe^{xy} + \ln x$

12. $z = e^{\ln xy} = xy$

$z_x = y, \quad z_y = x$

13. $f(x, y) = 4x^3 - 5xy^2 + y^3$

$f_x = 12x^2 - 5y^2$

$f_x(1, 2) = 12 - 20 = -8$

14. $z = 5x^4 - 3xy^2 + y^2$

$z_x = 20x^3 - 3y^2, \quad z_x(1, 2) = 20 - 12 = 8$

15. $z = x^2y - 3xy$

$z_x = 2xy - 3y, \quad z_y = x^2 - 3x$

  a. $z_{xx} = 2y$

  b. $z_{yy} = 0$

  c. $z_{xy} = 2x - 3$

  d. $z_{yx} = 2x - 3$

16. $z = 3x^3y^4 - \dfrac{x^2}{y^2}$

$z_x = 9x^2y^4 - \dfrac{2x}{y^2}, \quad z_y = 12x^3y^3 + \dfrac{2x^2}{y^3}$

  a. $z_{xx} = 18xy^4 - \dfrac{2}{y^2}$

  b. $z_{yy} = 36x^3y^2 - \dfrac{6x^2}{y^4}$

  c. $z_{xy} = 36x^2y^3 + \dfrac{4x}{y^3}$

  d. $z_{yx} = 36x^2y^3 + \dfrac{4x}{y^3}$

17. $z = x^2e^{y^2}$

$z_x = 2xe^{y^2}, \quad z_y = x^2e^{y^2} \cdot 2y$

  a. $z_{xx} = 2e^{y^2}$

  b. $z_{yy} = x^2e^{y^2} \cdot 2 + 2yx^2e^{y^2} \cdot 2y$

$= 2x^2e^{y^2} + 4x^2y^2e^{y^2}$

  c. $z_{xy} = 2xe^{y^2} \cdot 2y = 4xye^{y^2}$

  d. $z_{yx} = 4xye^{y^2}$

18. $z = \ln(xy+1)$

$$z_x = \frac{y}{xy+1}, \quad z_y = \frac{x}{xy+1}$$

a. $z_{xx} = \frac{-y^2}{(xy+1)^2}$

b. $z_{yy} = \frac{-x^2}{(xy+1)^2}$

c. $z_{xy} = \frac{(xy+1)-y\cdot x}{(xy+1)^2} = \frac{1}{(xy+1)^2}$

d. $z_{yx} = \frac{(xy+1)-x\cdot y}{(xy+1)^2} = \frac{1}{(xy+1)^2}$

19. $z = 16 - x^2 - xy - y^2 + 24y$

$z_x = -2x - y, \quad z_y = -x - 2y + 24$

$z_x = 0$ gives $y = -2x$. Using $z_y = 0$ and $y = -2x$ yields $x = -8$ and thus, $y = 16$.

Critical point at $(-8, 16)$.

$z_{xx} = -2, \quad z_{yy} = -2, \quad z_{xy} = -1$

$D = (-2)(-2) - (-1)^2 > 0$

Since $D > 0$ and $z_{xx} < 0$, there is a relative maximum of $z = 208$ at $(-8, 16)$.

20. $z = x^3 + y^3 - 3xy$

$z_x = 3x^2 - 3y, \quad z_y = 3y^2 - 3x$

$z_x = 0$ gives $y = x^2$. Using $z_y = 0$ and $y = x^2$

yields $x^4 - x = 0$ or $x(x^3 - 1) = 0$.

When $x = 0$, $y = 0$ and when $x = 1$, $y = 1$.

The critical points are $(0, 0)$ and $(1, 1)$.

$z_{xx} = 6x, \quad z_{yy} = 6y, \quad z_{xy} = -3$

At $(0, 0)$, $D = (6x)(6y) - (-3)^2 = -9 < 0$. The point $(0, 0, 0)$ is a saddle point.

At $(1, 1)$, $D = (6x)(6y) - (-3)^2 = 36 - 9 = 27 > 0$ with $z_{xx} > 0$. A relative minimum at $(1, 1, -1)$.

21. The objective function is

$F(x, y, \lambda) = 4x^2 + y^2 + \lambda(x + y - 10)$

$F_x = 8x + \lambda, \quad F_y = 2y + \lambda, \quad F_\lambda = x + y - 10$

$F_x = F_y = 0$ gives $y = 4x$. $y = 4x$ and $F_\lambda = 0$

gives $5x - 10 = 0$ or $x = 2$. So, $y = 8$.

The minimum is $z = 4(4) + 64 = 80$ at $(2, 8)$.

22. The objective function is

$F(x, y, \lambda) = x^4y^2 + \lambda(x + y - 9)$

$F_x = 4x^3y^2 + \lambda, \quad F_y = 2x^4y + \lambda,$

$F_\lambda = x + y - 9$

$F_x = F_y = 0$ gives $4x^3y^2 = 2x^4y$ or

$2x^3y(2y - x) = 0$ so $x = 0$, $y = 0$, or $x = 2y$.

$F_\lambda = 0$: If $x = 0$, then $y = 9$. If $y = 0$, then $x = 9$. If $x = 2y$, then $y = 3$ and $x = 6$.

At $(0, 9)$ $z = 0 \cdot 81 = 0$.

At $(9, 0)$, $z = 6561 \cdot 0 = 0$.

At $(6, 3)$, $z = 6^4 \cdot 3^2 = 11,664$.

The maximum value is 11,664 at $(6, 3)$.

23. At $x = 6$ and $y = 15$, $U = 6^2 \cdot 15 = 540$.

To retain the same value of U at $y = 60$, we have $540 = x^2 \cdot 60$ or $x^2 = 9$ or $x = 3$.

24. $xy = 1600$

When $x = 80$, $y = 20$.

25. a. $w = 2\sqrt{\dfrac{Th}{d}}, \quad w = 2\sqrt{\dfrac{293(16.8)}{5}} = 62.8$ km

b. At a surface temperature of 287 K, an altitude of 17.1 km, and a vertical temperature gradient of 4.9 K/km, the width of the region on the ground on either side where people hear the sonic boom is about 63.3 km.

c. $\dfrac{\partial f}{\partial h} = \dfrac{T}{d\sqrt{\frac{Th}{d}}} = \dfrac{\sqrt{T}}{\sqrt{dh}} = \sqrt{\dfrac{T}{dh}}$

At $(293, 16.8, 5)$, $\dfrac{\partial f}{\partial h} = \sqrt{\dfrac{293}{5(16.8)}} = 1.87$.

For the given T and d, a change of 1 km in the altitude will result in an increase of 1.87 km in the width of the sonic boom path on either side.

d. $\dfrac{\partial f}{\partial d} = 2\sqrt{Th} \cdot \left(-\dfrac{1}{2}d^{-3/2}\right) = -\dfrac{\sqrt{Th}}{(\sqrt{d})^3}$

At $(293, 16.8, 5)$,

$\dfrac{\partial f}{\partial d} = -\dfrac{\sqrt{293(16.8)}}{(\sqrt{5})^3} = -6.28$.

For the given T and h, a change of 1 K/km in d will result in a decrease of 6.28 km in the width of the sonic boom path on either side.

26. $C(x,y) = x^2\sqrt{y^2+13}$

   a. $C_x = 2x\sqrt{y^2+13}$

       $C_x(20,6) = 40\sqrt{36+13} = 280$

   b. $C_y = \dfrac{x^2 y}{\sqrt{y^2+13}}$

       $C_y(20,6) = \dfrac{400\cdot 6}{\sqrt{36+13}} = \dfrac{2400}{7}$

27. $Q = 80K^{1/4}L^{3/4}$, $K = 625$, $L = 4096$

   $\dfrac{\partial Q}{\partial K} = \dfrac{20L^{3/4}}{K^{3/4}}$, $\dfrac{\partial Q}{\partial L} = \dfrac{60K^{1/4}}{L^{1/4}}$

   For the given values $\dfrac{\partial Q}{\partial K} = 81.92$ and

   $\dfrac{\partial Q}{\partial L} = 37.5$ (hundreds).

   When work-hours are fixed at 4096, an increase of $1000 in expenditures results in an increase of 8192 units produced. When expenditures are fixed at $625,000 an increase of one work-hour results in an increase of 3750 units produced.

28. $q_A = 400 - 2p_A - 3p_B$

   $q_B = 300 - 5p_A - 6p_B$

   a. $\dfrac{\partial q_A}{\partial p_A} = -2$

   b. $\dfrac{\partial q_B}{\partial p_B} = -6$

   c. The two products are complementary since

       $\dfrac{\partial q_A}{\partial p_B}$ and $\dfrac{\partial q_B}{\partial p_A}$ are both negative.

29. $q_A = 800 - 40p_A - 2(p_B+1)^{-1}$

   $q_B = 1000 - 10(p_A+4)^{-1} - 30p_B$

   $\dfrac{\partial q_A}{\partial p_A} = -40$

   $\dfrac{\partial q_A}{\partial p_B} = \dfrac{2}{(p_B+1)^2}$

   $\dfrac{\partial q_B}{\partial p_A} = \dfrac{10}{(p_A+4)^2}$

   $\dfrac{\partial q_B}{\partial p_B} = -30$

   Competitive since $\dfrac{\partial q_A}{\partial p_B}$ and $\dfrac{\partial q_B}{\partial p_A}$ are both positive.

30. $P(x,y) = 40x + 80y - x^2 - y^2$

   $P_x = 40 - 2x$, $P_y = 80 - 2y$

   $P_x = 0$ when $x = 20$ and $P_y = 0$ when $y = 40$.

   Critical point at $(20,40)$.

   $P_{xx} = -2$, $P_{yy} = -2$, $P_{xy} = 0$

   $D = (-2)(-2) - (0)^2 > 0$ with $P_{xx} < 0$ means profit is maximized when $x = 20$ and $y = 40$.

31. The objective function is

   $F(x,y,\lambda) = x^2 y + \lambda(4x + 5y - 60)$

   $F_x = 2xy + 4\lambda$, $F_y = x^2 + 5\lambda$

   $F_\lambda = 4x + 5y - 60$

   $F_x = 0$ and $F_y = 0$ give $\dfrac{-xy}{2} = \dfrac{-x^2}{5}$ or

   $5xy - 2x^2 = 0$ or $x(5y - 2x) = 0$. Since x

   cannot be 0, we have $5y - 2x = 0$ or $y = \dfrac{2}{5}x$.

   So, $4x + 5\cdot\frac{2}{5}x - 60 = 0$ or $x = 10$. Then

   $y = \frac{2}{5}\cdot 10 = 4$.

32. **a.** The objective function is

       $F(x,y,\lambda) = 300x^{2/3}y^{1/3}$

                $+ \lambda(50x + 50y - 75,000)$

       $F_x = \dfrac{200y^{1/3}}{x^{1/3}}$, $F_y = \dfrac{100x^{2/3}}{y^{2/3}}$

       $\dfrac{\partial F}{\partial \lambda} = 50x + 50y - 75,000$

       $F_x = F_y = 0$ gives $\dfrac{4y^{1/3}}{x^{1/3}} = \dfrac{2x^{2/3}}{y^{2/3}}$ or $x = 2y$.

       $F_\lambda = 0$ and $x = 2y$ give $150y - 75,000 = 0$

       or $y = 500$, so $x = 1000$.

       Production is maximized at $x = 1000$,

       $y = 500$.

   b. $-\lambda = \dfrac{4y^{1/3}}{x^{1/3}} = \dfrac{4\sqrt[3]{500}}{\sqrt[3]{1000}} \approx 3.17$ means each

       additional dollar spent on production results in an additional 3.17 units produced.

   c.

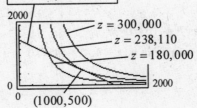

33. a. Use a graphing utility to obtain
$$\hat{y} = 0.127x + 8.926$$

b. If $x = 7000$, $\hat{y} = \$897.926$.

Chapter Test

1. $f(x,y) = \dfrac{2x+3y}{\sqrt{x^2 - y}}$

a. Domain =
$$\{(x,y): x \text{ and } y \text{ are real and } y < x^2\}$$

b. $f(-4, 12) = \dfrac{-8+36}{2} = 14$

2. $z = 5x - 9y^2 + 2(xy+1)^5$

$z_x = 5 + 10(xy+1)^4 (y) = 5 + 10y(xy+1)^4$

$z_y = -18y + 10(xy+1)^4 (x)$

$\quad = -18y + 10x(xy+1)^4$

$z_{xx} = 40y(xy+1)^3 (y) = 40y^2 (xy+1)^3$

$z_{yy} = -18 + 40x(xy+1)^3 (x)$

$\quad = -18 + 40x^2 (xy+1)^3$

$z_{xy} = z_{yx} = 10y \cdot 4(xy+1)^3 (x) + (xy+1)^4 (10)$

$\quad = 10(xy+1)^3 (5xy+1)$

3. $z = 6x^2 + x^2 y + y^2 - 4y + 9$

$z_x = 12x + 2xy = 2x(6+y)$

$z_y = x^2 + 2y - 4$

$z_x = 0$ if $x = 0$ or $y = -6$

If $x = 0$ and $z_y = 0$, then $y = 2$. If $y = -6$ and

$z_y = 0$, then $x^2 = 16$ or $x = \pm 4$.

Critical points are $(0,2)$, $(4,-6)$, and $(-4,-6)$.

$z_{xx} = 12 + 2y$, $z_{yy} = 2$, $z_{xy} = 2x$

$D = z_{xx} \cdot z_{yy} - (z_{xy})^2$

At $(0,2)$: $D = 16 \cdot 2 - 0 > 0$

At $(4,-6)$: $D = 0 \cdot 2 - 64 < 0$

At $(-4,-6)$: $D = 0 \cdot 2 - 64 < 0$

$D > 0$ and $z_{xx} > 0$ yield a relative minimum at $(0,2)$. $D < 0$ yields saddle points at $(4,-6)$ and $(-4,-6)$.

34. Use a graphing utility to obtain
$$q = -55.09p + 34,726.46.$$

4. $Q = 10K^{0.45}L^{0.55}$, $K = 10$, $L = 1590$

a. $Q = 10(10)^{0.45}(1590)^{0.55} = \$1625$ thousands

b. $Q_K = \dfrac{4.5L^{0.55}}{K^{0.55}}$, $Q_K = \dfrac{4.5(1590)^{0.55}}{10^{0.55}} = 73.11$

If labor hours remain constant and capital investment increases to $11,000$, the monthly production should increase $73.11 thousands.

c. $Q_L = \dfrac{5.5K^{0.45}}{L^{0.45}}$, $Q_L = \dfrac{5.5(10)^{0.45}}{(1590)^{0.45}} = 0.56$

If capital investment remains constant and labor hours increase to 1591, the monthly production should increase $0.56 thousands.

5. a. $f(94.5, 25, 7) = \$667.91$

When $94,500 is borrowed for 25 years at 7%, compounded monthly, the monthly payment is $667.91.

b. $\dfrac{\partial f}{\partial r}(94.5, 25, 7) = \$49.76$

If the rate increases from 7% to 8%, the monthly payment will increase $49.76. (The $94,500 and 25 years remain fixed.)

c. $\dfrac{\partial f}{\partial n}(94.5, 25, 7) < 0$ means that the monthly

payment decreases if the time to pay off the loan increases. (The $94,500 and 7% rate remain fixed.)

6. $f(x,y) = 2e^{x^2 y^2}$

$\dfrac{\partial f}{\partial y} = 2e^{x^2 y^2} (2x^2 y) = 4x^2 ye^{x^2 y^2}$

$\dfrac{\partial^2 f}{\partial x \partial y} = 4x^2 y(e^{x^2 y^2} \cdot 2xy^2) + e^{x^2 y^2} (8xy)$

$\quad = 8xye^{x^2 y^2} (1 + x^2 y^2)$

7. Find $\dfrac{\partial q_1}{\partial p_2}$ and $\dfrac{\partial q_2}{\partial p_1}$ and compare their signs.

Both positive means the products are competitive. Both negative means the products are complementary.

$\dfrac{\partial q_1}{\partial p_2} = -5$, $\dfrac{\partial q_2}{\partial p_1} = -4$

These products are complementary.

8. $P = 915x - 30x^2 - 45xy + 975y - 30y^2 - 3500$

$P_x = 915 - 60x - 45y = 15(61 - 4x - 3y)$

$P_y = -45x + 975 - 60y = 15(65 - 3x - 4y)$

Set P_x and P_y to 0 and solve.

$$3x + 4y = 65$$
$$4x + 3y = 61$$

The solution is $x = 7$ and $y = 11$.

$P_{xx} = -60$, $P_{yy} = -60$, $P_{xy} = -45$

$D = (-60)(-60) - (-45)^2 > 0$ and $P_{xx} < 0$ means

that P is a maximum at $(7, 11)$.

9. $U = x^3 y$ with budget constraint

$30x + 20y = 8000$.

The objective function is

$F(x, y, \lambda) = x^3 y + \lambda(30x + 20y - 8000)$

$\dfrac{\partial F}{\partial x} = 3x^2 y + 30\lambda$, $\dfrac{\partial F}{\partial y} = x^3 + 20\lambda$

$\dfrac{\partial F}{\partial \lambda} = 30x + 20y - 8000$

Set each partial to zero and solve for λ.

$\lambda = -\dfrac{x^2 y}{10}$ and $\lambda = -\dfrac{x^3}{20}$ yield $2x^2 y = x^3$ or

$x^2(x - 2y) = 0$, $x \neq 0$. Substituting $x = 2y$

into $\dfrac{\partial F}{\partial \lambda}$ gives $80y - 8000 = 0$ or $y = 100$.

Then $x = 200$. The function is maximized at

800,000,000 when $x = 200$ and $y = 100$.

10. a. $\sum x = 66, \sum y = 129,680, \sum x^2 = 506,$

$\sum xy = 795,260$

$b = \dfrac{66(129,680) - 12(795,260)}{(66)^2 - 12(506)} = 573.57$

$a = \dfrac{129,680 - 573.57(66)}{12} = 7652.05$

$\hat{y} = 7652.05 + 573.57x$

b. If $x = 14$,

$\hat{y} = 7652.05 + 573.57(14) = \$15,682$

c. The model would not make sense since it is a model for children and not adults.